SELECTED TOPICS IN
INFORMATION AND
CODING THEORY

Series on Coding Theory and Cryptology

Editors: Harald Niederreiter *(National University of Singapore, Singapore)* and
San Ling *(Nanyang Technological University, Singapore)*

Series on Coding Theory and Cryptology – Vol. 7

SELECTED TOPICS IN
INFORMATION AND
CODING THEORY

Editors

I. Woungang
Ryerson University, Canada

S. Misra
Indian Institute of Technology, Kharagpur, India

S. C. Misra
State University of New York at Buffalo, USA

 **World Scientific**

NEW JERSEY • LONDON • SINGAPORE • BEIJING • SHANGHAI • HONG KONG • TAIPEI • CHENNAI

Published by

World Scientific Publishing Co. Pte. Ltd.

5 Toh Tuck Link, Singapore 596224

USA office: 27 Warren Street, Suite 401-402, Hackensack, NJ 07601

UK office: 57 Shelton Street, Covent Garden, London WC2H 9HE

British Library Cataloguing-in-Publication Data
A catalogue record for this book is available from the British Library.

SELECTED TOPICS IN INFORMATION AND CODING THEORY
Series on Coding Theory and Cryptology — Vol. 7

ISBN-13 978-981-283-716-5
ISBN-10 981-283-716-7

Typeset by Stallion Press
Email: enquiries@stallionpress.com

Printed in Singapore by Mainland Press Pte Ltd.

JUN 2 2 2010

Dedicated to

Isaac's wife: Clarisse and sons: Clyde, Lenny, Kylian
Subhas's daughter: Devarati
Sudip's son: Devadeep

PREFACE

Overview and Goals

Information and Coding Theory research and applications are undergoing rapid advancements. The last few years have witnessed rapid advancements in Information and Coding Theory research and applications. This book provides a comprehensive guide to selected topics, both ongoing and emerging, in Information and Coding Theory. Consisting of contributions from well known and high profile researchers in their respective specialties, topics that are covered include applications of coding theory to computational complexity, algebraic combinatorics in coding theory, codes construction and existence, source coding, channel capacity, network coding, and few other selected topics in Information and Coding Theory research.

The book has been prepared keeping in mind that it needs to prove itself to be a valuable resource dealing with both the important core and the specialized issues in Information and Coding Theory. We hope that it will be a valuable reference for students, instructors, researchers, engineers, and industry practitioners in these fields. All of the chapters are integrated in a manner that renders the book as a supplementary reference volume and/or textbook for use in both undergraduate and graduate courses on Information and Coding Theory. Each chapter is of an expository, but also of a scholarly, tutorial, or survey style, on a particular topic within the scope of Information and Coding Theory.

Organization and Features

The book is organized into 15 chapters, each chapter written by topical area experts. These chapters are grouped into four parts.

Part 1 is devoted to the applications of coding theory to computational complexity, and is composed of three chapters: Chaps. 1–3. Chapter 1 discusses several theoretical methods for analyzing the linear complexity and related complexity measures and proposes several classes of interesting sequences with high linear complexity. Chapter 2 focuses on the construction of high coding gain lattices with low decoding complexity from good codes in larger dimensions, and proposes a possible lattice construction with high coding gain using turbo codes and Low Density Parity Check codes. Chapter 3 is dedicated to the issues of cooperative communication in wireless relay networks. Various constructions of the distributed space-time block codes with low maximum-likelihood decidability are surveyed, and new upper bounds on the maximum rate of certain classes of single-symbol decodable distributed space-time block codes are proposed.

Part 2 focuses on methods of algebraic combinatorics in coding theory, and methods of codes construction and existence. It is composed of four chapters: Chaps. 4–7. Chapter 4 discusses in-depth the interplay of coding theory and algebraic combinatorics, focusing on the interaction of codes with combinatorial designs. Chapter 5 discusses ways and results in which to define, construct, prove theorems, and analyze codes from group rings in general, using both zero-divisor and units within a group ring. The codes derived are described as either zero-divisor or unit-derived codes. Chapter 6 is a continuation of the work initiated in Chapter 5, by presenting a new algebraic group ring-based method for constructing codes with no short cycles in the check matrix, and a general algebraic method for constructing Low Density Parity Check codes with no short cycles. Chapter 7 presents the construction of some well-known classes of algebraic block codes, and discusses recent generalizations of quasi-cyclic codes, as well as some algebraic and combinatorial methods of obtaining new codes from existing ones.

Part 3 centers on source coding, channel capacity, and network coding issues. It is composed of three chapters: Chaps. 8–10. Chapter 8 introduces a new approach to estimation, prediction and hypothesis testing for time series based on ideas of universal coding or universal data compression. Chapter 9 presents a subjective approach to network coding, which is concerned with the deterministic multicast encoding of cyclic networks. This topic is presented at a level of detail that is not found elsewhere in the literature. Chapter 10 addresses the problem of transmission of several distributed sources over a multiple access channel with side information at the sources and the decoder, and proposes a joint source channel coding approach, which generalizes previous results available on the studied problem.

Part 4 addresses other selected topics in Information and Coding Theory, and is composed of five chapters; Chaps. 11–15. Chapter 11 presents a tutorial exposition of Low Density Parity Check codes. Chapter 12 focuses on some selected topics in the theory of variable length codes, including connections with codes for constrained channels and sources. Chapter 13 deals with decoding techniques and methods for finding the minimum distance of linear codes by means of Gröbner bases. Chapter 14 presents an overview of cooperative diversity, along with latest advances and open issues in this evolving field. In Chap. 15, algebraic coding theory is used as an alternative way to define secure cryptographic primitives.

We list below some of the important features of this book, which, we believe, would make it a valuable resource for our readers:

- This book is designed, in structure and content, to aid the learning process with the intention of making the book useful at all learning levels.
- Most of the chapters of the book are authored by prominent academicians/researchers, practitioners, in Information and Coding Theory that have been working with these topics for quite a few years now and have thorough understanding of the concepts.
- The authors of this book are distributed in a large number of countries and most of them are affiliated with institutions of worldwide repute. This gives this book an international flavor.
- Most of the chapters in this book have a distinct section providing direction for future research, which, particularly, targets researchers working in these areas. We believe that this section should provide insight to the researchers about some of the current research issues.
- The authors of each chapter have attempted to provide a comprehensive bibliography, which should greatly help the readers interested further to dig into the topics.
- Most of the chapters of this book have a separate section outlining thoughts for practitioners. We believe that this section in every chapter will be particularly useful for industry practitioners working directly with the practical aspects behind enabling these technologies in the field.
- All chapters, except one, provide a set of questions at the end that can help in assessing the understanding of the readers. In most chapters, solutions are provided to some of these questions.

- To make the book useful for pedagogical purposes, all chapters of the book have a corresponding set of presentation slides. The slides can be obtained as a supplementary resource by contacting the publisher, World Scientific, Singapore.

We have made attempts in all possible ways we could to make the different chapters of the book look as much coherent and synchronized as possible. However, it cannot be denied that due to the fact chapters were written by different authors, it was not fully possible to fully achieve this task. We believe that this is a limitation of most edited books of this sort.

Target Audience

The book is written by primarily targeting the student community. This includes the students of all levels — those getting introduced to these areas, those having an intermediate level of knowledge of the topics, and those who are already knowledgeable about many of the topics. To keep up with this goal, we have attempted to design the overall structure and content of the book in such a manner that makes it useful at all learning levels. To aid in the learning process, almost all chapters have a *set of questions* at the end of the chapter. Also, in order that teachers can use this book for classroom teaching, the book also comes with *presentation slides* and *sample solutions* to exercise questions, which are available as supplementary resources.

The secondary audience for this book is the research community, whether they are working in the academia or in the industry. To meet the specific needs to this audience group, certain chapters of the book provide directions for future research.

Finally, we have also taken into consideration the needs to those readers, typically from the industries, who have quest for getting insight into the practical significance of the topics, i.e. how the spectrum of knowledge and the ideas are relevant for real-life applications of coding and information theory.

Supplementary Resources

As mentioned earlier, the book comes with *presentation slides* for each chapter, which can be used for classroom instruction by teachers.

Teachers can contact the publisher, World Scientific, Singapore, to get access to these resources.

Acknowledgments

We are extremely thankful to the 25 authors of the 15 chapters of this book, who have worked very hard to bring this unique resource forward for help of the student, researcher, and practitioner community. The authors were very much interactive at all stages of preparation of the book from initial development of concept to finalization. We feel it is contextual to mention that as the individual chapters of this book are written by different authors, the responsibility of the contents of each of the chapters lies with the concerned authors.

We are also very much thankful to our colleagues in the World Scientific publishing and marketing teams, in particular, Ms. Kimberly Chua, Ms. Chelsea Chin, and Ms. Mei Kian, who tirelessly worked with us and guided us in the publication process. Special thanks also go to them for taking special interest in publishing this book, considering the current worldwide market needs for such a book.

Finally, we would like to thank our parents, Mr. J. Sime, Ms. C. Seupa, Prof. J.C. Misra, Ms. Shorasi Misra, our wives Clarisse, Satamita, and Sulagna, and our children Clyde, Lenny, Kylian, Babai, and Tultuli, for the continuous support and encouragement they offered during the course of this project.

Dr. Isaac Woungang
Toronto, ON, Canada

Dr. Subhas Chandra Misra
Kanpur, UP, India

Dr. Sudip Misra
Kharagpur, WB, India

CONTENTS

CONTRIBUTORS

Nuh Aydin
Department of Mathematics
Kenyon College, Gambier, OH 43022, USA
aydinn@kenyon.edu

Tsvetan Asamov
Department of Mathematics
Kenyon College, Gambier, OH 43022, USA
asamovt@kenyon.edu

Marie-Pierre Béal
Institut Gaspard-Monge (IGM)
Université Paris-Est
77454 Marne-la-Vallée Cedex 2, Paris, France
beal@univ-mlv.fr

Jean Berstel
Institut Gaspard-Monge (IGM), Université Paris-Est
77454 Marne-la-Vallée Cedex 2, Paris, France
berstel@univ-mlv.fr

Dominique Perrin
Institut Gaspard-Monge (IGM)
Université Paris-Est
77454 Marne-la-Vallée Cedex 2, Paris, France
dominique.perrin@esiee.fr

Brian H. Marcus
University of British Columbia
B.C., Vancouver, Canada
marcus@math.ubc.ca

Christophe Reutenauer
LaCIM, Université du Québec à Montréal
Montréal, Canada
reutenauer.christophe@uqam.ca

Paul H. Siegel
Department of Electrical and Computer Engineering
University of California at San Diego, San Diego, USA
psiegel@ucsd.edu

Boris Ryabko
Siberian State University of Telecommunications
    and Informatics and Institute of Computational Technology
    of Siberian Branch of Russian Academy of Science, Russia
boris@ryabko.net

Stanislav Bulygin
Department of Mathematics
University of Kaiserslautern
P.O. Box 3049, 67653 Kaiserslautern, Germany
bulygin@mathematik.uni-kl.de

Ruud Pellikaan
Department of Mathematics and Computing Science
Eindhoven University of Technology, P.O. Box 513
NL-5600 MB, Eindhoven, The Netherlands
g.r.pellikaan@tue.nl

Michael Huber
Institut für Mathematik, Technische Universität Berlin,
Straße des 17. Juni 136, D-10623, Berlin, Germany
mhuber@math.TU-Berlin.DE

Murat Uysal
Department of Electrical and Computer Engineering
University of Waterloo, Waterloo, Ontario, Canada, N2L3G1
muysal@ece.uwaterloo.ca

Muhammad Mehboob Fareed
Department of Electrical and Computer Engineering
University of Waterloo, Waterloo, Ontario, Canada
mmfareed@ece.uwaterloo.ca

Angela I. Barbero
Department of Applied Mathematics
University of Valladolid
47011 Valladolid, Spain
angbar@wmatem.eis.uva.es

Oyvind Ytrehus
Department of Informatics
University of Bergen, N-5020 Bergen, Norway
oyvind@ii.uib.no

G. Susinder Rajan
Department of Electrical Communication Engineering
Indian Institute of Science, Bangalore 560012, India
susinder@ece.iisc.ernet.in

B. Sundar Rajan
Department of Electrical Communication Engineering
Indian Institute of Science, Bangalore 560012, India
bsrajan@ece.iisc.ernet.in

Mohammd-Reza Sadeghi
Faculty of Mathematics and Computer Science
Amirkabir University of Technology Hafez Ave.
Tehran, Iran
msadeghi@aut.ac.ir

Vinod Sharma
Department of Electrical Communication Engineering
Indian Institute of Science, Bangalore 560012, India
vinod@ece.iisc.ernet.in

R. Rajesh
Department of Electrical Communication Engineering
Indian Institute of Science, Bangalore 560012, India
rajesh@pal.ece.iisc.ernet.in

Paul Hurley
IBM Research
Zurich Research Laboratory, Switzerland
pah@zurich.ibm.com

Ted Hurley
Department of Mathematics
National University of Ireland
Galway, Ireland
ted.hurley@nuigalway.ie

Pascal Véron
Institut de Mathématiques de Toulon
Université du Sud Toulon-Var
Toulon, France
veron@univ-tln.fr

Xudong Ma
Department of Electrical and Computer Engineering
University of Waterloo, Waterloo
Ontario N2L 3G1, Canada
x3ma@bbcr.uwaterloo.cay

Part I

APPLICATIONS OF CODING THEORY TO
COMPUTATIONAL COMPLEXITY

Chapter 1

LINEAR COMPLEXITY AND RELATED COMPLEXITY MEASURES

ARNE WINTERHOF

Johann Radon Institute for Computational and Applied Mathematics,
Austrian Academy of Sciences,
Altenbergerstr. 69, 4040 Linz, Austria
arne.winterhof@oeaw.ac.at

The linear complexity of a sequence is not only a measure for the unpredictability and thus suitability for cryptography but also of interest in information theory because of its close relation to the Kolmogorov complexity. However, in contrast to the Kolmogorov complexity the linear complexity is computable and so of practical significance.

It is also linked to coding theory. On the one hand, the linear complexity of a sequence can be estimated in terms of its correlation and there are strong ties between low correlation sequence design and the theory of error-correcting codes. On the other hand, the linear complexity can be calculated with the Berlekamp–Massey algorithm which was initially introduced for decoding BCH-codes.

This chapter surveys several mainly number theoretic methods for the theoretical analysis of the linear complexity and related complexity measures and describes several classes of particularly interesting sequences with high linear complexity.

1.1. Introduction

A sequence (s_n) of elements of the finite field $\mathbb{F}_q$ of q elements is called a *(homogeneous) linear recurring sequence of order k* if there exist $c_0, c_1, \ldots, c_{k-1}$ in $\mathbb{F}_q$, satisfying the *linear recurrence of order k over $\mathbb{F}_q$*;

$$s_{n+k} = c_{k-1}s_{n+k-1} + c_{k-2}s_{n+k-2} + \cdots + c_0 s_n, \quad n = 0, 1, \ldots. \qquad (1.1)$$

Now let (s_n) be a sequence over $\mathbb{F}_q$. One can associate to it a non-decreasing sequence $L(s_n, N)$ of non-negative integers as follows: The *linear complexity profile* of a sequence (s_n) over $\mathbb{F}_q$ is the sequence $L(s_n, N)$,

$N \geq 1$, where its Nth term is defined to be the smallest L such that a linear recurrence of order L over $\mathbb{F}_q$ can generate the first N terms of (s_n). We use the convention that $L(s_n, N) = 0$ if the first N elements of (s_n) are all zero and $L(s_n, N) = N$ if the first $N - 1$ elements of (s_n) are zero and $s_{N-1} \neq 0$. The value

$$L(s_n) = \sup_{N \geq 1} L(s_n, N),$$

is called the *linear complexity over* $\mathbb{F}_q$ of the sequence (s_n). For the linear complexity of any periodic sequence of period t one can easily verify that

$$L(s_n) = L(s_n, 2t) \leq t.$$

Linear complexity and linear complexity profile of a given sequence (as well as the linear recurrence defining it) can be determined by using the well-known Berlekamp–Massey algorithm, see Sec. 1.3. The algorithm is efficient for sequences with low linear complexity and hence such sequences can easily be predicted. One typical example is the so-called "linear generator"

$$s_{n+1} = as_n + b, \tag{1.2}$$

for $a, b \in \mathbb{F}_q, a \neq 0$, and some initial value $s_0 \in \mathbb{F}_q$, which satisfies $L(s_n) \leq 2$.

The expected values of linear complexity and linear complexity profile show that a "random" sequence should have $L(s_n, N)$ close to $\min\{N/2, t\}$ for all $N \geq 1$, see Sec. 1.4.

Two types of problems concerning linear complexity and linear complexity profile are of interest. One would like to construct sequences with high linear complexity (and possibly with other favorable properties). We illustrate such constructions. One would also like to find lower bounds for widely used sequences in order to judge whether it is reasonable to use them for cryptographic purposes. We present lower bounds and exact values of the linear complexity (profile) of many interesting sequences in Sec. 1.5.

Several other quality measures for sequences in view of different applications are closely related to linear complexity including the k-error linear complexity and the correlation measure of order k. We give an overview on these measures and their relations to linear complexity in Sec. 1.6.

1.2. Background

The *Kolmogorov complexity* is the central topic in *algorithmic information theory*. The Kolmogorov complexity of a binary sequence is, roughly speaking, the length of the shortest computer program that generates the sequence. The relationship between linear complexity and Kolmogorov complexity was studied in [5, 69]. Kolmogorov complexity and linear complexity are the same for almost all sequences over $\mathbb{F}_2$ of sufficiently (but only moderately) large length. In contrast to the linear complexity, the Kolmogorov complexity is in general not computable and so of no practical significance. The linear complexity (profile) is not only of theoretical interest, but can also be obtained algorithmically with the Berlekamp–Massey algorithm, see Sec. 1.3, which was initially introduced for decoding BCH-codes.

Mainly, the linear complexity (profile) is an important cryptographic characteristic of sequences (see the monographs and survey [9, 49, 50, 52, 56, 66]). A low linear complexity profile has turned out to be undesirable for cryptographical applications as stream ciphers.

Example 1.1 (Stream Cipher). We consider a message $m_0, m_1, \ldots$ represented as a sequence over $\mathbb{F}_q$. In a stream cipher each message symbol m_j is enciphered with an element x_j of another sequence $x_0, x_1, \ldots$ over $\mathbb{F}_q$, the key stream, by

$$c_j = m_j + x_j.$$

The cipher text $c_0, c_1, \ldots$ can be deciphered by subtracting the key stream

$$m_j = c_j - x_j.$$

The security of such a stream cipher depends on the unpredictability of the key stream. Since a sequence of small linear complexity is highly predictable, a high linear complexity of the sequence (x_n) is necessary (but not sufficient).

Sequences with low linear complexity are shown to be unsuitable for some applications using quasi-Monte Carlo methods as well (see [51–53, 66]). The following example describes a typical quasi-Monte-Carlo application.

Example 1.2 (Quasi-Monte-Carlo Calculation of π).

(1) Choose N pairs of a sequence (x_n) in $[0, 1)$

$$(x_n, x_{n+1}) \in [0, 1)^2, \quad n = 0, \ldots, N - 1.$$

(2) Count the number K of pairs (x_n, x_{n+1}) in the unit circle.

(3) Approximate π by $\frac{4K}{N}$.

Note that sequences in $[0, 1)$ can easily be derived from sequences over $\mathbb{F}_q$, see Sec. 1.7.

In [8] the linear complexity profile of a given binary sequence is estimated in terms of its *correlation measure of order k* which was introduced by Mauduit and Sárközy [38] and is closely related to its *autocorrelation*, see Sec. 1.6.4. There are strong ties between low correlation sequence design and the theory of error-correcting codes, see [27]. Good error-correcting codes very often correspond to a sequence with low correlation (and large linear complexity).

1.3. The Berlekamp–Massey Algorithm

The proof of the following Theorem contains the Berlekamp–Massey algorithm (see [4, 35] and also e.g. [30]).

Theorem 1.1. *If $L(s_n, N) > N/2$, then we have*

$$L(s_n, N + 1) = L(s_n, N).$$

If $L(s_n, N) \leq N/2$, then we have either

$$L(s_n, N + 1) = L(s_n, N),$$

or

$$L(s_n, N + 1) = N + 1 - L(s_n, N).$$

Proof. Put $L := L(s_n, N)$. Then there are $c_0, \ldots, c_{L-1} \in \mathbb{F}_q$ with

$$s_{n+L} = c_{L-1}s_{n+L-1} + \cdots + c_0 s_n, \quad 0 \leq n \leq N - L - 1.$$

If the same recurrence holds for $n = N - L$ as well then we have $L(s_n, N + 1) = L(s_n, N)$. Otherwise put

$$\lambda := s_N - c_{L-1}s_{N-1} - \cdots - c_0 s_{N-L} \neq 0.$$

Put $a_n := s_n$ for $n = 0, \ldots, N - 1$ and $a_N := s_N - \lambda$, so we have

$$N + 1 = L(s_n - a_n, N + 1) \leq L(s_n, N + 1) + L(-a_n, N + 1)$$
$$= L(s_n, N + 1) + L(a_n, N + 1) = L(s_n, N + 1) + L(s_n, N).$$

Hence we have

$$L(s_n, N+1) \geq \max(L(s_n, N), \quad N+1 - L(s_n, N)).$$

The equality is proven by induction. For $N = 1$, the equality is obvious and we assume $N > 1$.

If $L(s_n, N) = L(s_n, N-1) = \cdots = L(s_n, 1) = 0$, then we have $s_n = 0$ for $0 \leq n \leq N-1$. Since $s_N \neq 0$ we get $L(s_n, N+1) = N+1$.

If $L(s_n, N) = L(s_n, N-1) = \cdots = L(s_n, 1) = 1$, then is $s_{n+N} = s_N s_0^{-1} s_n$ the desired linear recurrence.

We may assume that there is $1 \leq M \leq N-1$ with

$$L(s_n, N) = L(s_n, N-1) = \cdots = L(s_n, M+1) > L(s_n, M).$$

By induction we have $L(s_n, M) = M + 1 - L$. Let

$$s_{n+M+1-L} = d_{M-L} s_{n+M-L} + \cdots + d_0 s_n, \quad 0 \leq n \leq L-2,$$

and put

$$\mu := s_M - d_{M-L} s_{M-1} - \cdots - d_0 s_{L-1} \neq 0.$$

If $L > N/2$ then

$$s_{n+L} = c_{L-1} s_{n+L-1} + \cdots + c_0 s_n$$
$$+ \lambda \mu^{-1}(s_{n+M-N+L} - d_{M-L} s_{n+M-N+L-1} - \cdots - d_0 s_{n-N+2L-1}),$$
$$0 \leq n \leq N-L,$$

is a linear recurrence of order L, and if $L \leq N/2$ then

$$s_{n+N+1-L} = c_{L-1} s_{n+N-L} + \cdots + c_0 s_{n+N-2L+1}$$
$$+ \lambda \mu^{-1}(s_{n+M-L+1} - d_{M-L} s_{n+M-L} - \cdots - d_0 s_n),$$
$$0 \leq n \leq L-1,$$

is a linear recurrence of length $N + 1 - L$ for the first $N + 1$ sequence elements. $\qquad\square$

The proof is constructive and provides an algorithm for the calculation of the linear complexity profile including the corresponding linear recurrences.

Example 1.3. Consider the finite sequence $(s_0, \ldots, s_9) = (1101011101)$ over $\mathbb{F}_2$. Then we have

N	$L(s_n, N)$	
1	1	$- - -$
2	1	$s_{n+1} = s_n$
3	2	$s_{n+2} = s_{n+1} + s_n$ or $s_{n+2} = 0$
4	2	$s_{n+2} = s_{n+1} + s_n$
5	3	$s_{n+3} = s_{n+1}$ or $s_{n+3} = s_{n+2} + s_n$
6	3	$s_{n+3} = s_{n+1}$
7	4	$s_{n+4} = s_{n+1} + s_n$ or $s_{n+4} = s_{n+3} + s_n$
8	4	$s_{n+4} = s_{n+1} + s_n$
9	5	$s_{n+5} = s_{n+4} + s_{n+3} + s_{n+2} + s_{n+1} + s_n$ or $s_{n+5} = s_{n+3} + s_{n+2}$
10	5	$s_{n+5} = s_{n+4} + s_{n+3} + s_{n+2} + s_{n+1} + s_n.$

1.4. The Expected Value of the Linear Complexity Profile

In this section, we show that for a "random" sequence (s_n) the value $L(s_n, N)$ is close to $N/2$.

Lemma 1.1. *If $L(s_n, N) \leq N/2$ then there is a unique linear recurrence of shortest length for the first N sequence elements of (s_n), i.e. for $L := L(s_n, N)$ the coefficients $c_0, \ldots, c_{L-1} \in \mathbb{F}_q$ in (1.1) are uniquely defined.*

Proof. Assume we had two different linear recurrences of the form (1.1) for the first N sequence elements of (s_n) with coefficients $c_0, \ldots, c_{L-1}$, respectively, $d_0, \ldots, d_{L-1}$. Put

$$k := \max\{j \mid c_j \neq d_j\},$$

such that $0 \leq k \leq L - 1$. Comparing the right hand sides in (1.1) yields

$$(c_0 - d_0)s_n + \cdots + (c_k - d_k)s_{n+k} = 0, \quad 0 \leq n \leq N - L - 1.$$

Since $c_k - d_k \neq 0$ this is a linear recurrence of order k for the first $N - (L - k)$ sequence elements of (s_n) and thus

$$L(s_n, N - (L - k)) \leq k. \tag{1.3}$$

Hence, $L(s_n, N - (L - k)) < L(s_n, N)$ and there exists a smallest positive index $j \leq L - k$ with $L(s_n, N - (L - k) + j) > L(s_n, N - (L - k))$. Applying the second part of Theorem 1.1 gives

$$L(s_n, N - (L - k) + j) = N - (L - k) + j - L(s_n, N - (L - k)).$$

From (1.3) and $L \leq N/2$ we get

$$L(s_n, N - (L - k) + j) \geq N - L + j \geq \frac{N}{2} + j.$$

Since $N - (L-k) + j \leq N$ we have $L(s_n, N) = L \geq N/2 + j$ in contradiction to $L \leq N/2$. $\qquad \square$

Let $A(N, L)$ be the number of finite sequences $s_0, \ldots, s_{N-1} \in \mathbb{F}_q$ of Length N with $L(s_n, N) = L$.

Lemma 1.2. *We have*

$$A(N, 0) = 1$$

and

$$A(N, L) = (q - 1)q^{\min(2L-1, 2N-2L)}$$

for $1 \leq L \leq N$.

Proof. We prove the result by induction. It is trivial for $N = 1$. We assume the assertion for N and derive the formula for $N + 1$.

First we consider the case $L := L(s_n, N + 1) \leq (N + 1)/2$.

By Theorem 1.1, we can have $L(s_n, N + 1) \leq (N + 1)/2$ only if $L(s_n, N+1) = L(s_n, N)$. By Lemma 1.1, we have a unique linear recurrence to the sequence $(s_0, \ldots, s_{N-1})$ with $L(s_n, N) = L$ which can be extended for exactly one choice of $s_N \in \mathbb{F}_q$. Hence we have

$$A(N + 1, L) = A(N, L),$$

and get the result by induction.

If $L > (N + 1)/2$ then we have by Theorem 1.1

$$L = L(s_n, N+1) = L(s_n, N) = \cdots = L(s_n, L+j) = L+j - L(s_n, L+j-1),$$

and thus $L(s_n, L + j - 1) = j$ with some $0 \leq j \leq N + 1 - L$. Each sequence (s_n) with $L(s_n, L + j - 1) = j$ corresponds to $(q - 1)q^{N-L+1-j}$ sequences with $L(s_n, N+1) = L$. Summation over j and induction provide

$$A(N+1, L)$$
$$= (q-1) \sum_{j=0}^{N+1-L} q^{N-L+1-j} A(L+j-1, j)$$
$$= (q-1)q^{N-L+1} + \sum_{j=1}^{N+1-L} (q-1)^2 q^{N-L+1-j} q^{2j-1},$$

and thus the assertion.   □

This Lemma was first proven by Gustavson [24]. The main result of this section can be found in [55, Proposition 4.2] for $q = 2$ and for arbitrary q in the unpublished work [62] of Smeets.

Theorem 1.2. *The expected value for* $L(s_n, N)$ *is*

$$\frac{1}{q^N} \sum_{L=0}^{N} A(N, L)L = \begin{cases} \dfrac{N}{2} + \dfrac{q}{(q+1)^2} - q^{-N} \dfrac{N(q+1) + q}{(q+1)^2} & \text{for even } N, \\[3mm] \dfrac{N}{2} + \dfrac{q^2 + 1}{2(q+1)^2} - q^{-N} \dfrac{N(q+1) + q}{(q+1)^2} & \text{for odd } N. \end{cases}$$

Proof. By the previous Lemma, we have

$$\sum_{L=1}^{N} A(N, L)L$$
$$= (q-1) \sum_{L=1}^{N} q^{\min(2L-1, 2N-2L)} L$$
$$= (q-1) \left(\sum_{L=1}^{\lfloor N/2 \rfloor} q^{2L-1} L + \sum_{L=\lfloor N/2 \rfloor + 1}^{N} q^{2N-2L} L \right)$$
$$= (q-1) \left(\sum_{L=1}^{\lfloor N/2 \rfloor} q^{2L-1} \sum_{k=1}^{L} 1 + \sum_{L=\lfloor N/2 \rfloor + 1}^{N} q^{2N-2L} \right.$$
$$\left. \times \left(\lfloor N/2 \rfloor + \sum_{k=\lfloor N/2 \rfloor + 1}^{L} 1 \right) \right)$$

$$= (q-1) \left(\sum_{k=1}^{\lfloor N/2 \rfloor} \sum_{L=k}^{\lfloor N/2 \rfloor} q^{2L-1} + \left\lfloor \frac{N}{2} \right\rfloor \sum_{L=\lfloor N/2 \rfloor+1}^{N} q^{2N-2L} \right.$$

$$\left. + \sum_{k=\lfloor N/2 \rfloor+1}^{N} \sum_{L=k}^{N} q^{2N-2L} \right)$$

$$= \sum_{k=1}^{\lfloor N/2 \rfloor} \frac{q^{2\lfloor N/2 \rfloor+2} - q^{2k}}{q^2+q} + \left\lfloor \frac{N}{2} \right\rfloor \frac{q^{2(N-\lfloor N/2 \rfloor)} - 1}{q+1}$$

$$+ \sum_{k=\lfloor N/2 \rfloor+1}^{N} \frac{q^{2(N-k+1)} - 1}{q+1}$$

$$= \left\lfloor \frac{N}{2} \right\rfloor \frac{q}{q+1} q^{2\lfloor N/2 \rfloor} - \frac{q}{q+1} \frac{q^{2\lfloor N/2 \rfloor} - 1}{q^2-1} + \left\lfloor \frac{N}{2} \right\rfloor \frac{q^{2(N-\lfloor N/2 \rfloor)} - 1}{q+1}$$

$$+ \frac{q^{2(N-\lfloor N/2 \rfloor+1)} - q^2}{(q+1)(q^2-1)} - \frac{N - \lfloor N/2 \rfloor}{q+1},$$

which implies the assertion. □

For results on the expected value of periodic sequences see [41].

1.5. Lower Bounds for Linear Complexity and Linear Complexity Profile

In this section, we describe some methods for determining or estimating the linear complexity (profile) and present results for several interesting classes of sequences.

1.5.1. *Explicit Non-linear Pseudorandom Numbers*

It is possible to express linear complexity in connection with various invariants of the sequences at hand.

In case of a q-periodic sequence (ξ_n) over $\mathbb{F}_q$, linear complexity is related to the degree of the polynomial $g(X) \in \mathbb{F}_q[X]$ representing the sequence (ξ_n). We recall that the polynomial $g(X)$ can be uniquely determined as follows: Consider a fixed ordered basis $\{\beta_1, \ldots, \beta_r\}$ of $\mathbb{F}_q$ over $\mathbb{F}_p$, and for $n = n_1 + n_2 p + \cdots + n_r p^{r-1}$ with $0 \leq n_k < p$, $1 \leq k \leq r$, order the elements of $\mathbb{F}_q$ as

$$\zeta_n = n_1 \beta_1 + n_2 \beta_2 + \cdots + n_r \beta_r.$$

Then $g(X)$ is the polynomial which satisfies $\deg g \leq q - 1$ and

$$\xi_n = g(\zeta_n), \quad 0 \leq n \leq q - 1. \tag{1.4}$$

When $q = p$ (and $\beta_1 = 1$) these sequences are called *explicit non-linear congruential generators* and we have

$$L(\xi_n) = \deg g + 1 \tag{1.5}$$

(for a proof, see [6, Theorem 8]). For a prime power q they are named *explicit non-linear digital generators*. In general (1.5) is not valid for $r \geq 2$. Meidl and Winterhof [43] showed, however, that the following inequalities hold

$$(\deg(g) + 1 + p - q)\frac{q}{p} \leq L(\xi_n) \leq (\deg(g) + 1)\frac{p}{q} + q - p.$$

For lower bounds on the linear complexity profile of (ξ_n) see Meidl and Winterhof [44].

A similar relation is valid for t-periodic sequences over $\mathbb{F}_q$ where t divides $q - 1$. For a t-periodic sequence (ω_n) one considers the unique polynomial $f \in \mathbb{F}_q[x]$ of degree at most $t - 1$, satisfying

$$\omega_n = f(\gamma^n), \quad n \geq 0,$$

for an element $\gamma \in \mathbb{F}_q$ of order t. In this case, $L(\omega_n)$ is equal to the number of non-zero coefficients of f (see [30]). Lower bounds for the linear complexity profile in some special cases are given by Meidl and Winterhof in [45]. For a general study of sequences with arbitrary periods see Massey and Serconek [36].

The following sequences exhibit a particularly nice behavior with respect to the linear complexity profile. The *explicit inversive congruential generator* (z_n) was introduced by Eichenauer–Herrmann in [15]. The sequence (z_n) in this case is produced by the relation

$$z_n = (an + b)^{p-2}, \quad n = 0, \ldots, p - 1, \quad z_{n+p} = z_n, \quad n \geq 0, \tag{1.6}$$

with $a, b \in \mathbb{F}_p$, $a \neq 0$, and $p \geq 5$. (The name stems from the fact that $n^{p-2} = n^{-1}$, $0 \neq n \in \mathbb{F}_p$.) It is shown in [44] that

$$L(z_n, N) \geq \begin{cases} (N-1)/3, & 1 \leq N \leq (3p-7)/2, \\ N - p + 2, & (3p-5)/2 \leq N \leq 2p - 3, \\ p - 1, & N \geq 2p - 2. \end{cases} \tag{1.7}$$

We provide the proof of a slightly weaker result.

Theorem 1.3. *Let (z_n) be as in (1.6), then*

$$L(z_n, N) \geq \min\left\{\frac{N-1}{3}, \frac{p-1}{2}\right\}, \quad N \geq 1.$$

Proof. Suppose (z_n) satisfies a linear recurrence relation of length L,

$$z_{n+L} = c_{L-1}z_{n+L-1} + \cdots + c_0 z_n, \quad 0 \leq n \leq N - L - 1, \quad (1.8)$$

with $c_0, \ldots, c_{L-1} \in \mathbb{F}_p$. We may assume $L \leq p - 1$. Put

$$C_L(N) = \{n; 0 \leq n \leq \min\{N - L, p\} - 1, \quad a(n+l) + b \neq 0, \ 0 \leq l \leq L\}.$$

Note that $card\{C_L(N)\} \geq \min\{p, N - L\} - (L + 1)$.

For $n \in C_L(N)$, the recurrence (1.8) is equivalent to

$$(a(n+L) + b)^{-1} = c_{L-1}(a(n+L-1)+b)^{-1} + \cdots + c_0(an+b)^{-1}.$$

Multiplication with

$$\prod_{j=0}^{L}(a(n+j)+b),$$

yields

$$\prod_{j=0}^{L-1}(a(n+j)+b) = \sum_{l=0}^{L-1} c_l \prod_{\substack{j=0 \\ j \neq l}}^{L}(a(n+j)+b),$$

for all $n \in C_L(N)$. Hence the polynomial

$$F(X) = -\prod_{j=0}^{L-1}(a(X+j)+b) + \sum_{l=0}^{L-1} c_l \prod_{\substack{j=0 \\ j \neq l}}^{L}(a(X+j)+b),$$

is of degree at most L and has at least $\min\{p, N - L\} - (L + 1)$ zeros. On the other hand,

$$F(-a^{-1}b - L) = -a^L \prod_{j=0}^{L-1}(j - L) \neq 0,$$

hence $F(X)$ is not the zero polynomial and we get

$$L \geq \deg(F) \geq \min\{p, N - L\} - (L + 1),$$

which implies the desired result. □

Analogs of (1.7) for *digital inversive generators*, i.e. for $r \geq 2$, are also given in [44]. For *t-periodic inversive generators*, where t is a divisor of $q-1$, see [45].

We mention one more explicit non-linear generator, namely the *quadratic exponential generator*, introduced by Gutierrez *et al.* [25]. Given an element $\vartheta \in \mathbb{F}_q^*$ we consider the sequence (q_n) where

$$q_n = \vartheta^{n^2}, \quad n = 0, 1, \ldots.$$

The lower bound

$$L(q_n, N) \geq \frac{\min\{N, t\}}{2}, \quad N \geq 1,$$

is obtained in [25]. Here the period t is at least $\tau/2$ where τ is the multiplicative order of ϑ.

1.5.2. *Recursive Non-linear Pseudorandom Numbers*

Given a polynomial $f(X) \in \mathbb{F}_p[X]$ of degree $d \geq 2$, the *non-linear congruential pseudorandom number generator* (u_n) is defined by the recurrence relation

$$u_{n+1} = f(u_n), \quad n \geq 0, \tag{1.9}$$

with some initial value $u_0 \in \mathbb{F}_p$. Obviously, the sequence (u_n) is eventually periodic with some period $t \leq p$. We assume it to be purely periodic.

The following lower bound on the linear complexity profile of a non-linear congruential generator is given in [25].

Theorem 1.4. *Let (u_n) be as in (1.9), where $f(X) \in \mathbb{F}_p[X]$ is of degree $d \geq 2$, then*

$$L(u_n, N) \geq \min\{\log_d(N - \lfloor \log_d N \rfloor), \log_d t\}, \quad N \geq 1.$$

Proof. Let us consider the following sequence of polynomials over $\mathbb{F}_p$:

$$F_0(X) = X, \quad F_i(X) = F_{i-1}(f(X)), \quad i = 1, 2, \ldots.$$

It is clear that $\deg(F_i) = d^i$ for every $i = 1, 2, \ldots$. Moreover $u_{n+j} = F_j(u_n)$ for any integers $n, j \geq 0$. Put $L = L(u_n, N)$ so that we have

$$u_{n+L} = \sum_{l=0}^{L-1} c_l u_{n+l}, \quad 0 \leq n \leq N - L - 1,$$

for some $c_0, \ldots, c_{L-1} \in \mathbb{F}_p$. Therefore, the polynomial

$$F(X) = -F_L(X) + \sum_{l=0}^{L-1} c_l F_l(X),$$

is of degree d^L and has at least $\min\{N - L, t\}$ zeros. Thus, $d^L \geq \min\{N - L, t\}$. Since otherwise the result is trivial, we may suppose $L \leq \lfloor \log_d N \rfloor$ and get $d^L \geq \min\{N - \lfloor \log_d N \rfloor, t\}$, which yields the assertion. $\qquad\square$

For some special classes of polynomials much better results are available, see [23, 25, 58]. For instance, in case of the largest possible period $t = p$ we have

$$L(u_n, N) \geq \min\{N - p + 1, p/d\}, \quad N \geq 1.$$

The *inversive (congruential) generator* (y_n) defined by

$$y_{n+1} = ay_n^{p-2} + b = \begin{cases} ay_n^{-1} + b & \text{if } y_n \neq 0, \\ b & \text{otherwise,} \end{cases} \quad n \geq 0, \qquad (1.10)$$

with $a, b, y_0 \in \mathbb{F}_p$, $a \neq 0$, has linear complexity profile

$$L(y_n, N) \geq \min\left\{ \frac{N-1}{3}, \frac{t-1}{2} \right\}, \quad N \geq 1. \qquad (1.11)$$

This sequence, introduced by Eichenauer and Lehn [14], has succeeded in drawing significant attention due to some of its enchanting properties. In terms of the linear complexity profile, the lower bound (1.11) shows that the inversive generator is almost optimal. The sequence (y_n) attains the largest possible period $t = p$ if, for instance, $X^2 - aX - b$ is a primitive polynomial over $\mathbb{F}_p$. See Flahive and Niederreiter [17] for a refinement of this result.

The *power generator* (p_n), defined as

$$p_{n+1} = p_n^e, \quad n \geq 0,$$

with some integer $e \geq 2$ and initial value $0 \neq p_0 \in \mathbb{F}_p$ satisfies

$$L(p_n, N) \geq \min\left\{ \frac{N^2}{4(p-1)}, \frac{t^2}{p-1} \right\}, \quad N \geq 1.$$

Results about the period length of (p_n) can be found in Friedlander *et al.* [19, 20].

The family of *Dickson polynomials* $D_e(X, a) \in \mathbb{F}_p[X]$ is defined by the following recurrence relation

$$D_e(X, a) = X D_{e-1}(X, a) - a D_{e-2}(X, a), \quad e = 2, 3, \ldots,$$

with initial values $D_0(X, a) = 2$, $D_1(X, a) = X$, where $a \in \mathbb{F}_p$. Obviously, the degree of D_e is e. It is easy to see that $D_e(X, 0) = X^e$, $e \geq 2$, which corresponds to the case of the power generator. In the special case that $a = 1$ the lower bound

$$L(u_n, N) \geq \frac{\min\{N^2, 4t^2\}}{16(p+1)} - (p+1)^{1/2}, \quad N \geq 1,$$

for a new class of non-linear congruential generators where $f(X) = D_e(X, 1)$ is proven by Aly and Winterhof [1]. Here the period t is a divisor of $p - 1$ or $p + 1$.

Another class of non-linear congruential pseudorandom number generators, where $f(X)$ is a Rédei function, is analyzed by Meidl and Winterhof [48]. Suppose that

$$r(X) = X^2 - \alpha X - \beta \in \mathbb{F}_p[X],$$

is an irreducible quadratic polynomial with the two different roots ξ and $\zeta = \xi^p$ in $\mathbb{F}_{p^2}$. We consider the polynomials $g_e(X)$ and $h_e(X) \in \mathbb{F}_p[X]$, which are uniquely defined by the equation

$$(X + \xi)^e = g_e(X) + h_e(X)\xi.$$

The *Rédei function* $f_e(X)$ of degree e is then given by

$$f_e(X) = \frac{g_e(X)}{h_e(X)}.$$

The Rédei function $f_e(X)$ is a permutation of $\mathbb{F}_p$ if and only if $\gcd(e, p + 1) = 1$, see Nöbauer [54]. For further background on Rédei functions we refer to [34, 54]. We consider generators (r_n) defined by

$$r_{n+1} = f_e(r_n), \quad n \geq 0,$$

with a Rédei permutation $f_e(X)$ and some initial element $u_0 \in \mathbb{F}_p$. The sequence (r_n) is periodic with period t, where t is a divisor of $\varphi(p + 1)$. As any mapping over $\mathbb{F}_p$, the Rédei permutation can be uniquely represented by a polynomial of degree at most $p - 1$ and, therefore, the sequence (r_n) belongs to the class of non-linear congruential pseudorandom number generators (1.9). In [48] the following lower bound on the linear complexity profile of the sequence (r_n) is obtained

$$L(r_n, N) \geq \frac{\min\{N^2, 4t^2\}}{20(p+1)^{3/2}}, \quad N \geq 2,$$

provided that $t \geq 2$.

The linear complexity profile of pseudorandom number generators over $\mathbb{F}_p$, defined by a recurrence relation of order $m \geq 1$ is studied in Topuzoğlu and Winterhof [65];

$$u_{n+1} = f(u_n, u_{n-1}, \ldots, u_{n-m+1}), \quad n = m-1, m, \ldots. \tag{1.12}$$

Here initial values $u_0, \ldots, u_{m-1}$ are in $\mathbb{F}_p$ and $f \in \mathbb{F}_p(X_1, \ldots, X_m)$ is a rational function in m variables over $\mathbb{F}_p$. The sequence (1.12) eventually becomes periodic with least period $t \leq p^m$. The fact that t can actually attain the value p^m gains non-linear generators of higher orders a particular interest. In case of a polynomial f, lower bounds for the linear complexity and linear complexity profile of higher order generators are given in [65].

A particular rational function f in (1.12) gives rise to a generalization of the inversive generator (1.10), as described below. Let (x_n) be the sequence over $\mathbb{F}_p$, defined by the linear recurring sequence of order $m + 1$;

$$x_{n+1} = a_0 x_n + a_1 x_{n-1} + \cdots + a_m x_{n-m}, \quad n \geq m,$$

with $a_0, a_1, \ldots, a_m \in \mathbb{F}_p$ and initial values $x_0, \ldots, x_m \in \mathbb{F}_p$. An increasing function $N(n)$ is defined by

$$N(0) = \min\{n \geq 0 : x_n \neq 0\},$$
$$N(n) = \min\{l \geq N(n-1) + 1 : x_l \neq 0\},$$

and the non-linear generator (z_n) is produced by

$$z_n = x_{N(n)+1} x_{N(n)}^{-1}, \quad n \geq 0$$

(see Eichenauer *et al.* [13]). It is easy to see that (z_n) satisfies

$$z_{n+1} = f(z_n, \ldots, z_{n-m+1}), \quad n \geq m-1,$$

whenever $z_n \cdots z_{n-m+1} \neq 0$ for the rational function

$$f(X_1, \ldots, X_m) = a_0 + a_1 X_1^{-1} + \cdots + a_m X_1^{-1} X_2^{-1} \cdots X_m^{-1}.$$

A sufficient condition for (z_n) to attain the maximal period length p^m is given in [13]. It is shown in [65] that the linear complexity profile $L(z_n, N)$ of (z_n) with the least period p^m satisfies

$$L(z_n, N) \geq \min\left(\left\lceil \frac{p-m}{m+1} \right\rceil p^{m-1} + 1, N - p^m + 1\right), \quad N \geq 1.$$

This result is in accordance with (1.11), i.e. the case $m = 1$.

1.5.3. *Legendre Sequence and Related Bit Sequences*

Let $p > 2$ be a prime. The *Legendre-sequence* (l_n) is defined by

$$l_n = \begin{cases} 1, & \left(\frac{n}{p}\right) = -1, \\ 0 & \text{otherwise}, \end{cases} \quad n \geq 0,$$

where $\left(\frac{\cdot}{p}\right)$ is the Legendre-symbol. Obviously, (l_n) is p-periodic. Results on the linear complexity of (l_n) can be found in [9, 67]. We give the proof here since the method is illustrative.

Theorem 1.5. *The linear complexity of the Legendre sequence is*

$$L(l_n) = \begin{cases} (p-1)/2, & p \equiv 1 \bmod 8, \\ p, & p \equiv 3 \bmod 8, \\ p-1, & p \equiv 5 \bmod 8, \\ (p+1)/2, & p \equiv 7 \bmod 8. \end{cases}$$

Proof. We start with the well-known relation

$$L(l_n) = p - \deg(\gcd(S(X), X^p - 1)),$$

where

$$S(X) = \sum_{n=0}^{p-1} l_n X^n,$$

(see, for example, [59, Lemma 8.2.1]), i.e. in order to determine the linear complexity it is sufficient to count the number of common zeros of $S(X)$ and $X^p - 1$ in the splitting field $\mathbb{F}$ of $X^p - 1$ over $\mathbb{F}_2$. Let $1 \neq \beta \in \mathbb{F}$ be a root of $X^p - 1$. For q with $\left(\frac{q}{p}\right) = 1$ we have

$$S(\beta^q) = \sum_{n=0}^{p-1} l_n \beta^{nq} = \sum_{\left(\frac{n}{p}\right)=-1} \beta^{nq} = \sum_{\left(\frac{n}{p}\right)=-1} \beta^n = S(\beta),$$

and for m with $\left(\frac{m}{p}\right) = -1$,

$$S(\beta^m) = \sum_{\left(\frac{n}{p}\right)=-1} \beta^{nm} = \sum_{\left(\frac{n}{p}\right)=1} \beta^n$$

$$= \sum_{n=1}^{p-1}(1+l_n)\beta^n = \frac{\beta^p - \beta}{\beta - 1} + S(\beta) = 1 + S(\beta).$$

Moreover, we have $S(\beta) \in \mathbb{F}_2$ if and only if $S(\beta)^2 = S(\beta^2) = S(\beta)$, i.e. $\left(\frac{2}{p}\right) = 1$ which is equivalent to $p \equiv \pm 1 \bmod 8$. Next we have

$$S(1) = \sum_{\left(\frac{n}{p}\right)=-1} 1 = \frac{p-1}{2} = \begin{cases} 0 & \text{if } p \equiv 1 \bmod 4, \\ 1 & \text{if } p \equiv 3 \bmod 4. \end{cases}$$

Let Q and N denote the sets of quadratic residues and non-residues modulo p, respectively. If $p \equiv \pm 1 \bmod 8$, then we have one of the following two cases: Either $S(\beta^q) = S(\beta^m) + 1 = 0$ for all $q \in Q$ and $m \in N$, or $S(\beta^m) = S(\beta^q) + 1 = 0$ for all $q \in Q$ and $m \in N$. Now the assertion is clear since $|Q| = |N| = (p-1)/2$. $\qquad\square$

The profile can be estimated using bounds on incomplete sums of Legendre symbols (cf. [59, Theorem 9.2]).

Theorem 1.6. *The linear complexity profile of the Legendre sequence satisfies*

$$L(l_n, N) > \frac{\min\{N, p\}}{1 + p^{1/2}(1 + \log p)} - 1, \quad N \geq 1.$$

Proof. Since $L(l_n, N) \geq L(l_n, p)$ for $N > p$ we may assume $N \leq p$. As usual, put $L = L(l_n, N)$ so that

$$l_{n+L} = c_{L-1}l_{n+L-1} + \cdots + c_0 l_n, \quad 0 \leq n \leq N - L - 1,$$

for some $c_0, \ldots, c_{L-1} \in \mathbb{F}_2$. Since $(-1)^{l_n} = \left(\frac{n}{p}\right)$, $1 \leq n \leq p-1$, with $c_L = 1$ we have

$$1 = (-1)^{\sum_{j=0}^{L} c_j l_{n+j}} = \left(\frac{\prod_{j=0}^{L}(n+j)^{c_j}}{p}\right), \quad 1 \leq n \leq N - L - 1,$$

and thus

$$N - L - 1 = \sum_{n=1}^{N-L-1} \left(\frac{\prod_{j=0}^{L}(n+j)^{c_j}}{p} \right).$$

The following bound for the right hand side of this equation

$$\left| \sum_{n=1}^{N-L-1} \left(\frac{\prod_{j=0}^{L}(n+j)^{c_j}}{p} \right) \right| < (L+1)p^{1/2}(1 + \log p), \qquad (1.13)$$

yields

$$N - (L+1) < (L+1)p^{1/2}(1 + \log p),$$

from which the assertion follows. The bound (1.13) can be proved as follows: For an integer $k \geq 2$ put $e_k(x) = \exp(2\pi i x/k)$. The relations below can be found in [68];

$$\sum_{a=0}^{k-1} e_k(au) = \begin{cases} 0, & u \not\equiv 0 \bmod k, \\ k, & u \equiv 0 \bmod k, \end{cases} \qquad (1.14)$$

$$\sum_{a=1}^{k-1} \left| \sum_{x=0}^{K-1} e_k(ax) \right| \leq k \log k, \quad 1 \leq K \leq k. \qquad (1.15)$$

The Weil bound, which we present in the following form (see [57, Theorems 2C and 2G]),

$$\left| \sum_{a=0}^{p-1} \chi(f(a))e_p(ax) \right| \leq \begin{cases} p^{1/2} \deg f, & 1 \leq x < p, \\ p^{1/2}(\deg f - 1), & x = 0, \end{cases} \qquad (1.16)$$

where χ denotes a non-trivial multiplicative character of $\mathbb{F}_p$ and $f \in \mathbb{F}_p[X]$ enables us to handle the complete hybrid character sum below. Application of Vinogradov's method (see [64]) with (1.14) and

$$f(X) = \prod_{j=0}^{L}(X + j)^{c_j},$$

gives

$$\left| \sum_{n=1}^{N-L-1} \left(\frac{f(n)}{p} \right) \right| = \frac{1}{p} \left| \sum_{x \in \mathbb{F}_p} \sum_{m \in \mathbb{F}_p} \left(\frac{f(m)}{p} \right) \sum_{n=1}^{N-L-1} e_p(x(n-m)) \right|$$

$$\leq \frac{1}{p} \sum_{x \in \mathbb{F}_p} \left| \sum_{m \in \mathbb{F}_p} \left(\frac{f(m) e_p(-xm)}{p} \right) \right| \left| \sum_{n=1}^{N-L-1} e_p(xn)) \right|$$

$$< (L+1) p^{1/2} (1 + \log p),$$

where we used that f is not a square (since $c_L = 1$) to apply (1.16) in the case $x = 0$. $\square$

For similar sequences, that are defined by the use of the quadratic character of arbitrary finite fields and the study of their linear complexity profiles, see [33, 42, 70].

Let γ be a primitive element and η be the quadratic character of the finite field $\mathbb{F}_q$ of odd characteristic. The *Sidelnikov sequence* (σ_n) is defined by

$$\sigma_n = \begin{cases} 1 & \text{if } \eta(\gamma^n + 1) = -1, \\ 0 & \text{otherwise,} \end{cases} \qquad n \geq 0.$$

In many cases one is able to determine the linear complexity $L(\sigma_n)$ over $\mathbb{F}_2$ exactly, see Meidl and Winterhof [47]. For example, if $(q-1)/2$ is an odd prime such that 2 is a primitive root modulo $(q-1)/2$, then (s_n) attains the largest possible linear complexity $L(\sigma_n) = q-1$. Moreover we have the lower bound, see [47],

$$L(\sigma_n, N) = \Omega \left(\frac{\min\{N, q\}}{q^{1/2} \log q} \right), \qquad N \geq 1,$$

where $f(n) = \Omega(g(n))$ means that $f(n) \geq cg(n)$ for all sufficiently large n and some constant $c > 0$. The linear complexity over $\mathbb{F}_p$ of this sequence has been estimated in Garaev *et al.* [22] by using bounds of character sums with middle binomial coefficients. For small values of p, the linear complexity can be evaluated explicitly.

Let p and q be two distinct odd primes. Put

$$Q = \{q, 2q, \ldots, (p-1)q\}, \quad Q_0 = Q \cup \{0\},$$

and

$$P = \{p, 2p, \ldots, (q-1)p\}.$$

The pq-periodic sequence (t_n) over $\mathbb{F}_2$, defined by

$$t_n = \begin{cases} 0 & \text{if } (n \bmod pq) \in Q_0, \\ 1 & \text{if } (n \bmod pq) \in P, \\ \left(1 - \left(\frac{n}{p}\right)\left(\frac{n}{q}\right)\right)/2 & \text{otherwise} \end{cases}$$

is called the *two-prime generator* (or *generalized cyclotomic sequence of order* 2) (see [7, 9]; Chapter 8.2). Under the restriction $\gcd(p-1, q-1) = 2$ it satisfies

$$L(t_n) = \begin{cases} pq - 1, & p \equiv 1 \bmod 8 \text{ and } q \equiv 3 \bmod 8 \\ & \text{or } p \equiv 5 \bmod 8 \text{ and } q \equiv 7 \bmod 8, \\ (p-1)q, & p \equiv 7 \bmod 8 \text{ and } q \equiv 3 \bmod 8 \\ & \text{or } p \equiv 3 \bmod 8 \text{ and } q \equiv 7 \bmod 8, \\ pq - p - q + 1, & p \equiv 7 \bmod 8 \text{ and } q \equiv 5 \bmod 8 \\ & \text{or } p \equiv 3 \bmod 8 \text{ and } q \equiv 1 \bmod 8, \\ (pq + p + q - 3)/2, & p \equiv 1 \bmod 8 \text{ and } q \equiv 7 \bmod 8 \\ & \text{or } p \equiv 5 \bmod 8 \text{ and } q \equiv 3 \bmod 8, \\ (p-1)(q-1)/2, & p \equiv 7 \bmod 8 \text{ and } q \equiv 1 \bmod 8 \\ & \text{or } p \equiv 3 \bmod 8 \text{ and } q \equiv 5 \bmod 8, \\ (p-1)(q+1)/2, & p \equiv 7 \bmod 8 \text{ and } q \equiv 7 \bmod 8 \\ & \text{or } p \equiv 3 \bmod 8 \text{ and } q \equiv 3 \bmod 8. \end{cases}$$

In the most important case when $|p - q|$ is small we have a lower bound on the linear complexity profile of order of magnitude

$$O(N^{1/2}(pq)^{-1/4}\log^{-1/2}(pq)),$$

for $2 \leq N < pq$, where $f(n) = O(g(n))$ is equivalent to $f(n) \leq cg(n)$ for all sufficiently large n and some constant $c > 0$.

1.5.4. *Elliptic Curve Generators*

We recall some definitions and basic facts about elliptic curves (see [32] or Chapter 5).

Let $p > 3$ be a prime and E be an elliptic curve over $\mathbb{F}_p$ of the form

$$Y^2 = X^3 + aX + b,$$

with coefficients $a, b \in \mathbb{F}_p$ such that $4a^3 + 27b^2 \neq 0$. The set $E(\mathbb{F}_p)$ of all $\mathbb{F}_p$-rational points on E forms an Abelian group where we denote addition by $\oplus$. The point O at infinity is the zero element of $E(\mathbb{F}_p)$. We recall the Hasse-Weil bound

$$|\#E(\mathbb{F}_p) - p - 1| \leq 2p^{1/2},$$

where $\#E(\mathbb{F}_p)$ is the number of $\mathbb{F}_p$-rational points, including O. For a given initial value $W_0 \in E(\mathbb{F}_p)$, a fixed point $G \in E(\mathbb{F}_p)$ of order t and a rational function $f \in \mathbb{F}_p(E)$ the *elliptic curve congruential generator* (with respect to f) is defined by $w_n = f(W_n)$, $n \geq 0$, where

$$W_n = G \oplus W_{n-1} = nG \oplus W_0, \quad n \geq 1.$$

Obviously, (w_n) is t-periodic. See [3, 28] and references therein for results on the properties of elliptic curve generators. For example, choosing the function $f(x, y) = x$, the work of Hess and Shparlinski [28] gives the following lower bound for the linear complexity profile:

$$L(w_n, N) \geq \min\{N/3, t/2\}, \quad N \geq 2.$$

Here we present an elementary proof of a slightly weaker result, see [66]. Let $x(Q)$ denote the first coordinate x of the point $Q = (x, y) \in E$.

Theorem 1.7. *Let (w_n) be the t-periodic sequence defined by*

$$w_n = x(nG), \quad 1 \leq n \leq t - 1, \tag{1.17}$$

with some $w_0 \in \mathbb{F}_p$ and $G \in E$ of order t. Then we have

$$L(w_n, N) \geq \frac{\min\{N, t/2\} - 3}{4}, \quad N \geq 2.$$

Proof. We may assume $N \leq t/2$ and $L(w_n, N) < t/2$. Put $nG = (x_n, y_n)$, $1 \leq n \leq t - 1$. Note that $x_k = x_m$ if and only if $k = m$ or $k = t - m$, $1 \leq k \leq t - 1$, and $y_k = 0$ if and only if t is even and $k = t/2$. Put $c_L = -1$

and assume that

$$\sum_{l=0}^{L} c_l w_{n+l} = 0, \quad L+1 \le n \le N-L-1,$$

or equivalently

$$\sum_{l=0}^{L} c_l w_{t-n-l} = 0, \quad L+1 \le n \le N-L-1.$$

Hence,

$$\sum_{l=0}^{L} c_l \frac{w_{n+l} + w_{t-n-l}}{2} = 0, \quad L+1 \le n \le N-L-1.$$

By the addition formulas for points on elliptic curves we have

$$x_{n+l} = \left(\frac{y_n - y_l}{x_n - x_l}\right)^2 - (x_n + x_l)$$

$$= \frac{x_l x_n^2 + (x_l^2 + a)x_n + ax_l + 2b - 2y_l y_n}{(x_n - x_l)^2}, \quad l+1 \le n \le t-l-1,$$

where we used $y_n^2 = x_n^3 + ax_n + b$. Similarly, we get

$$x_{t-n-l} = \frac{x_l x_n^2 + (x_l^2 + a)x_n + ax_l + 2b + 2y_l y_n}{(x_n - x_l)^2}, \quad l+1 \le n \le t-l-1,$$

and hence

$$\frac{x_{n+l} + x_{t-n-l}}{2} = \frac{x_l x_n^2 + (x_l^2 + a)x_n + ax_l + 2b}{(x_n - x_l)^2}, \quad l+1 \le n \le t-l-1.$$

So we get

$$\sum_{l=0}^{L} c_l \frac{x_l x_n^2 + (x_l^2 + a)x_n + ax_l + 2b}{(x_n - x_l)^2} = 0, \quad L+1 \le n \le N-L-1.$$

Clearing denominators we get

$$\sum_{l=0}^{L} c_l(x_l x_n^2 + (x_l^2 + a)x_n + ax_l + 2b) \prod_{\substack{j=0 \\ j \ne l}}^{L} (x_n - x_j)^2 = 0,$$

$$L+1 \le n \le N-L-1.$$

So the polynomial

$$F(X) = \sum_{l=0}^{L} c_l(x_l X^2 + (x_l^2 + a)X + ax_l + 2b) \prod_{\substack{j=0 \\ j \neq l}}^{L} (X - x_j)^2,$$

of degree at most $2(L+1)$ has at least $N - 2L - 1$ different zeros. Moreover, we have

$$F(x_L) = -2(x_L^3 + ax_L + b) \prod_{j=0}^{L-1} (x_L - x_j)^2 = -2y_L^2 \prod_{j=0}^{L-1} (x_L - x_j)^2 \neq 0.$$

Hence we get $2(L+1) \geq N - 2L - 1$ and the result follows. □

1.6. Related Measures

1.6.1. *Lattice Test*

In order to study the structural properties of a given periodic sequence (s_n) over $\mathbb{F}_q$, it is natural to consider the subspaces $\mathcal{L}(s_n, s)$ of $\mathbb{F}_q^s$ for $s \geq 1$, spanned by the vectors $\mathbf{s}_n - \mathbf{s}_0$, $n = 1, 2, \ldots$, where

$$\mathbf{s}_n = (s_n, s_{n+1}, \ldots, s_{n+s-1}), \quad n = 0, 1, \ldots.$$

We recall that (s_n) is said to pass the *s-dimensional lattice test* for some $s \geq 1$, if $\mathcal{L}(s_n, s) = \mathbb{F}_q^s$. It is obvious for example that the linear generator (1.2) can pass the s-dimensional lattice test at most for $s = 1$. On the other hand for $q = p$, the non-linear generator (1.4) passes the test for all $s \leq \deg g$ (see [51]). However, this test is well known to be unreliable since sequences, which pass the lattice test for large dimensions, yet having bad statistical properties are known [51].

Accordingly the notion of *lattice profile at N* is introduced by Dorfer and Winterhof [12]. For given $s \geq 1$ and $N \geq 2$, we say that (s_n) passes the *s-dimensional N-lattice test* if the subspace spanned by the vectors $\mathbf{s}_n - \mathbf{s}_0$, $1 \leq n \leq N - s$, is $\mathbb{F}_q^s$. The largest s for which (s_n) passes the s-dimensional N-lattice test is called the *lattice profile at N*, and is denoted by $S(s_n, N)$.

The lattice profile is closely related to the linear complexity profile, as the following result in [12] shows:

We have either

$$S(s_n, N) = \min\{L(s_n, N), N + 1 - L(s_n, N)\}$$

$$\text{or} \tag{1.18}$$

$$S(s_n, N) = \min\{L(s_n, N), N + 1 - L(s_n, N)\} - 1.$$

The results of Dorfer *et al.* [11] on the expected value of the lattice profile show that a "random" sequence should have $S(s_n, N)$ close to $\min\{N/2, t\}$.

1.6.2. *k-Error Linear Complexity*

We have remarked that a cryptographically strong sequence necessarily has a high linear complexity. It is also clear that the linear complexity of such a sequence should not decrease significantly when a small number of its terms are altered. The error linear complexity is introduced in connection with this observation [10, 63].

Let (s_n) be a sequence over $\mathbb{F}_q$, with period t. The *k-error linear complexity* $L_k(s_n)$ of (s_n) is defined as

$$L_k(s_n) = \min_{(y_n)} L(y_n),$$

where the minimum is taken over all t-periodic sequences (y_n) over $\mathbb{F}_q$, for which the Hamming distance of the vectors $(s_0, s_1, \ldots, s_{t-1})$ and $(y_0, y_1, \ldots, y_{t-1})$ is at most k.

One problem of interest here is to determine the minimum value k, for which $L_k(s_n) \leq L(s_n)$. This problem is tackled by Meidl [39], in case (s_n) is a bit sequence with period length p^n, where p is an odd prime and 2 is a primitive root modulo p^2. Meidl [39] also describes an algorithm to determine the k-error linear complexity that is based on an algorithm of [72]. Stronger results for p^n-periodic sequences over $\mathbb{F}_p$ have been recently obtained in Meidl [40].

In Klapper [31] an attack is discussed, where the idea is to decrease the linear complexity of a given sequence by considering it over a field which is different from the field where the sequence is naturally defined (and its high linear complexity is guaranteed). Although the result in Shparlinski and Winterhof [60] shows that this approach has very limited chance to succeed it is still important to analyze the (k-error) linear complexity of a sequence over different fields. Since Legendre sequences are constructed using properties of $\mathbb{F}_p$ it is somewhat natural to consider them not only over $\mathbb{F}_2$ but also over $\mathbb{F}_p$.

Here we give the proof of the following result on the k-error linear complexity over $\mathbb{F}_p$ of Legendre sequences, obtained by Aly and Winterhof in [2].

Theorem 1.8. *Let $L_k(l_n)$ denote the k-error linear complexity over $\mathbb{F}_p$ of the Legendre sequence (l_n). Then,*

$$L_k(l_n) = \begin{cases} p, & k = 0, \\ (p+1)/2, & 1 \le k \le (p-3)/2, \\ 0, & k \ge (p-1)/2. \end{cases}$$

Proof. Put

$$g_1(X) = \frac{1}{2}\left(X^{p-1} - X^{(p-1)/2}\right) \quad \text{and} \quad g_2(X) = \frac{1}{2}\left(1 - X^{(p-1)/2}\right).$$

Since $l_n = g_1(n)$ for $n \ge 0$ we get that the Legendre sequence (l_n) over $\mathbb{F}_p$ has linear complexity $L(l_n) = p$ by (1.5).

Consider now the p-periodic sequence (l'_n) defined by $l'_n = g_2(n)$, $n \ge 0$. Note that

$$g_1(n) = g_2(n), \quad 1 \le n \le p - 1,$$

and

$$L_k(l_n) \le L(l'_n) = \frac{p+1}{2}, \quad k \ge 1.$$

Assume now that $1 \le k \le (p-3)/2$. Let (s_n) be any sequence obtained from (l_n) by changing at most $(p-3)/2$ elements. Suppose that g is the polynomial in $\mathbb{F}_p[x]$ of degree at most $p-1$, which represents the sequence (s_n), i.e. $s_n = g(n)$, $n \ge 0$.

It is obvious that the sequences (s_n) and (l'_n) coincide for at least $p - 1 - k \ge (p+1)/2$ elements in a period. Hence, the polynomial $g(X) - g_2(X)$ has at least $(p+1)/2$ zeros, which implies that either $g(X) = g_2(X)$ or $\deg g \ge (p+1)/2$. Therefore, $L_k(l_n) = L(l'_n) = (p+1)/2$.

Finally, we remark that $L_k(l_n) = 0$ for $k \ge (p-1)/2$, since we have exactly $(p-1)/2$ non-zero elements in a period of (l_n) and the zero sequence of linear complexity 0 can be obtained by $(p-1)/2$ changes. □

Aly and Winterhof also give a lower bound for the k-error linear complexity over $\mathbb{F}_p$ of Sidelnikov sequences in the same paper,

$$L_k(\sigma_n) \ge \min\left(\left(\frac{p+1}{2}\right)^r - 1, \frac{q-1}{k+1} - \left(\frac{p+1}{2}\right)^r + 1\right).$$

For $k \ge (q-1)/2$ we have $L_k(\sigma_n) = 0$. The 1-error linear complexity over $\mathbb{F}_p$ of Sidelnikov sequences has recently be determined by Eun *et al.* in [16] to be

$$L_1(\sigma_n) = \left(\frac{p+1}{2}\right)^r - 1, \quad q > 3.$$

1.6.3. *Non-linear Complexity Profile*

We recall that the *non-linear complexity profile* $NL_m(s_n, N)$ of an infinite sequence (s_n) over $\mathbb{F}_q$ is the function, which is defined for every integer $N \geq 2$, as the smallest k such that a polynomial recurrence relation

$$s_{n+k} = \Psi(s_{n+k-1}, \ldots, s_n), \quad 0 \leq n \leq N - k - 1,$$

with a polynomial $\Psi(\lambda_1, \ldots, \lambda_k)$ over $\mathbb{F}_q$ of total degree at most m can generate the first N terms of (s_n). Note that generally speaking $NL_1(s_n, N) \neq L(s_n, N)$ because in the definition of $L(s_n, N)$ one can use only homogeneous linear polynomials. Obviously, we have

$$L(s_n, N) \geq NL_1(s_n, N) \geq NL_2(s_n, N) \geq \ldots.$$

See [25] for the presentation of results on the linear complexity profile of non-linear, inversive, and quadratic exponential generators in a more general form, namely in terms of lower bounds on the non-linear complexity profile.

1.6.4. *Autocorrelation and Related Distribution Measures for Binary Sequences*

One would expect that a periodic random sequence and a shift of it would have a low correlation. Autocorrelation measures the similarity between a sequence (s_n) of period t and its shifts by k positions, for $1 \leq k \leq t - 1$.

The *(periodic) autocorrelation* of a t-periodic binary sequence (s_n) is the function defined by

$$A(s_n, k) = \sum_{n=0}^{t-1} (-1)^{s_{n+k}+s_n}, \quad 1 \leq k \leq t - 1.$$

Obviously, a low autocorrelation is a desirable feature for pseudorandom sequences that are used in cryptographic systems. Local randomness of periodic sequences is also of importance cryptographically, since only small parts of the period are used for the generation of stream ciphers.

The *aperiodic autocorrelation* reflects local randomness and is defined by

$$\mathrm{AA}(s_n, k, u, v) = \sum_{n=u}^{v} (-1)^{s_{n+k}+s_n}, \quad 1 \le k \le t-1, \ 0 \le u < v \le p-1.$$

For the Legendre sequences, for example, $\mathrm{A}(l_n, k)$ can be immediately derived from the well-known formula, see e.g. [30]

$$\sum_{n=0}^{p-1} \left(\frac{n}{p}\right)\left(\frac{n+k}{p}\right) = -1, \quad 1 \le k \le p-1,$$

and the following bound on the aperiodic autocorrelation of Legendre sequences follows immediately from (1.13).

Theorem 1.9. *The (aperiodic) autocorrelation of the Legendre sequence satisfies*

$$\mathrm{A}(l_n, k) = \left(\frac{k}{p}\right)\left(1 + (-1)^{(p-1)/2}\right) - 1, \quad 1 \le k \le p-1,$$

$$|\mathrm{AA}(l_n, k, u, v)| \le 2p^{1/2}(1 + \log p) + 2, \quad 1 \le k \le p-1, \ 0 \le u \le v \le p-1.$$

For bounds on the aperiodic autocorrelation of extended Legendre sequences see [46]. For the aperiodic autocorrelation of Sidelnikov sequences see [61] and of the two-prime generator see [7].

In Mauduit and Sárközy [38] the *correlation measure of order k* of a binary sequence (s_n) is introduced as

$$C_k(s_n) = \max_{M,D} \left| \sum_{n=0}^{M=1} (-1)^{s_{n+d_1}} \cdots (-1)^{s_{n+d_k}} \right|, \quad k \ge 1,$$

where the maximum is taken over all $D = (d_1, d_2, \ldots, d_k)$ with non-negative integers $d_1 < d_2 < \cdots < d_k$ and M such that $M - 1 + d_k \le T - 1$. $C_2(s_n)$ is obviously bounded by the maximal absolute value of the aperiodic autocorrelation of (s_n). (We remark that some of our references deal actually with the corresponding sequences $s'_h = (-1)^{s_h}$ over $\{-1, 1\}$ with the adequate definition of the correlation measure.)

It is also shown in [38] that the Legendre sequence has small correlation measure up to rather high orders.

The following family of pseudorandom binary sequences is introduced in Gyarmati [26]. Let p be an odd prime and g be a primitive root modulo p.

Denote by ind n, the *discrete logarithm* of n to the base g, i.e. ind $n = j$ if $n = g^j$ with $1 \le j \le p - 1$. Let $f(X)$ be a polynomial of degree k modulo p. Then the finite sequence (e_n') is defined by

$$e_n' = \begin{cases} 1 & \text{if } 1 \le \text{ind } f(n) \le (p-1)/2, \\ -1 & \text{if } (p+1)/2 \le \text{ind } f(n) \le p-1 \text{ or } p \mid f(n), \end{cases} \quad 1 \le n \le p-1.$$

The correlation measure of the sequence (e_n) defined by $e_n' = (-1)^{e_n}$ is also analyzed in [26].

The sequence (k_n') of signs of Kloosterman sums is defined as follows;

$$k_n' = \begin{cases} 1 & \text{if } \displaystyle\sum_{j=1}^{p-1} \exp(2\pi i (j + nj^{-1})/p) > 0, \\ & \\ -1 & \text{if } \displaystyle\sum_{j=1}^{p-1} \exp(2\pi i (j + nj^{-1})/p) < 0, \end{cases} \quad 1 \le n \le p-1,$$

where j^{-1} is the inverse of j modulo p. Bounds on the correlation measure of order k of (k_n) defined by $k_n' = (-1)^{k_n}$ are given in Fouvry *et al.* [18].

Recently Brandstätter and Winterhof [8] have shown that the linear complexity profile of a given t-periodic sequence can be estimated in terms of its correlation measure;

$$L(s_n, N) \ge N - \max_{1 \le k \le L(s_n, N)+1} C_k(s_n), \quad 2 \le N \le t-1.$$

Hence, a lower bound on $L(s_n, N)$ can be obtained whenever an appropriate bound on $\max C_k(s_n)$ is known.

1.6.5. *Discrepancy*

Let (x_n) be a sequence in the unit interval $[0, 1)$. For $0 \le d_1 < \cdots < d_k < N$ we put

$$\mathbf{x}_n = \mathbf{x}_n(d_1, \ldots, d_k) = (x_{n+d_1}, \ldots, x_{n+d_k}), \quad 1 \le n \le N - d_k.$$

The *discrepancy* of the vectors $\mathbf{x}_1(d_1, \ldots, d_k), \ldots, \mathbf{x}_{N-d_k}(d_1, \ldots, d_k)$ is defined as

$$\sup_I \left| \frac{A(I, \mathbf{x}_1, \ldots, \mathbf{x}_{N-d_k})}{N - d_k} - V(I) \right|,$$

where the supremum is taken over all subintervals of $[0, 1)^k$, $V(I)$ is the volume of I and $A(I, \mathbf{x}_1, \ldots, \mathbf{x}_{N-d_k})$ is the number of points $\mathbf{x}_n$, $n = 1, \ldots, N - d_k$, in the interval I.

We can derive a binary sequence (e_n) from (x_n) by $e_n = 1$ if $0 \le x_n < 1/2$ and $e_n = 0$ otherwise.

In [37, Theorem 1] the correlation measure of order k of (e_n) is estimated in terms of the above discrepancy of vectors derived from the sequence (x_n). Hence, using the relation between linear complexity profile and correlation measure of (e_n) we can obtain (weak) linear complexity profile lower bounds for (e_n) from discrepancy upper bounds for (x_n).

1.7. Thoughts for Practitioners

Faster algorithms than the Berlekamp–Massey algorithm are known for sequences of particular periods [21, 71, 72].

The Legendre symbol needed for the generation of several sequences in Sec. 1.5.3 can be efficiently evaluated using the quadratic reciprocity law and its supplement.

Inversion is the most expensive operation in the generation of inversive generators. For fields $\mathbb{F}_p$ of prime order we can use the Euclidean algorithm. For fields $\mathbb{F}_{2^r}$ of characteristic 2 we recommend to use the Itoh-Tsujii algorithm [29] and an optimal normal basis representation [30].

For many practical applications, for example for qasi-Monte Carlo methods, we need sequences in the unit interval $[0, 1)$ (or any other interval) instead of sequences over finite fields. However, we can derive a sequence (x_n) over $[0, 1)$ from a sequence (ξ_n) over the finite field $\mathbb{F}_q$ in the following way. We fix a basis $\{\beta_1, \ldots, \beta_r\}$ of $\mathbb{F}_q$ over its prime field $\mathbb{F}_p$, i.e. $q = p^r$, and identify $\mathbb{F}_p$ with the integers $\{0, 1, \ldots, p-1\}$. Then we derive from the element

$$\xi_n = c_1\beta_1 + \cdots + c_r\beta_r, \quad c_1, \ldots, c_r \in \mathbb{F}_p,$$

an integer

$$y_n = c_r + c_{r-1}p + \cdots + c_0 p^{r-1} \in \{0, 1, \ldots, q-1\}.$$

The sequence (x_n) over $[0, 1)$ is obtained by

$$x_n = y_n/q.$$

1.8. Directions for Future Research

(1) Find more recursive non-linear generators for which substantial better lower bounds on the linear complexity profile can be proven.
(2) Find more classes of sequences over $[0, 1)$ for which the discrepancy with arbitrary lags $0 \leq d_1 < \cdots < d_k$ and thus the linear complexity of the corresponding binary sequence can be estimated.
(3) Extend the linear complexity profile lower bounds on the inversive generators of higher orders to arbitrary period.
(4) Analyze finer lattice tests with arbitrary lags.
(5) Find other quality measures which are related to linear complexity, e.g. the non-linearity of a Boolean function corresponding to a binary sequence.
(6) Prove results on other quality measures for the Sidelnikov sequence where analog results for the Legendre sequence are known, e.g. for the merit factor.

1.9. Conclusions

In this survey, we pointed to the strong ties between the cryptographic quality measure linear complexity and information theory and coding theory. We presented several lower bounds and exact values on the linear complexity (profile) of particular interesting sequences over a finite field using several illustrative methods. Finally, we mentioned other quality measures for sequences and their relations to linear complexity.

1.10. Questions

(1) Calculate the linear complexity profile of the finite sequence $(s_0, \ldots, s_9) = (1101011101)$ over $\mathbb{F}_2$ using the Berlekamp–Massey algorithm.
(2) Prove the following result which shows that lower bounds on the linear complexity profile provide upper bounds, as well.
Let (l_N) be a sequence with $l_1 \leq 0$ and $l_N \leq l_{N-1} + 1$ for $N \geq 2$. If

$$L(s_n, N) \geq l_N \quad \text{for } N \geq 2$$

then we have

$$L(s_n, N) \leq N - l_{N-1} \quad \text{for } N \geq 2.$$

Apply this result to get an upper bound on the linear complexity profile of the explicit inversive congruential generator of period p.

(3) Let g be an element of $\mathbb{F}_q$ of order t and $a, b \in \mathbb{F}_q \setminus \{0\}$. Prove a lower bound on the linear complexity profile of the sequence $z_n = (ag^n + b)^{q-2}$. Find conditions on a, b such that the bound is stronger than in the general case.

(4) Prove the formula for the exact value of the linear complexity of the two prime generator.

(5) Prove a lower bound on the linear complexity profile of the two-prime generator.

(6) Prove the relation between linear complexity profile and correlation measure of order k.

(7) Prove an upper bound on the correlation of order k of the Sidelnikov sequence and derive a lower bound on its linear complexity profile.

(8) Use the lower bound on the explicit inversive congruential generator of period p to derive an upper bound on the lattice profile $S(z_n, N)$.

(9) Find a sequence with large linear complexity but small 1-error linear complexity.

(10) Find an integer sequence with large linear complexity over $\mathbb{F}_p$, $p \geq 3$, but with small linear complexity over $\mathbb{F}_2$.

Solutions:

(1)

N	$L(s_n, N)$	
1	1	$- - -$
2	1	$s_{n+1} = s_n$
3	2	$s_{n+2} = s_{n+1} + s_n$ or $s_{n+2} = 0$
4	2	$s_{n+2} = s_{n+1} + s_n$
5	3	$s_{n+3} = s_{n+1}$ or $s_{n+3} = s_{n+2} + s_n$
6	3	$s_{n+3} = s_{n+1}$
7	4	$s_{n+4} = s_{n+1} + s_n$ or $s_{n+4} = s_{n+3} + s_n$
8	4	$s_{n+4} = s_{n+1} + s_n$
9	5	$s_{n+5} = s_{n+4} + s_{n+3} + s_{n+2} + s_{n+1} + s_n$ or $s_{n+5} = s_{n+3} + s_{n+2}$
10	5	$s_{n+5} = s_{n+4} + s_{n+3} + s_{n+2} + s_{n+1} + s_n$.

(2) For $N = 2$ we trivially have $L(s_n, N) \leq 2 \leq N - l_1$. For $N \geq 2$ by Theorem 1.1 we have either $L(s_n, N + 1) = L(s_n, N) \leq N - l_{N-1} \leq N + 1 - l_N$ by induction, where we used the condition $l_N \leq l_{N-1} + 1$, or $L(s_n, N + 1) = N + 1 - L(s_n, N) \leq N + 1 - l_N$. For the inversive generator (z_n) we have $l_N = \min\{(N-1)/3, (p-1)/2\}$ by Theorem 1.3 and get $L(z_n, N) \leq N - \min\{(N - 4)/3, (p - 3)/2\}$.

(3) As in Theorem 1.3, we can prove $L(z_n, N) \geq \min\{(N-1)/3, (t-1)/2\}$. We get the stronger bound $L(z_n, N) \geq \min\{N/2, t\}$ if $-a^{-1}b$ is not in the subgroup of $\mathbb{F}_q^*$ generated by g.

(4) See [9, Theorem 8.2.9].

(5) See [7].

(6) See [8].

(7) See [8].

(8) Use Exercise 2 and (1.18).

(9) A t-periodic sequence with exactly one non-zero entry has linear complexity t but 1-error linear complexity 0.

(10) A t-periodic sequence with one entry 2 and all other entries 0 has linear complexity t over $\mathbb{F}_p$, $p \geq 3$, but linear complexity 0 over $\mathbb{F}_2$.

1.11. Keywords

Linear complexity (profile)

For $N \geq 1$ the *linear complexity profile* $L(s_n, N)$ of a sequence (s_n) over $\mathbb{F}_q$ is the shortest length L of a linear recurrence relation

$$s_{n+L} = c_{L-1}s_{n+L-1} + \cdots + c_0 s_n, \quad 0 \leq n \leq N - L - 1,$$

over $\mathbb{F}_q$ satisfied by the first N sequence elements. The *linear complexity* $L(s_n)$ is defined by

$$L(s_n) := \sup_{N \geq 1} L(s_n, N).$$

k-error linear complexity

Let (s_n) be a sequence over $\mathbb{F}_q$, with period t. The *k-error linear complexity* $L_k(s_n)$ of (s_n) is defined as

$$L_k(s_n) := \min_{(y_n)} L(y_n),$$

where the minimum is taken over all t-periodic sequences (y_n) over $\mathbb{F}_q$, for which the Hamming distance of the vectors $(s_0, s_1, \ldots, s_{t-1})$ and $(y_0, y_1, \ldots, y_{t-1})$ is at most k.

Explicit non-linear congruential generator

For a prime p and a polynomial $f(X) \in \mathbb{F}_p[X]$ with $2 \leq \deg(f) \leq p-1$ the *explicit non-linear congruential generator* is the p-periodic sequence (x_n)

over $\mathbb{F}_p$ defined by

$$x_n = f(n), \quad n \geq 0.$$

Explicit inversive congruential generator

For $a, b \in \mathbb{F}_p$ with $a \neq 0$ the *explicit inversive congruential generator* (z_n) is defined by

$$z_n = (an + b)^{p-2}, \quad n \geq 0.$$

Recursive non-linear congruential generator

The *recursive non-linear congruential generator* (x_n) is defined by

$$x_{n+1} = f(x_n), \quad n \geq 0,$$

with some initial value $x_0 \in \mathbb{F}_p$ and a polynomial $f(X) \in \mathbb{F}_p[X]$ with $2 \leq \deg(f) \leq p - 1$.

Recursive inversive generator

For $a, b \in \mathbb{F}_p$ with $a \neq 0$ the *recursive inversive congruential generator* (z_n) is defined by

$$z_{n+1} = az_n^{p-2} + b, \quad n \geq 0,$$

with some initial value z_0.

Legendre sequence

The *Legendre sequence* (l_n) of period p is the sequence over $\mathbb{F}_2$ defined by $l_n = 1$ if $\left(\frac{n}{p}\right) = -1$, i.e. n is a quadratic non-residue modulo p and $l_n = 0$ otherwise.

Sidelnikov sequence

Let g be a primitive root modulo p. Then the *Sidelnikov sequence* (s_n) is the the $p - 1$ periodic sequence over $\mathbb{F}_2$ defined by $s_n = 1$ if $g^n + 1$ is a quadratic non-residue modulo p and $s_n = 0$ otherwise.

Two-prime generator

Let p and q be two distinct odd primes. Put

$$Q = \{q, 2q, \ldots, (p-1)q\}, \quad Q_0 = Q \cup \{0\},$$

and

$$P = \{p, 2p, \ldots, (q-1)p\}.$$

The pq-periodic sequence (t_n) over $\mathbb{F}_2$, defined by

$$t_n = \begin{cases} 0 & \text{if } (n \bmod pq) \in Q_0, \\ 1 & \text{if } (n \bmod pq) \in P, \\ \left(1 - \left(\frac{n}{p}\right)\left(\frac{n}{q}\right)\right)/2 & \text{otherwise}, \end{cases}$$

is called the *two-prime generator*.

Correlation measure of order k

The *correlation measure of order k* of a binary sequence (s_n) is introduced as

$$C_k(s_n) = \max_{M, D} \left| \sum_{n=1}^{M} (-1)^{s_{n+d_1}} \cdots (-1)^{s_{n+d_k}} \right|, \quad k \geq 1,$$

where the maximum is taken over all $D = (d_1, d_2, \ldots, d_k)$ with non-negative integers $d_1 < d_2 < \cdots < d_k$ and M such that $M - 1 + d_k \leq T - 1$.

References

1. H. Aly and A. Winterhof, On the linear complexity profile of nonlinear congruential pseudorandom number generators with Dickson polynomials, *Des. Codes Cryptogr.* **39**, 2 (2006), 155–162.
2. H. Aly and A. Winterhof, On the k-error linear complexity over $\mathbb{F}_p$ of Legendre and Sidelnikov sequences, *Des. Codes Cryptogr.* **40**, 3 (2006), 369–374.
3. P. H. T. Beelen and J. M. Doumen, Pseudorandom sequences from elliptic curves, *Finite Fields with Applications to Coding Theory, Cryptography and Related Areas*, Oaxaca, 2001 (Springer, Berlin, 2002), pp. 37–52.
4. E. R. Berlekamp, *Algebraic Coding Theory* (McGraw-Hill Book Co., New York-Toronto, Ont.-London, 1968).
5. T. Beth and Z. D. Dai, On the complexity of pseudo-random sequences — or: If you can describe a sequence it can't be random, *Advances in Cryptology — EUROCRYPT '89* (Houthalen, 1989), Lecture Notes in Computer Science (Springer, Berlin, 1990), Vol. 434, pp. 533–543.
6. S. R. Blackburn, T. Etzion and K. G. Paterson, Permutation polynomials, de Bruijn sequences, and linear complexity, *J. Combin. Theory Ser. A* **76**, 1 (1996), 55–82.
7. N. Brandstätter and A. Winterhof, Some notes on the two-prime generator, *IEEE Trans. Inform. Theory* **51** (2005), 3645–3647.

8. N. Brandstätter and A. Winterhof, Linear complexity profile of binary sequences with small correlation measure, *Period. Math. Hungar.* **52**, 2 (2006), 1–8.

9. T. W. Cusick, C. Ding and A. Renvall, *Stream Ciphers and Number Theory*, revised ed., North-Holland Mathematical Library, 66 (Elsevier Science B.V., Amsterdam, 2004).

10. C. Ding, G. Xiao and W. Shan, *The Stability Theory of Stream Ciphers*, Lecture Notes in Computer Science, (Springer-Verlag, Berlin, 1991), Vol. 561.

11. G. Dorfer, W. Meidl and A. Winterhof, Counting functions and expected values for the lattice profile at n, *Finite Fields Appl.* **10**, 4 (2004), 636–652.

12. G. Dorfer and A. Winterhof, Lattice structure and linear complexity profile of nonlinear pseudorandom number generators, *Appl. Algebra Engrg. Comm. Comput.* **13**, 6 (2003), 499–508.

13. J. Eichenauer, H. Grothe, J. Lehn and A. Topuzoğlu, A multiple recursive nonlinear congruential pseudo random number generator, *Manuscripta Math.* **59**, 3 (1987), 331–346.

14. J. Eichenauer and J. Lehn, A nonlinear congruential pseudorandom number generator, *Statist. Hefte* 4 (1986), 315–326.

15. J. Eichenauer-Herrmann, Statistical independence of a new class of inversive congruential pseudorandom numbers, *Math. Comp.* **60**, 201 (1993), 375–384.

16. Y.-C. Eun, H.-Y. Song and G. M. Kyureghyan, One-error linear complexity over $\mathbb{F}_p$ of Sidelnikov sequences, *Sequences and Their Applications SETA 2004*, Lecture Notes in Computer Science (Springer, Berlin, 2005), Vol. 3486, pp. 154–165.

17. M. Flahive and H. Niederreiter, On inversive congruential generators for pseudorandom numbers, *Finite Fields, Coding Theory, and Advances in Communications and Computing* (Las Vegas, NV, 1991), Lecture Notes in Pure and Applied Mathematics (Dekker, New York, 1993), Vol. 141, pp. 75–80.

18. É. Fouvry, P. Michel, J. Rivat and A. Sárközy, On the pseudorandomness of the signs of Kloosterman sums, *J. Aust. Math. Soc.* **77**, 3 (2004), 425–436.

19. J. B. Friedlander, C. Pomerance and I. E. Shparlinski, Period of the power generator and small values of Carmichael's function, *Math. Comp.* **70**, 236 (2001), 1591–1605.

20. J. B. Friedlander, C. Pomerance and I. E. Shparlinski, Corrigendum to: Period of the power generator and small values of Carmichael's function, *Math. Comp.* **71**, 240 (2002), 1803–1806.

21. R. A. Games and A. H. Chan, A fast algorithm for determining the complexity of a binary sequence with period 2^n, *IEEE Trans. Inform. Theory* **29**, 1 (1983), 144–146.

22. M. Z. Garaev, F. Luca, I. E. Shparlinski and A. Winterhof, On the lower bound of the linear complexity over $\mathbb{F}_p$ of Sidelnikov sequences, *IEEE Trans. Inform. Theory* **52**, 7 (2006), 3299–3304.

23. F. Griffin and I. E. Shparlinski, On the linear complexity profile of the power generator, *IEEE Trans. Inform. Theory* **46**, 6 (2000), 2159–2162.

24. F. G. Gustavson, Analysis of the Berlekamp–Massey linear feedback shift-register synthesis algorithm, *IBM J. Res. Develop.* **20**, 3 (1976), 204–212.

25. J. Gutierrez, I. E. Shparlinski and A. Winterhof, On the linear and nonlinear complexity profile of nonlinear pseudorandom number-generators, *IEEE Trans. Inform. Theory* **49**, 1 (2003), 60–64.

26. K. Gyarmati, On a family of pseudorandom binary sequences, *Period. Math. Hungar.* **49**, 2 (2004), 45–63.

27. T. Helleseth and P. V. Kumar, Sequences with low correlation, *Handbook of Coding Theory* (North-Holland, Amsterdam, 1998) Vol. I, II, pp. 1765–1853.

28. F. Hess and I. E. Shparlinski, On the linear complexity and multidimensional distribution of congruential generators over elliptic curves, *Des. Codes and Cryptogr.* **35**, 1 (2005), 111–117.

29. T. Itoh and S. Tsujii, A fast algorithm for computing multiplicative inverses in $GF(2^m)$ using normal bases, *Inform. and Comput.* **78**, 3 (1988), 171–177.

30. D. Jungnickel, Finite fields. Structure and arithmetics. Bibliographisches Institut, Mannheim, (1993).

31. A. Klapper, The vulnerability of geometric sequences based on fields of odd characteristic, *J. Cryptology* **7**, 1 (1994), 33–51.

32. N. Koblitz, *Algebraic Aspects of Cryptography.* (Springer-Verlag, Berlin Heidelberg, 1998).

33. S. Konyagin, T. Lange and I. Shparlinski, Linear complexity of the discrete logarithm, *Des. Codes Cryptogr.* **28**, 2 (2003), 135–146.

34. R. Lidl, G. L. Mullen and G. Turnwald, Dickson polynomials, *Pitman Monographs and Surveys in Pure and Applied Mathematics* (Longman Scientific & Technical, Harlow; copublished in the United States with John Wiley & Sons, Inc., New York, 1993.), Vol. 65.

35. J. L. Massey, Shift-register synthesis and BCH decoding, *IEEE Trans. Inform. Theory* **IT-15** (1969) 122–127.

36. J. L. Massey and S. Serconek, Linear complexity of periodic sequences: A general theory, *Advances in cryptology — CRYPTO '96* (Santa Barbara, CA), Lecture Notes in Computer Science (Springer, Berlin, 1996), Vol. 1109, pp. 358–371.

37. C. Mauduit, H. Niederreiter and A. Sárközy, On pseudorandom $[0,1)$ and binary sequences, *Publ. Math. Debrecen* **71**, 3–4 (2007), 305–324.

38. C. Mauduit and A. Sárközy, On finite pseudorandom binary sequences. I. Measure of pseudorandomness, the Legendre symbol, *Acta Arith.* **82**, 4 (1997), 365–377.

39. W. Meidl, How many bits have to be changed to decrease the linear complexity? *Des. Codes Cryptogr.* **33**, 2 (2004), 109–122.

40. W. Meidl, Linear complexity and k-error linear complexity for p^n-periodic sequences, *Coding, Cryptography and Combinatorics*, Progr. Comput. Sci. Appl. Logic (Birkhäuser, Basel, 2004.), Vol. 23, pp. 227–235.

41. W. Meidl and H. Niederreiter, On the expected value of the linear complexity and the k-error linear complexity of periodic sequences, *IEEE Trans. Inform. Theory* **48**, 11 (2002), 2817–2825.

42. W. Meidl and A. Winterhof, Lower bounds on the linear complexity of the discrete logarithm in finite fields, *IEEE Trans. Inform. Theory* **47**, 7 (2001), 2807–2811.

43. W. Meidl and A. Winterhof, Linear complexity and polynomial degree of a function over a finite field, *Finite Fields with Applications to Coding Theory, Cryptography and Related Areas*, Oaxaca, 2001 (Springer, Berlin, 2002), pp. 229–238.

44. W. Meidl and A. Winterhof, On the linear complexity profile of explicit nonlinear pseudorandom numbers, *Inform. Process. Lett.* **85**, 1 (2003), 13–18.

45. W. Meidl and A. Winterhof, On the linear complexity profile of some new explicit inversive pseudorandom numbers, *J. Complexity* **20**, 2–3 (2004), 350–355.

46. W. Meidl and A. Winterhof, On the autocorrelation of cyclotomic generators, *Finite Fields and Applications*, Lecture Notes in Computer Science (Springer, Berlin, 2004), Vol. 2948, 1–11.

47. W. Meidl and A. Winterhof, Some notes on the linear complexity of Sidelńikov-Lempel-Cohn-Eastman sequences, *Des. Codes Cryptogr.* **38**, 2 (2006), 159–178.

48. W. Meidl and A. Winterhof, On the linear complexity profile of nonlinear congruential pseudorandom number generators with Rdei functions, *Finite Fields Appl.* **13**, 3 (2007), 628–634.

49. A. J. Menezes, P. C. van Oorschot and S. A. Vanstone, *Handbook of Applied Cryptography*, With a foreword by R. L. Rivest. CRC Press Series on Discrete Mathematics and its Applications (CRC Press, Boca Raton, FL, 1997).

50. H. Niederreiter, Some computable complexity measures for binary sequences, *Sequences and their Applications*, Singapore, 1998, Springer Ser. Discrete Math. Theor. Comput. Sci. (Springer, London, 1999), pp. 67–78.

51. H. Niederreiter, Random number generation and quasi-Monte Carlo methods, *CBMS-NSF Regional Conference Series in Applied Mathematics, 63*, Society for Industrial and Applied Mathematics (SIAM), Philadelphia, PA, (1992).

52. H. Niederreiter, Linear complexity and related complexity measures for sequences, *Progress in Cryptology — INDOCRYPT 2003*, Lecture Notes in Computer Science (Springer, Berlin, 2003), Vol. 2904, pp. 1–17,

53. H. Niederreiter and I. E. Shparlinski, Recent advances in the theory of nonlinear pseudorandom number generators, *Monte Carlo and Quasi-Monte Carlo Methods*, 2000, Hong Kong, (Springer, Berlin, 2002), pp. 86–102.

54. R. Nöbauer, Rédei-Permutationen endlicher Körper, *Contributions to General Algebra*, Salzburg, 1986 (Hölder-Pichler-Tempsky, Vienna, 1987), Vol. 5, pp. 235–246.

55. R. A. Rueppel, in *Analysis and Design of Stream Ciphers*, With a foreword by J. L. Massey. Communications and Control Engineering Series (Springer-Verlag, Berlin, 1986).

56. R. A. Rueppel, Stream ciphers, *Contemporary Cryptology* (IEEE, New York, 1992), pp. 65–134.

57. W. M. Schmidt, Equations over finite fields, *An Elementary Approach*, Lecture Notes in Mathematics (Springer-Verlag, Berlin-New York, 1976), Vol. 536.

58. I. Shparlinski, On the linear complexity of the power generator, *Des. Codes Cryptogr.* **23**, 1 (2001), 5–10.

59. I. Shparlinski, Cryptographic applications of analytic number theory, *Complexity Lower Bounds and Pseudorandomness*, Progress in Computer Science and Applied Logic (Birkhäuser Verlag, Basel, 2003), Vol. 22.

60. I. E. Shparlinski and A. Winterhof, On the linear complexity of bounded integer sequences over different moduli, *Inform. Process. Lett.* **96**, 5 (2005), 175–177.

61. V. M. Sidel'nikov, Some k-valued pseudo-random sequences and nearly equidistant codes, *Prob. Inform. Transmission* **5**, 1 (1969), 12–16.; translated from Problemy Peredači Informacii, **5**, 1 (1969), 16–22 (Russian).

62. B. Smeets, The linear complexity profile and experimental results on a randomness test of sequences over the field $\mathbb{F}_q$, Preprint (1988).

63. M. Stamp and C. F. Martin, An algorithm for the k-error, linear complexity of binary sequences with period 2^n, *IEEE Trans. Inform. Theory* **39**, 4 (1993), 1398–1401.

64. A. Tietäväinen, Vinogradov's method and some applications, *Number Theory and Its Applications*, Ankara, 1996, Lecture Notes in Pure and Applied Mathematics (Dekker, New York, 1999), Vol. 204, pp. 261–282.

65. A. Topuzoğlu and A. Winterhof, On the linear complexity profile of nonlinear congruential pseudorandom number generators of higher orders, *Appl. Algebra Engrg. Comm. Comput.* **16**, 4 (2005), 219–228.

66. A. Topuzoğlu and A. Winterhof, Pseudorandom sequences, *Topics in Geometry, Coding Theory and Cryptography*, Algebr. Appl. (Springer, Dordrecht, 2007), Vol. 6, pp. 135–166.

67. R. J. Turyn, The linear generation of Legendre sequence, *J. Soc. Indust. Appl. Math.* **12** (1964) 115–116.

68. I. M. Vinogradov, *Elements of Number Theory*, Translated by S. Kravetz (Dover Publications, Inc., New York, 1954).

69. Y. Wang, Linear complexity versus pseudorandomness: on Beth and Dai's result, *Advances in Cryptology — ASIACRYPT'99* (Singapore), Lecture Notes in Computer Science (Springer, Berlin, 1999), Vol. 1716, pp. 288–298.

70. A. Winterhof, A note on the linear complexity profile of the discrete logarithm in finite fields, *Coding, Cryptography and Combinatorics*, Progr. Comput. Sci. Appl. Logic (Birkhäuser, Basel, 2004), Vol. 23, pp. 359–367.

71. G. Xiao and S. Wei, in Fast algorithms for determining the linear complexity of period sequences, ed., A. Menezes *et al.*, *Progress in Cryptology — INDOCRYPT 2002. Third International Conference on Cryptology in India*, Hyderabad, India (December 16–18, 2002), Proceedings. Berlin: Springer. Lecture Notes Computer Science 2551, pp. 12–21.

72. G. Xiao, S. Wei, K. Y. Lam and K. Imamura, A fast algorithm for determining the linear complexity of a sequence with period p^n over GF(q), *IEEE Trans. Inform. Theory* **46**, 6 (2000), 2203–2206.

Chapter 2

LATTICE AND CONSTRUCTION OF HIGH CODING GAIN LATTICES FROM CODES

MOHAMMD-REZA SADEGHI

Faculty of Mathematics and Computer Science,
Amirkabir University of Technology,
Hafez Ave., Tehran, Iran

In this study, we conduct a comprehensive investigation on lattices and several constructions of lattice from codes. These constructions include some known constructions such as, Construction A, B, C, D, and D′. For each construction we derive some corresponding lattice factors such as label groups, label codes, etc. Finally, we compare these constructions together to come up with a possible lattice construction with high coding gain using Turbo codes and low density parity check codes.

2.1. Introduction

The lattice version of the Gaussian channel coding problem is to find, for a given value of signal-to-noise-ratio (SNR) the n-dimensional lattice for which the error probability is minimized [9]. Shannon's fundamental results guarantee the existence of such lattice codes when the dimension increases. Some codes such as Turbo codes [6, 19] and low density parity check (LDPC) codes [16, 23] can have a remarkable performance under iterative decodings. Therefore, it is reasonable to investigate construction of high coding gain lattices with low decoding complexity from good codes in larger dimensions. In this study, first we give a detailed introduction to lattice, its relation with channel coding problem, and some related factors such as its trellis diagram and label code. These factors are used as a measure of decoding complexity [3, 25]. Then, we give some important lattice constructions from codes. These constructions include some known constructions such as, Construction A, B, C, D, and D′ [9]. For each Construction, we derive the corresponding lattice factors such as label groups, label codes, etc.

Both turbo codes [6, 19] and LDPC codes [16, 24] can have a remarkable performance under iterative decoding. Finally, we compare those constructions together to come up with a possible lattice construction with high coding gain using LDPC and Turbo codes.

2.2. Background

2.2.1. *Codes*

A *sequence space* W is defined as the Cartesian product of n sets of alphabet symbols G_i. A *block code* C is any subset of a sequence space $W = G_1 \times \cdots \times G_n$. The elements of C are called codewords, or blocks. If $C = \{\mathbf{c_1}, \ldots, \mathbf{c_M}\}$ has M codewords then C is called an (n, M)-code and is denoted by $C = [n, M]$. If all code symbols (coordinates) are drawn from the same alphabet $G = \{g_1, \ldots, g_q\}$, then the code C is called a *q-ary* (n, M)-*code over* G. In particular when $G = \{0, 1\}$ the code C is called a *binary code*.

A *group sequence space* W is a sequence space in which all symbol alphabets are groups G_i. Then $W = G_1 \times \cdots \times G_n$ is a direct product group under the componentwise group operation.

Definition 2.1. A *group code* C is a subgroup of a group sequence space W.

In other words, a group code is a set of codewords (sequences) that forms a group under the componentwise group operation. If all code symbols are drawn from a common group G, then $W = G^n$ and C will be called a *group code over* G.

Denote by $\mathbb{F}_q$ the Galois field of order q, where q is a prime or power of a prime. Then, a q-ary code C over the sequence space $\mathbb{F}_q^n$ is a **linear code** if and only if C forms a vector subspace over $\mathbb{F}_q$. The dimension of a linear code C is the dimension of the corresponding subspace. A linear q-ary code C with length n and dimension k has q^k codewords, it is called an (n, k)-code and is denoted by $C = [n, k]$. Let $\{\mathbf{g_1}, \ldots, \mathbf{g_k}\}$ be a basis for a k-dimensional code C, then the matrix B whose rows are g_i's, is called a *generator matrix B* for the code C.

Let $\mathbf{u} = (u_1, \ldots, u_n)$ and $\mathbf{v} = (v_1, \ldots, v_n)$ be vectors in the vector space V over the field F. The *inner product* $\langle \mathbf{u}, \mathbf{v} \rangle$ is defined as $\langle \mathbf{u}, \mathbf{v} \rangle = \sum_{i=1}^{n} u_i.v_i$. Two elements $\mathbf{u}$ and $\mathbf{v}$ are said to be *orthogonal* if $\langle \mathbf{u}, \mathbf{v} \rangle = 0$.

Let S be a subspace of the vector space V. Let S^* be the set of all vectors $\mathbf{v}$ in V such that for all $\mathbf{u} \in S$, $\langle \mathbf{u}, \mathbf{v} \rangle = 0$. Then S^* is a subspace and is called the *dual space* of S.

Definition 2.2. Let $C = [n, k]$ be a q-ary linear code. Then the dual space of C, which has dimension $n - k$, is called the *dual code* of C and is denoted by C^*.

Let $\{h_1, \ldots, h_{n-k}\}$ be a basis for C^*, then the *parity check matrix H* of C is defined by

$$\mathbf{H} = \begin{bmatrix} \mathbf{h_1} \\ \vdots \\ \mathbf{h_{n-k}} \end{bmatrix}.$$

It is trivial that a vector $c \in C$ if and only if $\mathbf{H}c^T = \mathbf{0}$. Therefore, the linear code C is the null space of the parity check matrix $\mathbf{H}$.

The *minimum distance $d_{\min}$* of a block code C is the minimum Hamming distance between all distinct pairs of codewords in C.

2.2.2. Lattices

Let $\mathbb{R}^m$ be the m-dimensional real space with scalar product $\langle ., . \rangle$ and standard basis $e_1 = (1, 0, \ldots, 0), \ldots, e_m = (0, \ldots, 0, 1)$. The Euclidian length is $\|\mathbf{x}\| = \langle \mathbf{x}, \mathbf{x} \rangle^{\frac{1}{2}}$.

A set Λ of vectors is called *discrete* if there exists a positive number β such that any two vectors of Λ have distance at least β.

Definition 2.3. An infinite discrete set $\Lambda \subseteq \mathbb{R}^m$ is called a **lattice** if Λ is a group under addition in $\mathbb{R}^m$. The elements of the lattice Λ are called points or vectors.

A **lattice** is a discrete additive subgroup of $\mathbb{R}^m$. The notation $d_{\min} = d_{\min}(\Lambda)$ is used to denote the length of the shortest non-zero vector of the lattice Λ. In fact $d_{\min}(\Lambda)$ refers to the **minimum distance** between lattice points.

It can be shown that every lattice is generated by the integer combinations of some linearly independent vectors $\mathbf{b_1}, \ldots, \mathbf{b_n} \in \mathbb{R}^m$ [27] i.e.

$$\Lambda = \{k_1 \mathbf{b}_1 + \cdots + k_n \mathbf{b}_n : k_1, \ldots, k_n \in \mathbb{Z}\}.$$

The set of vectors $\{\mathbf{b_1}, \ldots, \mathbf{b_n}\}$ is called a **lattice basis** and n is called the *dimension* or *rank* of Λ. Since $b_1 \ldots, b_n$ are linearly independent, $n \le m$. If $m = n$ the lattice Λ is called *full rank (dimension)*.

By a change of basis, we may transform an n-dimensional lattice Λ to be a subgroup of $\mathbb{R}^n$. Thus, without loss of generality, we may always assume that Λ has full rank.

Definition 2.4. Any subgroup of a lattice Λ is called a **sublattice** of Λ.

Example 2.5. The lattice $\mathbb{Z}^m$ generated by standard basis of $\mathbb{R}^m$ consists of all the vectors having integer components. It is called the integer lattice.

Definition 2.6. Let $\mathbf{b_i} = (b_{i1}, \ldots, b_{im})$, $i = 1, \ldots, n$ be the lattice basis vectors. The generator matrix of Λ is defined by

$$\mathbf{B} = \begin{bmatrix} \mathbf{b_1} \\ \vdots \\ \mathbf{b_n} \end{bmatrix} = \begin{bmatrix} b_{11} & \cdots & b_{1m} \\ \vdots & & \vdots \\ b_{n1} & \cdots & b_{nm} \end{bmatrix}.$$

With the above definition, we can write

$$\Lambda = \{\mathbf{v} = \mathbf{xB} : \mathbf{x} \in \mathbb{Z}^n\}.$$

Example 2.7. For $n \ge 0$, the lattice A_n is defined as

$$A_n = \left\{ (x_0, x_1, \ldots, x_n) \in \mathbb{Z}^{n+1} : \sum_{i=0}^{n} x_i = 0 \right\},$$

has an $n \times (n+1)$ generator matrix

$$\mathbf{B} = \begin{bmatrix} -1 & 1 & 0 & 0 & \cdots & 0 & 0 \\ 0 & -1 & 1 & 0 & \cdots & 0 & 0 \\ 0 & 0 & -1 & 1 & \cdots & 0 & 0 \\ . & . & . & . & \cdots & . & . \\ 0 & 0 & 0 & 0 & \cdots & -1 & 1 \end{bmatrix}.$$

We will refer to A_2 as the **Hexagonal** lattice.

The **Gram matrix** of Λ is defined as $A = \mathbf{BB}^T$. Two lattices are called *equivalent* or *similar* if one of them can be obtained from the other by (possibly) a rotation, a reflection, or a change of scale. Two generator matrices B and B' define equivalent lattices if and only if they are related by $B' = cUBM$ where c is a non-zero constant, U is a matrix with integer entries and determinant -1 or 1, and M is a real orthogonal matrix with $MM^T = I$.

Definition 2.8. A lattice is called **orthogonal (rectangular)** if it has a basis with mutually orthogonal vectors.

Definition 2.9. Given a full rank lattice Λ of $\mathbb{R}^n$, its **dual lattice** is defined as

$$\Lambda^* = \{\mathbf{x} \in \mathbb{R}^n : \langle \mathbf{x}, \mathbf{v} \rangle \in \mathbb{Z}, \text{for all } \mathbf{v} \in \Lambda\}.$$

Definition 2.10. The **fundamental parallelotope** of the lattice Λ is the set of points

$$\{\theta\mathbf{B} : \theta = (\theta_1, \ldots, \theta_n) \in \mathbb{R}^n, \, 0 \leq \theta_i < 1\}.$$

The fundamental parallelotope of a lattice is not unique since it depends on the choice of the lattice basis. In the case of full rank lattices the volume of the fundamental parallelotope is equal to $|\det \mathbf{B}|$, a number independent of the lattice basis. This number is called the **determinant** of Λ, and is denoted by $\det(\Lambda)$.

For non-full rank lattices, the volume of the fundamental parallelotope is not well defined since it has zero measure in $\mathbb{R}^m$. The determinant of Λ is then defined by using its Gram matrix.

Definition 2.11. Let $\mathbf{B}$ be the basis matrix of the lattice Λ, the **determinant (or volume) of** Λ is defined by the determinant of the Gram matrix of Λ, i.e.

$$\det(\Lambda) = \det(A)^{\frac{1}{2}} = \det(\mathbf{B}\mathbf{B}^T)^{\frac{1}{2}}.$$

The notation $\text{span}(\Lambda) = \text{span}\{\mathbf{b_1}, \ldots, \mathbf{b_n}\}$ denotes the real span of the generators $\{\mathbf{b_1}, \ldots, \mathbf{b_n}\}$, i.e. $\text{span}(\Lambda) = \{\alpha_1\mathbf{b_1} + \cdots + \alpha_n\mathbf{b_n} : (\alpha_1, \ldots, \alpha_n) \in \mathbb{R}^n\}$.

Definition 2.12. A **fundamental region** of a lattice Λ is defined as a subset of $span(\Lambda)$ which, when translated by lattice points, partitions $span(\Lambda)$ with just one lattice point in each copy.

Fundamental parallelotope is an example of a fundamental region, see Fig. 2.1.

Definition 2.13. For any point $\mathbf{p} \in \Lambda$ the **Voronoi cell** $\mathcal{V}(\mathbf{p})$ is defined by the set of those points of $\text{span}(\Lambda)$ that are at least as close to $\mathbf{p}$ as to

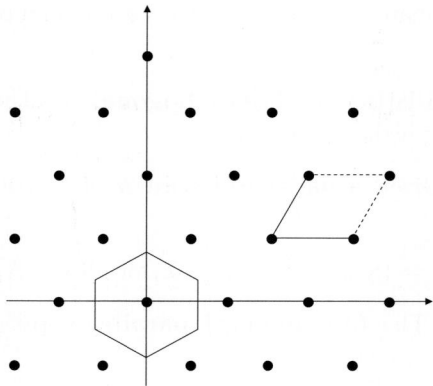

Fig. 2.1.   Voronoi cells and fundamental parallelotope.

any other point $\mathbf{q} \in \Lambda$, i.e. $\mathcal{V}(\mathbf{p}) = \{\mathbf{x} \in \mathrm{span}(\Lambda) : \|\mathbf{x} - \mathbf{p}\| \leq \|\mathbf{x} - \mathbf{q}\|$, for all $\mathbf{q} \in \Lambda\}$.

The translation of a Voronoi cell by a vector in Λ gives another Voronoi cell. A Voronoi cell is a fundamental region of the lattice Λ. The Voronoi cell for the origin is denoted by $\mathcal{V}$. If $0 \neq \mathbf{u} \in \Lambda$, then $\|\mathbf{x}\| = \|\mathbf{x} - \mathbf{u}\|$ defines a hyperplane half way between $\mathbf{0}$ and $\mathbf{u}$ of equation $\langle \mathbf{x}, \mathbf{u} \rangle = \|\mathbf{u}\|^2/2$. The points $\mathbf{p}$ of the lattice for which the hyperplane between $\mathbf{0}$ and $\mathbf{p}$ contains a facet of $\mathcal{V}$ are called *Voronoi-relevant points* and the hyperplanes are called *Voronoi-relevant hyperplanes*.

If Λ' is a sublattice of Λ such that Λ' is also full rank, then the index $|\Lambda/\Lambda'|$ is finite, and $\det(\Lambda') = |\Lambda/\Lambda'| \det(\Lambda)$. In fact since only one of every $|\Lambda/\Lambda'|$ points in Λ is in Λ', the volume of fundamental region of Λ' must be $|\Lambda/\Lambda'|$ times larger than that of Λ for its fundamental region to fill $\mathbb{R}^n$.

Definition 2.14. The **(fundamental) coding gain** (or Hermite parameter) of the lattice Λ is defined as

$$\gamma(\Lambda) = \frac{d_{\min}^2(\Lambda)}{(\det(\Lambda))^{\frac{2}{n}}}.$$

For any n, we have $\gamma(\mathbb{Z}^n) = 1$.

An n-dimensional constellation is any finite set of n-tuples or points in n-dimensional space. In particular, a signal constellation is the set of all possible signal points that are taken from a finite subset of points lying within a translate of a lattice. An uncoded system may be defined as the one that uses a constellation based on $\mathbb{Z}^n$. For example, PAM uses constellations based on $\mathbb{Z}$ and QAM [18] uses constellations based on $\mathbb{Z}^2$). Thus, the coding gain $\gamma(\Lambda)$ of an arbitrary lattice Λ may be considered to be the gain using constellations based on Λ over an uncoded system using constellations based on $\mathbb{Z}^n$ [11]. Therefore, the coding gain of a lattice is a useful measure to assess the performance of the lattice.

Geometrically, $\gamma(\Lambda)$ measures the increase in density of Λ over the baseline lattice $\mathbb{Z}$ or $\mathbb{Z}^n$. The coding gain is invariant to scaling, $\gamma(c\Lambda) = \gamma(\Lambda)$. More generally this quantity is invariant to any scaled orthogonal transformation T, $\gamma(T\Lambda) = \gamma(\Lambda)$, because $\det(T\Lambda) = |\det T| \det(\Lambda)$ and $d^2_{\min}(T\Lambda) = |\det T|^{2/n} d^2_{\min}(\Lambda)$.

For every n, the maximum value of the coding gain is attainable [17] and is called Hermite parameter. It is denoted by γ_n. The maximum value of γ_n is presently known only for $n = 1, 2, \ldots, 8$ [9].

2.2.2.1. *The Sphere Packing Problem*

A very old problem in mathematics asks to stack a large number of identical three-dimensional spheres in a very large box in the most efficient way, i.e. to maximize the number of spheres which can fit inside the box. Such arrangement of spheres is called a *sphere packing*. The spheres do not fill all the space in the box. The percentage of space occupied by the spheres is called the *packing density* and is denoted by Δ.

The above problem can be generalized to higher or lower dimensions.

Definition 2.15. Let Λ be an n-dimensional (full rank) lattice with minimum distance $d_{\min}$. The **packing radius** ρ of Λ is defined by

$$\rho = \frac{d_{\min}}{2}.$$

For an n-dimensional lattice Λ, assume that an n-dimensional sphere of radius ρ is centered at each lattice point. This arrangement of spheres is called a *lattice packing* or briefly a *packing* corresponding to the lattice Λ. We observe that the sphere of radius ρ is the largest sphere inscribed in the Voronoi cell of the lattice.

The **packing density** Δ of a lattice packing is

$$\Delta = \frac{\text{volume of one sphere}}{\text{volume of fundamental region}} = \frac{\rho^n V_n}{\det(\Lambda)} \leq 1,$$

where V_n is the volume of the n-dimensional sphere of radius 1. It is known that

$$V_n = \frac{\pi^{n/2}}{(n/2)!} = \begin{cases} \dfrac{\pi^k}{k!}, & n = 2k \\[2ex] \dfrac{2^n \pi^k k!}{n!}, & n = 2k+1 \end{cases} \tag{2.1}$$

where $(n/2)! = \Gamma(\frac{n}{2} + 1)$ and $\Gamma(t) = \int_0^\infty u^{t-1} e^{-u} du$ is Euler's gamma function.

We also define the *center density* $\delta(\Lambda)$ of the lattice packing as

$$\delta(\Lambda) = \frac{\Delta}{V_n}, \tag{2.2}$$

which gives the number of centers per unit volume and enable one to compare lattices in different dimensions. It is easy to see that

$$\gamma(\Lambda) = 4\delta(\Lambda)^{\frac{2}{n}}. \tag{2.3}$$

The density and the center density are independent of the scaling factor of the lattice. The optimal lattice sphere packings are known only up to dimension 8 [9].

Another factor of a sphere packing which will be used in the estimation of the probability of decoding error, is the kissing number. The number of spheres touching one sphere in a sphere packing is called the *kissing number* and is denoted by τ. For sphere packing the kissing number varies from one sphere to another one, but for lattice packing it is the same for all spheres (Fig. 2.2).

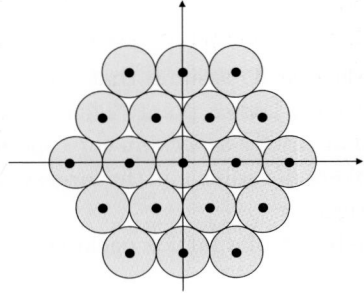

Fig. 2.2. The optimal two-dimensional lattice sphere packing.

2.3. The Channel Coding Problem

A source of information emits a sequence of messages (either continuous signals or discrete sets of symbols from a finite alphabet). These messages need to be transmitted over a channel impaired by noise in such a way that the probability of correct detection at the receiver side is maximized.

We assume that the channel impairments are only due to white Gaussian noise (Additive White Gaussian Noise (AWGN) [8] channel). The channel bandwidth is larger than the bandwidth of the transmitted messages so that no other type of distortion appears. AWGN can also be represented as a random vector $\mathbf{n} = (n_1, \ldots, n_n)$ in the Euclidian space whose components are independent and identically Gaussian distributed random variable with zero mean and variance σ^2. Let the vector $\mathbf{c}$ be transmitted, then the received vector $\mathbf{y}$ can be written as

$$\mathbf{y} = \mathbf{c} + \mathbf{n}.$$

The channel output $\mathbf{c} + \mathbf{n}$ is decoded to $\mathbf{c}$ under *maximum likelihood decoding* if and only if $\mathbf{c} + \mathbf{n}$ belongs to $\mathcal{V}(c)$, the Voronoi cell of $\mathbf{c}$, in the lattice Λ. Thus, the probability of correct decoding is given by

$$P_c = \frac{1}{(\sigma\sqrt{2\pi})^n} \int_{\mathcal{V}(c)} e^{-\frac{\|\mathbf{x}\|^2}{2\sigma^2}} \, d\mathbf{x}. \tag{2.4}$$

Using the fact that the Voronoi cells of a lattice are congruent, therefore P_e the probability of error under maximum likelihood decoding of Λ is the probability that a white Gaussian n-tuple $\mathbf{n}$ with variance σ^2 per dimension falls outside the Voronoi cell $\mathcal{V}$, i.e.

$$P_e = 1 - P_c = 1 - \frac{1}{(\sigma\sqrt{2\pi})^n} \int_{\mathcal{V}} e^{-\frac{\|\mathbf{x}\|^2}{2\sigma^2}} \, d\mathbf{x}. \tag{2.5}$$

Let C be a set of code points $\{c_1, \ldots, c_M\}$ and let $\mathcal{V}(c_k)$ be the Voronoi cell of c_k. Assuming that all code points are equally likely to be used, then the error probability for this code is

$$P_e = 1 - \frac{1}{M} \sum_{k=1}^{M} \frac{1}{(\sigma\sqrt{2\pi})^n} \int_{\mathcal{V}(c_k)} e^{-\frac{\|\mathbf{x}\|^2}{2\sigma^2}} \, d\mathbf{x}. \tag{2.6}$$

If all the Voronoi cells are congruent (i.e. $\mathcal{V}(c_k) = \mathcal{V}$) as for lattices, again we write

$$P_e = 1 - \frac{1}{(\sigma\sqrt{2\pi})^n} \int_{\mathcal{V}} e^{-\frac{\|\mathbf{x}\|^2}{2\sigma^2}} \, d\mathbf{x}. \tag{2.7}$$

For practical applications only a finite number M of points of the lattice within a bounded region containing the origin are used. This signal constellation is called a *lattice code* or *lattice constellation*. The error probability (2.7) is exact only for an infinite lattice. When a finite lattice constellation is considered, the assumption that the Voronoi cells are congruent is no longer valid. In fact the points laying on the edge of the constellation have unbounded Voronoi region while the inner points have lattice Voronoi regions. Since the unbounded regions contain the Voronoi cell of the lattice, Eq. (2.7) gives an upper bound to the error probability of a lattice code. The larger the number of points M of the lattice code the tighter is the bound.

The general channel coding problem for the Gaussian channel asks for the code C which minimizes P_e in (2.6) under the signal energy constraint

$$\|c_k\| \leq P = \text{constant} \quad \text{for } k = 1, \dots, M.$$

The lattice version of the Gaussian channel coding problem is to find, for a given value of σ, the n-dimensional lattice of determinant 1 for which (2.7) is minimized [9].

For a lattice with determinant $\det(\Lambda)$ we follow the suggestion of Forney [13] and Tarokh [34], and define the SNR as the following so that the error probability as a function of SNR remains independent of the lattice scaling factor.

Definition 2.16. Let $S_{\mathcal{V}}$ denote the n-dimensional sphere of radius r about the origin, having the same volume as $\mathcal{V}$. Define the normalized radius ρ_0 of $S_{\mathcal{V}}$ by the relation $r^2 = n\rho_0^2$. Then the lattice SNR is defined as

$$\text{SNR} = \frac{\rho_0^2}{\sigma^2}. \tag{2.8}$$

The volume of $S_{\mathcal{V}}$ is $V_n r^n = \text{vol}(\mathcal{V}) = \det(\Lambda)$, thus

$$\text{SNR} = \frac{\det(\Lambda)^{2/n}}{n\sigma^2 V_n^{2/n}}. \tag{2.9}$$

The difficulty in calculations involving the lattice coding problem lies mainly in the complexity of evaluation of the error probability of (2.7). To overcome this difficulty most investigators have replaced (2.7) by simpler expressions.

We assume that all the points are transmitted with equal probability. Due to the geometric uniformity of the lattice [12] it is enough to consider the probability of decoding a point different from $\mathbf{0}$, when $\mathbf{0}$ is transmitted. Let $\mathbf{y}$ be the received vector when $\mathbf{0}$ is transmitted. The components of $\mathbf{y}$ are Gaussian distributed random variable with zero mean and variance σ^2. Now let $\mathbf{p_i}$ be the Voronoi-relevant points around $\mathbf{0}$. Using the union bound [22] with all the Voronoi-relevant hyperplanes only, we obtain the upper bound [9]

$$
P_e \leq P\left(\bigcup_i \{\langle \mathbf{y}, \mathbf{p_i}\rangle \geq \|\mathbf{p_i}\|^2/2\}\right)
$$

$$
\leq \sum_i P\left(\bigcup_i \{\langle \mathbf{y}, \mathbf{p_i}\rangle \geq \|\mathbf{p_i}\|^2/2\}\right) = \sum_i \frac{1}{2}\mathrm{erfc}\left(\frac{\|\mathbf{p_i}\|/2}{\sqrt{2}\sigma}\right), \quad (2.10)
$$

where

$$
\mathrm{erfc}(x) = \frac{2}{\sqrt{\pi}}\int_x^\infty e^{-t^2}\,dt.
$$

In general this bound is very difficult to evaluate, since it is difficult to find all the Voronoi-relevant points.

Well-known upper and lower bounds [29] are given by

$$
P_e < P(\mathbf{y} \notin S_\rho) = 1 - \Gamma\left(\frac{n}{2}, \frac{\rho^2}{2\sigma^2}\right)\bigg/\Gamma\left(\frac{n}{2}\right) \quad (2.11)
$$

and

$$
P_e > P(\mathbf{y} \notin S_\Pi) = 1 - \Gamma\left(\frac{n}{2}, \frac{r^2}{2\sigma^2}\right)\bigg/\Gamma\left(\frac{n}{2}\right), \quad (2.12)
$$

where S_ρ and S_Π are n-dimensional spheres of radius ρ, and r, respectively, and

$$
\Gamma(a, x) = \int_0^x t^{a-1}e^{-t}\,dt,
$$

is the incomplete gamma function. By using the definitions of coding gain and SNR these bounds can be written as

$$
P_e < P(\mathbf{y} \notin S_\rho) = 1 - \Gamma\left(\frac{n}{2}, \frac{nV_n^{2/n}}{8}\gamma(\Lambda)\mathrm{SNR}\right)\bigg/\Gamma\left(\frac{n}{2}\right) \quad (2.13)
$$

and

$$
P_e > P(\mathbf{y} \notin S_\Pi) = 1 - \Gamma\left(\frac{n}{2}, \frac{n}{2}\mathrm{SNR}\right)\bigg/\Gamma\left(\frac{n}{2}\right). \quad (2.14)
$$

When SNR is large and $\sigma \to 0$, a simple estimation of the probability of error in (2.10) is given by [9]

$$P_e \approx \frac{\tau}{2} \text{erfc} \left(\frac{\rho}{\sigma\sqrt{2}} \right). \qquad (2.15)$$

This approximation is obtained from (2.10) by considering only the τ largest terms of the sum.

If we use the definition of the packing density and substitute $\gamma(\Lambda)$ from (2.2) and (2.3), we have

$$P_e \approx \frac{\tau}{2} \text{erfc} \left(\sqrt{\frac{n}{8} V_n^{2/n} \gamma(\Lambda).\text{SNR}} \right),$$

$$= \frac{\tau}{2} \text{erfc} \left(\sqrt{\frac{n}{2} \Delta^{2/n} \text{SNR}} \right).$$

These equations simply show the dependency of the probability of error on the lattice parameters, SNR, and the lattice packing problem.

2.4. Trellis Diagram and Label Code of Lattices

The **trellis** *of an n-dimensional lattice* [10, 14], Λ is an efficient graph representation of Λ in terms of the cosets of Λ', an n-dimensional orthogonal sublattice of Λ, in Λ . Therefore, it is natural to consider the construction of the Tanner graph for Λ based on Λ/Λ'. In this section, we describe the trellis diagram of lattices and their label code, and then apply them to the hexagonal lattice A_2y defined in Example 2.7. Consider a nested sequence of vector spaces

$$\{0\} \subseteq V_0 \subseteq V_1 \subseteq \cdots \subseteq V_n = \mathbb{R}^n,$$

where $\dim(V_i) = i$. Let W_i be the orthogonal complement of V_{i-1} in V_i, that is

$$V_i = V_{i-1} \oplus W_i, \quad 1 \leq i \leq n.$$

For any subspace V of $\mathbb{R}^n$, the **cross section** Λ_V of Λ is defined as the set of lattice points that lie in V

$$\Lambda_V \stackrel{\Delta}{=} \Lambda \cap V.$$

It can be seen that the cross sections Λ_{V_i} and Λ_{W_i} have a lattice structure.

We use the notation P_V for the projection operator onto the vector space V.

Definition 2.17. The ***state space*** $\Sigma_i(\Lambda)$ at level i, $0 \leq i \leq n$, is defined as the quotient group $P_{V_i}(\Lambda)/\Lambda_{V_i}$ and the ***label group*** $G_i(\Lambda)$ of Λ at trellis section i, $1 \leq i \leq n$ is defined as the quotient group $P_{W_i}(\Lambda)/\Lambda_{W_i}$, that is,

$$P_{V_i}(\Lambda)/\Lambda_{V_i} = \{P_{V_i}(\mathbf{x}) + \Lambda_{V_i} : \mathbf{x} \in \Lambda\},$$
$$P_{W_i}(\Lambda)/\Lambda_{W_i} = \{P_{W_i}(\mathbf{x}) + \Lambda_{W_i} : \mathbf{x} \in \Lambda\}.$$

The *trellis* T of the lattice Λ is defined as a directed graph whose nodes (states) at each level i, $0 \leq i \leq n$, are the elements of $\Sigma_i(\Lambda)$. Each edge between level $i - 1$ and i (in trellis section i) is labeled by the elements of $G_i(\Lambda)$. The set of all paths through the trellis diagram corresponds to Λ, i.e. for each lattice point $x \in \Lambda$, there is a path through T which starts from the initial state $\Sigma_0(\Lambda)$ and ends at the final state $\Sigma_n(\Lambda)$.

Definition 2.18. For every lattice point $\mathbf{x} \in \Lambda$, the *state sequence* $\sigma(\mathbf{x})$ and ***label sequence*** $\mathbf{g}(\mathbf{x})$ are defined

$$\sigma(\mathbf{x}) = (\sigma_0(\mathbf{x}), \ldots, \sigma_n(\mathbf{x})), \quad \mathbf{g}(\mathbf{x}) = (g_0(\mathbf{x}), \ldots, g_n(\mathbf{x})),$$

where

$$\sigma_i(\mathbf{x}) = P_{V_i}(\mathbf{x}) + \Lambda_{V_i}, \quad g_i(\mathbf{x}) = P_{W_i}(\mathbf{x}) + \Lambda_{W_i}.$$

It is clear that $\sigma_0(\mathbf{x}) = \Sigma_0(\Lambda), \sigma_n(\mathbf{x}) = \Sigma_n(\Lambda)$. For every $\mathbf{x} \in \Lambda$, the path passes through the state sequence $\sigma(\mathbf{x})$ and label sequence $\mathbf{g}(\mathbf{x})$.

Definition 2.19. The set of all possible state sequences

$$\boldsymbol{\sigma}(\Lambda) = \{\boldsymbol{\sigma}(\mathbf{x}) : \mathbf{x} \in \Lambda\},$$

is called the *state code* and the set of all possible label sequences

$$\mathbf{L} = \mathbf{g}(\Lambda) = \{\mathbf{g}(\mathbf{x}) : \mathbf{x} \in \Lambda\},$$

is called the ***label code***.

It can be shown that $\boldsymbol{\sigma}(\Lambda)$ and $\mathbf{g}(\Lambda)$ are isomorphic [15]. Their cardinality is denoted by $N(\Lambda)$, which is equal to the number of distinct paths in the trellis diagram of Λ.

In the above construction, the ordered system of subspaces $\mathrm{S} = \{W_i\}_{i=1}^n$ corresponding to the sequence $\{V_i\}_{i=1}^n$, is called (trellis) *coordinate system* of Λ.

The lattice Λ is said to have a *finite trellis* if there exists a trellis diagram for Λ with a finite number of states (or edges). It can be shown [3] that Λ has a finite trellis if and only if it has an n-dimensional orthogonal sublattice

$$\Lambda' = \Lambda_{W_1} \oplus \cdots \oplus \Lambda_{W_n}.$$

In this case

$$N(\Lambda) = \frac{\det \Lambda'}{\det \Lambda}.$$

Therefore the number of paths in the trellis T of Λ, which is equal to $|\mathbf{L}|$, is the number of cosets of Λ' in Λ. Hence the trellis of Λ is an efficient way of representing Λ as the union of cosets of Λ' in Λ.

Theorem 2.20. *Let Λ be an n-D lattice with finite Trellis. Then in any trellis coordinate system that Λ and Λ^* have finite trellises, the sizes of their label groups are the same, i.e. for all $1 \leq i \leq n$ we have:*

$$g_i(\Lambda) = g_i(\Lambda^*). \tag{2.16}$$

Proof. The proof is given in [3].   □

Suppose Λ has an n-dimensional orthogonal sublattice Λ' and Λ' has a set of basis vectors along the orthogonal subspaces $S = \{W_i\}_{i=1}^n$. To construct the Tanner graph for a lattice Λ we apply the Tanner graph construction for group codes to the label code $\mathbf{L}$ of Λ, which is an Abelian group code over the alphabet sequence space $\mathbf{G} = G_1 \times \cdots \times G_n$.

Let $|G_i| = g_i$ and let v_i be the shortest vector of Λ_{W_i}, i.e. $\Lambda_{W_i} = Zv_i$, then each element of G_i can be written of the form

$$\Lambda_{W_i} + j \det(P_{W_i}) \frac{v_i}{\|v_i\|},$$

for $j = 0, 1, \ldots, g_i - 1$. Then the map

$$\Phi : G_i \to Z_{g_i},$$

defined by

$$\Lambda_{W_i} + j \det P_{W_i}(\Lambda) \frac{v_i}{\|v_i\|} \mapsto j,$$

is an isomorphism between G_i and Z_{g_i}. Based on this isomorphism, we assume $\mathbf{G} = Z_{g_1} \times \cdots \times Z_{g_n}$.

In general, the elements of each label group G_i are of the form $(Z+a_j)v_i$ where

$$a_j = j \frac{\det(P_{W_i}(\Lambda))}{\det(\Lambda_{W_i})} \quad \text{for } j = 0, 1, \ldots, g_i - 1.$$

Example 2.21. Consider the hexagonal lattice A_2 [9] in Fig. 2.3 that is generated by

$$B = \begin{bmatrix} 1 & 0 \\ 1/2 & \sqrt{3}/2 \end{bmatrix}.$$

Here $V_0 = \{0\}$, $V_1 = \{(x,0) : x \in \mathbb{R}\}$, and $V_2 = \mathbb{R}^2$. Therefore, the trellis coordinate system is $S = \{W_1, W_2\}$ where

$$W_1 = V_1 = \langle(1,0)\rangle = \{(x,0) : x \in \mathbb{R}\}, \quad W_2 = \langle(0,1)\rangle = \{(0,y) : y \in \mathbb{R}\}.$$

The projections of the lattice points onto W_i and V_i are

$$P_{W_1}(\Lambda) = P_{V_1}(\Lambda) = \frac{1}{2}\mathbb{Z}, \quad P_{W_2}(\Lambda) = \frac{\sqrt{3}}{2}\mathbb{Z}.$$

It is clear that $P_{V_0} = \{0\}$ and $P_{V_2} = \Lambda$.

For the cross sections we have

$$\Lambda_{W_1} = \mathbb{Z}, \quad \Lambda_{W_2} = \sqrt{3}\mathbb{Z}.$$

The label groups are:

$$G_1(\Lambda) = \frac{\frac{1}{2}\mathbb{Z}}{\mathbb{Z}} = \left\{\mathbb{Z}, \frac{1}{2} + \mathbb{Z}\right\} \cong \mathbb{Z}_2,$$

$$G_2(\Lambda) = \frac{\frac{\sqrt{3}}{2}\mathbb{Z}}{\sqrt{3}\mathbb{Z}} = \left\{\sqrt{3}\mathbb{Z}, \frac{\sqrt{3}}{2} + \sqrt{3}\mathbb{Z}\right\} \cong \mathbb{Z}_2,$$

and for the label sequences we have

$$g_1(\mathbf{x}) = P_{W_1}(\mathbf{x}) + \Lambda_{W_1} = \frac{1}{2}n + \mathbb{Z}, \quad n \in \mathbb{Z},$$

$$g_2(\mathbf{x}) = P_{W_2}(\mathbf{x}) + \Lambda_{W_2} = \frac{\sqrt{3}}{2}n + \sqrt{3}\mathbb{Z}, \quad n \in \mathbb{Z}.$$

For n even, $g_1(\mathbf{x}) = \mathbb{Z} \leftrightarrow 0$ and $g_2(\mathbf{x}) = \sqrt{3}\mathbb{Z} \leftrightarrow 0$ and for n odd, $g_1(\mathbf{x}) = \frac{1}{2} + \mathbb{Z} \leftrightarrow 1$ and $g_2(\mathbf{x}) = \frac{\sqrt{3}}{2} + \sqrt{3}\mathbb{Z} \leftrightarrow 1$. Therefore, the label code $\mathbf{L}$ of the lattice is $\mathbf{L} = \{00, 11\}$. The trellis diagram is illustrated in Fig. 2.3.

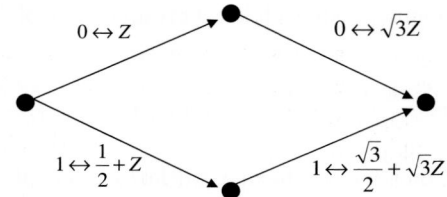

Fig. 2.3.   Trellis diagram of the hexagonal lattice A_2.

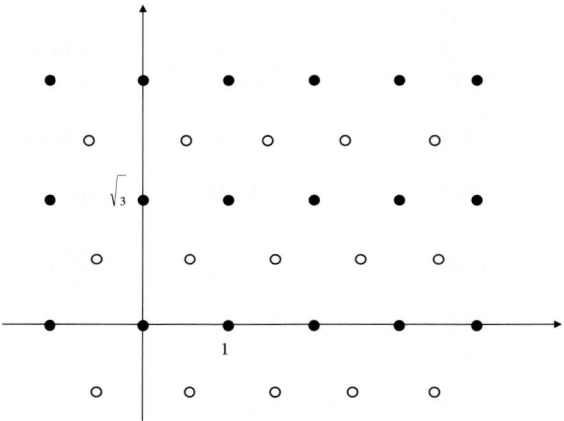

Fig. 2.4.   The hexagonal lattice A_2.

In Fig. 2.4 the orthogonal sublattice $\mathbb{Z} \oplus \sqrt{3}\mathbb{Z}$, denoted by black dots, corresponds to the top path. The other coset of this sublattice corresponds to the bottom path. The label code **L** of this lattice is **L** $= \{00, 11\}$ which is a repetition code and has check equation

$$x_1 + x_2 = 0 \mod 2.$$

2.5. Constructions of Lattices from Codes

2.5.1. *Construction A*

Let $C = [n, M, d_{\min}]$ be a (not necessarily linear) binary code of length n with M codewords. We construct Λ as follows: $x \in \mathbb{R}^n$ is in Λ if and only if the n-tuple $(x_1, \ldots, x_n)$ is congruent modulo 2 to a codeword of C, i.e

$$x \in \Lambda \text{ if and only if } x \equiv c \mod 2, \text{ for some } c \in C.$$

On the unit cube at origin $\{(x_1, \ldots, x_n) : 0 \leq x_i \leq 1, i = 1, \ldots, n\}$, the centers (lattice points) are the M codewords, each one marked on the vertices. All other centers are obtained by adding even integers to any of the coordinates of a codeword. Hence all the centers may be obtained by repeating a building block consisting of a $2 \times 2 \times \cdots \times 2$ cube with the unit cube in one corner.

It can be seen that when $C = [n, k, d_{\min}]$ is linear, $M = 2^k$, then Λ has a lattice structure. It is shown [9] that the packing radius and the center density of Λ are $\rho = \frac{1}{2} \min\{2, \sqrt{d_{\min}}\}$, and $\delta = 2^k \rho^n 2^{-n}$. Therefore, the minimum distance $d_{\min}(\Lambda)$ of Λ is

$$d_{\min}(\Lambda) = \min\{2, \sqrt{d_{\min}}\},$$

and its coding gain is

$$\gamma(\Lambda) = 4\delta^{2/n} = 4^{k/n} \frac{d_{\min}^2(\Lambda)}{4}.$$

Hence

$$\gamma(\Lambda) = \begin{cases} 4^{\frac{k}{n}}, & d_{\min} \geq 4, \\ \frac{d_{\min}^2}{4} 4^{\frac{k}{n}}, & d_{\min} < 4. \end{cases}$$

Figure 2.5 shows an example of the Construction A applied to the repetition code $C = \{00, 11\}$. We can imagine that the lattice is obtained by shifting the square (cube 2×2) over all the plane. The translation of the two points $(0,0)$ and $(1,1)$ appear once in every translated square and form a lattice. The lattice has basis vectors $v_1 = (1, 1)$ and $v_2 = (0, 2)$.

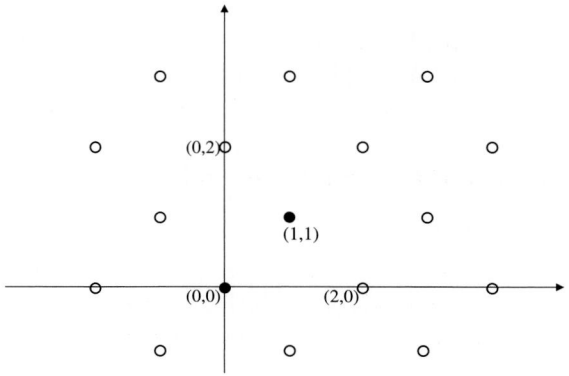

Fig. 2.5.   Construction A of the repetition code.

Construction A can be generalized to group codes as the following.

Definition 2.22. Let C be a group code over $G = \mathbb{Z}_{g_1} \times \cdots \times \mathbb{Z}_{g_n}$. Define Λ as

$$\Lambda = \mathbb{Z}^n \mathrm{diag}(g_i) + C,$$

where $\mathrm{diag}(g_i)$ is the diagonal matrix with diagonal $(g_1, \ldots, g_n)$. Clearly

$$\Lambda = \{(z_1 g_1 + c_1, \ldots, z_n g_n + c_n) : z_i \in \mathbb{Z}, \, c = (c_1, \ldots, c_n) \in C\}.$$

Theorem 2.23. *The set Λ has a lattice structure with*

$$d_{\min}(\Lambda) = \min\{g_{\min}, \sqrt{d_{\min}}\}, \quad \det \Lambda = \frac{\prod_{i=1}^{n} g_i}{|C|},$$

where $g_{\min} = \min\{g_1, \ldots, g_n\}$ and $d_{\min}$ is the minimum weight of C.

Proof. The code C has group structure, so for any two elements x_1 and x_2 in Λ, $x_1 + x_2$ is in Λ. Consider two distinct lattice points x_1, x_2 in Λ. If x_1 and x_2 are congruent to the same codeword, i.e. for some $c \in C$

$$x_1 = (g_1 z_1, \ldots, g_n z_n) + c, \quad x_2 = (g_1 z_1', \ldots, g_n z_n') + c,$$

then their distance apart is at least $g_{\min}$.

If x_1 and x_2 are congruent to different codewords then they differ by at least 1 in at least $d_{\min}$ places. Therefore, their distance apart is at least $\sqrt{d_{\min}}$. Hence

$$d_{\min}(\Lambda) = \min\{g_{\min}, \sqrt{d_{\min}}\}.$$

Let $\Lambda' = g_1 \mathbb{Z} \oplus \cdots \oplus g_n \mathbb{Z}$ then $\Lambda = \Lambda' + C$ where Λ' is the orthogonal sublattice of Λ with $\det \Lambda' = \prod_{i=1}^{n} g_i$. The lattice Λ is the union of $|C|$ cosets of Λ' in Λ, so

$$|C| = \frac{\det \Lambda'}{\det \Lambda}.$$

Therefore,

$$\det \Lambda = \frac{\prod_{i=1}^{n} g_i}{|C|}.$$

$\square$

Another generalization of Construction A is given in [2].

2.5.2. *Constructions B and C*

Construction B (Leech [21]). Let $C = [n, k, d_{\min} = 8]$ be a binary linear code with the property that the weight of each codeword is even. The lattice Λ is constructed in the following way: $x = (x_1, \ldots, x_n) \in \mathbb{R}^n$ is in Λ if and only if x is congruent (modulo 2) to a codeword of C and $\sum_{i=1}^n x_i$ is divisible by 4.

The number of centers is half that of Construction A, so

$$\delta = 2^k \rho^n 2^{-n-1}.$$

Here $\rho = \frac{1}{2}\sqrt{8}$ therefore the coding gain is

$$\gamma(\Lambda) = 4\delta^{\frac{2}{n}} = 2 \times 4^{\frac{k-1}{n}}.$$

Since $1 \leq k \leq n$ then $1 \leq \gamma(\Lambda) \leq 24^{(n-1)/n} < 8$. Hence for n large we cannot gain more than 8, while much denser lattices exist. Also we are restricted to the class of linear codes with $d_{\min} = 8$. Therefore for the codes with large length n, this construction cannot be successful.

Before introducing Construction C we comment on the coordinate array of a point. The coordinate array of a point $x = (x_1, \ldots, x_n)$ with integer coordinates is obtained by writing the binary expansion of the coordinates x_i in columns, beginning with the least significant digit. Complementary notation is used for negative numbers.

Let $C_\ell = [n, M_\ell, d_{\min}^\ell]$, $\ell = 0, 1, \ldots, a$, be a family of $a + 1$ codes (not necessarily linear) with M_ℓ codewords and $d_{\min}^\ell = \alpha 4^{a-\ell}$, where $\alpha = 1$ or 2. A sphere packing in $\mathbb{R}^n$ is given by the following construction.

Construction C (Leech [21]). A point x with integer coordinates is a center if and only if the 2^ℓth row of the coordinate array of x is in C_ℓ, for $\ell = 0, 1, \ldots, a$. It can be shown [9] that the radius of the spheres is $\rho = \sqrt{\alpha}2^{a-\ell}$ and the center density is

$$\delta = M_0 M_1 \cdots M_a \alpha^{\frac{2}{n}} 2^{-2n}.$$

Therefore we can show that if the codes are linear with $\dim(C_\ell) = k_\ell$, then

$$\delta = 2^{\sum_{\ell=0}^a k_\ell - 2n} \quad \text{if } \alpha = 1,$$
$$\delta = 2^{\sum_{\ell=0}^a k_\ell - \frac{3n}{2}} \quad \text{if } \alpha = 2.$$

Hence the coding gain is

$$\gamma = 4^{\sum_{\ell=0}^a \frac{k_\ell}{n} - 1} \quad \text{if } \alpha = 1,$$
$$\gamma = \frac{1}{2} 4^{\sum_{\ell=0}^a \frac{k_\ell}{n}} \quad \text{if } \alpha = 2.$$

In general a non-lattice packing is obtained, in fact the multiplication of a center by an integer may not be a center. A modification of this construction (Construction D) is given in the following section. This one always produces a lattice packing. Construction C is also restricted to a class of codes with given minimum distance.

2.5.3. *Construction D*

This construction, first given in [4], uses a nested family of binary codes to produce a lattice packing Λ in $\mathbb{R}^n$.

Let $\alpha = 1$ or 2 and let

$$C_0 \supseteq C_1 \supseteq \cdots \supseteq C_a,$$

be a family of $a + 1$ linear codes where $C_\ell = [n, k_\ell, d^\ell_{\min}]$ with $d^\ell_{\min} \geq \frac{4^\ell}{\alpha}$, for $\ell = 1, \ldots, a$ and C_0 be the $[n, n, 1]$ trivial code $\mathbb{F}_2^n$. Choose a basis $\{\mathbf{c}_1, \ldots, \mathbf{c}_n\}$ for $\mathbb{F}_2^n$ such that

$$C_\ell = \langle \mathbf{c}_1, \ldots, \mathbf{c}_{k_\ell} \rangle.$$

For any element $\mathbf{x} = (x_1, \ldots, x_n) \in \mathbb{F}_2^n$ and for $\ell = 1, \ldots, a$ consider

$$\frac{1}{2^{\ell-1}}\mathbf{x} = \left(\frac{x_1}{2^{\ell-1}}, \ldots, \frac{x_n}{2^{\ell-1}} \right),$$

as a vector in $\mathbb{R}^n$. Define $\Lambda \subseteq \mathbb{R}^n$ to consist of all vectors of the form

$$\mathbf{z} + \sum_{\ell=1}^{a} \sum_{j=1}^{k_\ell} \alpha_j^{(\ell)} \frac{1}{2^{\ell-1}} \mathbf{c}_j, \tag{2.17}$$

where $\mathbf{z} \in (2\mathbb{Z})^n$ and $\alpha_j^{(\ell)} = 0$ or 1.

When $a = \alpha = 1$, Construction D reduces to the linear case of Construction A. When $a = \alpha = 2$, C_1 is the $[n, n-1, 2]$ even weight code, C_2 has minimum distance 8 and the weight of every codeword of C_2 is a multiple of 4, then Construction D reduces to Construction B.

Theorem 2.24. *Suppose that C_i's are as above and b_t's for $1 \leq t \leq n-k_1$ are $n - k_1$ vectors of the form $(0, \ldots, 0, 2, 0, \ldots, 0)$. Define H to be*

$$H = [c_1, \ldots, c_{k_a}, \ldots, 2^{a-1}c_{k_2+1}, \ldots, 2^{a-1}c_{k_1}, 2^a b_1, \ldots, 2^a b_{n-k_1}],$$

then $x \in \Lambda^*$ if and only if $Hx^T \equiv 0 \mod 2^{a-1}$.

Proof. We see that $\Lambda^* = \{x \in R^n | \langle x, c \rangle \in Z, \ \forall c \in B_\Lambda \}$ where B_Λ is a basis for Λ. The set B_Λ contains vectors of the form $\frac{1}{2^{l-1}}c_j$, for $1 \leq l \leq a$

and $k_{l+1} + 1 \leq j \leq k_l$ plus $n - k_1$ vectors b_t. Then $x \in \Lambda^*$ if and only if $\langle x, \frac{1}{2^{l-1}} c_j \rangle \in Z$ and $\langle x, b_t \rangle \in Z$ for $1 \leq l \leq a$, $k_{l+1} + 1 \leq j \leq k_l$, and $1 \leq t \leq n - k_1$. By multiplying $\frac{1}{2^{l-1}} c_j$ appropriate powers of 2, we have:

$$x \in \Lambda^* \quad \text{if and only if} \quad \langle x, c_j \rangle \equiv 0, \langle x, e_t \rangle \equiv 0 \mod 2^{a-1} \qquad (2.18)$$

for $1 \leq l \leq a$, $k_{l+1} + 1 \leq j \leq k_l$, and $1 \leq t \leq n - k_1$.

So $x \in \Lambda^*$ if and only if (2.18) holds, this is equivalent to $Hx^T \equiv 0$ mod 2^{a-1}. $\qquad \square$

Theorem 2.25. *If Λ be a lattice constructed using Construction D. Let H be defined as above. For any $1 \leq j \leq n$ and for $k_{l+1} + 1 \leq s_j \leq k_l$, let s_j be the smallest number such that*

$$[H]_{s_j, j} \neq 0.$$

Then $(\Lambda^*)_{W_j} = 2^{l-1} Z$ where $W_j = \langle e_j \rangle$.

Proof. The proof is similar to the proof of the Theorem 2.34. $\qquad \square$

Corollary 2.26. *Let Λ be a lattice constructed using Construction D. Then the sizes of the alphabet sequences of the label code of Λ are powers of 2.*

Proof. Based on previous theorem, we have

$$g_i(\Lambda^*) = \frac{2^{l-1}}{2^{g(i)}}$$

where $g(i) \in \{0, \ldots, l - 1\}$. Now by Theorem 2.20 the proof is complete. $\qquad \square$

Example 2.27. Suppose that we have constructed lattice Λ by $C_1 = [8, 4, 4]$ (Hamming code H_8) using Construction D, then the generator matrix of Λ is

$$B = \begin{bmatrix} 1 & 1 & 1 & 1 & 1 & 1 & 1 & 1 \\ 0 & 1 & 0 & 1 & 0 & 1 & 0 & 1 \\ 0 & 0 & 1 & 1 & 0 & 0 & 1 & 1 \\ 0 & 0 & 0 & 0 & 1 & 1 & 1 & 1 \\ 0 & 0 & 0 & 2 & 0 & 0 & 0 & 0 \\ 0 & 0 & 0 & 0 & 0 & 2 & 0 & 0 \\ 0 & 0 & 0 & 0 & 0 & 0 & 2 & 0 \\ 0 & 0 & 0 & 0 & 0 & 0 & 0 & 2 \end{bmatrix}.$$

In order to find g_i's for $1 \leq i \leq 8$, first we compute Gram Schmidt coefficient matrix and its inverse

$$[\mu_{ij}]_B = \begin{bmatrix} 1 & 0 & 0 & 0 & 0 & 0 & 0 & 0 \\ \frac{1}{2} & 1 & 0 & 0 & 0 & 0 & 0 & 0 \\ \frac{1}{2} & 0 & 1 & 0 & 0 & 0 & 0 & 0 \\ \frac{1}{2} & 0 & 0 & 1 & 0 & 0 & 0 & 0 \\ \frac{1}{4} & \frac{1}{2} & \frac{1}{2} & \frac{-1}{2} & 1 & 0 & 0 & 0 \\ \frac{1}{4} & \frac{1}{2} & \frac{-1}{2} & \frac{1}{2} & 0 & 1 & 0 & 0 \\ \frac{1}{4} & \frac{-1}{2} & \frac{1}{2} & \frac{1}{2} & 0 & 0 & 1 & 0 \\ \frac{1}{4} & \frac{1}{2} & \frac{1}{2} & \frac{1}{2} & \frac{-1}{2} & \frac{-1}{2} & \frac{-1}{2} & 1 \end{bmatrix},$$

$$[\mu_{ij}]_B^{-1} = \begin{bmatrix} 1 & 0 & 0 & 0 & 0 & 0 & 0 & 0 \\ \frac{-1}{2} & 1 & 0 & 0 & 0 & 0 & 0 & 0 \\ \frac{-1}{2} & 0 & 1 & 0 & 0 & 0 & 0 & 0 \\ \frac{-1}{2} & 0 & 0 & 1 & 0 & 0 & 0 & 0 \\ 0 & \frac{-1}{2} & \frac{1}{2} & \frac{1}{2} & 1 & 0 & 0 & 0 \\ 0 & \frac{-1}{2} & \frac{1}{2} & \frac{-1}{2} & 0 & 1 & 0 & 0 \\ 0 & \frac{1}{2} & \frac{-1}{2} & \frac{-1}{2} & 0 & 0 & 1 & 0 \\ \frac{1}{2} & \frac{-3}{4} & \frac{-3}{4} & \frac{-3}{4} & \frac{1}{2} & \frac{1}{2} & \frac{1}{2} & 1 \end{bmatrix}.$$

So $\eta = (\eta_1, \ldots, \eta_8) = (\frac{1}{4}, \frac{1}{2}, \frac{1}{2}, \frac{1}{2}, \frac{1}{2}, \frac{1}{2}, \frac{1}{2}, 1)$ and $\alpha = (\alpha_1, \ldots, \alpha_8) = (1, 2, 2, 2, 2, 2, 2, 4)$ where η_i's and α_i's are computed by using the method given in [3]. Now we have $g_i = \frac{\alpha_i}{\eta_i} = 4$ for $1 \leq i \leq 8$.

Theorem 2.28. *Let Λ be a lattice constructed using Construction D, then its minimal norm is at least $4/\gamma$, its volume is*

$$det(\Lambda) = 4^{n - \sum_{i=1}^{a} k_i},$$

and hence center density,

$$\delta \geq \gamma^{-n/2} 2^{\sum_{i=1}^{a} k_i - n},$$

and coding gain

$$\gamma(\Lambda) \geq \alpha^{-1} 4^{\sum_{i=1}^{a} \frac{k_l}{n}}.$$

Proof. The proof is in [4].   □

Definition 2.29. The rth order *Reed–Muller code* $R(r, m)$ of length 2^m is formed by appending a zero-check digit[a] to the rth order punctured Reed–Muller code, for $1 \leq r \leq m - 2$. An rth order Reed–Muller code has parameters

$$R(r, m) = \left[n = 2^m, k = \sum_{i=0}^{r} \binom{m}{i}, d = 2^{m-r} \right].$$

The 0th and mth order Reed–Muller codes are the repetition and universal codes, respectively. It is shown [30] that the dual of a Reed–Muller code is a Reed–Muller code and

$$R^*(r, m) = R(m - r - 1, m). \tag{2.19}$$

Definition 2.30. Let $C_{a-r} = R(2r, m)$, where $R(2r, m)$ is the $2r$th Reed–Muller code of length $n = 2^m$. If m is even, take $\alpha = 1$ and $a = m/2$, while if m is odd, take $\alpha = 2$ and $a = (m - 1)/2$. The resulting lattice in called *Barnes–Wall lattice* [4] and is denoted by BW_n. In both cases the radius is $2^{(m-1)/2}$. Using properties of Reed–Muller codes, the density and coding gain of the corresponding lattice are

$$\delta = 2^{\frac{-5n}{4}} n^{\frac{n}{4}}, \quad \gamma(\Lambda) = \sqrt{\frac{n}{2}}.$$

Theorem 2.31. *Let Λ be a lattice constructed based on Construction D, then we have:*

$$d_{\min}(\Lambda) = \min \left\{ 2, d_{\min}^1, \frac{d_{\min}^2}{2}, \frac{d_{\min}^3}{2^2}, \dots, \frac{d_{\min}^a}{2^{a-1}} \right\},$$

where $d_{\min}(C_l)$ is the minimum distance of C_l.

Proof. Denote c^l to be the codeword with minimum weight in C_l for $1 \leq l \leq a$. There exist $\alpha_j^{(l)} \in \{0, 1\}$ such that $c^l = \sum_{j=1}^{k_l} \alpha_j^{(l)} c_j$. Since $\frac{1}{2^{l-1}} c^l$ is in the form of (2.17), then it belongs to Λ. So we have $d_{\min}(\Lambda) \leq \| \frac{1}{2^{l-1}} c^l \| = \frac{1}{2^{l-1}} d_{\min}(C_l)$. This means that $d_{\min}(\Lambda) \leq \min_{1 \leq l \leq a} \{ \frac{1}{2^{l-1}} d_{\min}(C_l) \}$. The vectors b_t's are in the lattice and this completes the proof. $\square$

---

[a]Given a codeword $(c_1, \dots, c_M)$, a zero-check digit is a symbol c_{M+1} for which $c_1 + \cdots + c_M + c_{M+1} = 0$.

2.5.4. *Construction D′*

As we saw Construction D converts a set of generators for a family of nested codes into a set of generators for a lattice. Construction D' [7] converts a set of parity checks defining a family of codes into congruences for a lattice.

Let $\alpha = 1$ or 2, and let

$$C_0 \supseteq C_1 \supseteq \cdots \supseteq C_a,$$

be a family of binary linear codes, where C_ℓ has parameters $[n, k_\ell, d^\ell_{\min}]$ such that

$$d^\ell_{\min} \geq \alpha 4^\ell \quad \ell = 0, 1, \ldots, a. \qquad (2.20)$$

Let $\mathbf{h}_1, \ldots, \mathbf{h}_n$ be linear independent vectors in $\mathbb{F}_2^n$ such that for $\ell = 0, 1, \ldots, a$ the code C_ℓ is defined by the $r_\ell = n - k_\ell$ parity check vectors $\mathbf{h}_1, \ldots, \mathbf{h}_{r_\ell}$ and let $r_{-1} = 0$. Consider vectors $\mathbf{h}_j$ as integral vectors in $\mathbb{R}^n$, with components 0 or 1. Define the new lattice Λ to consist of those $\mathbf{x} \in \mathbb{Z}^n$ that satisfy the congruences

$$\mathbf{h}_j.\mathbf{x} \equiv 0 \mod 2^{\ell+1}, \qquad (2.21)$$

for all $\ell = 0, 1, \ldots, a$ and $r_{a-\ell-1} + 1 \leq j \leq r_{a-\ell}$.

It is enough to see that the points satisfying (2.21) do form a lattice.

Theorem 2.32. *The minimal norm $d^2_{\min}(\Lambda)$ is at least $\alpha 4^a$, and*

$$\det \Lambda = 2^{\sum_{\ell=0}^a r_\ell}. \qquad (2.22)$$

Proof. The proof is given by Bos [7]. A more general result will be given in Theorem 2.37 □

By multiplying the modular equations by appropriate powers of 2, we can restate Construction D'. Indeed $\mathbf{x} \in \Lambda$ if

$$
\begin{array}{lll}
\mathbf{h}_j.\mathbf{x} \equiv 0 & \mod 2^{a+1} & 1 \leq j \leq r_0, \\
2\mathbf{h}_j.\mathbf{x} \equiv 0 & \mod 2^{a+1} & r_0 + 1 \leq j \leq r_1, \\
\quad \vdots & \quad \vdots & \quad \vdots \\
2^\ell \mathbf{h}_j.\mathbf{x} \equiv 0 & \mod 2^{a+1} & r_{\ell-1} + 1 \leq j \leq r_\ell, \\
\quad \vdots & \quad \vdots & \quad \vdots \\
2^{a-1} \mathbf{h}_j.\mathbf{x} \equiv 0 & \mod 2^{a+1} & r_{a-2} + 1 \leq j \leq r_{a-1}, \\
2^a \mathbf{h}_j.\mathbf{x} \equiv 0 & \mod 2^{a+1} & r_{a-1} + 1 \leq j \leq r_a.
\end{array}
\qquad (2.23)
$$

Therefore:

Lemma 2.33. *Let* $C_\ell = [n, k_\ell, d_{\min}^\ell]$, $\ell = 0, 1, \ldots, a$ *and* Λ *be as in the previous theorem. Let*

$$\widehat{H} = \{\mathbf{h}_1, \ldots, \mathbf{h}_{r_0}, 2\mathbf{h}_{r_0+1}, \ldots, 2\mathbf{h}_{r_1}, \ldots, 2^a\mathbf{h}_{r_{a-1}+1}, \ldots, 2^a\mathbf{h}_{r_a}\},$$

and let $\mathbf{H}$ *be the matrix whose rows are the elements of* $\widehat{H}$, *i.e.*

$$\mathbf{H} = [\mathbf{h}_1 \cdots \mathbf{h}_{r_0}\ 2\mathbf{h}_{r_0+1} \cdots 2\mathbf{h}_{r_1} \cdots 2^a\mathbf{h}_{r_{a-1}+1} \cdots 2^a\mathbf{h}_{r_a}]^{tr}.$$

Then $\mathbf{x} \in \mathbb{Z}^n$ *is in* Λ *if*

$$\mathbf{H}.\mathbf{x}^T \equiv \mathbf{0} \quad \mod 2^{a+1}. \tag{2.24}$$

These constitute the check equations of the lattice. Therefore the row vectors of $\mathbf{H}$, i.e. the elements of $\widehat{H}$, form the generators for the dual of the label code of Λ.

The following theorem [25] provides us a computational method of finding cross sections Λ_{W_j}, $j = 1, \ldots, n$.

Theorem 2.34 (Sadeghi *et al.* [25]). *Let H be defined as in the previous lemma. For any $1 \leq j \leq n$, for $r_{k-1} + 1 \leq s_j \leq r_k$, let s_j be the smallest number such that*

$$[H]_{s_j, j} \neq 0.$$

Let Λ_{W_j} be the cross section of Λ in the coordinate system $W_j = \langle e_j \rangle$. Then

$$\Lambda_{W_j} = 2^{a-k+1}\mathbb{Z}.$$

Proof. Let $\mathbf{x} \in \Lambda_{W_j}$, then $\mathbf{x} = (0, \ldots, 0, x_j, 0, \ldots, 0)$ where x_j is in the jth coordinate of $\mathbf{x}$. Therefore, $\mathbf{x} \in \Lambda$ and we have $2^k\mathbf{h}_{s_j} \cdot \mathbf{x} \equiv 0 \mod 2^{a+1}$. Thus $2^k x_j = q2^{a+1}$, for some $q \in \mathbb{Z}$. Hence $x_j = 2^{a-k+1}q$ and we can conclude that

$$\Lambda_{W_j} \subseteq 2^{a-k+1}\mathbb{Z}.$$

Now let $x_j \in 2^{a-k+1}\mathbb{Z}$, then $x_j = q2^{a-k+1}$, for some $q \in \mathbb{Z}$. Let $\mathbf{x} = (0, \cdots, 0, x_j, 0, \cdots, 0)$ where x_j is the jth coordinate of $\mathbf{x}$. In the following, we show that $\mathbf{x}$ is in Λ.

By the definition of s_j the jth coordinate of any $\mathbf{h}_i$ for which $i = 1, \ldots, s_j - 1$, is zero. So

$$\mathbf{x} \cdot \mathbf{h}_i = 0, \quad 1 \le i \le s_j - 1. \tag{2.25}$$

By multiplying both sides of the Eq. (2.25) by any integer m we have $m\mathbf{h}_i \cdot \mathbf{x} \equiv 0 \mod 2^{a+1}$, $1 \le i \le s_j - 1$. In particular, since $r_{k-1} \le s_j - 1$ then

$$2^\ell \mathbf{h}_i \cdot \mathbf{x} \equiv 0 \mod 2^{a+1}, \quad r_{\ell-1}+1 \le i \le r_\ell, \quad \ell = 0, 1, \ldots, k-1. \tag{2.26}$$

On the other hand $\mathbf{h}_i \cdot \mathbf{x} = 0$ or $\mathbf{h}_i \cdot \mathbf{x} = q2^{a-k+1}$, $s_j \le i \le r_a$, thus

$$2^k \mathbf{h}_i \cdot \mathbf{x} \equiv 0 \mod 2^{a+1}, \quad s_j \le i \le r_a. \tag{2.27}$$

We can multiply both sides of the Eq. (2.27) by any integer. Since $r_k \ge s_j$ then

$$2^\ell \mathbf{h}_i \cdot \mathbf{x} \equiv 0 \mod 2^{a+1}, \quad r_{\ell-1}+1 \le i \le r_\ell, \quad \ell = k+1, \ldots, a. \tag{2.28}$$

Equations (2.26) and (2.28) along with the fact that $x_j = q2^{a+1-k}$ imply that $\mathbf{x} \in \Lambda$. This complete the proof. $\qquad\square$

Corollary 2.35. *Let Λ be a lattice constructed using Construction D'. Then the sizes of the alphabet sequences of the label code of Λ are powers of 2. In fact for every $i = 1, \ldots, n$,*

$$g_i = \frac{2^{a+1}}{2^{g(i)}}, \tag{2.29}$$

for some $g(i) \in \{0, 1, \ldots, a\}$.

Corollary 2.36. *Suppose C_0 is a code for which $d_{\min}^0 \ge 2$, then*

$$\Lambda_{W_i} = 2^{a+1}\mathbb{Z}^n, \quad i = 1, \ldots, n. \tag{2.30}$$

Proof. If the jth column of the parity check matrix of C_0 was zero, then $(0, \ldots, 0, 1, 0, \ldots, 0)$, with 1 in its jth position, would be a codeword, contradicting the hypothesis. Thus every column of this matrix is a non-zero vector. Therefore in Theorem 2.34 $k = 0$ and so $\Lambda_{W_i} = 2^{a+1}\mathbb{Z}^n$. $\qquad\square$

By changing the scale and relabeling the codes, Construction D may be restated in such a way that the norm and determinant of the corresponding lattices agree. Then if C_i are Reed–Muller codes, the two lattices coincide. We illustrate this in the following example.

The following theorem generalizes Construction D' to any collection family of nested codes without the condition $d_{\min}^\ell \geq \alpha 4^\ell$. This theorem also relates the performance of the lattice to the performance of its underlying subcodes.

Theorem 2.37 (Sadeghi *et al.* [25]). *Let $C_0 \supseteq C_1 \supseteq \cdots \supseteq C_a$ be a family of linear binary codes with $C_\ell = [n, k_\ell, d_{\min}^\ell]$ and let $\{\mathbf{h}_1, \ldots, \mathbf{h}_n\}$ be linear independent vectors in $\mathbb{F}_2^n$ such that*

$$C_\ell^* = \langle \mathbf{h}_1, \ldots, \mathbf{h}_{r_\ell} \rangle, \quad \ell = 0, 1, \ldots, a.$$

Let Λ be the corresponding lattice given by Construction D'. Then for the minimal norm $d_{\min}^2(\Lambda)$ we have

$$\min\{d_{\min}^a, 4d_{\min}^{a-1}, 4^2 d_{\min}^{a-2}, \ldots, 4^a d_{\min}^0, 4^{a+1}\} \leq d_{\min}^2(\Lambda) \leq 4^{a+1}.$$

Proof. For any $\mathbf{x} \in \Lambda$ we can write the modular equations as

$$\mathbf{x} \cdot \mathbf{h}_j \equiv 0 \mod 2^{a-k+1}, \quad r_{k-1} \leq j \leq r_k,$$

for $k = 0, 1, \ldots, a$. Since $2^{a-k+1} \mid 2^{a-i+1}$, for $i = 0, 1, \ldots, k-1$, we have $\mathbf{x} \cdot \mathbf{h}_j \equiv 0 \mod 2^{a-k+1}$, $1 \leq j \leq r_k$. Define $\Lambda_k = \{\mathbf{x} \in \mathbb{Z}^n : \mathbf{x} \cdot \mathbf{h}_j \equiv 0 \mod 2^{a-k+1}, 1 \leq j \leq r_k\}$. Then each Λ_k is a lattice and by the definition of Λ we have

$$\Lambda = \bigcap_{k=0}^{a} \Lambda_k. \tag{2.31}$$

In the next step, we prove that if $\mathbf{x} \in \mathbb{Z}^n$ has at least one coordinate which is an odd number and

$$\mathbf{x} \cdot \mathbf{h}_j \equiv 0 \mod 2, \quad 1 \leq j \leq r_k, \tag{2.32}$$

then

$$\|\mathbf{x}\|^2 \geq d_{\min}^k. \tag{2.33}$$

To prove this, let $\mathbf{x} = \mathbf{c} + 2\mathbf{z}$, where $\mathbf{z} \in \mathbb{Z}^n$ and $\mathbf{c} = (c_1, \ldots, c_n)$, with $c_i = 0$ or 1. Since $\mathbf{x}$ has at least one odd coordinate, then $\mathbf{c} \neq 0$. Also $\mathbf{h}_j \cdot \mathbf{c} \equiv 0 \mod 2$, therefore $\mathbf{c} \in C_k$ and $\|\mathbf{c}\|^2 \geq d_{\min}^k$. Hence $\|\mathbf{x}\|^2 \geq d_{\min}^k$.

Now let $0 \neq \mathbf{x} \in \Lambda$. Since $\mathbf{x} \neq 0$ we can chose $k \geq 0$ such that $2^{-k}\mathbf{x} \in \mathbb{Z}^k$ and $2^{-k-1}\mathbf{x} \notin \mathbb{Z}^k$. If $k < a+1$ then $\frac{\mathbf{x}}{2^k} \cdot \mathbf{h}_j \equiv 0 \mod 2$, for $1 \leq j \leq r_{a-k}$. So $\|\mathbf{x}\|^2 \geq 4^k d_{\min}^{a-k}$. If $k \geq a+1$ then $\|\mathbf{x}\|^2 \geq (2^{a+1})^2$. Therefore,

$$\|\mathbf{x}\|^2 \geq d_{\min}^a \quad \text{or} \quad \|\mathbf{x}\|^2 \geq 4d_{\min}^{a-1} \quad \text{or} \cdots \text{or}$$

$$\|\mathbf{x}\|^2 \geq 4^a d_{\min}^0 \quad \text{or} \quad \|\mathbf{x}\|^2 \geq 4^{a+1}.$$

This complete the proof for the lower bound on $d_{\min}^2(\Lambda)$. It is trivial that the vector $2^{a+1}\mathbf{e_i} \in \Lambda$ which proves the upper bound on $d_{\min}^2(\Lambda)$. $\qquad \square$

Corollary 2.38. *The coding gain of Construction D' is*

$$\gamma(\Lambda) \geq \frac{\min\{d_{\min}^a, 4d_{\min}^{a-1}, 4^2 d_{\min}^{a-2}, \ldots, 4^a d_{\min}^0, 4^{a+1}\}}{4\sum_{\ell=0}^a \frac{r_\ell}{n}}.$$

Furthermore when $d_{\min}^\ell \geq 4^{\ell+1}$, $\ell = 0, 1, \ldots, a$, then

$$\gamma(\Lambda) = \frac{4^{a+1}}{4\sum_{\ell=0}^a \frac{r_\ell}{n}}. \tag{2.34}$$

Proof. It follows from the last theorem and the definition of coding gain. $\qquad \square$

We note that by using the bound that is given by Bos (Theorem 2.32) we have the following bound on coding gain

$$\gamma(\Lambda) \geq \frac{\alpha 4^a}{4\sum_{\ell=0}^a \frac{r_\ell}{n}}, \tag{2.35}$$

where $\alpha = 1$ or 2.

Corollary 2.39. *Let $C_0 = [n, n, 1]$ be the trivial code, then Construction D' produces a lattice with the same density (and coding gain) as Construction D does. In this case*

$$\delta \geq \alpha^{\frac{n}{2}} 2^{\sum_{\ell=1}^a k_\ell - n},$$

and coding gain

$$\gamma(\Lambda) \geq \alpha 4^{\sum_{\ell=1}^a \frac{k_\ell}{n}},$$

where $\alpha = 1, 2, 3$ or 4.

Proof. By using the fact that $r_0 = 0$, the proof is straightforward. □

Theorem 2.40. *Let $\Lambda_{D'}$ and Λ_A be the two lattices that are constructed from a code $C = [n, k]$ by Constructions D' and A, respectively. Then $\Lambda_{D'}$ and Λ_A coincide.*

Proof. Let $\{h_1, \ldots, h_r\}$ be the parity check vectors of a code C. Let $x \in \Lambda_{D'}$, then by using the Euclidian algorithm we have $\mathbf{x} = \mathbf{c} + 2\mathbf{z}$ where $\mathbf{c} = (c_1, \ldots, c_n)$, with $c_i = 0$ or 1 and $\mathbf{z} = (z_1, \ldots, z_n) \in \mathbb{Z}^n$. Since

$$\mathbf{x} \cdot h_j \equiv 0 \mod 2 \quad \text{and} \quad 2\mathbf{z} \cdot h_j \equiv 0 \mod 2, \quad j = 1, \ldots, r,$$

then $\mathbf{c} \cdot h_j \equiv 0 \mod 2$. Hence $\mathbf{c} \in C$.

Let $\mathbf{x} \in \Lambda_A$ then $\mathbf{x} = \mathbf{c} + 2\mathbf{z}$, where $\mathbf{z} \in Z^n$, $\mathbf{c} \in C$. We have

$$\mathbf{x} \cdot h_j = \mathbf{c} \cdot h_j + 2\mathbf{z}h_j \equiv 0 \mod 2,$$

and hence $\mathbf{x} \in \Lambda_{D'}$. □

Definition 2.41. A lattice Λ constructed based on Construction D' is called LDPC lattice if its parity check matrix is a sparse matrix.

It is trivial that if the underlying nested codes C_ℓ are LDPC codes then the corresponding lattice is an LDPC lattice and vice versa. Construction and decoding analysis of these type of lattices is given in [25].

2.6. Conclusions and Future Research

A detailed introduction to group codes, lattices are given. The relation between lattice and channel coding problem.

Several constructions have been investigated. These constructions are based on binary codes. Construction A is generalized to group codes. This generalization can be used to construct lattices using group codes, e.g. LDPC codes over a finite field $\mathbb{F}_q$ can be used to generate lattices.

Construction A cannot generate a lattice with coding gain more than 4: Construction B in limited to the class of codes with $d_{\min} = 8$. Construction C produces non-lattice packings.

Construction D and D' can have higher coding gain than the others. Turbo codes can be good candidate for both Constructions. The resulting lattice called **Turbo lattice** may be interesting in terms of its performance and its capacity achieving on AWGN channels.

LDPC codes identified by a set of parity check vector seem to be the best candidate to be used as underlying codes in Construction D'.

It would be of interest if one could show that for some of these constructions asymptotically the lattices achieve capacity on AWGN channel.

2.7. Questions

1. Let C be a linear $[8,4]$ code with $d_{\min} = 3$. Let λ be the lattice constructed using Construction A. What is the coding gain of λ?
2. *Barnes–Wall lattice* BW_{16}: Use Construction D with $n = 16$, $a = 2$, $\alpha = 1$ and Reed–Muller codes $C_0 = R(4,4), C_1 = R(2,4), C_3 = R(0,4)$. to generate BW_{16}. Find its generator matrix.
3. Find the determinant, center density and coding gain of BW_{16}.
4. Use Construction D' with Reed–Muller codes $C_0 = R(2,4)$ and $C_1 = R(0,4)$ to produce a 16-dimensional lattice denoted by BW'_{16}. Find its parity check matrix.
5. Find all projections, determinant, cross sections and sizes of alphabet sequences g_i's in the previous question.
6. Show that BW_{16} and BW'_{16} are coincide.
7. Let BW_{64} be the Barnes–Wall lattice generate by Construction D with $a = 2$, $\alpha = 1$ and Reed–Muller codes $C_0 = RM(6,6)$ $C_1 = RM(4,6)$ $C_2 = RM(2,6)$ $C_3 = RM(0,6)$. Find the volume and the coding gain of BW_{16} and explain its generator matrix.
8. Let $a = 2$ and $C_0 = \langle 1111, 0010, 0100 \rangle$, $C_1 = \langle 1111, 0010 \rangle$, and $C_2 = \langle 1111 \rangle$. Find the parity check matrix and the check equations of the lattice obtained from Construction D'.
9. Use Theorem 2.34 to find all cross sections of the lattice in the previous question.
10. By showing that the two vectors $(1,1,1,7)$ and $(7,3,1,1)$ belong to the lattice Λ in Question 7, find the sizes of alphabet sequence of the label code of Λ.

2.8. Answers

1. $d_{\min} < 4$ so $\gamma\Lambda = 4.5$.
2. The rows of the generator matrix are a basis for $C_0 = \mathbb{F}_2^{16}$. The first 11 rows are the generators of C_1 which are divided by 1 and the first row is the generator of C_2 which is divided by 2. The last five rows are

the part of generators of $\mathbb{F}_2^{16}$ that are not included in any set of code generators hence are multiplied by 2.

$$
\begin{bmatrix}
\frac12 & \frac12 & \frac12 & \frac12 & \frac12 & \frac12 & \frac12 & \frac12 & \frac12 & \frac12 & \frac12 & \frac12 & \frac12 & \frac12 & \frac12 & \frac12 \\
0 & 0 & 0 & 0 & 0 & 0 & 0 & 0 & 1 & 1 & 1 & 1 & 1 & 1 & 1 & 1 \\
0 & 0 & 0 & 0 & 1 & 1 & 1 & 1 & 0 & 0 & 0 & 0 & 1 & 1 & 1 & 1 \\
0 & 0 & 1 & 1 & 0 & 0 & 1 & 1 & 0 & 0 & 1 & 1 & 0 & 0 & 1 & 1 \\
0 & 1 & 0 & 1 & 0 & 1 & 0 & 1 & 0 & 1 & 0 & 1 & 0 & 1 & 0 & 1 \\
0 & 0 & 0 & 0 & 0 & 0 & 0 & 0 & 0 & 0 & 0 & 0 & 1 & 1 & 1 & 1 \\
0 & 0 & 0 & 0 & 0 & 0 & 0 & 0 & 0 & 1 & 0 & 1 & 1 & 0 & 1 & 1 \\
0 & 0 & 0 & 0 & 0 & 0 & 0 & 0 & 0 & 1 & 0 & 1 & 0 & 1 & 0 & 1 \\
0 & 0 & 0 & 0 & 0 & 0 & 1 & 1 & 0 & 0 & 0 & 0 & 0 & 0 & 1 & 1 \\
0 & 0 & 0 & 0 & 0 & 1 & 0 & 1 & 0 & 0 & 0 & 0 & 0 & 1 & 0 & 1 \\
0 & 0 & 0 & 1 & 0 & 0 & 0 & 1 & 0 & 0 & 0 & 1 & 0 & 0 & 0 & 1 \\
0 & 0 & 0 & 0 & 0 & 0 & 0 & 2 & 0 & 0 & 0 & 0 & 0 & 0 & 0 & 0 \\
0 & 0 & 0 & 0 & 0 & 0 & 0 & 0 & 0 & 0 & 0 & 0 & 0 & 0 & 2 & 0 \\
0 & 0 & 0 & 0 & 0 & 0 & 0 & 0 & 0 & 0 & 0 & 0 & 0 & 2 & 0 & 0 \\
0 & 0 & 0 & 0 & 0 & 0 & 0 & 0 & 0 & 0 & 0 & 2 & 0 & 0 & 0 & 0 \\
0 & 0 & 0 & 0 & 0 & 0 & 0 & 0 & 0 & 0 & 0 & 0 & 0 & 0 & 0 & 2
\end{bmatrix}.
$$

3. We have $k_0 = 16, k_1 = 11, k_2 = 1$, and $d_1 = 4, d_2 = 16$. So $\det \Lambda = 2^{16-(11+1)} = 16$, $\delta = 2^{-4}$, $\gamma(\Lambda) = \sqrt{2^3}$.

4. By using the properties of Reed–Muller codes we have

$$
C_0^* = R(1,4) \quad \text{and} \quad C_1^* = R(3,4).
$$

The code C_0^* consists of $r_0 = 5$ generators and C_1^* consists of $r_1 = 15$ generators. Put all 15 vectors of C_1^* in the rows of $\mathbf{H}$ in such a way that the first five rows of $\mathbf{H}$ are the generators of C_0^* and the next 11 generators are multiplied by 2. Therefore, the matrix H that defines the modular equations is

$$
\begin{bmatrix}
1 & 1 & 1 & 1 & 1 & 1 & 1 & 1 & 1 & 1 & 1 & 1 & 1 & 1 & 1 & 1 \\
0 & 0 & 0 & 0 & 0 & 0 & 0 & 0 & 1 & 1 & 1 & 1 & 1 & 1 & 1 & 1 \\
0 & 0 & 0 & 0 & 1 & 1 & 1 & 1 & 0 & 0 & 0 & 0 & 1 & 1 & 1 & 1 \\
0 & 0 & 1 & 1 & 0 & 0 & 1 & 1 & 0 & 0 & 1 & 1 & 0 & 0 & 1 & 1 \\
0 & 1 & 0 & 1 & 0 & 1 & 0 & 1 & 0 & 1 & 0 & 1 & 0 & 1 & 0 & 1 \\
0 & 0 & 0 & 0 & 0 & 0 & 0 & 0 & 0 & 0 & 0 & 0 & 2 & 2 & 2 & 2 \\
0 & 0 & 0 & 0 & 0 & 0 & 0 & 0 & 0 & 0 & 2 & 2 & 0 & 0 & 2 & 2 \\
0 & 0 & 0 & 0 & 0 & 0 & 0 & 0 & 0 & 2 & 0 & 2 & 0 & 2 & 0 & 2 \\
0 & 0 & 0 & 0 & 0 & 0 & 2 & 2 & 0 & 0 & 0 & 0 & 0 & 0 & 2 & 2 \\
0 & 0 & 0 & 0 & 0 & 2 & 0 & 2 & 0 & 0 & 0 & 0 & 0 & 2 & 0 & 2 \\
0 & 0 & 0 & 2 & 0 & 0 & 0 & 2 & 0 & 0 & 0 & 2 & 0 & 0 & 0 & 2 \\
0 & 0 & 0 & 0 & 0 & 0 & 0 & 2 & 0 & 0 & 0 & 0 & 0 & 0 & 0 & 2 \\
0 & 0 & 0 & 0 & 0 & 0 & 0 & 0 & 0 & 0 & 0 & 0 & 0 & 0 & 2 & 2 \\
0 & 0 & 0 & 0 & 0 & 0 & 0 & 0 & 0 & 0 & 0 & 0 & 0 & 2 & 0 & 2 \\
0 & 0 & 0 & 0 & 0 & 0 & 0 & 0 & 0 & 0 & 0 & 2 & 0 & 0 & 0 & 2
\end{bmatrix}.
$$

5. It is clear that the vector $(1,1,1,1,1,1,1,1,1,1,1,1,1,1,1,1)$ is in Λ, therefore $\det P_{W_i} = 1$, for $i = 1, 2, \ldots, 16$. Also by using the previous theorem we conclude that $\Lambda_{W_j} = 4$, $j = 1, \ldots, 16$. Therefore the sizes of alphabet sequence are $g_i = 4$, for $i = 1, 2, \ldots, 16$. Here $r_0 = 5$ and $r_1 = 15$ so $\det \Lambda = 2$ [20].

6. By multiplying the generator matrix of the Barnes–Wall lattice BW_{16} by 2, the two lattices coincide.

7. In this case we have $n = 64$, $k_0 = 64$, $k_1 = 57$, $k_2 = 22$, and $k_3 = 1$. Also we have $d_0 = 1$, $d_1 = 4$, $d_2 = 16$, and $d_3 = 64$. Volume of Λ is

$$\det(\Lambda) = 2^{64-57-22-1} = 2^{-16}.$$

The coding gain is $\sqrt{32}$. The first 57 rows of the generators of C_1 which divided by 1 set as first 57 rows of the generator matrix of Λ. Then divide 22 first row of that matrix (which are the generators of C_2) by 2. At the last divide by 2 the first row (which is the generators of C_3) of at hand matrix. The last seven rows are the part of generators of $\mathbb{F}_2^{64}$ that are not included in any set of code generators hence are multiplied by 2.

8. The dual codes are $C_0^* = \langle 1001 \rangle$, $C_1^* = \langle 1001, 0101 \rangle$, and $C_2^* = \langle 1001, 0101, 0011 \rangle$. Hence $h_1 = (1001)$, $h_2 = (0101)$, and $h_3 = (0011)$ and

$$H = \begin{bmatrix} 1 & 0 & 0 & 1 \\ 0 & 2 & 0 & 2 \\ 0 & 0 & 4 & 4 \end{bmatrix}.$$

Then $\mathbf{x} \in \mathbb{Z}^n$ belongs to Λ if and only if

$$H \cdot \mathbf{x}^T \equiv \mathbf{0} \mod 2^3.$$

In other words, $\mathbf{x} = (x_1, x_2, x_3, x_4) \in \Lambda$ if and only if

$$\begin{aligned} x_1 + x_4 &\equiv 0 \mod 8 \\ 2x_2 + 2x_4 &\equiv 0 \mod 8 \\ 4x_3 + 4x_4 &\equiv 0 \mod 8. \end{aligned}$$

9. In Theorem 2.34 we have $s_1 = 1$ ($k = 0$), $s_2 = 2$ ($k = 1$), $s_3 = 3$ ($k = 2$), $s_4 = 1$ ($k = 0$). Therefore, $\Lambda_{W_1} = 8\mathbb{Z}$, $\Lambda_{W_2} = 4\mathbb{Z}$, $\Lambda_{W_3} = 2\mathbb{Z}$, $\Lambda_{W_4} = 8\mathbb{Z}$.

10. $\det P_{W_j} = 1$, for $i = 1, \ldots, 4$. Hence the sizes of the alphabet sequence of the label code of Λ are

$$g_1 = 8, \quad g_2 = 4, \quad g_3 = 2, \quad g_4 = 8.$$

2.9. Keywords

Sequence space

A *sequence space* W is defined as the Cartesian product of n sets of alphabet symbols G_i.

(Block) code

A *block code* C is any subset of a sequence space $W = G_1 \times \cdots \times G_n$.

Group sequence space

Group sequence space W is a sequence space in which all symbol alphabets are groups G_i.

Group code

A *group code* C is a subgroup of a group sequence space W.

Linear code

A q-ary code C over the sequence space $\mathbb{F}_q^n$ is a linear code if C forms a vector subspace over $\mathbb{F}_q$.

Dual code

The dual space of a linear code C is called its dual code.

Lattice

A discrete subgroup of R^n.

Minimum distance of a lattice

The length of the shortest vector in a lattice.

Basis of a lattice

A set of independent lattice vectors that their integer linear combinations produce all lattice vectors.

Dual of a lattice

Given a full rank lattice Λ of $\mathbb{R}^n$, its dual lattice is defined as $\Lambda^* = \{\mathbf{x} \in \mathbb{R}^n : \langle \mathbf{x}, \mathbf{v} \rangle \in \mathbb{Z}, \text{for all } \mathbf{v} \in \Lambda\}$.

The fundamental parallelotope

The set of points $\{\boldsymbol{\theta}\mathbf{B} : \boldsymbol{\theta} = (\theta_1, \ldots, \theta_n) \in \mathbb{R}^n, 0 \leq \theta_i < 1\}$.

Volume of a lattice

Determinant (or volume) of Λ is defined by the determinant of the Gram matrix of Λ, i.e. $\det(\Lambda) = \det(A)^{\frac{1}{2}} = \det(\mathbf{B}\mathbf{B}^T)^{\frac{1}{2}}$.

Fundamental region

A fundamental region of a lattice Λ is defined as a subset of $span(\Lambda)$ which, when translated by lattice points, partitions $span(\Lambda)$ with just one lattice point in each copy.

Voronoi cell

For any point $\mathbf{p} \in \Lambda$ the Voronoi cell $\mathcal{V}(\mathbf{p})$ is defined by the set of those points of $span(\Lambda)$ that are at least as close to $\mathbf{p}$ as to any other point $\mathbf{q} \in \Lambda$.

Coding gain of a lattice

The (fundamental) coding gain (or Hermite parameter) of the lattice Λ is defined as $\gamma(\Lambda) = \frac{d_{\min}^2(\Lambda)}{(\det(\Lambda))^{\frac{2}{n}}}$.

Orthogonal lattice

A lattice is called orthogonal (rectangular) if it has a basis with mutually orthogonal vectors.

Packing density of a lattice

The packing density Δ of a lattice packing is

$$\Delta = \frac{\text{volume of one sphere}}{\text{volume of fundamental region}}.$$

Lattice SNR

Let $S_\mathcal{V}$ denote the n-dimensional sphere of radius r about the origin, having the same volume as $\mathcal{V}$. Define the normalized radius ρ_0 of $S_\mathcal{V}$ by the relation $r^2 = n\rho_0^2$. Then the lattice SNR is defined as $\text{SNR} = \frac{\rho_0^2}{\sigma^2} = \frac{\det(\Lambda)^{2/n}}{n\sigma^2 V_n^{2/n}}$.

Cross section of lattice

For any subspace V of $\mathbb{R}^n$, the *cross section* Λ_V of Λ is defined as the set of lattice points that lie in V.

Label group a lattice

The *Label group* $G_i(\Lambda)$ of Λ at trellis section i, $1 \le i \le n$ is defined as the quotient group $P_{W_i}(\Lambda)/\Lambda_{W_i}$, where P_{W_i} is the projection onto W_i, the ith coordinates system.

Label sequence of a lattice

For every lattice point $\mathbf{x} \in \Lambda$, the *label sequence* $\mathbf{g}(\mathbf{x})$ is defined as $\mathbf{g}(\mathbf{x}) = (g_0(\mathbf{x}), \ldots, g_n(\mathbf{x}))$, where $g_i(\mathbf{x}) = P_{W_i}(\mathbf{x}) + \Lambda_{W_i}$.

Label code of a lattice

The set of all possible label sequences is called the label code of a lattice.

References

1. E. Agrell, Closest point search in lattices, *IEEE Trans. Inform. Theory* **48** (2002), 2201–2214.
2. A. H. Banihashemi and F. R. Kschinschang, Tanner graphs for group codes and lattices: construction and complexity, *IEEE Trans. Inform. Theory* **47** (2001), 824–882.
3. A. H. Banihashemi and I. F. Blake, Trellis complexity and minimal trellis diagram of lattices, *IEEE Trans. Inform. Theory* **44** (1998), 1829–1847.
4. E. S. Barnes and N. J. A. Sloan, New lattice packings of spheres, *Can. J. Math.* **35** (1983), 117–130.
5. E. S. Barnes and G. E. Wall, Some extreme forms defined in terms of Abelian groups, *J. Australian Math. Soc.* **1** (1959), 47–63.
6. C. Berrou, A. Glavieux and P. Thitimajshima, Near shannon limit error-correcting coding and decoding: Turbo codes, *Proc. Int. Conf. on Communications*, 1993, 1064–1070.
7. A. Bos, J. H. Conway and N. J. A. Sloan, Further lattice packings in high dimensions, *Mathematica* **29** (1982), 171–180.
8. T. M. Cover and J. A. Thomas, *Elements of Information Theory* (John Wiley, New York, 1991).
9. J. H. Conway and N. J. A. Sloan, *Sphere Packings, Lattices and Groups*, 3rd edn. (Springer-Verlag, New York, 1998).
10. G. D. Forney Jr., Density/length profile and trellis complexity of lattices, *IEEE Trans. Inform. Theory* **40** (1988), 1753–1772.
11. G. D. Forney Jr., Coset codes. Part I: Introduction and geometrical classification, *IEEE Trans. Inform. Theory* **34** (1988), 1123–1151.
12. G. D. Forney Jr., Geometrically uniform codes", *IEEE Trans. Inform. Theory* **37**, 5 (1991), 1241–1260.
13. G. D. Forney Jr., Approaching AWGN channel capacity with coset codes and multilevel coset codes, *IEEE Inform. Theory, Proc. Int. Symp.*, 1997, 164–166.
14. G. D. Forney Jr., Modulation and coding for linear Gaussian channels, *IEEE Trans. Inform. Theory* **44** (1998), 2304–2414.
15. G. D. Forney Jr. and M. D. Trott, The dynamics of group codes: State spaces, trellis digrams and canonical encoders, *IEEE Trans. Inform. Theory* **IT-39**, 9 (1993), 1491–1513.
16. R. G. Gallager, Low-density parity-check codes, *IRE Trans Inform. Theory* **IT-8** (1962), 21–28.
17. P. M. Gruber and C. G. Lekkerkerker, *Geometry of Numbers*, 2nd edn. (Elsevier Science Publishers B.V., North-Holland, Amsterdam, 1987).
18. S. Haykin, *Communication Systems* (John Wiley, New York, 2001).
19. F. Jiang, E. Psota and L. C. Perez, The generator and parity-check matrices of turbo codes, *40th Annual Conference on Information Sciences and Systems*, 2006, 1451–1454.
20. R. Kannan, Improved algorithm on integer programming and related lattice problem, *Proc. 15th Annual ACM Symp. on Theory of Computing*, 1983, 193–206.

21. J. Leech, Some sphere packings in higher space, *Can. J. Math.* **16** (1964), 657–682.
22. A. Leon-Garcia, *Probability and Random Processes for Electrical Engineering*, 2nd edn. (Addison Wesley, New York, 1994).
23. D. J. C. MacKay, Good error correcting codes based on very sparse matrices, *IEEE Trans. Inform. Theory* **27**, 2 (1999), 399–431.
24. D. J. C. MacKay and R. M. Neal, Near Shannon limit performance of low density parity check codes, *Electron. Lett.* **32**, 18 (1996), 1645–1671.
25. M. R. Sadeghi, A. H. Banihashemi and D. Panario, Low-density parity-check lattices, construction and decoding alanysis, *IEEE Trans. Inform. Theory* **52** (2006), 4481–4495.
26. M. R. Sadeghi, Low density parity check lattices, PhD Thesis, Carleton University, Canada, 2003.
27. P. Samuel, *Algebraic Theory of Numbers* (Paris Hermann, 1971).
28. C. E. Shannon, A mathematical theory of communication, *The Bell Syst. Tech. J.* **27** (1948), 379–423.
29. C. E. Shannon, Probability of error for optimal codes in a Gaussian channel, *The Bell Syst. Tech. J.* **38**, 3 (1959).
30. S. B. Wicker, *Error Control Systems for Digital Communication and Storage* (Prentice Hall, New Jersey, 1995).
31. R. M. Tanner, A recursive approach to low complexity codes, *IEEE Trans. Inform. Theory* **27**, (1981), 533–547.
32. V. Tarokh, Minimal Tanner graphs for block codes and lattices, *Proc. 5th Can. Workshop on Inform. Theory*, Toronto, (1997), 9–10.
33. V. Tarokh and I. F. Blake, Trellis complexity vesus the coding gain of lattices I, *IEEE Trans. Inform. Theory* **42**, 6 (1996), 1796–1807.
34. V. Tarokh and A. Vardy, Universal bound on the performance of lattice codes, *IEEE Trans. Inform. Theory* **45** (1999), 670–681.
35. E. Viterbo, Tecniche matematiche computazionali per l'analisi ed il progtto di costellezioni a reticolo, PhD. thesis, Politecnico di Torino, Italy, 1995.
36. D. B. West, *Introduction To Graph Theory* (Prentice-Hall, Englewood Cliffs, NJ, 1996).
37. N. Wiberg, H.-A. Loeliger and R. Kotter, Codes and iterative decoding on graphs, *European Trans. on Telecom.* **6**, 5 (1995), 513–526.
38. S. B. Wicker, *Error Control Systems for Digital Communication and Storage* (Prentice Hall, New Jersey, 1995).

Chapter 3

DISTRIBUTED SPACE-TIME CODES WITH LOW ML DECODING COMPLEXITY

G. SUSINDER RAJAN* and B. SUNDAR RAJAN[†]

*Department of Electrical Communication Engineering,
Indian Institute of Science, Bangalore, India 560012*
susinder@ece.iisc.ernet.in
[†]*bsrajan@ece.iisc.ernet.in*

Cooperative communication in a wireless relay network consisting of a source node, a destination node, and several relay nodes equipped with half duplex single antenna transceivers is considered. Recently, for such a setting, it has been shown that diversity order equal to the number of relays can be achieved provided appropriate distributed space-time block codes (DSTBCs) are manufactured/used by the relay nodes, thus emulating a virtual multiple input multiple output (MIMO) system. This chapter deals with DSTBCs with low maximum likelihood (ML) decoding complexity. After a brief introduction and background to the study of STBCs and DSTBCs, a survey of STBC constructions for point to point MIMO systems that are amenable for low complexity ML decoding is presented. Then, DSTBC constructions achieving full cooperative diversity along with low ML decoding complexity are discussed separately for two cases: (i) the destination has perfect channel state information (CSI) of the source to relay as well as the relay to destination channel fading gains and no CSI is available at the source and relay nodes and (ii) the destination has full CSI and the relays have partial CSI corresponding to the phase component of the source to relay channel gains. For some of these cases, upper bounds on the maximum rate of single symbol ML decodable DSTBCs are also provided. These are counterparts of the well-known complex orthogonal designs for point to point MIMO systems that includes the Alamouti code as a special case.

3.1. Introduction

Communicating with multiple transmit and/or receive antennas helps to combat the detrimental effects of fading in wireless channels by providing additional diversity/reliability to the system. This diversity arises because the channel fading gain varies depending on the location of the antenna and thus well separated multiple transmit/receive antennas can bring in independent channel fading gains and hence more diversity or protection to the transmitted information. A typical multiple input multiple output (MIMO) communication system with N_T transmit antennas and N_R receive antennas is depicted in Fig. 3.1(a) where $\mathbf{H}_{j,k}$ denotes the channel fading gain between the jth transmit antenna and kth receive antenna. However, in some applications a wireless node/terminal may not have the required space for mounting multiple transmit/receive antennas. Cooperative communication is a promising technique which allows a set of geographically distributed users/terminals to share their single transmit antennas and collaboratively transmit a distributed code so as to achieve spatial diversity gains in such situations as well. The spatial diversity thus achieved is called as "cooperative diversity." It has been shown in [1, 2] that cooperative communication can lead to capacity gains and/or significant savings in transmit power for a required communication reliability. Hence, cooperative communication becomes attractive for applications such as wireless sensor networks, which essentially consist of small power constrained terminals. Consider the scenario of $R + 1$ users wanting to communicate using cooperative techniques to a common destination. Assume that each terminal is equipped with a single antenna and is half duplex constrained, i.e. it cannot transmit and receive simultaneously in the same frequency band. For the sake of simplicity, let us consider one of the users to be the source terminal without loss of generality and the remaining R users as relays, whose main job is to help the source communicate to the

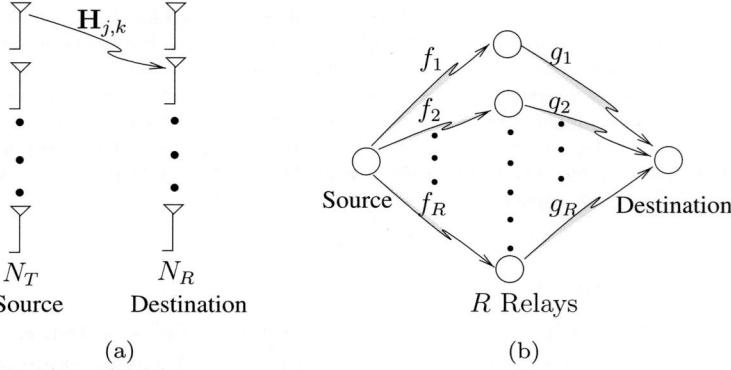

Fig. 3.1.   (a) Point to point MIMO system; (b) cooperative wireless relay network.

destination by relaying the information from the source to the destination. This setup is pictorially shown in Fig. 3.1(b). Mainly, there are two types of relaying methods discussed in the literature: (1) Amplify and forward [3–7], that is linearly process the received signals, amplify, and transmit the resulting signal to the destination and (2) Decode and forward [8], that is decode the received signal from the source, re-encode the decoded data, and transmit to the destination. In this chapter, we discuss only amplify and forward-based relaying protocols since it is easier to implement them in the small relay nodes and moreover, it does not require the knowledge of the channel fading gains at the relay nodes, which facilitates a possible extension to a completely non-coherent strategy [10–15]. Various protocols for cooperative communication have been discussed in the literature (see [3–9] and many more). One such important cooperative communication protocol is the distributed space-time coded protocol [5, 8] which relies on employing space-time block codes (STBCs), that is codes for point to point MIMO systems in a distributed fashion by making each relay transmit a column of a space-time codeword. It has been shown in [5] that a diversity order equal to the number of relays can be achieved provided appropriate distributed space-time block codes (DSTBCs) are employed. It is important to note that DSTBCs are required to have a certain special structure which is a constraint imposed by the protocol [5] and hence not all STBCs qualify as DSTBCs.

In this cooperative communication setting, the destination has to jointly decode the signals received from all the relays in order to decode the data transmitted by the source. In other words, the destination has to perform maximum likelihood (ML) detection of the DSTBC, whose complexity grows with increasing number of relays and/or transmission rate. There are suboptimal reception techniques like zero forcing, minimum mean squared error, etc. but full cooperative diversity is not guaranteed for such reception techniques. The sphere decoder [16] is an efficient ML decoder but even its complexity becomes unmanageable for large number of relays. Thus, the issue of ML decoding complexity gains significant importance especially in the context of cooperative communication in wireless sensor networks, wherein there are large number of cooperating small sensor nodes. It is well known that for point to point MIMO systems, STBCs from orthogonal designs (ODs) admit single complex symbol ML decoding and also offer full diversity for all complex constellations. Recently in [17–25], ODs have been extended to multi-symbol/multi-group ML decodable STBCs in order to improve the transmission rate in complex symbols per channel use as compared to ODs. Thus, it is of natural interest to study the counterpart single symbol ML decodable DSTBCs and multi-symbol/multi-group ML decodable DSTBCs.

3.1.1. *Organization of the Chapter*

Section 3.2 provides the necessary background material such as basic definitions and important theorems concerning STBCs for point to point MIMO systems. Section 3.3 gives a brief summary of the main results on multi-symbol ML decodable STBCs for point to point MIMO systems. In Sec. 3.4, DSTBC constructions with low ML decoding complexity for several scenarios are discussed. The remaining sections conclude this chapter and present some thoughts for the practitioners and directions for further research.

3.1.2. *Abbreviations*

STBC: Space-time block code
LSTD: Linear space-time design
bpcu: bits per channel use
SNR: Signal to noise ratio
OD: Orthogonal design
ROD: Real orthogonal design
UW: Unitary weight
NUW: Non-unitary weight
SSD: Single symbol decodable
CUW-SSD: Clifford unitary weight-SSD
MDC-QSTBC: Minimum decoding complexity quasi-orthogonal STBC
CIOD: Coordinate interleaved orthogonal design
PSSD: Proper SSD
TNU-SSD: Transformed non-unitary weight-SSD
QAM: Quadrature amplitude modulation
PSK: Phase shift keying
OFDM: Orthogonal frequency division multiplexing
PAPR: peak to average power ratio
AWGN: Additive white Gaussian noise
DOSTBC: Distributed orthogonal space time block code
S-PDSSDC: Semi-orthogonal precoded distributed single-symbol decodable
　　　　　　STBC

3.1.3. *Notation*

$\mathbb{R}$, $\mathbb{C}$, $\mathbb{N}$, and $\mathbb{Z}$ are used to represent the set of real numbers, complex numbers, natural numbers, and integers, respectively. The symbol i is used to denote $\sqrt{-1}$. For a complex number s, its real and imaginary parts are denoted by s_I and s_Q, respectively. Matrices and vectors are denoted by uppercase and lowercase bold letters, respectively. The $N \times N$ identity matrix will be denoted by $\mathbf{I_N}$. The conjugate, transpose, and conjugate transpose of a complex matrix $\mathbf{A}$ are denoted by $\mathbf{A}^*$, $\mathbf{A}^T$, and $\mathbf{A}^H$, respectively. The determinant of a square matrix $\mathbf{B}$ is denoted by $|\mathbf{B}|$. The functions $\| \cdot \|_F$, $\lceil \cdot \rceil$ denote the Frobenius norm and ceiling function, respectively.

3.2. Background

Consider a MIMO system with N_T transmit antennas and N_R receive antennas as shown in Fig. 3.1(a). Let T be the number of channel uses over which coding is done at the transmitter and the channel is assumed to be constant for the entire T channel uses. Then the equivalent baseband signal model is given by

$$\mathbf{Y} = \mathbf{XH} + \mathbf{N}, \tag{3.1}$$

where $\mathbf{Y}$ is the $T \times N_R$ received matrix, $\mathbf{H}$ is the channel matrix with its (j, k)th entry as $\mathbf{H}_{j,k}$, $\mathbf{X}$ is the $T \times N_T$ transmitted codeword, and $\mathbf{N}$ is the additive white Gaussian noise (AWGN) at the destination represented by a $T \times N_R$ matrix whose entries are independent and identically distributed complex Gaussian with zero mean and unit variance. Since coding is done over space (antennas) and time, the set of all possible transmitted codewords is called a "STBC" and denoted by $\mathscr{C}$.

Definition 3.1. An STBC $\mathscr{C}$ is a finite set of $T \times N_T$ complex matrices such that every element $\mathbf{X}$ of $\mathscr{C}$ can be expressed as:

$$\mathbf{X} = \sum_{j=1}^{K} x_j \mathbf{A_j}, \tag{3.2}$$

where $\mathbf{A_j}$, $j = 1, \ldots, K \in \mathbb{C}^{T \times N_T}$ are linearly independent complex matrices over $\mathbf{R}$ and x_j, $j = 1, \ldots, K \in \mathbb{R}$. Also $\mathrm{E}\{\|\mathbf{X}\|_F^2\} \leq PT$ where P is the transmit power constraint per channel use.

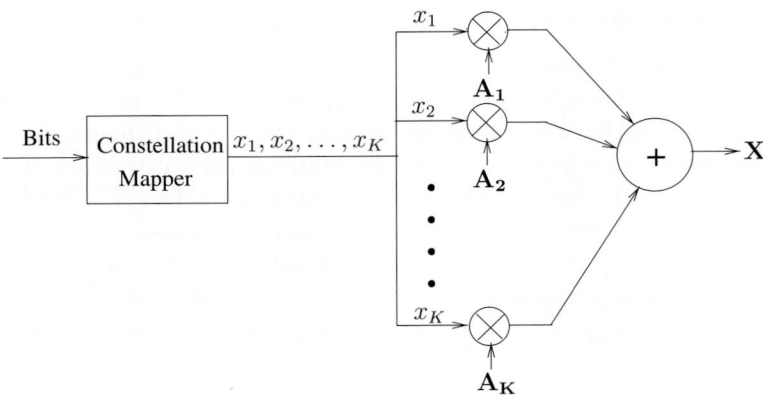

Fig. 3.2.　Encoding of STBCs.

The matrices $\mathbf{A_j}$, $j = 1, \ldots, K$ are called weight matrices or basis matrices or modulation matrices. The encoding of a STBC can be done using Eq. (3.2) by viewing the x_j's as real symbols and then allowing each one of them to take values from an appropriate subset of $\mathbb{R}$ depending on the data to be transmitted. This is pictorially illustrated in Fig. 3.2. Thus K real symbols are transmitted in T channel uses. Alternatively, we can pair two real symbols and think of it as a complex symbol and hence we can also view it as though $\frac{K}{2}$ complex symbols are transmitted in T channel uses. Viewing the x_j's as real symbols leads to the following definition of linear space-time design (LSTD) which will be useful later.

Definition 3.2. A $T \times N_T$ LSTD in K real symbols $x_1, x_2, \ldots, x_K$ is a $T \times N_T$ matrix which can be expressed as $\sum_{j=1}^{K} x_j \mathbf{A_j}$ for some linearly independent complex matrices $\mathbf{A_j}, j = 1, 2, \ldots, K$ over $\mathbb{R}$.

Definition 3.3. The rate of a STBC (as defined in Definition 3.1) is equal to $\frac{K}{2T}$ complex symbols per channel use.

Example 3.1. Consider the VBLAST or spatial multiplexing system [26] for 2 transmit antennas. The STBC corresponding to VBLAST is given by

$$\mathscr{C} = \left\{ \begin{bmatrix} s_1 \\ s_2 \end{bmatrix} \mid s_1, s_2 \in 4 - QAM \right\}$$

which has four real symbols s_{1I}, s_{1Q}, s_{2I}, s_{2Q}, and their corresponding weight matrices are:

$$\mathbf{A_1} = \begin{bmatrix} 1 \\ 0 \end{bmatrix}, \quad \mathbf{A_2} = \begin{bmatrix} i \\ 0 \end{bmatrix}, \quad \mathbf{A_3} = \begin{bmatrix} 0 \\ 1 \end{bmatrix}, \quad \text{and} \quad \mathbf{A_4} = \begin{bmatrix} 0 \\ i \end{bmatrix}.$$

Thus, the rate of VBLAST transmission scheme is 2 complex symbols per channel use.

Now ML detection of the transmitted codeword is equivalent to decoding the corresponding transmitted vector of real symbols $\mathbf{x} = [\,x_1 \ x_2 \ \ldots \ x_K\,]^T$ and is given as follows:

$$\hat{\mathbf{x}} = \arg\min \left\| \mathbf{Y} - \left(\sum_{j=1}^{K} x_j \mathbf{A_j} \right) H \right\|_F^2, \tag{3.3}$$

where the minimum is taken over all the possible values of the vector $\mathbf{x}$. Note that the ML decoding metric is a function $f(x_1, x_2, \ldots, x_K)$ which depends on the set of real symbols $x_1, x_2, \ldots, x_K$ and hence has to be computed $|\mathscr{C}|$ times for ML detection. Thus, ML detection of STBCs in general is computationally prohibitive for MIMO systems with large K. Suppose the STBC $\mathscr{C}$ is such that the function $f(x_1, x_2, \ldots, x_K)$ decomposes into g functions f_j, $j = 1, \ldots, g$ as follows:

$$f = f_1^2 + f_2^2 + \cdots + f_g^2 + e,$$

where the function e is independent of all the real symbols, the functions f_j, $j = 1, \ldots, g$ are mutually independent and each f_j is a function of at most $\frac{K}{g}$ real symbols. Then

$$\min\{f\} = \min\{f_1\} + \min\{f_2\} + \cdots + \min\{f_g\}. \tag{3.4}$$

Thus ML detection of $\mathbf{x}$ can be done by g independent parallel decoders, i.e. the jth decoder decodes only those real symbols which are in the argument of the function f_j. Thus, the total number of computations required for decoding reduces to $g|\mathscr{C}|^{\frac{1}{g}}$ if each f_j is a function of exactly $\frac{K}{g}$ real symbols. In essence, the ML decoding can be done independently for groups of real symbols rather than all the real symbols together.

Definition 3.4. A STBC $\mathscr{C}$ is said to be g-group ML decodable or $\frac{K}{g}$ real symbol ML decodable or $\frac{K}{2g}$ complex symbol ML decodable if the minimization of the ML decoding metric in Eq. (3.3) can be done as in Eq. (3.4).

Theorem 3.1. *A STBC $\mathscr{C}$ is g-group ML decodable if*

(1) *Whenever $\mathbf{A_j}$ and $\mathbf{A_k}$ correspond to real symbols belonging to different groups, we have*

$$\mathbf{A_j}^H \mathbf{A_k} + \mathbf{A_k}^H \mathbf{A_j} = \mathbf{0}. \tag{3.5}$$

(2) *The real symbols in one group take values independently of the real symbols in the other groups.*

Simple examples of multi-group ML decodable STBCs include the class of STBCs from ODs for which $g = K$.

Definition 3.5 (Taroka *et al.* [27]). A $T \times N_T$ OD $\mathbf{X}_{\mathrm{OD}}$ in K real symbols $x_1, x_2, \ldots, x_K$ is a LSTD which satisfies the following identity:

$$\mathbf{X}_{\mathrm{OD}}^H \mathbf{X}_{\mathrm{OD}} = \left(\sum_{j=1}^{K} x_j^2 \right) \mathbf{I_{N_T}}. \tag{3.6}$$

Theorem 3.2 (Liang [28]). *The set of matrices $\mathbf{A_j}$, $j = 1, \ldots, K$ of an OD $\mathbf{X}_{OD}$ satisfy the following two identities:*

$$\mathbf{A_j}^H \mathbf{A_j} = \mathbf{I_{N_T}}, \quad \forall j = 1, \ldots, K, \tag{3.7}$$

$$\mathbf{A_j}^H \mathbf{A_k} + \mathbf{A_k}^H \mathbf{A_j} = \mathbf{0}, \qquad 1 \le j < k \le K. \tag{3.8}$$

Definition 3.6. A STBC $\mathscr{C}$ is said to be fully diverse or achieve full diversity if $|(\mathbf{X_j} - \mathbf{X_k})^H (\mathbf{X_j} - \mathbf{X_k})| \ne 0 \; \forall \; \mathbf{X_j} \ne \mathbf{X_k} \in \mathscr{C}$.

For fully diverse STBCs, the pair wise error probability that a ML decoder decodes wrongly to a codeword $\mathbf{X_k}$ given that $\mathbf{X_j}$ was transmitted can be upper bounded for high signal to noise ratio (SNR) as $(|(\mathbf{X_j} - \mathbf{X_k})^H (\mathbf{X_j} - \mathbf{X_k})|^{\frac{1}{N_T}} P)^{-N_T N_R}$. The quantity $\min_{\mathbf{X_j} \ne \mathbf{X_k} \in \mathscr{C}} (|(\mathbf{X_j} - \mathbf{X_k})^H (\mathbf{X_j} - \mathbf{X_k})|^{\frac{1}{N_T}}$ is referred to as the "coding gain." Thus, for good error performance at high SNR, we need to construct fully diverse STBCs with large coding gain. Note that to achieve full diversity of $N_T N_R$, it is necessary that $T \ge N_T$. Hence, we shall refer to STBCs with $T = N_T$ as minimum delay STBCs or square STBCs.

STBCs can be easily obtained from ODs by specifying a signal set from which the real symbols $x_j, j = 1, \ldots, K$ take values from. Such STBCs are fully diverse for arbitrarily chosen signal sets because by Eq. (3.6), $(\mathbf{X_j} - \mathbf{X_k})^H(\mathbf{X_j} - \mathbf{X_k})$ is an identity matrix multiplied by a non-zero scale factor and hence $|(\mathbf{X_j} - \mathbf{X_k})^H(\mathbf{X_j} - \mathbf{X_k})| \neq 0 \,\forall\, \mathbf{X_j} \neq \mathbf{X_k} \in \mathscr{C}$.

Moreover, since ODs satisfy Eq. (3.8), by Theorem 3.1 we have that STBCs obtained from ODs are single real symbol ML decodable if all the K real symbols $x_j, j = 1, \dots, K$ are allowed to take values independently.

Example 3.2. Consider the Alamouti 2×2 OD in 2 complex symbols given by $\begin{bmatrix} s_1 & -s_2^* \\ s_2 & s_1^* \end{bmatrix}$. There are four real symbols in the Alamouti OD given by $s_{1I}, s_{1Q}, s_{2I}, s_{2Q}$, and their associated weight matrices are:

$$\mathbf{A_1} = \begin{bmatrix} 1 & 0 \\ 0 & 1 \end{bmatrix}, \quad \mathbf{A_2} = \begin{bmatrix} i & 0 \\ 0 & -i \end{bmatrix}, \quad \mathbf{A_3} = \begin{bmatrix} 0 & -1 \\ 1 & 0 \end{bmatrix}, \quad \text{and} \quad \mathbf{A_4} = \begin{bmatrix} 0 & i \\ i & 0 \end{bmatrix},$$

respectively. It can be checked that the Alamouti OD satisfies Eqs. (3.6)–(3.8).

If we allow the complex symbols s_1, s_2 to take values independently from 16-QAM constellation, then we get a rate 1, single real symbol ML decodable, full diversity STBC since 16-QAM is a Cartesian product of two 4-PAM signal sets and thus $s_{1I}, s_{1Q}, s_{2I}, s_{2Q}$ take values independently. ML decoding of the associated 4-real symbols can be done as follows:

$$\hat{s}_{1I} = \arg\min_{s_{1I} \in 4-\mathrm{PAM}} \|\mathbf{Y} - s_{1I}\mathbf{A_1}\mathbf{H}\|_F^2,$$
$$\hat{s}_{1Q} = \arg\min_{s_{1Q} \in 4-\mathrm{PAM}} \|\mathbf{Y} - s_{1Q}\mathbf{A_2}\mathbf{H}\|_F^2,$$
$$\hat{s}_{2I} = \arg\min_{s_{2I} \in 4-\mathrm{PAM}} \|\mathbf{Y} - s_{2I}\mathbf{A_3}\mathbf{H}\|_F^2,$$
$$\hat{s}_{2Q} = \arg\min_{s_{2Q} \in 4-\mathrm{PAM}} \|\mathbf{Y} - s_{2Q}\mathbf{A_4}\mathbf{H}\|_F^2.$$

On the contrary if s_1, s_2 are allowed to take values independently from a 16-PSK constellation, then we get a rate 1, single complex symbol ML decodable, full diversity STBC since the real and imaginary parts of s_1 and s_2 have to take values jointly in the 16-PSK case. ML decoding of the complex symbols s_1, s_2 can be done as follows:

$$\hat{s}_1 = \arg\min_{s_1 \in 16-\mathrm{PSK}} \|\mathbf{Y} - (s_{1I}\mathbf{A_1} + s_{1Q}\mathbf{A_2})\mathbf{H}\|_F^2,$$
$$\hat{s}_2 = \arg\min_{s_2 \in 16-\mathrm{PSK}} \|\mathbf{Y} - (s_{2I}\mathbf{A_3} + s_{2Q}\mathbf{A_4})\mathbf{H}\|_F^2.$$

The signal sets of 16-QAM and 16-PSK are shown in Fig. 3.3 and it can be observed that in the case of 16-PSK the real and imaginary parts cannot take values independently.

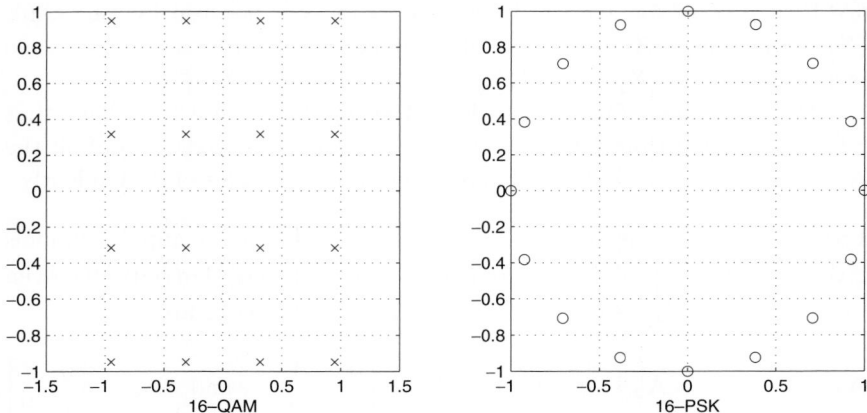

Fig. 3.3.   16-QAM and 16-PSK constellations.

Example 3.2 illustrates how signal sets play an important role in controlling the ML decoding complexity.

3.3. Multi-Group ML Decodable STBCs for Point to Point MIMO Systems

In this section, we briefly summarize the results available on low ML decodable STBCs for point to point MIMO systems. To the best of our knowledge, such a summary is not available in the literature. We begin with the well known result of the variation of the maximum rate of ODs with respect to the increase in the number of transmit antennas.

Theorem 3.3 (Tirkkonen and Hottinen [29]). *The maximum rate of a square OD for* $N_T = 2^a(2b + 1)$, $a, b \in \mathbb{N}$ *transmit antennas is* $\frac{a+1}{2^a(2b+1)}$ *complex symbols per channel use.*

The above theorem states that the rate of square ODs falls rapidly with increasing number of transmit antennas. For example, the maximum rate of an OD for 4 transmit antennas is $\frac{3}{4}$. The transmission rate of single complex symbol ML decodable STBCs can be improved to $\frac{a}{2^{a-1}(2b+1)}$ by relaxing one of the two identities (Eqs. (3.7) and (3.8)) satisfied by ODs [17, 18]. This introduces a classification of STBCs into unitary weight (UW)-STBCs and non-unitary weight (NUW)-STBCs. UW-STBCs are those which satisfy Eq. (3.7) and NUW-STBCs are those which do not satisfy Eq. (3.7). Let us see examples of both the cases below.

Example 3.3. Consider the 4×4 coordinate interleaved orthogonal design (CIOD) given by

$$\mathbf{X}_{\mathrm{CIOD}} = \begin{bmatrix} x_1 + ix_2 & -x_3 + ix_4 & 0 & 0 \\ x_3 + ix_4 & x_1 - ix_2 & 0 & 0 \\ 0 & 0 & x_5 + ix_6 & -x_7 + ix_8 \\ 0 & 0 & x_7 + ix_8 & x_5 - ix_6 \end{bmatrix}.$$

It has totally eight real symbol (rate $= 1$ complex symbols per channel use) and their corresponding weight matrices are:

$$\mathbf{A_1} = \begin{bmatrix} 1 & 0 & 0 & 0 \\ 0 & 1 & 0 & 0 \\ 0 & 0 & 0 & 0 \\ 0 & 0 & 0 & 0 \end{bmatrix}, \quad \mathbf{A_2} = \begin{bmatrix} i & 0 & 0 & 0 \\ 0 & -i & 0 & 0 \\ 0 & 0 & 0 & 0 \\ 0 & 0 & 0 & 0 \end{bmatrix},$$

$$\mathbf{A_3} = \begin{bmatrix} 0 & -1 & 0 & 0 \\ 1 & 0 & 0 & 0 \\ 0 & 0 & 0 & 0 \\ 0 & 0 & 0 & 0 \end{bmatrix}, \quad \mathbf{A_4} = \begin{bmatrix} 0 & i & 0 & 0 \\ i & 0 & 0 & 0 \\ 0 & 0 & 0 & 0 \\ 0 & 0 & 0 & 0 \end{bmatrix},$$

$$\mathbf{A_5} = \begin{bmatrix} 0 & 0 & 0 & 0 \\ 0 & 0 & 0 & 0 \\ 0 & 0 & 1 & 0 \\ 0 & 0 & 0 & 1 \end{bmatrix}, \quad \mathbf{A_6} = \begin{bmatrix} 0 & 0 & 0 & 0 \\ 0 & 0 & 0 & 0 \\ 0 & 0 & i & 0 \\ 0 & 0 & 0 & -i \end{bmatrix},$$

$$\mathbf{A_7} = \begin{bmatrix} 0 & 0 & 0 & 0 \\ 0 & 0 & 0 & 0 \\ 0 & 0 & 0 & -1 \\ 0 & 0 & 1 & 0 \end{bmatrix}, \quad \text{and} \quad \mathbf{A_8} = \begin{bmatrix} 0 & 0 & 0 & 0 \\ 0 & 0 & 0 & 0 \\ 0 & 0 & 0 & i \\ 0 & 0 & i & 0 \end{bmatrix}.$$

It is clear that CIOD belongs to the class of NUW-STBCs. However, all the eight weight matrices satisfy (3.8). We have $|\Delta \mathbf{X}_{\mathrm{CIOD}}^{H} \Delta \mathbf{X}_{\mathrm{CIOD}}| = (\sum_{j=1}^{4} \Delta x_j^2)^2 (\sum_{j=5}^{8} \Delta x_j^2)^2$, where the notation Δ denotes the codeword difference matrix. Thus, STBCs from CIODs are not fully diverse for arbitrary signal sets. However, if the employed signal set has non-zero coordinate product distance, that is $\Delta x_j \Delta x_{j+4} \neq 0, \forall\, j = 1, \ldots, 4$ then full diversity is guaranteed. This constraint on the signal set can be easily met, for example, by letting the real symbols x_j and x_{j+4} take values jointly from an appropriately rotated QAM constellation [17]. Note that constellation rotation entangles the real symbols. Single complex symbol ML decoding is done as follows:

$$(\hat{x}_j, \hat{x}_{j+4}) = \arg\min \| \mathbf{Y} - (x_j \mathbf{A_j} + x_{j+4} \mathbf{A_{j+4}}) \mathbf{H} \|_F^2, \quad \forall\, j = 1, \ldots, 4,$$

where the above minimum is computed over all possible constellation points.

Example 3.4. Consider
$\begin{bmatrix} x_1 + ix_3 & -x_5 + ix_7 & x_2 + ix_4 & -x_6 + ix_8 \\ x_5 + ix_7 & x_1 - ix_3 & x_6 + ix_8 & x_2 - ix_4 \\ x_2 + ix_4 & -x_6 + ix_8 & x_1 + ix_3 & -x_5 + ix_7 \\ x_6 + ix_8 & x_2 - ix_4 & x_5 + ix_7 & x_1 - ix_3 \end{bmatrix}$. It has

eight real symbols (rate = 1) and their corresponding weight matrices are:

$$\mathbf{A_1} = \begin{bmatrix} 1 & 0 & 0 & 0 \\ 0 & 1 & 0 & 0 \\ 0 & 0 & 1 & 0 \\ 0 & 0 & 0 & 1 \end{bmatrix}, \quad \mathbf{A_2} = \begin{bmatrix} 0 & 0 & 1 & 0 \\ 0 & 0 & 0 & 1 \\ 1 & 0 & 0 & 0 \\ 0 & 1 & 0 & 0 \end{bmatrix},$$

$$\mathbf{A_3} = \begin{bmatrix} i & 0 & 0 & 0 \\ 0 & -i & 0 & 0 \\ 0 & 0 & i & 0 \\ 0 & 0 & 0 & -i \end{bmatrix}, \quad \mathbf{A_4} = \begin{bmatrix} 0 & 0 & i & 0 \\ 0 & 0 & 0 & -i \\ i & 0 & 0 & 0 \\ 0 & -i & 0 & 0 \end{bmatrix},$$

$$\mathbf{A_5} = \begin{bmatrix} 0 & -1 & 0 & 0 \\ 1 & 0 & 0 & 0 \\ 0 & 0 & 0 & -1 \\ 0 & 0 & 1 & 0 \end{bmatrix}, \quad \mathbf{A_6} = \begin{bmatrix} 0 & 0 & 0 & -1 \\ 0 & 0 & 1 & 0 \\ 0 & -1 & 0 & 0 \\ 1 & 0 & 0 & 0 \end{bmatrix},$$

$$\mathbf{A_7} = \begin{bmatrix} 0 & i & 0 & 0 \\ i & 0 & 0 & 0 \\ 0 & 0 & 0 & i \\ 0 & 0 & i & 0 \end{bmatrix}, \quad \text{and} \quad \mathbf{A_8} = \begin{bmatrix} 0 & 0 & 0 & i \\ 0 & 0 & i & 0 \\ 0 & i & 0 & 0 \\ i & 0 & 0 & 0 \end{bmatrix}.$$

Note that all the weight matrices are clearly unitary but they do not satisfy (3.8). Let us consider the grouping of real symbols: $\{x_1, x_2\}$, $\{x_3, x_4\}$, $\{x_5, x_6\}$, and $\{x_7, x_8\}$. For this grouping of real symbols, it can be checked that the associated weight matrices satisfy the conditions for single complex symbol decoding in Eq. (3.5). Now similar to the case of CIODs, if the real symbols x_j, x_{j+1} are allowed to take values from an appropriately rotated

QAM constellation a fully diverse single complex symbol ML decodable UW-STBC can be obtained.

Note from the above two examples that STBCs which are not obtained from ODs fail to offer full diversity for arbitrary signal sets since they do not satisfy Eq. (3.6). The maximum rate of general single complex symbol ML decodable STBCs is an open problem. The inter-relationship between the various single symbol decodable (SSD) STBCs [18] is illustrated pictorially in Fig. 3.4.

To further improve the transmission rate of SSD codes, multi-symbol/multi-group ML decodable STBCs have been studied in the literature. A summary of multi-symbol ML decodable square STBC constructions in the literature is presented in Table 3.1. Note that all the multi-symbol ML decodable square STBC constructions are available either for even or power of two number of transmit antennas. For the other cases, we drop appropriate number of columns from a square STBC constructed for the nearest power of two or the nearest even number as the case may be. Hence, for odd number of transmit antennas, non-square STBCs are usually preferred due to their higher transmission rate.

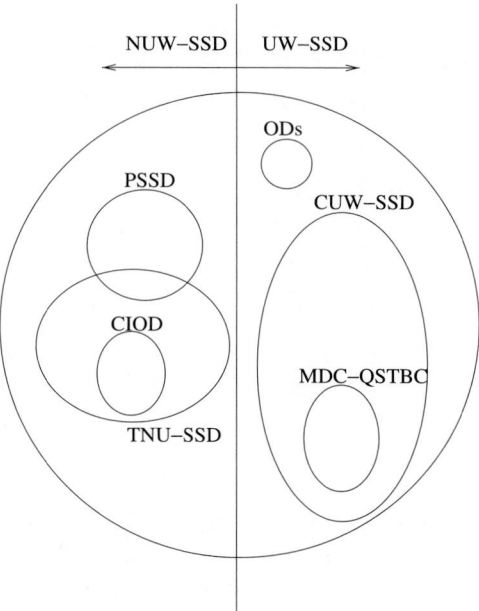

Fig. 3.4. Inter-relationship between single symbol ML decodable STBCs for point to point MIMO systems.

Table 3.1. Brief summary of multi-symbol ML decodable square STBCs in literature.

Ref.	UW/NUW	Tx. antennas N_T	Maximum rate	Signal set	Decoding complexity	Full diversity
ODs [27–28]	UW	$2^a(2b+1)$	$\dfrac{a+1}{2^a(2b+1)}$	QAM	Single real symbol	Yes
				Other complex constellations	Single complex symbol	Yes
Square CIOD [17]	NUW	$2^a(2b+1),\ a>1$	$\dfrac{a}{2^{a-1}(2b+1)}$	31.7175° rotated QAM	Single complex symbol	Yes
MDC-QSTBC [19]	UW	$2^a(2b+1),\ a>1$	$\dfrac{a}{2^{a-1}(2b+1)}$	13.29° rotated QAM	Single complex symbol	Yes
Optimal MDC-QSTBC [20]	UW/NUW	$2^a(2b+1),\ a>1$	$\dfrac{a}{2^{a-1}(2b+1)}$	Linear transformation of QAM symbols	Single complex symbol	Yes
Clifford algebras [18]	UW	2^a	$\dfrac{a}{2^{a-1}}$	Rotated QAM	Single complex symbol	Yes
ABBA m-times on OD [31, 30]	UW	Even	equal to underlying OD	Not specified	m complex symbol	No
ABBA-QSTBC [32, 34, 36]	UW	$2^a(2b+1),\ a>1$	$\dfrac{a}{2^{a-1}(2b+1)}$	Rotated QAM	Double complex symbol	Yes
ABBA-CIOD [21]	NUW	8	1	Rotated QAM	Double complex symbol	Yes

Table 3.1. (*Continued*)

Ref.	UW/NUW	Tx. antennas N_T	Maximum rate	Signal set	Decoding complexity	Full diversity
4Gp-QSTBC [25]	UW	2^a	1	Rotated $\mathbb{Z}^{\frac{N_T}{2}}$ lattice	Four group decodable	Yes
PCIOD [45]	NUW	Even	1	Rotated $\mathbb{Z}^{\frac{N_T}{2}}$ lattice	Four group decodable	Yes
4Gp-SAST [23]	NUW	Even	1	Rotated $\mathbb{Z}^{\frac{N_T}{2}}$ lattice	Four group decodable	Yes
g-Group tensor product construction [22]	UW	Even	$\dfrac{g}{2^{(\lfloor \frac{g-1}{2}\rfloor+1)}}$	Rotated $\mathbb{Z}^{\frac{2N_T}{g}}$ lattice	g-Group decodable	Yes
g-Group extended Clifford algebra [46]	UW	2^a	$\dfrac{g}{2^{(\lfloor \frac{g-1}{2}\rfloor+1)}}$	Rotated $\mathbb{Z}^{\frac{2N_T}{g}}$ lattice	g-Group decodable	Yes
QSTBC [35, 33, 38]	UW	2^a	1	Rotated QAM	Two group decodable	Yes
SAST [37]	UW	Even	1	Rotated QAM	Two group decodable	Yes
High rate QSTBC [39]	UW	4	$\dfrac{5}{4}$	Not specified	Two group decodable	No

3.4. Distributed Space-Time Block Code Constructions

In this section, we briefly describe the amplify and forward-based distributed space-time coding strategy for wireless relay networks in [5]. Consider a wireless relay network with a source node, R relay nodes, and a destination node, each of them equipped with half duplex single antenna transceivers as shown in Fig. 3.1(b). The wireless channel between any two nodes is assumed to be quasi-static, flat fading, and is modeled by a complex Gaussian random variable with mean zero and unit variance. Let f_j denote the channel fading gain between the source and the jth relay node and let g_j denote the channel fading gain between the jth relay node and the destination. All the terminals are assumed to be synchronized at the symbol level as well as at the carrier frequency level. The source node is assumed to have no channel state information (CSI), i.e. absence of knowledge of $f_j, g_j, j = 1, 2, \ldots, R$ and the destination is assumed to have perfect CSI of all the channels (also known as the coherent case). We consider the following two different scenarios and discuss low ML decoding complexity DSTBC constructions for each of these cases:

(1) The relays do not have any CSI.
(2) The relays have partial CSI. Specifically, the relays know the phase of the channel fading gains from the source to the relays.

3.4.1. Coherent Case Without CSI at Relays

In this case, complete CSI is assumed to be available at the destination node, whereas the relay nodes are assumed to have no CSI. For this case, we present two broad classes of DSTBCs: (i) Four-group ML decodable DSTBCs and (ii) Single complex symbol ML decodable DSTBCs. Let P denote the average total power per channel use spent by the source node and the R relay nodes together. The transmission of information from source node to destination node takes place in two phases which comprises totally of $T_1 + T_2$ channel uses during which the channel fading gains $f_j, g_j, j = 1, 2, \ldots, R$ are assumed to be constant. In the first phase comprising of T_1 channel uses, the source transmits a T_1 length vector $\mathbf{s} \in \mathbb{C}^{T_1}$ satisfying $\|\mathbf{s}\|_F^2 \leq T_1$ using a fraction π_1 of the total power P. All the R relay nodes receive a faded noise corrupted version of the source node's transmission. The vector $\mathbf{r_j}$ received by the jth relay node is given by:

$$\mathbf{r_j} = \sqrt{\pi_1 P} f_j \mathbf{s} + \mathbf{v_j},$$

where $\mathbf{v_j}$ is the vector representing the AWGN at the jth relay node.

In the second phase, which comprises of T_2 channel uses, all the R relay nodes are scheduled to transmit together a linearly transformed version of their respective received vectors consuming a fraction π_2 of the total power P. The power allocation factors π_1 and π_2 are chosen to satisfy $\pi_1 P T_1 + \pi_2 P R T_2 = P(T_1 + T_2)$. To be precise, the jth relay node is equipped with a $T_2 \times T_1$ complex matrix $\mathbf{B_j}$ satisfying $\|\mathbf{B_j}\|_F^2 \leq T_2$ called "relay matrix" and it transmits vector $\mathbf{t_j}$ given by

$$\mathbf{t_j} = \sqrt{\frac{\pi_2 P}{\pi_1 P + 1}} \mathbf{B_j} \tilde{\mathbf{r}}_\mathbf{j},$$

where the notation $\tilde{\mathbf{x}}$ denotes either $\mathbf{x}$ or $\mathbf{x}^*$. Then,

$$\mathbf{t_j} = \sqrt{\frac{\pi_1 \pi_2 P^2}{\pi_1 P + 1}} \tilde{f}_j \mathbf{B_j} \tilde{\mathbf{s}} + \sqrt{\frac{\pi_2 P}{\pi_1 P + 1}} \mathbf{B_j} \tilde{\mathbf{v}}_\mathbf{j}. \tag{3.9}$$

Note from Eq. (3.9), that amplify and forward operation at the relay node causes it to transmit a linearly transformed version of its additive noise also to the destination node. This is a drawback of amplify and forward protocols. During the second phase, the destination node receives the signals transmitted by the relay nodes and it is given by

$$\mathbf{y} = \sum_{j=1}^{R} g_j \mathbf{t_j} + \mathbf{w} = \sqrt{\frac{\pi_1 \pi_2 P^2}{\pi_1 P + 1}} \left(\sum_{j=1}^{R} \tilde{f}_j g_j \mathbf{B_j} \tilde{\mathbf{s}} \right)$$

$$+ \sqrt{\frac{\pi_2 P}{\pi_1 P + 1}} \left(\sum_{j=1}^{R} g_j \mathbf{B_j} \tilde{\mathbf{v}}_\mathbf{j} \right) + \mathbf{w},$$

where the vector $\mathbf{w}$ denotes the AWGN at the destination node. As for classical point to point MIMO systems, the received vector $\mathbf{y}$ can also be expressed in a similar form as follows:

$$\mathbf{y} = \sqrt{\frac{\pi_1 \pi_2 P^2}{\pi_1 P + 1}} \mathbf{X} \mathbf{h} + \mathbf{n},$$

where

$$\mathbf{X} = \begin{bmatrix} \mathbf{B_1} \tilde{\mathbf{s}} & \mathbf{B_2} \tilde{\mathbf{s}} & \cdots & \mathbf{B_R} \tilde{\mathbf{s}} \end{bmatrix}, \quad \mathbf{h} = \begin{bmatrix} \tilde{f}_1 g_1 & \tilde{f}_2 g_2 & \cdots & \tilde{f}_R g_R \end{bmatrix}^T \tag{3.10}$$

and

$$\mathbf{n} = \sqrt{\frac{\pi_2 P}{\pi_1 P + 1}} \left(\sum_{j=1}^{R} g_j \mathbf{B_j} \tilde{\mathbf{v}}_\mathbf{j} \right) + \mathbf{w}.$$

Comparing Eq. (3.10) to the point to point MIMO case (Eq. (3.1)), we observe that for the distributed case, the codeword $\mathbf{X}$ is constrained to have a special structure and moreover the entries of the equivalent channel matrix $\mathbf{h}$ are a product of complex Gaussian random variables. Another important difference is that the equivalent noise vector seen by the destination $\mathbf{n}$ does not necessarily have a scaled identity matrix as its covariance matrix. If $\boldsymbol{\Gamma}$ denotes the covariance matrix of $\mathbf{n}$, then

$$\boldsymbol{\Gamma} = \mathrm{E}[\mathbf{nn}^H] = \mathbf{I_{T_2}} + \frac{\pi_2 P}{\pi_1 P + 1} \left(\sum_{j=1}^{R} |g_j|^2 \mathbf{B_j} \mathbf{B_j}^H \right). \qquad (3.11)$$

Thus, the covariance matrix of $\mathbf{n}$ depends on the set of relay matrices. The set of all possible codewords $\mathbf{X}$ is called as a "DSTBC" denoted by $\mathscr{C}$ since the columns of $\mathbf{X}$ are transmitted by geographically distributed nodes. Supposing the entries of the vector $\mathbf{s}$ are viewed as complex symbols, then observe that the matrix $\mathbf{X}$ has the property that any column of X has linear functions of either only the complex symbols or only their complex conjugates. We refer to this special property of DSTBCs as "conjugate linearity" property. Note that it is not necessary for classical STBCs to satisfy this property and in fact all square ODs [28, 29] for more than 2 transmit antennas fail to satisfy the conjugate linearity property. We define the counterpart rate of a DSTBC to be $\frac{T_1}{T_2}$ complex symbols per channel use. But, it is important to note that from the viewpoint of the source, the effective transmission rate is only $\frac{T_1}{T_1+T_2}$ complex symbols per channel use. However, it is easy to show that maximizing $\frac{T_1}{T_2}$ is equivalent to maximizing $\frac{T_1}{T_1+T_2}$.

The ML decoder for DSTBCs thus becomes

$$\hat{\mathbf{s}} = \arg \min_{\mathbf{X} \in \mathscr{C}} \left\| \boldsymbol{\Gamma}^{-\frac{1}{2}} \left(\mathbf{y} - \sqrt{\frac{\pi_1 \pi_2 P^2}{\pi_1 P + 1}} \mathbf{Xh} \right) \right\|_F^2, \qquad (3.12)$$

whose complexity is clearly exponential in the transmission rate of the source. Hence, the problem of constructing DSTBCs with low ML decoding complexity gains significant importance especially for large transmission rates and/or large number of relays. Thus, it is important to study the counterpart single symbol ML decodable DSTBCs and multi-group ML decodable DSTBCs.

Theorem 3.4 (Jing and Hassibi [5]; Susinder Rajan and Sundar Rajan [44, 48]). *For the power allocation* $\pi_1 = 1$, $\pi_2 = \frac{1}{R}$, *if* $\mathbf{B_j B_j}^H$ *is a diagonal matrix for all* $j = 1, 2, \cdots, R$ *and* $\mathrm{rank}(\mathbf{X_j} - \mathbf{X_k}) = R, \forall \mathbf{X_j} \neq \mathbf{X_k} \in \mathscr{C}$ *then, the DSTBC* $\mathscr{C}$ *achieves full cooperative diversity equal to* R *under ML decoding.*

Theorem 3.4 shows that the classical rank criterion applicable for point to point MIMO systems continues to hold for achieving full cooperative diversity in the distributed case also. Before proceeding to construct multi-group ML decodable DSTBCs we need to first identify the conditions for multi-group ML decoding of DSTBCs.

Theorem 3.5 (Susinder Rajan and Sundar Rajan [48]). *A DSTBC* $\mathscr{C}$ *having the conjugate linearity property is g-group ML decodable if and only if the following two conditions are satisfied.*

(1) *The associated basis matrices* $\mathbf{A_j}, j = 1, \ldots, K$ *of* $\mathscr{C}$ *satisfy*

$$\mathbf{A_j^H \Gamma^{-1} A_k} + \mathbf{A_k^H \Gamma^{-1} A_j} = \mathbf{0}, \tag{3.13}$$

whenever $\mathbf{A_j}$ *and* $\mathbf{A_k}$ *belong to different groups.*
(2) *The real symbols in one group take values independently of the real symbols in the other groups.*

Constructing multi-group ML decodable DSTBCs is much more challenging than multi-group ML decodable STBCs since DSTBCs need to satisfy (1) conjugate linearity property (2) row orthogonal relay matrices, and (3) Eq. (3.13) for appropriate number of groups g. Full diversity DSTBCs with rate one were constructed in [41–43]. In [43], a rate one, full diversity, double complex symbol ML decodable DSTBC was presented for four relays. Full diversity, rate one, two group ML decodable DSTBCs were constructed in [40] using division algebras. Later in [45, 47, 48], three constructions of rate one, full diversity, four group ML decodable DSTBCs were constructed. On similar lines, constructions and maximal rate of full diversity, single complex symbol ML decodable DSTBCs were addressed in [52, 53]. However, it is an important open problem to construct full diversity, maximal rate g-group ML decodable DSTBCs for arbitrary g. In the following, we briefly present some of the constructions of four group ML decodable DSTBCs and single complex symbol ML decodable DSTBCs in the literature.

3.4.1.1. *Four group ML decodable DSTBCs*

Construction of PCIOD:

For an even number of relays R, the $R \times R$ Precoded CIOD (PCIOD) $\mathbf{X}_{\text{PCIOD}}$ is given by

$$
\text{diag} \left\{ \begin{bmatrix} x_1 + ix_2 & -x_3 + ix_4 \\ x_3 + ix_4 & x_1 - ix_2 \end{bmatrix}, \ldots, \begin{bmatrix} x_k + ix_{k+1} & -x_{k+2} + ix_{k+3} \\ x_{k+2} + ix_{k+3} & x_k - ix_{k+1} \end{bmatrix}, \ldots, \right.
$$
$$
\left. \begin{bmatrix} x_{2R-3} + ix_{2R-2} & -x_{2R-1} + ix_{2R} \\ x_{2R-1} + ix_{2R} & x_{2R-3} - ix_{2R-2} \end{bmatrix} \right\}.
$$

$$(3.14)$$

The corresponding vector $\mathbf{s}$ to be transmitted by the source in the first phase is given by $[\, x_1 + ix_2 \; x_3 + ix_4 \ldots x_{2R-1} + ix_{2R} \,]^T$.

If R is odd, then construct a PCIOD for $R+1$ relays and drop the last column to get a $(R+1) \times R$ DSTBC which can be employed for R relays.

Note that $\mathbf{X}_{\text{PCIOD}}$ is a block diagonal matrix with replicas of the Alamouti STBC as the diagonal subblocks. Since the Alamouti STBC is conjugate linear, PCIOD also has the conjugate linearity property which is necessary for application as a DSTBC. Also, it is easy to check that the relay matrices of PCIOD have orthogonal rows as required by Theorem 3.4. There are totally $2R$ real variables $x_1, x_2, \ldots, x_{2R}$ in $\mathbf{X}_{\text{PCIOD}}$. It can be verified that the corresponding noise covariance matrix $\mathbf{\Gamma}$ for PCIOD is a block diagonal matrix with 2×2 scaled identity matrices as the diagonal subblocks. This special structure of $\mathbf{\Gamma}$ results in the $2R$ basis matrices $A_j, \; j = 1, 2, \ldots, 2R$ of $\mathbf{X}_{\text{PCIOD}}$ to satisfy

$$\mathbf{A}_j^H \mathbf{\Gamma}^{-1} \mathbf{A}_k + \mathbf{A}_k^H \mathbf{\Gamma}^{-1} \mathbf{A}_j = 0, \quad \forall \, j \neq k.$$

which is a condition for multi-group ML decoding (see Eq. (3.13)). However, the following expression shows that PCIOD is not fully diverse for arbitrary signal sets.

$$
|(\Delta \mathbf{X}_{\text{PCIOD}})^H (\Delta \mathbf{X}_{\text{PCIOD}})|
$$
$$
= \left(\sum_{j=1}^{4} \Delta x_j^2 \right)^2 \cdots \left(\sum_{j=k}^{k+3} \Delta x_j^2 \right)^2 \cdots \left(\sum_{j=2R-3}^{2R} \Delta x_j^2 \right)^2 .
$$

Thus, constellation precoding (entangling of some real symbols) is necessary to achieve full diversity. Precoding is to be done in the following manner. The $2R$ real variables are first partitioned into four groups as follows:

- First group: $\{x_{1+4k} | k = 0, 1, \ldots, \frac{R}{2} - 1\}$
- Second group: $\{x_{2+4k} | k = 0, 1, \ldots, \frac{R}{2} - 1\}$
- Third group: $\{x_{3+4k} | k = 0, 1, \ldots, \frac{R}{2} - 1\}$
- Fourth group: $\{x_{4+4k} | k = 0, 1, \ldots, \frac{R}{2} - 1\}$.

There are $\frac{R}{2}$ real variables in each group. It has been shown in [48] that

$$|(\Delta \mathbf{X}_{\text{PCIOD}})^H (\Delta \mathbf{X}_{\text{PCIOD}})| \geq \min_{j=1,\ldots,4} \left\{ \left(\prod_{k=0}^{\frac{R}{2}-1} \Delta x_{j+4k} \right)^2 \right\},$$

thus guaranteeing full diversity if $\prod_{k=0}^{\frac{R}{2}-1} \Delta x_{j+4k} \neq 0, \ \forall j = 1, 2, 3, 4$, i.e. product distance is not equal to zero. This can be ensured by letting the real symbols in each group take values from a rotated $\mathbb{Z}^{\frac{R}{2}}$ lattice constellation which is designed to maximize the minimum product distance. Algebraic number theory provides effective means to construct rotated $\mathbb{Z}^n$ lattices with large minimum product distance [49, 50] for any $n \in \mathbb{N}$ and the corresponding lattice generator matrices can be explicitly obtained from [50] for $1 \leq n \leq 30$.

The following example illustrates the construction procedure for $R = 6$.

Example 3.5. The PCIOD for six relays is as shown below:

$$\mathbf{X}_{\text{PCIOD}} = \begin{bmatrix} x_1 + ix_2 & -x_3 + ix_4 & 0 & 0 & 0 & 0 \\ x_3 + ix_4 & x_1 - ix_2 & 0 & 0 & 0 & 0 \\ 0 & 0 & x_5 + ix_6 & -x_7 + ix_8 & 0 & 0 \\ 0 & 0 & x_7 + ix_8 & x_5 - ix_6 & 0 & 0 \\ 0 & 0 & 0 & 0 & x_9 + ix_{10} & -x_{11} + ix_{12} \\ 0 & 0 & 0 & 0 & x_{11} + ix_{12} & x_9 - ix_{10} \end{bmatrix},$$

(3.15)

where

$$\begin{bmatrix} x_j \\ x_{j+4} \\ x_{j+8} \end{bmatrix} = \mathcal{G} \begin{bmatrix} y_j \\ y_{j+4} \\ y_{j+8} \end{bmatrix}, \quad j = 1, \ldots, 4,$$

and the vectors $\begin{bmatrix} y_j \\ y_{j+4} \\ y_{j+8} \end{bmatrix}$, $j = 1, 2, 3, 4$ take values from a subset (the cardinality of this subset is chosen based on the required transmission rate in bpcu) of $\mathbb{Z}^3$. The 3×3 lattice generator matrix $\mathcal{G}$ can be taken from [50]. The vector transmitted by the source is $\mathbf{s} = [x_1 + ix_2 \ x_3 + ix_4 \ x_5 + ix_6 \ x_7 + ix_8 \ x_9 + ix_{10} \ x_{11} + ix_{12}]^T$. At the destination, ML decoding of the symbols $\{x_j, x_{j+4}, x_{j+8}\}$ has to be done jointly for each $j = 1, 2, 3, 4$ separately. Thus, the resulting DSTBC is 4-group ML decodable or 3-real symbol ML decodable.

Example 3.6. If $R = 5$ relays, simply drop the last column of $\mathbf{X}_{\text{PCIOD}}$ in Eq. (3.15) which will result in a full diversity, 4-group ML decodable DSTBC for a five relay network.

Thus, the simple construction of PCIOD using Alamouti STBCs on the diagonal along with appropriate constellation precoding provides full diversity, four group ML decodable DSTBCs for arbitrary number of relays. Note that for the case of $R = 4$, PCIOD coincides with CIOD [17] and hence PCIOD is a generalization of CIOD. It is important to observe that PCIODs have NUW matrices as well as non-unitary relay matrices. Amidst many advantages, PCIODs have a drawback that they possess a large number of zeros entries. This leads to a large peak to average power ratio (PAPR) at the relays, which is undesirable since it demands the use of larger power amplifiers. Thus it is desirable to have alternate constructions of four-group ML decodable DSTBCs with unitary relay matrices and UW matrices to avoid the large PAPR problem. The following two constructions from extended Clifford algebras [47] solve this problem for power of two number of relays. For more details on the algebraic construction procedure, we refer the readers to [48].

Construction I from extended Clifford algebras: In order to describe the constructions from extended Clifford algebras, we need to introduce the ABBA construction [30, 31] and the doubling construction [40].

ABBA construction [30, 31]: Given a $M \times M$ LSTD $\mathbf{A}$, a $2M \times 2M$ LSTD $\mathbf{D}$ is constructed as

$$\mathbf{D} = \begin{bmatrix} \mathbf{A} & \mathbf{B} \\ \mathbf{B} & \mathbf{A} \end{bmatrix}, \qquad (3.16)$$

where the LSTD $\mathbf{B}$ is obtained from the LSTD $\mathbf{A}$ by introducing a new labeling to the associated real symbols of LSTD $\mathbf{A}$. The following example illustrates the ABBA construction.

Example 3.7. Let the given LSTD be the Alamouti LSTD. On applying the ABBA construction, we get a 4×4 LSTD given by

$$
\mathbf{D} = \begin{bmatrix}
x_1 + ix_2 & -x_3 + ix_4 & x_5 + ix_6 & -x_7 + ix_8 \\
x_3 + ix_4 & x_1 - ix_2 & x_7 + ix_8 & x_5 - ix_6 \\
x_5 + ix_6 & -x_7 + ix_8 & x_1 + ix_2 & -x_3 + ix_4 \\
x_7 + ix_8 & x_5 - ix_6 & x_3 + ix_4 & x_1 - ix_2
\end{bmatrix}.
$$

Doubling construction [40], Given a $M \times M$ LSTD $\mathbf{A}$, a $2M \times 2M$ LSTD $\mathbf{D}$ is constructed as:

$$
\mathbf{D} = \begin{bmatrix} \mathbf{A} & -\mathbf{B}^H \\ \mathbf{B} & \mathbf{A}^H \end{bmatrix}. \tag{3.17}
$$

Now construction I from extended Clifford algebras can be described in the following steps for $R = 2^a$, $a \in \mathbb{N}$ relays:

Step 1: Starting with the linear design $[s_1]$ in complex variable s_1, keep applying ABBA construction iteratively on it till a $2^{a-1} \times 2^{a-1}$ LSTD $\mathbf{D}$ is obtained.

Step 2: Apply doubling construction on $\mathbf{D}$ to get the required LSTD.

If $s_1, s_2, \ldots, s_R$ are the complex symbols in the resulting LSTD, then the corresponding vector $\mathbf{s}$ to be transmitted by the source in the first phase is given by $[s_1 \ s_2 \ \cdots \ s_R]^T$. It has been shown in [47] that four group ML decoding can be done and the associated grouping of the $2R$ real symbols into four groups is given by:

- First group: $\{s_{1I}, s_{2I}, \ldots, s_{\frac{R}{2}I}\}$.
- Second group: $\{s_{1Q}, s_{2Q}, \ldots, s_{\frac{R}{2}Q}\}$.
- Third group: $\{s_{(\frac{R}{2}+1)I}, s_{(\frac{R}{2}+2)I}, \ldots, s_{RI}\}$.
- Fourth group: $\{s_{(\frac{R}{2}+1)Q}, s_{(\frac{R}{2}+2)Q}, \ldots, s_{RQ}\}$.

The properties of conjugate linearity, UW matrices and unitary relay matrices can be easily proven [48]. Full diversity is achieved by these DSTBCs if the real symbols in each group take values from appropriately rotated $\mathbb{Z}^{\frac{R}{2}}$ lattice constellations. The following example illustrates the construction procedure.

Example 3.8. Let $R = 8$ relays. Then the LSTD as per construction I from extended Clifford algebras is given by

$$
\mathbf{X} = \begin{bmatrix}
s_1 & s_2 & s_3 & s_4 & -s_5^* & -s_6^* & -s_7^* & -s_8^* \\
s_2 & s_1 & s_4 & s_3 & -s_6^* & -s_5^* & -s_8^* & -s_7^* \\
s_3 & s_4 & s_1 & s_2 & -s_7^* & -s_8^* & -s_5^* & -s_6^* \\
s_4 & s_3 & s_2 & s_1 & -s_8^* & -s_7^* & -s_6^* & -s_5^* \\
s_5 & s_6 & s_7 & s_8 & s_1^* & s_2^* & s_3^* & s_4^* \\
s_6 & s_5 & s_8 & s_7 & s_2^* & s_1^* & s_4^* & s_3^* \\
s_7 & s_8 & s_5 & s_6 & s_3^* & s_4^* & s_1^* & s_2^* \\
s_8 & s_7 & s_6 & s_5 & s_4^* & s_3^* & s_2^* & s_1^*
\end{bmatrix}.
$$

Let $s = [\, s_1 \ s_2 \ s_3 \ s_4 \ s_5 \ s_6 \ s_7 \ s_8 \,]^T$. Then, it is easy to observe that the above LSTD is conjugate linear and has UW matrices, unitary relay matrices. The associated four groups of real symbols for decoding purposes is given by: $\{s_{1I}, s_{2I}, s_{3I}, s_{4I}\}$, $\{s_{1Q}, s_{2Q}, s_{3Q}, s_{4Q}\}$, $\{s_{5I}, s_{6I}, s_{7I}, s_{8I}\}$, and $\{s_{5Q}, s_{6Q}, s_{7Q}, s_{8Q}\}$. Let us compute the multi-dimensional signal sets which lead to full diversity. Note that $\mathbf{X}$ has the form of $\begin{bmatrix} \mathbf{A} & -\mathbf{B}^H \\ \mathbf{B} & \mathbf{A}^H \end{bmatrix}$ and it can be checked that $|(\Delta\mathbf{X})^H(\Delta\mathbf{X})| = \begin{bmatrix} \Delta\mathbf{A}^H\Delta\mathbf{A} + \Delta\mathbf{B}^H\Delta\mathbf{B} & \mathbf{0} \\ \mathbf{0} & \Delta\mathbf{A}\Delta\mathbf{A}^H + \Delta\mathbf{B}\Delta\mathbf{B}^H \end{bmatrix}$, which then implies that $|(\Delta\mathbf{X})^H(\Delta\mathbf{X})| \geq \min(|\Delta\mathbf{A}|^2, |\Delta\mathbf{B}|^2)$. Hence, it is sufficient if we ensure that either $|\Delta\mathbf{A}| \neq 0$ or $|\Delta\mathbf{B}| \neq 0$. For this example,

$$
|\Delta\mathbf{A}| = (\Delta s_2 + \Delta s_1 - \Delta s_4 - \Delta s_3)(\Delta s_2 + \Delta s_1 + \Delta s_4 + \Delta s_3)
$$
$$
(-\Delta s_2 + \Delta s_1 + \Delta s_4 - \Delta s_3)(-\Delta s_2 + \Delta s_1 - \Delta s_4 + \Delta s_3).
$$

Thus, full diversity can be ensured if every term in the product is not zero. We can ensure that the real part of every term in the product is non-zero in the following manner (a similar technique can be used for imaginary terms as well). Let

$$
T = \begin{bmatrix}
1 & 1 & -1 & -1 \\
1 & 1 & 1 & 1 \\
1 & -1 & -1 & 1 \\
1 & -1 & 1 & -1
\end{bmatrix}
$$

and define

$$
\mathbf{t} = \begin{bmatrix} y_1 \\ y_2 \\ y_3 \\ y_4 \end{bmatrix} = \mathbf{T} \begin{bmatrix} s_{1I} \\ s_{2I} \\ s_{3I} \\ s_{4I} \end{bmatrix}.
$$

Thus, $|\Delta\mathbf{A}| \neq 0$ if $(\Delta y_1)(\Delta y_2)(\Delta y_3)(\Delta y_4) \neq 0$ which is nothing but product distance. This can be satisfied if we let $\mathbf{t}$ take values from a subset of a lattice $\mathbb{J}\mathbb{Z}^4$, where the lattice generator matrix $\mathbb{J}$ is taken from [50]. Thus, we can let $\begin{bmatrix} s_{1I} \\ s_{2I} \\ s_{3I} \\ s_{4I} \end{bmatrix} = \mathbf{T}^{-1}\mathbf{t}$. In a similar manner, we can ensure that $|\Delta\mathbf{B} \neq 0|$ as well.

Construction II from extended Clifford algebras: Construction II from extended Clifford algebras starts with the following LSTD for four relays.

$$\mathbf{A} = \begin{bmatrix} s_1 & -s_2^* & -s_3^* & -s_4 \\ s_2 & s_1^* & -s_4^* & s_3 \\ s_3 & s_4^* & s_1^* & -s_2 \\ s_4 & -s_3^* & s_2^* & s_1 \end{bmatrix}.$$

For $R = 2^a$, $a > 2$, apply ABBA construction on the above LSTD $\mathbf{A}$ iteratively. The grouping of real symbols into four groups is done as follows:

- First group: $\{s_{1I}, s_{5I}, \ldots, s_{(R-3)I}, s_{4Q}, s_{8Q}, \ldots, s_{RQ}\}$.
- Second group: $\{s_{1Q}, s_{5Q}, \ldots, s_{(R-3)Q}, s_{4I}, s_{8I}, \ldots, s_{RI}\}$.
- Third group: $\{s_{3Q}, s_{7Q}, \ldots, s_{(R-1)Q}, s_{2I}, s_{6I}, \ldots, s_{(R-2)I}\}$.
- Fourth group: $\{s_{3I}, s_{7I}, \ldots, s_{(R-1)I}, s_{2Q}, s_{6Q}, \ldots, s_{(R-2)Q}\}$.

Full diversity can be obtained in a similar way as for DSTBCs from construction I (see Exercise 2).

3.4.1.2. *Single complex symbol ML decodable DSTBCs*

The counterpart of ODs and single complex symbol ML decodable STBCs for the distributed case have been studied in [52, 53] and we briefly present the important results here. Towards studying distributed orthogonal STBCs (DOSTBCs) and single complex symbol ML decodable DSTBCs, let us consider a more generalized setup wherein, the relays are allowed to linearly transform the received vector as well as its conjugate and transmit the resulting sum. To be precise, the jth relay can transmit

$$\mathbf{t_j} = \sqrt{\frac{\pi_2 P}{\pi_1 P + 1}}(\mathbf{B_j}\mathbf{r_j} + \mathbf{C_j}\mathbf{r_j}^*)$$

$$= \sqrt{\frac{\pi_1\pi_2 P^2}{\pi_1 P + 1}}\left(\mathbf{B_j}f_j\mathbf{s} + \mathbf{C_j}f_j^*\mathbf{s}^*\right) + \sqrt{\frac{\pi_2 P}{\pi_1 P + 1}}\sum_{j=1}^{R} g_j(\mathbf{B_j}\mathbf{v_j} + \mathbf{C_j}\mathbf{v_j}^*),$$

where $\mathbf{B_j}, \mathbf{C_j} \in \mathbb{C}^{T_2 \times T_1}$ satisfying $\|\mathbf{B_j}\|_F^2 + \|\mathbf{C_j}\|_F^2 \leq T_2$. As before, P denotes the average total power spent by the source and relays. The received vector $\mathbf{y}$ can then be expressed as follows:

$$\mathbf{y} = \sqrt{\frac{\pi_1 \pi_2 P^2}{\pi_1 P + 1}} \mathbf{Xh} + \mathbf{n},$$

where

$$\mathbf{X} = \left[\mathbf{B_1} f_1 \mathbf{s} + \mathbf{C_1} f_1^* \mathbf{s}^* \ \ \mathbf{B_2} f_2 \mathbf{s} + \mathbf{C_2} f_2^* \mathbf{s}^* \ \cdots \ \mathbf{B_R} f_R \mathbf{s} + \mathbf{C_R} f_R^* \mathbf{s}^* \right], \tag{3.18}$$
$$\mathbf{h} = \left[g_1 \ g_2 \ \cdots g_R \right]^T,$$

and

$$\mathbf{n} = \sqrt{\frac{\pi_2 P}{\pi_1 P + 1}} \left(\sum_{j=1}^{R} g_j \left(\mathbf{B_j v_j} + \mathbf{C_j v_j^*} \right) \right) + \mathbf{w}.$$

Note the difference between Eq. (3.18) and Eq. (3.10) that the channel fading gains are now considered as part of $\mathbf{X}$. This simplifies the discussion as well as construction of DOSTBCs. Also, note that the covariance matrix of the equivalent noise vector $\mathbf{n}$ now becomes

$$\mathbf{\Gamma} = \mathbf{I_{T_2}} + \frac{\pi_2 P}{\pi_1 P + 1} \left(\sum_{j=1}^{R} |\mathbf{g_j}|^2 \left(\mathbf{B_j B_j}^H + \mathbf{C_j C_j}^H \right) \right). \tag{3.19}$$

The ML decoder remains the same as in Eq. (3.12) with $\mathbf{X}$ and $\mathbf{h}$ as given by Eq. (3.18).

Definition 3.7 (Yi and Kim [52]). A $T_2 \times R$ matrix $\mathbf{X}$ is called a DOSTBC in complex variables $s_1, s_2, \ldots, s_N$ if the following two conditions are satisfied:

(1) The entries of $\mathbf{X}$ are $0, \pm f_j s_k, \pm f_j^* s_k^*$, $1 \leq j \leq R$, $1 \leq k \leq N$ or multiples of these indeterminates by i.
(2) The matrix $\mathbf{X}$ satisfies the following equality

$$\mathbf{X}^H \mathbf{\Gamma}^{-1} \mathbf{X} = \sum_{j=1}^{R} |f_j|^2 \mathbf{D_j}, \tag{3.20}$$

where $\mathbf{D_j} = \text{diag}\{|f_1|^2 D_{j,1}, \ldots, |f_R|^2 D_{j,R}\}$ and $D_{j,k}$ are non-zero for $k = 1, \ldots, R$.

Using Eq. (3.20) and Theorem 3.5, it is easy to show that DOSTBCs are single complex symbol ML decodable and furthermore achieve full diversity when arbitrary complex constellations are employed for the complex symbols $s_1, s_2, \ldots, s_N$. To employ a DOSTBC $\mathbf{X}$ in N complex symbols, in the first phase the source transmits $\mathbf{s} = [\, s_1 \; s_2 \; \cdots \; s_N \,]^T$ and hence $T_1 = N$. The problem is thus to construct DOSTBCs which maximize the rate $\frac{N}{T_2}$.

Theorem 3.6 (Yi and Kim [55]). *The rate of a DOSTBC in N complex variables for R relays satisfies the following inequality:*

$$\frac{N}{T_2} \le \frac{N}{\lceil \frac{NR}{2} \rceil}.$$

The above theorem gives an upper bound on the rate of DOSTBCs but constructions achieving the above upper bound are not yet known. The authors in [52] restrict to a subclass of DOSTBCs which have diagonal covariance matrix $\mathbf{\Gamma}$. Towards that end, we define a matrix to be *column-monomial (row-monomial)* if there is at most one non-zero entry on every column (row) of it.

Theorem 3.7 (Yi and Kim [52]). *The covariance matrix $\mathbf{\Gamma}$ in Eq. (3.19) is a diagonal matrix if and only if $\mathbf{B_j}$ and $\mathbf{C_j}$ are column monomial for $j = 1, 2, \ldots, R$.*

Definition 3.8 (Yi and Kim [52]). *A $T_2 \times R$ matrix is called a column-monomial DOSTBC in complex variables $s_1, s_2, \ldots, s_N$ if it is a DOSTBC whose associated matrices $\mathbf{B_j}$ and $\mathbf{C_j}$, $j = 1, 2, \ldots, R$ are all column-monomial.*

Theorem 3.8 (Yi and Kim [52]). *If a column-monomial DOSTBC $\mathbf{X}$ in variables $s_1, s_2, \ldots, s_N$ exists, its rate satisfies the following inequality:*

$$\frac{N}{T_2} \le \begin{cases} \dfrac{1}{m}, & \text{when } N = 2l, R = 2m, \\[2mm] \dfrac{2l+1}{2lm+2m}, & \text{when } N = 2l+1, R = 2m, \\[2mm] \dfrac{1}{m+1}, & \text{when } N = 2l, R = 2m+1, \\[2mm] \min\left(\dfrac{2l+1}{2lm+2m+l+1}, \right. & \text{when } N = 2l+1, R = 2m+1, \\[2mm] \left. \dfrac{2l+1}{2lm+2l+m+1} \right) \end{cases}$$

$$(3.21)$$

where, $N \ge 1$, $R \ge 1$, and l, m are some positive integers.

Systematic construction methods of column-monomial DOSTBCs achieving the upper bound in Theorem 3.8 are provided in [52] when the number of relays (R) and/or the number of information-bearing symbols (N) are even.

Example 3.9. For $R = 5$ and $N = 5$, the construction in [52] results in

$$
\mathbf{X} =
\begin{bmatrix}
f_1 s_2 & f_2^* s_3^* & 0 & 0 & 0 \\
-f_1 s_3 & f_2^* s_2^* & 0 & 0 & 0 \\
f_1 s_4 & f_2^* s_5^* & 0 & 0 & 0 \\
-f_1 s_5 & f_2^* s_4^* & 0 & 0 & 0 \\
0 & 0 & f_3 s_1 & f_4^* s_3^* & 0 \\
0 & 0 & -f_3 s_3 & f_4^* s_1^* & 0 \\
0 & 0 & f_3 s_4 & f_4^* s_5^* & 0 \\
0 & 0 & -f_3 s_5 & f_4^* s_4^* & 0 \\
f_1^* s_1^* & 0 & 0 & 0 & f_5 s_5 \\
f_1^* s_5^* & 0 & 0 & 0 & -f_5 s_1 \\
0 & 0 & f_3^* s_2^* & 0 & f_5 s_4 \\
0 & 0 & f_3^* s_4^* & 0 & -f_5 s_2 \\
0 & f_2 s_1 & 0 & 0 & f_3^* s_3^* \\
0 & -f_2 s_3 & 0 & 0 & f_5^* s_1^* \\
0 & 0 & 0 & f_4 s_2 & 0
\end{bmatrix},
$$

whose rate achieves the upper bound of $\frac{1}{3}$ as per Theorem 3.8.

Recently in [53], precoding of the information vector **s** and its conjugate using matrices $\mathbf{P}, \mathbf{Q}$ before transmission by the source has been considered. Using this precoding technique, it has been shown in [53] that the bounds in Theorem 3.8 can be approximately doubled for a class of single complex symbol ML decodable DSTBCs called as Semi-orthogonal Precoded Distributed Single-Symbol Decodable STBCs (S-PDSSDCs). Explicit constructions of S-PDSSDCs achieving the upper bounds are also provided for certain number of relays in [53]. The DSTBCs constructed in [53] admit single complex symbol ML decoding and achieve full diversity for appropriately rotated QAM constellations only. But it is important to note that the DOSTBCs in [52] admit single real symbol ML decoding for QAM constellations and moreover achieve full diversity for any complex constellation.

3.4.2. *Coherent Case with Partial CSI at Relays*

In this subsection, we consider the coherent case with the additional assumption that the relay nodes have partial CSI of the source to relay channel gains, i.e. the jth relay node is assumed to know the phase component of the source to relay channel gain f_j. In this scenario, if the channel fading gain $f_j = |f_j|e^{i\phi}$, then the jth relay can pre-multiply the received vector r_j by $e^{-i\phi}$ to compensate for the rotational effect of the fading channel. Thus, without loss of generality we can consider the channel fading gains f_j, $j = 1, 2, \ldots, R$ to be modeled as real Rayleigh random variables with unit variance. This scenario of partial CSI at relays has been studied in [43, 51, 54, 55] and it has been shown that it leads to several benefits compared to the coherent case. The important difference arises in the signal model because f_j is now a real random variable. We assume that the relays linearly transform the received vector as well as its conjugate. To be precise, the jth relay transmits

$$\mathbf{t_j} = \sqrt{\frac{\pi_2 P}{\pi_1 P + 1}}(\mathbf{B_j r_j} + \mathbf{C_j r_j}^*),$$

where $\mathbf{B_j}, \mathbf{C_j} \in \mathbb{C}^{T_2 \times T_1}$ which satisfy $\|\mathbf{B_j}\|_F^2 + \|\mathbf{C_j}\|_F^2 \leq T_2$. Since f_j is a real random variable, we have

$$\mathbf{t_j} = \sqrt{\frac{\pi_1 \pi_2 P^2}{\pi_1 P + 1}} f_j (\mathbf{B_j s} + \mathbf{C_j s}^*) + \sqrt{\frac{\pi_2 P}{\pi_1 P + 1}}(\mathbf{B_j v_j} + \mathbf{C_j v_j}^*),$$

where $\mathbf{v_j}$ is the AWGN at the jth relay node as before. Thus, we get

$$\mathbf{y} = \sqrt{\frac{\pi_1 \pi_2 P^2}{\pi_1 P + 1}} \mathbf{Xh} + \mathbf{n},$$

where

$$\mathbf{X} = [\mathbf{B_1 s} + \mathbf{C_1 s}^* \ \ \mathbf{B_2 s} + \mathbf{C_2 s}^* \ \cdots \ \mathbf{B_R s} + \mathbf{C_R s}^*], \tag{3.22}$$

$$\mathbf{h} = [f_1 g_1 \ \ f_2 g_2 \ \cdots \ f_R g_R]^T,$$

and

$$\mathbf{n} = \sqrt{\frac{\pi_2 P}{\pi_1 P + 1}} \left(\sum_{j=1}^{R} g_j (\mathbf{B_j v_j} + \mathbf{C_j v_j}^*)\right) + \mathbf{w}.$$

Note from Eq. (3.22) that linear dispersion STBCs [56] can be used in this scenario as opposed to only conjugate linear DSTBCs (Eq. (3.10)) for the coherent case without CSI at relays. Similarly note that the channel fading gains f_j, $j = 1, 2, \ldots, R$ do not appear in the matrix $\mathbf{X}$ in Eq. (3.22) as was the case in Eq. (3.18). It can be proven [43] that the rank criterion for achieving full diversity ($|\Delta \mathbf{X} \neq 0|$) holds true for this partial CSI case as well. As for the coherent case without CSI at relays, single symbol ML decodable DSTBCs for the coherent case with partial CSI at relays have been studied independently in [54, 55]. We briefly present some of the important results in the sequel.

Definition 3.9 (Yi and Kim [55]). A $T_2 \times R$ matrix $\mathbf{X}$ (as in Eq. (3.22)) is called a column-monomial DOSTBC-CPI in complex variables $s_1, s_2, \ldots, s_N$ if the following two conditions are satisfied:

(1) The entries of $\mathbf{X}$ are $0, \pm s_k, \pm s_k^*$, $1 \leq k \leq N$ or multiples of these indeterminates by i.
(2) The matrix $\mathbf{X}$ satisfies the following equality:

$$\mathbf{X}^H \mathbf{\Gamma}^{-1} \mathbf{X} = \sum_{j=1}^{R} |f_j|^2 \mathbf{D_j},$$

where $\mathbf{D_j} = \text{diag}\{|f_1|^2 D_{j,1}, \ldots, |f_R|^2 D_{j,R}\}$ and $D_{j,k}$ are nonzero for $k = 1, \ldots, R$. Furthermore its associated matrices $\mathbf{B_j}, \mathbf{C_j}$, $j = 1, 2, \ldots, R$ are all column-monomial.

Similar to DOSTBCs, it is easy to show that column-monomial DOSTBCs-CPI are single symbol ML decodable and also achieve full diversity for arbitrary complex constellations.

Theorem 3.9 (Yi and Kim [55]). *For $R > 2$, the rate of column-monomial DOSTBC-CPI satisfies the following inequality:*

$$\frac{N}{T_2} \leq \frac{1}{2}.$$

Theorem 3.10 (Sreedhar et al. [55] Yi and Kim [54]). *Let $\mathbf{G}$ be a $p \times R$ Real Orthogonal Design (ROD) with rate one constructed as in [27]. Then the rate $\frac{1}{2}$ complex OD constructed using the stacking construction [27]* as: $\mathbf{X} = \begin{bmatrix} \mathbf{G} \\ \mathbf{G}^* \end{bmatrix}$ *is a column-monomial DOSTBC-CPI which achieves the upper bound* $\frac{1}{2}$ *on the rate.*

Example 3.10. For $R = 4$, the column-monomial DOSTBC-CPI with rate $\frac{1}{2}$ is given by

$$
\mathbf{X} = \begin{bmatrix}
s_1 & s_2 & s_3 & s_4 \\
-s_2 & s_1 & -s_4 & s_3 \\
-s_3 & s_4 & s_1 & -s_2 \\
-s_4 & -s_3 & s_2 & s_1 \\
s_1^* & s_2^* & s_3^* & s_4^* \\
-s_2^* & s_1^* & -s_4^* & s_3^* \\
-s_3^* & s_4^* & s_1^* & -s_2^* \\
-s_4^* & -s_3^* & s_2^* & s_1^*
\end{bmatrix}.
$$

3.5. Thoughts for Practitioners

STBCs with low ML decoding complexity is very high on the agenda of standard making bodies such as IEEE 802.16e, etc. For example, a modified version of the CIOD [17] is part of the IEEE 802.16e standard for three transmit antenna systems. In recent years, cooperative communication and relaying mechanisms are also investigated for futuristic IEEE standards such as IEEE 802.16j, etc. (see [62]). Thus, DSTBCs with low ML decoding complexity as well as good performance will attract wide attention especially when it comes to practical applications.

In practice, there are many other issues apart from decoding complexity which have not been considered in this chapter. For example, perfect symbol as well as carrier frequency synchronization was assumed among the geographically distributed relay nodes, which in reality is difficult to achieve. To combat the problem of symbol asynchronism, asynchronous DSTBCs which achieve full cooperative diversity have been studied in [48, 57–59]. Most of these works fail to address decoding complexity issues. In [48, 57, 58], OFDM has been used along with certain structured DSTBCs to achieve full asynchronous cooperative diversity along with low ML decoding complexity. Recently [61] shows that the Alamouti DSTBC combined with OFDM can combat both symbol asynchronism as well as carrier frequency offsets at the relay nodes for a 2 relay network. Similarly partner selection and coordination, design of training symbols, and estimation of CSI are few of the several other important practical issues. A simple training and channel estimation method has been recently proposed in [60].

3.6. Conclusions

A survey of multi-symbol ML decodable STBCs for point to point MIMO systems in the literature was presented. The various tradeoffs involving rate, decoding complexity, and diversity of a STBC were discussed and few examples were given to highlight the important aspects. Then, the parallel developments in DSTBCs with low ML decoding complexity were reviewed and the new issues, problems, opportunities that arise due to the distributed cooperative setup were briefly touched upon. This area of cooperative communication and distributed coding offers lot of new opportunities as well as brings in new issues. In this chapter, we have attempted to provide a survey as well as give a brief overview of this developing area.

3.7. Directions for Future Research

Below, we list few of the important open problems and directions for future research.

(1) The maximum rate of single symbol ML decodable DSTBCs was presented in Theorem 3.8. What is the maximum rate of general λ-symbol ML decodable DSTBCs and how to construct DSTBCs with those rates?

(2) For the case of no CSI at any of the relay nodes, it has been shown in [15] that differential encoding can be combined with the four group ML decodable DSTBCs from extended Clifford algebras to obtain full diversity, four group decodable distributed differential STBCs albeit with a suboptimal receiver. Constructing full diversity, single symbol decodable distributed differential STBCs is an interesting direction for further research.

(3) In this chapter, we have considered only single antennas at all the nodes. Full diversity DSTBCs for relay networks with multiple antennas has been addressed in [64]. Are there full diversity, DSTBC constructions with low ML decoding complexity for relay networks with multiple antennas?

(4) If CSI is available at the source and/or relays, then is it possible to combine beamforming along with DSTBCs with low ML decoding

complexity? Beamforming in relay networks has been addressed recently in [63].

(5) How to construct full diversity, single symbol ML decodable DSTBCs for asynchronous relay networks? Full diversity, single complex symbol ML decodable DSTBCs are available in the literature [48, 57, 58] only for asynchronous networks with 2 and 4 relay nodes.

(6) All the DSTBC constructions with low ML decoding complexity presented in this chapter are essentially applicable only for cooperative protocols built upon the Jing-Hassibi protocol described in [5]. However, there are several other cooperative protocols discussed in the literature [3, 4, 7] some of which are targeted towards optimally trading off diversity and multiplexing gains. Constructing codes with low ML decoding complexity for the other cooperative protocols is also worth investigating.

(7) [65] has exposed other means (apart from the conditions in Theorem 3.1) to reduce the ML decoding complexity of STBCs and few high rate STBCs have also been discussed. The counterpart of these techniques may be of use to construct high rate, full diversity DSTBCs with reduced ML decoding complexity.

3.8. Keywords

MIMO	A communication system with multiple transmit and multiple receive antennas.
PAM	Pulse amplitude modulation.
QAM	Quadrature amplitude modulation.
PSK	Phase shift keying.
PAPR	Ratio of the peak to average power.
AWGN	Additive white Gaussian noise.
SNR	Ratio of the signal power to noise power.
OFDM	Orthogonal frequency division multiplexing.
STBC	A coding technique for MIMO systems in which coding of data is done in blocks.
DSTBC	An effective STBC which is generated by geographically distributed nodes.
Diversity	Negative of the asymptotic slope of the error probability when plotted on a log-log scale versus SNR.
Coding gain	Effective gain in SNR due to coding.

3.9. Exercise

(1) Consider the Alamouti LSTD $\begin{bmatrix} s_1 & -s_2^* \\ s_2 & s_1^* \end{bmatrix}$, where s_1, s_2 are complex variables.

   (a) If a STBC $\mathscr{C}$ is obtained from the Alamouti LSTD by choosing 64 QAM as the signal constellation for s_1 and s_2, then what is the rate of $\mathscr{C}$ in bpcu?

   (b) Choose an appropriate signal constellation for s_1 and s_2 in the Alamouti LSTD such that the resulting STBC $\mathscr{C}$ consists of 64 unitary matrices.

(2) Let $\mathbf{A}$ be an $N_T \times N_T$ OD in real variables $x_1, \ldots, x_K$ and let $\mathbf{B} = (x_{K+1} + ix_{K+2})\mathbf{I_{N_T}}$. Prove that $\mathbf{D} = \begin{bmatrix} \mathbf{A} & -\mathbf{B}^H \\ \mathbf{B} & \mathbf{A}^H \end{bmatrix}$ obtained using the doubling construction is an $2N_T \times 2N_T$ OD in real variables $x_1, x_2, \ldots, x_{K+2}$.

(3) What is the rate of the LSTD $\mathbf{X} = \begin{bmatrix} s_1 + \frac{1}{\sqrt{2}}s_2 & s_3 + \frac{1}{\sqrt{2}}s_2 \\ -s_3 - \frac{1}{\sqrt{2}}s_2 & s_1 + \frac{1}{\sqrt{2}}s_2 \end{bmatrix}$, where s_1, s_2, s_3 denote complex variables?

(4) Are SSD codes necessarily ODs?

(5) Suppose that $\mathbf{X} = x_1\mathbf{A_1} + x_2\mathbf{A_2} + x_3\mathbf{A_3} + x_4\mathbf{A_4}$ where x_j, $j = 1, 2, 3, 4$ are real variables, is single complex symbol ML decodable with the partitioning: $\{x_1, x_2\}, \{x_3, x_4\}$. Let $\mathbf{\bar{A}_1} = \mathbf{c_{11}A_1} + \mathbf{c_{12}A_2}$, $\mathbf{\bar{A}_2} = \mathbf{c_{21}A_1} + \mathbf{c_{22}A_2}$, $\mathbf{\bar{A}_3} = \mathbf{c_{31}A_3} + \mathbf{c_{32}A_4}$, $\mathbf{\bar{A}_4} = \mathbf{c_{41}A_3} + \mathbf{c_{42}A_4}$, where $c_{j,k}$, $1 \le j \le 4, 1 \le k \le 2$ are some real constants. Then show that $\mathbf{\bar{X}} = x_1\mathbf{\bar{A}_1} + x_2\mathbf{\bar{A}_2} + x_3\mathbf{\bar{A}_3} + x_4\mathbf{\bar{A}_4}$ is also single complex symbol ML decodable for the same partitioning of real symbols. Infer that linear transformation of weight matrices when constrained to weight matrices within the same group does not disturb the ML decoding complexity.

(6) Construct a rate one, full diversity, single complex symbol ML decodable STBC for four transmit antennas using Construction II from extended Clifford algebras. List down the corresponding weight matrices and write down the single complex symbol ML decision rule.

(7) Consider the diagonal LSTD $\mathbf{X} = \begin{bmatrix} s_1 & 0 \\ 0 & s_2 \end{bmatrix}$, where s_1, s_2 are complex variables.

   (a) Prove that the given LSTD is two group decodable for the following partitioning of real symbols: $\{s_{1I}, s_{2I}\}, \{s_{1Q}, s_{2Q}\}$.

(b) Use the partitioning of real symbols specified in part (a) and provide full diversity signal sets for the given LSTD to result in a two group ML decodable STBC.

(c) Are the weight matrices of the STBC constructed in part (b) unitary?

(d) Generalize the ideas above to give a construction of rate one, full diversity two group ML decodable STBCs from diagonal LSTDs for arbitrary number of transmit antennas N_T.

(8) Consider the rate $\frac{3}{4}$ OD given by $\begin{bmatrix} s_1 & -s_2^* & -s_3^* & 0 \\ s_2 & s_1^* & 0 & -s_3^* \\ s_3 & 0 & s_1^* & s_2^* \\ 0 & s_3 & -s_2 & s_1 \end{bmatrix}$, where s_1, s_2, s_3
are complex variables. In a distributed setting, if the source node transmits $\mathbf{s} = \begin{bmatrix} s_1 & s_2 & s_3 \end{bmatrix}^T$ in the first phase and if the relay nodes implement the given OD, show that the destination will not see a DOSTBC.

(9) Construct a rate one, full diversity four group ML decodable DSTBC for a three relay network.

(10) In a 2 relay network, if the relay nodes employ

$$\mathbf{B_1} = \begin{bmatrix} 1 & 0 \\ 0 & 1 \end{bmatrix}, \quad \mathbf{C_1} = 0, \quad \mathbf{B_2} = 0, \quad \mathbf{C_2} = \begin{bmatrix} \frac{1}{\sqrt{2}} & 0 \\ -\frac{1}{\sqrt{2}} & 1 \end{bmatrix}$$

as relay matrices, show that the noise covariance matrix Γ is not a diagonal matrix.

3.10. Sample Solutions to Exercise Problems

(1) (a) $|\mathscr{C}| = 64^2 = 2^{12}$. Thus rate $= \frac{\log_2 |\mathscr{C}|}{2} = \frac{\log_2 2^{12}}{2} = 6\,\text{bpcu}$.

(b) Let s_1 and s_2 take values from 8-PSK constellation scaled by $\frac{1}{\sqrt{2}}$, i.e. $s_1, s_2 \in \{\frac{1}{\sqrt{2}}e^{-\frac{i2\pi k}{8}}, \ k = 0, 1, \ldots, 7\}$. $|\mathscr{C}| = 8^2 = 64$. If $\mathbf{X}$ is a codeword, then $\mathbf{X}^H\mathbf{X} = (|s_1|^2 + |s_2|^2)\mathbf{I_2} = \mathbf{I_2}$.

(2) We know that $\mathbf{A}^H\mathbf{A} = (\sum_{j=1}^{K} x_j^2)\mathbf{I_{N_T}}$ and $\mathbf{B}^H\mathbf{B} = (x_{K+1}^2 + x_{K+2}^2)\mathbf{I_{N_T}}$. Compute $\mathbf{D}^H\mathbf{D}$ and substitute for $\mathbf{A}^H\mathbf{A}$ and $\mathbf{B}^H\mathbf{B}$ to show that $\mathbf{D}^H\mathbf{D} = (\sum_{j=1}^{K+2} x_j^2)\mathbf{I_{2N_T}}$.

(3) The six weight matrices of the given LSTD are first found. Then it can be checked that only 4 of the 6 weight matrices are linearly independent. Thus, the rate of the given LSTD is $\frac{4}{4} = 1$ complex symbols per channel use.

(4) No. The weight matrices of SSD codes need not be unitary. Moreover, ODs are single real symbol ML decodable whereas in SSD codes, some of the weight matrices need not satisfy the orthogonality condition (see Theorem 3.1).

(5) It is given that $\mathbf{A_1}^H\mathbf{A_3} + \mathbf{A_3}^H\mathbf{A_1} = 0$, $\mathbf{A_1}^H\mathbf{A_4} + \mathbf{A_4}^H\mathbf{A_1} = 0$, $\mathbf{A_2}^H\mathbf{A_3} + \mathbf{A_3}^H\mathbf{A_2} = 0$, $\mathbf{A_2}^H\mathbf{A_4} + \mathbf{A_4}^H\mathbf{A_2} = 0$. Check that $\bar{\mathbf{A}}_1^H\bar{\mathbf{A}}_3 + \bar{\mathbf{A}}_3^H\bar{\mathbf{A}}_1 = 0$, $\bar{\mathbf{A}}_1^H\bar{\mathbf{A}}_4 + \bar{\mathbf{A}}_4^H\bar{\mathbf{A}}_1 = 0$, $\bar{\mathbf{A}}_2^H\bar{\mathbf{A}}_3 + \bar{\mathbf{A}}_3^H\bar{\mathbf{A}}_2 = 0$, $\bar{\mathbf{A}}_2^H\bar{\mathbf{A}}_4 + \bar{\mathbf{A}}_4^H\bar{\mathbf{A}}_2 = 0$.

(6) The required STBC is given by $\begin{bmatrix} s_1 & -s_2^* & -s_3^* & -s_4 \\ s_2 & s_1^* & -s_4^* & s_3 \\ s_3 & s_4^* & s_1^* & -s_2 \\ s_4 & -s_3^* & s_2^* & s_1 \end{bmatrix}$, where $\{s_{1I}, s_{4Q}\}$, $\{s_{1Q}, s_{4I}\}$, $\{s_{3Q}, s_{2I}\}$, $\{s_{2Q}, s_{3I}\}$ are allowed to take values independently from 13.29° rotated QAM constellation. It can be checked that full diversity is achieved when $\Delta s_{1I} \neq \pm\Delta s_{4Q}$, $\Delta s_{4I} \neq \pm\Delta s_{1Q}$, $\Delta s_{2I} \neq \pm\Delta s_{3Q}$, $\Delta s_{3I} \neq \pm\Delta s_{2Q}$, and rotating QAM constellation by 13.29° satisfies this. Let $\mathbf{A_{jI}}$, $\mathbf{A_{jQ}}$ denote the weight matrices of s_{jI} and s_{jQ}, respectively. Single complex symbol ML decoding is done as follows:

$$(\hat{s}_{1I}, \hat{s}_{4Q}) = \arg\min \|\mathbf{Y} - (s_{1I}\mathbf{A_{1I}} + s_{4Q}\mathbf{A_{4Q}})\mathbf{H}\|_F^2$$

$$(\hat{s}_{1Q}, \hat{s}_{4I}) = \arg\min \|\mathbf{Y} - (s_{1Q}\mathbf{A_{1Q}} + s_{4I}\mathbf{A_{4I}})\mathbf{H}\|_F^2$$

$$(\hat{s}_{3Q}, \hat{s}_{2I}) = \arg\min \|\mathbf{Y} - (s_{2I}\mathbf{A_{2I}} + s_{3Q}\mathbf{A_{3Q}})\mathbf{H}\|_F^2$$

$$(\hat{s}_{3I}, \hat{s}_{2Q}) = \arg\min \|\mathbf{Y} - (s_{3I}\mathbf{A_{3I}} + s_{2Q}\mathbf{A_{2Q}})\mathbf{H}\|_F^2$$

(7) (a) The associated weight matrices are $\mathbf{A_{1I}} = \begin{bmatrix} 1 & 0 \\ 0 & 0 \end{bmatrix}$, $\mathbf{A_{1Q}} = \begin{bmatrix} i & 0 \\ 0 & 0 \end{bmatrix}$, $\mathbf{A_{2I}} = \begin{bmatrix} 0 & 0 \\ 0 & 1 \end{bmatrix}$, $\mathbf{A_{2Q}} = \begin{bmatrix} 0 & 0 \\ 0 & i \end{bmatrix}$. Check that the necessary and sufficient condition for two group decoding, i.e. $\mathbf{A_{jI}^H A_{kI}} + \mathbf{A_{kI}^H A_{jI}} = 0$ and $\mathbf{A_{jI}^H A_{kQ}} + \mathbf{A_{kQ}^H A_{jI}} = 0$ are satisfied for $j \neq k$.

(b) $|\Delta X| = (\Delta s_1)(\Delta s_2)$ which is not equal to zero if $(\Delta s_{1I})(\Delta s_{2I}) \neq 0$ or $(\Delta s_{1Q})(\Delta s_{2Q}) \neq 0$. This can be ensured by letting $\{s_{1I}, s_{2I}\}$, $\{s_{1Q}, s_{2Q}\}$ take values from a rotated $\mathbb{Z}^2$ lattice constellation which is designed to maximize the minimum product distance [49, 50].

(c) The weight matrices are non-unitary.

(d) For N_T transmit antennas, a rate one, full diversity, two group ML decodable STBC is given by $\mathbf{X} = \text{diag}\{s_1, s_2, \ldots, s_{N_T}\}$, where $\{s_{1I}, s_{2I}, \ldots, s_{N_TI}\}$, $\{s_{1Q}, s_{2Q}, \ldots, s_{N_TQ}\}$ take values from a rotated $\mathbb{Z}^{N_T}$ lattice constellation designed to maximize minimum product distance [49, 50].

(8) Compute $\mathbf{X}_{OD}{}^H\boldsymbol{\Gamma}^{-1}\mathbf{X}_{OD}$ and show that it is not a diagonal matrix and hence by Definition 3.7, $\mathbf{X}_{OD}$ is not a DOSTBC.

(9) The required DSTBC is obtained from PCIODs and is given by

$$\mathbf{X}_{PCIOD} = \begin{bmatrix} s_1 & -s_2^* & 0 \\ s_2 & s_1^* & 0 \\ 0 & 0 & s_3 \\ 0 & 0 & s_4 \end{bmatrix}, \text{ where, } \{s_{1I}, s_{3I}\},\ \{s_{2I}, s_{4I}\},\ \{s_{1Q}, s_{3Q}\},$$

$\{s_{2Q}, s_{4Q}\}$ take values from $31.7175°$ rotated QAM constellation. The information vector transmitted by the source in the first phase is given by $\mathbf{s} = \begin{bmatrix} s_1 & s_2 & s_3 s_4 \end{bmatrix}^T$.

(10) By Theorem 3.7, the noise covariance matrix $\boldsymbol{\Gamma}$ is not a diagonal matrix since $\mathbf{C_2}$ is not column-monomial.

References

1. A. Sendonaris, E. Erkip and B. Aazhang, User cooperation diversitypart I: System description, *IEEE Trans. Comm.* **51**, 11 (2003), 1927–1938.
2. A. Sendonaris, E. Erkip and B.m Aazhang, User cooperation diversitypart I: Implementation aspects and performance analysis, *IEEE Trans. Comm.* **51**, 11 (2003), 1939–1948.
3. K. Azarian, H. El Gamal and P. Schniter, On the achievable diversity-multiplexing tradeoff in half-duplex cooperative channels, *IEEE Trans. Inform. Theory* **51**, 12 (2005), 4152–4172.
4. S. Yang and J.-C. Belfiore, Towards the optimal amplify-and-forward cooperative diversity scheme, *IEEE Trans. Inform. Theory* **53**, 9 (2007), 3114–3126.
5. Y. Jing and B. Hassibi, Distributed space-time coding in wireless relay networks, *IEEE Trans. Wireless Commun.* **5**, 12 (2006), 3524–3536.
6. P. Elia, K. Vinodh, M. Anand and P. Vijay Kumar, D-MG tradeoff and optimal codes for a class of AF and DF cooperative communication protocols, *IEEE Trans. Inform. Theory* (to appear in).
7. S. O. Gharan, A. Bayesteh and A. K. Khandani, Optimum diversity-multiplexing tradeoff in the multiple relays network, Technical Report UW-E&CE#2007-19, University of Waterloo, (11 April, 2007). Available in arXiv:0709.4506.
8. J. N. Laneman and G. W. Wornell, Distributed space-time-coded protocols for exploiting cooperative diversity in wireless networks, *IEEE Trans. Inform. Theory* **49**, 10 (2003), 2415–2425.
9. S. Yiu, R. Schober and L. Lampe, Distributed space-time block coding, *IEEE Trans. Commun.* **54**, 7 (2006), 1195–1206.
10. T. Kiran and B. Sundar Rajan, Partially-coherent distributed space-time codes with differential encoder and decoder, *IEEE J. Select. Areas Commun.* **25**, 2 (2007), 426–433.

11. Y. Jing and H. Jafarkhani, Distributed differential space-time coding for wireless relay networks, *IEEE Trans. Commun.* (to appear in).

12. F. Oggier and B. Hassibi, A coding strategy for wireless networks with no channel information, *Forty-Fourth Annual Allerton Conference on Communication, Control and Computing*, (September 27–29, 2006).

13. F. Oggier, Cyclic distributed space-time codes for wireless relay networks with no channel information, *IEEE Trans. Inform. Theory.* Available online http://www.systems.caltech.edu/~frederique/submitDSTCnoncoh.pdf.

14. F. Oggier and E. Lequeu, Differential distributed space-time coding based on Cayley codes, *Proc. IEEE Int. Symp. on Information Theory 2008*, Toronto, Canada, (6–11 July, 2008), 2548–2552.

15. G. Susinder Rajan and B. Sundar Rajan, Algebraic distributed differential space-time codes with low decoding complexity, *IEEE Trans. Wireless Commun.* **7**, 10 (2008), 3962–3971.

16. O. Damen, A. Chkeif and J.-C. Belfiore, Lattice code decoder for space-time codes, *IEEE Commun. Lett.* **4**, 5 (2000).

17. M. Z. A. Khan and B. Sundar Rajan, Single-symbol maximum-likelihood decodable linear STBCs, *IEEE Trans. Inform. Theory* **52**, 5 (2006), 2062–2091.

18. S. Karmakar and B. Sundar Rajan, Minimum-decoding complexity, maximum-rate space-time block codes from clifford algebras, *Proc. IEEE Intnl. Symp. Inform. Theory*, Seattle, (9–14 July, 2006), pp. 788–792.

19. C. Yuen, Y. L. Guan and T. T. Tjhung, Quasi-orthogonal STBC with minimum decoding complexity, *IEEE Trans. Wireless Commun.* **4**, 5 (2005), 2089–2094.

20. H. Wang, D. Wang and X. G. Xia, Optimal quasi-orthogonal space-time block codes with minimum decoding complexity, *IEEE Trans. Inform. Theory.* Available online http://www.ece.udel.edu/~xxia/paper_wwx.pdf.

21. Z. A. Khan, B. Sundar Rajan and M. H. Lee, On single-symbol and double-symbo decodable STBCs, *Proc. IEEE Intnl. Symp. Inform. Theory*, Yokohama, Japan, (June 29–July 4), 2003, p. 127.

22. S. Karmakar and B. Sundar Rajan, Multi-group decodable STBCs from clifford algebras, *Proc. IEEE Information Theory Workshop*, Chengdu, China, (October 22–26, 2006), pp. 448–452.

23. D. N. Dao, C. Yuen, C. Tellambura, Y. L. Guan and T. T. Tjhung, Four-group decodable space-time block codes, *IEEE Trans. Signal Process.* **56**, 1 (2008), 424–430.

24. C. Yuen, Y. L. Guan and T. T. Tjhung, *Quasi-Orthogonal Space-Time Block Code*, (Imperial College Press).

25. C. Yuen, Y. L. Guan and T. T. Tjhung, A class of four-group quasi-orthogonal space-time block code achieving full rate and full diversity for any number of antennas, *Proc. IEEE Personal, Indoor and Mobile Radio Communications Symposium*, Vol. 1, pp. 92–96, Berlin, Germany, September 11–14, (2005).

26. P. W. Wolniansky, G. J. Foschini, G. D. Golden and R. A. Valenzuela, V-BLAST: An architecture for realizing very high data rates over the

rich-scattering wireless channel, *URSI Int. Symp. on Signals, Systems, and Electronics*, (29 September–2 October 1998), pp. 295–300.

27. V. Tarokh, H. Jafarkhani and A. R. Calderbank, Space-time block codes from orthogonal designs, *IEEE Trans. Inform. Theory* **45**, 5 (1999), 1456–1467.

28. X.-B. Liang, Orthogonal designs with maximal rates, *IEEE Trans. Inform. Theory* **49**, 10 (2003), pp. 2468–2503.

29. O. Tirkkonen and A. Hottinen, Square-matrix embeddable space-time block codes for complex signal constellations, *IEEE Trans. Inform. Theory* **48**, 2 (2002), 384–395.

30. H. Jafarkhani, A quasi-Orthogonal space-time block code, *IEEE Trans. Commun.* **49**, 1 (2001), 1–4.

31. O. Tirkkonen, A. Boariu and A. Hottinen, Minimal non-orthogonality rate 1 space-time block code for 3+ Tx antennas, *Proc. IEEE Int. Symp. on Spread-Spectrum Tech. and Appl.*, New Jersey, (Sept. 6–8, 2000), pp. 429–432.

32. O. Tirkkonen, Optimizing space-time block codes by constellation rotations, *Proc. 2nd Finnish Wireless Communications Workshop*, Tampere, Finland, (October 2001).

33. C. Yuen, Y. L. Guan and T. T. Tjhung, Full-rate full-diversity STBC with constellation rotation, *Proc. IEEE Vehicular Technology Conference*, 22–25 (April, 2003), pp. 296–300.

34. W. Su and X. Gen Xia, Signal constellations for quasi-orthogonal space-time block codes with full diversity, *IEEE Trans. Inform. Theory* **50**, 10 (2004), pp. 2331–2347.

35. N. Sharma and C. B. Papadias, Improved quasi-orthogonal codes through constellation rotation, *IEEE Trans. Commun.* **51**, 3 (2003), 332–335.

36. N. Sharma and C. B. Papadias, Full-rate full-diversity linear quasiorthogonal space-time codes for any number of transmit antennas, *EURASIP J. Appl. Signal Process.* **9** (2004), 1246–1256.

37. D. N. Dao and C. Tellambura, Capacity-approaching semi-orthogonal space-time block codes, *Proc. IEEE Global Telecommunications Conference*, St. Louis, MO, USA, (28 November-2 December 2005).

38. L. Xian and H. Liu, Rate-one space-time block codes with full diversity, *IEEE Trans. Commun.* **53** (2005), 1986–1990.

39. C. Yuen, Y. L. Guan and T. T. Tjhung, On the search for high-rate quasi-orthogonal space-time block code, *Int. J. Wireless Inform. Network* **13**, 4 (2006), 329–340.

40. T. Kiran and Sundar Rajan, B. Distributed space-time codes with reduced decoding complexity, *Proc. IEEE Intnl. Symp. Inform. Theory*, Seattle, (9–14 July, 2006), pp. 542–546.

41. F. Oggier and B. Hassibi, An algebraic family of distributed space-time codes for wireless relay networks, *Proc. IEEE Intnl. Symp. Inform. Theory*, Seattle, (9–14 July, 2006), pp. 538–541.

42. P. Elia, F. Oggier and P. Vijay Kumar, Asymptotically optimal cooperative wireless networks with reduced signaling complexity, *IEEE J. Select. Areas Commun.* **25**, 2 (2007), 258–267.

43. Y. Jing and H. Jafarkhani, Using Orthogonal and Quasi-orthogonal designs in wireless relay networks, *IEEE Trans. Inform. Theory* **53**, 11 (2007), 4106–4118.

44. G. Susinder Rajan and B. Sundar Rajan, A non-orthogonal distributed space-time protocol, Part-I: Signal model and design criteria, *Proc. IEEE International Workshop in Information Theory*, Chengdu, China, (22–26 October, 2006), pp. 385–389.

45. G. Susinder Rajan and B. Sundar Rajan, A non-orthogonal distributed space-time coded protocol, Part-II: Code construction and DM-G tradeoff, *Proc. IEEE Information Theory Workshop*, Chengdu, China, (22–26 October, 2006), pp. 488–492.

46. G. Susinder Rajan and B. Sundar Rajan, STBCs from representation of extended clifford algebras, *Proc. IEEE Intnl. Symp. Inform. Theory*, Nice, France, (24–29 June, 2007), pp. 1626–1630.

47. G. Susinder Rajan and B. Sundar Rajan, Algebraic distributed space-time codes with low ML decoding complexity, *Proc. IEEE Intnl. Symp. Inform. Theory*, Nice, France (24–29 June, 2007), pp. 1516–1520.

48. G. Susinder Rajan and B. Sundar Rajan, Multi-group ML decodable collocated and distributed space time block codes, *IEEE Trans. Inform. Theory*. Available in arXiv: 0712.2384.

49. E. Bayer-Fluckiger, F. Oggier and E. Viterbo, New algebraic constructions of rotated $\mathbb{Z}^n$-lattice constellations for the Rayleigh fading channel, *IEEE Trans. Inform. Theory* **50**, 4 (2004), 702–714.

50. Full Diversity Rotations, http://www1.tlc.polito.it/~viterbo/rotations/rotations.html.

51. G. Susinder Rajan and B. Sundar Rajan, Distributed space-time codes for cooperative networks with partial CSI, *Proc. IEEE Wireless Communications and Networking Conference*, Hong Kong, (11–15 March, 2007).

52. Z. Yi and I.-M. Kim, Single-symbol ML decodable distributed STBCs for cooperative networks, *IEEE Trans. Inform. Theory* **53**, 8 (2007), 2977–2985.

53. J. Harshan and B. Sundar Rajan, High rate single-symbol decodable precoded DSTBCs for cooperative networks, Technical Report TR-PME-2007-08 of DRDO-IISc Programme on Advanced Research in Mathematical Engineering. Available in arxiv:0708.4214.

54. D. Sreedhar, A. Chockalingam and B. Sundar Rajan, Single-symbol ML decodable distributed STBCs for partially-coherent cooperative networks, *IEEE Trans. Inform. Theory*. Available in arXiv:0708.3019.

55. Z. Yi and I.-M. Kim, The impact of noise correlation and channel phase information on the data-rate of the single-symbol ML decodable distributed STBCs, *IEEE Trans. Inform. Theory*. Available in arXiv:0708.3378.

56. B. Hassibi and B. M. Hochwald, High-rate codes that are linear in space and time, *IEEE Trans. Inform. Theory* **48**, 7 (2002), 1804–1824.

57. G. Susinder Rajan and B. Sundar Rajan, OFDM based distributed space time coding for asynchronous relay networks, *IEEE Int. Conf. Communications*, Beijing, China, (19–23 May, 2008).

58. Z. Li and X. G. Xia, A simple alamouti space-time transmission scheme for asynchronous cooperative systems, *IEEE Signal Process. Lett.* **14**, 11 (2007), 804–807.

59. X. Guo and X. G. Xia, A distributed space-time coding in asynchronous wireless relay networks, *IEEE Trans. Wireless Commun.* **7**, 5 (2008), 1812–1816.

60. G. Susinder Rajan and B. Sundar Rajan, Leveraging coherent distributed space time codes for non-coherent communication in relay networks via training, *IEEE Trans. Wireless Commun.* **8**, 2 (2009), 683–688.

61. Z. Li and X.-G. Xia, An alamouti coded OFDM transmission for cooperative systems robust to both timing errors and frequency offsets, *IEEE Trans. Wireless Commun.* **7**, 5 (2008), 1839–1844.

62. The IEEE 802.16 Working Group on Broadband Wireless Access Standards, http://www.ieee802.org/16/.

63. Y. Jing and H. Jafarkhani, Network beamforming with channel mean and covariance at relays, Technical Report, (2007). Available online https:// webfiles.uci.edu/yjing/www/papers/NetworkBF-MC-JSAC-report.pdf.

64. F. Oggier and B. Hassibi, An algebraic coding scheme for wireless relay networks with multiple-antenna nodes, *IEEE Trans. Signal Process.* Available online http://www.systems.caltech.edu/˜frederique/revisedDSTC.ps.

65. E. Biglieri, Y. Hong and E. Viterbo, On fast-decodable space-time block codes, *IEEE Trans. Inform. Theory.* Available online in arXiv:cs.IT/0708. 2804v1.

Part II

METHODS OF ALGEBRAIC COMBINATORICS IN CODING THEORY/CODES CONSTRUCTION AND EXISTENCE

Chapter 4

CODING THEORY
AND ALGEBRAIC COMBINATORICS

MICHAEL HUBER

Institut für Mathematik, Technische Universität Berlin,
Straße des 17. Juni 136, D-10623 Berlin, Germany
mhuber@math.tu-berlin.de

This chapter introduces and elaborates on the fruitful interplay of coding theory and algebraic combinatorics, with most of the focus on the interaction of codes with combinatorial designs, finite geometries, simple groups, sphere packings, kissing numbers, lattices, and association schemes. In particular, special interest is devoted to the relationship between codes and combinatorial designs. We describe and recapitulate important results in the development of the state-of-the-art. In addition, we give illustrative examples and constructions, and highlight recent advances. Finally, we provide a collection of significant open problems and challenges concerning future research.

4.1. Introduction

The classical publications "A mathematical theory of communication" by Shannon [1] and "Error detecting and error correcting codes" by Hamming [2] gave birth to the twin disciplines of information theory and coding theory. Since their inceptions the interactions of information and coding theory with many mathematical branches have continually deepened. This is in particular true for the close connection between coding theory and algebraic combinatorics.

This chapter introduces and elaborates on this fruitful interplay of coding theory and algebraic combinatorics, with most of the focus on the interaction of codes with combinatorial designs, finite geometries, simple groups, sphere packings, kissing numbers, lattices, and association schemes. In particular, special interest is devoted to the relationship between codes and combinatorial designs. Since we do not assume the reader is familiar with the theory of combinatorial designs, an accessible and reasonably self-contained exposition is provided. Subsequently, we describe and recapitulate important results in the development of the state of the art, provide illustrative examples and constructions, and highlight recent advances. Furthermore, we give a collection of significant open problems and challenges concerning future research.

The chapter is organized as follows. In Sec. 4.2, we give a brief account of basic notions of algebraic coding theory. Section 4.3 consists of the main part of the chapter: After an introduction to finite projective planes and combinatorial designs, a subsection on basic connections between codes and combinatorial designs follows. The next subsection is on perfect codes and designs, and addresses further related concepts. Section 4.3.4 deals with the classical Assmus–Mattson theorem and various analogues. A subsection on codes and finite geometries follows the discussion on the non-existence of a projective plane of order 10. In Sec. 4.3.6, interrelations between the Golay codes, the Mathieu–Witt designs, and the Mathieu groups are studied. Section 4.3.7 deals with the Golay codes and the Leech lattice, as well as recent milestones concerning kissing numbers and sphere packings. The last topic of this section considers codes and association schemes. The chapter concludes with sections on directions for further research as well as conclusions and exercises.

4.2. Background

For our further purposes, we give a short account of basic notions of algebraic coding theory. For additional information on the subject of algebraic coding theory, the reader is referred to [3–13]. For some historical notes on its origins, see [14, 16 (Chap. 1)] as well as [15] for a historical survey on coding theory and information theory.

We denote by $\mathbb{F}^n$ the set of all n-tuples from a q-symbol alphabet. If q is a prime power, we take the finite field $\mathbb{F} = \mathbb{F}_q$ with q elements, and interpret $\mathbb{F}^n$ as an n-dimensional vector space $\mathbb{F}_q^n$ over $\mathbb{F}_q$. The elements of $\mathbb{F}^n$ are called *vectors* (or *words*) and will be denoted by bold symbols.

The (*Hamming*) *distance* between two codewords $\mathbf{x}, \mathbf{y} \in \mathbb{F}^n$ is defined by the number of coordinate positions in which they differ, i.e.

$$d(\mathbf{x}, \mathbf{y}) := |\{i \mid 1 \leq i \leq n, \ x_i \neq y_i\}|.$$

The *weight* $w(\mathbf{x})$ of a codeword $\mathbf{x}$ is defined by

$$w(\mathbf{x}) := d(\mathbf{x}, \mathbf{0}),$$

whenever 0 is an element of $\mathbb{F}$.

A subset $C \subseteq \mathbb{F}^n$ is called a (*q-ary*) *code* of *length* n (*binary* if $q = 2$, *ternary* if $q = 3$). The elements of C are called *codewords*. A *linear code* (or $[n, k]$ *code*) over the field $\mathbb{F}_q$ is a k-dimensional linear subspace C of the vector space $\mathbb{F}_q^n$. We note that large parts of coding theory are concerned with linear codes. In particular, as many combinatorial configurations can be described by their incidence matrices, coding theorists have started in the early 1960s to consider as codes the vector spaces spanned by the rows of the respective incidence matrices over some given field.

The *minimum distance d* of a code C is defined as

$$d := \min \{d(\mathbf{x}, \mathbf{y}) \mid \mathbf{x}, \mathbf{y} \in C, \ \mathbf{x} \neq \mathbf{y}\}.$$

Clearly, the minimum distance of a linear code is equal to its *minimum weight*, i.e. the minimum of the weights of all non-zero codewords. An $[n, k, d]$ code is an $[n, k]$ code with minimum distance d.

The minimum distance of a (not necessarily linear) code C determines the error-correcting capability of C: If $d = 2e + 1$, then C is called an *e-error-correcting code*. Defining by

$$S_e(\mathbf{x}) := \{\mathbf{y} \in \mathbb{F}^n \mid d(\mathbf{x}, \mathbf{y}) \leq e\}$$

the *sphere* (or *ball*) of radius e around a codeword $\mathbf{x}$ of C, this implies that the spheres of radius e around distinct codewords are disjoint.

Counting the number of codewords in a sphere of radius e yields to the subsequent *sphere packing* (or *Hamming*) *bound*.

Theorem 4.1. *Let C be a q-ary code of length n and minimum distance $d = 2e + 1$. Then*

$$|C| \cdot \sum_{i=0}^{e} \binom{n}{i} (q-1)^i \leq q^n.$$

If equality holds, then C is called a *perfect code*. Equivalently, C is perfect if the spheres of radius e around all codewords cover the whole space $\mathbb{F}^n$. Certainly, perfect codes are combinatorially interesting objects, however, they are extremely rare.

We will call two codes *(permutation) equivalent* if one is obtained from the other by applying a fixed permutation to the coordinate positions for all codewords. A *generator matrix* G for an $[n, k]$ code C is a $(k \times n)$-matrix for which the rows are a basis of C. We say that G is in *standard form* if $G = (I_k, P)$, where I_k is the $(k \times k)$ identity matrix.

For an $[n, k]$ code C, let

$$C^{\perp} := \{\mathbf{x} \in \mathbb{F}_q^n \mid \forall_{\mathbf{y} \in C}[\langle \mathbf{x}, \mathbf{y} \rangle = 0]\}$$

denote the *dual code* of C, where $\langle \mathbf{x}, \mathbf{y} \rangle$ is the standard inner (or dot) product in $\mathbb{F}_q^n$. The code $C^{\perp}$ is an $[n, n - k]$ code. If H is a generator matrix for $C^{\perp}$, then clearly

$$C = \{\mathbf{x} \in \mathbb{F}_q^n \mid \mathbf{x}H^T = \mathbf{0}\},$$

and H is called a *parity check matrix* for the code C. If $G = (I_k, P)$ is a generator matrix of C, then $H = (-P^T, I_{n-k})$ is a parity check matrix of C. A code C is called *self-dual* if $C = C^{\perp}$. If $C \subset C^{\perp}$, then C is called *self-orthogonal*.

If C is a linear code of length n over $\mathbb{F}_q$, then

$$\overline{C} := \left\{ (c_1, \ldots, c_n, c_{n+1}) \mid (c_1, \ldots, c_n) \in C, \sum_{i=1}^{n+1} c_i = 0 \right\}$$

defines the *extended code* corresponding to C. The symbol c_{n+1} is called the *overall parity check symbol*. Conversely, C is the *punctured* (or *shortened*) code of $\overline{C}$.

The *weight distribution* of a linear code C of length n is the sequence $\{A_i\}_{i=0}^n$, where A_i denotes the number of codewords in C of weight i. The polynomial

$$A(x) := \sum_{i=0}^n A_i x^i$$

is the *weight enumerator* of C.

The weight enumerators of a linear code and its dual code are related, as shown by the following theorem, which is one of the most important results in the theory of error-correcting codes.

Theorem 4.2 (MacWilliams [16]). *Let C be an $[n, k]$ code over $\mathbb{F}_q$ with weight enumerator $A(x)$ and let $A^{\perp}(x)$ be the weight enumerator of the dual code $C^{\perp}$. Then*

$$A^{\perp}(x) = q^{-k}(1 + (q - 1)x)^n A\left(\frac{1 - x}{1 + (q - 1)x}\right).$$

We note that the concept of the weight enumerator can be generalized to non-linear codes (so-called *distance enumerator*, cf. [17, 18] and Sec. 4.3.8).

An $[n, k]$ code C over $\mathbb{F}_q$ is called *cyclic* if

$$\forall_{(c_0, c_1, \ldots, c_{n-1}) \in C}[(c_{n-1}, c_0, \ldots, c_{n-2}) \in C],$$

i.e. any cyclic shift of a codeword is again a codeword. We adopt the usual convention for cyclic codes that n and q are coprime. Using the isomorphism

$$(a_0, a_1, \ldots, a_{n-1}) \rightleftarrows a_0 + a_1 x + \cdots + a_{n-1} x^{n-1}$$

between $\mathbb{F}_q^n$ and the residue class ring $\mathbb{F}_q[x]/(x^n - 1)$, it follows that a cyclic code corresponds to an ideal in $\mathbb{F}_q[x]/(x^n - 1)$.

4.3. Thoughts for Practitioners

In the following, we introduce and elaborate on the fruitful interplay of coding theory and algebraic combinatorics, with most of the focus on the interaction of codes with combinatorial designs, finite geometries, simple groups, sphere packings, kissing numbers, lattices, and association schemes. In particular, special interest is devoted to the relationship between codes and combinatorial designs. We give an accessible and reasonably self-contained exposition in the first subsection as we do not assume the reader is familiar with the theory of combinatorial designs. In what follows, we describe and recapitulate important results in the development of the state-of-the-art. In addition, we give illustrative examples and constructions, and highlight recent achievements.

4.3.1. *Introduction to Finite Projective Planes and Combinatorial Designs*

Combinatorial design theory is a subject of considerable interest in discrete mathematics. We give in this subsection an introduction to the topic, with emphasis on the construction of some important designs. For a more general treatment of combinatorial designs, the reader is referred to [19–24]. In particular, [19, 21] provide encyclopedias on key results.

Besides coding theory, there are many interesting connections of design theory to other fields. We mention in our context especially its links to finite geometries [25], incidence geometry [26], group theory [27–30], graph theory [4, 31], cryptography [32–34], as well as classification algorithms [35]. In addition to that, we recommend [22, 36–39] for the reader interested in the broad area of combinatorics in general.

We start by introducing several notions.

Definition 4.1. A *projective plane of order n* is a pair of *points* and *lines* such that the following properties hold:

(i) any two distinct points are on a unique line,
(ii) any two distinct lines intersect in a unique point,
(iii) there exists a *quadrangle*, i.e. four points no three of which are on a common line,
(iv) there are $n+1$ points on each line, $n+1$ lines through each point, and the total number of points, respectively lines, is $n^2 + n + 1$.

It follows easily from (i), (ii), and (iii) that the number of points on a line is a constant. When setting this constant equal to $n + 1$, then (iv) is a consequence of (i) and (iii).

Combinatorial designs can be regarded as generalizations of projective planes:

Definition 4.2. For positive integers $t \leq k \leq v$ and λ, we define a *t-design*, or more precisely a *t-(v, k, λ) design*, to be a pair $\mathcal{D} = (X, \mathcal{B})$, where X is a finite set of *points*, and $\mathcal{B}$ a set of k-element subsets of X called *blocks*, with the property that any t points are contained in precisely λ blocks.

We will denote points by lower-case and blocks by upper-case Latin letters. Via convention, we set $v := |X|$ and $b := |\mathcal{B}|$. Throughout this chapter, "repeated blocks" are not allowed, that is, the same k-element subset of points may not occur twice as a block. If $t < k < v$ holds, then we speak of a *non-trivial t-design*.

Designs may be represented algebraically in terms of incidence matrices: Let $\mathcal{D} = (X, \mathcal{B})$ be a t-design, and let the points be labeled $\{x_1, \ldots, x_v\}$ and the blocks be labeled $\{B_1, \ldots, B_b\}$. Then, the $(b \times v)$-matrix $A = (a_{ij})$ $(1 \leq i \leq b, 1 \leq j \leq v)$ defined by

$$a_{ij} := \begin{cases} 1, & \text{if } x_j \in B_i \\ 0, & \text{otherwise} \end{cases}$$

is called an *incidence matrix* of $\mathcal{D}$. Clearly, A depends on the respective labeling, however, it is unique up to column and row permutation.

If $\mathcal{D}_1 = (X_1, \mathcal{B}_1)$ and $\mathcal{D}_2 = (X_2, \mathcal{B}_2)$ are two t-designs, then a bijective map $\alpha : X_1 \to X_2$ is called an *isomorphism* of $\mathcal{D}_1$ onto $\mathcal{D}_2$, if

$$B \in \mathcal{B}_1 \Leftrightarrow \alpha(B) \in \mathcal{B}_2.$$

In this case, the designs $\mathcal{D}_1$ and $\mathcal{D}_2$ are *isomorphic*. An isomorphism of a design $\mathcal{D}$ onto itself is called an *automorphism* of $\mathcal{D}$. Evidently, the set of all automorphisms of a design $\mathcal{D}$ form a group under composition, the *full automorphism group* of $\mathcal{D}$. Any subgroup of it will be called *an automorphism group* of $\mathcal{D}$.

If $\mathcal{D} = (X, \mathcal{B})$ is a t-(v, k, λ) design with $t \geq 2$, and $x \in X$ arbitrary, then the *derived design with respect to* x is $\mathcal{D}_x = (X_x, \mathcal{B}_x)$, where $X_x = X \backslash \{x\}$, $\mathcal{B}_x = \{B \backslash \{x\} \mid x \in B \in \mathcal{B}\}$. In this case, $\mathcal{D}$ is also called an *extension* of $\mathcal{D}_x$. Obviously, $\mathcal{D}_x$ is a $(t-1)$-$(v-1, k-1, \lambda)$ design. The *complementary design* $\overline{\mathcal{D}}$ is obtained by replacing each block of $\mathcal{D}$ by its complement.

For historical reasons, a t-(v, k, λ) design with $\lambda = 1$ is called a *Steiner t-design*. Sometimes this is also known as a *Steiner system* if the parameter t is clearly given from the context.

The special case of a Steiner design with parameters $t = 2$ and $k = 3$ is called a *Steiner triple system of order* v (briefly $STS(v)$). The question regarding their existence was posed in the classical "Combinatorische Aufgabe" (1853) of the nineteenth century geometer Jakob Steiner [40]:

> "Welche Zahl, N, von Elementen hat die Eigenschaft, dass sich die Elemente so zu dreien ordnen lassen, dass je zwei in einer, aber nur in einer Verbindung vorkommen?"

However, there had been earlier work on these particular designs going back to, in particular, J. Plücker, W. S. B. Woolhouse, and most notably T. P. Kirkman. For an account on the early history of designs, see [21 (Chap. 1.2), 41].

A Steiner design with parameters $t = 3$ and $k = 4$ is called a *Steiner quadruple system of order* v (briefly $SQS(v)$).

If a 2-design has equally many points and blocks, i.e. $v = b$, then we speak of a *square design* (as its incidence matrix is square). By tradition, square designs are often called *symmetric designs*, although here the term does not imply any symmetry of the design. For more on these interesting designs, see, e.g. [42].

We give some illustrative examples of finite projective planes and combinatorial designs. We assume that q is always a prime power.

Example 4.1. Let us choose as point set

$$X = \{1, 2, 3, 4, 5, 6, 7\}$$

and as block set

$$\mathcal{B} = \{\{1, 2, 4\}, \{2, 3, 5\}, \{3, 4, 6\}, \{4, 5, 7\}, \{1, 5, 6\}, \{2, 6, 7\}, \{1, 3, 7\}\}.$$

This gives a 2-$(7, 3, 1)$ design, the well-known *Fano plane*, the smallest design arising from a projective geometry, which is unique up to isomorphism. We give the usual representation of this projective plane of order 2 as shown in Fig. 4.1:

Example 4.2. We take as point set

$$X = \{1, 2, 3, 4, 5, 6, 7, 8, 9\}$$

and as block set

$$\mathcal{B} = \{\{1, 2, 3\}, \{4, 5, 6\}, \{7, 8, 9\}, \{1, 4, 7\}, \{2, 5, 8\}, \{3, 6, 9\},$$

$$\{1, 5, 9\}, \{2, 6, 7\}, \{3, 4, 8\}, \{1, 6, 8\}, \{2, 4, 9\}, \{3, 5, 7\}\}.$$

This gives a 2-$(9, 3, 1)$ design, the smallest non-trivial design arising from an affine geometry, which is again unique up to isomorphism. This affine plane of order 3 can be constructed from the array

$$\begin{array}{ccc} 1 & 2 & 3 \\ 4 & 5 & 6 \\ 7 & 8 & 9 \end{array}$$

as shown in Fig. 4.2.

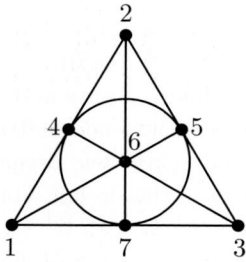

Fig. 4.1.   Fano plane.

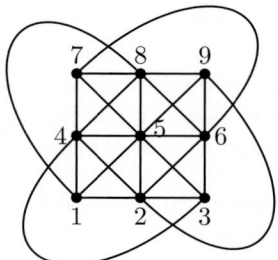

Fig. 4.2.   Affine plane of order 3.

More generally, we obtain:

Example 4.3. We choose as point set X the set of 1-dimensional subspaces of a vector space $V = V(d, q)$ of dimension $d \geq 3$ over $\mathbb{F}_q$. As block set $\mathcal{B}$ we take the set of 2-dimensional subspaces of V. Then there are $v = (q^d - 1)/(q - 1)$ points and each block $B \in \mathcal{B}$ contains $k = q + 1$ points. Since obviously any two 1-dimensional subspaces span a single 2-dimensional subspace, any two distinct points are contained in a unique block. Thus, the *projective space* $PG(d - 1, q)$ is an example of a 2-$(\frac{q^d - 1}{q - 1}, q + 1, 1)$ design. For $d = 3$, the particular designs are *projective planes of order* q, which are square designs. More generally, for any fixed i with $1 \leq i \leq d - 2$, the points and i-dimensional subspaces of $PG(d - 1, q)$ (i.e. the $(i + 1)$-dimensional subspaces of V) yield a 2-design.

Example 4.4. We take as point set X the set of elements of a vector space $V = V(d, q)$ of dimension $d \geq 2$ over $\mathbb{F}_q$. As block set $\mathcal{B}$ we choose the set of affine lines of V (i.e. the translates of 1-dimensional subspaces of V). Then there are $v = q^d$ points and each block $B \in \mathcal{B}$ contains $k = q$ points. As clearly any two distinct points lie on exactly one line, they are contained in a unique block. Hence, we obtain the *affine space* $AG(d, q)$ as an example of a 2-$(q^d, q, 1)$ design. When $d = 2$, these designs are *affine planes of order* q. More generally, for any fixed i with $1 \leq i \leq d - 1$, the points and i-dimensional subspaces of $AG(d, q)$ form a 2-design.

Remark 4.1. It is well-established that both affine and projective planes of order n exist whenever n is a prime power. The conjecture that no such planes exist with orders other than prime powers is unresolved so far. The classical result of Bruck and Ryser [43] still remains the only general statement: If $n \equiv 1$ or 2 (mod 4) and n is not equal to the sum of two squares of integers, then n does not occur as the order of a finite projective plane. The smallest integer that is not a prime power and not covered by the Bruck–Ryser theorem is 10. Using substantial computer analysis, Lam *et al.* [44] proved the non-existence of a projective plane of order 10 (cf. Remark 4.10). The next smallest number to consider is 12, for which neither a positive nor a negative answer has been proved.

Needless to mention that — apart from the existence problem — the question on the number of different isomorphism types (when existent) is fundamental. There are, for example, precisely four non-isomorphic projective planes of order 9. For a further discussion, in particular of the rich history of affine and projective planes, we refer to [25, 45–49].

Example 4.5. We take as points the vertices of a 3-dimensional cube. As illustrated in Fig. 4.3, we can choose three types of blocks:

  (i) a face (six of these),
 (ii) two opposite edges (six of these), and
(iii) an inscribed regular tetrahedron (two of these).

This gives a 3-$(8, 4, 1)$ design, which is unique up to isomorphism.

We have more generally:

Example 4.6. In $AG(d, q)$ any three distinct points define a plane unless they are collinear (that is, lie on the same line). If the underlying field is $\mathbb{F}_2$, then the lines contain only two points and hence any three points cannot be collinear. Therefore, the points and planes of the affine space $AG(d, 2)$ form a 3-$(2^d, 4, 1)$ design. More generally, for any fixed i with $2 \leq i \leq d-1$, the points and i-dimensional subspaces of $AG(d, 2)$ form a 3-design.

Example 4.7. The unique 2-$(9, 3, 1)$ design whose points and blocks are the points and lines of the affine plane $AG(2, 3)$ can be extended precisely three times to the following designs which are also unique up to isomorphism: the 3-$(10, 4, 1)$ design, which is the Möbius plane of order

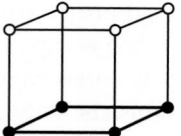 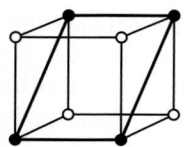 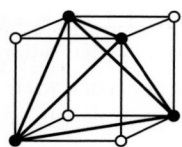

Fig. 4.3.   Steiner quadruple system of order 8.

3 with $P\Gamma L(2,9)$ as full automorphism group, and the two *Mathieu–Witt designs* 4-(11,5,1) and 5-(12,6,1) with the sporadic Mathieu groups M_{11} and M_{12} as point 4-transitive and point 5-transitive full automorphism groups, respectively.

To construct the "large" Mathieu–Witt designs one starts with the 2-(21,5,1) design whose points and blocks are the points and lines of the projective plane $PG(2,4)$. This can be extended also exactly three times to the following unique designs: the *Mathieu–Witt design* 3-(22,6,1) with $\text{Aut}(M_{22})$ as point 3-transitive full automorphism group as well as the *Mathieu–Witt designs* 4-(23,7,1) and 5-(24,8,1) with M_{23} and M_{24} as point 4-transitive and point 5-transitive full automorphism groups, respectively.

The five Mathieu groups were the first sporadic simple groups and were discovered by Mathieu [50, 51] over 100 years ago. They are the only finite 4- and 5-transitive permutation groups apart from the symmetric or alternating groups. The Steiner designs associated with the Mathieu groups were first constructed by both Carmichael [28] and Witt [52], and their uniqueness established up to isomorphism by Witt [53]. From the meanwhile various alternative constructions, we mention especially those of Lüneburg [54] and Aschbacher [55, (Chap. 6)]. However, the easiest way to construct and prove uniqueness of the Mathieu–Witt designs is via coding theory, using the related *binary* and *ternary Golay codes* (see Sec. 4.3.6).

Remark 4.2. By classifying Steiner designs which admit automorphism groups with sufficiently strong symmetry properties, specific characterizations of the Mathieu–Witt designs with their related Mathieu groups were obtained (see, e.g. [56–61] and [62 (Chap. 5)] for a survey).

Remark 4.3. We mention that, in general, for $t = 2$ and 3, there are many infinite classes of Steiner t-designs, but for $t = 4$ and 5 only a finite number are known. Although Teirlinck [63] has shown that non-trivial t-designs exist for all values of t, no Steiner t-designs have been constructed for $t \geq 6$ so far.

In what follows, we need some helpful combinatorial tools:

A standard combinatorial double counting argument gives the following assertions.

Lemma 4.1. *Let $\mathcal{D} = (X, \mathcal{B})$ be a t-(v, k, λ) design. For a positive integer $s \leq t$, let $S \subseteq X$ with $|S| = s$. Then the total number λ_s of blocks containing* *all the points of S is given by*

$$\lambda_s = \lambda \frac{\binom{v-s}{t-s}}{\binom{k-s}{t-s}}.$$

In particular, for $t \geq 2$, a t-(v, k, λ) design is also an s-(v, k, λ_s) design.

For historical reasons, it is customary to set $r := \lambda_1$ to be the total number of blocks containing a given point (referring to the "replication number" from statistical design of experiments, one of the origins of designs theory).

Lemma 4.2. *Let $\mathcal{D} = (X, \mathcal{B})$ be a t-(v, k, λ) design. Then the following holds:*

(a) $bk = vr$.
(b) $\binom{v}{t}\lambda = b\binom{k}{t}$.
(c) $r(k - 1) = \lambda_2(v - 1)$ for $t \geq 2$.

Since in Lemma 4.1 each λ_s must be an integer, we have moreover the subsequent necessary arithmetic conditions.

Lemma 4.3. *Let $\mathcal{D} = (X, \mathcal{B})$ be a t-(v, k, λ) design. Then*

$$\lambda \binom{v - s}{t - s} \equiv 0 \left(\mathrm{mod} \ \binom{k - s}{t - s} \right)$$

for each positive integer $s \leq t$.

The following theorem is an important result in the theory of designs, generally known as *Fisher's Inequality*.

Theorem 4.3 (Fisher [64]). *If $\mathcal{D} = (X, \mathcal{B})$ is a non-trivial t-(v, k, λ) design with $t \geq 2$, then we have $b \geq v$, that is, there are at least as many blocks as points in $\mathcal{D}$.*

We remark that equality holds exactly for square designs when $t = 2$. Obviously, the equality $v = b$ implies $r = k$ by Lemma 4.2(a).

4.3.2. *Basic Connections Between Codes and Combinatorial Designs*

There is a rich and fruitful interplay between coding theory and design theory. In particular, many t-designs have been found in the last decades by considering the codewords of fixed weight in some special, often linear codes. As we will see in the sequel, these codes typically exhibit a high degree of regularity. There is a good amount of literature [4, 7, 13, 31, 65–72] discussing to some extent in more detail various relations between codes and designs.

For a codeword $x \in \mathbb{F}^n$, the set

$$\text{supp}(x) := \{i \mid x_i \neq 0\}$$

of all coordinate positions with non-zero coordinates is called the *support* of x. We shall often form a t-design of a code in the following way: Given a (usually linear) code of length n, which contains the zero vector, and non-zero weight w, we choose as point set X the set of n coordinate positions of the code and as block set $\mathcal{B}$ the supports of all codewords of weight w.

Since we do not allow repeated blocks, clearly only distinct representatives of supports for codewords with the same supports are taken in the non-binary case.

We give some elementary examples.

Example 4.8. The matrix

$$G = \begin{pmatrix} 1 & 1 & 0 & 1 & 0 & 0 & 0 \\ 0 & 1 & 1 & 0 & 1 & 0 & 0 \\ 0 & 0 & 1 & 1 & 0 & 1 & 0 \\ 0 & 0 & 0 & 1 & 1 & 0 & 1 \end{pmatrix}$$

is a generator matrix of a binary $[7, 4, 3]$ Hamming code, which is the smallest non-trivial Hamming code (see also Example 4.12). This code is a perfect single-error-correcting code with weight distribution $A_0 = A_7 = 1$, $A_3 = A_4 = 7$. The seven codewords of weight 3 are precisely the seven rows of the incidence matrix

$$\begin{pmatrix} 1 & 1 & 0 & 1 & 0 & 0 & 0 \\ 0 & 1 & 1 & 0 & 1 & 0 & 0 \\ 0 & 0 & 1 & 1 & 0 & 1 & 0 \\ 0 & 0 & 0 & 1 & 1 & 0 & 1 \\ 1 & 0 & 0 & 0 & 1 & 1 & 0 \\ 0 & 1 & 0 & 0 & 0 & 1 & 1 \\ 1 & 0 & 1 & 0 & 0 & 0 & 1 \end{pmatrix}$$

of the Fano plane $PG(2,2)$ of Fig. 4.1. The supports of the seven codewords of weight 4 yield the complementary 2-$(7,4,2)$ design, i.e. the biplane of order 2.

Example 4.9. The matrix $(I_4, J_4 - I_4)$, where J_4 denotes the (4×4) all-one matrix, generates the extended binary $[8,4,4]$ Hamming code. This code is self-dual and has weight distribution $A_0 = A_8 = 1$, $A_4 = 14$. As any two codewords of weight 4 have distance at least 4, they have at most two 1's in common, and hence no codeword of weight 3 can appear as a subword of more than one codeword. On the other hand, there are $\binom{8}{3} = 56$ words of weight 3 and each codeword of weight 4 has four subwords of weight 3. Hence each codeword of weight 3 is a subword of exactly one codeword of weight 4. Therefore, the supports of the fourteen codewords of weight 4 form a 3-$(8,4,1)$ design, which is the unique $SQS(8)$ (cf. Example 4.5).

We give also a basic example of a non-linear code constructed from design theory.

Example 4.10. We take the rows of an incidence matrix of the (unique) Hadamard 2-$(11,5,2)$ design, and adjoin the all-one codeword. Then, the 12 codewords have mutual distance 6, and if we delete a coordinate, we get a binary non-linear code of length 10 and minimum distance 5.

For a detailed description of the connection between non-linear codes and design theory as well as the application of design theory in the area of (majority-logic) decoding, the reader is referred to [13, 71, 72].

Using highly transitive permutation groups, a further construction of designs from codes can be described (see, e.g. [31]).

Theorem 4.4. *Let C be a code which admits an automorphism group acting t-homogeneously (in particular, t-transitively) on the set of coordinates. Then the supports of the codewords of any non-zero weight form a t-design.*

Example 4.11. The r-th order *Reed–Muller (RM) code* $\mathrm{RM}(r,m)$ of length 2^m is a binary $[2^m, \sum_{i=0}^{r} \binom{m}{i}, 2^{m-r}]$ code with its codewords the value-vectors of all Boolean functions in m variables of degree at most r. These codes were first considered by Muller [73] and Reed [74] in 1954. The dual of the Reed–Muller code $\mathrm{RM}(r,m)$ is $\mathrm{RM}(m-r-1,m)$. Clearly, the extended binary $[8,4,4]$ Hamming code in Example 4.9 is $\mathrm{RM}(1,3)$.

Alternatively, a codeword in $\mathrm{RM}(r,m)$ can be viewed as the sum of characteristic functions of subspaces of dimension at least $m-r$ of the affine space $AG(m, 2)$. Thus, the full automorphism group of $\mathrm{RM}(r, m)$ contains the 3-transitive group $AGL(m, 2)$ of all affine transformations, and hence the codewords of any fixed non-zero weight yield a 3-design.

4.3.3. *Perfect Codes and Designs*

The interplay between coding theory and combinatorial designs is most evidently seen in the relationship between perfect codes and t-designs.

Theorem 4.5 (Assmus and Mattson [75]). *A linear e-error-correcting code of length n over $\mathbb{F}_q$ is perfect if and only if the supports of the codewords of minimum weight $d = 2e + 1$ form an $(e + 1)$-$(n, d, (q - 1)^e)$ design.*

The question

> "Does every Steiner triple system on n points extend to a Steiner quadruple system on $n + 1$ points?"

which goes also back to Steiner [40], is still unresolved in general. However, in terms of binary e-error-correcting codes, there is a positive answer.

Theorem 4.6 (Assmus and Mattson [75]). *Let C be a (not necessarily linear) binary e-error correcting code of length n, which contains the zero vector. Then C is perfect if and only if the supports of the codewords of minimum weight $d = 2e + 1$ form a Steiner $(e + 1)$-$(n, d, 1)$ design. Moreover, the supports of the minimum codewords in the extended code $\overline{C}$ form a Steiner $(e + 2)$-$(n + 1, d + 1, 1)$ design.*

We have seen in Examples 4.8 and 4.9 that the supports of the seven codewords of weight 3 in the binary $[3, 4, 7]$ Hamming code form an $STS(7)$, while the supports of the 14 codewords of weight 4 in the extended $[8, 4, 4]$ Hamming code yield an $SQS(8)$. In view of the above theorems, we get more generally:

Example 4.12. Let $n := (q^m - 1)/(q - 1)$. We consider a $(m \times n)$-matrix H over $\mathbb{F}_q$ such that no two columns of H are linearly dependent. Then H clearly is a parity check matrix of an $[n, n - m, 3]$ code, which is the *Hamming code* over $\mathbb{F}_q$. The number of its codewords is q^{n-m}, and for any codeword $\mathbf{x}$, we have $S_1(\mathbf{x}) = 1 + n(q - 1) = q^m$. Hence, by the sphere packing bound (Theorem 4.1), this code is perfect, and the supports of codewords of minimum weight 3 form a 2-$(n, 3, q - 1)$ design. Furthermore, in a binary $[2^m - 1, 2^m - 1 - m, 3]$ Hamming code the supports of codewords of weight 3 form a $STS(2^m-1)$, and the supports of the codewords of weight 4 in the extended code yield a $SQS(2^m)$.

Note. The Hamming codes were developed by Hamming [2] in the mid 1940s, who was employed at Bell Laboratories, and addressed a need for error correction in his work on the primitive computers of the time. We remark that the extended binary $[2^m, 2^m - m - 1, 4]$ Hamming code is the Reed–Muller code $RM(m-2, m)$.

Example 4.13. The *binary Golay code* is a $[23, 12, 7]$ code, while the *ternary Golay code* is a $[11, 6, 5]$ code. For both codes, the parameters imply equality in the sphere packing bound, and hence these codes are perfect. We will discuss later various constructions of these some of the most famous codes (see Example 4.14 and Construction 4.12). By the above theorems, the supports of codewords of minimum weight 7 in the binary $[23, 12, 7]$ Golay code form a Steiner 4-$(23, 7, 1)$ design, and the supports of the codewords of weight 8 in the extended binary $[24, 12, 8]$ Golay code give a Steiner 5-$(24, 8, 1)$ design. The supports of codewords of minimum weight 5 in the ternary $[11, 6, 5]$ Golay code yield a 3-$(11, 5, 4)$ design. It can be shown (e.g. via Theorem 4.4) that this is indeed a Steiner 4-$(11, 5, 1)$ design. We will see in Example 4.15 that the supports of the codewords of weight 6 in the extended ternary $[12, 6, 6]$ Golay code give a Steiner 5-$(12, 6, 1)$ design; thus the above results are not best possible.

Note. The Golay codes were discovered by Golay [76] in 1949 in the process of extending Hamming's construction. They have numerous practical real-world applications, e.g. the use of the extended binary Golay code in the Voyager spacecraft program during the early 1980s or in contemporary standard automatic link establishment (ALE) in high frequency (HF) data communication for forward error correction (FEC).

Remark 4.4. It is easily seen from their construction that the Hamming codes are unique (up to equivalence). It was shown by Pless [77] that this is also true for the Golay codes. Moreover, the binary and ternary Golay codes are the only non-trivial perfect e-error-correcting codes with $e > 1$ over any field $\mathbb{F}_q$. Using integral roots of the Lloyd polynomial, this remarkable fact was proven by Tietäväinen [78] and van Lint [79], and independently by Zinov'ev and Leont'ev [80]. Best [81] and Hong [82] extended this result to arbitrary alphabets for $e > 2$. For a thorough account of perfect codes, we refer to [83, 84 (Chap. 11)].

As trivial perfect codes can only form trivial designs, we have (up to equivalence) a complete list of non-trivial linear perfect codes with their associated designs:

Code	Code parameters		Design parameters
Hamming code	$[\frac{q^m-1}{q-1}, \frac{q^m-1}{q-1} - m, 3]$	q any prime power	$2\text{-}\left(\frac{q^m-1}{q-1}, 3, q-1\right)$
Binary Golay code	$[23, 12, 7]$	$q = 2$	$4\text{-}(23, 7, 1)$
Ternary Golay code	$[11, 6, 5]$	$q = 3$	$4\text{-}(11, 5, 1)$

There are various constructions of non-linear single-error-correcting perfect codes. For more details, see, e.g. [9, 13, 71, 72, 85] and references therein. However, a classification of these codes seems out of reach at present, although some progress has been made recently, see, for instance [86–88].

Remark 4.5. The long-standing question whether every Steiner triple system of order $2^m - 1$ occurs in a perfect code has been answered recently in the negative. Relying on the classification [89] of the Steiner quadruple systems of order 16, it was shown in [90] that the unique anti-Pasch Steiner triple system of order 15 provides a counterexample.

Remark 4.6. Due to the close relationship between perfect codes and some of the most interesting designs, several natural extensions of perfect codes have been examined in this respect: *Nearly perfect codes* [91], and the more general class of *uniformly packed codes* [92, 93], were studied extensively and eventually lead to t-designs. Van Tilborg [94] showed that e-error correcting uniformly packed codes do not exist for $e > 3$, and classified those for $e \leq 3$. For more details, see [4, 10, 13, 94]. The concept of *diameter perfect codes* [95, 96] is related particularly to Steiner designs. For further generalizations of perfect codes, see e.g. [84 (Chap. 11), 13 (Chap. 6)].

4.3.4. *The Assmus-Mattson Theorem and Analogues*

We consider in this subsection one of the most fundamental results in the interplay of coding theory and design theory. We start by introducing two important classes of codes.

Let q be an odd prime power. We define a function χ (the so-called *Legendre–symbol*) on $\mathbb{F}_q$ by

$$\chi(x) := \begin{cases} 0, & \text{if } x = 0 \\ 1, & \text{if } x \text{ is a non-zero square} \\ -1, & \text{otherwise.} \end{cases}$$

We note that χ is a character on the multiplicative group of $\mathbb{F}_q$. Using the elements of $\mathbb{F}_q$ as row and column labels a_i and a_j ($0 \leq i,j < q$), respectively, a matrix $Q = (q_{ij})$ of order q can be defined by

$$q_{ij} := \chi(a_j - a_i). \tag{4.1}$$

If q is a prime, then Q is a circulant matrix. We call a matrix

$$C_{q+1} := \begin{pmatrix} 0 & 1 & \cdots & 1 \\ \chi(-1) & & & \\ \vdots & & Q & \\ \chi(-1) & & & \end{pmatrix}$$

of order $q + 1$ a *Paley matrix*. These matrices were constructed by R. A. Paley in 1933 and are a specific type of conference matrices, which have their origin in the application to conference telephone circuits.

Construction 4.7. Let n be an odd prime, and q be a *quadratic residue* (mod n), i.e. $q^{(n-1)/2} \equiv 1$ (mod n). The *quadratic residue code* (or *QR code*) of length n over $\mathbb{F}_q$ is an $[n, (n+1)/2]$ code with minimum weight $d \geq \sqrt{n}$ (so-called *square root bound*). It can be generated by the $(0,1)$-circulant matrix of order n with top row the incidence vector of the non-zero quadratic residues (mod n). These codes are a special class of cyclic codes and were first constructed by A. M. Gleason in 1964. For $n \equiv 3$ (mod 4), the extended quadratic residue code is self-dual. We note for the important binary case that q is a quadratic residue (mod n) if and only if $n \equiv \pm 1$ (mod 8).

Note. By a theorem of A. M. Gleason and E. Prange, the full automorphism group of an extended quadratic residue code of length n contains the group $PSL(2, n)$ of all linear fractional transformations whose determinant is a non-zero square.

Example 4.14. The binary $[7, 4, 3]$ Hamming code is a quadratic residue code of length 7 over $\mathbb{F}_2$. The binary $[23, 12, 7]$ Golay code is a quadratic residue code of length 23 over $\mathbb{F}_2$, while the ternary $[11, 6, 5]$ Golay code is a quadratic residue code of length 11 over $\mathbb{F}_3$.

Construction 4.8. For $q \equiv -1 \pmod{6}$ a prime power, the *Pless symmetry code* $\mathrm{Sym}_{2(q+1)}$ of dimension $q + 1$ is a ternary $[2(q + 1), q + 1]$ code with generator matrix $G_{2(q+1)} := (I_{q+1}, C_{q+1})$, where C_{q+1} is a Paley matrix. Since $C_{q+1}C_{q+1}^T = -I_{q+1}$ (over $\mathbb{F}_3$) for $q \equiv -1 \pmod{3}$, the code $\mathrm{Sym}_{2(q+1)}$ is self-dual. This infinite family of cyclic codes were introduced by Pless [97, 98] in 1972. We note that the first symmetry code S_{12} is equivalent to the extended $[12, 6, 6]$ Golay code.

The celebrated Assmus–Mattson theorem gives a sufficient condition for the codewords of constant weight in a linear code to form a t-design.

Theorem 4.9 (Assmus and Mattson [99]). *Let C be an $[n, k, d]$ code over $\mathbb{F}_q$ and $C^\perp$ be the $[n, n - k, e]$ dual code. Let n_0 be the largest integer such that $n_0 - \frac{n_0 + q - 2}{q - 1} < d$, and define m_0 similarly for the dual code $C^\perp$, whereas if $q = 2$, we assume that $n_0 = m_0 = n$. For some integer t with $0 < t < d$, let us suppose that there are at most $d - t$ non-zero weights w in $C^\perp$ with $w \leq n - t$. Then, for any weight v with $d \leq v \leq n_0$, the supports of codewords of weight v in C form a t-design. Furthermore, for any weight w with $e \leq w \leq \min\{n - t, m_0\}$, the support of the codewords w in $C^\perp$ also form a t-design.*

The proof of the theorem involves a clever use of the MacWilliams relations (Theorem 4.2). Along with these, Lemma 4.1 and the immediate observation that codewords of weight less than n_0 with the same support must be scalar multiples of each other, form the basis of the proof (for a detailed proof, see, e.g. [4 (Chap. 14)]).

Remark 4.7. Until this result by Assmus, Jr. and Mattson, Jr. in 1969, only very few 5-designs were known: the Mathieu–Witt designs 5-$(12, 6, 1)$ and 5-$(24, 8, 1)$, the 5-$(24, 8, 48)$ design formed by the codewords of weight 12 (the *dodecads*) in the extended binary Golay code, as well as 5-$(12, 6, 2)$ and 5-$(24, 8, 2)$ designs which have been found without using coding theory. However, by using the Assmus–Mattson theorem, it was possible to find a number of new 5-designs. In particular, the theorem is most useful when the dual code has relatively few non-zero weights. Nevertheless, it has not been possible to detect t-designs for $t > 5$ by the Assmus–Mattson theorem.

We illustrate in the following examples some applications of the theorem.

Example 4.15. The extended binary $[24, 12, 8]$ Golay code is self-dual (cf. Construction 4.7) and has codewords of weight $0, 8, 12, 16$, and 24 in view of Theorem 4.2. For $t = 5$, we obtain the Steiner $5\text{-}(24, 8, 1)$ design as in Example 4.13. In the self-dual extended ternary $[12, 6, 6]$ Golay code all codewords are divisible by 3, and hence for $t = 5$, the supports of the codewords of weight 6 form a Steiner $5\text{-}(12, 6, 1)$ design.

Example 4.16. The extended quadratic residue code of length 48 over $\mathbb{F}_2$ is self-dual with minimum distance 12. By Theorem 4.2, it has codewords of weight $0, 12, 16, 20, 24, 28, 32, 36$, and 48. For $t = 5$, each of the values $v = 12, 16, 20$, or 24 yields a different 5-design and its complementary design.

Example 4.17. The Pless symmetry code Sym_{36} of dimension 18 is self-dual (cf. Construction 4.8) with minimum distance 12. The supports of codewords of weight $12, 15, 18$, and 21 yield 5-designs together with their complementary designs.

Remark 4.8. We give an overview of the state of knowledge concerning codes over $\mathbb{F}_q$ with their associated 5-designs (cf. also the tables in [13 (Chap. 16), 65, 71, 72]). In fact, these codes are all self-dual. Trivial designs as well as complementary designs are omitted.

Code	Code parameters		Design parameters	Ref.
Extended cyclic code	$[18, 9, 8]$	$q = 4$	$5\text{-}(18, 8, 6)$	[100]
			$5\text{-}(18, 10, 180)$	
Extended binary Golay code	$[24, 12, 8]$	$q = 2$	$5\text{-}(24, 8, 1)$	[101]
			$5\text{-}(24, 12, 48)$	
Extended ternary Golay code	$[12, 6, 6]$	$q = 3$	$5\text{-}(12, 6, 1)$	
Lifted Golay code over $\mathbb{Z}_4$	$[24, 12]$	$\mathbb{Z}_4$	$5\text{-}(24, 10, 36)$	[102, 103]
			$5\text{-}(24, 11, 336)$	[102]
			$5\text{-}(24, 12, 1584)$	[102]
			$5\text{-}(24, 12, 1632)$	[102]
Extended quadric residue codes	$[24, 12, 9]$	$q = 3$	$5\text{-}(24, 9, 6)$	[99, 65]
			$5\text{-}(24, 12, 576)$	
			$5\text{-}(24, 15, 8580)$	
	$[30, 15, 12]$	$q = 4$	$5\text{-}(30, 12, 220)$	[99, 65]
			$5\text{-}(30, 14, 5390)$	
			$5\text{-}(30, 16, 123000)$	
	$[48, 24, 12]$	$q = 2$	$5\text{-}(48, 12, 8)$	[99, 65]
			$5\text{-}(48, 16, 1365)$	
			$5\text{-}(48, 20, 36176)$	
			$5\text{-}(48, 24, 190680)$	

(*Continued*)

(*Continued*)

Code	Code parameters		Design parameters	Ref.
	$[48, 24, 15]$	$q = 3$	5-$(48, 12, 364)$	[99, 65]
			5-$(48, 18, 50456)$	
			5-$(48, 21, 2957388)$	
			5-$(48, 24, 71307600)$	
			5-$(48, 27, 749999640)$	
	$[60, 30, 18]$	$q = 3$	5-$(60, 18, 3060)$	[99, 65]
			5-$(60, 21, 449820)$	
			5-$(60, 24, 34337160)$	
			5-$(60, 27, 1271766600)$	
			5-$(60, 30, 24140500956)$	
			5-$(60, 33, 239329029060)$	
Pless symmetry codes	$[24, 12, 9]$	$q = 3$	5-$(24, 9, 6)$	[98]
			5-$(24, 12, 576)$	
			5-$(24, 15, 8580)$	
	$[36, 18, 12]$	$q = 3$	5-$(36, 12, 45)$	[98]
			5-$(36, 15, 5577)$	
			5-$(36, 18, 209685)$	
			5-$(36, 21, 2438973)$	
	$[48, 24, 15]$	$q = 3$	5-$(48, 12, 364)$	[98]
			5-$(48, 18, 50456)$	
			5-$(48, 21, 2957388)$	
			5-$(48, 24, 71307600)$	
			5-$(48, 27, 749999640)$	
	$[60, 30, 18]$	$q = 3$	5-$(60, 18, 3060)$	[97, 98]
			5-$(60, 21, 449820)$	
			5-$(60, 24, 34337160)$	
			5-$(60, 27, 1271766600)$	
			5-$(60, 30, 24140500956)$	
			5-$(60, 33, 239329029060)$	

Note. The lifted Golay code over $\mathbb{Z}_4$ is defined in [104] as the extended Hensel lifted quadric residue code of length 24. The supports of the codewords of Hamming weight 10 in the lifted Golay code and certain extremal double circulant Type II codes of length 24 yield (non-isomorphic) 5-$(24, 10, 36)$ designs. We further note that the quadratic residue codes and the Pless symmetry codes listed in the table with the same parameters are not equivalent as shown in [98] by inspecting specific elements of the automorphism group $PSL(2, q)$.

Remark 4.9. The concept of the weight enumerator can be generalized to non-linear codes (so-called *distance enumerator*), which leads to an analog of the MacWilliams relations as well as to similar results to the Assmus–Mattson theorem for non-linear codes (see [17, 18, 105] and Sec. 4.3.8).

The question whether there is an analogous result to the Assmus–Mattson theorem for codes over $\mathbb{Z}_4$ proposed in [102] was answered in the affirmative in [106]. Further generalizations of the Assmus–Mattson theorem are known, see in particular [107–113].

4.3.5. *Codes and Finite Geometries*

Let A be an incidence matrix of a projective plane $PG(2, n)$ of order n. When we consider the subspace C of $\mathbb{F}_2^{n^2+n+1}$ spanned by the rows of A, we obtain for odd n only the $[n^2 + n + 1, n^2 + n, 2]$ code consisting of all codewords of even weight. The case for even n is more interesting, in particular if $n \equiv 2 \pmod 4$.

Theorem 4.10. *For $n \equiv 2 \pmod 4$, the rows of an incidence matrix of a projective plane $PG(2, n)$ of order n generate a binary code C of dimension $(n^2 + n + 2)/2$, and the extended code $\overline{C}$ is self-dual.*

In a projective plane $PG(2, n)$ of even order n, there exist sets of $n + 2$ points, no three of which are collinear, and which are called *hyperovals* (sometimes just *ovals*, cf. [46]). This gives furthermore

Theorem 4.11. *The code C has minimum weight $n + 1$. Moreover, the codewords of minimum weight correspond to the lines and those of weight $n + 2$ to the hyperovals of $PG(2, n)$.*

Remark 4.10. The above two theorems arose in the context of the examination of the existence of a projective plane of order 10 (cf. Remark 4.1; for detailed proofs see, e.g. [4 (Chap. 13)]). Assuming the existence of such a plane, the obtained properties of the corresponding code led to very extensive computer searches. For example, in an early crucial step, it was shown [114] that this code could not have codewords of weight 15. On the various attempts to attack the problem and the final verification of the non-existence, we refer to [44, 115, 116] as well as [22 (Chap. 17), 35 (Chap. 12)].

Note. We note that at present the Fano plane is the only known projective plane with order $n \equiv 2 \pmod 4$.

For further accounts on codes and finite geometries, the reader is referred to [66 (Chaps. 5 and 6), 3, 4, 67, 116–119] as well as [120] from a more group-theoretical perspective and [121] with an emphasis on quadratic forms over $\mathbb{F}_2$.

4.3.6. *Golay Codes, Mathieu–Witt Designs, and Mathieu Groups*

We highlight some of the remarkable and natural interrelations between the Golay codes, the Mathieu–Witt designs, and the Mathieu groups.

There are various different constructions for the Golay codes besides the description as quadratic residue codes in Example 4.14. We briefly illustrate some exemplary constructions. For further details and more constructions, we refer to [122, 13 (Chap. 20), 4 (Chap. 11), 123 (Chap. 11)].

Construction 4.12.

- Starting with the zero vector in $\mathbb{F}_2^{24}$, a linear code of length 24 can be obtained by successively taking the lexicographically least binary codeword which has not been used and which has distance at least 8 to any predecessor. At the end of this process, we have 4096 codewords which form the extended binary Golay code. This construction is due to Conway and Sloane [124].

- Let A be an incidence matrix of the (unique) 2-$(11, 6, 3)$ design. Then $G := (I_{12}, P)$ with

$$
P := \begin{pmatrix} 0 & 1 & \cdots & 1 \\ 1 & & & \\ \vdots & & A & \\ 1 & & & \end{pmatrix}
$$

  is a (12×24)-matrix in which each row (except the top row) has eight 1's, and generates the extended binary Golay code.

- Let N be a (12×12)-adjacency matrix of the graph formed by the vertices and edges of the regular icosahedron. Then $G := (I_{12}, J_{12} - N)$ is a generator matrix for the extended binary Golay code.

- We recall that $\mathbb{F}_4 = \{0, 1, \omega, \omega^2\}$ is the field of four elements with $\omega^2 = \omega + 1$. The *hexacode* is the $[6, 3, 4]$ code over $\mathbb{F}_4$ generated by the matrix $G := (I_3, P)$ with

$$
P := \begin{pmatrix} 1 & \omega^2 & \omega \\ 1 & \omega & \omega^2 \\ 1 & 1 & 1 \end{pmatrix}.
$$

The extended binary Golay code can be defined by identifying each codeword with a binary (4×6)-matrix M (with rows $\mathbf{m}_0, \mathbf{m}_1, \mathbf{m}_2, \mathbf{m}_3$), satisfying

(i) each column of M has the same parity as the first row $\mathbf{m}_0$,

(ii) the sum $\mathbf{m}_1 + \omega\mathbf{m}_2 + \omega^2\mathbf{m}_3$ lies in the hexacode.

This description is essentially equivalent to the computational tool *MOG (Miracle Octad Generator)* of Curtis [125]. The construction via the hexacode is by Conway, see, e.g. [123 (Chap. 11)].

- Let Q be the circulant matrix of order 5 defined by Eq. (4.1). Then $G := (I_6, P)$, where P is the matrix Q bordered on top with a row of 1's, is a generator matrix of the ternary Golay code.

Remark 4.11. Referring to Example 4.7, we note that the automorphism groups of the Golay codes are isomorphic to the particular Mathieu groups, as was first pointed out in [101, 126]. Moreover, the Golay codes are related in a particularly deep and interesting way to a larger family of sporadic finite simple groups (cf., e.g. [55]).

Remark 4.12. We have seen in Example 4.13 that the supports of the codewords of weight 8 in the extended binary [24, 12, 8] Golay code form a Steiner 5-(24, 8, 1) design. The uniqueness of the large Mathieu–Witt design (up to isomorphism) can be established easily via coding theory (cf. Example 4.7). The main part is to show that any binary code of 4096 codewords, including the zero vector, of length 24 and minimum distance 8, is linear and can be determined uniquely (up to equivalence). For further details, in particular for a uniqueness proof of the small Mathieu–Witt designs, see, e.g. [70, 122, 4 (Chap. 11)].

4.3.7. *Golay Codes, Leech Lattice, Kissing Numbers, and Sphere Packings*

Sphere packings closely connect mathematics and information theory via the sampling theorem as observed by Shannon [1] in his classical article of 1948. Rephrased in a more geometric language, this can be expressed as follows:

> "Nearly equal signals are represented by neighboring points, so to keep the signals distinct, Shannon represents them by n-dimensional 'billiard balls', and is therefore led to ask: what is the best way to pack 'billiard balls' in n dimensions?" [127]

One of the most remarkable lattices, the *Leech lattice* in $\mathbb{R}^{24}$, plays a crucial role in classical sphere packings. We recall that a *lattice* in $\mathbb{R}^n$

is a discrete subgroup of $\mathbb{R}^n$ of rank n. The extended binary Golay code led to the discovery by Leech [128] of the 24-dimensional Euclidean lattice named after him. There are various constructions besides the usual ones from the binary and ternary Golay codes in the meantime, see, e.g. [123, 129 (Chap. 24)]. We outline some of the fundamental connections between sphere packings and the Leech lattice.

The *kissing number problem* deals with the maximal number τ_n of equal size non-overlapping spheres in the n-dimensional Euclidean space $\mathbb{R}^n$ that can touch a given sphere of the same size. Only a few of these numbers are actually known. For dimensions $n = 1, 2, 3$, the classical solutions are: $\tau_1 = 2$, $\tau_2 = 6$, $\tau_3 = 12$. The number τ_3 was the subject of a famous controversy between Isaac Newton and David Gregory in 1694, and was finally verified only in 1953 by Schütte and van der Waerden [130]. Using an approach initiated by Delsarte [18, 131] in the early 1970s, which gives linear programming upper bounds for binary error-correcting codes and for spherical codes [132] (cf. Sec. 4.3.8), Odlyzko and Sloane [133], and independently Levenshtein [134], proved that $\tau_8 = 240$ and $\tau_{24} = 196{,}560$. These exact solutions are the number of non-zero vectors of minimal length in the root lattice E_8 and in the Leech lattice, respectively. By extending and improving Delsarte's method, Musin [135] verified in 2003 that $\tau_4 = 24$, which is the number of non-zero vectors of minimal length in the root lattice D_4.

The *sphere packing problem* asks for the maximal density of a packing of equal size non-overlapping spheres in the n-dimensional Euclidean space $\mathbb{R}^n$. A sphere packing is called a *lattice packing* if the centers of the spheres form a lattice in $\mathbb{R}^n$. The Leech lattice is the unique densest lattice packing (up to scaling and isometries) in $\mathbb{R}^{24}$, as was shown by Cohn and Kumar [136, 137] recently in 2004, again by a modification of Delsarte's method. Moreover, they showed that the density of any sphere packing in $\mathbb{R}^{24}$ cannot exceed the one given by the Leech lattice by a factor of more than $1 + 1.65 \times 10^{-30}$ (via a computer calculation). The proof is based on the work [138] by Cohn and Elkies in 2003 in which linear programming bounds for the sphere packing problem are introduced and new upper bounds on the density of sphere packings in $\mathbb{R}^n$ with dimension $n \leq 36$ are proven.

For further details on the kissing number problem and the sphere packing problem, see [123 (Chap. 1), 127, 139, 14], as well as the survey articles [140–142]. For an on-line database on lattices, see [143].

4.3.8. *Codes and Association Schemes*

Any finite non-empty subset of the unit sphere S^{n-1} in the n-dimensional Euclidean space $\mathbb{R}^n$ is called a *spherical code*. These codes have many practical applications, e.g. in the design of signals for data transmission and storage. As a special class of spherical codes, *spherical designs* were introduced by Delsarte *et al.* [132] in 1977 as analogs on S^{n-1} of the classical combinatorial designs. For example, in S^2 the tetrahedron is a spherical 2-design; the octahedron and the cube are spherical 3-designs, and the icosahedron and the dodecahedron are spherical 5-designs. In order to obtain the linear programming upper bound mentioned in the previous subsection, Krawtchouk polynomials were involved in the case of binary error-correcting codes and Gegenbauer polynomials in the case of spherical codes.

However, Delsarte's approach was indeed much more general and far-reaching. He developed for association schemes, which have their origin in the statistical theory of design of experiments, a theory to unify many of the objects we have been addressing in this chapter. We give a formal definition of *association schemes* in the sense of Delsarte [18] as well as introduce the *Hamming* and the *Johnson schemes* as important examples of the two fundamental classes of *P-polynomial* and *Q-polynomial* association schemes.

Definition 4.3. A *d-class association scheme* is a finite point set X together with $d+1$ relations R_i on X, satisfying

(i) $\{R_0, R_1, \ldots, R_d\}$ is a partition of $X \times X$,
(ii) $R_0 = \{(x, x) \mid x \in X\}$,
(iii) for each i with $0 \leq i \leq d$, there exists a j with $0 \leq j \leq d$ such that $(x, y) \in R_i$ implies $(y, x) \in R_j$,
(iv) for any $(x, y) \in R_k$, the number p_{ij}^k of points $z \in X$ with $(x, z) \in R_i$ and $(z, y) \in R_j$ depends only on i, j, and k,
(v) $p_{ij}^k = p_{ji}^k$ for all i, j and k.

The numbers p_{ij}^k are called the *intersection numbers* of the association scheme. Two points $x, y \in X$ are called *i-th associates*, if $\{x, y\} \in R_i$.

Example 4.18. The *Hamming scheme* $H(n, q)$ has as point set X the set $\mathbb{F}^n$ of all n-tuples from a q-symbol alphabet; two n-tuples are *i*-th associates if their Hamming distance is i. The *Johnson scheme* $J(v, k)$, with $k \leq \frac{1}{2}v$, has as point set X the set of all k-element subsets of a set of size v; two k-element subset S_1, S_2 are i-th associates if $|S_1 \cap S_2| = k - i$.

Delsarte introduced the Hamming and Johnson schemes as settings for the classical concept of error-correcting codes and combinatorial t-designs, respectively. In this manner, certain results become formally dual, like the sphere packing bound (Theorem 4.1) and Fisher's inequality (Theorem 4.3).

For a more extended treatment of association schemes, the reader is referred to [144–148, 4 (Chap. 17), 13 (Chap. 21)], and in particular to [149, 150] with an emphasis on the close connection between coding theory and associations schemes. For a survey on spherical designs, see [21 (Chap. VI.54)].

4.4. Directions for Further Research

We present in this section a collection of significant open problems and challenges concerning future research.

Problem 4.1 (cf. [40]). Does every Steiner triple system on n points extend to a Steiner quadruple system on $n + 1$ points?

Problem 4.2 Does there exist any non-trivial Steiner 6-design?

Problem 4.3 (cf. [13 (p. 180)]). Find all non-linear single-error-correcting perfect codes over $\mathbb{F}_q$.

Problem 4.4 (cf. [6 (p. 106)]). Characterize codes where all codewords of the same weight (or of minimum weight) form a non-trivial design.

Problem 4.5 (cf. [6 (p. 116)]). Find a proof of the non-existence of a projective plane of order 10 without the help of a computer or with an easily reproducible computer program.

Problem 4.6 Does there exist any finite projective plane of order 12, or of any other order that is neither a prime power nor covered by the Bruck–Ryser theorem (cf. Remark 4.1)?

Problem 4.7 Does the root lattice D_4 give the unique kissing number configuration in $\mathbb{R}^4$?

Problem 4.8 Solve the kissing number problem in n dimensions for any $n > 4$ apart from $n = 8$ and 24. For presently known lower and upper bounds, we refer to [151, 152], respectively. Also any improvements of these bounds would be desirable.

Problem 4.9 (cf. [138, Conjecture 8.1]). Verify the conjecture that the Leech lattice is the unique densest sphere packing in $\mathbb{R}^{24}$.

4.5. Conclusions

Over the last sixty years a substantial amount of research has been inspired by the various interactions of coding theory and algebraic combinatorics. The fruitful interplay often reveals the high degree of regularity of both the codes and the combinatorial structures. This has lead to a vivid area of research connecting closely mathematics with information and coding theory. The emerging methods can be applied sometimes surprisingly effectively, e.g. in view of the recent advances on kissing numbers and sphere packings.

A further development of this beautiful interplay as well as its application to concrete problems would be desirable, certainly also in view of the various still open and long-standing problems.

4.6. Exercises

(1) Verify (numerically) that the Steiner quadruple system $SQS(8)$ of order 8 (cf. Example 4.5) has 14 blocks, and that the Mathieu–Witt design 5-$(24, 8, 1)$ (cf. Example 4.7) has 759 blocks.

(2) What are the parameters of the 2-design consisting of the points and hyperplanes (i.e. the $(d-2)$-dimensional projective subspaces) of the projective space $PG(d-1, q)$?

(3) Does there exist a self-dual $[8, 4]$ code over the finite field $\mathbb{F}_2$?

(4) Show that the ternary $[11, 6, 5]$ Golay code has 132 codewords of weight 5.

(5) Compute the weight distribution of the binary $[23, 12, 7]$ Golay code.

(6) Show that any binary code of 4096 codewords, including the zero vector, of length 24 and minimum distance 8 is linear.

(7) Give a proof for the sphere packing bound (cf. Theorem 4.1).

(8) Give a proof for Fisher's Inequality (cf. Theorem 4.3).

(9) Show that a binary code generated by the rows of an incidence matrix of any projective plane $PG(2, n)$ of even order n has dimension at most $(n^2 + n + 2)/2$ (cf. Theorem 4.10).

(10) (Todd's Lemma). In the Mathieu–Witt design 5-$(24, 8, 1)$, if B_1 and B_2 are blocks (*octads*) meeting in four points, then $B_1 + B_2$ is also a block.

4.7. Solutions

ad (1): By Lemma 4.2(b), we have to calculate $b = \frac{8 \cdot 7 \cdot 6}{4 \cdot 3 \cdot 2} = 14$ in the case of the Steiner quadruple system $SQS(8)$, and $b = \frac{24 \cdot 23 \cdot 22 \cdot 21 \cdot 20}{8 \cdot 7 \cdot 6 \cdot 5 \cdot 4} = 759$ in the case of the Mathieu–Witt design 5-$(24, 8, 1)$.

ad (2): Starting from Example 4.3, we obtain via counting arguments (or by using the transitivity properties of the general linear group) that the points and hyperplanes of the projective space $PG(d - 1, q)$ form a 2-$\left(\frac{q^d - 1}{q - 1}, \frac{q^{d-1} - 1}{q - 1}, \frac{q^{d-2} - 1}{q - 1} \right)$ design.

ad (3): Yes, the extended binary $[8, 4, 4]$ Hamming code is self-dual (cf. Example 4.9).

ad (4): Since the ternary $[11, 6, 5]$ Golay code is perfect (cf. Example 4.13), every word of weight 3 in $\mathbb{F}_3^{11}$ has distance 2 to a codeword of weight 5. Thus $A_5 = 2^3 \cdot \binom{11}{3} / \binom{5}{2} = 132$.

ad (5): The binary $[23, 12, 7]$ Golay code contains the zero vector and is perfect. This determines the weight distribution as follows $A_0 = A_{23} = 1$, $A_7 = A_{16} = 253$, $A_8 = A_{15} = 506$, $A_{11} = A_{12} = 1288$.

ad (6): Let C denote a binary code of 4096 codewords, including the zero vector, of length 24 and minimum distance 8. Deleting any coordinate leads to a code which has the same weight distribution as the code given in Exercise (5). Hence, the code C only has codewords of weight $0, 8, 12, 16$, and 24. This is still true if the code C is translated by any codeword (i.e. $C + \mathbf{x}$ for any $\mathbf{x} \in C$). Thus, the distances between pairs of codewords are also $0, 8, 12, 16$, and 24. Therefore, the standard inner product $\langle \mathbf{x}, \mathbf{y} \rangle$ vanishes for any two codewords $\mathbf{x}, \mathbf{y} \in C$, and hence C is self-orthogonal. For cardinality reasons, we conclude that C is self-dual and hence in particular linear.

ad (7): The sum $\sum_{i=0}^{e} \binom{n}{i} (q - 1)^i$ counts the number of words in a sphere of radius e. As the spheres of radius e about distinct codewords are disjoint, we obtain $|C| \cdot \sum_{i=0}^{e} \binom{n}{i} (q - 1)^i$ words. Clearly, this number cannot exceed the total number q^n of words, and the claim follows.

ad (8): As a non-trivial t-design with $t \geq 2$ is also a non-trivial 2-design by Lemma 4.1, it is sufficient to prove the assertion for an arbitrary non-trivial 2-(v, k, λ) design $\mathcal{D}$. Let A be an incidence matrix of $\mathcal{D}$ as defined in Sec. 4.3.1. Clearly, the (i, k)-th entry

$$(AA^t)_{ik} = \sum_{j=1}^{b} (A)_{ij}(A^t)_{jk} = \sum_{j=1}^{b} a_{ij}a_{kj}$$

of the $(v \times v)$-matrix AA^t is the total number of blocks containing both x_i and x_k, and is thus equal to r if $i = k$, and to λ if $i \neq k$. Hence

$$AA^t = (r - \lambda)I + \lambda J,$$

where I denotes the $(v \times v)$-unit matrix and J the $(v \times v)$-matrix with all entries equal to 1. Using elementary row and column operations, it follows easily that

$$\det(AA^t) = rk(r - \lambda)^{v-1}.$$

Thus AA^t is non-singular (i.e. its determinant is non-zero) as $r = \lambda$ would imply $v = k$ by Lemma 4.1, yielding that the design is trivial. Therefore, the matrix AA^t has rank$(A) = v$. But, if $b < v$, then rank$(A) \leq b < v$, and thus rank$(AA^t) < v$, a contradiction. It follows that $b \geq v$, proving the claim.

ad (9): Let C denote a binary code generated by the rows of an incidence matrix of $PG(2, n)$. By assumption n is even, and hence the extended code $\overline{C}$ must be self-orthogonal. Therefore, the dimension of C is at most $n^2 + n + 2/2$.

ad (10): For given blocks $B_1 = \{01, 02, 03, 04, 05, 06, 07, 08\}$ and $B_2 = \{01, 02, 03, 04, 09, 10, 11, 12\}$ in the Mathieu–Witt design 5-$(24, 8, 1)$, let us assume that $B_1 + B_2$ is not a block. The block B_3 which contains $\{05, 06, 07, 08, 09\}$ must contain just one more point of B_2, say $B_3 = \{05, 06, 07, 08, 09, 10, 13, 14\}$. Similarly, $B_4 = \{05, 06, 07, 08, 11, 12, 15, 16\}$ is the block containing $\{05, 06, 07, 08, 11\}$. But hence, it is impossible to find a block which contains $\{05, 06, 07, 09, 11\}$ and intersects with B_i, $1 \leq i \leq 4$, in 0, 2 or 4 points. Therefore, we obtain a contradiction as there must be a block containing any five points by Definition 4.2.

4.8. Keywords

Error-correcting codes, combinatorial designs, perfect codes and related concepts, Assmus–Mattson theorem and analogues, projective geometries, nonexistence of a projective plane of order 10, Golay codes, Leech lattice, kissing numbers, sphere packings, spherical codes, association schemes.

Acknowledgment

The author gratefully acknowledges the support by the Deutsche Forschungsgemeinschaft (DFG).

References

1. C. E. Shannon, A mathematical theory of communication, *Bell Syst. Tech. J.* **27** (1948), 379–423 and 623–656.
2. R. W. Hamming, Error detecting and error correcting codes, *Bell Syst. Tech. J.* **29** (1950), 147–160.
3. E. R. Berlekamp, *Algebraic Coding Theory* (McGraw-Hill, New York, 1968); [Revised edition: Aegean Park Press 1984].
4. P. J. Cameron and J. H. van Lint, *Designs, Graphs, Codes and Their Links* (Cambridge University Press, Cambridge, 1991).
5. R. Hill, *A First Course in Coding Theory* (Clarendon Press, Oxford, 1986).
6. W. C. Huffman and V. Pless, (eds.) *Handbook of Coding Theory*, Vols. I and II (North-Holland, Amsterdam, New York, Oxford, 1998).
7. W. C. Huffman and V. Pless, *Fundamentals of Error-Correcting Codes* (Cambridge University Press, Cambridge, 2003).
8. A. Betten, M. Braun, H. Fripertinger, A. Kerber, A. Kohnert and A. Wassermann, *Error-Correcting Linear Codes* (Springer, Berlin, Heidelberg, New York, 2006).
9. J. H. van Lint, Codes, in *Handbook of Combinatorics*, (eds.) R. L. Graham, M. Grötschel and L. Lovász, Vol. I (North-Holland, Amsterdam, New York, Oxford, 1995), pp. 773–807.
10. J. H. van Lint, *Introduction to Coding Theory*, 3rd edn. (Springer, Berlin, Heidelberg, New York, 1999).
11. W. W. Peterson and E. J. Weldon Jr., *Error-Correcting Codes*, 2nd edn. (MIT Press, Cambridge, 1972).
12. R. Roth, *Introduction to Coding Theory* (Cambridge University Press, Cambridge, 2006).
13. F. J. MacWilliams and N. J. A. Sloane, *The Theory of Error-Correcting Codes* (North-Holland, Amsterdam, New York, Oxford, 1977); 12. impression 2006.
14. T. M. Thompson, *From Error-Correcting Codes through Sphere Packings to Simple Groups*, Carus Math. Monograph, Vol. 21 (1983).

15. A. R. Calderbank, The art of signaling: Fifty years of coding theory, *IEEE Trans. Inform. Theor.* **44** (1998), 2561–2595.
16. F. J. MacWilliams, A theorem on the distribution of weights in a systematic code, *Bell Syst. Tech. J.* **42** (1963), 79–94.
17. F. J. MacWilliams, N. J. A. Sloane and J.-M. Goethals, The MacWilliams identities for non-linear codes, *Bell Syst. Tech. J.* **51** (1972), 803–819.
18. P. Delsarte, An algebraic approach to the association schemes of coding theory, *Philips Res. Rep. Suppl.* **10** (1973).
19. T. Beth, D. Jungnickel and H. Lenz, *Design Theory*, Vols. I and II, *Encyclopedia of Math. and Its Applications* **69/78**, (Cambridge University Press, Cambridge, 1999).
20. P. J. Cameron, *Parallelisms of Complete Designs* (Cambridge University Press, Cambridge, 1976).
21. C. J. Colbourn and J. H. Dinitz, (eds.) *Handbook of Combinatorial Designs*, 2nd edn. (CRC Press, Boca Raton, 2006).
22. M. Hall Jr., *Combinatorial Theory*, 2nd edn. (J. Wiley, New York, 1986).
23. D. R. Hughes and F. C. Piper, *Design Theory* (Cambridge University Press, Cambridge, 1985).
24. D. R. Stinson, *Combinatorial Designs: Constructions and Analysis* (Springer, Berlin, Heidelberg, New York, 2004).
25. P. Dembowski, *Finite Geometries* (Springer, Berlin, Heidelberg, New York, 1968); [Reprint 1997].
26. F. Buekenhout, (ed.), *Handbook of Incidence Geometry* (North-Holland, Amsterdam, New York, Oxford, 1995).
27. P. J. Cameron, *Permutation Groups* (Cambridge University Press, Cambridge, 1999).
28. R. D. Carmichael, *Introduction to the Theory of Groups of Finite Order* (Ginn, Boston, 1937); [Reprint: Dover Publications, New York, 1956].
29. J. D. Dixon and B. Mortimer, *Permutation Groups* (Springer, Berlin, Heidelberg, New York, 1996).
30. H. Wielandt, *Finite Permutation Groups* (Academic Press, New York, 1964).
31. V. D. Tonchev, *Combinatorial Configurations: Designs, Codes, Graphs* (Longman, Harlow, 1988).
32. D. Pei, *Authentication Codes and Combinatorial Designs* (CRC Press, Boca Raton, 2006).
33. D. R. Stinson, Combinatorial designs and cryptography, in *Surveys in Combinatorics, 1993*, (ed.) K. Walker (Cambridge University Press, Cambridge, 1993), pp. 257–287.
34. C. J. Colbourn, J. H. Dinitz and D. R. Stinson, Applications of combinatorial designs to communications, cryptography, and networking, in *Surveys in Combinatorics, 1999*, (eds.) J. D. Lamb and D. A. Preece (Cambridge University Press, Cambridge, 1999), pp. 37–100.
35. P. Kaski and P. R. J. Östergård, *Classification Algorithms for Codes and Designs* (Springer, Berlin, Heidelberg, New York, 2006).
36. P. J. Cameron, *Combinatorics: Topics, Techniques, Algorithms* (Cambridge University Press, Cambridge, 1994); [Reprint 1996].

37. J. H. van Lint and R. M. Wilson, *A Course in Combinatorics*, 2nd edn. (Cambridge University Press, Cambridge, 2001).
38. H. J. Ryser, ed. *Combinatorial Mathematics* (Math. Assoc. Amer., Buffalo, NY, 1963).
39. R. L. Graham, M. Grötschel and L. Lovász, (eds.) *Handbook of Combinatorics*, Vols. I and II (North-Holland, Amsterdam, New York, Oxford, 1995).
40. J. Steiner, Combinatorische Aufgabe, *J. Reine Angew. Math.* **45** (1853), 181–182.
41. R. J. Wilson, The early history of block designs, *Rend. Sem. Mat. Messina Ser. II* **9** (2003), 267–276.
42. Y. J. Ionin and M. S. Shrikhande, *Combinatorics of Symmetric Designs* (Cambridge University Press, Cambridge, 2006).
43. R. H. Bruck and H. J. Ryser, The non-existence of certain finite projective planes, *Canad. J. Math.* **1** (1949), 88–93.
44. C. W. H. Lam, L. Thiel and S. Swiercz, The non-existence of finite projective planes of order 10, *Canad. J. Math.* **41** (1989), 1117–1123.
45. A. Beutelspacher, Projective planes, in *Handbook of Incidence Geometry*, (ed.) F. Buekenhout (North-Holland, Amsterdam, New York, Oxford, 1995), pp. 101–136.
46. J. W. P. Hirschfeld, *Projective Geometries over Finite Fields*, 2nd edn. (Clarendon Press, Oxford, 1998).
47. D. R. Hughes and F. C. Piper, *Projective Planes*, 2nd edn. (Springer, Berlin, Heidelberg, New York, 1982).
48. H. Lüneburg, *Translation Planes* (Springer, Berlin, Heidelberg, New York, 1980).
49. G. Pickert, *Projektive Ebenen*, 2nd edn. (Springer, Berlin, Heidelberg, New York, 1975).
50. E. Mathieu, Mémoire sur l'étude des fonctions de plusieurs quantitiés, *J. Math. Pures Appl.* **6** (1861), 241–323.
51. E. Mathieu, Sur la fonction cinq fois transitive de 24 quantités, *J. Math. Pures Appl.* **18** (1873), 25–46.
52. E. Witt, Die 5-fach transitiven Gruppen von Mathieu, *Abh. Math. Sem. Univ. Hamburg* **12** (1938), 256–264.
53. E. Witt, Über Steinersche Systeme, *Abh. Math. Sem. University Hamburg* **12** (1938), 265–275.
54. H. Lüneburg, *Transitive Erweiterungen endlicher Permutationsgruppen* (Springer, Berlin, Heidelberg, New York, 1969).
55. M. Aschbacher, *Sporadic Groups* (Cambridge University Press, Cambridge, 1994).
56. J. Tits, Sur les systèmes de Steiner associés aux trois "grands" groupes de Mathieu, *Rendic. Math.* **23** (1964), 166–184.
57. H. Lüneburg, Fahnenhomogene Quadrupelsysteme, *Math. Z.* **89** (1965), 82–90.
58. M. Huber, Classification of flag-transitive Steiner quadruple systems, *J. Combin. Theor. Ser. A* **94** (2001), 180–190.

59. M. Huber, The classification of flag-transitive Steiner 3-designs, *Adv. Geom.* **5** (2005), 195–221.

60. M. Huber, The classification of flag-transitive Steiner 4-designs, *J. Algebr. Comb.* **26** (2007), 183–207.

61. M. Huber, A census of highly symmetric combinatorial designs, *J. Algebr. Comb.* **26** (2007), 453–476.

62. M. Huber, *Flag-transitive Steiner Designs* (Birkhäuser, Basel, Berlin, Boston, 2009).

63. L. Teirlinck, Non-trivial t-designs without repeated blocks exist for all t, *Discrete Math.* **65** (1987), 301–311.

64. R. A. Fisher, An examination of the different possible solutions of a problem in incomplete blocks, *Ann. Eugenics* **10** (1940), 52–75.

65. E. F. Assmus Jr. and H. F. Mattson Jr., Coding and combinatorics, *SIAM Rev.* **16** (1974), 349–388.

66. E. F. Assmus Jr. and J. D. Key, *Designs and their Codes* (Cambridge University Press, Cambridge, 1993).

67. E. F. Assmus Jr. and J. D. Key, Designs and codes: An update, *Des. Codes Cryptography* **9** (1996), 7–27.

68. I. F. Blake, Codes and designs, *Math. Mag.* **52** (1979), 81–95.

69. J. H. van Lint, Codes and designs, in *Higher Combinatorics: Proc. NATO Adv. Study Inst. (Berlin)*, (ed.) M. Aigner (Reidel, Dordrecht, Boston, 1977), pp. 241–256.

70. J. H. van Lint, Codes and combinatorial designs, in *Proc. Marshall Hall conference on Coding Theory, Design Theory, Group Theory (Burlington, VT)*, (eds.) D. Jungnickel and S. A. Vanstone (J. Wiley, New York, 1993), pp. 31–39.

71. V. D. Tonchev, Codes and designs, in *Handbook of Coding Theory*, Vol. II, (eds.) W. C. Huffman and V. Pless (North-Holland, Amsterdam, New York, Oxford, 1998), pp. 1229–1267.

72. V. D. Tonchev, Codes, in *Handbook of Combinatorial Designs*, 2nd edn. (eds.) C. J. Colbourn and J. H. Dinitz (CRC Press, Boca Raton, 2006), pp. 677–702.

73. D. E. Muller, Application of boolean algebra to switching circuit design and to error correction, *IEEE Trans. Comp.* **3** (1954), 6–12.

74. I. S. Reed, A class of multiple-error-correcting codes and the decoding scheme, *IEEE Trans. Inform. Theor.* **4** (1954), 38–49.

75. E. F. Assmus Jr. and H. F. Mattson Jr., On tactical configurations and error-correcting codes, *J. Combin. Theor. Ser. A* **2** (1967), 243–257.

76. M. J. E. Golay, Notes on digital coding, *Proc. IRE* **37** (1949), 657.

77. V. Pless, On the uniqueness of the Golay codes, *J. Combin. Theor. Ser. A* **5** (1968), 215–228.

78. A. Tietäväinen, On the nonexistence of perfect codes over finite fields, *SIAM J. Appl. Math.* **24** (1973), 88–96.

79. J. H. van Lint, Nonexistence theorems for perfect error correcting codes, in *Computers in Algebraic Number Theory*, Vol. IV, (eds.) G. Birkhoff and M. Hall Jr. (SIAM-AMS Proc., Providence, RI, 1971), pp. 89–95.

80. V. A. Zinov'ev and V. K. Leont'ev, The nonexistence of perfect codes over Galois fields, *Probl. Control and Inform. Theor.* **2** (1973), 123–132.
81. M. R. Best, A contribution to the nonexistence of perfect codes, PhD thesis, University of Amsterdam, 1982.
82. Y. Hong, On the nonexistence of unknown perfect 6- and 8-codes in Hamming schemes $H(n, q)$ with q arbitrary, *Osaka J. Math.* **21** (1984), 687–700.
83. J. H. van Lint, A survey of perfect codes, *Rocky Mountain J. Math.* **5** (1975), 189–224.
84. G. Cohen, I. Honkala, S. Litsyn and A. Lobstein, *Covering Codes* (North-Holland, Amsterdam, New York, Oxford, 1997).
85. A. M. Romanov, A survey of methods for constructing nonlinear perfect binary codes, *Diskretn. Anal. Issled. Oper. Ser. 1* **13** (2006), 60–88.
86. S. V. Avgustinovich, O. Heden and F. I. Solov'eva, The classification of some perfect codes, *Des. Codes Cryptography* **31** (2004), 313–318.
87. O. Heden and M. Hessler, On the classification of perfect codes: Side class structures, *Des. Codes Cryptography* **40** (2006), 319–333.
88. K. T. Phelps, J. Rifà and M. Villanueva, Kernels and p-kernels of p^r-ary 1-perfect codes, *Des. Codes Cryptography* **37** (2005), 243–261.
89. P. Kaski, P. R. J. Östergård and O. Pottonen, The Steiner quadruple systems of order 16, *J. Combin. Theor. Ser. A* **113** (2006), 1764–1770.
90. P. R. J. Östergård and O. Pottonen, There exist Steiner triple systems of order 15 that do not occur in a perfect binary one-error-correcting code, *J. Combin. Des.* **15** (2007), 465–468.
91. J.-M. Goethals and S. Snover, Nearly perfect binary codes, *Discrete Math.* **2** (1972), 65–88.
92. N. V. Semakov, V. A. Zinov'ev and G. V. Zaitsev, Uniformly packed codes, *Probl. Peredaci Inform.* **7** (1971), 38–50.
93. J.-M. Goethals and H. C. A. van Tilborg, Uniformly packed codes, *Philips Res. Reports* **30** (1975), 9–36.
94. H. C. A. van Tilborg, Uniformly packed codes, PhD thesis, Technology University Eindhoven, 1976.
95. R. Ahlswede, H. K. Aydinian, and L. H. Khachatrian, On perfect codes and related concepts, *Des. Codes Cryptography* **22** (2001), 221–237.
96. M. Schwartz and T. Etzion, Codes and anticodes in the Grassman graph, *J. Combin. Theor. Ser. A* **97** (2002), 27–42.
97. V. Pless, The weight of the symmetry code for $p = 29$ and the 5-designs obtained therein, *Ann. New York Acad. Sci.* **175** (1970), 310–313.
98. V. Pless, Symmetry codes over $GF(3)$ and new five-designs, *J. Combin. Theor. Ser. A* **12** (1972), 119–142.
99. E. F. Assmus Jr. and H. F. Mattson Jr., New 5-designs, *J. Combin. Theor. Ser. A* **6** (1969), 122–151.
100. F. J. MacWilliams, A. M. Odlyzko, N. J. A. Sloane and H. N. Ward, Self-dual codes over $GF(4)$, *J. Combin. Theor. Ser. A* **25** (1978), 288–318.
101. L. J. Paige, A note on the Mathieu groups, *Canad. J. Math.* **9** (1956), 15–18.

102. M. Harada, New 5-designs constructed from the lifted Golay code over $\mathbb{Z}_4$, *J. Combin. Des.* **6** (1998), 225–229.

103. T. A. Gulliver and M. Harada, Extremal double circulant type II codes over $\mathbb{Z}_4$ and construction of 5-(24, 10, 36) designs, *Discrete Math.* **194** (1999), 129–137.

104. A. Bonnecaze, P. Solé and A. R. Calderbank, Quaternary quadratic residue codes and unimodular lattices, *IEEE Trans. Inform. Theor.* **41** (1995), 366–377.

105. P. Delsarte, Four fundamental parameters of a code and their combinatorial significance, *Inform. Contr.* **23** (1973), 407–438.

106. K. Tanabe, A criterion for designs in $\mathbb{Z}_4$-codes on the symmetrized weight enumerator, *Des. Codes Cryptography* **30** (2003), 169–185.

107. A. R. Calderbank, P. Delsarte and N. J. A. Sloane, A strengthening of the Assmus-Mattson theorem, *IEEE Trans. Inform. Theor.* **37** (1991), 1261–1268.

108. G. T. Kennedy and V. Pless, On designs and formally self-dual codes, *Des. Codes Cryptography* **4** (1994), 43–55.

109. J. Simonis, MacWilliams identities and coordinate partitions, *Linear Algebra Appl.* **216** (1995), 81–91.

110. C. Bachoc, On harmonic weight enumerators of binary codes, *Des. Codes Cryptography* **18** (1999), 11–28.

111. J.-L. Kim and V. Pless, Designs in additive codes over $GF(4)$, *Des. Codes Cryptography* **30** (2003), 187–199.

112. D.-J. Shin, P. V. Kumar and T. Helleseth, An Assmus–Mattson-type approach for identifying 3-designs from linear codes over $\mathbb{Z}_4$, *Des. Codes Cryptography* **31** (2004), 75–92.

113. T. Britz and K. Shiromoto, Designs from subcode supports of linear codes, *Des. Codes Cryptography* **46** (2008), 175–189.

114. F. J. MacWilliams, N. J. A. Sloane and J. G. Thompson, On the existence of a projective plane of order 10, *J. Combin. Theor. Ser. A* **14** (1973), 66–78.

115. C. W. H. Lam, The search for a finite projective plane of order 10, *Amer. Math. Monthly* **98** (1991), 305–318.

116. P. J. Cameron, Finite geometries, in *Handbook of Combinatorics*, Vol. I, (eds.) R. L. Graham, M. Grötschel, and L. Lovász (North-Holland, Amsterdam, New York, Oxford, 1995), pp. 647–692.

117. E. F. Assmus Jr. and J. D. Key, Polynomial codes and finite geometries, in *Handbook of Coding Theory*, Vol. II, (eds.) W. C. Huffman and V. Pless (North-Holland, Amsterdam, New York, Oxford, 1998), pp. 1269–1343.

118. J. A. Thas, Finite geometries, varieties and codes, *Doc. Math. J. DMV, Extra Vol. ICM III*, (1998), pp. 397–408.

119. L. Storme, Projective geometry and coding theory, COM^{2}MAC Lect. Note Series **9**. Combin. Comput. Math., Center Pohang Univ. of Science and Technology, (2003).

120. C. Hering, On codes and projective designs. Technical Report 344, Kyoto Univ., Math. Research Inst. Seminar Notes, (1979).

121. P. J. Cameron, Finite geometry and coding theory, Socrates Course Notes, (1999). Published electronically at http://dwispc8.vub.ac.be/Potenza/lectnotes.html.

122. T. Beth and D. Jungnickel, Mathieu groups, Witt designs, and Golay codes, in *Geometries and Groups*, (eds.) M. Aigner and D. Jungnickel (Springer, Berlin, Heidelberg, New York, 1981), pp. 157–179.

123. J. H. Conway and N. J. A. Sloane, *Sphere Packings, Lattices and Groups*, 3rd edn. (Springer, Berlin, Heidelberg, New York, 1998).

124. J. H. Conway and N. J. A. Sloane, Lexicographic codes: Error-correcting codes from game theory, *IEEE Trans. Inform. Theor.* **32** (1986), 337–348.

125. R. T. Curtis, A new combinatorial approach to M_{24}, *Math. Proc. Cambridge Philos. Soc.* **79** (1976), 25–42.

126. E. F. Assmus Jr. and H. F. Mattson Jr., Perfect codes and the Mathieu groups, *Arch. Math.* **17** (1966), 121–135.

127. N. J. A. Sloane, The sphere packing problem, *Doc. Math. J. DMV, Extra Vol. ICM III*, (1998), pp. 387–396.

128. J. Leech, Some sphere packings in higher space, *Canad. J. Math.* **16** (1964), 657–682.

129. J. H. Conway and N. J. A. Sloane, Twenty-three constructions for the Leech lattice, *Proc. Roy. Soc. Lond. Ser. A* **381** (1982), 275–283.

130. K. Schütte and B. L. van der Waerden, Das Problem der dreizehn Kugeln, *Math. Ann.* **125** (1953), 325–334.

131. P. Delsarte, Bounds for unrestricted codes, by linear programming, *Philips Res. Rep.* **27** (1972), 272–289.

132. P. Delsarte, J.-M. Goethals and J. Seidel, Spherical codes and designs, *Geom. Dedicata* **6** (1977), 363–388.

133. A. M. Odlyzko and N. J. A. Sloane, New bounds on the number of unit spheres that can touch a unit sphere in n dimensions, *J. Comb. Theor. Ser. A* **26** (1979), 210–214.

134. V. I. Levenshtein, On bounds for packings in n-dimensional Euclidean space, *Sov. Math. Doklady* **20** (1979), 417-421.

135. O. R. Musin, The kissing number in four dimensions, *Ann. Math.* (to appear). Preprint published electronically at http://arxiv.org/abs/math.MG/0309430.

136. H. Cohn and A. Kumar, The densest lattice in twenty-four dimensions, *Elect. Res. Announc. Amer. Math. Soc.* **10** (2004), 58–67.

137. H. Cohn and A. Kumar, Optimality and uniqueness of the Leech lattice among lattices, *Ann. Math.* (to appear). Preprint published electronically at http://arxiv.org/abs/math.MG/0403263.

138. H. Cohn and N. D. Elkies, New upper bounds on sphere packings, *Ann. Math.* **157** (2003), 689–714.

139. O. R. Musin, An extension of Delsarte's method. The kissing problem in three and four dimensions, in *Proc. COE Workshop on Sphere Packings (Kyushu Univ. 2004)*, (2005), pp. 1–25.

140. N. D. Elkies, Lattices, linear codes, and invariants, *Notices Amer. Math. Soc.* **47** (2000), 1238–1245 and 1382–1391.

141. F. Pfender and G. M. Ziegler, Kissing numbers, sphere packings and some unexpected proofs, *Notices Amer. Math. Soc.* **51** (2004), 873-883.

142. K. Bezdek, Sphere packings revisited, *Europ. J. Comb.* **27** (2006), 864–883.

143. G. Nebe and N. J. A. Sloane, A catalogue of lattices. Published electronically at http://www.research.att.com/njas/lattices/.

144. R. C. Bose and T. Shimamoto, Classification and analysis of partially balanced incomplete block designs with two associate classes, *J. Amer. Statist. Assoc.* **47** (1952), 151–184.

145. R. C. Bose and D. M. Mesner, On linear associative algebras corresponding to association schemes of partially balanced designs, *Ann. Math. Statist.* **30** (1959), 21–38.

146. E. Bannai and T. Ito, *Algebraic Combinatorics I: Association Schemes* (Benjamin, New York, 1984).

147. A. E. Brouwer, A. M. Cohen and A. Neumaier, *Distance-Regular Graphs* (Springer, Berlin, Heidelberg, New York, 1989).

148. A. E. Brouwer and W. H. Haemers, *Association Schemes*, in *Handbook of Combinatorics*, Vol. I, (eds.) R. L. Graham, M. Grötschel, and L. Lovász, (North-Holland, Amsterdam, New York, Oxford, 1995), pp. 747–771.

149. P. Camion, Codes and association schemes: Basic properties of association schemes relevant to coding, in *Handbook of Coding Theory*, Vol. II, (eds.) W. C. Huffman and V. Pless (North-Holland, Amsterdam, New York, Oxford, 1998), pp. 1441–1567.

150. P. Delsarte and V. Levensthein, Association schemes and coding theory, *IEEE Trans. Inform. Theor.* **44** (1998), 2477–2504.

151. G. Nebe and N. J. A. Sloane, Table of the highest kissing numbers presently known. Published electronically at http://www.research.att.com/njas/lattices/.

152. C. Bachoc and F. Vallentin, New upper bounds for kissing numbers from semidefinite programming. Preprint published electronically at http://arxiv.org/abs/math.MG/0608426.

Chapter 5

BLOCK CODES FROM MATRIX AND GROUP RINGS

PAUL HURLEY* and TED HURLEY†

*IBM Research, Zurich Research Laboratory, Switzerland

†Department of Mathematics,
National University of Ireland, Galway, Ireland
*pah@zurich.ibm.com
†ted.hurley@nuigalway.ie

In this chapter, the algebra of groups ring and matrix rings is used to construct and analyze systems of zero-divisor and unit-derived codes. These codes are more general than codes from ideals (e.g. cyclic codes) in group rings. They expand the space of linear block codes, offering additional flexibility in terms of desired properties as algebraic formulations, while also have readily available generator and check matrices.

A primer is presented in the necessary algebra, showing how group rings and certain rings of matrices can be used interchangeably. Then it is shown how the codes may be derived, showing particular cases such as self-dual codes and codes from dihedral group rings.

5.1. Introduction

Any (n, r) linear block code can be described using a generator matrix G and a check matrix H that are related by $GH^{\mathrm{T}} = 0$. The generator matrix G is $r \times n$ with rank r, while the check matrix H is $(n-r) \times n$ with rank $n-r$.

Thus G and H are in a sense *zero-divisors*. In certain cases, the code is an *ideal* within some algebraic structure. For example, a cyclic code is obtained from zero-divisors in the cyclic group ring and is also an ideal therein.

Algebras have *units* as well as zero-divisors, and *modules* in addition to ideals. The codes described in this treatise are in general not ideals, thus expanding the space of block codes. Ways to construct block codes from group rings in general using both zero-divisors and units are described. The codes derived are either *zero-divisor* or *unit-derived* codes.

Group rings, as well as being subalgebras of matrix algebras, have a rich structure of their own. Thus, properties of codes such as self-duality and low density parity check (LPDC) are easy to describe in terms of group ring concepts, enabling codes of particular types and with precise, provable, properties to be defined and analyzed from group ring structures.

Section 5.2 describes the algebra used throughout: the concept of a ring, algebra, zero-divisor, unit, group ring, ideal, etc., with illustrative examples. In particular, the relationship between group rings and rings of matrices is established. This relationship enables the immediate production of generator and check matrices and allows the power of matrix algebra to be exploited in conjunction with the algebra of group rings.

Next, the basic notion of a code from a group ring encoding, the two different types — zero-divisor and unit-derived — are described in Sec. 5.3. In each case the generator and check matrices may be immediately obtained. The cases when group ring encodings are and are not ideals are examined, and the relationship with cyclic codes presented.

Derivations of particular types and examples of codes and their properties are presented next. The natural first choice for a non-commutative group ring is the dihedral group ring, one previously examined from the standpoint of ideals. The zero-divisor and unit-derived codes from this group ring and their associated matrices are described in Sec. 5.4, followed by a description of self-dual codes from group ring elements in Sec. 5.5.

In the next chapter of this book, LDPC codes and convolutional codes from the general group ring method are presented.

5.2. Algebraic Background

It is assumed that the reader is familiar with the concept of a *group* although no advanced theory is required.

A ring R is a structure with two binary operations $+$ and $\cdot$ such that

(1) $(R, +)$ is an abelian group,
(2) $(R, \cdot)$ is an associative binary system; that is:
- $a \cdot b \in R, \, \forall \, a, b \in R$
- $(a \cdot b) \cdot c = a \cdot (b \cdot c), \, \forall \, a, b, c \in R$
(3) Multiplication $(\cdot)$ distributes over addition $(+)$; that is $\forall \, a, b, c \in R$:
- $a \cdot (b + c) = a \cdot b + a \cdot c$
- $(b + c) \cdot a = b \cdot a + c \cdot a$

Some authors do not require that $(R, \cdot)$ satisfy the associative law and then those which do are referred to as *associate rings*.

0_R, or simply 0 when the context is understood, denotes the zero element of $(R, +)$.

A *ring with identity* is a ring R such that there exists $1_R \in R$ with $a \cdot 1_R = a = 1_R \cdot a, \forall a \in R$. When the context is clear, often 1 is written for the identity 1_R of R.

The symbol $\cdot$ is usually omitted in which case the multiplication is denoted by the juxtaposition ab.

A commutative ring is a ring R in which $ab = ba, \forall a, b \in R$.

Suppose now R is a ring with identity 1. Then $a \in R$ is said to be *invertible*, or to be a *unit*, in R if and only if there exists a $b \in R$ with $ab = 1 = ba$. The set of invertible elements in a ring R forms a group which is referred to as the *group of units of R* and is denoted by $U(R)$. Thus $U(R)$ should be read as *units of R*.

A ring R (with identity) is said to be a *division ring* if and only if every non-zero element in R has an inverse, that is, if and only if $U(R) = R - \{0\}$. A commutative division ring is a *field*, that is a field is a commutative ring (with identity) in which every non-zero element has an inverse.

5.2.1. *Examples*

(1) The integers $\mathbb{Z}$ is a commutative ring with identity and $U(\mathbb{Z}) = \{1, -1\}$.

(2) The natural numbers $\mathbb{N}$ is not a ring.

(3) The set $\mathbb{Z}_n$ of integers modulo n with addition and multiplication defined modulo n is a ring. This is an example of a finite ring. When $n = p$ is a prime, then $\mathbb{Z}_p$ is a (finite) field.

(4) The rationals, $\mathbb{Q}$, the reals $\mathbb{R}$ and the complex numbers $\mathbb{C}$ are all examples of fields.

(5) The set $R_{n \times n}$ of $n \times n$ matrices with coefficients from a ring R is also a ring. If R has an identity then so does $R_{n \times n}$. When R is a field, $U(R_{n \times n})$ consists of those elements in $R_{n \times n}$ with non-zero determinant.

(6) The set $R_{n \times m}$ of $n \times m$ matrices with coefficients from a ring R with $m \neq n$ does not form a ring as multiplication for such a system does not exist.

(7) $R[x]$, the set of polynomials with variable x and coefficients from a ring R, is a ring.

(8) $R(x)$, the set of power series with variable x and coefficients from a ring R, is a ring.

(9) $R(z, z^{-1})$, the set of Laurent series with variable z and coefficients from a ring R, is a ring. Now $R(x, x^{-1}) = \{\sum_{i=-\infty}^{\infty} \alpha_i z^i \mid \alpha_i \in R\}$.

(10) $R[x, x^{-1}]$, the set of Laurent series with variable x of finite support and coefficients from a ring R, is also a ring. Here "finite support" means that only a finite number of the coefficients in the Laurent series are non-zero.

(11) An important example used below is the concept of a *group ring*. Let R be a ring and G a group. Then the group ring RG is the set of all $\sum_{g \in G} \alpha_g g$ where $\alpha_g \in R$ and only a finite number of the α_g may be non-zero. Further details are given below.

5.2.2. *Zero-Divisors*

In coding theory, zero-divisors are important. A non-zero element u in the ring R is said to be a *zero-divisor* in R if there exists a non-zero $v \in R$ such that $uv = 0$. This is the concept of a left zero-divisor; the concept of a right zero-divisor is similar. In many cases if u is a left zero-divisor then it is also a right zero-divisor but it may happen that $uv = 0$ with $vu \neq 0$ but there still exists a $w \neq 0$ such that $wu = 0$.

5.2.2.1. *Examples*

(1) Suppose $n \in \mathbb{N}$ is not prime so that $n = rs$, with $r, s > 1$. Then $rs = 0$ in $\mathbb{Z}_n$ so that r, s are zero-divisors.

(2) A field F has no zero-divisors.

(3) $A \in F_{n \times n}$ for a field F is a zero-divisor if and only if $A \neq 0_{n \times n}$ and $\det(A) = 0$.

5.2.3. *Subrings and Ideals*

Let R be a ring. A subset S of R which is also a ring (relative to the same operations $(+, \cdot)$ of R) is said to be a *subring* of R.

If S is a subring of R and $rs \in S, \forall r \in R, \forall s \in S$ then S is said to be a *left ideal* in R. If S is a subring of R and $sr \in S, \forall r \in R, \forall s \in S$ then S is said to be a *right ideal* in R.

If S is both a left and right ideal in R then S is said to be a two-sided ideal in R or simply an *ideal* in R. For commutative rings the concepts of left-, right-, and two-sided are equivalent.

5.2.4. *Vector Spaces and Modules*

It is assumed that the reader is familiar with the concept of a vector space over a field, including linear (in)dependence and basis. Familiarity is also assumed with the matrix concepts, such as *row-space, column-space,* and properties, such that *row-rank = column-rank = rank*, descriptions of which may be found in any book on linear algebra (for example [1]).

A vector space V is defined over a field F. The concept of a module over a ring R is similar and when the ring R is a field a module over R is the same as a vector space — the *scalars* in a module are elements of a ring rather than a field.

Thus a module M over a ring R is an abelian group $(M, +)$ together with an operation $R \times M \to M$ usually written juxtaposition such that $\forall r, s \in R, \forall m, n \in M$:

(1) $r(m + n) = rm + rn$,
(2) $(r + s)m = rm + sm$,
(3) $(rs)m = r(sm)$

If in addition R has an identity 1_R then it is insisted that $1_R x = x$ $\forall x \in M$.

5.2.5. *Algebra*

The word "algebra" is used widely to mean many different things. Here it is restricted to mean a system which is a module over R with a multiplication which makes it into a ring. A *subalgebra* of an algebra $\mathbb{A}$ is a subset of $\mathbb{A}$ which is itself an algebra under the same operations as in $\mathbb{A}$.

5.2.5.1. *Examples of Algebras*

- $\mathbb{C} = \{a + ib \,|\, a, b \in \mathbb{R}\}$ is the set of complex numbers. This algebra is defined once the multiplication on i has been defined; this is $i^2 = -1$. A basis for the vector space $\mathbb{C}$ over $\mathbb{R}$ is $\{1, i\}$.

    $\mathbb{C}$ is represented by 2×2 matrices of the form $\left\{ \begin{pmatrix} a & b \\ -b & a \end{pmatrix} \right\}$, with $a, b \in \mathbb{R}$.

- $\mathbb{H} = \{a + ib + jc + kd \,|\, a, b, c, d \in \mathbb{R}\}$ is the set of *quaternions* of Hamilton. A basis for the vector space over $\mathbb{R}$ is $\{1, i, j, k\}$.

    The algebra is defined once the multiplications on $i, j,$ and k have been defined. The multiplication is $i^2 = j^2 = k^2 = -1, ij = k = -ji$,

$jk = i = -kj$, and $ki = j = -ik$. The quaternions also have a matrix representation: $\left\{ \begin{pmatrix} a & b & c & d \\ -b & a & -d & c \\ -c & d & a & -b \\ -d & -c & b & a \end{pmatrix} \right\}$ with $a, b, c, d \in \mathbb{R}$.

- $\{a_1 e_1 + a_2 e_2 + \cdots + a_n e_n \mid a_i \in \mathbb{R}\}$ is a *hypercomplex system* and an algebra is defined once the multiplications on the e_i are defined.
- $F_{n \times n}$ is the algebra of $n \times n$ (square) matrices with coefficients from the field F.
- $F[x]$ is the algebra of polynomials with coefficients from F. This algebra has *infinite dimensions* as a vector space.
- $R(z, z^{-1})$ is the algebra of Laurent series over R with variable z.
- $R[z, z^{-1}]$ is the algebra of Laurent series of finite support over R with variable z. This is a subalgebra of $R(z, z^{-1})$.

5.2.5.2. *Subalgebras of Matrices*

The $n \times n$ matrices over F, $F_{n \times n}$, is an algebra and within this algebra are many rather useful (sub)algebras.

(1) The set of *circulant matrices* $n \times n$ matrices over F is a subalgebra of $F_{n \times n}$. $\left\{ \begin{pmatrix} a & b & c \\ c & a & b \\ b & c & a \end{pmatrix} \right\}$ is a subalgebra of $F_{3 \times 3}$.

More generally, $\left\{ \begin{pmatrix} a_0 & a_1 & \cdots & a_{n-1} \\ a_{n-1} & a_0 & \cdots & a_{n-2} \\ \vdots & \vdots & \vdots & \vdots \\ a_1 & a_2 & \cdots & a_0 \end{pmatrix} \right\}$ is a subalgebra of $F_{n \times n}$.

It is necessary to verify that the sum (easy) and product (not so easy) of circulant matrices is a circulant matrix. It is also true that circulant matrices commute. For further information on circulant matrices see [2].

(2) Consider circulant-by-circulant matrices as, for example, $\left\{ \begin{pmatrix} A & B & C \\ C & A & B \\ B & C & A \end{pmatrix} \right\}$ where A, B and C are circulant matrices of the same size $m \times m$ over R.

These form a subalgebras of $R_{3m \times 3m}$ and also commute with one another.

(3) Matrices of the form $\begin{pmatrix} A & B \\ B^{\mathrm{T}} & A^{\mathrm{T}} \end{pmatrix}$, where A and B are circulant matrices of size $m \times m$ over R, are a subalgebra of $R_{2m \times 2m}$. These matrices form not circulant-by-circulant and do not commute in general.

5.2.6. *Group Rings: More Algebras*

The most natural multiplication is group multiplication. Cayley noticed that the group multiplication could be used to define the multiplication on an algebra which is now called the *group ring* of a group.

For R a ring and G a group, the *group ring RG* is defined by:

$$RG = \left\{ \sum_{g \in G} \alpha_g g \mid \alpha_g \in R \right\},$$

in which only a finite number of the α_g allowed to be non-zero.

Thus the *basis* for the group ring over R consists of the *elements of the group G* and the group multiplication is sufficient, together with the distributive law, to define multiplication in the group ring. When R is a field, RG is often called a *group algebra*. However, group rings where the ring is not a field are useful and indeed, as shown in the next chapter, a group ring where the ring is actually a group ring itself is proving useful in constructing convolutional codes.

The group ring RG can thus be considered as the module over R with basis consisting of the elements of G and with a multiplication determined by the *convolutional type multiplication* of the elements of G.

The convolution used in digital signal processing is exactly the group ring multiplication when the group is the simplest of all, the cyclic group — finite cyclic gives the circulant convolution while the infinite cyclic yields the convolution used in digital filters.

The connection between group representation theory and group rings, as the structure theory of algebras, was widely recognised after the influential paper by Noether [3] and then by papers of Noether and Brauer.

Many survey papers and books as well as a huge number of research articles have appeared on group rings over the years. Particularly worth mentioning is the recent book by Milies and Sehgal "Introduction to group rings" [4] as well as previous books by Passman and Sehgal.

5.2.6.1. *Examples*

(1) $\mathbb{Z}_2 C_3$ consists of $\alpha_0 + \alpha_1 g + \alpha_2 g^2$ where $\alpha_i \in \{0, 1\}$. It is 3-dimensional and has 2^3 elements. This algebra is the same as (isomorphic to) the 3×3 circulant matrices over $\mathbb{Z}_2$.

(2) $\mathbb{R}(C_2 \times C_2)$ consists of $\alpha_0 + \alpha_1 a + \alpha_2 b + \alpha_3 ab$ with $\alpha_i \in \mathbb{R}$ and $C_2 \times C_2$ is the Klein 4-group generated by a, b each of order 2.

In fact this group ring is the same as (isomorphic to) the set of

$$\text{matrices} \left\{ \begin{pmatrix} \alpha_0 & \alpha_1 & \alpha_2 & \alpha_3 \\ \alpha_1 & \alpha_0 & \alpha_3 & \alpha_2 \\ \alpha_2 & \alpha_3 & \alpha_0 & \alpha_1 \\ \alpha_3 & \alpha_2 & \alpha_1 & \alpha_0 \end{pmatrix} \right\} \text{ with } \alpha_i \in R.$$

5.2.7. *Zero-Divisors and Units in Matrix and Group Rings*

Zero-divisors and units in an algebra turn out to be useful in coding theory. A zero-divisor in a ring is an element $u \neq 0$ such that there exists $v \neq 0$ with $uv = 0$. Up to now codes have been mainly constructed from zero-divisors. But *units* also come into play and show extreme promise. A unit is an element u such that there exists v with $uv = 1 = vu$.

For example, matrices have lots of zero-divisors (singular matrices), and lots of units (non-singular matrices). Group rings are fertile sources of zero-divisors and units, with the advantage of a rich structure.

5.2.7.1. *Zero-Divisor Example*

Let

$$U = \begin{pmatrix} 1 & 1 & 0 & 1 & 0 & 0 & 0 \\ 0 & 1 & 1 & 0 & 1 & 0 & 0 \\ 0 & 0 & 1 & 1 & 0 & 1 & 0 \\ 0 & 0 & 0 & 1 & 1 & 0 & 1 \\ 1 & 0 & 0 & 0 & 1 & 1 & 0 \\ 0 & 1 & 0 & 0 & 0 & 1 & 1 \\ 1 & 0 & 1 & 0 & 0 & 0 & 1 \end{pmatrix} \quad \text{and} \quad V = \begin{pmatrix} 1 & 1 & 1 & 0 & 1 & 0 & 0 \\ 0 & 1 & 1 & 1 & 0 & 1 & 0 \\ 0 & 0 & 1 & 1 & 1 & 0 & 1 \\ 1 & 0 & 0 & 1 & 1 & 1 & 0 \\ 0 & 1 & 0 & 0 & 1 & 1 & 1 \\ 1 & 0 & 1 & 0 & 0 & 1 & 1 \end{pmatrix}.$$

Let $\mathbb{Z}_2$ denote the integers modulo 2, which is the same as the field of two elements $\mathbb{F}_2$ — also written $\mathrm{GF}(2)$.

It is easy to check that $UV = 0$ over $\mathbb{Z}_2$. The last 3 rows of U are dependent on the first 4 rows of U. The last 4 rows of V are dependent on the first 3 rows of V.

Thus consider

$$G = \begin{pmatrix} 1 & 1 & 0 & 1 & 0 & 0 & 0 \\ 0 & 1 & 1 & 0 & 1 & 0 & 0 \\ 0 & 0 & 1 & 1 & 0 & 1 & 0 \\ 0 & 0 & 0 & 1 & 1 & 0 & 1 \end{pmatrix} \quad \text{and} \quad H = \begin{pmatrix} 1 & 1 & 1 & 0 & 1 & 0 & 0 \\ 0 & 1 & 1 & 1 & 0 & 1 & 0 \\ 0 & 0 & 1 & 1 & 1 & 0 & 1 \end{pmatrix}.$$

Then $GH^{\mathrm{T}} = 0$ and thus a linear code, the first Hamming code, is obtained.

The U and V zero-divisors come from the group ring of the cyclic group of order 7. In Z_2C_7, $(1 + g + g^3)(1 + g + g^2 + g^4) = 0$; say $uv = 0$.

Now form the circulant matrices with first rows obtained from u and v, where the columns of the matrices are thought of as being numbered by the elements of C_7.

$$\begin{pmatrix} 1 & g & g^2 & g^3 & g^4 & g^5 & g^6 \\ \hline 1 & 1 & 0 & 1 & 0 & 0 & 0 \\ 0 & 1 & 1 & 0 & 1 & 0 & 0 \\ \vdots & \vdots & \vdots & \vdots & \vdots & \vdots & \vdots \end{pmatrix}$$

This gives the U as above and similarly the V as above is obtained.

5.2.8. *Group Rings and Matrices*

Let $\{g_1, g_2, \ldots, g_n\}$ be a fixed listing of the elements of G. The RG-*matrix* of $w = \sum_{i=1}^{n} \alpha_{g_i} g_i \in RG$, is in $R_{n \times n}$, the ring of $n \times n$ matrices over R, and defined as,

$$M(RG, w) = \begin{pmatrix} \alpha_{g_1^{-1} g_1} & \alpha_{g_1^{-1} g_2} & \cdots & \alpha_{g_1^{-1} g_n} \\ \alpha_{g_2^{-1} g_1} & \alpha_{g_2^{-1} g_2} & \cdots & \alpha_{g_2^{-1} g_n} \\ \vdots & \vdots & \vdots & \vdots \\ \alpha_{g_n^{-1} g_1} & \alpha_{g_n^{-1} g_2} & \cdots & \alpha_{g_n^{-1} g_n} \end{pmatrix}. \tag{5.1}$$

The first column is, in essence, labeled by g_1, the second by g_2, etc.; the first row is in essence labeled by g_1^{-1}, the second by g_2^{-1}, etc. Each row and each column is a permutation, determined by the group multiplication, of the initial row.

Only finite groups are presented, but many of the results hold for infinite groups with corresponding infinite RG-matrices. A group ring RG

is isomorphic to a ring of *RG-matrices* over R, a subring of $R_{n \times n}$, and the ring of $n \times n$ matrices over R.

Theorem 5.1. *Given a listing of the elements of a group G of order n there is a bijective ring homomorphism $\sigma : w \mapsto M(RG, w)$ between RG and the $n \times n$ RG-matrices over R.*

A proof of this is given in [5]. This result means that the group ring and the ring of matrices are interchangeable. One can thus exploit results from matrix algebra and group rings as needed.

For every $u \in RG$, its *RG-matrix* $\sigma(u)$ is denoted by the corresponding capital letter U.

Another useful result is that in a group algebra, every element must be a unit or zero-divisor, and there is a method to determine which.

Theorem 5.2. *Let R be a field. A non-zero $u \in RG$ is a zero-divisor if and only if $\det(\sigma(u)) = 0$, and otherwise a unit.*

Again the proof is given in [5] along with further details. Thus, when R is a field, an element $u \in RG$ is a zero-divisor if and only if rank $U < n$ and is a unit if and only if rank $U = n$. Additionally, it can easily be shown that a finite ring with identity contains only zero-divisors and units.

The isomorphism between group rings and the *RG-matrices* allows the generator and check matrices for the group ring codes, to be defined in Sec. 5.3, to be immediately derived from the group ring description.

5.2.9. *Examples of RG Matrices*

Details on these examples of *RG-matrices* appear in [5]. In the cyclic group ring case the matrices are the circulant matrices [2]. All cyclic codes can be generated from singular circulant matrices.

The *RG*-matrix types which turn up as isomorphic to certain group rings include Toeplitz-type matrices, Walsh–Toeplitz matrices, and circulant matrices, Toeplitz combined with Hankel-type matrices and block-type circulant matrices.

In the general finite abelian group case, RG is isomorphic to certain block circulant matrices that, when R is commutative, commute and are normal. In the case of an elementary abelian 2-group of rank m and order 2^m, where the matrix size is $2^m \times 2^m$, the *RG*-matrices are the Walsh–Toeblitz matrices over R.

In the case of the dihedral group the RG-matrices are of the form $\begin{pmatrix} A & B \\ B & A \end{pmatrix}$ where A is a circulant matrix and B is a reverse circulant matrix. These reverse circulant matrices are of *Hankel-type* as they bear the same relationships to Hankel matrices as circulant matrices bear to Toeplitz matrices.[a] See [6] for further details.

5.2.10. *Element Properties*

One has the concept of support and weight of an element.

Definition 5.1. The *support* of an element $u = \sum_{g \in G} \alpha_g g$ in a group ring is the set of all group elements for which the coefficients are non-zero, i.e. $\mathrm{supp}(u) = \{g \in G : \alpha_g \neq 0\}$.

Definition 5.2. The *weight* of an element u, written $\mathrm{wt}(u)$, in a group ring is the number of non-zero coefficients it has, i.e. $\mathrm{wt}(u) = |\mathrm{supp}(u)|$.

Many concepts and properties of matrices turn out to have useful equivalents in the group ring context. These are inherent to the group ring itself, existing independent of any group listing chosen to establish the isomorphism, while maintaining consistency with their equivalent matrix evocation. First the transpose of a group ring element is described.

Definition 5.3. The *transpose* of an element $u = \sum_{g \in G} \alpha_g g$ in RG is $u^{\mathrm{T}} = \sum_{g \in G} \alpha_g g^{-1}$, or equivalently, $u^{\mathrm{T}} = \sum_{g \in G} \alpha_{g^{-1}} g$.

This is consistent with the matrix definition of transpose. For a given listing $G = \{g_1, \ldots, g_n\}$, let U be the RG-matrix of u. The entry (i, j) of the RG-matrix of u^{T} is $\alpha_{(g_i^{-1} g_j)^{-1}} = \alpha_{g_j^{-1} g_i}$, so U^{T} is the RG-matrix of u^{T}.

The transpose u^{T} has also been called the canonical antiautomorphism of u [7], denoted $\bar{u}$. When dealing with the cyclic groups, it has been referred to as a transpose [8], and it is associated with the reciprocal polynomial.

Definition 5.4. $u \in RG$ is *symmetric* if and only if $u^{\mathrm{T}} = u$.

Clearly, the definition is consistent as u is symmetric if and only if U is a symmetric matrix.

---

[a]The Toeplitz $n \times n$ matrices may be embedded in the $(2n-1) \times (2n-1)$ circulant matrices; the Hankel $n \times n$ matrices may be embedded similarly in the $(2n-1) \times (2n-1)$ reverse circulant/Hankel-type matrices

5.2.11. *Some Notation*

We denote $Wu = \{xu : x \in W\}$ and $uW = \{ux : x \in W\}$.

The notation $\underline{x}$ is used to indicate that $\underline{x}$ is a vector as opposed to an element of the group ring. For $\underline{x} = (\alpha_1, \alpha_2, \ldots, \alpha_n) \in R^n$, the mapping $\zeta(\underline{x}) = \sum_{i=1}^n \alpha_i g_i = x$ is an element in RG according to a given listing of G. $\zeta^{-1}(x)$ denotes the inverse map.

5.3. Codes from Group Rings

Most existing linear codes in use are cyclic codes. These include such important codes as BCH, Reed–Solomon, Golay, and Hamming codes.

All cyclic codes can be obtained from zero-divisors in a cyclic group ring. Matrices used in cyclic codes and some others are circulant matrices or are derived from circulant matrices in the sense that certain rows of a circulant matrix are used for a generator matrix G and certain columns of a circulant matrix H are used for a check matrix, with $GH^{\mathrm{T}} = 0$.

Circulant matrices are simply the group ring elements of the cyclic group under the injection of this group ring into the ring of matrices. By looking at group rings in general and their zero-divisors and units, new codes and new systems of codes can be obtained which encompass the previous class as a special case.

The definition of code from a group ring encoding is now given and the two different types, zero-divisor and unit-identified. The concept of these types of codes originally appeared in [9, 10].

Let RG be the group ring of the group G over the ring R. A listing of the elements of G is given by $G = \{g_1, g_2, \ldots, g_n\}$. Suppose W is a submodule of RG, and $u \in RG$ is given.

Definition 5.5. Let $x \in W$. A *group ring encoding* is a mapping $f : W \to RG$, such that $f(x) = xu$ or $f(x) = ux$. In the latter case, f is a *left group ring encoding*. In the former, it is a *right group ring encoding*.

A code $\mathcal{C}$ *derived from a group ring encoding* is then the image of a group ring encoding, namely for a given $u \in RG$, $\mathcal{C} = \{ux : x \in W\}$ or $\mathcal{C} = \{xu : x \in W\}$.

Multiplication need not necessarily commute in a group ring. Allowing non-commutative groups enables the construction of non-commutative codes.

Definition 5.6. If $xu = ux$ for all x then the code $\{xu : x \in W\}$ is said to be *commutative*; and otherwise *non-commutative*.

When u is a zero-divisor, it generates a *zero-divisor code* and when it is a unit, it generates a *unit-derived code*.

There is no restriction on the ring R. It can be a field, but the techniques we describe are more general, and enable codes over other rings such as the integers $\mathbb{Z}$, rings of matrices or others.

In practice, the submodule W has dimension $r < n$. It can have the basis $\{g_1, g_2, \ldots, g_r\}$. Other submodules also turn out to be useful, e.g. as generated by $\{g_{k_1}, g_{k_2}, \ldots, g_{k_t}\}$ with $1 \le t < n$ where $\{k_1, k_2, \ldots, k_t\}$ is a subset of $\{1, 2, \ldots, n\}$.

For unit-derived codes, there is complete freedom in the choice of W (and hence r). Zero-divisor codes, as we show in Sec. 5.3.2, have restrictions placed on what W can be in order for a one-to-one map from the code back to W to exist. When RG is finite and has an identity, only zero-divisors and units are contained in RG. This is also true when R is a field by Theorem 5.2.

5.3.1. *Codes from Units*

Let u be a unit in RG, where G is of order n and listed $G = \{g_1, g_2, \ldots, g_n\}$. Let W be a submodule of RG generated (as an R-module) by r group elements $S = \{g_{k_1}, g_{k_2}, \ldots, g_{k_r}\}$ with $r < n$.

The unit-derived code is $\mathcal{C} = \{ux : x \in W\}$ or $\mathcal{C} = \{xu : x \in W\}$. The code is thus constructed from a unit u, a submodule W and, when RG does not commute, a choice over left or right encoding. Assume in what follows that the encoding is on the right ($x \mapsto xu$). The left-encoding case $x \mapsto ux$ is similar, following, with minor adjustments, the same general procedure.

Now c is a codeword (i.e. in $\mathcal{C}$) if and only if $cu^{-1} \in W$, namely, *if and only if the coefficients of $G \backslash S$ in cu^{-1} are zero*. Notice that multiplying a codeword by the inverse of the unit recovers the original input.

A unit-derived code can also be considered a mapping from R^r to R^n. First, map a vector $\underline{x} = (\alpha_1, \alpha_2, \ldots, \alpha_r) \in R^r$ by $\lambda_W(\underline{x}) = \sum_{i=1}^{r} \alpha_i g_{k_i}$ to an element $x \in W$. Then a codeword $xu \in \mathcal{C}$ is obtained which may be written $xu = \sum_{i=1}^{n} \beta_i g_i$. This gives an encoding $\underline{x} \mapsto (\beta_1, \beta_2, \ldots, \beta_n)$ which is a map from $R^r \to R^n$.

Each unit-derived code is also associated with an equivalent code, called the *matrix-generated code* $\mathcal{D}$. This is a code from R^r to R^n and has an $r \times n$

generator matrix A extracted from the RG-matrix U, and a check matrix extracted from V. If A is such a generating matrix, then $\mathcal{D} = \{\underline{x}A : \underline{x} \in R^r\}$. The distinction between codes $\mathcal{C}$ and $\mathcal{D}$ is of convenience. They are equivalent, exhibiting the same properties. This procedure is conceptual and any practical implementation works on producing $\mathcal{D}$ or $\mathcal{C}$ directly, depending on what is desired.

5.3.1.1. *Generator and Check Matrices*

The check and generator matrices that result from a unit-derived code arise as follows. Suppose $uu^{-1} = 1$ in the group ring and let U and U^{-1} respectively be the corresponding $n \times n$ RG-matrices.

First, consider W to be the submodule generated by $\{g_1, g_2, \ldots, g_r\}$ with $r < n$ (i.e. has as basis the first r elements in the chosen listing of G). Later the case when W has a general basis of group elements is dealt with. An element in W is thus of the form $x = \sum_{i=1}^{r} \alpha_i g_i$.

Divide $U = \left(\begin{smallmatrix} A \\ B \end{smallmatrix} \right)$ into block matrices where A is $r \times n$ and B is $(n-r) \times n$. Similarly, let $U^{-1} = (C \; D)$ where C is $n \times r$ and D is $n \times (n - r)$.

Now $AD = 0$ as $UU^{-1} = I$. It is easy to see that A is a generator matrix for the matrix-generated code. It can be shown that D^{T} is a check matrix.

Theorem 5.3. *Let $\underline{y} \in R^n$ and $\mathcal{D} = \{\underline{x}A : \underline{x} \in R^r\}$. Then $\underline{y} \in \mathcal{D}$ if and only if $\underline{y}D = 0$.*

Proof. If $\underline{y} = \underline{x}A$ for some $\underline{x} \in R^r$, then clearly $\underline{y}D = 0$. If, on the other hand, $\underline{y}D = 0$,

$$\underline{y} = \underline{y}U^{-1}U = \underline{y}\,(C \; D) \begin{pmatrix} A \\ B \end{pmatrix} = (\underline{y}C \; \underline{y}D)\,(A \; B) = (\underline{y}C \; 0) \begin{pmatrix} A \\ B \end{pmatrix} = \underline{y}CA.$$

Now $\underline{y}C$ is in R^r and $\underline{y} = \underline{y}U^{-1}U = \underline{x}A$ for some $\underline{x} \in R^r$ as required. $\square$

So, D^{T} is a check matrix for the matrix-generated code $\mathcal{D}$: $\underline{y}$ is a codeword if and only if $D^{\mathrm{T}}\underline{y}^{\mathrm{T}} = 0$ if and only if $\underline{y}D = 0$. The $r \times n$ generator matrix A and $(n - r) \times n$ check D^{T} produced from this unit and submodule have full allowable rank, r and $n - r$, respectively.

Units in group rings result in non-singular matrices, enabling the construction of codes from units. Any non-singular matrix could also produce a code by the above arguments, although of course one could not exploit any underlying algebraic structure of a group ring.

When W is generated by a general basis $S = \{g_{k_1}, g_{k_2}, \ldots, g_{k_r}\}$, the generator and check matrices are obtained by "extracting" from and "adding" to certain rows and columns from U and U^{-1}. A generator matrix results from the $r \times n$ matrix consisting of the $k_1, k_2, \ldots, k_r$ rows of U. Additionally, let D be the $(n - r) \times n$ matrix obtained by deleting the $k_1, k_2, \ldots, k_r$ columns of V. Then D^{T} is a check matrix.

5.3.1.2. *Constructing Unit-Derived Codes*

The generator and check matrices for the matrix-generated code $\mathcal{D}$ are immediate from the construction. However, working with the unit-derived code $\mathcal{C}$ itself can be advantageous. For example, using the group ring check conditions directly may be the best method for decoding.

In summary, unit-derived code of length n and dimension r can be constructed quite freely as follows. Choose a group G of order n and a ring R over which the code will be defined. Typically, R is a field but this is not a requirement; codes over the integers, rings of matrices or other rings are also useful.

Find a unit u in the group ring RG and its inverse u^{-1}. As previously mentioned, if R is a field or RG of finite order, every element in RG is either a zero-divisor or a unit. When R is a field, there is a straightforward algorithm to determine which. Generation of units is therefore not difficult.

Any basis for a submodule W consisting of r group elements will generate a code, e.g. the first r elements $\{g_1, g_2, \ldots, g_r\}$ according to a listing of G.

It can prove advantageous to choose another, appropriate, basis, to say increase the minimum distance of the code or optimize some other criteria. This freedom of basis nicely leads to the concept of an optimal one for a given unit $u \in RG$ and dimension r — a so-called best-basis:

$$\underset{S \subset G, |S| = r}{\arg\max} \; \min_{x \in W(S)} \; \mathrm{wt}(xu)$$

where $W(S)$ denotes the submodule generated by S and $\mathrm{wt}(y)$ the number of non-zero coefficients of y.

This flexibility in choice of r and W and the full ranks obtained are major advantages of a unit-derived code over one derived from a zero-divisor.

5.3.1.3. Derivation of Units

The fresh approach of deriving codes from units in the group ring RG offers promise as group rings are a rich source of units, and there is sizeable flexibility as outlined. Units exist and are known in RG, where R can be any ring and not just a field, and from these codes of different types can be constructed. Once a unit is known there is still a choice on the submodule/dimension for the code and codes of different dimensions may be obtained from a particular unit.

To fully describe a unit-derived code in terms of generator and check conditions a unit and its inverse are needed. The inverse may be known from the algebra; general formulae for certain units, and their inverses, in group rings are known. Please refer [4] and the references therein. In the cases of group rings over cyclic groups it is worth noting that the Euclidean algorithm, a fast algorithm, may be used to obtain an inverse as $RG \cong R[x]/\langle x^n - 1 \rangle$. Computer algebra packages such as GAP and MAGMA which are particularly useful for handling groups, may also be used to find units in group rings; for example, there exist functions such as DihedralGroup() and DirectProduct() which return dihedral groups and direct products of groups, respectively.

5.3.1.4. Dual and Orthogonal Codes

The dual of a code from a group ring encoding is $\mathcal{C}^\perp = \{y \in RG : \langle ux, y \rangle = 0, \forall x \in W\}$. Recall the concept of the transpose of a group ring element (Definition 5.3). We now show that the dual of a unit-derived code can be generated from $(u^{-1})^{\mathrm{T}}$.

Theorem 5.4. *Let W be a submodule with basis of group elements $S \subset G$ and $W^\perp$ be the submodule with basis $G \backslash S$. Let $u \in RG$ be a unit such that $uu^{-1} = 1$. Then the dual code of $\mathcal{C} = \{xu : x \in W\}$ is $\mathcal{C}^\perp = \{x(u^{-1})^{\mathrm{T}} : x \in W^\perp\}$.*

Proof. Let $z \neq 0$ be an element in RG. We need to show that $\langle xu, z \rangle = 0$, $\forall x \in W$ if and only if $zu^{\mathrm{T}} \in W^\perp$.

Note that $\langle xu, y(u^{-1})^{\mathrm{T}} \rangle = \langle x, y \rangle$. Thus, if $zu^{\mathrm{T}} \in W^\perp$, then, for all $x \in W$, $\langle xu, z \rangle = \langle x, zu^{\mathrm{T}} \rangle = 0$.

Conversely, if $zu^{\mathrm{T}} \in W$, pick a $g \in S$ that has a non-zero coefficient γ in zu^{T}. Then, $\langle gu, z \rangle = \langle g, zu^{\mathrm{T}} \rangle = \gamma \neq 0$. $\qquad\square$

Strict equivalence of a unit-derived code and its dual, whereby $\mathcal{C} = \mathcal{C}^\perp$, requires that for all $x \in W$, $xuu^{\mathrm{T}} \in W^\perp$, which imposes an impractical restriction. However, it is natural to say that a unit-derived code is *self-dual* if $\mathcal{C}$ and $\mathcal{C}^\perp$ are equivalent codes, or equivalently, that the resultant matrix-generated codes $\mathcal{D}$ and $\mathcal{D}^\perp$ are equal.

Definition 5.7. An unit $u \in RG$ is *orthogonal* if and only if its inverse is u^T (i.e. $uu^T = 1$).

It is easy to see that the RG-matrix from an orthogonal unit u is an orthogonal matrix.

5.3.2. *Codes from Zero-Divisors*

Let us now look at code construction from zero-divisors for a given group ring RG. Assume G is of order n with listing $\{g_1, g_2, \ldots, g_n\}$. The code will be of length n and its dimension will depend on the choice of the submodule W. The presentation will first deal with details from the group ring, and then incorporate the matrix algebra relationship.

Let u be a zero-divisor in RG, namely that $uv = 0$ for some non-zero $v \in RG$. Let W be a submodule of RG with basis of group elements $S \subseteq G$.

As defined, a resultant zero-divisor code is $\mathcal{C} = \{ux : x \in W\} = uW$ or $\mathcal{C} = \{xu : x \in W\} = Wu$. The code is thus constructed from a zero-divisor u, a submodule W and, for RG non-commutative, a choice over left or right encoding. We shall describe the case of right-encoding, that is $\mathcal{C} = Wu$; the left-encoding case is similar.

We say that u is a *generator element* of the code $\mathcal{C} = Wu$ relative to the submodule W. It is of course possible that $\mathcal{C}$ has another generator element and indeed may also be defined in terms of a different submodule W.

The case when $Wu = RGu$ is the particular traditional case where the code is a left ideal — see Sec. 5.3.3 for a more complete discussion on this. This is the case where rank U has the same rank or dimension as Wu.

When u is a zero-divisor then there is an element $v \neq 0$ with $uv = 0$ and thus $y \in \mathcal{C}$ satisfies $yv = 0$.

Definition 5.8. $v \in RG$ is said to be a (left) *check element* for a zero-divisor code $\mathcal{C}$ when $y \in \mathcal{C}$ if and only if $vy = 0$. We can then write $\mathcal{C} = \{y \in RG : vy = 0\}$.

We shall show that, given a zero-divisor u and the code $\mathcal{C}$, there is a set $v_1, v_2, \ldots, v_t$ of elements in RG such that $y \in \mathcal{C}$ if and only if $yv_i = 0$ for $1 \leq i \leq t$.

Zero-divisor codes with a single check element are particularly useful and exist in many cases.

We have defined codes in RG as generated by a zero-divisor and relative to a submodule W. We note that in addition to using a zero-divisor as a generator, codes can also be constructed by using a zero-divisor instead directly as a check element, regardless of whether it has a single generating element or not.

Definition 5.9. Suppose T is a submodule of RG. Define $T_v = \{x \in T \mid xv = 0\}$ and say T_v is the *check zero-divisor code* relative to T.

Note that T_v is a submodule of RG and in the case where $T = RG$ we have that T_v is actually a left ideal. It only makes sense to consider the case where v is a zero-divisor in which case $T_v \neq 0$. In some cases this code will have a single generator matrix but it will always be possible to describe a set of generator elements.

For the particular choice of module, we now restrict our attention to the case when R is a field. Some of the results hold over integral domains and also for rings in general.

Definition 5.10. A set of group ring elements $T \subset RG$ is *linearly independent* if, for $\alpha_x \in R$, $\sum_{x \in T} \alpha_x x = 0$ only when $\alpha_x = 0$ for all $x \in T$. Otherwise, T is *linearly dependent*.

We define rank(T) to be the maximum number of linearly independent elements of T. Thus rank$(T) = |T|$ if and only if T is linearly independent.

Note that a zero-divisor code $\mathcal{C} = Wu$, where W is generated by S, is the submodule of RG consisting of all elements of the form $\sum_{g \in S} \alpha_g gu$. The dimension of this submodule is thus rank(Su).

If Su is linearly dependent then there exists a subset $S'u$ of Su which is linearly independent and generates the same module as Su.[b] Then let W' to be the submodue of W generated by S' and then the code $\mathcal{C} = Wu = W'u$, and $S'u$ is linearly independent.

The maximum dimension a code for a given zero-divisor u is $r = $ rank(Gu).

The zero-divisor codes are thus (n, k) codes for where $k = $ rank(Su) and $k \leq r = $ rank(Gu). As pointed out, for a given u and W it is always

---

[b]It is here that we require R to be a field.

possible to find a submodule W' of W (which may be W itself) such that W' is generated by S' with $S'u$ linearly independent and $Wu = W'u$. One way of finding S' inside S is explained below using the matrix U of u and finding an appropriate basis for the matrix consisting of the relevant rows of U corresponding to the elements of Su.

A way of finding an (n, t) zero-divisor code is to consider U and find t linearly independent rows $i_1, i_2, \ldots, i_t$ of U. Then $S = g_{i_1}, g_{i_r 2}, \ldots, g_{i_t}$ is such that Su is linearly independent and generates an (n, t) code. The case $t = \text{rank}(U)$ is the code RGu and can be obtained from considering (any) $t = \text{rank}(U)$ linearly independent rows of U. The codes with $t < \text{rank}(Su)$ can be considered as "shortened" codes. Their generator and check matrices are easily obtained by methods described below.

Notice Su linearly independent is equivalent to saying that W contains no zero-divisors of u.

For G the cyclic group $C_n = \{1, g, g^2, \ldots, g^{n-1}\}$, let r be the first value such that $\{u, gu, g^2u, \ldots, g^tu\}$ is linearly dependent. Then it can easily be shown that r is the $\text{rank}(Gu)$, and any basis $S \subset \{1, g, g^2, \ldots, g^{r-1}\}$ can be chosen. Shortened cyclic codes are obtained by choosing a subset of this set S.

5.3.2.1. *Equivalent Codes in R^n*

The established relationship between the group ring and matrices in Sec. 5.2.8 enables us to express the code in terms of matrices, and to derive generator and check matrices for an equivalent code. More precisely, we shall define two equivalent codes in the module R^n, the second of which has generator/check matrices of the usual form.

Let U be the RG-matrix of u, and W be a submodule with basis $S = \{g_{i_1}, g_{i_2}, \ldots, g_{i_r}\}$ such that Su is linearly independent.

As previously stated, $C = Wu$ is a code defined on RG. A (n, r) code $\mathcal{E}$ can be defined from R^r to R^n as follows. Let $\underline{w} = (\alpha_1, \alpha_2, \ldots, \alpha_r) \in R^r$ be the vector to be encoded. Using the basis S, we write $\underline{w}$ as $\underline{x} \in R^n$ with α_j in position i_j for $1 \leq i \leq r$ and zero everywhere else.

$\underline{x}$ can then be mapped to an element in W by $x = \zeta(\underline{x}) = \sum_{j=1}^{r} \alpha_j g_{i_j}$ and a codeword $xu \in C$ equated with a codeword in $\mathcal{E}$ given by $\zeta^{-1}(xu) = \underline{x}U$.

In summary, we obtain the (n, r) code $\mathcal{E} = \{\underline{x}U : \underline{x} \in R^r\}$ in R^n equivalent to C. Considering codewords as $\underline{x}U \in \mathcal{E}$ where $\underline{x} \in R^n$ as described proves, as we will show, convenient for analyzis purposes.

For any $n \times n$ matrix A let $\underline{a}_1, \underline{a}_2, \ldots, \underline{a}_n$ denote the rows of A in order.

Generator for matrix-generated code. We can derive a generator matrix A for a code from R^r to R^n called the *matrix-generated code*, and given by $\mathcal{D} = \{\underline{x}A : x \in R^r\}$. It is equivalent to codes $\mathcal{C}$ and $\mathcal{E}$.

Note that codewords in $\mathcal{E}$ consist of linear combinations of the rows $\underline{u}_{i_1}, \underline{u}_{i_2}, \ldots, \underline{u}_{i_r}$ of U. Let A be the $r \times n$ matrix consisting of the $i_1, \ldots, i_r$ rows of U. We will show in Lemma 5.2 that the rows of A are linearly independent if and only if Su is.

Linear independence. We now tie up the relationship between linear (in)dependent rows of the RG-matrix U and the linear (in)dependence of the set Su. Specifically, it is established that the rows $\underline{u}_{i_1}, \underline{u}_{i_2}, \ldots, \underline{u}_{i_r}$ are linearly independent if and only if Su are, and that rank $U = \text{rank}(Gu)$.

Theorem 5.5. *Suppose U has rank t. Let $S \subset G$ be a set of group elements such that $|S| = t + 1$. Then Su is linearly dependent.*

Proof. Let the rows of U be $\underline{u}_1, \underline{u}_2, \ldots, \underline{u}_n$ in order. Suppose $Su = \{g_{j_1}u, g_{j_2}u, \ldots, g_{j_{t+1}}u\}$.

Any $t+1$ rows of U are dependent so there exists $\alpha_{j_1}, \alpha_{j_2}, \ldots, \alpha_{j_{t+1}}$ not all zero such that $\sum_{k=1}^{t+1} \alpha_{j_k} \underline{u}_{j_k} = 0_{1 \times n}$. Let A be the RG-matrix with first row having α_{j_k} in the j_kth position for $1 \leq k \leq t + 1$ and zeros elsewhere.

Then A is the RG matrix corresponding to the group ring element $a = \alpha_{j_1} g_{j_1} + \alpha_{j_2} g_{j_2} + \cdots + \alpha_{j_{t+1}} g_{i_{t+1}}$. Also AU is an RG-matrix whose first row consists of zeros and hence $AU = 0_{n \times n}$. Thus $au = 0$ and therefore $\{g_{i_1}u, g_{i_2}u, \ldots, g_{i_{r+1}}u\}$ is linearly dependent as required. $\square$

Then from the last theorem it follows that we may take S to have r elements such that $r \leq \text{rank}\, U$. For if $r > \text{rank}\, U$ then Su is generated by r elements $S'u$ (where $S' \subset S$), and the code is given by $\mathcal{C} = W'u$ where W' is the module generated by S'.

Alternatively, assuming U has rank $\geq r$, one can choose or find r linearly independent rows $\underline{u}_{i_1}, \underline{u}_{i_2}, \ldots, \underline{u}_{i_r}$ of U and then construct S by reference to these, let $S = \{g_{i_1}, g_{i_2}, \ldots, g_{i_r}\}$ and then Su is linearly independent.

Define G_j to be the RG-matrix corresponding to the group element $g_j \in G$ — this is consistent with the notation for the RG-matrix corresponding to g_j. Then G_j is the matrix whose first row has a 1 in the jth position and zeros elsewhere. It is then clear that $G_j U$ is the RG-matrix with first row $\underline{u}_j$.

Lemma 5.1. *Suppose $\underline{u}_1, \underline{u}_2, \ldots, \underline{u}_s$ are the first rows (or first columns) of the RG-matrices $U_1, U_2, \ldots, U_s$ respectively. Then $\alpha_1 \underline{u}_1 + \alpha_2 \underline{u}_2 + \cdots + \alpha_s \underline{u}_s = 0$ if and only if $\alpha_1 U_1 + \alpha_2 U_2 + \cdots + \alpha_s U_s = 0$.*

Proof. Suppose $\alpha_1 \underline{u}_1 + \alpha_2 \underline{u}_2 + \cdots + \alpha_s \underline{u}_s = 0$. Let $U = \alpha_1 U_1 + \alpha_2 U_2 + \cdots + \alpha_s U_s$. Then U is an RG-matrix whose first row consists of zeros and hence $U = 0$.

On the other hand it is clear that if $\alpha_1 U_1 + \alpha_2 U_2 + \cdots + \alpha_s U_s = 0$ then $\alpha_1 \underline{u}_1 + \alpha_2 \underline{u}_2 + \cdots + \alpha_s \underline{u}_s = 0$. □

Lemma 5.2. $\{g_{i_1} u, g_{i_2} u, \ldots, g_{i_r} u\}$ *is linearly independent if and only if* $\{\underline{u}_{i_1}, \underline{u}_{i_2}, \ldots, \underline{u}_{i_r}\}$ *is linearly independent.*

Proof. Follows immediately from Lemma 5.1. □

Suppose then $\operatorname{rank} U = r$ and that $\{\underline{u}_{i_1}, \underline{u}_{i_2}, \ldots, \underline{u}_{i_r}\}$ were linearly independent. Then by Lemma 5.2, Su is linearly independent and it is also clear that $\mathcal{C} = RGu$, the right ideal generated by u.

5.3.2.2. *Check Elements and Matrices*

For now, assume the code under question is (n, r) where $r = \operatorname{rank} U$. Afterwards, we generalize to obtain check conditions for (n, k) codes where $k < \operatorname{rank} U$.

Clearly $cv = 0$ for any codeword c. The most convenient situation is when the code $\mathcal{C}$ (and thus for codes $\mathcal{D}, \mathcal{E}$) has a (single) check element, i.e. that $y \in C$ is a codeword if and only if $yv = 0$. Equivalently, V checks $\mathcal{E}$ provided $\underline{y} \in \mathcal{E}$ if and only if $\underline{y} V = 0$ if and only if $YV = 0$, where $\underline{y}$ is the first row of Y. We now examine these requirements for a check element.

Definition 5.11. For a zero-divisor u with $\operatorname{rank} U = r$, say u is a *principal zero-divisor* if and only if there exists a $v \in RG$ such that $uv = 0$ and $\operatorname{rank} V = n - r$.

This is the situation for example when RG is a principal ideal domain as for example when G is a cyclic group — see Sec. 5.3.3 for the relationship between them and codes from zero-divisors in the case where the group is the cyclic group.

It is also possible in other cases that for a given zero-divisor u there is a v with $uv = 0$ and $\operatorname{rank} U + \operatorname{rank} V = n$; for example if $u^2 = 0$ or $uu^T = 0$ and $\operatorname{rank} U = \frac{n}{2}$, in which case also $\operatorname{rank} U^T = \frac{n}{2}$.

Suppose that $uv = 0$ and rank $V = n - r$. Then y is a codeword if and only if $yv = 0$ if and only if $YV = 0$. This is not immediately obvious and depends on the fact that U and V are RG-matrices; the proof, in stages, is shown.

Lemma 5.3. *Let y be the first row of an RG-matrix Y. Suppose also V is an RG-matrix. Then $YV = 0$ if and only if $\underline{y}V = 0$.*

Proof. Suppose $\underline{y}V = 0$. Then YV is an RG-matrix with first row consisting of zeros. Hence $YV = 0$. On the other hand if $YV = 0$ then clearly $\underline{y}V = 0$. $\square$

The proof for the following theorem is contained in [10].

Theorem 5.6. *Let $\mathcal{C} = \{xu : x \in W\}$ where W is generated by S such that Su is linearly independent and $|S| = \operatorname{rank} U = r$. Suppose further that $uv = 0$ in the group ring RG so that rank $V = n - r$. Then y is a codeword if and only if $yv = 0$.*

Corollary 5.1. *$\mathcal{C} = \{xu : x \in W\}$ has a single check element if and only if u is a principal zero-divisor.*

Corollary 5.2. *$y \in \mathcal{E}$ if and only if $\underline{y}V = 0$ if and only if $YV = 0$ where Y is the RG-matrix with first row $\underline{y}$.*

General check conditions. Define the null-space of U to be $\operatorname{Ker}(U) = \{\underline{x} : U\underline{x} = 0\}$ where $\underline{x}$ is an $n \times 1$ vector. Since U has rank r, the dimension of $\operatorname{Ker}(U)$ is $n - r$. Let $\underline{v}_1, \underline{v}_2, \ldots, \underline{v}_{n-r}$ be a basis for $\operatorname{Ker}(U)$; these $\underline{v}_i$ are $n \times 1$ column vectors. Let V_i be the RG-matrix with first column $\underline{v}_i$. Then clearly $UV_i = 0$ for $1 \le i \le n - r$ since UV_i is the RG-matrix with first row consisting of zeros and hence must be zero. Hence $uv_i = 0$ where v_i is the group ring element corresponding to the RG-matrix V_i.

Note that the null-space of U is easily and quickly obtained using linear operations on the rows of U. The basis for the null-space may be read off from the *row-reduced echelon* form of U, which also puts the generator in standard form. This is also very useful in producing a check matrix for the corresponding encoding $R^r \to R^n$ — see Theorem 5.8.[c]

Thus if y is a codeword then $yv_i = 0$ for $1 \le i \le n - r$. The following theorem may be proved along similar lines to Theorem 5.6. Its proof is omitted.

---

[c]The rows of the row-reduced echelon form generate the code.

Theorem 5.7. *Suppose u is a zero-divisor, rank $U = r$ and W is generated by S with r elements such that Su is linearly independent. Let v_i be defined as above. Then $y \in C$ if and only if $yv_i = 0$ for all $i = 1, \ldots, n - r$.*

Corollary 5.3. *Y is a codeword if and only if $YV_i = 0$.*

Not all the v_i are needed — just enough so that the corresponding matrices V_i contain a basis for the null-space. In many cases a particular V_i of rank $n - r$ can be found. The check conditions for the code $\mathcal{E}$ follow:

Theorem 5.8. *Let $\hat{V} = (v_1, v_2, \ldots, v_{n-r}, 0, 0, \ldots, 0)$ be the $n \times n$ matrix with first $n - r$ columns consisting of v_i in order and then r columns with zeros. Then $\underline{y} \in \mathcal{E}$ if and only if $\underline{y}\hat{V} = 0$.*

Proof. Clearly if $\underline{y} \in \mathcal{E}$ then $\underline{y}\hat{V} = 0$. Suppose then $\underline{y}\hat{V} = 0$. Define $\mathrm{Ker}(\hat{V}) = \{t \in R^n : t\hat{V} = 0\}$. Since $\hat{V}$ has dimension $n - r$, the dimension of $\mathrm{Ker}(\hat{V})$ is r. Now each row of U is in $\mathrm{Ker}(\hat{V})$ since $U\hat{V} = 0$. Since U also has rank r this implies the rows of U generate $\mathrm{Ker}(\hat{V})$. Hence $\underline{y}$ is a linear combination of the rows of U. The rows $u_{i_1}, u_{i_2}, \ldots, u_{i_r}$ are linearly independent and hence are a basis for the row space of U which has dimension r. Hence $\underline{y} = \alpha_1 u_{i_1} + \alpha_2 u_{i_2} + \cdots + \alpha_r u_{i_r}$ and is thus a codeword. $\qquad\square$

Generator from a check element. The argument above may also be used to obtain a generator when we use a zero-divisor $v \in RG$ to act as a check element and produce the code $T_v = \{y \in T : yv = 0\}$, regardless of whether the code has a single generating element or not.

Take the case $T = RG$. Suppose the resultant RG-matrix V has rank $n - r$, then $n - r$ of the rows of V are linearly independent and the other rows of V are linear combinations of these. Thus the code may be considered a (n, r) code with check matrix of size $(n - r) \times n$.

Define $\mathrm{Ker}(V) = \{\underline{x} : \underline{x}V = 0\}$. Then $\mathrm{Ker}(V)$ has rank r. Let $\underline{u}_1, \underline{u}_2, \ldots, \underline{u}_r$ be a basis for $\mathrm{Ker}(V)$. Form the matrix $\hat{U}$ with rows $\underline{u}_i$. Then we get the following:

Theorem 5.9. *$\hat{U}$ is a generator matrix of the code $\mathcal{E}$.*

Proof. The proof is similar to that of Theorem 5.8. $\qquad\square$

Check matrices when the dimension is less than the rank. We now take the case where W is the submodule generated by S such that $|S| = s < r = \mathrm{rank}\, U$. This generates an (n, s) code.

One way to create a check matrix for $\mathcal{D}$ would be to apply standard row operations to obtain a basis for the null-space of the generator A. However, it can also be obtained from the RG-matrices U and V by adding certain $r - s$ vectors to V as explained below.

Let V_{n-r} denote a submatrix of V consisting of $n - r$ linearly independent columns.

Consider the indexing set $T = \{k_1, k_2, \ldots, k_s\}$ ($1 \le k_1 < k_2 < \cdots < k_s \le n$) which defines $S = \{g_{k_1}, \ldots, g_{k_s}\}$. Extend the set T to a set of linearly independent rows $R = \{k_1, k_2, \ldots, k_s, w_1, \ldots, w_{r-s}\}$ of U.

Let U_r be the matrix formed from R with the rows in order. Then U_r has rank r and size $r \times n$. There exists an $n \times r$ matrix C such that $U_r C = I_r$.

Delete the $k_1, k_2, \ldots, k_s$ columns of C to get an $n \times (r-s)$ matrix, which we call C_{r-s}. We now add this C_{n-r} matrix to V_{n-r} to get the matrix D. This D then has rank $n - s$ and satisfies $U_r D = 0$. It follows that $\underline{y} \in \mathcal{D}$ if and only if $D^{\mathrm{T}} \underline{y}^{\mathrm{T}} = 0$.

Thus D^{T} is a check matrix for $\mathcal{D}$, obtained by adding certain $r - s$ columns from C, the right inverse of U_r, to the matrix V_{n-r}.

5.3.2.3. *Dual Codes*

By definition, a dual of a code $\mathcal{C}$ considered as vectors over R^n is its orthogonal complement, namely $\mathcal{C}^{\perp} = \{v \in R^n : \langle v, c \rangle = 0, \forall c \in \mathcal{C}\}$.

Let $x, y \in RG$. The inner (or dot) product, is given by term-by-term multiplication of the elements in the ring R, namely $\langle x, y \rangle = \sum_{g \in G} \alpha_g \beta_g$ where $x = \sum_{g \in G} \alpha_g$ and $y = \sum_{g \in G} \beta_g$.

Thus, the dual of a code from a group ring encoding is $\mathcal{C}^{\perp} = \{y \in RG : \langle ux, y \rangle = 0, \forall x \in W\}$. It can easily be shown that the dual of a zero-divisor code has a easy form, summarized in the following theorem (proved in [10]).

Theorem 5.10. *Let $u, v \in RG$ such that $uv = 0$. Let U and V be the RG-matrices of u and v, respectively, such that $\operatorname{rank} U = r$ and $\operatorname{rank} V = n - r$. Let W be a submodule over a basis $S \subset G$ of dimension r such that Su is linearly independent and $W^{\perp}$ denote the submodule over basis $G \backslash S$. Then the code $\mathcal{C} = \{xu : x \in W\}$ has dual code $\mathcal{C}^{\perp} = \{xv^T : x \in W^{\perp}\} = \{y \in RG : yu^T = 0\}$.*

This is consistent with cyclic codes whereby the dual for a code with generator u and check v is usually expressed as having the reciprocal (polynomial) $g^{n-r} v^{\mathrm{T}}$ (where the $\deg(v) = n - r$) as generator [11]. Using this as generator with W having basis $\{1, g, \ldots, g^r\}$ yields the same code as generator v^{T} with submodule $W^{\perp}$.

Of course, for the matrix-generated code $\mathcal{D}$, one may obtain a code equivalent to its dual by interchanging the generator and check matrices; see for example [12].

5.3.3. *Relationship with Cyclic Codes*

The cyclic codes are now shown to be exactly zero-divisor codes in group rings on cyclic groups for special cases of the module W. Also noteworthy is that Reed–Muller codes are extended cyclic codes and have been shown to be associated with the group ring of the elementary abelian 2-group [13].

For a given polynomial $h(g)$ over a ring R, let $r(g)$, be the polynomial of minimal degree such that $h(g)r(g) \equiv 0 \mod (g^n - 1)$ (whenever it exists). Then, a cyclic code over R is generated by $h(g)$ with corresponding check polynomial $r(g)$.

The group ring RC_n of the cyclic group C_n over a ring R is isomorphic to $R[g]/\langle g^n - 1 \rangle$. This is a well-known result (e.g. [11]). Cyclic codes of degree n are given by zero-divisors in RC_n and the check matrices the counterpart of the zero-divisor.

Theorem 5.11. *Let $h(g) = \alpha_0 1 + \alpha_1 g + \cdots + \alpha_r g^r$. Then $h(g)$ is the generator polynomial of a cyclic code of length n if and only if the RG-matrix $M(RC_n, h(g)) = \sigma(h(g))$ is a zero-divisor in $R_{n \times n}$.*

The check matrix of the code is given by the polynomial or group ring element $r(g)$ such that $h(g)r(g) \equiv 0 \pmod{g^n - 1}$.

Proof. If $h(g)$ is a zero-divisor in $R(G)$ then $h(g)r(g) = l(g)(g^n - 1)$ as polynomials.

The code generated by $h(g)$ is the same as the one generated by $d(g) = \gcd(h(g), g^n - 1)$ and then $d(g)p(g) = g^n - 1$.

$M(RG, d(g))$ gives the generator matrix of the code and $M(RG, p(g))$ the check matrix. $\qquad \square$

Cyclic codes in standard notation are presented using non-square matrices. However, the matrices can be made square by considering them as group ring elements.

In cyclic codes, the matrix representation of u is used and the mapping is given as $x \mapsto xu$ where $x \in R^r$ and $u \in R_{r \times n}$. Consider x as an element in R^n (by adding 0's) and the element u as an element in $R_{n \times n}$ which is the group ring completion of the matrix u. Then the mapping is $x \mapsto xu$ in matrix form and is equivalent to the previous mapping.

See also Sec. 5.3.4 below on ideals in the cyclic group ring.

5.3.4. *When are They Ideals?*

Here we discuss conditions under which codes as defined in Sec. 5.3 are ideals. It transpires that unit-derived codes as defined in Sec. 5.3.1 are never ideals and that zero-divisor codes as defined in Sec. 5.3.2 are ideals only in very special cases.

Definition 5.12. *I is said to be an* ideal *in the ring H if I is a subring such that (i) $hi \in I$, $\forall h \in H, \forall i \in I$ and (ii) $ih \in I$, $\forall h \in H, i \in I$. Such an ideal is often referred to as a* two-sided *ideal. I is said to be a* left ideal *in H if I is a subring such that $hi \in I$, $\forall i \in I, \forall h \in H$. I is said to be a* right ideal *in H if I is a subring of H such that $ih \in I$, $\forall i \in I, \forall h \in H$.*

In the case of a commutative ring there is no distinction between left, right, and two-sided ideals.

The cyclic codes are ideals in the group ring on the cyclic group but for example the quasi-cyclic and shortened cyclic codes are not ideals. Our intention here is to clarify the situation with respect to codes in group rings as defined in Sec. 5.3.

Recall that such a code is either Wu or uW where W is a submodule of RG and $u \in RG$. We consider the right encoding Wu; the other one is similar. Assume that $G = \{g_1, g_2, \ldots, g_n\}$ and here consider cases where the ring R is a field.

Assume also that W is generated by $S = \{g_{i_1}, g_{i_2}, \ldots, g_{i_r}\}$. The code is then $\mathcal{C} = \{xu : x \in W\}$ and is generated by $Su = \{g_{i_1}u, g_{i_2}u, \ldots, g_{i_r}u\}$.

We may assume as explained in Sec. 5.3.2 that $Su = \{g_{i_1}u, g_{i_2}u, \ldots, g_{i_r}u\}$ is linearly independent.

The rows of an $n \times n$ matrix U are designated in order by $\underline{u}_1, \underline{u}_2, \ldots, \underline{u}_n$.

Define G_j to be the RG-matrix corresponding to the group element $g_j \in G$. Thus G_j is the matrix whose first row has a 1 in the jth position and zeros elsewhere and this first row determines the RG-matrix G_j. It then follows G_jU is the RG-matrix with first row $\underline{u}_j$.

Lemma 5.4. *Suppose $g_{i_1}u, g_{i_2}u, \ldots, g_{i_r}u$ is linearly independent. Then* rank $U \geq r$.

Proof. Since $g_{i_1}u, g_{i_2}u, \ldots, g_{i_r}u$ is linearly independent so is $G_{i_1}U, G_{i_2}U, \ldots, G_{i_r}U$. Thus $\underline{u}_{i_1}, \underline{u}_{i_2}, \ldots, \underline{u}_{i_r}$ is linearly independent and hence U contains r linearly independent rows. □

Theorem 5.12. *Let C be the code Wu with W generated by $S = \{g_{i_1}, g_{i_2}, \ldots, g_{i_r}\}$ such that $Su = \{g_{i_1}u, g_{i_2}u, \ldots, g_{i_r}u\}$ is linearly independent. Then C is a left ideal if and only if* $\operatorname{rank} U = r$.

Please refer to [10] for the proof. Note that when C is a left ideal, then $C = Wu = RGu$.

If u is a unit then $\operatorname{rank} U = n$. Hence C is not an ideal in a unit-derived code — the module W is never equal to all of RG in our definition of unit-derived code.

Let I be a left ideal in a ring R. Similar remarks apply to right and two-sided ideals. If I contains a unit then clearly I contains the identity and hence $I = R$. Now I is said to be a *proper* ideal if $I \neq 0$ and $I \neq R$. Here again we can see that a unit-derived code is never an ideal as this has the form Wu where u is a unit and thus Wu has a unit.

Suppose I is a proper left ideal of a group ring RG where R is a field. Then I is a subspace of RG and so is generated as an R-module by $u_1, u_2, \ldots, u_s$ with $u_i \in RG$. Since I is a proper ideal none of the u_i can be units and so by Theorem 5.2 each u_i is a zero-divisor. Also no linear combination of the u_i can be a unit as I does not contain any unit. Thus:

Theorem 5.13. *I is a proper ideal in a group ring RG, with R a field, if and only if I is generated as a module by $u_1, u_2, \ldots, u_s$ where each u_i is a zero-divisor and no linear combination of the u_i is a unit.*

A left ideal I is *principal* in RG if and only if it has the form RGu for an element $u \in RG$. Thus proper principal ideals in the group ring over a field are of the form RGu where u is a zero-divisor. These are the particular zero-divisor codes where the rank of U equals the number of elements in the generating set S of W where Su is linearly independent.

Ideals in cyclic group rings. Because of its particular nature as a polynomial type ring, it is easy to show that all ideals in the cyclic group ring are principal.

Lemma 5.5. *Every ideal in the group ring of the cyclic group is principal.*

Proof. Let I be an ideal in the group ring of the cyclic group RG where G is generated by g. Choose $f(g)$ in I of minimal degree in g. Let $x(g) \in I$.

Then by division algorithm to polynomials, $x(g) = q(g)f(g) + r(g)$ where $r(g) = 0$ or $r(g)$ has degree less than $f(g)$. Now $r(g) \in I$ since $x(g) \in I$ and $f(g) \in I$. Since $f(g)$ is of minimal degree in I this implies $r(g) = 0$ and $x(g) = q(g)f(g)$. Hence I is a principal ideal generated by $f(g)$. $\qquad\square$

Suppose now we have a zero-divisor u in the cyclic group ring RG and that $\operatorname{rank} U = r$. In order for u to be a principal zero-divisor we require an element $v \in RG$ such that $uv = 0$ and $\operatorname{rank} V = n - r$. Such a v always exists in the cyclic group ring. Choose v to be an element of least degree such that $uv = 0$. Here in fact what we are doing is choosing a generator for the annihilator of u which is a principal ideal.

Theorem 5.14. *Suppose u is a zero-divisor in the cyclic group ring RG with G of order n. Let v be an element of least degree such that $uv = 0$. If $\operatorname{rank} U = r$ then $\operatorname{rank} V = n - r$.*

Proof. If $uv_1 = 0$ then by division algorithm, $v_1 = qv + r$ where $r = 0$ or $\deg(r) < \deg(v)$. Then multiplying through by u, and noting that elements commute, we see that $ur = 0$. Since v is of least degree such that $uv = 0$ this implies that $r = 0$ and hence $v_1 = qv$.

We now show that the null-space of U is generated by the rows of V. The null-space of U is the set of all vectors $\underline{x}$ such that $U\underline{x} = 0$.

Suppose now $\underline{x}$ is a vector in the null-space of U, so that $U\underline{x} = 0$. Let T be the RG-matrix with first row $\underline{x}$. Now $\underline{x}$ determines T. Then UT is an RG-matrix with first row 0 and hence $UT = 0$. Therefore $ut = 0$ in the group ring. Hence $t = qv$ for some $q \in RG$. Therefore $T = QV$. Hence the rows of T are linear combinations of the rows of V. In particular $\underline{x}$ is a linear combination of the rows of V. Thus the rows of V generate the null-space of U. Since by linear algebra the null-space of U has dimension $n - r$ it follows that V has rank $n - r$. $\qquad\square$

Suppose u is a zero-divisor in the group ring RG, with G generated by g, and that as *polynomials* $uv = g^n - 1$,[d] then v as an element of RG has the property that it is of minimal degree such that $uv = 0$. It follows that $\operatorname{rank} U + \operatorname{rank} V = n$.

Corollary 5.4. *If $uv = g^n - 1$ then $\operatorname{rank} U + \operatorname{rank} V = n$.*

---

[d]Cyclic codes are defined using such elements.

5.4. Dihedral Codes

Previous cyclic codes, as already noted are zero-divisor codes in the cyclic group ring. The first natural series of non-abelian groups are the dihedral groups.

There is some potential from codes that arise from the dihedral group. Investigation into codes from ideals in the dihedral group algebra started as early as 1969 [14], and a more recent result [15] showed that there exist a random ideal in $Z_2 D_{2n}$ for infinitely many n, such that the resultant code of rate $1/2$ is "good".

The dihedral group D_{2n} of order $2n$ is given $D_{2n} = \langle a, b : a^2, b^n, ab = b^{-1}a \rangle$. Further information on the dihedral group can be obtained in a standard book on abstract algebra, such as [16].

There are a number of possible listings of the elements of D_{2n}, of which $D_{2n} = \{1, b, b^2, \ldots, b^{n-1}, a, ab, ab^2, \ldots, ab^{n-1}\}$ proves convenient. An element $u \in RD_{2n}$ can be written

$$u = \sum_{i=0}^{n-1} \alpha_i b^i + \sum_{i=0}^{n-1} \beta_i ab^i,$$

where $\alpha_i, \beta_i \in R$. The associated RD_{2n}-matrix U is then,

$$U = \left(\begin{array}{ccccc|ccccc} \alpha_0 & \alpha_1 & \alpha_2 & \cdots & \alpha_{n-1} & \beta_0 & \beta_1 & \beta_2 & \cdots & \beta_{n-1} \\ \alpha_{n-1} & \alpha_0 & \alpha_1 & \cdots & \alpha_{n-2} & \beta_1 & \beta_2 & \beta_3 & \cdots & \beta_0 \\ \vdots & \vdots & \vdots & \vdots & \vdots & \vdots & \vdots & \vdots & \vdots & \vdots \\ \alpha_1 & \alpha_2 & \alpha_3 & \cdots & \alpha_0 & \beta_{n-1} & \beta_0 & \beta_1 & \cdots & \beta_{n-2} \\ \hline \beta_0 & \beta_1 & \beta_2 & \cdots & \beta_{n-1} & \alpha_0 & \alpha_1 & \alpha_2 & \cdots & \alpha_{n-1} \\ \beta_1 & \beta_2 & \beta_3 & \cdots & \beta_0 & \alpha_{n-1} & \alpha_0 & \alpha_1 & \cdots & \alpha_{n-2} \\ \vdots & \vdots & \vdots & \vdots & \vdots & \vdots & \vdots & \vdots & \vdots & \vdots \\ \beta_{n-1} & \beta_0 & \beta_1 & \cdots & \beta_{n-2} & \alpha_1 & \alpha_2 & \alpha_3 & \cdots & \alpha_0 \end{array} \right).$$

This can be written $U = \begin{pmatrix} A & B \\ B & A \end{pmatrix}$, where A is circulant and B is a reverse circulant as each row is a circulant shift to the left of the one previous. Interestingly, in a non-group ring context, reverse circulants have appeared before in codes [17].

From this, units and zero-divisors in such a group ring may be classified.

Any multiplication xy in RD_{2n} or, equivalently, using the corresponding RG-matrices can be done with low-complexity using existing techniques on circulant matrices (such as fast Fourier transform methods). A is already circulant, and the operation xB can be done by calculating xB' where B' is the "flip" of B.

For a zero-divisor $u \in RD_{2n}$ where R is a field, finding a basis S for the submodule W can be done by a simple algorithm. Pick, in order, elements from the set $\{1, b, b^2, \ldots, b^{n-1}\}$ until the first k such that $\{u, bu, b^2u, \ldots, b^ku\}$ is linearly dependent, or else they are all linearly independent. Then add elements, in order, from $\{a, ab, ab^2, \ldots, ab^{n-1}\}$ until the combined set $\{u, bu, \ldots, b^{k-1}u, au, abu, \ldots, ab^l\}$ is linearly dependent. This can be shown to give maximum rank. An equivalent process can be performed using the RG-matrix U directly.

5.4.1. *Some Dihedral Zero-Divisor Codes*

Consider the following example of codes over D_{2n} built up from elements in C_n. Let $u \in RC_n$ be a zero-divisor such that $uv = 0$ where C_n has generating element b. From it a zero-divisor of the form $u + axu \in D_{2n}$ can be constructed for any $x \in C_n$ with $(u + axu)(v + ayv^{\mathrm{T}}) = 0$ for any $y \in C_n$. For simplicity, consider the case $x = y = 1$. Additionally, any element $u \in \mathbb{Z}_2C_n$ (including units) will produce a zero-divisor in $\mathbb{Z}_2D_{2n}$ of the form $u + au$.

This simple construction can produce surprisingly decent codes. Consider $u = 1 + b^2 + b^5 \in \mathbb{Z}_2C_7$, which produces the Hamming $(7, 4, 3)$ code. The zero-divisor $u + au$ has a corresponding RG-matrix with rank 7 and produces a $(14, 7, 4)$ code which is the best possible for this length and dimension. Similarly $u = 1 + b^2 + b^3 + b^9 + b^{10} + b^{11}$ produces a $(24, 11, 8)$ code — also the best distance possible for $(24, 11)$ binary code. For $u \in \mathbb{Z}_2C_{31}$ given by $1 + g + g^6 + g^9 + g^{10} + g^{14} + g^{15} + g^{16} + g^{17} + g^{19} + g^{20} + g^{21} + g^{22} + g^{23} + g^{25} + g^{27}$ yields a $(62, 30, 12)$ code over $\mathbb{Z}_2D_{62}$ which has the same distance as the best-known $(62, 30)$ code (although codes with better distance may exist).

5.5. Self-Dual Codes

Self-dual codes can be formed in a group ring RG as follows. Suppose $|G| = m = 2q$ and that $u \in RG$ has the following properties:

(1) $u^2 = 0$.
(2) $u = u^{\mathrm{T}}$.
(3) $\operatorname{rank} U = \operatorname{rank} u = q$.

Then u generates a self-dual code as follows. Consider $G = \{g_1, g_2, \ldots, g_m\}$. Let $S = \{g_{i_1}, g_{i_2}, \ldots, g_{i_q}\}$ be chosen so that Su is linearly independent. As pointed out in Sec. 5.3.2.3, such a set always exists since rank $u = q$. In most cases for a natural ordering the set $S = \{g_1, g_2, \ldots, g_q\}$, the first q elements of G, is such that Su is linearly independent and in any case by reordering the elements of G it may be assumed that S consists of the first q elements.

The self-dual code is then $\mathcal{C} = Wu$ where W is the R-module generated by S. A matrix version is obtained by applying the isomorphism $\phi : RG \to R_{n \times n}$.

Now, suppose that $\mathcal{C}$ is a code obtained with RG-matrices K, L satisfying $KL = 0$ with rank $K + $ rank $L = n = |G|$. The code is generated by K and "checked" by L. This is the case for example with all cyclic codes which are zero-divisor group ring codes of the cyclic group ring. More precisely the matrix code is $\mathcal{C} = \alpha K$ where α has length equal to $r = $ rank K, and the first r rows of K are linearly independent. Then $y \in C$ if and only if $yK = 0$ if and only if $L^T y^T = 0$. Thus L^T is the check matrix in the usual notation.

The code is self-dual if and only if $K = L^T$. If $k, l \in RG$ are the elements corresponding to K, L respectively, it can be seen that the conditions for a self-dual code translates in the group ring setting to finding an element $k \in RG$ such that $kk^T = 0$ and rank $k = $ rank $k^T = 1/2|G|$. Further if $k = k^T$ (i.e. k and K are symmetric) then the code is obtained from a group ring element k with $k^2 = 0$ and rank $k = $ rank $K = 1/2|G|$.

Thus in many situations, including (symmetric) cyclic codes, self-dual codes are obtained. It is more difficult to obtain elements k such that $kk^T = 0$ with rank $k = $ rank $k^T = 1/2|G|$ but all self-dual are obtainable this way when derived from group rings; in particular cyclic self-dual codes and many other group ring self-dual codes come about this way.

Isodual codes. If the symmetric condition $u^T = u$ is omitted then *isodual codes* are obtained. An isodual code is a code equivalent to its dual. In this case, $u^2 = 0$, rank $u = q = $ rank u^T. The check matrix is U^T as opposed to U in the self-dual case. However the group ring code determined by u is equivalent to the group ring code determined by u^T; note from [5] that u^T is the element obtained by interchanging the coefficients of g and g^{-1} for every $g \in G$ in the expression for u and that if U is the matrix of u then U^T is the matrix of u^T in the isomorphism between the group ring and the ring of matrices.

There is more freedom in the choice of u if it is not required that u be symmetric. Thus isodual codes may be obtained in some cases with higher distance than strictly self-dual codes.

In other cases also it is possible to obtain codes which contain codes equivalent to its dual. These should possibly be called *isodual-containing codes*.

5.6. Exercises and Suggested Research

5.6.1. *Exercises*

(1) List all 16 elements in $\mathbb{Z}_2 C_4$ and identify them as either zero-divisors or units. Determine the rank of each zero-divisor. Construct the full-rank zero-divisor codes for elements $1 + g$ and $1 + g^2$ and determine each code's distance.

Construct a $(4, 2)$ and a $(4, 3)$ code from the unit $1 + g + g^2$. Determine the distance of each code.

(2) Show that every element $u \in \mathbb{Z}_2 C_n$ with even support (i.e. $\text{wt}(u)$ is even) is a zero-divisor (hint: show $1 + g$ divides u).

(3) Show that $1 + g + g^3$ generates the Hamming $(7, 4, 3)$ code in $\mathbb{Z}_2 C_7$, using the module generated by $\{1, g, g^2, g^3\}$. Does it generate the Hamming code using the module generated by $\{1, g, g^2, g^4\}$?

Show that $1 + g + g^3$ is a unit in $\mathbb{Z}_2 C_5$ and that it generates a $(5, 3, 2)$ code when using a module generated by $\{1, g, g^3\}$ but only a $(5, 3, 1)$ code when using a module generated by $\{1, g, g^2\}$.

(4) Show that, for every $\mathbb{Z}_2 C_n$, $\text{rank}(1 + g) = n - 1$ where g generates C_n. What ubiquitous code is the code $\{(1 + g)x : x \in \mathbb{Z}_2 C_n\}$?

(5) Identify each of 16 elements in $\mathbb{Z}_2(C_2 \times C_2)$ as either zero-divisors or units. Determine the rank of each zero-divisor.

Consider each C_2 as generated by a_1 and a_2, respectively. Construct the RG-matrix for $u = 1 + a_1 a_2$. Determine the distance of the $(4, 2)$ zero-divisor code that results from using the u with the module generated by $\{1, a_1\}$.

Determine the maximum distance of the possible $(4, 2)$ unit-derived codes generated by $1 + a_2 + a_1 a_2$ (i.e. find the best-distance over all possible bases for the submodule).

(6) Consider the group ring $\mathbb{Z}_2(C_2 \times C_2 \times C_2)$ with a_1, a_2, a_3 as generators for each C_2, respectively. Show that $1 + a_1 + a_1 a_2 a_3$ is a self-check unit (i.e. of form $u^2 = 1$) and determine the distance of the best $(8, 4)$ code that results from it.

Prove that $1 + a_1 + a_2 + a_1a_2a_3$ is a zero-divisor that generates an $(8, 4, 4)$ self-dual code.

(7) Consider the dihedral group ring $\mathbb{Z}_2 D_8$ where $D_8 = \langle a, b \,|\, a^2 = 1, b^4 = 1, ab = b^{-1}a \rangle$. Show that $1 + b + b^2 + ab^2 + ab^3$ is a unit, and from it construct a $(8, 3, 4)$ code (best distance possible for a linear code).

(8) In $\mathbb{Z}_2 C_{2n}$ with C_{2n} generated by g, show that $u = 1 + g^n$ satisfies $u^2 = uu^{\mathrm{T}} = 0$ and that it has rank n. What is the distance of this self-dual code?

Show that $u = g^i + g^{n-i} + g^{n+i} + g^{2n-i}$ generates a $(2n, n)$ self-dual code.

(9) In $\mathbb{Z}_2 D_{2n}$, let $\hat{b} = b + b^2 + \cdots + b^{n-1}$. Show that $u = 1 + a\hat{b}$ defines a zero-divisor self-dual code.

Show that for $n = 3$, a $(6, 3, 3)$ code results and that for $n \geq 4$ the distance is always 4.

(10) Show that $u = 1 + g^2 + g^3 + g^7 + g^8$ satisfies $u^2 = uu^{\mathrm{T}} = 1$ in $\mathbb{Z}_2 C_{10}$.

Determine the distance of the $(10, 5)$ code using u with the module generated by $\{1, g, g^2, g^3, g^4, g^5\}$.

Let $u_i = g^i + g^{n-i} + g^{n+i} + g^{2n-i}$ in $\mathbb{Z}_2 C_{2n}$. Show that $a = 1 + u$ is an orthogonal unit.

In $\mathbb{Z}_2 C_{14}$, show that $u = 1 + g^2 + g^5 + g^9 + g^{12}$ generates a $(14, 7, 4)$ self-dual code (the best possible distance for a $(14, 7)$ code).

(11) Show that $w = a + a^2 + a^3$ in $\mathbb{Z}_2 C_4$, where C_4 is generated by a, is symmetric and satisfies $w^2 = 1$. Prove that the code defined using the first two rows of the matrix of w is a $(4,2,2)$ code. Let $u = 1 + hw$ in $\mathbb{Z}_2(C_4 \times C_2)$ where C_2 is generated by h. Prove that $u^2 = 0$, that u is symmetric and has rank 4. Define the resulting $(8,4)$ self-dual code and show it has distance 4.

(12) Let $w = (a + a^2 + a^3)(b + b^2 + b^3)$ in $\mathbb{Z}_2(C_4 \times C_4)$ with the C_4 generated by a and b, respectively. Show w is symmetric and satisfies $w^2 = 1$. Construct $(16,8),(16,12)$ codes from w and determine their distances. Let $u = 1 + hw$ in $\mathbb{Z}_2(C_4 \times C_4 \times C_2)$ where C_2 is generated by h. Show $u^2 = 0$, rank $u = 16$ and that u generates a $(32,16)$ self-dual code whose distance is 6.

(13) Let $w = a + a^2 + a^3 + a^4 + a^5$ in $\mathbb{Z}_2 C_6$ with C_6 generated by a. Show that w is symmetric and that $w^2 = 1$. Construct $(6,3),(6,4)$ codes from w and determine their distances. Let $u = 1 + hw$ in $\mathbb{Z}_2(C_6 \times C_2)$ with C_2 generated by h. Show w is symmetric and satisfies $w^2 = 1$. Construct a $(12,6)$ self-dual code from w and show that its distance is 4.

(14) Let $w = (a + a^2 + a^3 + a^4 + a^5)(b + b^2 + b^3 + b^4 + b^5)$ in $\mathbb{Z}_2(C_6 \times C_6)$ with the C_6 generated by a and b, respectively. Show w is symmetric and satisfies $w^2 = 1$. Construct (36,18) and (36,24) codes from w and determine their distances. Let $u = 1 + hw$ in $\mathbb{Z}_2(C_6 \times C_6 \times C_2)$ where C_2 is generated by h. Show $u^2 = 0$, $\operatorname{rank} u = 16$ and that u generates a $(36, 18)$ self-dual code whose distance is 8.

(15) Let $w = a + a^8 + a^{15}$ in the group ring $\mathbb{Z}_2 C_{16}$ with C_{16} generated by a. Show $u^8 = 1$ and $u^i \neq 1$ for $1 \leq i < 8$. Set $u = 1 + hu$ in the group ring $\mathbb{Z}_2(C_{16} \times C_2)$ where C_2 is generated by h. Show $u^8 = 0$. Find the matrix of u and show it has rank 28. From this derive a (32,28) code of rate 7/8 and show that its distance is 2. What is the check matrix of this code? Show that the dual of the code is contained in the code.

(16) Let $w = a + a^{16} + a^{31}$ in the group ring $\mathbb{Z}_2 C_{32}$ with C_{32} generated by a. Show $u^{16} = 1$ and $u^i \neq 1$ for $1 \leq i < 16$. Set $u = 1 + hu$ in the group ring $\mathbb{Z}_2(C_{32} \times C_2)$ where C_2 is generated by h. Show $u^{16} = 0$. Find the matrix of u and show it has rank 60. From this derive a (64,60) code of rate 15/16 and show that is distance is 2. What is the check matrix of this code? Show that the dual of the code is contained in the code.

(17) Consider $w = a + a^3 + a^{16} + a^{29} + a^{31}$ in $\mathbb{Z}_2 C_{32}$ with C_{32} generated by a. Show $w^8 = 1$ and that $w^i \neq 1$ for $1 \leq i < 8$. Let $u = 1 + hw$ in the group ring $\mathbb{Z}_2(C_{32} \times C_2)$ with C_2 generated by h. Show u is symmetric and that $u^8 = 0$. Find the rank r of u and construct a $(32, r)$ code from u and determine its distance.

(18) Suppose w is symmetric and satisfies $w^2 = 1$ in FG, where F is a field of characteristic 2. Consider $u = 1 + hw$, in $F(G \times C_2)$, where C_2 is generated by h. Show that $u^2 = 1$, $\operatorname{rank} u = |G|$ and that u generates a self-dual code.

(19) Suppose w is symmetric and satisfies $w^3 = 1$ in FG, where F is a field of characteristic 3. Consider $u = 2 + hw$, in $F(G \times C_3)$, where C_3 is generated by h. Show that $u^3 = 0$, and $\operatorname{rank} u = 2|G|$. Show that u generates a $(3|G|, 2|G|)$ code. Show that $\operatorname{rank} u^2 = |G|$ and that $(u^T)^2 = u^2$ gives the check matrix of the code and prove that the dual of the code is contained within the code.

(20) (Computer required) Find elements u, v in the group ring $\mathbb{Z}_2 C_{10}$, of weight at least 3, such that $uv = 1$. Construct the circulant matrices U, V corresponding to u, v respectively. From U construct a generator matrix for (i) a $(10, 5)$ code, (ii) a $(10, 7)$ code, (iii) a $(10, 3)$ code and derive directly from V the check matrices of these codes.

Determine the distances of these codes.

(21) (Computer needed) Find units in u, v in the group ring $\mathbb{Z}_2 C_{20}$ with weight at least 3. Construct the circulant matrices U, V, corresponding to u, v respectively. From U construct a generator matrix for (i) a $(20, 10)$ code, (ii) a $(20, 15)$ code, (iii) a $(20, 5)$ code and derive directly from V the check matrices of these codes.

Determine the distances of these codes.

(22) (Computer needed) Find a unit in a $\mathbb{Z}_2 C_{1000}$ with weight at least 3. From this unit construct generator and check matrices for $(1000, 500)$ and $(1000, 750)$ codes.

How many different $(1000, 500)$ codes may be constructed from this unit?

5.6.2. *Research Problems*

(1) Construct a unit of weight 3 in $\mathbb{Z}_2 C_{100}$. Using this unit from which the check matrices are derived construct 10 random (100,50) codes and 10 random (100,75) codes. Simulate the performance of these codes using MATLAB Communications Toolbox or similar package.

(2) Construct a unit of weight 5 in $\mathbb{Z}_2 C_{1000}$. Using this unit from which the check matrices are derived construct 10 random (1000,500) codes and 10 random (1000,750) codes. Simulate the performance of these codes using MATLAB Communications Toolbox or similar package.

(3) In general, find a method to construct a unit of a given support x in $\mathbb{Z}_2 C_n$ for a given n. Using this unit from which the check matrices are derived construct 10 random (n, r) codes and 10 random (n, s) codes. Simulate the performance of these codes using MATLAB Communications Toolbox or similar package.

(4) Classify the zero-divisors and units in dihedral group ring, and obtain methods to ascertain the distance of the resulting codes.

(5) Determine fundamentally how unit-derived codes in cyclic group rings perform in relationship to cyclic codes.

(6) Find a way to efficiently solve the optimization problem of determining the best-basis for a given unit (i.e. one with the best distance property) in either particular group ring cases or more generally.

References

1. P. J. Olver and C. Shakiban, *Applied Linear Algebra* (Pearson Prentice Hall, 2006).
2. P. J. Davis, *Circulant Matrices* (John Wiley and Sons, Inc., New York, 1979).

3. E. Noether, Hypercomplexe Grössen und Darstellingtheorie, *M. Zeit.* **30** (1929) 641–692.

4. C. P. Milies and S. K. Sehgal, *An Introduction to Group Rings* (Klumer, Dordrecht/Boston/London, 2002).

5. T. Hurley, Group rings and rings of matrices, *Int. J. Pure Appl. Math.* **31**, 3 (2006), 319–335.

6. R. Gray, Toeplitz and circulant matrices: A review, *Foundations and Trends in Communications and Information Theory.* **2**, 3 (2005), 155–329, URL http://ee.stanford.edu/ gray/toeplitz.pdf.

7. H. N. Ward, Quadratic residue codes and divisibility, in *Handbook of Coding Theory*, Chap. 9 (Elsevier, Amsterdam, 1998), pp. 827–870.

8. F. J. MacWilliams, Orthogonal circulant matrices over finite fields, and how to find them, *J. Comb. Theor.* **10** (1971) 1–17.

9. P. Hurley and T. Hurley, Module codes in group rings, *Proc. IEEE Int. Symp. on Information Theory (ISIT)*, (2007).

10. P. Hurley and T. Hurley, Codes from zero-divisors and units in group rings, *Int. J. Inform. and Coding Theory* **1**, 1 (2009), 57–87.

11. R. E. Blahut, *Algebraic Codes for Data Transmission* (Cambridge University Press, 2003).

12. F. J. MacWilliams and N. Sloane, *The Theory of Error-Correcting Codes* (North-Holland Amsterdam/London/New York/Tokyo, 1998).

13. J. E. F. Assmus, On Berman's characterization of the Reed–Muller codes, *J. Stat. Plan. Infer.* **56** (1996) 17–21.

14. F. J. MacWilliams, Codes and ideals in group algebras, *Comb. Math. Appl.* (1969) 312–328.

15. L. M. Bazzi and S. K. Mitter, Some constructions of codes from group actions, *Preprint. Under Submission* (2003).

16. D. S. Dummit and R. M. Foote, *Abstract Algebra*, 2nd edn. (John Wiley and Sons, 1999).

17. W. C. Huffman and V. Pless, *Fundamentals of Error-Correcting Codes* (Cambridge University Press, 2003).

Chapter 6

LDPC AND CONVOLUTIONAL CODES FROM MATRIX AND GROUP RINGS

PAUL HURLEY* and TED HURLEY[†]

*IBM Research, Zurich Research Laboratory, Switzerland
[†]Department of Mathematics,
National University of Ireland, Galway, Ireland
*pah@zurich.ibm.com
[†]ted.hurley@nuigalway.ie

In this chapter, low density parity check (LDPC) codes and convolutional codes are constructed and analyzed using matrix and group rings. It is shown that LDPC codes may be constructed using units or zero-divisors of small support in group rings. From the algebra it is possible to identify where short cycles occur in the matrix of a group ring element thereby allowing the construction, directly and algebraically, of LDPC codes with no short cycles. It is then also possible to construction units of small support in group rings with no short cycles at all in their matrices, thus allowing a huge choice of LDPC codes with no short cycles which may be produced from a single unit element. A general method is given for constructing codes from units in abstract systems. Applying the general method to the system of group rings with their rich algebraic structure allows the construction and analysis of series of convolutional codes. Convolutional codes are constructed and analyzed within group rings in the the infinite cyclic group over rings which are themselves group rings.

6.1. Introduction

Low density parity check (LDPC) codes and convolutional codes are hot topics and compete with one another for applications. Algebraic methods for convolutional codes exist for small memory but seem not to exist otherwise, see [1]. Although LDPC codes are usually produced by randomized techniques, there has been recent activity in the area of algebraic constructions [2–4].

Just as, see [5], the cyclic group ring is fundamental for the algebraic construction and analysis of linear codes so also general group rings can be fundamental building blocks for the algebraic construction and analysis of LDPC and convolutional codes.

The group ring method for constructing LDPC codes as described here simply requires a group ring element of small support. "Small support" means that only a small number, compared to the size, of the coefficients of the group ring element are non-zero. From the algebra of group rings it is then possible to determine where the short cycles can occur in the matrix of the group ring element, thus enabling LDPC codes with no short cycles to be constructed. Indeed it is then easy to construct group ring elements of small support which have no short cycles at all in their matrices thereby allowing a huge choice of LDPC codes with no short cycles which may be constructed from a particular group ring element. Examples and simulations of performances of LDPC codes produced by this method and comparisons of these with known LDPC codes and standards may be obtained in [6].

Having no short cycles in the check matrix of an LDPC code has been shown to dramatically improve the performance of the code. Short cycles in an LDPC code deteriorate the performance of the decoding algorithms and having no short cycles may in effect increase the distances of such codes.

The group ring method for LDPC and other codes needs only a relatively few initial parameters as input and can re-create the matrix line-by-line without the need to store the whole structure in memory. Thus, in addition these methods have applications to systems with low power and low storage requirements.

In Sec. 6.2, a method is derived to construct group ring elements so that their corresponding matrices are low density and have no short cycles in their (Tanner) graphs. This is then applied to the construction of LDPC codes with no short cycles, algebraically and directly, from group rings.

Although the word "graph" is used, no knowledge of graph theory is assumed or necessary as is explained in Sec. 6.2.

A method is devised in Sec. 6.3 for constructing convolutional codes using units in certain algebraic systems. From this, a general method is devised from which convolutional codes are readily constructed and analyzed using group rings. A number of applications of the general methods are given — detailed descriptions on these may be found in Secs. 6.1, 6.3.1, to this Sec. 6.3.

Exercises and suggested research areas are given in Sec. 6.4.

Basic algebra requirements are an understanding of the concepts of *units, zero-divisors,* and occasionally *nilpotent elements* in rings. The group theory requirements are also fairly basic and no deep properties of group rings are required except perhaps for Sec. 6.3.18 which is independent. Details on basic ring theory may be obtained in the previous chapter and more details on group rings may be obtained in [7].

Generally, RG denotes the group ring of the group G over the ring R. R may be a field, but is not necessarily so; in the convolutional code constructions, for example, R is sometimes taken to be a group ring itself. The group ring FG is often called a *group algebra* when F is a field.

For a group ring RG where $|G| = n$ there is an injection, see [8], $RG \to R_{n \times n}$ mapping the group ring into the ring of $n \times n$ matrices over R. Note that R can be any ring in this injection and is not confined to a field. This allows codes constructed in RG to be translated into codes constructed in matrices over R.

C_n denotes the cyclic group of order n and C_∞ denotes the infinite cyclic group. Both of these are generated by a single element and this generator is often denoted by a lower case letter such as g. $H \times K$ denotes the direct product of groups H, K so that for example $C_n \times C_2$ denotes the direct product of C_n and C_2.

Cyclic and related codes are formed and analyzed within FC_n, the *group ring* of the *cyclic* group C_n over a *field* or *ring* F — see, for example, [5]. Convolutional codes here are constructed and analyzed within $(FG)C_\infty$, the *group ring* of the infinite *cyclic* group C_∞ over the (group) *ring* FG.

LDPC codes here are constructed using units and/or zero-divisors within a general group ring FG where G can be any group; in many applications so far $G = C_n \times C_m$. What is required to produce an LDPC code is that one of the group ring elements u, v in an equation such as $uv = 0$ or $uv = 1$ has small support (has a small number of non-zero coefficients) compared to the size of the group G.

6.2. LDPC Codes

A LDPC code is a code where the check matrix has only a small number, compared to its length, of non-zero elements in each row and column. See [9] for further details.

In the group ring set-up being LDPC is equivalent to say that the check group ring element has small support compared to the length of the code.

As mentioned in Sec. 6.1, most LDPC codes have been produced by randomized techniques, but there has been recent activity in the area of algebraic constructions [2–4]. The group ring encoding framework is a useful tool in this direction, as it seems that some of these constructions are implicitly working in a group ring. Some advantages of the group ring method are also mentioned in the introduction and it may also be noticed that this method gives immediate generator as well as check matrices. One could also envisage a hybrid whereby a random construction is performed within the parameters of an algebraic construction.

For any matrix $H = (h_{ij})$, the Tanner graph [10] of H is a bipartite graph $K = K_1 \cup K_2$ where K_1 has one vertex for each row of H and K_2 has one vertex for column in H and there is an edge between two vertices i, j exactly when $h_{ij} \neq 0$. A short cycle in the (Tanner) graph of a matrix is a cycle of length 4.

Thus, a matrix has no short cycles in its graph if and only the intersection of positions in which two columns have non-zero values is at most 1. This definition is used when considering the absence or otherwise of short cycles and thus no deep graph theory is involved.

An algebraic group ring method is presented here for constructing codes with no short cycles in the check matrix. When applied to check elements in the group ring with small support this gives a general algebraic method for constructing LDPC codes with no short cycles.

Examples of performances of LDPC codes produced with this method and comparisons of these with performances of known codes and standards may be found in [6].

Codes are obtained from either zero-divisors or units in the group ring RG. Corresponding matrix versions are obtained by mapping these group ring elements to a ring of matrices. If, for example, the check group ring element v has short support then its corresponding matrix V has only a small number of elements in each row and column and is thus low density.

LDPC codes have been produced randomly and the short cycles are eliminated as the construction of the code takes place.

Here, it is determined precisely where the short cycles can occur in the matrix of a group ring element in terms of the group elements which appear with non-zero coefficients in this group ring element. It is then easy to construct group ring elements which will have no short cycles anywhere in its matrix. Using group ring elements with small support and with no short cycles, LDPC codes with no short cycles can be constructed algebraically.

To construct LDPC codes with no short cycles it is then a matter of making sure that the group ring check element, of low density, used for the construction has no short cycles at all in its matrix or else if it has then more care must be taken when selecting the columns to be used for the construction so as to avoid these short cycles. It is obviously easier to ensure there are no short cycles if the original (check) element used to construct the code has no short cycles at all in its matrix; this can be ensured by Theorem 6.1.

One advantage of codes derived from units is that there is a huge choice as to which rows (or columns) are used to form the check or generator matrix for a code. In the case where the original unit group ring element is low density and has no short cycles in its matrix there is then a big choice as to which columns (or rows) are taken from this unit for the check matrix and each choice gives an LDPC code with no short cycles. For example, from a unit of size 1000×1000 with no short cycles and low density and it is required to derive a 500×1000 code then there are $\binom{1000}{500}$, which is of the order of 2^{500}, choices and each code derived is low density and has no short cycles. Furthermore, since any set of rows or columns in a unit (non-singular) matrix is linearly independent this means that the codes are all different. The potential to construct a huge number of codes from a single unit all performing well has obvious advantages.

6.2.1. *How to Avoid Short Cycles Algebraically*

A short cycle in the graph of a matrix is a cycle of length 4.

We now give necessary and sufficient conditions on the group ring element u in terms of the group elements that occur in its expression so that its corresponding matrix U has no short cycles in its graph. Mathematical proofs are provided.

The conditions also determine when and where short cycles can occur in a group ring matrix U constructed from u. It is then possible to avoid these short cycles when constructing codes from U. It should also be noted that the unit-derived method for constructing codes has complete freedom as to which module W to choose and thus W can be chosen so as to avoid short cycles. Choice of the module W determines the generator and check matrices.

The general group ring results are given initially. Special cases, as for example when the group is cyclic or abelian, are derived and these are easier to describe.

6.2.2. *Collections of Differences, Special Case*

Collection of differences are usually defined with respect to a set a non-negative integers, see, for example, [11]. These collections of differences are special cases of collections of group differences, to be defined in Sec. 6.2.4, when the group in question is a cyclic group. The integer definition is recapped here and used to show examples of the general definition.

Let $S = \{i_1, i_2, \ldots, i_r\}$ be a set of non-negative unequal integers and n an integer with $n > i_j$ for all $j = 1, 2, \ldots, r$.

Then the *collection of cyclic differences of* S mod n is defined by $DS(n) = \{i_j - i_k \mod n \mid 1 \leq j, k \leq r, j \neq k\}$. This collection has possibly repeated elements.

For example, if $S = \{1, 3, 7, 8\}$ and $n = 12$ then

$$
DS(12) = \left\{ \begin{array}{ccc} 2 & 6 & 7 \\ & 4 & 5 \\ & & 1 \\ 10 & 6 & 5 \\ & 8 & 7 \\ & & 11 \end{array} \right\} = \{2, 6, 7, 4, 5, 1, 10, 6, 5, 8, 7, 11\}.
$$

In this case $6, 7, 5$ occur twice.

If $|S| = r$ then counting repeats $|DS(n)| = r(r-1)$.

6.2.3. *Cyclic Differences*

Consider the group ring RC_n where C_n is generated by g. Suppose $u = \alpha_{i_1} g^{i_1} + \alpha_{i_2} g^{i_2} + \cdots + \alpha_{i_r} g^{i_r} \in RC_n$ with $\alpha_{i_j} \neq 0$.

For each g^i, g^j in u (with non-zero coefficients) form $g^i g^{-j}, g^j g^{-i}$, and set $DS(u)$ to be the collection of all such $g^i g^{-j}, g^j g^{-i}$.

Set $S = \{i_1, i_2, \ldots, i_r\}$ and define the collection of cyclic differences $DS(n)$. It is clear that $DS(n)$ and $DS(u)$ are equivalent, the only difference being in the notation used.

The proof of the following theorem is a direct corollary of the more general Theorem 6.2.

Theorem 6.1. *U has no 4-cycles in its graph if and only if $DS(u)$ has no repeated elements.*

6.2.3.1. *Example*

Set $u = 1 + g + g^3 + g^7$ in $\mathbb{Z}_2 C_{15}$. The collection of differences is formed from $\{0, 1, 3, 7\}$ and is $DS(u) = \{1, 3, 7, 2, 6, 4, 14, 12, 8, 13, 9, 11\}$ which has no repeats. Thus the matrix formed from u, which is circulant in this case, has no short cycles.

Set $u = 1+g+g^3+g^7$ in $\mathbb{Z}_2 C_{13}$. The collection of differences formed from $\{0, 1, 3, 7\}$ is $\{1, 3, 7, 2, 6, 4, 12, 10, 6, 11, 7, 9\}$ and has repeats $6, 7$. Thus, matrix formed from u has short cycles — but we can identify where they occur.

6.2.4. *Collection of Differences of a Group Ring Element*

Let RG denote the group ring of the group G over the ring R. Let G be listed by $G = \{g_1, g_2, \ldots, g_n\}$ and suppose $u = \sum_{i=1}^{n} \alpha_i g_i$ in RG.

For each (distinct) pair g_i, g_j occurring in u with non-zero coefficients, form the (group) *differences* $g_i g_j^{-1}, g_j g_i^{-1}$. Then *the collection of differences of u, $DS(u)$, consists of all such differences.* Thus:

$$DS(u) = \{g_i g_j^{-1}, g_j g_i^{-1} \mid g_i \in G, g_j \in G, i \neq j, \alpha_i \neq 0, \alpha_j \neq 0\}.$$

Note that the collection of difference of u consists of group elements and for each $g, h \in G, g \neq h$ both gh^{-1} and its inverse hg^{-1} are formed and included in the collection.

Theorem 6.2. *The matrix U has no short cycles in its graph if and only if the $DS(u)$ has no repeated (group) elements.*

Proof. The rows of U correspond in order to $ug_i, i = 1, \ldots, n$. Then U has a 4-cycle

$\Leftrightarrow$ for some $i \neq j$ and some $k \neq l$, the coefficients of g_m, g_l, in ug_i and ug_j are non-zero.

$\Leftrightarrow$

$ug_i = \cdots + \alpha g_k + \beta g_l + \cdots$
and
$ug_j = \cdots + \alpha_1 g_k + \beta_1 g_l + \cdots$

$\Leftrightarrow$

$u = \cdots + \alpha g_k g_i^{-1} + \beta g_l g_i^{-1} + \cdots$
and
$u = \cdots + \alpha_1 g_k g_j^{-1} + \beta_1 g_l g_j^{-1} + \cdots .$

$\Leftrightarrow$

$DS(u)$ contains both $g_k g_i^{-1} g_i g_l^{-1} = g_k g_l^{-1}$ and $g_k g_j^{-1} g_l^{-1} = g_k g_l^{-1}$.

This happens if and only if $DS(u)$ has a repeated element.   □

6.2.5. *Repeated Elements*

Suppose now u is such that $DS(u)$ has repeated elements.

Hence $u = \cdots + \alpha_m g_m + \alpha_r g_r + \alpha_p g_p + \alpha_q g_q + \cdots$, where the displayed α_i are not zero, so that $g_m g_r^{-1} = g_p g_q^{-1}$. The elements causing a short cycle are displayed and note that the elements g_m, g_r, g_p, g_q are not necessarily in the order of the listing of G.

Since we are interested in the graph of the element and thus in the non-zero coefficients, replace a non-zero coefficient by the coefficient 1. Thus, write $u = \cdots + g_m + g_r + g_p + g_q + \cdots$ so that $g_m g_r^{-1} = g_p g_q^{-1}$.

Include the case where one p, q could be one of m, r in which case it should not be listed in the expression for u.

Then

$$u g_m^{-1} g_p = \cdots + g_p + g_r g_m^{-1} g_p + \cdots = \cdots + g_p + g_q + \cdots$$

and

$$u g_p^{-1} g_m = \cdots + g_m + g_q g_p^{-1} g_m = \cdots + g_m + g_r + \cdots .$$

(Note that $u g_m^{-1} g_p = u g_r^{-1} g_q$ and $u g_p^{-1} g_m = u g_q^{-1} g_r$.)

Thus to avoid short cycles, if using the row determined by g_i then delete the rows determined by $g_i g_m^{-1} g_p$ and $g_i g_p^{-1} g_m$.

6.2.6. *Special Cases*

The special case when $G = C_n$ was dealt with in Sec. 6.2.3.

Let $G = C_n \times C_m$ be the direct product of cyclic groups C_n, C_m generated by g, h, respectively.

These groups have been particularly useful in practice, see [6]. List the elements of G by

$$G = \{1, g, g^2, \ldots, g^{n-1}, h, hg, hg^2, \ldots, hg^{n-1}, \ldots,$$
$$h^{m-1}, h^{m-1} g, \ldots, h^{m-1} g^{n-1}\}.$$

Every element $u \in RG$ relative to this listing has the form $u = a_0 + h a_1 + \cdots + h^{m-1} a_{m-1}$ with each $a_i \in C_n$. The collection of differences of u is easy to determine and elements with no repeats in their collection of differences are thus easy to construct.

Relative to this listing the matrix of an element in RG is a circulant-by-circulant matrix of size $mn \times mn$ [8].

Another particularly useful group which is relatively easy to work with is the dihedral group D_{2n} given by $\langle a, b \,|\, a^2 = 1, b^n = 1, ab = b^{-1}a \rangle$ [7]. This group is non-abelian for $n \geq 3$. Every element u in RD_{2n} may be written as $u = f(b) + ag(b)$ with $f(b), g(b) \in RC_n$ where C_n is generated by b. The collection of differences of u is easy to determine. The corresponding matrix U of u has the form $\begin{pmatrix} A & B \\ B & A \end{pmatrix}$ where A is circulant and B is reverse circulant [8]. This gives non-commutative matrices and non-commutative codes.

6.3. Convolutional Codes

6.3.1. *Introduction*

Methods are presented for constructing and analyzing convolutional codes using units in Laurent series of finite support over matrix rings. By considering group rings as matrix rings, convolutional codes are constructed and analyzed from units in Laurent series over group rings. Thus, convolutional codes are constructed in group rings RG where the ring R is itself a group ring. Matrix versions may be directly obtained from the group ring description.

Group rings have a rich structure and this rich structure is exploited here just as, for example, the rich structure of the cyclic group ring is exploited in the study and construction of cyclic linear codes.

The lack of algebraic methods for constructing convolutional codes is mentioned a number of times in the oft-quoted source [1]. Many of the existing convolutional codes have been found by computer search and thus of necessity have relatively short memories.

Using algebraic methods here, the range of convolutional codes available is expanded and series of convolutional codes are derived. Free distances and codes to a prescribed free distances are also derived.

Some known convolutional codes can be obtained in this manner thus giving an algebraic setting for these.

The methods here are fairly general and use properties of group rings and their embedding into matrix rings. Zero-divisors and units in group rings enable the construction of units in certain polynomial rings and/or group rings over these group rings from which convolutional codes can be constructed. Properties of the convolutional codes can be studied and derived from properties of group rings. In many instances, the free distances can be calculated algebraically and convolutional codes to a specified free distance, as, for example, in Theorem 6.4 or Theorem 6.9, can be constructed. Some theorems on free distances are proved.

The following applications of the general method are included. These in themselves constitute new constructions and new methods for convolutional codes. Here C_∞ denotes the infinite cyclic group, C_n is the cyclic group of order n, and F is a field.

(1) Infinite series of $(2, 1)$ convolutional codes are constructed in Sec. 6.3.12 using $(FC_2)C_\infty$. Free distances are calculated algebraically.
(2) Given a linear cyclic code $\mathbb{CC}$ with $d = \min(d_1, d_2)$ where d_1 is the minimum distance of $\mathbb{CC}$ and d_2 is the distance of the dual of $\mathbb{CC}$, the generator polynomial f of $\mathbb{CC}$ is mimicked in $(FC_2)C_\infty$ to construct convolutional $(2, 1)$ codes of minimum free distance $d + 2$.
(3) Rate $\frac{3}{4}$ and higher rate convolutional codes are constructed and free distances calculated.
(4) Convolutional codes over a field F of characteristic p for any prime p using nilpotent elements in the field FG are constructed where G is a group whose order is divisible by p.
(5) *Hamming type* convolutional codes are constructed and lower bounds on free distances obtained.
(6) Convolutional codes are constructed using idempotents in group rings. These are particularly used in cases where the characteristic of the field does not divide the order of the group; *characters* of groups and *character tables* come into play in constructing these convolutional codes.

The convolutional codes in (1) and (2) above are reminiscent of cyclic codes with properties of cyclic codes but also being convolutional have the advantage of a "memory." Cyclic codes are obtained in the group ring FC_n for a ring (often a field) F and (finite) cyclic group C_n. Now replace, in FC_n, the ring F by the ring FC_2 and the finite group C_n by the infinite cyclic group C_∞ to obtain the group ring $(FC_2)C_\infty$; this group ring is then used to define and analyze the convolutional codes.

To obtain codes convolutional codes over fields of characteristic p one would use the group ring $(FC_p)C_\infty$ where p is the characteristic of F, or in general use $(FG)C_\infty$ where the characteristic of F divides the order of G.

The "Hamming-type" codes mentioned in (5) above use the group ring $(\mathbb{Z}_2(C_4 \times C_2))C_\infty$ where $C_4 \times C_2$ denotes the direct product of C_4 and C_2.

6.3.2. *Algebraic Description of Convolutional Codes*

Background on general algebra and group rings may be obtained in [7].

$R[z]$ denotes the polynomial ring with coefficients from a ring R and $R_{r \times n}$ denotes the ring of $r \times n$ matrices with coefficients from R. R^n is used to denote $R_{1 \times n}$ and thus $R^n = \{(r_1, r_2, \ldots, r_n) : r_i \in R\}$.

It is easy to verify that $R_{r \times n}[z] \cong R[z]_{r \times n}$.

$R[z, z^{-1}]$ is used to denote the set of Laurent series of finite support in z with coefficients from R. *Finite support* means that only a finite number of the coefficients are non-zero. It is clear that $R[z, z^{-1}] \cong RC_\infty$, where C_∞ denotes the infinite cyclic group. Note that elements in group rings have finite support.

Note also that $R[z] \cong T$ where T is the algebra of *non-negative elements* in RC_∞, i.e. the algebra of elements $w = \sum_{i=0}^{\infty} \alpha_i g^i$, in RC_∞, where C_∞ is generated by g.

If $\mathbb{F}$ is an integral domain then $\mathbb{F}[z]$ has no zero-divisors and only trivial units — the units of $\mathbb{F}[z]$ are the units of $\mathbb{F}$.

See [1] and/or [5] for basic information on convolutional codes, and algebraic descriptions are described therein. The (equivalent) extremely useful algebraic description as described in [12] is given below.

A convolutional code $\mathcal{CC}$ of length n and dimension k is a direct summand of $\mathbb{F}[z]^n$ of rank k. Here $\mathbb{F}[z]$ is the polynomial ring over $\mathbb{F}$ and $\mathbb{F}[z]^n = \{(v_1, v_2, \ldots, v_n) : v_i \in \mathbb{F}[z]\}$.

Suppose V is a submodule of $\mathbb{F}[z]^n$ and that $\{v_1, \ldots, v_r\} \subset \mathbb{F}[z]^n$ forms a generating set for V. Then

$$V = \text{Image } M = \{uM : u \in \mathbb{F}[z]^r\} \text{ where } M = \begin{bmatrix} v_1 \\ \vdots \\ v_r \end{bmatrix} \in \mathbb{F}[z]_{r \times n}.$$

This M is called a *generating matrix* of V.

A generating matrix $G \in \mathbb{F}[z]_{r \times n}$ having rank r is called a *generator* or *encoder matrix* of $\mathcal{CC}$.

A matrix $H \in \mathbb{F}[z]_{n \times (n-k)}$ satisfying $\mathcal{CC} = \ker H = \{v \in \mathbb{F}[z]^n : vH = 0\}$ is said to be a *control matrix* of the code $\mathcal{CC}$.

6.3.3. *Convolutional Codes from Units*

Let R be a ring which is a subring of the ring of matrices $F_{n \times n}$. In particular the group ring FG is a subring of $F_{n \times n}$, where $n = |G|$, by an explicit embedding given in [8]. There is no restriction on F in general but it is assumed to be a field here; however, many of the results will hold more generally. *Units* and *zero-divisors* in any ring are defined in the usual manner.

Construct R-convolutional codes as follows:

6.3.4. *General Construction*

Suppose $f(z, z^{-1}), g(z, z^{-1}) \in R[z, z^{-1}]$ satisfy $f(z, z^{-1})g(z, z^{-1}) = 1$.

Consider (compatible) block decompositions of f, g by

$$f(z, z^{-1}) = \begin{pmatrix} f_1(z, z^{-1}) \\ f_2(z, z^{-1}) \end{pmatrix},$$

$$g(z, z^{-1}) = \big(g_1(z, z^{-1}), \quad g_2(z, z^{-1}) \big),$$

where $f_1(z, z^{-1})$ is an $r \times n$ matrix, $f_2(z, z^{-1})$ is an $(n - r) \times n$ matrix, $g_1(z, z^{-1})$ is an $n \times r$ matrix and $g_2(z, z^{-1})$ is an $n \times (n - r)$ matrix.

Then

$$\begin{pmatrix} f_1(z, z^{-1}) \\ f_2(z, z^{-1}) \end{pmatrix} \times \big(g_1(z, z^{-1}), \quad g_2(z, z^{-1}) \big) = 1.$$

Thus

$$\begin{pmatrix} f_1 g_1 & f_1 g_2 \\ f_2 g_1 & f_2 g_2 \end{pmatrix} = 1.$$

From this it follows that

$$f_1(z, z^{-1})g_1(z, z^{-1}) = I_{r \times r},$$
$$f_1(z, z^{-1})g_2(z, z^{-1}) = 0_{r \times (n-r)},$$
$$f_2(z, z^{-1})g_1(z, z^{-1}) = 0_{(n-r) \times r},$$
$$f_2(z, z^{-1})g_2(z, z^{-1}) = I_{(n-r) \times (n-r)}.$$

Thus $f_1(z, z^{-1})$ is taken as the generator or encoder matrix of an (n, r) convolutional codes and $g_2(z, z^{-1})$ is then the check or control matrix for the code. It is seen in particular that $f_1(z, z^{-1})$, (and $f_2(z, z^{-1})$) have right finite support inverses and thus by Theorem 6.6 of [1], the generator matrix f_1 is non-catastrophic.

Given $f(z, z^{-1})g(z, z^{-1}) = 1$ by the general method of unit-derived code as described in [13] *any* rows of $f(z, z^{-1})$ can be used to construct a generator matrix of a convolutional code and the control matrix may be obtained directly from $g(z, z^{-1})$. If rows $\{j_1, j_2, \ldots, j_r\}$ are chosen from $f(z, z^{-1})$ then an encoding $F^r[z] \to F^n[z]$ is obtained with generator matrix consisting of these r rows of $f(z)$ and check/control matrix is obtained by deleting the $\{j_1, j_2, \ldots, j_r\}$ columns of $g(z)$. This can also be seen by permuting the rows of f appropriately and using the particular case as described above.

6.3.5. *Polynomial Case*

This construction holds in particular when one or both of f, g are polynomials.

It is known that convolutional codes have polynomial generator matrices but the more general case as described in Sec. 6.3.4 is often more suitable for initially constructing and analyzing the codes. It is easy to obtain polynomial generator and control matrices from Laurent matrices of finite support.

For clarity, a detailed analysis of this polynomial case is now given.

Suppose $f(z)g(z) = 1$ in $R[z]$. Essentially, then the encoder matrix is obtained from $f(z)$ and the decoder or control matrix is obtained from $g(z)$.

Now $f(z) = (f_{i,j}(z))$ is an $n \times n$ matrix with entries $f_{i,j}(z) \in F[z]$. Similarly $g(z) = (g_{i,j}(z))$ is an $n \times n$ matrix over $F[z]$. Suppose $r[z] \in F[z]^r$ and consider $r[z]$ as an element of $\mathbb{F}[z]^n$ (by adding zeros to the end of it). Then define a mapping $\gamma : F[z]^r \to F[z]^n$ by $\gamma : r(z) \mapsto r(z)f(z)$. The code $\mathcal{CC}$ is the image of γ. Since $r[z]$ has zeros in its last $(n - r)$ entries as a member of $F[z]^n$, this means that *the generator matrix is the first r rows* of $f(z)$ which is an $r \times n$ matrix over $F[z]$. Since $f(z)$ is invertible, this generator matrix has rank r and is thus the encoder matrix which we denote by $G(z)$. For this polynomial case, $G(z)$ is a basic generator matrix — see A.1 Theorem in [1].

$w(z) \in \mathbb{F}[z]^n$ is a codeword if and only if $w(z)g(z)$ is in $\mathbb{F}[z]^r$, that is, if and only if the final $(n - r)$ entries of $w(z)g(z)$ are all 0. Suppose $w(z) = (\alpha_1(z), \alpha_2(z), \ldots, \alpha_n(z))$. Then this condition is that

$$(\alpha_1(z), \alpha_2(z), \ldots, \alpha_n(z)) * \begin{pmatrix} g_{1,r+1}(z) & g_{1,r+2}(z) & \cdots & g_{1,n}(z) \\ g_{2,r+1}(z) & g_{2,r+2}(z) & \cdots & g_{2,n}(z) \\ \vdots & \vdots & \vdots & \vdots \\ g_{n,r+1}(z) & g_{n,r+2}(z) & \cdots & g_{n,n}(z) \end{pmatrix} = 0.$$

The check or *control matrix* $H(z)$ of the code is thus:

$$
\begin{pmatrix}
g_{1,r+1}(z) & g_{1,r+2}(z) & \cdots & g_{1,n}(z) \\
g_{2,r+1}(z) & g_{2,r+2}(z) & \cdots & g_{2,n}(z) \\
\vdots & \vdots & \vdots & \vdots \\
g_{n,r+1}(z) & g_{n,r+2}(z) & \cdots & g_{n,n}(z)
\end{pmatrix}.
$$

This has size $n \times (n-r)$ and is the matrix consisting of the last $(n-r)$ columns of $g(z)$ or in other words the matrix obtained by deleting the first r columns of $g(z)$.

Since $f(z), g(z)$ are units, it is automatic that $\operatorname{rank} G(z) = r$ and $\operatorname{rank} H(z) = n - r$.

6.3.6. *Particular Useful Case*

Suppose $f(z)g(z) = z^t$ in $R[z]$. It is easy to obtain such f, g and a general method for producing these is given later. Then $f(z)(g(z)/z^t) = 1$. Now $g(z)/z^t$ involves negative powers of z but has finite support. The encoder matrix is obtained from $f(z)$ and the decoder or control matrix is obtained from $g(z)/z^t$ using the method as formulated in Sec. 6.3.4. It is also possible to consider $(f(z)/z^i)(g(z)/z^j) = 1$ with $i+j = t$ and to derive the generator matrix from $f(z)/z^i$ and the check/control matrix from $g(z)/z^j$.

The control matrix contains negative powers of z, but it is easy to obtain a polynomial control matrix from this.

In these cases, it is possible to multiply by the unit z^{-t} and stay within $R[z, z^{-1}]$. Other elements in $R[z, z^{-1}]$ have inverses, but with infinite support and thus outside $R[z, z^{-1}]$.

6.3.7. *Group Ring Convolutional Codes*

In the constructions of Sec. 6.3.4, R is any subring of $F_{n\times n}$.

The group ring $R = FG$ with $|G| = n$ is a subring of $F_{n\times n}$ using an explicit injection from FG to a subring of $F_{n\times n}$ as found for example in [8].

Thus the methods of Sec. 6.3.4 may be used to define convolutional codes by considering $R[z, z^{-1}] \cong RC_\infty$, which is the group ring over C_∞ with coefficients from the group ring $R = FG$.

To obtain units in $R[z, z^{-1}]$ (which includes $R[z]$) we are lead to consider zero-divisors and units in $R = FG$.

$R = FG$ is a rich source of zero-divisors and units, and consequently $R[z, z^{-1}]$ is a rich source of units. There are methods available for constructing units and zero-divisors in FG. What is required are units in $R[z]$, where $R = FG$, a group ring, and these can be obtained by the use of zero-divisors and units in R as coefficients of the powers of z.

FG also has a rich structure in which to analyze any resulting code. When F is a field, every non-zero element of FG is either a unit or a zero-divisor.

In what follows bear in mind that in $R[z, z^{-1}]$ it is desirable that R has zero-divisors and units, as when R is a group ring.

6.3.8. *Equations in the Infinite Cyclic Group Ring*

Suppose then $\sum_{i=-m}^{n} \alpha_i z^i \times \sum_{j=-m}^{n} \beta_j z^j = 1$ in the group ring $RC_\infty = R[z, z^{-1}]$ with $\alpha_i \in R$ and C_∞ generated by z. By multiplying through by a power of z this is then $\sum_{i=0}^{n} \alpha_i z^i \times \sum_{j=-m}^{n} \beta_j z^j = 1$.

The case with $m = 0$ gives polynomials over z. Here we have $\sum_{i=0}^{n} \alpha_i z^i \times \sum_{i=0}^{t} \beta_j z^j = 1$ where $\alpha_n \neq 0, \beta_t \neq 0$ and looking at the coefficient of z^0 it is clear that we must also have $\alpha_0 \neq 0, \beta_0 \neq 0$. This can be considered an an equation in RC_∞ with non-negative powers. Solutions may be used to construct convolutional codes.

By looking at the highest and lowest coefficients we then have that $\alpha_0 \times \beta_0 = 1$ and $\alpha_n \times \beta_t = 0$. Thus in particular α_0 is a unit with inverse β_0 and α_n, β_t are zero-divisors.

Solutions of the general equation $\sum_{i=0}^{n} \alpha_i z^i \times \sum_{j=-m}^{n} \beta_j z^j = 1$ can also be used to form convolutional codes and polynomial generator matrices may be derived from these.

6.3.9. *Convolutional Codes from Nilpotent Elements*

The following two Lemmas are useful in constructing new classes of convolutional codes. Their proofs are straight-forward. The second is a generalization of the first but it is useful for later references to state each separately.

Lemma 6.1. *Let* $R = FG$ *be the group ring of a group* G *over a field* F *with characteristic 2. Suppose* $\alpha_i \in R$ *commute. Let* $w = \sum_{i=0}^{n} \alpha_i z^i \in RC_\infty$. *Then* $w^2 = 1$ *if and only if* $\alpha_0^2 = 1, \alpha_i^2 = 0, i > 0$.

Lemma 6.2. *Let $R = FG$ be the group ring of a group G over a field F with characteristic 2. Suppose $\alpha_i \in R$ commute. Let $w = \sum_{i=0}^{n} \alpha_i z^i \in RC_\infty$. Then $w^2 = z^{2t}$ if and only if $\alpha_i^2 = 0, i \neq t$ and $\alpha_t^2 = 1$.*

Now construct convolutional codes by the following general method: Find elements α_i with $\alpha_i^2 = 0$ and units u with $u^2 = 1$ in the group ring R. Form units in $R[z]$ or $R[z, z^{-1}]$ using Lemma 6.1 or Lemma 6.2. From these units, convolutional codes are defined using the methods described in Sec. 6.3.4.

6.3.10. *Examples of Group Ring Convolutional Codes*

6.3.10.1. *Example 1*

Consider now $\alpha_0 = a + a^2 + a^3$ and for $i > 0$ define $\alpha_i = a + a^3$ or $\alpha_i = 0$ in the group ring $R = \mathbb{Z}_2 C_4$. Then $\alpha_0^2 = 1$ and $\alpha_i^2 = 0, i > 0$. We could also take $\alpha_i = 1 + a^2$.

Define $w(z) = \sum_{i=0}^{n} \alpha_i z^i$ in RC_∞. By Lemma 6.1, $w^2 = 1$.

The matrix corresponding to α_0 is $\begin{pmatrix} 0 & 1 & 1 & 1 \\ 1 & 0 & 1 & 1 \\ 1 & 1 & 0 & 1 \\ 1 & 1 & 1 & 0 \end{pmatrix}$ and the matrix

corresponding to $\alpha_i, i \neq 0$ is $\begin{pmatrix} 0 & 1 & 0 & 1 \\ 1 & 0 & 1 & 0 \\ 0 & 1 & 0 & 1 \\ 1 & 0 & 1 & 0 \end{pmatrix}$ or else is the zero matrix.

Now specify that the first two rows of w give the generator matrix and from this it follows that the last two columns of w is a control matrix.

This gives the following generator matrix:

$$G = \begin{pmatrix} 0 & 1 & 1 & 1 \\ 1 & 0 & 1 & 1 \end{pmatrix} + \delta_1 \begin{pmatrix} 0 & 1 & 0 & 1 \\ 1 & 0 & 1 & 0 \end{pmatrix} z + \delta_2 \begin{pmatrix} 0 & 1 & 0 & 1 \\ 1 & 0 & 1 & 0 \end{pmatrix} z^2$$

$$+ \cdots + \delta_n \begin{pmatrix} 0 & 1 & 0 & 1 \\ 1 & 0 & 1 & 0 \end{pmatrix} z^n$$

where $\delta_i = 1$ when $\alpha_i \neq 0$ and $\delta_i = 0$ when $\alpha_i = 0$.

The control matrix is

$$H = \begin{pmatrix} 1 & 1 \\ 1 & 1 \\ 0 & 1 \\ 1 & 0 \end{pmatrix} + \delta_1 \begin{pmatrix} 0 & 1 \\ 1 & 0 \\ 0 & 1 \\ 1 & 0 \end{pmatrix} z + \delta_2 \begin{pmatrix} 0 & 1 \\ 1 & 0 \\ 0 & 1 \\ 1 & 0 \end{pmatrix} z^2 + \cdots + \delta_n \begin{pmatrix} 0 & 1 \\ 1 & 0 \\ 0 & 1 \\ 1 & 0 \end{pmatrix} z^n.$$

The code has length 4 and dimension 2. The free distance is at least 6 for any $n \geq 2$ and in many cases it will be larger. Polynomials used for generating cyclic linear codes suitably converted to polynomials in $R[z]$ prove particularly useful and amenable — see for example Sec. 6.3.12.

6.3.10.2. *Particular Example*

The $(4, 2)$ convolutional code with generator and check matrices as follows has free distance 8.

$$
G = \begin{pmatrix} 0 & 1 & 1 & 1 \\ 1 & 0 & 1 & 1 \end{pmatrix} + \begin{pmatrix} 0 & 1 & 0 & 1 \\ 1 & 0 & 1 & 0 \end{pmatrix} z + \begin{pmatrix} 0 & 1 & 0 & 1 \\ 1 & 0 & 1 & 0 \end{pmatrix} z^3
$$
$$
+ \begin{pmatrix} 0 & 1 & 0 & 1 \\ 1 & 0 & 1 & 0 \end{pmatrix} z^4
$$

$$
H = \begin{pmatrix} 1 & 1 \\ 1 & 1 \\ 0 & 1 \\ 1 & 0 \end{pmatrix} + \begin{pmatrix} 0 & 1 \\ 1 & 0 \\ 0 & 1 \\ 1 & 0 \end{pmatrix} z + \begin{pmatrix} 0 & 1 \\ 1 & 0 \\ 0 & 1 \\ 1 & 0 \end{pmatrix} z^3 + \begin{pmatrix} 0 & 1 \\ 1 & 0 \\ 0 & 1 \\ 1 & 0 \end{pmatrix} z^4.
$$

6.3.11. **Direct Products: Turbo-Effect**

Examples of convolutional codes formed using α_i with $\alpha_i^2 = 0$ in FG have been produced. Consider now $F(G \times H)$ and let $w = \beta \times \alpha_i$ for *any* $\beta \in FH$. Then $w^2 = \beta^2 \alpha_i^2 = 0$. This expands enormously the range of available elements whose square is zero. Note also that over a field of characteristic 2 if $\alpha^2 = 0 = \gamma^2$ then $(\alpha + \gamma)^2 = 0$.

For example in $\mathbb{Z}_2 C_2$ the element $1 + a$ was used where C_2 generated by a. Then in $\mathbb{Z}_2(G \times C_2)$ consider $\alpha = \beta(1 + a)$ for any $\beta \in \mathbb{Z}_2 G$. Then $\alpha^2 = 0$.

A simple example of this is $\mathbb{Z}_2(C_2 \times C_2)$ where $\alpha = (1+a)b+(1+b)a = a + b$. The matrix of $a + b$ is $\begin{pmatrix} A & B \\ B & A \end{pmatrix}$ where $A = \begin{pmatrix} 0 & 1 \\ 1 & 0 \end{pmatrix}$ and $B = \begin{pmatrix} 1 & 0 \\ 0 & 1 \end{pmatrix}$. In forming $(4, 2)$ convolutional codes we would only use the top half of the matrices, i.e. $P = \begin{pmatrix} 0 & 1 & 1 & 0 \\ 1 & 0 & 0 & 1 \end{pmatrix}$. Note that in this encoding the vector (γ, δ) is mapped to $(\gamma, \delta)P = (\delta \quad \gamma | \gamma \quad \delta)$. This is like an interweaving of two codes.

To get a permutation effect, use the direct product with S_n, the permutation or symmetric group on n letters.

6.3.12. *(2,1) Codes*

Examples of $(2,1)$ optimal codes up to degree 10 may be obtained in [1]. Some of these can be reproduced algebraically and properties derived using the methods developed here.

Further new $(2,1)$ convolutional codes and series of convolutional $(2,1)$ are constructed in this section as an application of the general methods described above. The free distances can often be determined algebraically and codes to a prescribed free distance can be constructed by using Theorem 6.4.

Let F be a field of characteristic 2 and $R = FC_2$, where C_2 is generated by a. Consider elements $\alpha_i \in R$, $i > 0$, where either $\alpha_i = 1 + a$ or $\alpha_i = 0$. Then $\alpha_i^2 = 0$.

Let $\alpha_0 = 1$ in R and define $w = \alpha_1 + \alpha_0 z + \alpha_2 z^2 + \cdots + \alpha_n z^n$. Then $w^2 = z^2$ and hence $w \times (w/z^2) = 1$. Thus w can be used to define a $(2,1)$ convolutional code.

More generally let t be an integer, $0 \le t \le n$, and define $w = \sum_{i=0}^{n} \beta_i z^i$ where $\beta_i = \alpha_i, i \ne t$, $\beta_t = 1$. Then $w^2 = z^{2t}$ gives that $w \times (w/z^{2t}) = 1$. Thus w can be used to define a convolutional $(2,1)$ code. The case $\alpha_0 = \beta_1$ is a special case.

Now determine the code by choosing the first row of the matrix of w to be the generator/encoder matrix and then the last column of w/z^{2t} is the control matrix.

The matrix of α_i is $\begin{pmatrix} 1 & 1 \\ 1 & 1 \end{pmatrix}$ when $\alpha_i = 1+a$ and is the zero 2×2 matrix when $\alpha_i = 0$.

Define $\delta_i = 1$ when $\alpha_i \ne 0$ and $i \ne t$; $\delta_i = 0$ when $\alpha_i = 0$ and $i \ne t$; and define $\delta_t(1,1)$ to be $(1,0)$.

Then the encoder matrix of the code is $G = (1,1) + \delta_1(1,1)z + \delta_2(1,1)z^2 + \cdots + \delta_n(1,1)z^n$ and with $H = \begin{pmatrix} 1 \\ 1 \end{pmatrix} + \delta_1\begin{pmatrix} 1 \\ 0 \end{pmatrix}z + \delta_2\begin{pmatrix} 1 \\ 1 \end{pmatrix}z^2 + \cdots + \delta_n\begin{pmatrix} 1 \\ 1 \end{pmatrix}z^n$, the control matrix is H/z^{2t}.

The generator matrix G obtained in this way is non-catastrophic as it has a right finite weight inverse — see Theorems 6.3 and 6.6 in [1].

For $n = 2$ we get as an example the code with the generator matrix $G = (1,1) + (1,0)z + (1,1)z^2$. This code has free distance 5 which is optimal. It is precisely the $(2,1,2,5)$ code as described in [1], page 1085.

Proposition 6.1. *G has free distance 5.*

Proof. Consider $\sum_{i=0}^{t} \beta_i z^i G$, with $\beta_i \in \mathbb{Z}_2$ and $\beta_t \neq 0$. In determining free distance we may consider $\beta_0 \neq 0$. The coefficients of $z^0 = 1$ and z^{t+2} are $(1,1)$, and also $(1,0)$ occurs in the expression for at least one other coefficient. Thus the free distance is $2 + 2 + 1$ which is attained by G. $\square$

The above proof illustrates a general method for proving free distance or getting a lower bound on the free distance. For example wherever $(1,0)$ appears in a sum making up a coefficient it will contribute a distance of at least 1 as the other non-zero coefficients. Consider a sum of $(1,1)$. This will add to $(0,0)$ or to $(1,1)$. Now adding $(1,0)$ on to either $(0,0)$ or $(1,1)$ will give $(1,0)$ or $(0,1)$ which will contribute a distance of 1.

The check matrix for this code is

$$\frac{\begin{pmatrix}1\\1\end{pmatrix} + \begin{pmatrix}0\\1\end{pmatrix} z + \begin{pmatrix}1\\1\end{pmatrix} z^2}{z^2} = \begin{pmatrix}1\\1\end{pmatrix} + \begin{pmatrix}0\\1\end{pmatrix} z^{-1} + \begin{pmatrix}1\\1\end{pmatrix} z^{-2}.$$

For $n \geq 3$, it may be verified directly by similar algebraic methods that the free distance is at least 6. Appropriate choices of the α_i will give bigger free distances. See Theorem 6.4.

For $n = 3$ and $\delta_2 = 1 = \delta_3$ a $(2,1,3,6)$ convolutional code is obtained which is also optimal. Thus a degree 3 optimal distance 6 is given by the encoder matrix $G = (1,1) + (1,0)z + (1,1)z^2 + (1,1)z^3$ and the control matrix is $H/z^2 = \begin{pmatrix}1\\1\end{pmatrix}/z^2 + \begin{pmatrix}1\\0\end{pmatrix}/z + \begin{pmatrix}1\\1\end{pmatrix} + \begin{pmatrix}1\\1\end{pmatrix}z$. It is clear that H is also a control matrix.

The next case is $(2,1,4)$ of degree 4. The optimal distance of one of these is 7. Consider $w = \alpha_1 + \alpha_0 z + \alpha_1 z^3 + \alpha_1 z^4$, where $\alpha_1 = 1 + a$ and $\alpha_0 = 1$ in $\mathbb{Z}_2 C_2$. Then $w^2 = z^2$ and thus w gives the encoder matrix and w/z^2 gives the check matrix. The encoder matrix is $G = (1,1) + (1,0)z + (1,1)z^3 + (1,1)z^4$. Call this code $\mathcal{C}$.

Proposition 6.2. *The free distance of $\mathcal{C}$ is 7.*

Proof. Consider $(\sum_{i=0}^{t} \beta_i z^i)G$, with $\beta_i \in \mathbb{Z}_2$. In determining free distance we may consider $\beta_0 \neq 0$ and $\beta_t \neq 0$. The coefficients of $z^0 (= 1)$ and z^{t+4} are both $(1,1)$. If there are more than two non-zero β_i in the sum then $(1,0)$ occurs in at least three coefficients giving a distance of $2 + 2 + 3 = 7$ at least. It is now necessary to consider the case when there are just two β_i in the sum. It is easy to see then that at least three of the coefficients of z^i are $(1, 1)$, and $(1, 0)$ or $(0, 1)$ is a coefficient of another. Thus the free distance is 7. □

Consider the next few degrees. Let $\alpha = 1 + a, \alpha_0 = 1$ in FC_2 where F has characteristic 2.

(1) deg 5: $w = \alpha + \alpha_0 z + \alpha z^3 + \alpha z^4 + \alpha z^5$; gives a free distance of 8.
(2) deg 6: $w = \alpha + \alpha z^2 + \alpha z^3 + \alpha_0 z^4 + \alpha z^5 + \alpha z^6$. This gives a free distance of 10.
(3) Consider for example the following degree 12 element.

$$w = \alpha + \alpha z^2 + \alpha z^4 + \alpha z^5 + \alpha z^6 + \alpha_0 z^9 + \alpha z^{10} + \alpha z^{11} + \alpha z^{12}.$$

Note that this resembles the polynomial used for the Golay $(23, 12)$ code — see e.g. [5], page 119. The difference is that a z^{12} has been added and the coefficient of z^9 appears with coefficient α_0 and not 0 as in the Golay code. It is possible to play around with this by placing α_0 as the coefficient of other powers of z in w.

We thus study the best performance of convolutional codes derived from $w = \sum_{i=0}^{t} \alpha_i z^i$ where some $\alpha_t = 1 \in FC_2$, and all the other α_i are either 0 or else $1 + a$ in FC_2. Try to choose the α_i as one would for a linear cyclic code so as to maximize the (free) distance.

The set-up indicates we should look at existing cyclic codes and form convolutional codes by mimicking the generating polynomials for the cyclic codes.

6.3.12.1. *From Cyclic Codes to Convolutional Codes*

FC_n denotes the group ring of the cyclic group C_n over the field F and suppose C_n is generated by g.

A cyclic code of length n over F is given by a zero-divisor in this group ring. Suppose then the cyclic code $\mathcal{C}$ is generated by $f(g) \in FC_n$ and that $h(g)f(g) = 0$, where $h(g)$ is of minimal degree such that $h(g)f(g) = 0$. If $r(g)f(g) = 0$ then $r(g) = s(g)h(g)$. Also $h(g)$ is the generator polynomial for the dual code $\hat{\mathcal{C}}$ of $\mathcal{C}$.

Assume now F has characteristic 2. Suppose then that C is an (n, k, d_1), with rank k and distance d_1, and that $\hat{\mathcal{C}}$ is an $(n, n - k, d_2)$ code, with rank $n - k$ and distance d_2. Let $d = \min(d_1, d_2)$.

Suppose $f(g) = \sum_{i=0}^{r} \epsilon_i g^i$, with $\epsilon_i \in F, (\epsilon_r \neq 0)$, is a generating polynomial for $\mathcal{C}$. In $f(g)$, assume $\epsilon_0 \neq 0$.

The degree of f is of course less than n. The number of non-zero coefficients in $f(g)$ is the weight of $f(g)$.

Consider the polynomial $f(z)$ in $F[z]$. This has degree $< n$ and weight $\geq d_1$ as does $f(g)$.

Consider $p(g) = (\sum_{i=0}^{t} \alpha_i g^i) f(g) = \sum_{i=0}^{n-1} \beta_i g^i$ in FC_n and $w(z) = (\sum_{i=0}^{t} \alpha_i z^i) f(z) = \sum_{i=0}^{s} \gamma_i z^i$ in $F[z]$. Note that it is not assumed that $t < n$.

The following lemma is straight-forward but fundamental.

Lemma 6.3. *For* $0 \leq j \leq n-1$ *the sum of the coefficients of* $z^j, z^{j+n}, z^{j+2n}, \ldots$ *in* $w(z)$ *is* β_j.

Corollary 6.1. *If* $\sum_{i=0}^{n-1} \beta_i g^i$ *has weight* $\geq d$, *then* $\sum_{i=0}^{n-1} \gamma_i z^i$ *has weight* $\geq d$.

Proof. Suppose in $\sum_{i=0}^{n-1} \beta_i g^i$ that $\beta_j \neq 0$. Then the sum of the coefficients of $z^j, z^{j+n}, \ldots$ in $\sum_{i=0}^{s} \gamma_i z^i$ is non-zero. Thus the coefficient of at least one of $z^j, z^{j+n}, \ldots$ in $\sum_{i=0}^{s} \gamma_i z^i$ is non-zero.

Hence the weight of $\sum_{i=0}^{s} \gamma_i z^i \geq d$. $\qquad\square$

Lemma 6.4. *Let* $w(z) = \sum_{i=0}^{t} \alpha_i z^i$ *and suppose for* $0 \leq j < n$ *the first non-zero coefficient in the sequence* $z^j, z^{j+n}, \ldots,$ *is* z^{j+kn}. *Then in* $w(z)(1 + z^n)$ *the first non-zero coefficient in the sequence* $z^j, z^{j+n}, \ldots$ *is also* z^{j+kn}.

Theorem 6.3. *Consider* $t(z) = \sum_{i=0}^{t} \alpha_i z^i$ *and let* $w(z) = t(z) f(z) = (\sum_{i=0}^{t} \alpha_i z^i) f(z)$. *Then either* $w(z)$ *has weight* $\geq d_1$ *or else* $t(g) = s(g)h(g)$ *in which case* $t(z)$ *has weight* $\geq d_2$.

Proof. As $f(g)$ has rank k, $(\sum_{i=0}^{t} \alpha_i g^i) f(g) = (\sum_{i=0}^{k-1} \delta_i g^i) f(g)$.

Case 1: $\sum_{i=0}^{k-1} \delta_i g^i \neq 0$.

In this case $\sum_{i=0}^{k-1} \delta_i g^i \neq 0$ has weight > 0. Then $(\sum_{i=0}^{k-1} \alpha_i g^i) f(g)$ has weight $\geq d_1$ as $\mathcal{C}$ has distance $\geq d_1$. Hence by Corollary 6.1 $w(z)$ has weight $\geq d_1$.

Case 2: $\sum_{i=0}^{k-1} \delta_i g^i = 0$.

Then $(\sum_{i=0}^{t} \alpha_i g^i) f(g) = 0$. Hence $\sum_{i=0}^{t} \alpha_i g^i = \alpha(g)h(g)$ for some polynomial $\alpha(g) \in FC_n$.

Case 2(1): $\sum_{i=0}^{t} \alpha_i g^i \neq 0$.

Then $\sum_{i=0}^{t} \alpha_i g^i = \alpha(g)h(g)$ has weight $\geq d_2$ as $\hat{\mathcal{C}}$ has distance d_2. Also in this case $\sum_{i=0}^{t} \alpha_i z^i$ has weight $\geq d_2$ for if $\sum_{i=0}^{t} \alpha_i z^i$ has weight $< d_2$ then $\sum_{i=0}^{t} \alpha_i g^i$ has weight $< d_2$.

Case 2(2): Suppose now $\sum_{i=0}^{t} \alpha_i g^i = 0$.

Then $\sum_{i=0}^{t} \alpha_i z^i = s(z)(1 + z^n)^k$ where $1 + z^n$ does not divide $s(z)$.

Consider $w(z) = s(z)f(z)(1 + z^n)^k$. Now $s(z)$ is such that $s(g) = \sum_{i=0}^{q} \omega_i g^i \neq 0 \in FC_n$.

Thus as in case 1 or case 2(1), $s(z)f(z) = \sum_{i=0}^{t} \gamma_i z^i$ has weight $\geq d$ and is such that for d of the j, $0 \leq j < n$, one of the coefficients in $z^j, z^{j+n}, \ldots$ is non-zero.

Then by Lemma 6.4 $\alpha(z)f(z)(1 + z^n)^k$ has weight $\geq d$.

Hence weight of $w(z)$ is this case is also $\geq d$. $\qquad\square$

Let $\beta_i = \epsilon_i(1 + a)$ in the group ring FC_2 with C_2 generated by a and form $f_1(z) = \sum_{i=1}^{r} \beta_i z^i$. Then $\beta_j = 0$ in FC_2 if and only if $\epsilon_j = 0$ in F.

Choose a $\beta_t \neq 0$ in f_1 and replace this β_t in $f_1(z)$ by $1 \in FC_2$ to form $f_0(z) = \sum_{i=0}^{r} \alpha_i z^i$. Then $\alpha_j = 0$ if and only if $\epsilon_j = 0$.

So $f_0(z) = \sum_{i=0}^{r} \alpha_i z^i$ where $\alpha_i = \epsilon_i(1 + a)$ except when $i = t$ in which case $\alpha_t = \epsilon_t 1 = \epsilon_t$, where 1 denotes the identity of FC_2.

Then $f_0(z)^2 = z^{2t}$ and thus $f_0(z) \times (f_0(z)/z^{2t}) = 1$. We now use $f_0(z)$ to generate a convolutional code by taking just the first rows of the α_i. Thus, the generating matrix is $\hat{f}(z) = \sum_{i=0}^{r} \hat{\alpha}_i z^i$ where $\hat{\alpha}_i$ is the first row of α_i. If $\alpha_i \neq 0$ and $i \neq t$ then $\hat{\alpha}_i = (1, 1)$, and $\hat{\alpha}_t = (1, 0)$.

Lemma 6.5. *Let* $w = \sum_{i=1}^{n} \alpha_i(1, 1) + \alpha(1, 0)$ *with* $\alpha \neq 0$. *Then at least one component of* w *is not zero.*

Proof. Now $w = (\sum_{i=1}^{n} \alpha_i + \alpha, \sum_{i=1}^{n} \alpha_i)$. Since $\alpha \neq 0$ it is clear that one component of w is not zero. $\qquad\square$

Assume now $\alpha_t \neq \alpha_0$ and $\alpha_t \neq \alpha_r$; i.e. ϵ_t is not the first or last coefficient of $f(z)$. (A similar result holds if it does occur in one of these positions, but the free distance derived is possibly less by 1.)

Theorem 6.4. *Let* CC *denote the convolutional code with generator matrix* $\hat{f}(z)$. *Then the free distance of* CC *is at least* $d + 2$.

Proof. Consider $w(z) = (\sum_{i=0}^{t} \beta_i z^i)\hat{f}$ and we wish to show that its weight is $\geq d + 2$. Here $\beta_i \in F$. Thus, we need to show that the sum of the weights of the coefficients of z in $w(z)$ is $\geq d + 2$. In calculating the weight of w we can assume $\beta_0 \neq 0$ and we also naturally assume $\beta_t \neq 0$.

Let $w_1(z) = \sum_{i=0}^{t} \beta_i z^i$. The weight of w_1, supp(w_1), is the number of non-zero β_i. Suppose then supp(w_1) $\geq d$. Then in $w(z)$, α_t appears with the coefficient of z^i, for at least d different i with $0 < i < t + r$. Also the coefficient of $1 = z^0$ is $\beta_0(1,1)$ and the coefficient of z^{t+r} is $\beta_t(1,1)$ and each of these have distance 2. Then by Lemma 6.5, $w(z)$ has free distance at least $d_2 + 2$.

Now weight of $w(z) = (\sum_{i=0}^{t} \beta_i z^i)\hat{f}$ is the same as the weight of $w_1(z) = (\sum_{i=0}^{t} \beta_i z^i)f(z)$. By Theorem 6.3 $w_1(z)$ has weight $\geq d_1$ or else $\sum_{i=0}^{t} \beta_i z^i$ has weight $\geq d_2$. In the first case $w(z)$ has free distance $\geq d_1 + 2$ and in the second case, as already shown, w has free distance $\geq d_2 + 2$. Thus w has free distance $\geq d + 2$ as required. □

The free distance may be bigger than $d + 2$; an upper bound is $2d - 1$. The free distance also depends on where the invertible element is placed in the expression for $f(z)$.

It is worth noting that if the weight of the input element is $\geq t$ then the free distance is at least $t + 2$; this may be seen from the proof of Theorem 6.4. Thus, it is possible to avoid short distance codewords by ensuring that the input elements have sufficient weight — this could be done by, for example, taking the complement of any element with small weight.

The best choice for $\mathcal{C}$ is possibly a self-dual code as in this case $d_1 = d_2 = d$.

These convolutional codes can be considered to be self-dual type convolutional codes in the sense that $f(z)$ determines the generator matrix and $f(z)/z^{2t}$ determines the control matrix.

6.3.12.2. $(2m, 1)$ *Codes*

Section 6.3.12 can be generalized to produce convolutional codes of smaller rate $(2m, 1)$ but with much bigger free distance. Essentially the free distance is multiplied by m over that obtained for similar $(2, 1)$ codes.

The group to consider is C_{2m} generated by a. Assume m is odd although similar results may be obtained when m is even. Let $\alpha = 1 + a + a^2 + \cdots + a^{2m-1}$ and $\alpha_0 = 1 + a^2 + \cdots + a^{2m-2}$. Then $\alpha^2 = 0$ and $\alpha_0^2 = 1$ as α_0 has odd weight.

Define as before $f(z) = \sum_{i=1}^{r} \alpha_i z^i$ where now $\alpha_i = \beta_i \alpha$ in $\mathbb{Z}_2 C_{2m}$. Replace some $\alpha_i \neq 0$, say α_t, by α_0.

Then $f(z)^2 = z^{2t}$ and $f(z)(f(z)/z^{2t}) = 1$. Thus use $f(z)$ to define a convolutional code $\mathcal{C}$ by taking the first row of the α_i.

For example $G(z) = (1, 1, 1, 1, 1, 1) + (1, 0, 1, 0, 1, 0)z + (1, 1, 1, 1, 1, 1)z^2$ defines a $(6, 1)$ convolutional code which has free distance 15. $G(z) = (1, 1, 1, 1, 1, 1) + (1, 0, 1, 0, 1, 0)z + (1, 1, 1, 1, 1, 1)z^3 + (1, 1, 1, 1, 1, 1)z^4$ defines a convolutional code which has free distance 21.

A theorem similar to Theorem 6.4 may also be proved. Let $f(g)$ denote the generator matrix of a cyclic code with distance d_1 and whose dual code has distance d_2. Let $d = \min(d_1, d_2)$ and let $\mathcal{CC}$ denote the convolutional code obtained from $f(z)$ where the coefficients of $f(g)$ have been replaced by α_i in all, but one coefficient which has been replaced by α_0 and the first row of each coefficient is used. Assume in the following theorem that α_0 is not the first or last coefficient.

Theorem 6.5. *The free distance of $\mathcal{CC}$ is at least $md + 2m$.*

6.3.13. *Higher Rate Convolutional Codes*

The methods of Sec. 6.3.12 can also be generalized to produce higher rate convolutional codes.

Consider achieving a rate of $3/4$. The rate of $1/2$ was achieved using FC_2.

In C_4 generated by a define $\alpha = 1 + a$ and $\alpha_0 = 1$. Then $\alpha^4 = 0$ and α can be used to define a (linear) code of rate $3/4$ and distance 2 in FC_4. Now α has matrix

$$\begin{pmatrix} 1 & 1 & 0 & 0 \\ 0 & 1 & 1 & 0 \\ 0 & 0 & 1 & 1 \\ 1 & 0 & 0 & 1 \end{pmatrix}$$

and the first three rows of this

$$A = \begin{pmatrix} 1 & 1 & 0 & 0 \\ 0 & 1 & 1 & 0 \\ 0 & 0 & 1 & 1 \end{pmatrix}$$

generates a $(4, 3, 2)$ code.

The matrix of α_0 is $I_{4 \times 4}$, the identity 4×4 matrix, and let B denote the first three rows of $I_{4 \times 4}$.

Lemma 6.6. *Let $\underline{x} \neq 0$ be a 1×3 vector. Then $\underline{x}(A + B)$ is not the zero vector and thus $\underline{x}(A + B)$ has distance at least 1.*

Proof. Now $(\alpha + 1)^4 = \alpha^4 + 1 = 1$ and so $(\alpha + 1)$ is a non-singular matrix. Thus in particular the first three rows of the matrix of $(\alpha + 1)$ are linearly independent. The first three rows of $\alpha + 1$ precisely constitutes the matrix $A + B$. Thus $\underline{x}(A + B)$ is not the zero vector.

Another way to look at this is that $\alpha + 1 = a$ but it is useful to look at the more general way in Lemma 6.6 for further developments. $\qquad\square$

Corollary 6.2. *If* $\underline{x}A + \underline{y}B = \underline{0}$ *then* $\underline{x} \neq \underline{y}$.

Form convolutional $(4, 3)$ codes as follows.

Let $f(z) = \sum_{i=0}^{n} \alpha_i z^i$ where $\alpha_i = \alpha$ or $\alpha_i = 0$ except for $\alpha_t = 1$ for some $t, 1 < t \leq n$. We could also use $\alpha_1 = \alpha_t = 1$, but this generally gives smaller distance codes.

Then $f(z)^4 = z^{4t}$ and so $f(z) \times (f(z)^3/z^{4t}) = 1$. Thus use $f(z)$ to generate the code and $(f(z)^3/z^{4t})$ to check/control the code. Take the first three rows of the matrix of $f(z)$ to generate a $(4, 3)$ code and delete the last three columns $(f(z)^3/z^{4t})$ to form the control matrix.

Thus $G(z) = \sum_{i=0}^{n} \hat{\alpha}_i z^i$ is the generator matrix where $\hat{\alpha}_i$ is the first three rows of the matrix of α_i.

In Sec. 6.3.12, we had the situation that when α_0 occurred in any coefficient then it contributed a distance of 1, so that when the weight of G is s then α_0 will contribute a free distance of s. Here we use the fact that if α_0 occurs then it will contribute a distance of at least 1 unless its coefficient equals the sum of the coefficients in the other non-zero α_i which occur with it in the same coefficient of z^j.

6.3.14. *Examples*

The generator matrix

$$G = \begin{pmatrix} 1 & 1 & 0 & 0 \\ 0 & 1 & 1 & 0 \\ 0 & 0 & 1 & 1 \end{pmatrix} + \begin{pmatrix} 1 & 0 & 0 & 0 \\ 0 & 1 & 0 & 0 \\ 0 & 0 & 1 & 0 \end{pmatrix} z + \begin{pmatrix} 1 & 1 & 0 & 0 \\ 0 & 1 & 1 & 0 \\ 0 & 0 & 1 & 1 \end{pmatrix} z^2$$

defines a $(4, 3)$ convolutional code. It may be shown that its free distance is 5. The proof is similar to the proof of Theorem 6.1 but also using Lemma 6.6.

The check matrix for the code is easy to write out.

Consider $n = 3$, and

$$G = \begin{pmatrix} 1 & 1 & 0 & 0 \\ 0 & 1 & 1 & 0 \\ 0 & 0 & 1 & 1 \end{pmatrix} + \begin{pmatrix} 1 & 0 & 0 & 0 \\ 0 & 1 & 0 & 0 \\ 0 & 0 & 1 & 0 \end{pmatrix} z + \begin{pmatrix} 1 & 1 & 0 & 0 \\ 0 & 1 & 1 & 0 \\ 0 & 0 & 1 & 1 \end{pmatrix} z^2$$

$$+ \begin{pmatrix} 1 & 1 & 0 & 0 \\ 0 & 1 & 1 & 0 \\ 0 & 0 & 1 & 1 \end{pmatrix} z^3$$

This is a $(4, 3)$ convolutional code and its free distance is 6.
The next example is

$$G = \begin{pmatrix} 1 & 1 & 0 & 0 \\ 0 & 1 & 1 & 0 \\ 0 & 0 & 1 & 1 \end{pmatrix} + \begin{pmatrix} 1 & 0 & 0 & 0 \\ 0 & 1 & 0 & 0 \\ 0 & 0 & 1 & 0 \end{pmatrix} z + \begin{pmatrix} 1 & 1 & 0 & 0 \\ 0 & 1 & 1 & 0 \\ 0 & 0 & 1 & 1 \end{pmatrix} z^3$$

$$+ \begin{pmatrix} 1 & 1 & 0 & 0 \\ 0 & 1 & 1 & 0 \\ 0 & 0 & 1 & 1 \end{pmatrix} z^4$$

This has free distance 7. This may be proved similar to Theorem 6.2 using Lemma 6.6.

It is then possible to proceed as in Sec. 6.3.12 to investigate further degrees (memories) with rate $3/4$.

6.3.15. *Polynomial Generator and Control Matrices*

In cases where a polynomial generator *and* polynomial right inverse for this generator are required, insist that $\alpha_0 = 1$. This gives slightly less free distance but is interesting in itself.

For example, consider the encoder matrix $G = (1, 0) + \delta_1(1, 1)z + \cdots + \delta_n(1, 1)z^n$ and the control matrix is

$$H = \begin{pmatrix} 1 \\ 0 \end{pmatrix} + \delta_1 \begin{pmatrix} 1 \\ 1 \end{pmatrix} z + \cdots + \delta_n \begin{pmatrix} 1 \\ 1 \end{pmatrix} z^n.$$

Here $\delta_i = 0$ or $\delta_i = 1$.

This code has free distance 4 for $n = 2$. For $n \geq 2$, the free distance will depend on the choice of the δ_i. As already noted, the choices where the z-polynomial corresponds to a known cyclic code polynomial deserve particular attention.

We may also increase the size of the field as, for example, as follows.

Consider now $R = GF(4)C_2$, the group ring of the cyclic group of order 2 over the field of 4 elements. Define $\alpha_0 = w + w^2 g$, $\alpha_1 = w + wg$, $\alpha_2 = w^2 + w^2 g$, where w is the primitive element in $GF(4)$ which satisfies $w^2 + w + 1 = 0, w^3 = 1$. Then $\alpha_0^2 = w^2 + w^4 = w^2 + w = 1$ and $\alpha_1^2 = \alpha_2^2 = 0$. Thus $w = \alpha_0 + \alpha_1 z + \alpha_2 z^2$ satisfies $w^2 = 1$ and can be used to define a convolutional code of length 2 and dimension 1. The encoder matrix is then $G = (w, w^2) + \delta_1(w, w)z + \delta_2(w^2, w^2)z^2 + \cdots + \delta_n(w^i, w^i)z^n$ and the control matrix is

$$H = \begin{pmatrix} w^2 \\ w \end{pmatrix} + \delta_1 \begin{pmatrix} w \\ w \end{pmatrix} z + \cdots + \delta_n \begin{pmatrix} w^i \\ w^i \end{pmatrix} z^n.$$

The *degree* of a convolutional code with encoder matrix $G(z)$ is defined to be the maximal degree of the full $k \times k$ size minors of $G(z)$ where k is the dimension; see [5]. The maximum free distance of a length 2, dimension one, degree δ code over any field is by [14], $2\delta + 2$.

Consider the case $n = 2$ and let the code be denoted by $\mathcal{C}$. The encoder matrix is $G = (w, w^2) + (w, w)z + (w^2, w^2)z^2$. The degree of this code is $\delta = 2$ since the dimension is 1. Let $G' = (1, w) + (1, 1)z + (w, w)z^2$ so that $wG' = G$.

Theorem 6.6. *The free distance of $\mathcal{C}$ is 6 and so is thus a maximum distance separable convolutional code.*

Proof. Consider combinations $(\alpha_0 + \alpha_1 z + \cdots + \alpha_t z^t)G$ and we wish to show that this has (free) distance 6. We may assume $\alpha_0 \neq 0$. It is clear when $t = 0$ that w has a distance of 6 and so in particular a distance of 6 is attained. Since also w is a factor of G we may now consider the minimum distance of $w = (\alpha_0 + \alpha_1 z + \cdots + \alpha_t z^t)G'$ with $\alpha_0 \neq 0, \alpha_t \neq 0$, and $t > 0$. The coefficient of z^0 is $\alpha_0(1, w)$; the coefficient of z^{t+2} is $\alpha_t(w, w)$, the coefficient of z^{t+1} is $\alpha_t(1, 1) + \alpha_{t-1}(w, w)$ and the coefficient of z^t is $\alpha_t(1, w) + \alpha_{t-1}(1, 1) + \alpha_{t-2}(w, w)$ when $t \geq 2$ and the coefficient of z is $\alpha_1(1, w) + \alpha_0(1, 1)$ and this is also the case when $t = 1$.

Case $t \geq 2$: If $\alpha_t \neq \alpha_{t-1}w$ then the coefficient of z^{t+1} has distance 2 giving a distance of 6 with 2 coming from each of the coefficients of z^0, z^{t+1}, z^{t+2}. If $\alpha_t = \alpha_{t-1}w$ the coefficient of z^t is $\alpha_{t-1}(w + 1, w^2 + 1) +$

$\alpha_{t-1}(\omega, \omega)$; in any case this has distance ≥ 1. Also the coefficient of z has distance ≥ 1. Thus, the total distance is at least $2 + 1 + 1 + 2 = 6$.

Case $t = 1$. If $\alpha_0 \omega \neq \alpha_1$ then the coefficient of z^2 has distance 2 and thus get a distance of $2 + 2 + 2 = 6$ for the coefficients of z^0, z^2, z^3. If $\alpha_0 \omega = \alpha_1$ then the coefficient of z is $\alpha_1(1, \omega) + \alpha_0(1, 1) = \alpha_0(\omega + 1, \omega^2 + 1)$ which has distance 2. Thus also we get a distance of $2 + 2 + 2 = 6$ from coefficients of z^0, z, z^3.

Note that the proof depends on the fact that $\{1, \omega\}$ is linearly independent in $GF(4)$.   □

6.3.16. *General Rank Considerations*

Let $w(z) = \sum_{i=0}^{t} \alpha_i z^i$ where $\alpha_i^2 = 0, i \neq t, \alpha_t^2 = 1$ with the α_i in some group ring RG. Suppose the α_i commute and that R has characteristic 2. Then $w(z)^2 = z^{2t}$.

Consider the ranks of the non-zero α_i in deciding which rows of w to choose with which to construct the convolutional code. For example, if the non-zero α_i satisfy rank $\alpha_i = 1/2|G| = m$ we choose the matrix with just half the rows of the matrix of each α_i.

Many good codes may be produced this way.

It is possible to have more than one α_t satisfying $\alpha_t^2 = 1$ in $w(z)$ but then the generator matrix produced can be catastrophic, although a valid code may still be defined.

6.3.16.1. *Example*

Let $u = 1 + h(a + a^2 + a^3)$ in $\mathbb{Z}_2(C_4 \times C_2)$. Then $u^2 = 0$ and rank $u = 4$. Define $w = u + z + uz^2$. Then $w^2 = z^2$ and w is used to define a $(8, 4)$ convolutional code. The generator matrix is

$$G = (I, B) + (I, 0)z + (I, B)z^2 \quad \text{where } B = \begin{pmatrix} 0 & 1 & 1 & 1 \\ 1 & 0 & 1 & 1 \\ 1 & 1 & 0 & 1 \\ 1 & 1 & 1 & 0 \end{pmatrix}.$$

Now (I, B) has distance 4. Any combination of $(I, B), (I, 0)$ has distance 1 at least as B is non-singular. Thus consider $(\sum_{i=0}^{t} \beta_i z^t)G$. The highest and lowest power of z has distance 4 and there is a power of z in between which has distance 1 so altogether we get a free distance of 9. The degree of the code is 8.

This can be extended. It can also be extended by finding higher dimensional u with $u^2 = 0$. See Sec. 6.3.21 for further development of these ideas.

6.3.17. *Higher Rate Convolutional Codes Using Nilpotent Elements*

So far we have used α_i with $\alpha_i^2 = 0$ and this generally give rate $1/2$ convolutional codes. We now look at elements α with $\alpha^4 = 0$ with which to produce convolutional rate $3/4$ codes.

See Sec. 6.3.13 for some preliminary examples on these.

Suppose then $w = \sum_{i=0}^{n} \alpha_i z^i$ in $\mathbb{F}G$ where $\alpha_i^4 = 0, i \neq t$ and $\alpha_t^4 = 1, 1 \leq t \leq n$. Suppose also $\mathbb{F}$ has characteristic 2 and that the α_i commute. Then $w^4 = z^{4t}$. Thus, w is used to generate a $3/4$ rate convolutional code by taking the first $3/4$ of the rows of the α_i; then w^3/z^{4t} will be the control matrix using the last $1/4$ of the columns of the α_i.

6.3.17.1. *Example*

Consider $\alpha = a + a^7 \in \mathbb{Z}_2 C_8$. Then $\alpha_i^4 = 0$ and α generates an $(8, 6, 2)$ linear cyclic code — this is the best distance for a linear $(8, 6)$ code. Now construct convolutional codes similar to the construction of the $(2, 1)$ codes.

An element $\alpha_0 \in \mathbb{Z}_2 C_8$ such that $\alpha_0^4 = 1$ is needed. There are a number of choices including $\alpha_0 = 1, \alpha_0 = 1 + a + a^3, \alpha_0 = 1 + a + a^7$. Choose α_0 so that the first three rows of the matrix of α_0 generates a linear code of largest distance. It is easy to verify that the first three rows of $\alpha_0 = 1 + a + a^3$ generates a linear code of distance 2.

- $w = \alpha + \alpha_0 z$. This gives a $(8, 6)$ code of free distance 4. The "degree" in the convolutional sense is 6.
- $w = \alpha + \alpha_0 z + \alpha z^2$. This is a $(8, 6)$ convolutional code of free distance 6. The "degree" here is 12.
- $w = \alpha + \alpha_0 z + \alpha z^2 + \alpha z^3$ gives an $(8, 6)$ code of free distance 6.
- $w = \alpha + \alpha_0 z + \alpha z^3 + \alpha z^4$ gives an $(8, 6)$ code of free distance 8.
- Polynomial degree 5: $w = \alpha + \alpha_0 z + \alpha z^3 + \alpha z^4 + \alpha z^5$. The free distance has to be determined.
- Polynomial degree 6: $w = \alpha + \alpha z^2 + \alpha z^3 + \alpha_0 z^4 + \alpha z^5 + \alpha z^6$. This should give a free distance of at least 10.
- As for the $(2, 1)$ convolutional codes in Sec. 6.3.12, by mimicking the polynomials used to generate cyclic codes, it should be possible to get $(8, 6)$ convolutional codes with increasing free distance.

6.3.18. *Using Idempotents to Generate Convolutional Codes*

Let FG be the group ring over a field F. In applications it may be necessary to require that $\operatorname{char} F \nmid |G|$. It may also be necessary to require that F contain a primitive nth root of unity. See [7] for further details.

Let $\{e_1, e_2, \ldots, e_k\}$ be a complete family of orthogonal idempotents in FG. Such sets always exist when $\operatorname{char} F \nmid |G|$.

Thus:

  (i) $e_i \neq 0$ and $e_i^2 = e_i$, $1 \leq i \leq k$.
  (ii) If $i \neq j$ then $e_i e_j = 0$.
  (iii) $1 = e_1 + e_2 + \cdots + e_k$.

Here 1 is used for the identity of FG.

Theorem 6.7. Let $f(z) = \sum_{i=0}^{k} \pm e_i z^{t_i}$. Then $f(z)f(z^{-1}) = 1$.

Proof. Since $e_1, e_2, \ldots, e_k$ is a set of orthogonal primitive idempotents, $f(z)f(z^{-1}) = e_1^2 + e_2^2 + \cdots + e_k^2 = 1$. $\qquad\square$

The result in Theorem 6.7 can be considered as an equation in RC_∞ with $R = FG$.

To now construct convolutional codes, decide on the rank r and then use the first r rows of the matrices of the e_i in Theorem 6.7. The control matrix is obtained from $f(z^{-1})$ by deleting the last r columns of the e_i.

If the e_i have rank $\geq k$ and for some i rank $e_i = k$ then it is probably best to take the $r = k$ for the rank of the convolutional code, although other cases also have uses depending on the application in mind.

6.3.18.1. *Idempotents in Group Rings*

Orthogonal sets of idempotents may be obtained in group rings from the conjugacy classes and character tables, see e.g. [7].

In the ring of matrices define e_{ii} to be the matrix with 1 in the ith diagonal and zeros elsewhere. Then $e_{11}, e_{22}, \ldots, e_{nn}$ is a complete set of orthogonal idempotents and can be used to define such $f(z)$. These in a sense are trivial but can be useful and can also be combined with others.

Notice also that a product $\Pi(z)$ of matrices $f(z)$ satisfying $f(z)f(z^{-1}) = 1$ also satisfies $\Pi(z)\Pi(z^{-1}) = 1$ and this product can also be used to define convolutional codes.

To construct convolutional codes from idempotents proceed as follows:

- Find sets of orthogonal idempotents.
- Decide on the $f(z)$ to be used with each set.
- Take the product of the $f(z)$.
- Decide on the rate.
- Convert these idempotents into matrices as per the isomorphism between the group ring and a ring of matrices.
- Construct the codes as in Sec. 6.3.4.

Group rings are a rich source of complete sets of orthogonal idempotents. This brings us into character theory and representation theory in group rings. Group rings and their relationships to representation theory was greatly enhanced by the fundamental work of Emmy Noether [15].

Orthogonal sets over the rationals and other fields are also obtainable.

The Computer Algebra packages GAP and Magma can construct character tables and conjugacy classes from which complete sets of orthogonal idempotents may be obtained.

6.3.18.2. *Example 1*

Consider CC_2 where C_2 is generated by a. Define $e_1 = \frac{1}{2}(1 + a)$ and $e_2 = 1 - e_1 = \frac{1}{2}(1 - a)$. This gives $f(z) = e_1 + e_2 z^t$ or $f(z) = e_2 + e_1 z^t$ for various t. Products of these could also be used but in this case we get another of the same form by a power of z.

6.3.18.3. *Cyclic*

The orthogonal idempotents and character table of the cyclic group are well known and are closely related to the Fourier matrix.

This gives for example in C_4, $e_1 = \frac{1}{4}(1 + a + a^2 + a^3)$, $e_2 = \frac{1}{4}(1 + wa + w^2 a^2 + w^3 a^3)$, $e_3 = \frac{1}{4}(1 - a + a^2 - a^3)$, $e_4 = \frac{1}{4}(1 + w^3 a + w^2 a^2 + wa^3)$ from which 4×4 matrices with degree 4 in z may be constructed, where w is a primitive 4th root of unity. Notice in this case that $w^2 = -1$.

Let $f(z) = e_1 + e_2 z + e_3 z^2 + e_4 z^3$. Then $f(z)f(z^{-1}) = 1$. We take the first row of the matrices to give the following generator matrix for a $(4, 1, 3)$ convolutional code:

$$G(z) = \frac{1}{4}\{(1, 1, 1, 1) + (1, w, -1, -w)z + (1, -1, 1, -1)z^2$$
$$+ (1, -w, -1, w)z^3\}.$$

It is easy to check that a combination of any one, two, or three of the vectors $(1, 1, 1, 1), (1, \omega, -1, -\omega), (1, -1, 1, -1), (1, -\omega, -1, \omega)$, which are the rows of the Fourier matrix, has distance at least 2 and a combination of all four of them has distance 1. From this, it is easy to show that the code has free distance 14 — any combination of more than one will have 4 at each end and three in the middle with distance at least 2. This gives a $(4, 1, 3, 14)$ convolutional codes which is optimal — see [14].

We can combine the e_i to get real sets of orthogonal idempotents. Note that it is enough to combine the conjugacy classes of g and g^{-1} in order to get real sets of orthogonal idempotents.

In this case then we get

$$\hat{e}_1 = e_1 = \frac{1}{4}(1 + a + a^2 + a^3), \quad \hat{e}_2 = e_2 + e_4 = \frac{1}{2}(1 - a^2),$$

$$\hat{e}_3 = e_3 = \frac{1}{4}(1 - a + a^2 - a^3),$$

which can then be used to construct real convolutional codes.

Then $G(z) = \frac{1}{4}\{(1, 1, 1, 1) + 2(2, 0, -2, 0)z + (1, -1, 1, -1)z^2\}$ gives a $(4, 1, 2)$ convolutional code. Its free distance is 10 which is also optimal.

Using $C_2 \times C_2$ gives different matrices. Here the set of orthogonal idempotents consists of $e_1 = \frac{1}{4}(1 + a + b + ab), e_2 = \frac{1}{4}(1 - a + b - ab), e_3 = \frac{1}{4}(1 - a - b + ab), e_4 = \frac{1}{4}(1 + a - b - ab)$ and the matrices derived are all real.

This gives $G(z) = \frac{1}{4}\{(1, 1, 1, 1) + (1, -1, 1, -1)z + (1, -1, -1, 1)z^2 + (1, 1, -1, -1)z^3\}$. Its free distance also seems to be 14.

6.3.18.4. *Symmetric Group*

The orthogonal idempotents of the symmetric group are well understood and are real.

We present an example here from S_3, the symmetric group on three letters.

Now $S_3 = \{1, (1, 2), (1, 3), (2, 3), (1, 2, 3), (1, 3, 2)\}$ where these are cycles. We also use this listing of S_3 when constructing matrices.

There are three conjugacy classes: $K_1 = \{1\}$; $K_2 = \{(1, 2), (1, 3), (2, 3)\}$; and $K_3 = \{(1, 2, 3), (1, 3, 2)\}$.

Define

$$\hat{e}_1 = 1 + (1, 2) + (1, 3) + (2, 3) + (1, 2, 3) + (1, 3, 2),$$
$$\hat{e}_2 = 1 - \{(1, 2) + (1, 3) + (2, 3)\} + (1, 2, 3) + (1, 3, 2),$$
$$\hat{e}_3 = 2 - \{(1, 2, 3) + (1, 3, 2)\},$$

and $e_1 = \frac{1}{6}\hat{e}_1; e_2 = \frac{1}{6}\hat{e}_2; e_3 = \frac{1}{3}\hat{e}_3$. Then $\{e_1, e_2, e_3\}$ form a complete orthogonal set of idempotents and may be used to construct convolutional codes.

The G-matrix 1 (see [8]) is

$$
\begin{pmatrix}
1 & (12) & (13) & (23) & (123) & (132) \\
(12) & 1 & (132) & (123) & (23) & (13) \\
(13) & (123) & 1 & (132) & (12) & (23) \\
(23) & (132) & (123) & 1 & (13) & (12) \\
(132) & (23) & (12) & (13) & 1 & (123) \\
(123) & (13) & (23) & (21) & (132) & 1
\end{pmatrix}.
$$

Thus the matrices of e_1, e_2, e_3 are, respectively,

$$
E_1 = \frac{1}{6}
\begin{pmatrix}
1 & 1 & 1 & 1 & 1 & 1 \\
1 & 1 & 1 & 1 & 1 & 1 \\
1 & 1 & 1 & 1 & 1 & 1 \\
1 & 1 & 1 & 1 & 1 & 1 \\
1 & 1 & 1 & 1 & 1 & 1 \\
1 & 1 & 1 & 1 & 1 & 1
\end{pmatrix},
$$

$$
E_2 = \frac{1}{6}
\begin{pmatrix}
1 & -1 & -1 & -1 & 1 & 1 \\
-1 & 1 & 1 & 1 & -1 & -1 \\
-1 & 1 & 1 & 1 & -1 & -1 \\
-1 & 1 & 1 & 1 & -1 & -1 \\
1 & -1 & -1 & -1 & 1 & 1 \\
1 & -1 & -1 & -1 & 1 & 1
\end{pmatrix},
$$

and

$$
E_3 = \frac{1}{3}
\begin{pmatrix}
2 & 0 & 0 & 0 & -1 & -1 \\
0 & 2 & -1 & -1 & 0 & 0 \\
0 & -1 & 2 & -1 & 0 & 0 \\
0 & -1 & -1 & 2 & 0 & 0 \\
-1 & 0 & 0 & 0 & 2 & -1 \\
-1 & 0 & 0 & 0 & -1 & 2
\end{pmatrix}.
$$

Note that e_1, e_2 have rank 1 and that e_3 has rank 2.

6.3.19. *Convolutional Codes Over Fields of Other Characteristic*

Convolutional codes over fields of arbitrary characteristic, and not just characteristic 2, may also be constructed using the general method of Sec. 6.3.4. The following theorem is similar to Lemma 6.2.

Lemma 6.7. *Let $R = FG$ be the group ring of a group G over a field F with characteristic p. Suppose $\alpha_i \in R$ commute and $\gamma_i \in F$. Let $w = \sum_{i=0}^{n} \alpha_i \gamma_i z^i \in RC_\infty$. Then $w^p = \gamma_t^p z^{pt}$ if and only if $\alpha_i^p = 0, i \neq t$ and $\alpha_t^p = 1$.*

The situation with $\gamma_t = 1$ is easiest to deal with and is not a great restriction.

Suppose now $|G| = n$. Construct convolutional codes as follows. Find elements α_i with $\alpha_i^p = 0$ and units u with $u^p = 1$ in the group ring R. Then use Lemma 6.7 to form units in $R[z]$. Thus get $f(z)^p = \gamma_t^p z^{pt}$ and hence $f(z) \times f(z)^{p-1}/(\gamma_t^p z^{pt}) = 1$. From these units, convolutional codes are defined as in Sec. 6.3.4.

Thus $f(z)$ may be used to define a convolutional code. Choose the first r rows of the α_i, considered as matrices, to define a (n, r) convolutional code. The generator matrix is $\hat{f}(z) = \sum_{i=0}^{n} \hat{\alpha}_i \gamma_i z^i$ where $\hat{\alpha}_i$ denotes the first r rows of the matrix of α_i.

It is necessary to decide which rows of the matrix to choose in defining the convolutional code. This is usually best decided by considering the ranks of the non-zero α_i.

6.3.19.1. *Examples in Characteristic 3*

Suppose then F has characteristic 3 and consider $F(C_3 \times C_3)$ where the C_3 are generated, respectively, by g, h.

Define $\alpha = 1 + h(1 + g)$. Then $\alpha^3 = 0$. Define $\alpha_0 = 2 + 2h$. Then $\alpha_0^3 = 1$. The matrix of α is

$$P = \begin{pmatrix} I & B & 0 \\ 0 & I & B \\ B & 0 & I \end{pmatrix}$$

where I is the identity 3×3 matrix, 0 is the zero 3×3 matrix and

$$B = \begin{pmatrix} 1 & 1 & 0 \\ 0 & 1 & 1 \\ 1 & 0 & 1 \end{pmatrix}.$$

By row (block) operations P is equivalent to $\begin{pmatrix} I & 0 & -B^2 \\ 0 & I & B \\ 0 & 0 & 0 \end{pmatrix}$. Thus P has rank 6 and the matrix

$$Q = \begin{pmatrix} I & 0 & -B^2 \\ 0 & I & B \end{pmatrix}$$

defines a block $(9, 6)$ code which indeed has distance 3.

Now define $\alpha_t = \alpha_0$ for some $0 < t < n$ and choose $\alpha_i = 0$ or $\alpha_i = \alpha$ for $i \neq t$. Define $f(z) = \sum_{i=0}^n \alpha_i z^i$. Then by Lemma 6.7, $f(z)^3 = z^{3t}$ and hence $f(z) \times (f(z)^2 / z^{3t}) = 1$. Choose the first 6 rows of the α_i in $f(z)$ to define the code and thus we get a $(9, 6)$ convolutional code. The generator matrix is $\hat{f}(z) = \sum_{i=0}^n \hat{\alpha}_i z^i$ where $\hat{\alpha}_i$ denotes the first 6 rows of α_i, considered as a matrix. The control matrix is obtained from $f(z)^2 / z^{3t}$ using the last three columns of the α_i.

Lemma 6.8. $\underline{x}\hat{\alpha}_i + \underline{y}\hat{\alpha}_0$ *has distance at least 1 for 1×6 vectors $\underline{x}, \underline{y}$ with $\underline{y} \neq \underline{0}$.*

6.3.19.2. *Specific Examples for Characteristic 3*

Define $f(z) = \alpha + \alpha_0 z + \alpha z^2$. Then $\hat{f}(z) = \hat{\alpha} + \hat{\alpha}_0 z + \hat{\alpha} z^2$ is a convolutional $(9, 6)$ code of free distance 8.

Define $f(z) = \alpha + \alpha_0 z + \alpha z^3 + \alpha z^4$. Then $\hat{f}(z) = \hat{\alpha} + \hat{\alpha}_0 z + \hat{\alpha} z^3 + \hat{\alpha} z^4$ defines a $(9, 6)$ convolutional code which has free distance 11.

6.3.19.3. *A General Result for Characteristic 3*

A result similar to Theorem 6.4 can also be proved. Suppose now $\mathcal{C}$ is a cyclic (n, k, d_1) code over the field F of characteristic 3. Suppose also that the dual of $\mathcal{C}$, denoted $\hat{\mathcal{C}}$, is an $(n, n - k, d_2)$ code.

Let $d = \min(d_1, d_2)$. Suppose $f(g) = \sum_{i=0}^r \beta_i g^i$, with $\beta_i \in F, (\beta_r \neq 0)$, is a generating polynomial for $\mathcal{C}$. In $f(g)$, assume $\beta_0 \neq 0$.

Consider $f(z) = \sum_{i=1}^r \alpha_i z^i$ where now $\alpha_i = \beta_i \alpha$ with α as above in $F(C_3 \times C_3)$. Note that if $\beta_i = 0$ then $\alpha_i = 0$. Replace some $\alpha_i \neq 0$, say α_t, by α_0 (considered as members of $F(C_3 \times C_3)$).

So assume $f(z) = \sum_{i=0}^r \alpha_i z^i$ with this $\alpha_t = \alpha_0$ and other $\alpha_i = \beta_i \alpha$ (for $i \neq t$).

Then $f(z)^3 = \beta_t^3 z^{3t}$ giving that $f(z) \times (f(z)^2 / (\beta_t^3 z^{3t}) = 1$. We now use $f(z)$ to generate a convolutional code by taking the first six rows of the α_i.

Thus the generating matrix is $\hat{f}(z) = \sum_{i=0}^{r} \hat{\alpha}_i \beta_i z^i$ where $\hat{\alpha}_i$ consists of the first six rows of α for $i \neq t$ and $\hat{\alpha}_t$ consists of the first six rows of α_0.

For the following theorem assume the invertible element α_0 does not occur in the first or the last position of f.

Theorem 6.8. *Let* CC *denote the convolutional code with generator matrix* $\hat{f}$. *Then the free distance of* CC *is at least* $d + 4$.

Note that this code CC has rate $\frac{6}{9} = \frac{2}{3}$. By considering fields of characteristic p it should be possible to obtain good convolutional codes of rate $\frac{p-1}{p}$ by this method.

6.3.20. *Nilpotent Type*

Many group rings R have elements α such that $\alpha^n = 0$ (and $\alpha^r \neq 0, r < n$). These can be exploited to produce convolutional codes.

Consider $\mathbb{F}C_{14}$ where $\mathbb{F}$ has characteristic 2. Let $w_0 = 1 + g^5 + g^6 + g^{12} + g^{13}, w_1 = 1 + g^2 + g^5 + g^7 + g^9 + g^{12}, w_2 = 1 + g + g^3 + g^7 + g^8 + g^{10}$ and define $p = w_0 + w_1 z + w_2 z^2$. Then $p^2 = 1$. Since $w_i^2 = 0$ for $i \geq 1$, consider a rate of $\frac{1}{2}$. Thus, consider the convolutional code with encoder matrix obtained from the first seven rows of p and then the control matrix is obtained from the last seven columns of p.

Now consider $\mathbb{Z}_2 C_8$ generated by g. Define $u = \alpha_0 + (1 + g^4)z + (1 + g^2)z^2 + (1 + g)z^3$ where $\alpha_0^8 = 1$. There are a number of choices for α_0, e.g. $\alpha_0 = 1 + g + g^3$.

Then $u^2 = \alpha_0^2 + (1 + g^8)z + (1 + g^4)z^2 + (1 + g^2)z^6 = \alpha_0^2 + (1 + g^4)z^2 + (1 + g^2)z^6$, $u^4 = \alpha_0^4 + (1 + g^4)z^{12}$, and $u^8 = 1$.

Then u can be used to define a convolutional code. Now $1 + g^4$ has rank 4 so for best results make it an $(8, 4)$ convolutional code by taking the first four rows of the matrices of u.

This is an $(8, 4, 9)$ convolutional code with degree/memory $\delta = 6$.

The rate could be increased but this would reduce the contribution from $1 + g^4$ matrix to distance essentially 0 as it has rank $= 4$. This would give a $(8, 6, 7)$ convolutional code.

To go further, consider $\mathbb{Z}_2 C_{16}$ etc. Here use degree 6 or 3 as the largest power of z and it is then possible to get a $(16, 8, 9)$ convolutional code. As $\text{rank}(1 + g^4) = 8$ it is probably possible to construct a $(16, 12, 9)$ but details have not been worked out.

These are binary codes. Going to bigger fields should give better distances.

6.3.21. *Hamming Type*

Set $R = \mathbb{Z}_2(C_4 \times C_2)$. Suppose C_4 is generated by a and C_2 is generated by h. Consider $\alpha_0 = 1 + h(1 + a^2)$ and $\alpha_i = 1 + h(a + a^2 + a^3)$ or $\alpha_i = 0$ for $i > 0$. Then $\alpha_0^2 = 1$ and $\alpha_i^2 = 0$. Define $w(z) = \sum_{i=0}^{n} \alpha_i z^i$ in RC_∞. By Lemma 6.1, $w^2 = 1$.

Let

$$A = \begin{pmatrix} 1 & 0 & 1 & 0 \\ 0 & 1 & 0 & 1 \\ 1 & 0 & 1 & 0 \\ 0 & 1 & 0 & 1 \end{pmatrix}, \quad B = \begin{pmatrix} 0 & 1 & 1 & 1 \\ 1 & 0 & 1 & 1 \\ 1 & 1 & 0 & 1 \\ 1 & 1 & 1 & 0 \end{pmatrix}$$

and I is the identity 4×4 matrix. The matrix corresponding to α_0 is then $\begin{pmatrix} I & A \\ A & I \end{pmatrix}$ and the matrix corresponding to α_i, $i \neq 0$, is either $\begin{pmatrix} I & B \\ B & I \end{pmatrix}$ or the zero matrix.

Now specify that the first four rows of w formulate the generator matrix of a code and then the last four columns of w formulate the control matrix. This gives a convolutional code of length 8 and dimension 4. It is easy to transform the resulting code into a systematic code.

The generator matrix is

$$G(z) = (I, A) + \delta_1(I, B)z + \delta_2(I, B)z^2 + \cdots + \delta_n(I, B)z^n, \text{ where } \delta_i \in \{0, 1\}.$$

The control matrix is

$$H(z) = \begin{pmatrix} A \\ I \end{pmatrix} + \delta_1 \begin{pmatrix} B \\ I \end{pmatrix} z + \delta_2 \begin{pmatrix} B \\ I \end{pmatrix} z^2 + \cdots + \delta_n \begin{pmatrix} B \\ I \end{pmatrix} z^n.$$

The (I, A) may be moved to the coefficient of any z^i in which case the (natural) control matrix will need to be divided by a power of z to get the true control matrix.

This convolutional code may be considered as a Hamming type convolutional code as (I, B) is a generator matrix of the Hamming $(8, 4)$ code.

For $n = 1$ the free distance turns out to be 6; this can be proved in a similar manner to Theorem 6.6.

An example of this type is $G(z) = (I, B) + (I, A)z + (I, B)z^2$ with control matrix $H(z)/z^2$ where $H(z) = \begin{pmatrix} B \\ I \end{pmatrix} + \begin{pmatrix} A \\ I \end{pmatrix} z + \begin{pmatrix} B \\ I \end{pmatrix} z^2$ has free distance 10.

6.3.21.1. *From Cyclic to Hamming Type*

For $n \geq 2$, proceed as previously to define the polynomials by reference to corresponding cyclic linear polynomials. This will give convolutional codes of this type which increasing free distance. Note that (I, A) has distance 2, (I, B) (the Hamming Code) has distance 4, any combination of (I, A) and (I, B) has distance ≥ 1.

The following may be proved in a similar manner to Theorem 6.4.

Suppose now $\mathcal{C}$ is a cyclic (n, k, d_1) code over the field F of characteristic 2 and that the dual of $\mathcal{C}$, $\hat{\mathcal{C}}$, is an $(n, n - k, d_2)$ code. Let $d = \min(d_1, d_2)$.

Assume $f(g) = \sum_{i=1}^{r} \beta_i g^i$ is a generator polynomial for $\mathcal{C}$. In $f(g)$, it is possible to arrange that $\beta_0 \neq 0$ and naturally assume that $\beta_r \neq 0$. Define $f(z) = \sum_{i=1}^{r} \alpha_i z^i$ with the $\alpha_i = \beta_i \alpha_i$, $i \neq t$ and $\alpha_t = \alpha_0$.

Then $f(z)^2 = z^{2t}$ giving $f(z) \times f(z)/z^{2t} = 1$. Now use $f(z)$ to generate a convolutional code $\mathcal{CC}$ by taking just the first four rows of the α_i. Thus the generating matrix is $G = \sum_{i=0}^{r} \hat{\alpha}_i z^i$ where $\hat{\alpha}_i$ consists of the first four rows of the matrix of α_i.

Theorem 6.9. *$\mathcal{CC}$ has free distance at least $d + 8$.*

6.4. Exercises and Suggested Research

6.4.1. *Exercises*

6.4.1.1. *LDPC and Collections of Differences*

(1) Determine the collection of differences of $v = 1 + g + g^3$ in $\mathbb{Z}_2 C_7$. Is v a unit or a zero-divisor in $\mathbb{Z}_2 C_7$? What is the rank of the matrix V corresponding to v? Construct a $(7, 3)$ code with no short cycles in its check matrix from V and determine its distance. Is this code familiar?

(2) Determine the collection of differences of $v = 1 + g + g^3$ in $\mathbb{Z}_2 C_8$. Is v a unit or a zero-divisor in $\mathbb{Z}_2 C_8$? What is the rank of the matrix V of v? Construct an $(8, 4)$ code with no short cycles in its check matrix from V and determine its distance.

(3) Determine the collection of differences of $v = 1 + g + g^3 + g^7$ in $\mathbb{Z}_2 C_{12}$. Show that v has short cycles in its matrix V and determine where these occur. Is v a unit? What is the rank of V? Construct a $(12, 6)$ code from the matrix V which has no short cycles. Determine the distance of this code.

Determine the collection of differences of $v = 1 + g + g^3 + g^7$ in $\mathbb{Z}_2 C_{15}$. Show that the matrix V of v has no short cycles. Is v a unit? What is the rank of V? Construct $(15, 8), (15, 10), (15, 5)$ codes using columns from V and determine their distances.

(4) Consider $v = 1 + g + g^3 + g^7 + g^{12}$. Construct the collection of differences of v in (1) $\mathbb{Z}_2C_{20}$; (2) $\mathbb{Z}_2C_{25}$.

   Is v a unit in $\mathbb{Z}_2C_{25}$? What is its rank in this group ring? (The rank of a group ring element is the rank of its corresponding matrix.)

   Construct a $(25, 12)$ code in $\mathbb{Z}_2C_{25}$ from v with no short cycles and determine its distance.

(5) Let $v = 1 + g + g^3 + g^7 + g^{12}$. Show that the matrix of v has no short cycles in $\mathbb{Z}_2C_n$ if and only if $n \geq 25$.

(6) Consider $v = 1 + g + g^3 + g^7 + g^{12} + g^{20}$ in $\mathbb{Z}_2C_n$. Determine the values of n for which v has no short cycles in its matrix V.

(7) Construct a 20×20 unit circulant matrix over $\mathbb{Z}_2$ (derived from the group ring $\mathbb{Z}_2C_{20}$) with weight ≥ 3 with no short cycles. From this derive a $(20, 10)$ code with no short cycles and determine the distance of this code.

(8) Find a unit element v with weight ≥ 3 in $\mathbb{Z}_2(C_2 \times C_2 \times C_2 \times C_2)$. Determine the collection of differences of this unit element. Is it possible to find a unit with weight ≥ 2 in $\mathbb{Z}_2(C_2 \times C_2 \times C_2 \times C_2)$ which has no short cycles in its matrix?

(9) Show that the square of every element in $\mathbb{Z}_2C_2^n$ is either 0 or 1 according to whether the element has even or odd weight. Hence classify the units and zero-divisors in $\mathbb{Z}_2C_2^n$. Show that every element in $\mathbb{Z}_2C_2^n$ of weight ≥ 2 has short cycles in its matrix.

(10) Computer required: Construct a 100×100 low density circulant unit matrix with no cycles of length 4 derived from a unit of weight ≥ 5 in the group ring $\mathbb{Z}_2C_{100}$. From this derive a $(100, 50)$ LDPC code with no short cycles. Use MATLAB or other package to simulate the performance of the code you have constructed.

(11) Suppose C_9 is generated by g and C_2 is generated by h. Consider $v = 1 + g + g^3 + h(g + g^5)$ in $\mathbb{Z}_2(C_9 \times C_2)$. Determine the collection of differences of v. Use the matrix of v to construct an $(18, 9)$ code.

(12) Let $D_{18} = \langle a, b \mid a^2 = 1, b^9 = 1, ab = b^{-1}a \rangle$ be the dihedral group of order 18. Let $v = 1 + b + b^3 + a(b + b^5)$ in $\mathbb{Z}_2D_{18}$. Determine the collection of differences of v. Use the matrix of v to construct an $(18, 9)$ code.

6.4.1.2. *Convolutional Exercises*

(1) Show in $\mathbb{Z}_2C_7$ that $(1 + g + g^3)(1 + g + g^2 + g^4) = 0$. Here $f(g) = 1 + g + g^3$ generates the Hamming $(7, 4, 3)$ code and the dual is a $(7, 3, 4)$ code generated by $h(g) = 1 + g + g^2 + g^4$. Consider then, as in Sec. 6.3.12,

$f(z) = (1 + a) + z + (1 + a)z^3$ with coefficients inside the group ring $\mathbb{Z}_2 C_2$ where C_2 is generated by a. Take the first row of the matrices in $f(z)$ to generate a $(2, 1)$ convolutional code $\mathcal{C}$. What does Theorem 6.4 say about the free distance of $\mathcal{C}$? Calculate the free distance of $\mathcal{C}$ directly.

(2) Consider similar to the last exercise $(2, 1)$ convolutional codes obtained from $h(z) = (1 + a) + z + (1 + a)z^2 + (1 + a)z^4$ and from $h_1(z) = (1 + a) + (1 + a)z + z^2 + (1 + a)z^4$. Calculate their free distances.

(3) Use the generating polynomial for the Reed Solomon $(255, 233, 32)$ cyclic code to generate a convolutional $(2, 1)$ code by the method of Sec. 6.3.12. Give a lower bound to the free distance of the code using Theorem 6.4.

(4) Use the generating polynomial for the Reed Solomon $(255, 233, 32)$ cyclic code to generate a convolutional $(8, 4)$ code by the method of Sec. 6.3.21. Give a lower bound to the free distance of the code using Theorem 6.9.

(5) Use the polynomial $f(z) = 1 + z + z^3$ which generates the Hamming cyclic $(7, 4, 3)$ code to construct a convolutional $(8, 4)$ code as in Sec. 6.3.21. Find a lower bound to the free distance of the code from Theorem 6.9. What is the exact free distance of the code?

(6) Use the polynomial $f(z) = 1 + z + z^2 + z^4$ which generates the Hamming cyclic $(7, 3, 4)$ code to construct a convolutional $(8, 4)$ code as in Sec. 6.3.21. Find a lower bound to the free distance of the code from Theorem 6.9. What is the exact free distance of the code?

(7) Consider $e_0 = 1/2(1 + a), e_1 = 1/2(1 - a)$ in the group ring FC_2 with C_2 generated by a and F a field not of characteristic 2. Show $e_0^2 = e_0, e_1^2 = e_1, e_0 e_1 = 0, e_0 + e_1 = 1$. Consider $f(z) = e_0 + e_1 z$ and show $f(z)f(z^{-1}) = 1$. Use the first row of the matrix of $f(z)$ to construct a $(2, 1)$ convolutional code. What is the control matrix? Show that this code has free distance 4.

(8) Consider $e_0 = 1/n(1 + a + a^2 + \cdots + a^{n-1}), e_1 = 1 - e_0$ in the group ring FC_n with C_n generated by a and F a field not of characteristic dividing n. Show $e_0^2 = e_0, e_1^2 = e_1, e_0 e_1 = 0, e_0 + e_1 = 1$. Consider $f(z) = e_0 + e_1 z$ and show $f(z)f(z^{-1}) = 1$. Suppose now $n = 2m$. Use the first m rows of the matrix of $f(z)$ to construct a $(2m, m)$ convolutional code. What is the control matrix? What is the free distance of this code?

(9) Consider $u = 1 + h(1 + g + g^2 + g^3)$ in $F(C_5 \times C_5)$ with the C_5 generated by h, g, respectively, and F a field of characteristic 5. Show $u^5 = 0$.

Show also that the matrix U of u has rank 20. Thus u generates a $(25, 20)$ linear code. Find the distance of this code.

Let $v = 3 + h(1 + g + g^2)$. Show $v^5 = 1$. Let $f(z) = u + vz + uz^2$. Show $f(z)^5 = z^5$. Define a $(25, 20)$ convolutional code by taking as generator matrix the first 20 rows of $f(z)$ when considered as a matrix. What is the control matrix? Investigate the free distance of this convolutional code.

(10) Find elements $u, v \in F(C_7 \times C_7)$ for which $u^7 = 0$ and $v^7 = 1$. Find also a u with $u^7 = 0$ and such that the matrix U of u has rank 42. Hence, construct a $(49, 42)$ linear code over F and determine its distance.

Construct a convolutional $(49, 42)$ code by considering $f(z) = u + vz + uz^2$ and showing $f(z)^7 = z^7$.

6.4.2. *Research Investigations*

6.4.2.1. *LDPC Related*

(1) Use the group ring $\mathbb{Z}_2(C_{128} \times C_4)$ to construct a low density unit (invertible) matrix of weight ≥ 3 with no short cycles.

Select 10 random LDPC rate $1/2$ codes with no short cycles from the unit and simulate their performances using MATLAB or other.

(2) Let D_{32} denotes the dihedral group of order 32. Find the units and zero-divisors in $\mathbb{Z}_2 D_{32}$. Determine which of these do not have short cycles in their matrices and investigate codes derived from these.

(3) Find the units and zero-divisors in $\mathbb{Z}_2 D_{64}$, where D_{64} denotes the dihedral group of order 64. Determine which of these do not have short cycles in their matrices and investigate codes derived from these. Which of the units and zero-divisors could be classified as "low density"?

(4) Let p be a prime. Show the the pth power of every element in $\mathbb{Z}_p C_p^n$ is the sum of its coefficients. (Note that, by Fermat's little theorem, for an integer a, $a^p \equiv a \mod p$.) Hence or otherwise determine the zero-divisors and units in $\mathbb{Z}_p C_p^n$. For $p > 2$, and for sufficiently large n, find those units in $\mathbb{Z}_p C_p^n$ of weight ≥ 3 which have no short cycles in their matrices. From such units which are also low density, investigate the resulting LDPC codes.

(5) Investigate and determine units with small support but of weight ≥ 3 in $\mathbb{Z}_2 C_n$ which have no short cycles in their matrices for various (large) n. Investigate the resulting LDPC matrices which could be formed.

(6) Investigate and determine units with small support, but of weight ≥ 3, in $\mathbb{Z}_2(C_n \times C_2)$ which have no short cycles in their matrices for various (large) n. Investigate the resulting LDPC codes which could be formed.

(7) Investigate units with small support, but of weight ≥ 3, in $\mathbb{Z}_2(C_n \times C_4)$ which have no short cycles in their matrices for various (large) n. Investigate the resulting LDPC codes which could be formed.

(8) Investigate and determine units with small support, but of weight ≥ 3, in $\mathbb{Z}_2 D_{2n}$ which have no short cycles in their matrices for various (large) n. Investigate the resulting LDPC code which could be formed. Here D_{2n} denotes the dihedral group of order $2n$.

6.4.2.2. *Convolutional Related*

(1) Find a polynomial generator for some large cyclic code. Construct a corresponding convolutional $(2,1)$ code by the methods of Sec. 6.3.12. Give a lower bound for the free distance of the convolutional code by using Theorem 6.4. Ascertain the exact free distance of the convolutional code constructed.

(2) Find a polynomial generator for some large cyclic code. Construct a corresponding convolutional $(8,4)$ code by the methods of Sec. 6.3.21. Give a lower bound for the free distance of the convolutional code by using Theorem 6.9. Ascertain the exact free distance of the convolutional code constructed.

(3) Use the group ring method to construct rate $7/8$ convolutional codes by for example considering elements $a \in FC_{16}$ such that $a^{16} = 0$ as suggested in Sec. 6.3.13. Calculate the free distances of the codes constructed.

(4) Investigate convolutional $(2,1)$ or $(8,4)$ codes and their free distances using variations on the generating polynomial for the Golay cyclic codes by the methods described in Sec. 6.3.12 or Sec. 6.3.21.

(5) Use the methods of Sec. 6.3.19 to construct convolutional codes over fields of characteristic 5 and 7. Determine the free distances, or at least lower bounds for the free distances, of the codes constructed.

(6) Use generating polynomials for BCH codes to construct $(2,1)$ or $(8,4)$ convolutional codes by the method of Sec. 6.3.12 or Sec. 6.3.21 and investigate their free distances.

References

1. R. J. McEliece, The algebraic theory of convolutional codes, *Handbook of Coding Theory*, Vol. 1 (Elsevier, 1999).
2. H. Tan, J. Xu, Y. Kou, S. Lin and K. A. S. Abdel-Ghaffar, On algebraic construction of Gallager and circulant low-density parity-check codes, *IEEE Trans. Inform. Theory* (2004) 1269–1279.

3. O. Milenkovic, I. Djordjevic and B. Vasic, Block-circulant low-density parity-check codes for optical communication systems, *IEEE J. Selected Topics in Quantum Electron.* **10**, 2 (2004) 294–299.

4. R. Tanner, D. Sridhara, A. Sridharan, T. Fuja and C. Daniel Jr., LDPC block and convolutional codes based on circulant matrices, *IEEE Trans. Inform. Theory* (2004) 2966–2984.

5. R. E. Blahut, *Algebraic Codes for Data Transmission* (Cambridge University Press, 2003).

6. T. Hurley, P. McEvoy and J. Wenus, Algebraic constructions of LDPC codes with no short cycles.

7. C. P. Milies and S. K. Sehgal, *An Introduction to Group Rings* (Klumer, Dordrecht/Boston/London, 2002).

8. T. Hurley, Group rings and rings of matrices, *Inter. J. Pure Appl. Math.* **31**, 3 (2006) 319–335.

9. D. J. C. MacKay, G. Mitchison and P. L. McFadden, Sparse-graph codes for quantum error-correction, *IEEE Trans. Inform. Theory* **50**, 10 (2004).

10. R. Tanner, A recursive approach to low complexity codes, *IEEE Trans. Inform. Theory* **27**, (1981) 533–547.

11. J. van Lint and R. Wilson, *A Course in Combinatorics* (Cambridge University Press, 2001).

12. H. Gluesing-Luerssen and W. Schmale, On cyclic convolutional codes, *Acta Applicandae Mathematicae* **82**, (2004) 183–237.

13. P. Hurley and T. Hurley, Codes from zero-divisors and units in group rings, *Available at arXiv:0710.5893*.

14. J. Rosenthal and R. Smarandache, Maximum distance separable convolutional codes, *Appl. Algebra Engrg. Comm. Comput.* **10**, 1 (1999) 15–37.

15. E. Noether, Hypercomplexe Grössen und Darstellingtheorie, *M. Zeit.* **30**, (1929) 641–692.

Chapter 7

SEARCH FOR GOOD LINEAR CODES IN THE CLASS OF QUASI-CYCLIC AND RELATED CODES

NUH AYDIN* and TSVETAN ASAMOV†

*Department of Mathematics, Kenyon College,
Gambier, OH, 43022, USA*
*aydinn@kenyon.edu
†asamovt@kenyon.edu

This chapter gives an introduction to algebraic coding theory and a survey of constructions of some of the well-known classes of algebraic block codes such as cyclic codes, BCH codes, Reed–Solomon codes, Hamming codes, quadratic residue codes, and quasi-cyclic (QC) codes. It then describes some recent generalizations of QC codes and open problems related to them. Also discussed in this chapter are elementary bounds on the parameters of a linear code, the main problem of algebraic coding theory, and some algebraic and combinatorial methods of obtaining new codes from existing codes. It also includes a section on codes over $\mathbb{Z}_4$, integers modulo 4, due to increased attention given to these codes recently. Moreover, a recently created database of best-known codes over $\mathbb{Z}_4$ is introduced.

Keywords: Code constructions; bounds on codes; cyclic codes; quasi-cyclic codes; quaternary codes; best-known codes; database of codes.

7.1. Introduction and Basic Definitions

Coding theory is concerned with reliability of communication over noisy channels. Error-correcting codes are used in a wide range of communication systems from deep space communication to quality of sound in compact disks and wireless phones.

The basic principle of coding theory is to employ redundancy to recover original messages even if errors occur during the transmission. Redundancy is naturally used in human languages. It is built into natural languages in many ways [67]. One of the ways the redundancy is manifest in human languages is the fact that not every possible string of symbols is a valid word

239

in a language. Humans as well as computers can use this fact to detect and even correct the errors in communication. For example, suppose you see the word "mistaky" in a text. An error can be detected and corrected even in the absence of surrounding context using the maximum likelihood principle: Among the valid words in the language, "mistake" is the closest one to the received string.

Using the same basic principles, we can formulate the basic notions of coding theory in a mathematical way.

Definition 7.1. A code C of length n over an alphabet F is a subset of $F^n = \{(a_1, a_2, \ldots, a_n) : a_i \in F, 1 \leq i \leq n\}$.

Note the analogy with the set of valid words in a language. Codewords in F^n can be likened to valid words in the language. Not every possible string of symbols is a valid word in a language, likewise not every vector in F^n is a codeword (except for the trivial and quite useless code of $C = F^n$). Throughout this chapter, vectors will be represented by bold face letters such as $\mathbf{u} \in F^n$.

The alphabet F of a code is a finite set, the most important case being a finite field. In this case, we often consider subsets of F^n that are vector subspaces. Such codes are called *linear codes*.

Example 7.1. Let $C_1 := \{u_1 = 1200, u_2 = 0102\}$ and $C_2 := \{v_1 = 00000, v_2 = 10110, v_3 = 11001, v_4 = 01111\}$. Then C_1 is a ternary code of length 4 and C_2 is a binary code of length 5. It can easily be verified that C_1 is not a linear code, whereas C_2 is. The dimension of C_2 is 2. These two specific codes will be referred to a few times in this section.

A fundamental concept in coding theory is distance.

Definition 7.2. For two vectors $\mathbf{u} = (u_1, u_2, \ldots, u_n)$, $\mathbf{v} = (v_1, v_2, \ldots, v_n)$ in F^n the Hamming distance between them is denoted and defined by $d(\mathbf{u}, \mathbf{v}) = |\{i : u_i \neq v_i\}|$, the number of positions in which $\mathbf{u}$ and $\mathbf{v}$ differ. (For a set A, $|A|$ denotes the number of elements in A.) For a code C, we define the *minimum distance* of C to be $\min\{d(\mathbf{u}, \mathbf{v}) : \mathbf{u}, \mathbf{v} \in C, \mathbf{u} \neq \mathbf{v}\}$.

Example 7.2. For the codes C_1 and C_2 defined above, we have $d(u_1, u_2) = 3$ and $d(v_1, v_2) = 4$. The minimum distance of both C_1 and C_2 is 3.

Exercise 7.1.1. Show that the *Hamming distance* defines a metric on F^n, i.e. for all $\mathbf{u}, \mathbf{v}, \mathbf{w} \in F^n$ it satisfies the following properties: (i) $d(\mathbf{u}, \mathbf{v}) \geq 0$,

(ii) $d(\mathbf{u}, \mathbf{v}) = 0 \Leftrightarrow \mathbf{u} = \mathbf{v}$, (iii) $d(\mathbf{u}, \mathbf{v}) = d(\mathbf{v}, \mathbf{u})$, (iv) $d(\mathbf{u}, \mathbf{v}) \leq d(\mathbf{u}, \mathbf{w}) + d(\mathbf{w}, \mathbf{v})$.

The minimum distance of a code determines its error-detecting and correcting capabilities. We like to have codes with large minimum distances so that few changes will not turn a codeword into another. At the same time, we want a code to contain as many codewords as possible so that we can transmit many different messages. Not surprisingly, these are two conflicting goals. There are trade offs between the two objectives. This is one of the main problems in coding theory. We will address this question more carefully later in this chapter.

Definition 7.3. We say that a code C is *e-error detecting*, if $\mathbf{v} \in F^n$ is such that $d(\mathbf{v}, \mathbf{u}) \leq e$ for some $\mathbf{u} \in C$ then either $\mathbf{v} = \mathbf{u}$ or $\mathbf{v} \notin C$.

Intuitively, this means that a set of at most e changes on a codeword does not produce another codeword. In the example above, the single change in the word "mistake" did not lead to another valid word. Is this true for every word in English language, i.e. is the English language single error detecting? For the codes C_1 or C_2, if you take any codeword and if you introduce two errors (i.e. if you change two coordinates) you do not end up with another codeword.

Definition 7.4. We say that a code C is *e-error correcting*, if for all $\mathbf{v} \in F^n$ such that $d(\mathbf{v}, \mathbf{u_1}) \leq e$ and $d(\mathbf{v}, \mathbf{u_2}) \leq e$ for some $\mathbf{u_1}, \mathbf{u_2} \in C$ we have $\mathbf{u_1} = \mathbf{u_2}$.

This means that a vector in F^n cannot be within a *Hamming distance e* of more than one codeword. Try to verify that both C_1 and C_2 are *1-error-correcting* codes by taking arbitrary vectors (of appropriate length over the relevant alphabet) and checking this property.

Now we can state the precise relationship between the minimum distance of a code and its error detecting and correcting capability.

Theorem 7.1. Let C be a code with minimum distance d. Then C is a $t = d - 1$ error-detecting code and $e = \lfloor \frac{d-1}{2} \rfloor$ error-correcting code, where $\lfloor x \rfloor$ denotes the greatest integer $\leq x$.

Exercise 7.1.2. Prove this theorem using the properties of the Hamming distance.

Determining the minimum distance of a code is an important and in general a difficult problem in coding theory. For a code of size M, there are $\binom{M}{2} = \frac{M(M-1)}{2} = O(M^2)$ distinct pairs to consider to find the minimum distance. For *linear codes*, we get an improvement. We first need to introduce the concept of Hamming weight.

Definition 7.5. For a vector $\mathbf{u} \in F^n$, the Hamming weight $w_H(\mathbf{u})$ of $\mathbf{u}$ is defined to be $|\{i : u_i \neq 0\}|$, the number of non-zero components of $\mathbf{u}$. For a code $C \subseteq F^n$, the *minimum Hamming weight* of C is $\min\{w_H(\mathbf{u}) : \mathbf{u} \in C, \mathbf{u} \neq \mathbf{0}\}$.

For the code C_1 both codewords have weight 2, for C_2 the weights are 0,3,3,4. Note that the *Hamming distance* and the Hamming weight are related by $d(\mathbf{u}, \mathbf{v}) = w_H(\mathbf{u} - \mathbf{v})$ if the alphabet is an additive group. For linear codes, the minimum distance is the same as the minimum weight.

Lemma 7.1. *Let C be a linear code. Then the minimum distance of C is equal to the minimum weight of C.*

Exercise 7.1.3. Show that the minimum distance of C_2 is the same as the minimum weight of C_2, but that is not the case for C_1. Then, prove this lemma.

Therefore, to compute the minimum distance of a linear code of size M, in the worst case one needs to consider $M - 1 = O(M)$ vectors, instead of $O(M^2)$ vectors.

We also note that the dimension of a linear code determines its size.

Exercise 7.1.4. Let C be a *linear code* of dimension k over a finite field of order q. Show that $|C| = q^k$.

For a linear code, the most important parameters are the length, the dimension, and the minimum distance. If a linear code C over $\mathbb{F}_q$, the finite field with q elements, has the values n, k, and d for the length, the dimension, and the minimum distance, respectively, it is referred to as an $[n, k, d]_q$-code. In the case of a non-linear code, we use the notation $(n, M, d)_q$ where M is the size of the code. Hence, we say that C_1 is a $(4, 2, 3)_3$-code and C_2 is a $[5, 2, 3]_2$-code.

Let C be an $[n, k, d]_q$ linear code. Since C is a vector subspace of $\mathbb{F}_q^n$, every basis of C has k elements. A $k' \times n$ $(k' \geq k)$ matrix G whose row space is equal to C is called a *generator matrix* for C. A generator matrix of the form $G = (I_k | A)$, where I_k denotes the $k \times k$ identity matrix, is said to be in the standard form. For the code C_2 a generator matrix is

$G_2 = \begin{pmatrix} 1 & 0 & 1 & 1 & 0 \\ 1 & 1 & 0 & 0 & 1 \end{pmatrix}$. After some elementary row operations it can be put into the standard form: $G_2 = \begin{pmatrix} 1 & 0 & 1 & 1 & 0 \\ 0 & 1 & 1 & 1 & 1 \end{pmatrix}$.

The inner product of two codewords $\mathbf{u}$ and $\mathbf{v}$ in $V := \mathbb{F}_q^n$ is defined in the usual way

$$\langle \mathbf{u}, \mathbf{v} \rangle := \sum_{i=1}^{n} v_i u_i.$$

The *dual* or *orthogonal code* $C^\perp$ of an $[n, k]_q$ linear code C is defined by

$$C^\perp := \{ \mathbf{v} \in V \; : \; \langle \mathbf{u}, \mathbf{v} \rangle = 0 \text{ for all } \mathbf{u} \in C \}.$$

It is easily verified that $C^\perp$ is a vector space of V of dimension $n - k$, i.e. an $[n, n-k]_q$ code. Let C be an $[n, k, d]_q$ code and let $G = (I_k | A)$ be a generator matrix of C. Let $H = (-A^{\mathrm{T}} | I_{n-k})$, where superscript T stands for the transpose, then

$$GH^{\mathrm{T}} = (I_k | A) \begin{pmatrix} -A \\ I_{n-k} \end{pmatrix} = -A + A = 0.$$

Thus, the rows of H are orthogonal to the rows of G, and since rank$(H) = n - k$, H is a generator matrix for $C^\perp$. The matrix H is a *parity check matrix* for C. More generally, a parity check matrix for a linear code C is a matrix whose row space is $C^\perp$, equivalently, a matrix whose null space is C. A linear code C is determined by either a generator matrix or a parity check matrix. For C_2, a parity check matrix is $H_2 = \begin{pmatrix} 1 & 1 & 1 & 0 & 0 \\ 1 & 1 & 0 & 1 & 0 \\ 0 & 1 & 0 & 0 & 1 \end{pmatrix}$. The parity check matrix of a linear code has the following important property:

Lemma 7.2. *Let C be a code with a parity check matrix H such that any set of $d - 1$ columns of H is linearly independent and there is a set of d columns of H that is linearly dependent. Then, the minimum distance of C is d.*

Exercise 7.1.5. Prove this lemma and use it to verify that the minimum distance of C_2 is 3.

The *Hamming weight enumerator*, $W_C^H(x, y)$, of a linear code C of length n is defined as

$$W_C^H(x, y) = \sum_{u \in C} x^{n - w(u)} y^{w(u)} = \sum_{i=0}^{n} A_i x^{n-i} y^i,$$

where $A_i := |\{\mathbf{u} \in C : w(\mathbf{u}) = i\}|$, the number of codewords in C with weight equal to i. The weight enumerator of the code C_2 is $W = x^5 + 2x^2y^3 + xy^4$.

One of the classical theorems of coding theory is MacWilliam's identity, which relates the weight enumerators of a code and its dual.

Theorem 7.2 (Macwilliams and Sloane [53]). *The relationship between the Hamming weight enumerators of a q-ary linear code C and its dual $C^\perp$ is given by*

$$W_{C^\perp}^H(x,y) = \frac{1}{|C|} W_C^H(x + (q-1)y, x - y).$$

Definition 7.6. The map that sends $W_C^H(x,y)$ to $\frac{1}{|C|} W_C^H(x+(q-1)y, x - y)$ is called the *MacWilliams transform*.

The MacWilliams transform of the weight enumerator W of C_2 is $x^5 + 2x^3y^2 + 4x^2y^3 + xy^4$. Therefore, the minimum distance of $C_2^\perp$ is 2 and it is a $[5,3,2]_2$-code. Note that the sum of the coefficients of the weight enumerator is the total number of codewords.

Definition 7.7. A linear code C is called *self-orthogonal* if $C \subseteq C^\perp$ and *self-dual* if $C = C^\perp$. A code C (not necessarily linear) is called *formally self-dual* if its *Hamming weight enumerator* coincides with its MacWilliams transform.

It is clear that a self-dual code is also formally self-dual. Another important notion in coding theory is equivalence of codes.

Definition 7.8 (Macwilliams and Sloane [53]). Let C_1 and C_2 be codes of length n over $\mathbb{F}_q$. We say that C_1 and C_2 are *equivalent* if there are n permutations $\pi_0, \pi_1, \ldots, \pi_{n-1}$ of $\mathbb{F}_q$ and a permutation σ of n coordinate positions such that

$$\text{If } (c_0, \ldots, c_{n-1}) \in C_1 \quad \text{then} \quad \sigma\big(\pi_0(c_0), \ldots, \pi_{n-1}(c_{n-1})\big) \in C_2.$$

For linear codes only those π_i's, which are the compositions of a scalar multiplication with a field automorphism, are allowed. The scalar multiple may vary for each coordinate, but the field automorphism must be the same.

There are some important special cases: when all π_i's are identity permutations, we say that C_1 and C_2 are *permutation equivalent* and when each π_i is a multiplication by a non-zero scalar, C_1 and C_2 are said to be *scalar multiple equivalent* or *monomially equivalent*. For *prime fields* such as $\mathbb{Z}_p$, integers modulo a prime p (also denoted by $\mathbb{F}_p$ or $GF(p)$), there are no non-trivial field automorphisms. Equivalent (linear) codes have the same weight enumerator, in particular they have the same minimum distance.

7.1.1. *Main Problem of Coding Theory*

One of the central problems of algebraic coding theory is to determine the best possible values of the parameters of a code and to explicitly construct codes with those parameters. There is an online table [33] of best-known linear codes over the finite fields of size ≤ 9. Additionally, there is a table of best-known (non-linear) binary codes [50].

To formulate the main problem for linear codes, we can first choose the alphabet size q, then fix two of the parameters and ask for the optimal value of the other. For example, fixing n and k, we ask for the largest value of d. This value is denoted by $d_q(n, k)$. Similarly, we can fix n and d, and try to maximize k, or the size of the code. The maximum size of such a code is denoted by $A_q(n, d)$. Or, fix k and d, and try to minimize n. This minimum value is denoted by $n_q(k, d)$. There are numerous bounds on the parameters of a code. We review some of the most elementary bounds in the following section. Others can be found in standard books in coding theory such as [53]. The problem of determining the values $n_q(k, d)$ (or $d_q(n, k)$) has been a central problem in coding theory. Many papers in the literature deal with this problem. In general, the optimal values are not determined except for small values of k, or when $n - k$ is small. A printed (but not up to date) table is available in [17] with online and up to date version at [33]. Among others, some of the cases where the problem is solved are $n_2(k, d), k \leq 8$ in [14], $d_3(n, k), k \leq 6$ (not all cases determined) in [15], $n_4(5, d)$ for many cases in [16], some cases of $n_2(9, d)$ in [27], and many cases of $n_5(k, d)$ for $k = 3, 4$ in [13, 44]. A survey on the subject can be found in [41].

There are various algebraic methods to construct codes with good parameters. In the following sections, we will describe some of these methods. We first review some necessary abstract algebra in Sec. 7.3, then introduce some of the well-known constructions in Sec. 7.4. There are also many ways to combine existing codes to produce new codes. Some of these methods are described in the following section. In Sec. 7.5, we pay special attention to the class of quasi-cyclic (QC) and related codes, which have proven to be promising toward a solution to the main problem of the coding theory. Section 7.6 focuses on the computationally difficult problem of determining the minimum distance of a code. We devote (Sec. 7.7) to the codes over $\mathbb{Z}_4$, the integers modulo 4, due to increased attention to those codes in recent years. This section also introduces a recently created database of best-known codes over $\mathbb{Z}_4$. We list some open problems in the area of QC and related codes in Sec. 7.8.

7.2. Some Elementary Constructions and Elementary Bounds on Codes

7.2.1. Some Elementary Constructions

Extending a code: Given a code C, there are many ways of obtaining longer codes by adding coordinates to C. The most common way to extend a linear code is by adding an overall parity check. If C is an $[n, k, d]_q$-code, the extended code $\hat{C}$ is defined by

$$\hat{C} = \left\{ (c_0, c_1, \ldots, c_{n-1}, c_n) : (c_0, c_1, \ldots, c_{n-1}) \in C, \sum_{k=0}^{n} c_k = 0 \right\},$$

and it is an $[n+1, k, \hat{d}]$-code where $\hat{d} = d$ or $d+1$.

Puncturing a code: Puncturing a code is opposite of extending where one fixed coordinate position is deleted from all codewords. If C is an $(n, M, d)_q$-code with $d \geq 2$, then the code C^* obtained by puncturing C once has parameters $(n-1, M, d^*)$ where $d^* = d$ or $d-1$. If C is a linear code with parameters $[n, k, d]_q$ then C^* has parameters $[n-1, k, d^*]_q$, $d^* = d$ or $d-1$.

Shortening a code: Let C be an $[n, k, d]_q$-code. Fix a position i and let C_i be the set of all codewords that have 0 at position i. Then C_i is a subcode of C. If we delete the coordinate i from the set of vectors in C_i then the resulting code is called a shortened code of C. The parameters of the shortened code are $[n-1, k-1, d]_q$.

Direct sum: Given two linear codes C_1, C_2 over F_q with parameters $[n_i, k_i, d_i]$, $i = 1, 2$, then their direct sum is the code given by $C_1 \oplus C_2 = \{(\mathbf{u}, \mathbf{v}) : \mathbf{u} \in C_1, \mathbf{v} \in C_2\}$ ($(\mathbf{u}, \mathbf{v})$ denotes the concatenation of the vectors $\mathbf{u}$ and $\mathbf{v}$). Then the parameters of the direct sum code are $[n_1 + n_2, k_1 + k_2, \min\{d_1, d_2\}]$. Moreover, if G_i and H_i are generator and parity check matrices of C_i, respectively, then a generator matrix of $C_1 \oplus C_2$ is $\begin{pmatrix} G_1 & 0 \\ 0 & G_2 \end{pmatrix}$ and a parity check matrix is $\begin{pmatrix} H_1 & 0 \\ 0 & H_2 \end{pmatrix}$.

The $(\mathbf{u}|\mathbf{u} + \mathbf{v})$ *construction:* Another way to combine two linear codes C_1, C_2 of the same length over the same field $\mathbb{F}_q$ to obtain a new code of double length is through the $(\mathbf{u}|\mathbf{u} + \mathbf{v})$ construction, which is defined as $C_3 = \{(\mathbf{u}, \mathbf{u} + \mathbf{v}) : \mathbf{u} \in C_1, \mathbf{v} \in C_2\}$. It is not hard to show that the parameters of C_3 are $[2n, k_1 + k_2, \min\{2d_1, d_2\}]$, where the parameters of C_i are $[n, k_i, d_i]$. It is also not difficult to show that the generator and parity check matrices of C_3 are $\begin{pmatrix} G_1 & G_1 \\ 0 & G_2 \end{pmatrix}$ and $\begin{pmatrix} H_1 & 0 \\ -H_2 & H_2 \end{pmatrix}$, respectively.

The $(\mathbf{u}|\mathbf{u} + a\mathbf{v}|\mathbf{u} + \mathbf{v} + \mathbf{w})$ *construction:* Given three codes $C_1[n_1, k_1, d_1]$, $C_2[n_2, k_2, d_2], C_3[n_3, k_3, d_3]$ over the same field $\mathbb{F}_q$, and $a \in \mathbb{F}_q$ we can generate a new code of the form $C_4 = \{(\mathbf{u}, \mathbf{u} + a\mathbf{v}, \mathbf{u} + \mathbf{v} + \mathbf{w}) : \mathbf{u} \in C_1, \mathbf{v} \in C_2, \mathbf{w} \in C_3\}$. When $a = -1$, the parameters of C_4 are $[n + \max\{n_1, n_2\} + \max\{n_1, n_2, n_3\}, k_1 + k_2 + k_3, \min\{3d_1, 2d_2, d_3\}]$. Moreover, C_4 has a generator matrix of the form $\begin{pmatrix} G_1 & G_1 & G_1 \\ 0 & aG_2 & G_2 \\ 0 & 0 & G_3 \end{pmatrix}$.

Construction X: Once again, consider three codes $C_1[n_1, k_1, d_1]$, $C_2[n_2, k_2, d_2]$, $C_3[n_3, k_3, d_3]$ over a field $\mathbb{F}_q$. If $k_1 = k_2 + k_3$ and C_2 is a subcode of C_1, implying $n_1 = n_2$ and $k_1 \geq k_2$, then we can split C_2 into a union of cosets of C_1 and append a different word from C_3 to each of the cosets. Thus, we end up with a code with parameters $[n_1 + n_3, k_1, d \geq \min\{d_2, d_1 + d_3\}]$ and a generator matrix given by $\begin{pmatrix} G_{12} & G_3 \\ G_2 & 0 \end{pmatrix}$. Here G_2 and G_3 are the generator matrices of, respectively, C_2 and C_3, while G_{12} is such that G_{12} and G_2 together generate C_1.

7.2.2. *Some Bounds on Codes*

There are many bounds on the parameters of a code. Here, we give two most elementary bounds: the singleton bound that is related to MDS codes and the sphere packing bound that is related to perfect codes. More bounds can be found in books on the subject such as [9, 42, 53, 64, 74].

Theorem 7.3 (The Singleton Bound). $A_q(n, d) \leq q^{n-d+1}$.

Proof. Let C be a q-ary (n, M, d)-code. If we remove the last (or any) $d-1$ coordinate positions from each codeword in C, the resulting M words are still distinct. Since those have length $n - d + 1$, we have $M \leq q^{n-d+1}$. □

The singleton bound implies that any $[n, k, d]_q$ code must satisfy $q^k \leq q^{n-d+1}$ or equivalently $d \leq n - k + 1$. A linear code for which the equality holds in this bound is called a *maximum distance separable code*, or *MDS code*. It is known that the dual of an MDS code is also MDS. So is a shortening. MDS codes are known to exist for small lengths. In fact for any prime power q and any dimension k, $1 \leq k \leq q+1$, there exists a $[q+1, k, q-k+1]_q$ MDS code (hence for any smaller length). It is conjectured that no longer non-trivial MDS codes exist except for $n = q+2$ for q even and $k = 3$ or $k = q-1$ [53]. The conjecture has been proven in some cases, but the general case is still open. See [77] as an example of a case for which the conjecture is proven.

Theorem 7.4 (The Sphere Packing Bound or the Hamming Bound).

$$A_q(n, d) \leq \frac{q^n}{\sum_{j=0}^{t} \binom{n}{j}(q-1)^j}, \quad where \ t = \left\lfloor \frac{d-1}{2} \right\rfloor.$$

The proof of this bound is based on the observation that the "spheres" of radius t around codewords are disjoint.

An $(n, M, 2t+1)_q$-code C is said to be *perfect* (or *t-perfect*) if the balls $B_t(\mathbf{c}) = \{\mathbf{x} \in F^n : d(\mathbf{x}, \mathbf{c}) \leq t\}$ of radius t around codewords are disjoint and cover the space F^n (here F is the alphabet of size q for the code, it is not necessarily a field), i.e.

$$\bigcup_{\mathbf{c} \in C} B_t(\mathbf{c}) = F^n.$$

For perfect codes the equality holds in the sphere packing bound. Perfect codes are rather rare. There has been intensive search to classify all perfect codes or to discover new ones. After much effort, the classification of all perfect codes is nearly complete. The result on the classification of all perfect codes can be found in 42 (p. 49), or with more details in [74 (Chap. 7)].

7.3. Some Background in Abstract Algebra

In algebraic coding theory, finite fields and polynomials over finite fields are very important. In this section, we review some of the basic facts about these objects. Owing to space considerations, we skip background information on such topics as rings, ideals, and Euclidean domains. For more details, the reader is referred to the books such as [30, 48, 64, 65]. Most of the definitions, theorems, and examples in this section (and its sub-sections) can be found in [48, 64] as well as the proofs of the theorems.

7.3.1. *Polynomials*

A fundamental theorem about polynomials over fields is the following theorem known as the *division algorithm*.

Theorem 7.5. *The polynomial ring $F[x]$ over a field F is a Euclidean domain with $\sigma(p(x)) = \deg(p(x))$, with the well-known division algorithm of polynomials: Given $f, g \in F[x]$ with $g \neq 0$, there exist unique polynomials $q, r \in F[x]$ such that $f = q \cdot g + r$, where $r = 0$ or $\deg(r) < \deg(g)$. Thus, $F[x]$ is a principal ideal domain, and hence a unique factorization domain.*

Example 7.3. Let $f(x) = 3x^4 + x^3 + 2x^2 + 1 \in \mathbb{Z}_5[x], g(x) = x^2 + 4x + 2 \in \mathbb{Z}_5[x]$. Then, $f = q \cdot g + r$ with $q(x) = 3x^2 + 4x, r(x) = 2x + 1$ and $\deg(r) < \deg(g)$.

Definition 7.9. Given two polynomials $f, g \in F[x]$, $g \neq 0$ we say that g *divides* f (also denoted by $g|f$) if there exists a polynomial $p \in F[x]$ such that $f = p \cdot g$. This is equivalent to saying that the remainder in the division algorithm is 0 when f is divided by g.

Definition 7.10. A non-zero, non-constant polynomial $f \in F[x]$ is said to be *irreducible* over F if whenever $f = p \cdot q$ for some polynomials $p, q \in F[x]$ then either p or q is a constant polynomial.

Irreducible polynomials are very important in finite field theory.

Definition 7.11. A polynomial d is called a *greatest common divisor* of polynomials f and g if

(i) $d|f$ and $d|g$,
(ii) whenever $p|f$ and $p|g$, $p|d$ as well.

A greatest common divisor is unique up to a constant multiple. If f, g are either integers or polynomials, the notation (f, g) is commonly used to denote their greatest common divisor. Thus, the notation $(f, g) = 1$ means f and g are relatively prime, where f and g could be either integers or polynomials.

An element $a \in F$ is called a *root* (or a *zero*) of a polynomial $f \in F[x]$ if $f(a) = 0$. It is well known (follows from Theorem 7.5) that an element $a \in F$ is a root of a polynomial $f \in F[x]$ if and only if $(x - a)|f(x)$. For polynomials of degree 2 or 3 existence of roots is equivalent to reducibility. This result again follows from the division algorithm.

Lemma 7.3. *A polynomial of degree 2 or 3 over a field F is irreducible in $F[x]$ if and only if it has no roots in F.*

Exercise 7.3.1. Prove this lemma, and give an example to show that it is not true for polynomials of higher degrees.

If $(x - a)^k | f(x)$ for $k > 1$ then a is called a *multiple root* of f. The largest integer r such that $(x - a)^r | f(x)$ but $(x - a)^{r+1} \nmid f(x)$ is called the multiplicity of a. If $r = 1$ then a is called a *simple root*. The notion of *derivative*, defined purely algebraically, is useful for determining multiplicity of roots. The derivative of a polynomial $f(x) = a_0 + a_1 x + a_2 x^2 + \cdots + a_n x^n \in F[x]$ is defined by $f'(x) = a_1 + 2a_2 x + \cdots + na_n x^{n-1} \in F[x]$. Then, it is not difficult to show that the usual laws of derivatives hold in polynomial rings. We can also use derivatives to detect multiple roots.

Proposition 7.1. *An element $a \in F$ is a multiple root of $f(x) \in F[x]$ if and only if it is a root of both $f(x)$ and $f'(x)$.*

7.3.2. *Field Extensions*

In working with polynomials over fields, we often need to consider larger fields to find the roots. Therefore, a discussion of field extensions is needed. If a field F contains a subset K that happens to be a field by itself with the induced operations, then K is called a *subfield* of F, and F is called an *extension (field)* of K. If $K \neq F$, then K is a *proper subfield* of F.

If K is a subfield of a finite field $\mathbb{F}_p$ with p elements where p is a prime, then K must contain 0 and 1 and, by closure under addition, all the other elements of $\mathbb{F}_p$. Therefore, $K = \mathbb{F}_p$ and $\mathbb{F}_p$ contains no proper subfields. A field containing no proper subfields is called a *prime field*. Any finite field of order p, p prime, is a prime field. Another example of a prime field is the field $\mathbb{Q}$ of rational numbers, which has characteristic 0. It turns out that these are the only prime fields. Any field F contains a prime field that is isomorphic to either $\mathbb{F}_p$ or $\mathbb{Q}$ depending on whether the characteristic of F is a prime p or 0.

A common way of obtaining extension fields is by *adjoining* a set of elements S to a given field K from a larger field F that contains K. The smallest field that contains both K and S is denoted by $K(S)$. For a finite set $S = \{\alpha_1, \ldots, \alpha_n\}$ the notation $K(\alpha_1, \ldots, \alpha_n)$ is common. If $S = \{\alpha\}$ is a singleton set then $E = K(\alpha)$ is called a *simple extension* of K and α is called a *generating element* of E over K.

Elements of a larger field that are roots of (non-zero) polynomials over subfields have a special place in field theory. For fields $K \subseteq F$ and $\alpha \in F$, if there exist $a_i \in K$, $0 \leq i \leq n$, not all equal to 0, such that $a_n\alpha^n + a_{n-1}\alpha^{n-1} + \cdots + a_1\alpha + a_0 = 0$ then α is said to be *algebraic* over K. For example, $\sqrt{2} \in \mathbb{R}$ is algebraic over $\mathbb{Q}$ but π is not. An extension E of K is called algebraic over K (or an *algebraic extension* of K) if every element of E is algebraic over K.

The set $I = \{f \in K[x] : f(\alpha) = 0\}$ of polynomials that has an algebraic element $\alpha \in K \subseteq F$ as root forms an ideal in $K[x]$. Since $K[x]$ is a principal ideal domain, there is a unique monic polynomial $g \in K[x]$ such that $I = \langle g \rangle$, ideal generated by g. This polynomial g is irreducible and it is called the *minimal polynomial* of α over K. The *degree* of α is defined as the degree of g. The minimal polynomial has the property that for any polynomial f over K, $f(\alpha) = 0$ if and only if $g|f$.

It is sometimes useful to regard an extension field F as a vector space over its subfield K. If F, considered as a vector space over K, is finite dimensional, then F is called a *finite extension* of K. The dimension of F over K is called the *degree* of F over K and is denoted by $[F : K]$.

The following are some of the standard results for extension fields.

Lemma 7.4. *If F is a finite extension of K and E is a finite extension of F, then E is a finite extension of K and $[E : K] = [E : F][F : K]$.*

Lemma 7.5. *Every finite extension is an algebraic extension.*

Lemma 7.6. *Let $\alpha \in F$ be algebraic of degree n over K and let p be the minimal polynomial of α over K. Then*

(1) *$K(\alpha)$ is isomorphic to $K[x]/\langle p \rangle$.*
(2) *$[K(\alpha) : K] = n = \deg p(x)$ and $\{1, \alpha, \ldots, \alpha^{n-1}\}$ is a basis of $K(\alpha)$ over K.*
(3) *Every $\beta \in K(\alpha)$ is algebraic over K and its degree over K is a divisor of n.*

Lemma 7.7. *Let α, β be two roots of an irreducible polynomial f over K. Then $K(\alpha)$ and $K(\beta)$ are isomorphic by an isomorphism that maps α to β and keeps the elements of K fixed.*

Given a polynomial f over a field K, it is often the case that K does not contain all (or any of) the roots of f and we need to consider extension fields of K. The smallest field F that contains all the roots of f is called the *splitting field* of f over K. In $F[x]$ f can be written as a product of linear factors, that is, there exist elements $\alpha_1, \alpha_2, \ldots, \alpha_n \in F$ such that $f(x) = a(x-\alpha_1)(x-\alpha_2)\cdots(x-\alpha_n)$ where $a \in K$ is the leading coefficient of f. It is also true that $F = K(\alpha_1, \alpha_2, \ldots, \alpha_n)$. It is well known that splitting fields exist and they are unique (up to isomorphism). The splitting field of a polynomial f over a field K is a finite, therefore algebraic, extension of F since it is obtained from K by adjoining finitely many elements.

7.3.3. *Structure of Finite Fields*

Finite fields play a central role in algebraic coding theory. For linear codes, finite fields have been traditionally used as the alphabet of a code. More recently, linear codes over rings have also gained considerable interest. In this section, we give a description of the basic properties of finite fields.

The field $\mathbb{Z}_p$, the integers modulo p for a prime p, is the most familiar example of a finite field; however, there are many other finite fields as well. The fields $\mathbb{Z}_p$ play an important role in general field theory since every field of characteristic p must contain an isomorphic copy of $\mathbb{Z}_p$. The most fundamental properties of finite fields are given by the following theorems.

Theorem 7.6. *Let F be a finite field. Then F has p^n elements, where prime p is the characteristic of F and n is the degree of F over its prime subfield $\mathbb{Z}_p$.*

Lemma 7.8. *In a finite field F with $q = p^n$ elements every $a \in F$ satisfies $a^q = a$. Therefore, the polynomial $x^q - x$ factors in $F[x]$ as*

$$x^q - x = \prod_{a \in F}(x - a).$$

Consequently, F is the splitting field *of $x^q - x$ over $\mathbb{Z}_p$.*

Here is the main characterization of finite fields:

Theorem 7.7. *For every prime p and every positive integer n, there exists a finite field with p^n elements. Any finite field with $q = p^n$ elements is isomorphic to the splitting field of $x^q - x$ over $\mathbb{Z}_p$.*

This theorem provides a justification for speaking of *the* finite field (or *the* Galois field) with q elements, or *the* finite field of order q. We shall denote this field by $\mathbb{F}_q$ or $\mathrm{GF}(q)$. In particular, for a prime p the notations $\mathbb{F}_p$ and $\mathbb{Z}_p$ are interchangeable.

The subfields of a given finite field are uniquely determined. Given a finite field $\mathbb{F}_q$, $q = p^n$, there is a unique subfield of order p^m, for each $m|n$.

For a finite field $\mathbb{F}_q$, we denote by $\mathbb{F}_q^*$ the multiplicative group of non-zero elements of $\mathbb{F}_q$. It is well known that $\mathbb{F}_q^*$ is a cyclic group. A generator of the cyclic group $\mathbb{F}_q^*$ is called a *primitive element* of $\mathbb{F}_q$. The existence of primitive elements implies that every finite field is a simple algebraic extension of its prime subfield, which in turn implies that for any positive integer n and every prime p, there exists an irreducible polynomial of degree n in $\mathbb{F}_p[x]$. Irreducible polynomials having primitive elements as their roots are given special names.

Definition 7.12. Let α be a primitive element of $\mathbb{F}_{q^n}$. The minimal polynomial of α over $\mathbb{F}_q$ is called a *primitive polynomial* for $\mathbb{F}_{q^n}$ over $\mathbb{F}_q$.

7.3.4. *Roots of Irreducible Polynomials*

Irreducible polynomials and their roots are important for constructing finite fields. They are also important for the construction of certain algebraic codes. In this section, we summarize important facts about the set of roots of an irreducible polynomial.

Lemma 7.9. *Let $f \in \mathbb{F}_q[x]$ be an irreducible polynomial and let α be a root of f in an extension of $\mathbb{F}_q$. Then for a polynomial $h \in \mathbb{F}_q[x]$, $h(\alpha) = 0$ if and only if $f|h$.*

One of the important properties of the roots of a polynomial $p(x)$ over a finite field $\mathbb{F}_q$ is that if α is a root of p, then so is α^q. This follows from the fact that $a^q = a$ and $(a + b)^q = a^q + b^q$ in $\mathbb{F}_q$. It also follows that $\alpha, \alpha^q, \alpha^{q^2}, \ldots$ are all roots of $p(x)$. If $p(x)$ is an irreducible polynomial of degree m, then a root α of $p(x)$ lies in $\mathbb{F}_q(\alpha) = \mathbb{F}_{q^m}$, and $\mathbb{F}_{q^m}$ is the smallest such field. Moreover, $\{\alpha, \alpha^q, \alpha^{q^2}, \ldots, \alpha^{q^{m-1}}\}$ is the set all roots of $p(x)$. This means that the splitting field of a polynomial of degree m over $\mathbb{F}_q$ is $\mathbb{F}_{q^m}$. The elements $\alpha, \alpha^q, \alpha^{q^2}, \ldots, \alpha^{q^{m-1}} \in \mathbb{F}_{q^m}$ are called *conjugates* of α with respect to $\mathbb{F}_q$. The conjugates of $\alpha \in \mathbb{F}_q^*$ with respect to any subfield of $\mathbb{F}_q$ have the same order in the group $\mathbb{F}_q^*$. Therefore, we can make the following definition.

Definition 7.13. The multiplicative order of any root of an irreducible polynomial $f(x) \in \mathbb{F}_q[x]$ in its splitting field is called the *order* of $f(x)$.

It follows that all the conjugates of a primitive element are also primitive. As an example, let us consider $\alpha \in \mathbb{F}_{16}$, where α is a root of the irreducible polynomial $f(x) = x^4 + x + 1$ over $\mathbb{F}_2$. The conjugates of α with respect to $\mathbb{F}_2$ are $\alpha, \alpha^2, \alpha^4 = \alpha + 1$, and $\alpha^8 = \alpha^2 + 1$, each of them being a primitive element of $\mathbb{F}_{16}$. The conjugates of α with respect to $\mathbb{F}_4$ are α and $\alpha^4 = \alpha + 1$.

On the basis of previous results, we can compute minimal polynomials as in the following lemma.

Lemma 7.10. *Let $\alpha \in \mathbb{F}_{q^m}$. Then the minimal polynomial of α over $\mathbb{F}_q$ is $m_\alpha(x) = (x - \alpha)(x - \alpha^q) \cdots (x - \alpha^{q^{d-1}}) \in \mathbb{F}_q[x]$ where d is the smallest positive integer such that $\alpha^{q^d} = \alpha$.*

Example 7.4. Let $p(x) = x^4 + x^3 + 1 \in \mathbb{F}_2$. It can be verified that $p(x)$ is a primitive polynomial for $\mathbb{F}_{16}$ over $\mathbb{F}_2$. Let α be a root of $p(x)$ in $\mathbb{F}_{16}$ (hence a primitive element of $\mathbb{F}_{16}$). Let us compute the minimal polynomials of all of the elements of $\mathbb{F}_{16}$ over $\mathbb{F}_2$. Let $[\beta]$ denote the set of conjugates of $\beta \in \mathbb{F}_{16}$ with respect to $\mathbb{F}_2$. Then,

$$[\alpha] = \{\alpha, \alpha^2, \alpha^4, \alpha^8\},$$
$$[\alpha^3] = \{\alpha^3, \alpha^6, \alpha^{12}, \alpha^{24} = \alpha^9\},$$
$$[\alpha^5] = \{\alpha^5, \alpha^{10}\},$$
$$[\alpha^7] = \{\alpha^7, \alpha^{14}, \alpha^{28} = \alpha^{13}, \alpha^{26} = \alpha^{11}\}.$$

Each conjugacy class $[\alpha^i]$ has the same minimal polynomial m_i. For example, the minimal polynomial of α^3 is $m_3(x) = (x - \alpha^3)(x - \alpha^6)(x - \alpha^9)(x - \alpha^{12}) = x^4 + x^3 + x^2 + x + 1$, and $m_3(x) = m_6(x) = m_9(x) = m_{12}(x)$. We can compute other minimal polynomials similarly and obtain

$$m_5(x) = m_{10}(x) = x^2 + x + 1,$$
$$m_7(x) = m_{11} = m_{13}(x) = m_{14}(x) = x^4 + x + 1.$$

7.3.5. *Roots of Unity*

The polynomial $x^n - 1$ over $\mathbb{F}_q$ is very important in algebraic coding theory due to its connections with cyclic codes. In this section, we review results on this polynomial, its roots, and its factorization.

First, we observe that if $(n, q) \neq 1$, then we can write $n = mp^k$ where $(m, q) = 1$ and $p = \text{char}(\mathbb{F}_q)$. Then, $x^n - 1 = x^{mp^k} - 1 = (x^m - 1)^{p^k}$. Therefore, we will always assume that $(n, q) = 1$.

Let $\mathbb{F}_{q^m}$ be the splitting field of $x^n - 1$ over $\mathbb{F}_q$. Since $(x^n - 1)' = nx^{n-1}$ is relatively prime with $x^n - 1$, the polynomial $x^n - 1$ does not have multiple roots. Thus, $x^n - 1$ has n distinct roots in $\mathbb{F}_{q^m}$. The roots of $x^n - 1$ in $\mathbb{F}_{q^m}$ are called nth *roots of unity over* $\mathbb{F}_q$. The set W_n of nth roots of unity over $\mathbb{F}_q$ has a nice algebraic structure.

Lemma 7.11. *When* $(n, q) = 1$, W_n *is a cyclic group, a cyclic subgroup of the multiplicative group* $\mathbb{F}_{q^m}^*$ *for a suitable* $m \in \mathbb{Z}$.

An nth root of unity Ω over $\mathbb{F}_q$ of order n, which is a generator of the cyclic group W_n, is called a *primitive nth root of unity* over $\mathbb{F}_q$. We can determine m (the smallest integer such that $\omega \in \mathbb{F}_{q^m}$) in terms of n and q. Let $\omega \in W_n$ be a primitive nth root of unity. Since ω has order n, we have
$$\omega \in \mathbb{F}_{q^r} \Leftrightarrow \omega^{q^r} = \omega \Leftrightarrow \omega^{q^r - 1} = 1 \Leftrightarrow n|(q^r - 1).$$

Since m is the smallest integer for which $\omega \in \mathbb{F}_{q^m}$, we have the following result.

Lemma 7.12 (Roman [64]). *If* $\mathbb{F}_{q^m}$ *is the splitting field of* $x^n - 1$ *over* $\mathbb{F}_q$, *then* m *is the smallest positive integer for which* $n|(q^m - 1)$, *that is,* m *is the smallest positive integer for which* $q^m \equiv 1 \bmod n$. *This integer* m *is called the* order *of* $q \bmod n$, *which is denoted by* $o_n(q)$.

7.3.6. *Factorization of* $x^n - 1$

The factorization of the polynomial $x^n - 1$ over a finite field is very important for the study of cyclic codes. For $(n, q) = 1$, the polynomial $x^n - 1$ over $\mathbb{F}_q$ has no multiple factors and can be factored using the fact that it has n distinct roots. It is therefore the product of the distinct minimal polynomials. Let α be a primitive element of $\mathbb{F}_{q^m}$, where $m = o_n(q)$. Then, we know that $\omega = \alpha^{(q^m-1)/n}$ is a primitive nth root of unity. Therefore, the roots of $x^n - 1$ are given by $\omega, \omega^2, \ldots, \omega^{n-1}, \omega^n = 1$. We need to determine the minimal polynomials for these roots and take the product of distinct ones.

For $0 \le i \le n - 1$, the conjugates of ω^i are $\omega^i, \omega^{iq}, \omega^{iq^2}, \ldots, \omega^{iq^{d-1}}$ where d is the smallest positive integer such that $\omega^{iq^d} = \omega^i$. Since
$$\omega^{iq^d} = \omega^i \Leftrightarrow \omega^{iq^d - i} = 1 \Leftrightarrow n|(iq^d - i) \Leftrightarrow iq^d \equiv i \quad \bmod n,$$
the minimal polynomial for ω^i (and its conjugates) is
$$m_i(x) = (x - \omega^i)(x - \omega^{iq}) \cdots \left(x - \omega^{iq^{d-1}}\right).$$

The set of exponents of ω in the last product is called ith *cyclotomic coset* of q modulo n. These sets can be defined independently of a primitive nth root of unity. In fact the relation $\sim$ defined on $\mathbb{Z}_n = \{0, 1, 2, \ldots, n-1\}$ by $i \sim j$ if and only if $j \equiv iq^r \bmod n$ for some integer r is an equivalence relation and the equivalence classes are exactly the cyclotomic cosets of q modulo n. There is a one-to-one correspondence between irreducible factors of $x^n - 1$ over $\mathbb{F}_q$ and cyclotomic cosets of q modulo n; every irreducible factor of degree k corresponds to a cyclotomic coset of size k and k must divide m. We now illustrate this factorization with an example.

Example 7.5. Let $q = 2, n = 15$ and consider the polynomial $f(x) = x^{15} - 1 = x^{15} + 1$ over $\mathbb{F}_2$. Since $m = o_{15}(2) = 4$, the splitting field of $f(x)$ over $\mathbb{F}_2$ is $\mathbb{F}_{16}$. The polynomial $p(x) = x^4 + x^3 + 1$ is primitive over $\mathbb{F}_2$. Let α be a root of $p(x)$, then it is a primitive element of $\mathbb{F}_{16}$ and happens to be a primitive 15th root of unity. Hence, the roots of $x^{15} - 1$ are $1, \alpha, \alpha^2, \ldots, \alpha^{14}$. We already computed the minimal polynomials in this case. Therefore, we obtain the factorization $x^{15} - 1 = (x+1)(x^4 + x + 1)(x^4 + x^3 + x^2 + x + 1)$ $(x^2 + x + 1)(x^4 + x^3 + 1)$ over $\mathbb{F}_2$.

Exercise 7.3.2. Obtain a factorization of $x^{11} - 1$ over $\mathbb{F}_3$ using cyclotomic cosets and an irreducible polynomial of degree 5.

7.4. Some Classes of Linear Codes

In this section, we will review some of the most fundamental and standard classes of algebraic codes. The material in this section can be found in most standard books on coding theory such as [64].

7.4.1. *Cyclic Codes*

Cyclic codes are very important for both theoretical and practical purposes. Their nice structure facilitates their implementation in practice. On the other hand, they establish a fundamental link between coding theory and algebra. We begin with the definition of a cyclic code.

Let $\mathbf{v} = (v_0, v_1, \ldots, v_{n-1})$ be a vector in $V := \mathbb{F}_q^n$. We may associate with vector $\mathbf{v} \in V$ a polynomial in $\mathbb{F}_q[x]$ as follows:

$$\phi : \mathbf{v} = (v_0, v_1, \ldots, v_{n-1}) \to v(x) = v_0 + v_1 x + \cdots + v_{n-1} x^{n-1}.$$

The map ϕ is a vector space isomorphism from V onto the subspace $\phi(V)$ of $\mathbb{F}_q[x]$. Given this map, we can identify $\phi(V)$ with V without an explicit reference to ϕ. Hence we will think of the vectors in V as polynomials of degree $< n$.

Definition 7.14. A linear code is *cyclic* if it is invariant under (*right*) cyclic shift, i.e. $(c_0, c_1, \ldots, c_{n-1}) \in C$ implies $(c_{n-1}, c_0, c_1, \ldots, c_{n-2}) \in C$.

Viewing a codeword $\mathbf{c}$ as a polynomial $c(x)$ in a cyclic code C implies that $xc(x) \bmod (x^n - 1) \in C$. Thus, a linear code C is cyclic if and only if C is an ideal of the factor ring

$$R_n = \frac{\mathbb{F}_q[x]}{\langle x^n - 1 \rangle}.$$

This relation links algebra to coding theory and enables us to use the algebraic structure of ideals in order to better understand the cyclic codes. From algebra, we know that $\mathbb{F}_q[x]$ is a principal ideal domain and $\mathbb{F}_q[x]/\langle f(x) \rangle$ is a principal ideal ring. Below are the most basic facts about the structure of cyclic codes.

Theorem 7.8 (Roman [64]). *Let C be an ideal in R_n, i.e. a cyclic code of length n.*

(1) *There is a unique monic polynomial $g(x) \in R_n$ of minimum degree, which generates C, i.e. $C = \langle g(x) \rangle$. This polynomial is called the generator polynomial of C. (The generator polynomial is usually not the only polynomial that generates C. The next lemma characterizes all the polynomials that generate C.)*

(2) $g(x) | x^n - 1$.

(3) *If $\deg(g(x)) = r$, then C has dimension $n - r$. In fact,*

$$C = \langle g(x) \rangle = \{r(x)g(x) : \deg(r(x)) < n - r\}.$$

(4) *If $g(x) = g_0 + g_1 x + \cdots + g_r x^r$, then $g_0 \neq 0$ and C has a generator matrix of the form*

$$\begin{bmatrix} g_0 & g_1 & g_2 & \cdots & g_r & 0 & 0 & 0 & \cdots & 0 \\ 0 & g_0 & g_1 & g_2 & \cdots & g_r & 0 & 0 & \cdots & 0 \\ 0 & 0 & g_0 & g_1 & \cdots & & g_r & 0 & & 0 \\ \vdots & & & & & & & \cdots & & 0 \\ 0 & 0 & 0 & \cdots & 0 & g_0 & g_1 & g_2 & \cdots & g_r \end{bmatrix},$$

where each row is a right cyclic shift of the previous row.

Lemma 7.13 (Macwilliams and Sloane [53]). *Let C be a cyclic code of length n with the least degree generator polynomial $g(x)$. Then,*

$$C = \langle f(x)g(x)\rangle$$

if and only if $(f(x), h(x)) = 1$, where $h(x) = (x^n - 1)/g(x)$, called the check polynomial of C.

Proof. $\Rightarrow$: Suppose $\langle g(x)\rangle = \langle f(x)g(x)\rangle$. Then $g(x) = g(x)f(x)t(x)$ for some $t(x) \in \mathbb{F}_q[x]$. Since $(g(x), h(x)) = 1$, there are polynomials $A(x), B(x)$ in $\mathbb{F}_q[x]$ such that $A(x)g(x) + B(x)h(x) = 1$. Replacing $g(x)$ with $g(x)f(x)t(x)$, we have $(A(x)g(x)t(x)) \cdot f(x) + B(x)h(x) = 1$. Therefore, $(f(x), h(x)) = 1$.
$\Leftarrow$: It is clear that $\langle f(x)g(x)\rangle \subseteq \langle g(x)\rangle$. Since $(f(x), h(x)) = 1$, there exist $s(x), t(x) \in \mathbb{F}_q[x]$ such that $s(x)f(x) + t(x)h(x) = 1$. Hence, $s(x)f(x)g(x) + t(x)(x^n - 1) = g(x)$. Reducing mod $x^n - 1$ we get, $s(x)f(x)g(x) = g(x)$. Thus, $g(x) \in \langle f(x)g(x)\rangle$, and $\langle g(x)\rangle \in \langle f(x)g(x)\rangle$. $\square$

There is an alternative way of describing cyclic codes. Every cyclic code of length n has a unique, monic generator polynomial of degree $\leq n$ that divides $x^n - 1$. Therefore, to find all the cyclic codes one needs to factor $x^n - 1$. If the factorization is $x^n - 1 = m_1(x)m_2(x) \cdots m_t(x)$, then there are a total of 2^t distinct factors of $x^n - 1$, hence 2^t cyclic codes. If α is a root of some $m_i(x)$ in some extension of $\mathbb{F}_q$, then m_i is the minimal polynomial of α over $\mathbb{F}_q$. Hence, for any $f(x) \in \mathbb{F}_q[x]$, $f(\alpha) = 0$ if and only if $m_i(x)|f(x)$. Therefore, we can specify C through the roots of its generator polynomial. If $g(x) = q_1(x) \cdots q_r(x)$, product of some irreducible factors of $x^n - 1$, then $\langle g(x)\rangle = \{f(x) \in R_n : f(\beta_1) = f(\beta_2) = \cdots = f(\beta_r) = 0\}$ where β_i is a root of $q_i(x)$. Note that the every element in the set $Z = \{\beta_i : 1 \leq i \leq r\}$ is an nth root of unity and therefore is a power of the primitive nth root of unity, say ω, over $\mathbb{F}_q$. The set Z is called the *zero set of the code C* and uniquely identifies it.

7.4.2. BCH Codes

A very important class of cyclic codes is BCH codes, discovered by Bose, Ray-Chaudhuri, and Hocquenghem. They are defined by specifying the roots of a cyclic code.

Definition 7.15. Let $q, n, b, d \in \mathbb{N}$, where q is a prime power, $(n, q) = 1$, and $2 \leq d \leq n$. Let ω be a primitive nth root of unity over $\mathbb{F}_q$ (we know that ω lies in $\mathbb{F}_{q^m}$ where $m = ord_n(q))$, m_i be the minimal polynomial of ω^i and $Z := \{b, b+1, \ldots, b+d-2\}$. Then the *BCH code* $C \subseteq \mathbb{F}_q^n$ of *designed distance* d is a cyclic code of length n over $\mathbb{F}_q$ defined by the following equivalent conditions:

(i) $\mathbf{v} \in C$ if and only if $\mathbf{v}(\omega^i) = 0$ for all $i \in Z$.

(ii) The polynomial $lcm\{m_i : i \in Z\}$ is the least degree (monic) *generator polynomial* of C.

(iii) A parity check matrix of C is the matrix

$$
\begin{bmatrix}
1 & \omega^b & \omega^{2b} & \cdots & \omega^{b(n-1)} \\
1 & \omega^{b+1} & \omega^{2(b+1)} & \cdots & \omega^{(b+1)(n-1)} \\
\cdots & \cdots & \cdots & \cdots & \cdots \\
1 & \omega^{b+d-2} & \omega^{2(b+d-2)} & \cdots & \omega^{(b+d-2)(n-1)}
\end{bmatrix}.
$$

Remark. If $b = 1$ in the last definition the resulting BCH code is called a *narrow-sense BCH code*. If $n = q^m - 1$, the BCH code is called *primitive*.

A well-known result about BCH codes is the BCH bound.

Theorem 7.9 (BCH bound). *A BCH code of designed distance d defined by 7.15 has minimum distance $\geq d$.*

The BCH bound can be proven by applying Lemma 7.2 on the parity check matrix. Alternatively, it can be proven using Mattson–Solomon polynomials [64].

A practical example for the use of BCH codes is the European and trans-Atlantic information communication system, which has been using such codes for many years [49]. The message symbols are of length 231 and the generator polynomial is of degree 24 so that $231 + 24 = 255 = 2^8 - 1$ is the length of the codewords. The code detects at least 6 errors and its failure (incorrect decoding) probability is one in sixteen million.

Exercise 7.4.1. Determine the parameters of the binary, narrow-sense BCH code of length 15, and *designed distance* 5.

7.4.3. *Reed Solomon Codes*

A special case of BCH codes is *Reed–Solomon codes* (or *RS codes* in short), which are defined as narrow sense BCH codes of designed distance d and of length $n = q - 1$ over $\mathbb{F}_q$. Hence $m = 1$ and $\mathbb{F}_q$ posses a primitive nth root of unity. The generator polynomial of least degree for an RS code is

$$g(x) = \prod_{i=1}^{d-1} (x - \omega^i),$$

where ω is a primitive element of $\mathbb{F}_q$. It turns out that an RS code has parameters $[n, k, n - k + 1]_q$, where $n = q - 1$, i.e., they are MDS codes.

RS codes are used to insure high sound quality of compact discs.

There is an alternative description of RS codes that motivates the construction of algebraic geometry codes: Let $1 \leq n \leq q$, $1 \leq k \leq n$ and let

$$P_k = \{f(x) \in \mathbb{F}_q[x] : \deg f(x) < k\}.$$

First, choose n distinct elements $\alpha_1, \alpha_2, \ldots, \alpha_n \in \mathbb{F}_q$, then define a Reed–Solomon code by

$$GRS_q(n, k) = \{(f(\alpha_1), f(\alpha_2), \ldots, f(\alpha_n)) : f(x) \in P_k\}.$$

It can easily be verified that the code $GRS_q(n, k)$ is a linear code with the parameters $[n, k, n - k + 1]_q$ over $\mathbb{F}_q$.

Exercise 7.4.2. Show that an RS code of length $n = q$ defined by the second method is equivalent to a cyclic code.

7.4.4. *Hamming Codes*

Hamming codes are an important class of linear codes. The binary Hamming code with parameters $[7, 4, 3]$ was one of the first codes designed and used in practice by Hamming [39]. In general, they are defined via a *parity check matrix* over any finite field $\mathbb{F}_q$. First, choose a positive integer r. Let H be a matrix whose columns consist of all vectors of length r over $\mathbb{F}_q$ whose first non-zero entry is 1. What are the parameters of the Hamming code? A counting argument shows that there are $1 + q + q^2 + \cdots + q^{r-1} = \frac{q^r - 1}{q - 1}$ such vectors, therefore, H is an r-by-n matrix, where $n = \frac{q^r - 1}{q - 1}$. What is the rank of H? It is easy to see that after a permutation of columns (if necessary), H can be put into the form $(I_r | H')$ where I_r is the identity matrix of order r. Therefore, the rank of H is r, and its nullity, which is the dimension of the Hamming code, is $n - r$. Finally, we want to determine the minimum distance of the Hamming code. One can easily show that no two columns are linearly independent (i.e. no column is a scalar multiple of another), and there exist three columns that are linearly dependent.

Hence, by Lemma 7.2 the minimum distance is 3, and the parameters of the Hamming code are $[n, n - r, 3]_q$, where $n = \frac{q^r - 1}{q - 1}$. Given these parameters, it is easy to show that the Hamming codes are perfect. They are an infinite family of perfect codes. Besides the Hamming codes, the only linear perfect codes are the Golay codes. Moreover, all binary Hamming codes are cyclic [64]. More generally, Hamming codes are equivalent to cyclic codes when $(r, q - 1) = 1$ [64].

7.4.5. *Quadratic Residue Codes*

Quadratic residue (QR) codes are also a special class of cyclic codes. To define them, we first need to introduce the concept of a QR. Let p be an odd prime. An integer a such that $(a, p) = 1$ is called a QR mod p if the equation $x^2 \equiv a \bmod p$ has a solution. Otherwise, a is called a quadratic non-residue. The set of QR's and non-residues mod p are denoted by QR and NR, respectively. For example, for $p = 23$, $QR = \{1, 2, 3, 4, 6, 8, 9, 12, 13, 16, 18\}$ and NR, is the rest of the non-zero integers modulo 23. It is well known that of the $p - 1$ non-zero integers mod p, exactly half of them are in QR and the other half are in NR. It is also easy to show that if $x, y \in QR$, then $xy \in QR$, and if $x, y \in NR$, then $xy \in QR$. On the other hand, if $x \in QR$ and $y \in NR$ then $xy \in NR$.

Let p be an odd prime, and let q be a prime that is a QR mod p. From the closure properties of QR and NR, it follows that whenever an element from a cyclotomic coset $cl_i = \{i, iq, iq^2, \dots\}$ of $q \bmod p$ is in QR, the entire set cl_i is contained in QR. The same is true for NR. Therefore, the QR and NR are unions of cyclotomic cosets of $q \bmod p$. For such primes p and q, let ω be a primitive pth root of unity over $\mathbb{F}_q$, and let $q(x) = \prod_{r \in QR}(x - \omega^r)$ and $n(x) = \prod_{s \in NR}(x - \omega^s)$. Then, since QR and NR are unions of cyclotomic cosets, the polynomials $q(x)$ and $n(x)$ are in $\mathbb{F}_q[x]$. Moreover, $x^p - 1 = (x - 1)q(x)n(x)$. The q-ary cyclic codes generated by $Q(p) = \langle q(x) \rangle$, $\overline{Q(p)} = \langle (x - 1)q(x) \rangle$, $N(p) = \langle n(x) \rangle$, $\overline{N(p)} = \langle (x - 1)n(x) \rangle$ are called QR codes. Clearly, $Q(p) \supseteq \overline{Q(p)}$, and $N(p) \supseteq \overline{N(p)}$. It is also clear that dim $Q(p) = $ dim $N(p) = p - \deg(q(x)) = \frac{p+1}{2}$. It can be shown that the codes $Q(p)$ and $N(p)$ are equivalent; hence, they have the same minimum distance d. The square root bound [53] states that $d^2 \geq p$. Furthermore, if $p = 4m - 1$ then $d^2 - d + 1 \geq p$.

Example 7.6. Let $q = 2$, and $p = 23$. Then q is a QR mod 23 ($5^2 \equiv 2 \bmod 23$). In general, 2 is a QR mod p (p odd prime) if and only if $p \equiv \pm 1 \bmod 8$. Then we know that $x^{23} - 1 = (x - 1)q(x)n(x)$ over $\mathbb{F}_2$, where

$\deg(q(x)) = \deg(n(x)) = 11$. The resulting QR codes $Q(p)$ and $N(p)$ have dimension 12. Moreover, by the square root bound, the minimum distance is at least 6. The actual minimum distance turns out to be 7. This is the famous binary Golay code with parameters $[23, 12, 7]$, denoted by G_{23}. Its extension is a $[24, 12, 8]$-code, G_{24}. These two codes have some fascinating properties. G_{23} is the only binary, multiple error-correcting, perfect code. G_{24} leads to a unique combinatorial design called a Steiner triple system (a 5-design) [58]. It also gives a construction of a densest lattice in dimension 24, called the Leech lattice [72]. Moreover, both G_{23} and G_{24} led to the discovery of some new simple groups [72].

Similarly, the ternary Golay code G_{11} is also a QR residue code with parameters $[11, 6, 5]_3$. It is well known that the Golay codes are the only multiple error-correcting perfect codes (up to equivalence) [58].

Exercise 7.4.3. Show that the Golay codes G_{23}, G_{11} are perfect.

7.5. Constacyclic and Quasi-Twisted Codes

This section is largely from [7], parts of it reprinted with kind permission of Springer Science and Business Media.

7.5.1. *Constacyclic Codes*

There are several generalizations of cyclic codes. One immediate generalizaton is the class of *constacyclic codes*. Let $a \in \mathbb{F}_q^* := \mathbb{F}_q - \{0\}$. A linear code of length n over $\mathbb{F}_q$ is called *constacyclic* if it is invariant under the constacyclic shift:

$$(c_0, c_1, \ldots, c_{n-1}) \mapsto (ac_{n-1}, c_0, \ldots, c_{n-2}).$$

Note that in the case $a = 1$ we recover cyclic codes. When $a = -1$, they are called *negacyclic* codes. Most of the results about cyclic codes are also true for constacyclic codes. These are summarized in the following proposition. Recall the identification of words (vectors) of $\mathbb{F}_q^n$ with polynomials of degree $\leq n - 1$.

Lemma 7.14.

(i) *Constacyclic codes are precisely the ideals in the ring* $\frac{\mathbb{F}_q[x]}{\langle x^n - a \rangle}$.

(ii) *The ring* $\frac{\mathbb{F}_q[x]}{\langle x^n - a \rangle}$ *is a principal ideal ring and for a conctacyclic code C there exists a polynomial $g(x)$ (called the generator polynomial) of smallest degree such that $C = \langle g(x) \rangle$ where $g(x)|(x^n - a)$ and $\dim(C) = n - \deg(g(x))$.*

(iii) *If* $g(x) = g_0 + g_1 x + \ldots + g_r x^r$, *then a generator matrix for C is*

$$
G = \begin{pmatrix}
g_0 & g_1 & g_2 & \cdots & g_r & 0 & \cdots & 0 \\
0 & g_0 & g_1 & \cdots & g_{r-1} & g_r & \cdots & 0 \\
\cdots & \cdots & \cdots & \cdots & \cdots & \cdots & \cdots & \cdots \\
0 & \cdots & 0 & g_0 & g_1 & g_2 & \cdots & g_r
\end{pmatrix},
$$

where each row of G is a constacyclic shift *of the previous one.*

Proof. Everything is proved as in the cyclic case. □

Similar to cyclic codes, a constacyclic code can be specified through the roots of its generator polynomial. In studying cyclic codes, the factorization of $x^n - 1$ was crucial. Now, we are interested in factorizing $x^n - a$ over $\mathbb{F}_q$. Before looking at this factorization, we remark that in certain cases constacyclic codes are equivalent to cyclic codes.

Lemma 7.15. *If* $\mathbb{F}_q$ *contains an* nth *root* δ *of a, then a constacyclic code of length* n *is equivalent to a cyclic code of length* n.

The following lemma tells us exactly when an element $a \in \mathbb{F}_q$ has an nth root in $\mathbb{F}_q$.

Lemma 7.16 (Roman [64]). *Let* $a = \alpha^i$ *where* α *is a primitive element of* $\mathbb{F}_q$. *Then the equation* $x^n = a$ *has a solution in* $\mathbb{F}_q$ *if and only if* $(n, q-1)|i$, *where* $(n, q-1)$ *denotes the greatest common divisor of the integers* n *and* $q - 1$.

Proof. The equation $x^n = a$ has a solution $x = \alpha^j \Leftrightarrow \alpha^{nj} = \alpha^i$
$\Leftrightarrow \alpha^{nj-i} = 1$
$\Leftrightarrow (q-1)|(nj - i)$
$\Leftrightarrow i = nj + r(q-1)$ for some integers r, j.
$\Leftrightarrow (n, q-1)|i$ □

Hence, in our investigation of constacyclic codes, we are going to consider the case $(n, q-1) \nmid i$.

7.5.2. Factorization of $x^n - a$ and a BCH bound

Let $a \in \mathbb{F}_q^*$ be such that it does not have an nth root in $\mathbb{F}_q$. We also assume that $(n, q) = 1$ so that the polynomial $x^n - a$ does not have multiple roots. The roots of $x^n - a$ are $\delta, \delta\zeta, \delta\zeta^2, \ldots, \delta\zeta^{n-1}$ where ζ is a primitive nth root of unity and $\delta^n = a$. Then ζ lies in F_{q^m} where $m = \text{ord}_q(n)$.

By assumption $\delta \notin \mathbb{F}_q$. Since $\delta^n = a$, $\delta^{nr} = a^r = 1$, where r is the order of a in the multiplicative group $\mathbb{F}_q^*$, which is equal to $\frac{q-1}{(i,q-1)}$, $a = \alpha^i$ and α is a primitive element of $\mathbb{F}_q$. Hence, δ is an nrth root of 1. Therefore, $\delta \in \mathbb{F}_{q^s}$ where $s = \mathrm{ord}_q(nr)$. Now, $q^s - 1 \equiv 0 \bmod nr \Rightarrow q^s - 1 \equiv 0 \bmod n$. This implies that $m | s$. Consequently, $\mathbb{F}_{q^m} \subseteq \mathbb{F}_{q^s}$. Hence, the field $\mathbb{F}_{q^s}$ contains both ζ and δ and we may take $\delta = w^t$ and $\zeta = w^{rt}$ where w is a primitive element of $\mathbb{F}_{q^s}$ (hence a primitive $(q^s - 1)$th root of unity) and $q^s - 1 = ntr$, for some integer t. Hence, $\zeta = \delta^r$ and $x^n - a$ factors are as follows:

$$x^n - a = \prod_{i=0}^{n-1}(x - \delta\zeta^i) = \prod_{i=0}^{n-1}\left(x - w^{t(1+ir)}\right) = \prod_{i=0}^{n-1}(x - \delta^{1+ir}).$$

Each irreducible factor of $x^n - a$ corresponds to a cyclotomic coset modulo nr (*not* modulo n), i.e. the degree of each irreducible factor is the same as size of a cyclotomic coset modulo nr. Since all the roots of $x^n - a$ are nrth roots of unity, we have that $(x^n - a)|(x^{nr} - 1)$ also, $(x^{nr} - 1)|(x^{n(q-1)} - 1)|(x^{q^s-1} - 1)$.

Example 7.7. Let $q = 5$ and $n = 6$ and let us consider the polynomial $x^6 - 3$ over $\mathbb{F}_5$ (hence constacyclic codes of length 6 over $\mathbb{F}_5$ with $a = 3$). A primitive element of $\mathbb{F}_5$ is 2, $3 = 2^3$ in $\mathbb{F}_5$, order of 3 in $\mathbb{F}_5$ is 4 and $(n, q-1) = (6, 4) = 2 \nmid 3$ so that there is no 6th root of 3 in $\mathbb{F}_5$. According to the discussion above,

$$x^6 - 3 = \prod_{i=0}^{5}(x - \delta^{4i+1}) = (x^2 + 3x + 3)(x^2 + 2x + 3)(x^2 + 3),$$

where δ is a primitive $6 \cdot 4 = 24$th root of unity. The powers of δ that appear in this factorization are $1, 5, 9, 13, 17, 21$ and these are precisely union of three (the same as the number of irreducible factors over $\mathbb{F}_5$) cyclotomic cosets modulo 24: $cl_1 = \{1, 5\}$, $cl_9 = \{9, 21\}$, $cl_{13} = \{13, 17\}$. On the other hand, $x^{24} - 1$ and $x^6 - 1$ factors over $\mathbb{F}_5$ are as follows:

$$\begin{aligned}
x^{24} - 1 =\ & (x^2 + 3x + 3)(x^2 + 2x + 3)(x^2 + 3)(x^2 + 4x + 1)(x^2 + x + 2)\\
& \times (x^2 + 2x + 4)(x^2 + x + 1)(x^2 + 4x + 2)(x^2 + 3x + 4)(x^2 + 2)\\
& \times (x + 3)(x + 4)(x + 2)(x + 1),\\
x^6 - 1 =\ & (x^2 + 4x + 1)(x^2 + x + 1)(x + 1)(x + 4).
\end{aligned}$$

The factors of $x^6 - 1$ correspond to the following cyclotomic cosets modulo 24: $cl_0 = \{0\}$, $cl_4 = \{4, 20\}$, $cl_8 = \{8, 16\}$, $cl_{12} = \{12\}$, which are obtained by shifting the cosets corresponding to $x^6 - 3$ by 1.

7.5.3. BCH Bound for Constacyclic Codes

Lemma 7.17. *Let C be a constacyclic code of length n over $\mathbb{F}_q$ and let the generator polynomial $g(x)$ have the elements $\{\delta\zeta^i : 1 \leq i \leq d - 1\}$ among its roots. Then the minimum distance of $C \geq d$.*

Proof. Consider the constacyclic code C of length n with generator polynomial $g(x)|(x^n - a)$ having $\{\delta\zeta, \delta\zeta^2, \ldots, \delta\zeta^{d-1}\} = \{\delta^{r+1}, \delta^{2r+1}, \ldots, \delta^{(d-1)r+1}\}$ among its roots. The corresponding cyclic code $\psi(C)$ generated by $g(\delta x)|(x^n - 1)$ has the elements $\zeta, \zeta^2, \ldots, \zeta^{d-1}$ among the roots. By the BCH bound, the minimum distance of $\psi(C) \geq d$. Therefore, $d(C) \geq d$ as well. $\qquad\square$

We now give an example of a constacyclic code that is optimal.

Example 7.8. Let $q = 3$ and $n = 28$ and consider constacyclic codes of length 28 over $\mathbb{F}_3$ with $a = 2$. We remark that the condition $(n, q - 1) \nmid i$ implies that we should consider only even lengths over $\mathbb{F}_3$. We find that $r = 2$ and therefore $(x^{28} - 2)|(x^{56} - 1)$. The factorization of $x^{28} - 2$ over $\mathbb{F}_3$ is as follows:

$$x^{28} - 2 = \prod_{i=0}^{27}(x - \delta\zeta^i) = \prod_{i=0}^{27}(x - \delta^{2i+1})$$

$$= (x^6 + 2x^4 + x^3 + x^2 + 2)(x^6 + 2x^5 + 2x + 2)(x^2 + x + 2)$$
$$\times (x^6 + x^5 + x + 2)(x^6 + 2x^4 + 2x^3 + x^2 + 2)(x^2 + 2x + 2),$$

where δ is a primitive 56th root of 1 and $\zeta = \delta^2$ is a primitive 28th root of 1 over $\mathbb{F}_3$. The exponents of δ in this factorization are exactly odd integers modulo 56 and they are partitioned into following cyclotomic cosets:

$$\{1, 3, 9, 19, 25, 27\}, \quad \{5, 13, 15, 23, 39, 45\}, \quad \{7, 21\},$$
$$\{11, 17, 33, 41, 43, 51\}, \quad \{29, 31, 37, 47, 53, 55\}, \quad \{35, 49\}.$$

Let $g(x)$ be the polynomial of smallest degree which contains $\delta^i, i = 7, 11, 29, 35$ among its roots. Then,

$$g(x) = x^{20} + 2x^{19} + x^{17} + 2x^{16} + 2x^{13} + 2x^{12} + 2x^{11} + x^{10} + x^9 + 2x^8$$
$$+ x^7 + 2x^4 + 2x^3 + x + 1$$

and the consecutive powers α^i, $14 \leq i \leq 27$ of α, are also among the zeros of $g(x)$. Therefore, by the BCH bound for constacyclic codes, the constacyclic code of length 28 generated by $g(x)$ has minimum distance at least 15 (and its dimension is 8). It turns out that these are the parameters of an *optimal* linear code over $\mathbb{F}_3$ of length 28 and dimension 8 [33].

Exercise 7.5.1. Construct a constacyclic code of length 26, dimension 16, and minimum distance 8 over $\mathbb{F}_5$ with constant $a = 2$ or 3. Note that such a code would be optimal.

7.5.4. *Quasi-Twisted Codes*

The class of quasi-twisted (QT) and their special case of QC codes are a generalization of constacyclic (cyclic) codes and they have been shown to be promising toward solving the main problem in coding theory: to construct codes with the best possible parameters. A large number of new codes in these classes have been discovered in recent years. Often, computers are employed in finding these codes. For a sample of results in this area, see [6–8, 22, 23, 34, 38, 68] among others.

Let $n = lm$ where $l, m \in \mathbb{N}$, $a \in \mathbb{F}_q^*$ and define $\mu_{a,l} : C \to \mathbb{F}_q^n$ by $\mu_{a,l}((c_0, \ldots, c_{n-1})) = (a \cdot c_{0-l}, \ldots, a \cdot c_{(l-1)-l}, c_{l-l}, \ldots, c_{n-l-1})$, where the subscripts are taken modulo n.

Definition 7.16. A linear code C is called l-QT if $\mu_{a,l}(C) = C$.

In words, a constacyclic shift of a codeword by l positions is still a codeword. Some of the most important classes of codes can be realized as special cases of QT codes. For example, the case $a = 1$ gives QC codes, $l = 1$ gives constacyclic codes (also known as pseudocyclic codes), $l = 1$ and $a = 1$ yields cyclic codes.

7.5.5. *Structure of 1-Generator QT Codes*

An l-QT code over $\mathbb{F}_q$ of length $n = ml$ can be viewed as an $\mathbb{F}_q[x]/\langle x^m - a \rangle$ submodule of $(\mathbb{F}_q[x]/\langle x^m - a \rangle)^l$ (after a permutation of the coordinates). Then an r-generator QT code is spanned by r elements of $(\mathbb{F}_q[x]/\langle x^m - a \rangle)^l$. In this chapter, we restrict ourselves to 1-generator QT codes. Let

$$
G_0 = \begin{bmatrix}
g_0 & g_1 & g_2 & \cdots & g_{m-1} \\
ag_{m-1} & g_0 & g_1 & \cdots & g_{m-2} \\
ag_{m-2} & ag_{m-1} & g_0 & \cdots & g_{m-3} \\
\vdots & \vdots & \vdots & & \vdots \\
ag_1 & ag_2 & ag_3 & \cdots & g_0
\end{bmatrix}_{m \times m}.
$$

An $(m \times m)$ matrix of the type G_0 is called a twistulant matrix of order m or simply a twistulant matrix. It is shown in [7] that the generator matrices of QT codes can be transformed into blocks of twistulant matrices by a suitable permutation of columns. Therefore, generator matrices of an r-generator and 1-generator QT codes can be assumed to be in the following forms:

$$\begin{bmatrix} G_{11} & G_{12} & \cdots & G_{1l} \\ G_{21} & G_{22} & \cdots & G_{2l} \\ \vdots & \vdots & & \vdots \\ G_{r1} & G_{r2} & \cdots & G_{rl} \end{bmatrix}_{rm \times n} , \quad \text{and} \quad \begin{bmatrix} G_1 \ G_2 \ \dots \ G_l \end{bmatrix}_{m \times n},$$

respectively, where each G_{ij} (or G_k) is a twistulant matrix of the form 7.5.5.

Most of the work in the literature is concerned with 1-generator QC or QT codes. Often, computer searches with heuristic search algorithms are employed [24, 37] to find new codes. A number of papers focus on rate $1/l$ and related QC codes [35, 38]. More recently, a different search algorithm was devised for a certain type of 1-generator QC [68] and QT [7] codes inspired by the work in [43]. This method produced a number of new codes over most small finite fields for which a database of best-known codes is available [5, 7, 22, 23, 25, 69, 70]. The method is described in detail in the rest of this section.

Let $1 \le i \le l$. For fixed i consider the following ith *restriction map* on an l-QT code C of length $n = ml$:

$$\prod_i : \mathbb{F}_q^n \to \mathbb{F}_q^m$$

$$(c_0, c_1, \dots, c_{(ml-1)}) \to (c_{(i-1)m}, c_{(1+(i-1)m)}, \dots, c_{(m-1+(i-1)m)}).$$

In view of the structure of QT codes described above, $\Pi_i(C)$ is a constacyclic code for all i. This yields the following theorem.

Theorem 7.10. *Let C be a 1-generator l-QT code over $\mathbb{F}_q$ of length $n = ml$. Then, a generator $\mathbf{g}(\mathbf{x}) \in (\mathbb{F}_q[x]/\langle x^m - a \rangle)^l$ of C has the following form:*

$$\mathbf{g}(\mathbf{x}) = (f_1(x)g_1(x), f_2(x)g_2(x), \dots, f_l(x)g_l(x)),$$

where $g_i(x)|(x^m - a)$ and $(f_i(x), (x^m - a)/g_i(x)) = 1$ for all $1 \le i \le l$.

Proof. Since $\Pi_i(C)$ is a constacyclic code for every i we have the result. $\square$

The following theorem plays an important role in some search methods mentioned above.

Theorem 7.11. *Let C be a 1-generator l-QT code of length $n = ml$ with a generator of the form:*

$$\mathbf{g(x)} = (f_1(x)g(x), f_2(x)g(x), \ldots, f_l(x)g(x)),$$

where $g(x)|(x^m - a), g(x), f_i(x) \in \mathbb{F}_q[x]/\langle x^m - a \rangle$, and $(f_i(x), h(x)) = 1$, $h(x) = \frac{x^m - a}{g(x)}$ for all $1 \le i \le l$. Then $l \cdot (d+1) \le d(C)$, where $\{\delta\zeta^i : s \le i \le s + (d-1)\}$ are among the zeros of $g(x)$ for some integers s, d $(d > 0)$ and dimension of C is equal to $n - \deg(g(x))$.

Next, we present two examples that illustrate how the results in this section and the last theorem in particular is used in designing a computer search algorithm to discover new linear codes.

Example 7.9. This example presents a ternary QT code that has the best-known parameters among all linear codes with the same length and dimension [7]. Let $q = 3$, $m = 40$, and $a = 2$ and consider constacyclic codes of length 40 over $\mathbb{F}_3$. The order of $2 \bmod 3$ is 2 and $x^{40} - 2$ factors over $\mathbb{F}_3$ as

$$x^{40} - 2 = \prod_{i=0}^{39}(x - \delta^{2i+1}).$$

The exponents of δ (a primitive 80th root of 1) are odd integers $\bmod\, 80$ and the corresponding powers of ζ (a primitive 40th root of 1) are broken into the following cyclotomic cosets $\bmod\, 40$:

$$\{0, 1, 4, 13\}, \{2, 7, 22, 27\}, \{3, 10, 14, 31\}, \{5, 9, 16, 28\},$$
$$\{6, 15, 18, 19\}, \{8, 25, 29, 36\}, \{11, 23, 30, 34\}, \{12, 17, 32, 37\},$$
$$\{20, 21, 24, 33\} \{26, 35, 38, 39\}.$$

Let $h(x)$ be the polynomial corresponding to cyclotomic cosets containing 0,3 and 12 and let

$$g(x) = \frac{x^{40} - 2}{h(x)}$$
$$= x^{28} + 2x^{27} + 2x^{25} + x^{24} + 2x^{23} + x^{21} + 2x^{20} + x^{19} + x^{18} + 2x^{17}$$
$$+ 2x^{15} + x^{14} + x^{13} + 2x^{11} + x^8 + 2x^7 + 2x^5 + x^3 + x^2 + 2.$$

Then, $g(x)$ has degree 28 and contains consecutive powers $18 \le i \le 30$ of δ among its roots. Therefore, the constacyclic code of length 40 generated by $g(x)$ has dimension 12 and minimum distance ≥ 14 and a QT code of the form (gf_1, gf_2, gf_3), where $(f_i, h) = 1$ has length 120, dimension 12 and minimum distance at least 42. Let $f_1 = x^{10} + x^9 + x^8 + x^2$, $f_2 = 2x^{10} + x^9 + x^6 + x$, and $f_3 = 2x^{11} + x^{10} + x^9 + x^8 + x^5 + x^4 + x^2 + 2x$ (found by a computer search). Then f_i's satisfy $(f_i, h) = 1$. The QT code with these generators has actual minimum distance of 66, three units larger than the previously best-known linear code over $\mathbb{F}_3$ with parameters $[120, 12, 63]$ (see [33]. The weight enumerator of this code is as follows:

$$0^1 66^{4000} 69^{15120} 72^{35200} 75^{77728} 78^{108000} 81^{122160}$$

$$84^{97120} 87^{47520} 90^{18832} 93^{5040} 96^{720},$$

where the bases are the weights and the exponents are the number of codewords of the given weight.

Example 7.10. This example presents a best-known code over $\mathbb{F}_4$ that is QT [5]. We represent the elements of $\mathbb{F}_4$ by $\{0, 1, a, b\}$, where $b = a^2 = a+1$. Let $g(x) = \frac{x^{39}-a}{h(x)}$ where $h(x) = (x^6 + ax^5 + x^4 + ax^3 + x + b)(x^6 + x^5 + ax^3 + x^2 + x + b)$. Then $g(x)$ generates a quaternary constacyclic code with parameters $[39, 12, 18]$. According to [33], these are the parameters of the best-known code for this length and dimension. Searching over the codes with a generator of the form $(g(x), g(x)f_1(x))$, we find, by the help of a computer, that if we choose $f_1(x) = x + bx^3 + ax^7 + bx^9 + bx^{10} + x^{11}$, then we obtain a $[78, 12, 44]_4$-code. This turns out to be a best-known code. The weight enumerator of this code is

$$0^1 44^{6786} 46^{24921} 48^{103194} 50^{321750} 52^{816075} 54^{1695096} 56^{2737215} 58^{3417453}$$

$$60^{3298464} 62^{2414529} 64^{1301391} 66^{491400} 68^{124371} 70^{21294} 72^{3159} 74^{117}.$$

Finally, we would like to remark that most of the search over QC/QT codes have been among 1-generator codes. There are few papers in the literature that report new codes from multiple generator QC codes. Two such examples are [21, 36].

7.6. Thoughts for Practitioners

7.6.1. *Computing Minimum Distance of a Linear Code*

The problem of finding the minimum weight of a general binary linear code was conjectured to be NP-complete by Berlekamp *et al.* [10]. Carey and Johnson, among others, repeatedly called for the resolution of the conjecture [31, 45, 46]. The increased interest in the topic resulted in the proved hardness of a number of related problems [1, 2, 18, 26, 51, 71] over the years. Yet, the original conjecture remained open for almost two decades. In 1997, employing a polynomial transformation from maximum-likelihood decoding to minimum distance, Vardy showed that finding the minimum distance of a linear code over a fixed finite field is NP-complete [75, 76]. More recently, Dumer *et al.* showed that the minimum distance of a linear code over a finite field cannot be approximated to within any constant factor in random polynomial (RP) time, unless RP equals NP [28]. Furthermore, the last result was translated to prove the hardness of approximating the minimum distance within an additive error that is linear in the block length of the code [29].

For a $[n, k, d]_q$ linear code, computing the minimum weight via complete codeword enumeration involves finding the minimum weight for $(q^k - 1)$ codewords of length n. This is computationally infeasible even for small values of the parameters. The fastest algorithm for finding the minimum weight of a linear code over a finite field is based on an unpublished work by Brouwer that was later improved by Zimmermann.

7.6.1.1. *The Brouwer–Zimmermann Algorithm for Linear Codes*

The work of Zimmermann was only published in German [19] but English summaries are available [11, 73, 79]. In determining the minimum weight of a $[n, k, d]_q$ linear code, the algorithm employs mutually disjoint(or partially disjoint) information sets and partial codeword enumeration to compute an upper bound d_u on d, while keeping track of a lower bound d_l of d that grows linearly with the number of disjoint information sets. Termination is reached when $d_l \geq d_u$. For a $[n, k, d]_q$ cyclic code having an information set formed by k consecutive columns, there always exist $\lfloor n/k \rfloor$ mutually disjoint information sets. Moreover, the generator matrix corresponding to a single information set is sufficient for codeword enumeration for all $\lfloor n/k \rfloor$ generator matrices [11]. Thus, for some cyclic codes, a faster growth of d_l can be achieved using a single information set. This result can be extended to the case of constacyclic codes as well.

Often times, one is interested in finding the minimum weight of a linear code but only if it is within a certain range. In this case, the termination point of the algorithm can be adjusted. The MAGMA algebra system [12] supports such a feature via the functions VerifyMinimumDistanceUpperBound(C, d) and VerifyMinimumDistanceLowerBound (C, d).

The authors have devised a recent combinatorial search algorithm that makes an extensive use of this functionality [3]. The algorithm examines a large number of codes in the search space and employs the VerifyMinimumDistanceUpperBound() function to quickly discard codes with minimum distance below a prescribed value.

As far as parallel computing is concerned, an easy parallelization of the Brouwer–Zimmermann algorithm has been implemented by van Dijk *et al.* [73]. Their approach is based on the revolving door algorithm by Nijenhuis and Wilf, and a combinatorial result by of Lüneburg [52] and Knuth [47]. Furthermore, the autoson program by McKay can be used to distribute the work over a network of Unix workstations [54].

Finally, for certain codes of high rate $R = k/n > 1/2$, finding the weight distribution of the dual code and using the MacWilliams transform might prove faster than running the Brouwer–Zimmermann Algorithm.

7.7. Codes over $\mathbb{Z}_4$ and Database of $\mathbb{Z}_4$ Codes

7.7.1. *Codes Over* $\mathbb{Z}_4$

After the discovery of good binary non-linear codes from linear codes over $\mathbb{Z}_4$, the ring of integers modulo 4 (sometimes called "quaternary codes") [40, 55] there has been intensive research on this class of codes. A *code* C of length n over $\mathbb{Z}_4$ is a subset of $\mathbb{Z}_4^n$. C is a *linear code* over $\mathbb{Z}_4$ if it is an additive subgroup of $\mathbb{Z}_4^n$, hence a submodule of $\mathbb{Z}_4^n$. An element of C is called a *codeword* and a *generator matrix* is a matrix whose rows generate C. The *Hamming weight* $w_H(x)$ of a vector $x = (x_1, x_2, \ldots, x_n)$ in $\mathbb{Z}_4^n$ is the number of components $x_i \neq 0$. The *Lee weight* $w_L(x)$ of a vector x is $\sum_{i=1}^{n} \min\{|x_i|, |4 - x_i|\}$. The Hamming and Lee distances $d_H(x, y)$ and $d_L(x, y)$ between two vectors x and y are $w_H(x - y)$ and $w_L(x - y)$, respectively. The minimum Hamming and Lee weights, d_H and d_L, of C are the smallest Hamming and Lee weights, respectively, amongst all non-zero codewords of C.

The *Gray map* $\phi : \mathbb{Z}_4^n \to \mathbb{Z}_2^{2n}$ is the coordinate-wise extension of the function from $\mathbb{Z}_4$ to $\mathbb{Z}_2^2$ defined by $0 \to (0,0), 1 \to (1,0), 2 \to (1,1), 3 \to (0,1)$. The image $\phi(C)$, of a linear code C over $\mathbb{Z}_4$ of length n by the Gray map, is a (in general non-linear) binary code of length $2n$. The Gray map is an isometry from $(\mathbb{Z}_4^n, w_L)$ to $(\mathbb{Z}_2^{2n}, w_H)$. Therefore, the minimum Hamming weight of $\phi(C)$ is equal to the minimum Lee weight of C.

The *dual code* $C^{\perp}$ of C is defined as $\{x \in \mathbb{Z}_4^n \mid x \cdot y = 0, \forall\, y \in C\}$, where $x \cdot y$ is the standard inner product of x and y. C is *self-orthogonal* if $C \subseteq C^{\perp}$ and C is *self-dual* if $C = C^{\perp}$.

Two codes are said to be *equivalent* if one can be obtained from the other by permuting the coordinates and (if necessary) changing the signs of certain coordinates. Codes differing by only a permutation of coordinates are called *permutation-equivalent*. Any linear code C over $\mathbb{Z}_4$ is permutation-equivalent to a code with generator matrix G of the form

$$ G = \begin{bmatrix} I_{k_1} & A_1 & B_1 + 2B_2 \\ 0 & 2I_{k_2} & 2A_2 \end{bmatrix}, $$

where A_1, A_2, B_1, and B_2 are matrices with entries 0 or 1 and I_k is the identity matrix of order k. Such a code has size $4^{k_1}2^{k_2}$. The code is a free module if and only if $k_2 = 0$. If C has length n and minimum Lee weight d_L, the code is referred to as an $[n, 4^{k_1}2^{k_2}, d_L]$-code.

7.7.2. *A Database of $\mathbb{Z}_4$ Codes*

Cyclic codes, QC codes and QT codes over $\mathbb{Z}_4$ are studied in [6, 8] and many new codes are discovered whose Gray images have better parameters than best-known binary linear codes. Among other results in this area, two new non-linear binary codes have been constructed using $\mathbb{Z}_4$ linear codes and their binary images. One of the codes has binary parameters $(64, 2^{37}, 12)$ [20]. Another code has binary parameters $(92, 2^{24}, 28)$ [8]. The latter code is QC over $\mathbb{Z}_4$ and its generator polynomial is related to the generator polynomial of the binary Golay code G_{23}. Moreover, many $\mathbb{Z}_4$ codes have been discovered whose Gray images have better parameters than the comparable binary linear codes (such codes are called "good codes") [4, 6, 8]. Since the Gray image of a $\mathbb{Z}_4$ linear code is most often non-linear, it is appropriate to compare their parameters with the codes in [50]. However, the database in [50] is very limited and often does not extend to the parameters of interest. Despite extensive research on codes over $\mathbb{Z}_4$, there has been no database of best-known $\mathbb{Z}_4$ codes. Recently, such a database has been created by the authors. It is available online at http://Z4codes.info.

While the Hamming distance has been the dominant metric in the field case, researchers have explored different distance functions for codes over $\mathbb{Z}_4$. Most researchers have focused their work on the Lee distance but

Euclidean and Hamming metrics have also been considered. In order to deal with the presence of multiple distance functions, we had to adopt a list, rather than the typical tabular structure for our database. Moreover, we chose not to overwrite existing codes when an improved code has been found but only add the new code to the list. Finally, for the sake of flexibility and convenience, we have decided to provide the willing researchers with editing privileges that would allow them to upload their new results instantly.

7.8. Directions for Future Research

In this section, we list a few open problems in algebraic coding theory, related to the material that is discussed in this chapter. These problems appear in [5], portions reprinted, with permission, from [5] ©2007 IEEE.

7.8.1. *QCT Codes*

This subsection reviews a generalization of QT codes (hence of QC codes as well), called QCT codes that are first introduced in [5], and investigates their structural properties. It then presents open problems in this class.

Let $a_1, a_2, \ldots, a_l$ be non-zero constants (not necessarily distinct) in $\mathbb{F}_q$. A linear code of length $n = ml$ will be called a QCT code if it is invariant under the following shift:

$$(c_1, c_2, \ldots, c_{(m-1)l}, c_{(m-1)l+1}, \ldots, c_{ml})$$
$$\rightarrow (a_1 c_{ml}, a_2 c_{ml-1}, \ldots, a_l c_{(m-1)l+1}, c_1, \ldots, c_{(m-1)l+1}).$$

We remark that if all the constants are equal then we obtain a QT code, if they are all equal to 1 then we obtain a QC code. If $l = 1$ then we obtain constacyclic and cyclic codes as special cases.

As in the case of a QT code, it is easy to see that after a suitable permutation of the coordinate positions, a generator matrix of a QCT code can be put into blocks of twistulant matrices (each block involving a possibly different constant).

To illustrate this construction, we present two examples of QCT codes, which are better than the best-known QC or QT codes over $\mathbb{Z}_4$ [5].
(i) A $[6, 2, 6]$ code generated by

$$G = \begin{bmatrix} 1 & 3 & 0 & 1 & 1 & 2 \\ 1 & 1 & 1 & 0 & 2 & 1 \end{bmatrix}.$$

This code has minimum Lee weight $d_L = 6$, while the best QC or QT code has $d_L = 5$ [6]. In addition, this code is *self-orthogonal*, and the best-known QT self-orthogonal code only has $d_L = 4$ [32].

(ii) An $[8, 4, 6]$ code generated by

$$
G = \begin{bmatrix}
0 & 0 & 1 & 2 & 0 & 1 & 1 & 1 \\
2 & 0 & 0 & 1 & 3 & 0 & 1 & 1 \\
1 & 2 & 0 & 0 & 3 & 3 & 0 & 1 \\
0 & 1 & 2 & 0 & 3 & 3 & 3 & 0
\end{bmatrix}.
$$

This is a *self-dual* code, and the best QC or QT code with length 8 and dimension 4 has $d_L = 4$ [8, 32]. Note that the Gray map image of this code is the Nordstrom–Robinson code [53]. Thus, this construction provides a new simple description of this code.

In addition to the codes above, many hundreds of QCT codes over $\mathbb{Z}_4$ and over finite fields have been found, which have the same parameters as the best-known codes. Therefore, it is likely that this class of codes contains some new codes.

7.8.2. *Algebraic Properties of QCT Codes*

Now we like to investigate the algebraic structure of QCT codes. Let $a_i \in \mathbb{F}_q^*$, $R_i = \frac{\mathbb{F}_q[x]}{\langle x^m - a_i \rangle}$, $1 \le i \le l$ and $R = R_1 \times R_2 \times \cdots \times R_l$. A QCT code C, after a suitable permutation of coordinates, can be regarded as an $\mathbb{F}_q[x]$-module of R. We say that C is *s-generated* if it is generated by s elements. Since each block (of length m) of a QCT code is actually a constacyclic code, we have the following result.

Lemma 7.18. *An s-generated QCT Code C has generators of the form* $\{g_1(x), g_2(x), \ldots, g_s(x)\}$ *where*

- $g_j(x) = (g_{j1}(x), g_{j2}(x), \ldots, g_{jl}(x))$;
- $g_{ji}(x) = f_{ji}(x)g_i(x)$ *for some* $g_i(x) \mid x^m - a_i$, $f_{ji}(x) \in R_i$ *and* $(f_{ji}, h_i) = 1$, *where* $x^m - a_i = g_i(x)h_i(x)$.

Again, we will focus on the 1-generator case. As a corollary we have that a 1-generator QCT code is generated by an element of the form

$$
g(x) = (f_1(x)g_1(x), f_2(x)g_2(x), \ldots, f_l(x)g_l(x)),
$$

where $f_i(x), g_i(x) \in R_i$ and $g_i(x) \mid (x^m - a_i)$. Moreover, we can show that f_i and g_i can be chosen so that $(f_i(x), h_i(x)) = 1$ where $h_i(x) = \frac{x^m - a_i}{g_i(x)}$.

For two polynomials f and g, we denote their greatest common divisor by (f, g) and their least common multiple by $[f, g]$.

Next we consider bounds on the parameters of a QCT code.

Theorem 7.12. *Let C be a 1-generator QCT code generated by an element of the form described above.*

(1) $\dim(C) = \deg([h_1, h_2, \ldots, h_l])$,
(2) $d(C) \geq \min\{d_i : 1 \leq i \leq l\}$ *where d_i is the minimum distance of the ith constacyclic block and $d(C)$ is the minimum distance of C.*

Proof. Let $h = [h_1, h_2, \ldots, h_l]$, then clearly, $h(x)g(x) = 0$, which implies that $\dim(C) \leq \deg(h)$. On the other hand, if $f(x)g(x) = 0$ then $f(x)f_i(x)g_i(x) = 0$ in R_i, for $1 \leq i \leq l$. This implies that $h_i(x) \mid f(x)f_i(x)$. Since $(h_i(x), f_i(x)) = 1$, $h_i(x) \mid f(x)$ for $1 \leq i \leq l$. Hence $h(x) \mid f(x)$. This shows $\dim(C) \geq \deg(h)$ and the assertion on the dimension. The statement on the minimum distance is rather obvious. $\qquad\square$

Example 7.11. Let $q = 4$ and let a be a root of the polynomial $x^2 + x + 1 \in \mathbb{F}_2[x]$ so that $\mathbb{F}_q = \mathbb{F}_2(a)$. Next choose $a_1 = 1, a_2 = a$, and $m = 11$, $g_1(x) = x^5 + ax^4 + x^3 + x^2 + bx + 1$ and $g_2(x) = x^5 + ax^4 + ax^3 + x^2 + x + a$. Then g_1 and g_2 divide $x^{11} - 1$ and $x^{11} - a$ over $\mathbb{F}_q$ (respectively) and they generate cyclic and constacyclic codes with parameters $[11, 6, 5]_q$. A code with these parameters is optimal [33]. Now we consider the QCT code generated by $\langle g_1, g_2 \rangle$. In this case, $h_1 = \frac{x^{11}-1}{g_1}$ and $h_2 = \frac{x^{11}-a}{g_2}$ are relatively prime so that $[h_1, h_2] = h_1 h_2$; hence, the dimension is 12. The minimum distance of this QCT code is 5, which shows that the lower bound on the minimum distance is attained. Thus, we obtain a quaternary $[22, 12, 5]$ code. According to [33], there exists a quaternary $[22, 12, 7]$ code.

Generalizing from this example, we can say more about the dimension and minimum distance of QCT codes in the special case when all the constants are distinct. If $a_1 \neq a_2$, then $x^m - a_1$ and $x^m - a_2$ are relatively prime. If $x^m - a_1 = g_1 h_1$ and $x^m - a_2 = g_2 h_2$ then $(h_1, h_2) = 1$ (as well as $(g_1, g_2) = 1$) so that $[h_1, h_2] = h_1 h_2$. Then the QCT code C generated by $g = \langle g_1, g_2 \rangle$ has dimension $k_1 + k_2$ where k_1, k_2 are, respectively, the dimensions of the constacyclic codes generated by g_1 and g_2. We also claim that in this case the minimum distance is actually equal to $\min\{d_1, d_2\}$, where d_i is the minimum distance of the constacyclic code generated by g_i. To see this, consider $\{th_2 g = (th_2 g_1, 0) : t \in \mathbb{F}_q[x], t \neq 0, \deg(t) < \deg(h_1)\}$. Since $\langle g_1 \rangle = \langle h_2 g_1 \rangle$ (because $(h_1, h_2) = 1$), we see that there is a codeword of weight d_1 in C. Similarly, one can show that C contains a codeword of weight d_2. The same argument can be applied to any l when $a_1, a_2, \ldots, a_l$ are all distinct. This shows that the minimum distance of such a QCT code is not very high. However, there is a way to impose a restriction so that a better bound on the minimum distance is obtained.

Theorem 7.13. *Let C be a 1-generator QCT code generated by, i.e. $\mathbb{F}_q$-span of $g(x) = (f_1(x)g_1(x), f_2(x)g_2(x), \ldots, f_l(x)g_l(x))$ with the conditions on the $f_i's$ and $g_i's$ as described before. Let $h = \min\{\deg(h_i) : 1 \leq i \leq l\}$. Then the subcode C' generated by $g(x), xg(x), x^2g(x), \ldots, x^{h-1}g(x)$ has dimension h and minimum distance $\geq d_1 + d_2 + \cdots + d_l$ where d_i is the minimum distance of the code $\langle g_i \rangle$.*

Example 7.12. Let $q = 5, m = 13, l = 3, a_1 = 1, a_2 = 2, a_3 = 4$, $g_1 = (x^4 + x^3 + 4x^2 + x + 1)(x + 4)$, $g_2 = (x^4 + 4x^3 + 4x^2 + x + 1)(x + 3)$, $g_3 = (x^4 + 2x^3 + 2x^2 + 1)(x + 1)$ where $g_1 \mid (x^{13} - 1)$, $g_2 \mid (x^{13} - 2)$, and $g_3 \mid (x^{13} - 4)$ over $\mathbb{F}_5$. The constacyclic codes $\langle g_1 \rangle$, $\langle g_2 \rangle$ $\langle g_3 \rangle$ all have parameters $[13, 8, 4]$ and they are optimal. The subcode of $\langle f_1g_1, f_2g_2, f_3g_3 \rangle$ given in the last theorem has length 39, dimension 8 and minimum distance ≥ 12. However, when we choose $f_1 = x^7$, $f_2 = x^7 + 2x^6 + 2x^5$ and $f_3 = 3x^6 + x + 2$ the resulting code is a $[39, 8, 21]$ code. This example shows that the actual minimum distance in this construction may be significantly larger than the lower bound promised by the theorem. This code is not the best-known code; however, according to [33], there is a $[39, 8, 23]$ code.

7.8.3. Open Problems

Open problem I: Let C be a cyclic (or constacyclic) code of length n. How should $a(x)$ be chosen so that the minimum distance of the code $\{|u(x)|a(x)u(x)(\bmod x^n - 1)| : u(x) \in C\}$ is as large as possible? Is there a difference between the field version and the ring version of this problem?

The practical evidence from searches over 1-generator QC and QT codes shows that in many cases we do get very large minimum distances. However, to the best of our knowledge, no explanation has been provided for any specific properties of the polynomials that achieve these large minimum distances (one obvious restriction on $a(x)$ is that it be relatively prime to the complement of the canonical generator). Also, we have not noticed any explicit connection with good QT codes and this problem.

This problem can also be expressed in the following alternative, combinatorial way: Consider a 1-generator QT code C_T with a generator of the form (g, gf) where $x^m - a = gh$ and $(f, h) = 1$. Since g and fg generate the same cyclic or constacyclic code C, C_T is obtained from C by listing the codewords of C in a certain order, then listing them in another order and taking the juxtaposition. Each choice of f corresponds to an ordering of C. What would be a good ordering that preserves the linearity of the code and gives a large minimum distance?

Open problem II: Naturally, Open Problem I can be stated for 1-generator QCT codes and their subclass described above.

Open problem III: Find an analogue of Theorem 7.11 for multi-generator QC codes.

7.9. Conclusion

Algebraic coding theory is a huge subject now. Despite much work on it, the main problem of coding theory is still a challenging yet promising area of research, looking for creative approaches. In this chapter, we present a selected subset of topics from the field with interesting results and related open problems. We give special attention to a promising class of codes (QC codes and their generalizations). Moreover, a new database of $\mathbb{Z}_4$ codes is introduced. The reader is referred to the cited references for more on the subject.

7.10. Key Concepts in the Chapter

(1) *A linear code* of length n over a field F is a vector subspace of F^n.
(2) *A self-dual code* C is a linear code whose dual $C^\perp$ code is equal to itself, where the dual code is defined by

$$C^\perp := \{\mathbf{v} \in F^n \,|\, \langle \mathbf{u}, \mathbf{v} \rangle = 0 \text{ for all } \mathbf{u} \in C\}.$$

(3) *A parity check matrix* of a linear code C is a matrix whose null space is C.
(4) *A generator matrix* of a linear code C is a matrix whose row space is C.
(5) *Hamming distance* between $\mathbf{u} = (u_1, u_2, \ldots, u_n), \mathbf{v} = (v_1, v_2, \ldots, v_n)$ in F^n is $d(\mathbf{u}, \mathbf{v}) = |\{i : u_i \neq v_i\}|$, the number of positions in which $\mathbf{u}$ and $\mathbf{v}$ differ.
(6) *A perfect code* is a code for which the sphere packing bound is attained with equality. If the code parameters are $(n, M, d)_q$ then

$M \cdot \sum_{j=0}^{t} \binom{n}{j}(q-1)^j = q^n$, where $t = \lfloor \frac{d-1}{2} \rfloor$. For a linear code $[n, k, d]_q$ the equality becomes $\sum_{j=0}^{t} \binom{n}{j}(q-1)^j = q^{n-k}$.

(7) *An MDS code* (maximum distance separable code) is a code for which the singleton bound is attained. For a linear code $[n, k, d]$ this means $d = n - k + 1$. For a non-linear code with parameters $(n, M, d)_q$, it means $M = q^{n-d+1}$.

(8) *A cyclic code* C is a linear code, which is closed under cyclic shifts, i.e. if $(u_0, u_1, \ldots, u_{n-1}) \in C$, then $(u_{n-1}, u_0, \ldots, u_{n-2}) \in C$. A cyclic code of length n over $\mathbb{F}_q$ is an ideal in the ring $\frac{\mathbb{F}_q[x]}{\langle x^n - 1 \rangle}$.

(9) *A quasi-twisted code* (more precisely an l-quasi-twisted code) is a linear code of length $n = ml$ that is closed under the quasi-twisted shifts by l-positions, i.e. if $(c_0, c_1, \ldots, c_{n-1}) \in C$ then $(a \cdot c_{0-l}, \ldots, a \cdot c_{(l-1)-l}, c_{l-l}, \ldots, c_{n-l-1}) \in C$ where the subscripts are taken as $\mod n$.

(10) *An nth root of unity* over $\mathbb{F}_q$ is an element in an extension field of $\mathbb{F}_q$ that is a root of $x^n - 1$. A primitive nth root of unity ω is an nth root of unity such that $\omega^k \neq 1$ for any $k < n$.

(11) *A quadratic residue* $\mod p$ is an integer a such that a is relatively prime with p and the equation $x^2 \equiv a \mod p$ has a solution in $\mathbb{Z}_p$.

7.11. Solution to Exercises

(1) Most of these properties follow from the definition and the observation that the minimum distance between two vectors $u = (u_1, u_2, \ldots, u_n)$ and $v = (v_1, v_2, \ldots, v_n)$ is $d(u, v) = \sum_{i=1}^{n} d(u_i, v_i)$. Hence, you can argue component-wise. For example, show that the triangle inequality holds for each component (i.e. $d(u_i, v_i) \leq d(u_i, w_i) + d(w_i, v_i)$) by considering cases where $d(u_i, v_i) = 0$ or 1, then summing up all inequalities for all components.

(2) If the minimum distance is d, $d - 1$ or fewer changes cannot change a codeword into another. Similarly, if there are at most $\lfloor \frac{d-1}{2} \rfloor$ then the resulting vector (after errors) will remain closest to the original vector. (One can use the triangle inequality to formally prove this.)

(3) For C_2 the minimum distance and minimum weight are both equal to 3. For C_1, the minimum weight is 2 but the minimum distance is 3. Let C be a linear code. Consider the sets $D = \{d(\mathbf{u}, \mathbf{v}) : \mathbf{u}, \mathbf{v} \in C, \mathbf{u} \neq \mathbf{v}\}$ and $W = \{w_H(\mathbf{u}) : \mathbf{u} \in C, \mathbf{u} \neq \mathbf{0}\}$. Using linearity we can show that every number in the set S appears in W and *vice versa*. Then the minimums of the sets S and W are equal. Let $\mathbf{u}, \mathbf{v} \in C$ such that

$\mathbf{u} \neq \mathbf{v}$. Let $\mathbf{w} = \mathbf{u} - \mathbf{v}$, then $w_H(\mathbf{w}) = d(\mathbf{u}, \mathbf{v})$, and $\mathbf{w} \in C$ (by linearity of C), and $\mathbf{w} \neq \mathbf{0}$. Conversely, every weight $w_H(\mathbf{u})$, $\mathbf{u} \neq \mathbf{0}$ can be written as the distance $d(\mathbf{u}, \mathbf{0})$.

(4) Let C be a linear code of dimension k. Then it has a basis $\{v_1, v_2, \ldots, v_k\}$ with k elements. Every element of C has a unique representation as a linear combination $a_1 v_1 + a_2 v_2 + \cdots + a_k v_k$, where each a_i has q possible values. Therefore, the total number of such linear combinations is q^k.

(5) The proof of this lemma is based on the following observation: Let H be a $k \times n$ matrix with columns $h_1, h_2, \ldots, h_n$ hence, write H as $H = [h_1, \ldots, h_n]$. Let $v = (v_1, \ldots, v_n)$ be a vector, then the product $H \cdot v$ gives the linear combination $v_1 h_1 + v_2 h_2 + \cdots + v_n h_n$. We proceed by contradiction. Suppose there is a non-zero vector v of weight less than d in C. Let v have non-zero components at positions $i_1, i_2, \ldots, i_r$ where $0 < r < d$. Then, we have $H \cdot v = 0$ hence $v_{i_1} h_{i_1} + v_{i_2} h_{i_2} + \cdots + v_{i_r} h_{i_r} = 0$. This means that the set $h_{i_1}, h_{i_2}, \ldots, h_{i_r}$ of $r < d$ columns is linearly dependent, but this contradicts the assumption. Hence there is no codeword of weight less than d. On the other hand, existence of a set of s linearly dependent columns implies the existence of a codeword of weight d. To show that the minimum distance of C_2 is 3, note that there is no duplication among the columns of H_2; hence, no two columns are linearly dependent, and there is a set of three columns that is linearly dependent (e.g. first, second, and last).

(6) $\Rightarrow$: Let f be a polynomial of degree 2 or 3 over a field F and suppose f is irreducible, yet it has a root a in F. Then, $(x - a)|f$, and so $f(x) = (x-a)g(x)$ for some polynomial g where $\deg(g(x)) = \deg(f(x)) - 1 \geq 1$. This means that f is reducible, contradicting the hypothesis.

$\Leftarrow$: Suppose f has no roots in f yet it is reducible over F. Then $f(x) = r(x)s(x)$ where both r and s have degree ≥ 1. Since degree of f is 2 or 3, one of r and s must have degree 1, i.e. must be a linear polynomial of the form $ax + b$. However, every linear polynomial over a field has a root. Hence, either r or s will have a root, and that root will also be a root of f. This is a contradiction again, completing the proof.

For a counterexample to the theorem for polynomials of degree 4 or higher, consider $(x^2 + 1)^2$ over $\mathbb{Z}_3$ or over reals. It is reducible (obviously) yet has no root in either field.

(7) Note $m = o_{11}(3) = 5$. Therefore, the splitting field of $x^{11} - 1$ over $\mathbb{F}_3$ is $\mathbb{F}_{3^5}$. Hence, we need a primitive polynomial of order 5. We are given

that $f(x) = x^5 + 2x + 1$ is one such polynomial. Let α be a root of $f(x)$. Then it is a primitive element of $\mathbb{F}_{3^5}$ and $\omega = \alpha^{242/11} = \alpha^{22}$ is a 11th root of unity over $\mathbb{F}_3$. To find the irreducible factors of degree 5 of $x^{11} - 1$ over $\mathbb{F}_3$, we compute the minimal polynomials

$$m_1(x) = (x - \omega)(x - \omega^3)(x - \omega^4)(x - \omega^5)(x - \omega^9)$$
$$= x^5 + x^4 + 2x^3 + x^2 + 2$$
$$m_2(x) = (x - \omega^2)(x - \omega^6)(x - \omega^7)(x - \omega^8)(x - \omega^{10})$$
$$= x^5 + 2x^3 + x^2 + 2x + 2$$

Therefore,

$$x^{11} - 1 = (x - 1)(x^5 + x^4 + 2x^3 + x^2 + 2)(x^5 + 2x^3 + x^2 + 2x + 2).$$

(8) The parameters are $[15, 7, 5]$. Note that this is an optimal code. By looking at the cyclotomic cosets from Example 1.2, we see that the generator polynomial must have degree 8; hence, the dimension of the code is 7. From the table of best-known codes, the minimum distance cannot be more than 5. Since we know that it is at least 5, it is therefore exactly 5.

(9) First, performing a permutation if necessary, we can assume that $\alpha_i = \alpha^i$ for $0 \le i \le n - 1$ where α is a primitive element of $\mathbb{F}_q$. Given a codeword $(f(\alpha^0), f(\alpha), f(\alpha^2), \ldots, f(\alpha^{n-1}))$ for some $f(x) \in P_k$, we need to show that its cyclic shift $(f(\alpha^{n-1}), f(\alpha^0), \ldots, f(\alpha^{n-2}))$ is also a codeword. This means that $(f(\alpha^{n-1}), f(\alpha^0), \ldots, f(\alpha^{n-2})) = (g(\alpha^0), g(\alpha^1), \ldots, g(\alpha^{n-1}))$ for some $g(x) \in P_k$. It is easy to verify that the polynomial $g(x) = f(\alpha^{-1}x)$ satisfies this condition.

(10) A code with parameters $[n, k, 2t + 1]$ is perfect if the equality $q^{n-k} = \sum_{j=0}^{t} \binom{n}{j}(q - 1)^j$ holds. For the binary Golay code G_{23} this means $2^{11} = \sum_{j=0}^{3} \binom{23}{j} = 1 + 23 + 253 + 1771$, which holds true. For the ternary Golay code G_{11}, this means $3^5 = \sum_{j=0}^{2} \binom{11}{j} 2^j = 1 + 11 \cdot 2 + 4 \cdot \frac{11 \cdot 10}{2}$, which also holds true.

(11) Let $q = 5$, $n = 26$, and $a = 2$ (or 3). Therefore, we are considering constacyclic codes of length 26 over $\mathbb{F}_5$ with $a = 2$ or 3. The order, r, of 2 in $\mathbb{F}_5^*$ is 4. Hence, we have

$$x^{26} - 2 = \prod_{i=0}^{25}(x - \delta^{4i+1}),$$

where δ is a primitive $26 \cdot 4 = 104$th root of unity. The exponents

$$\{1, 5, 9, 13, 17, 21, 25, 29, 33, 37, 41, 45, 49, 53, 57, 61, 65, 69, 73, 77, 81,$$
$$85, 89, 93, 97, 101\}$$

of δ in this factorization are partitioned into the following cyclotomic cosets:

$$\{1, 5, 21, 25\}, \ \{9, 17, 45, 85\}, \ \{13, 65\}, \ \{29, 41, 89, 101\}$$
$$\{33, 61, 69, 97\}, \ \{37, 49, 81, 93\}, \ \{53, 57, 73, 77\}$$

and the corresponding powers of ζ, where ζ is a primitive 26th root of 1, are

$$\{0, 1, 5, 6\}, \ \{2, 4, 11, 21\}, \ \{3, 16\}, \ \{7, 10, 22, 25\}, \ \{8, 15, 17, 24\}$$
$$\{9, 12, 20, 23\} \ \{13, 14, 18, 19\}.$$

Let $g(x)$ be the polynomial of smallest degree, which has ζ^i, $13 \leq i \leq 19$ among its roots (hence the roots of $g(x)$ are precisely ζ^i where i runs through cyclotomic cosets containing 3, 8, and 13). Then $g(x) | (x^{26} - 2)$ and $\deg g(x) = 8$. Hence, the constacyclic code generated by $g(x)$ has length 26, dimension 16, and minimum distance ≥ 8. According to the linear codes table, these are the *optimal* parameters. (Consequently, minimum distance is exactly 8.)

References

1. S. Arora, L. Babai, J. Stern and Z. Sweedyk, The hardness of approximate optima in lattices, codes, and systems of linear equations, *Proc. 34th Annual Symp. the Foundation of Computer Science*, Palo Alto, CA, (1993), pp. 724–733.
2. S. Arora, L. Babai, J. Stern and Z. Sweedyk, The hardness of approximate optima in lattices, codes, and systems of linear equations, *J. Comput. Syst. Sci.* **54**, 2 (1997), 317–331.
3. T. Asamov and N. Aydin, A search algorithm for linear codes: Progressive dimension growth, *Des. Codes Cryptogr.* **45**, 2 (2007), 213–217.
4. N. Aydin and T. Asamov, Database of $\mathbb{Z}_4$ codes, preprint.
5. N. Aydin, T. Asamov and T. A. Gulliver, Some open problems on quasi-twisted and related code constructions and good quaternary codes, *Proc. 2007 IEEE Int. Symp. Information Theory*, (2007), pp. 856–860.
6. N. Aydin and T. A. Gulliver, Some good cyclic and quasi-twisted $\mathbb{Z}_4$-linear codes, Ars *Combin.*, to appear.
7. N. Aydin, I. Siap and D. K. Ray-Chaudhuri, The structure of 1-generator quasi-twisted codes and new linear codes, *Des. Codes Cryptogr.* **24** (2001), 313–326.

8. N. Aydin and D. K. Ray-Chaudhuri, Quasi-cyclic codes over $\mathbb{Z}_4$ and some new binary codes, *IEEE Trans. Inform. Theory* **48** (2002), 2065–2069.

9. E. R. Berlekamp, *Algebraic Coding Theory* (McGraw-Hill, New York, 1968).

10. E. R. Berlekamp, R. J. McEliece and H C A. van Tilborg, On the inherent intractability of certain coding problems, *IEEE Trans. Inform. Theory* **24** (1978), 384–386.

11. W. Bosma and J. Cannon, *Discovering Mathematics with Magma: Reducing the Abstract to Concrete* (Springer, Berlin, 2006).

12. W. Bosma, J. Cannon and C. Playoust, The Magma algebra system I: The user language, *J. Symbol. Comput.* **24** (1997), 235–265.

13. I. Bouyukliev, S. Kapralov, T. Maruta and M. Fukui, Optimal linear codes of dimension 4 over $\mathbb{F}_5$, *IEEE Trans. Inform. Theory* **43** (1997), 308–313.

14. I. Bouyukliev, D. B. Jaffe and V. Vavrek, The smallest length of eight-dimensional binary linear codes with prescribed minimum distance, *IEEE Trans. Inform. Theory* **46** (2000), 1539–1544.

15. I. Bouyukliev and J. Simonis, Some new results for optimal ternary linear codes, *IEEE Trans. Inform. Theory* **48** (2002), 981–985.

16. I. Bouyukliev, M.Grassl and Z. Varbanov, New bounds for $n_4(k,d)$ and classification of some optimal codes over GF(4), *Disc. Math.* **281** (2004), 43–66, 981–985.

17. A. E. Brouwer, Bounds on the size of linear codes, *Handbook of Coding Theory*, eds. V. Pless and W. C. Huffmann, Elsevier, Amsterdam, The Netherlands (1998).

18. J. Bruck and M. Naor, The hardness of decoding linear codes with preprocessing, *IEEE Trans. Inform. Theory* **36** (1990), 381–385.

19. A. Betten, H. Fripertinger, A. Kerber, A. Wassermann and K.-H Zimmermann, *CodierungstheorieKonstruktion und Anwendung Linearer Codes* (Springer-Verlag, Heidelberg, Germany 1998).

20. A. R. Calderbank and G. McGuire, Construction of a $(64, 2^{37}, 12)$ code via Galois rings, *Des. Codes Cryptogr.* **10** (1997), 157–165.

21. E. Z. Chen, New constructions of a family of 2-generator quasi-cyclic two-weight codes and related codes, *Proc. 2007 IEEE Int. Symp. Information Theory*, (2007), pp. 861–864.

22. R. Daskalov and P. Hristov, New quasi-twisted degenerate ternary linear codes, *IEEE Trans. Inform. Theory* **49** (2003), 2259–2263.

23. R. Daskalov and P. Hristov, New binary one-generator quasi-cyclic codes, *IEEE Trans. Inform. Theory* **49** (2003), 3001–3005.

24. R. Daskalov and T. A. Gulliver, New quasi-twisted quaternary linear codes, *IEEE Trans. Inform. Theory* **46** (2000), 2642–2643.

25. R. Daskalov, E. Metodieva and P. Khristov, New bounds for the minimal distance of linear codes over GF(9) (Russian), *Problemy Peredachi Informatsii* **40**, 1 (2004), 15–26; translation in *Probl. Inf. Transm.* **40**, 1 (2004), 13–24.

26. P. Diaconis and R. L. Graham, The Radon transform on $\mathbb{Z}_2^k$, *Pacific J. Math.* **118** (1985), 176–185.

27. S. Dodunekov, S. Guritman and J. Simonis, Some new results on the minimum length of binary linear codes of dimension nine, *IEEE Trans. Inform. Theory* **45** (2003), 3001–3005.

28. I. Dumer, D. Micciancio and M. Sudan, Hardness of approximating the minimum distance of a linear code, *Proc. 40th Annual Symp. Foundations of Computer Science (FOCS)*, (1999), 475–484.

29. I. Dumer, D. Micciancio and M. Sudan, Hardness of approximating the minimum distance of a linear code, *IEEE Trans. Inform. Theory* **49** (2003), 22–37.

30. D. S. Dummit and R. M. Foote, *Abstract Algebra* (Princeton Hall, 1991).

31. M. R. Garey and D. S. Johnson, *Computers and Intractability: A Guide to the Theory of NP-Completeness* (Freeman, San Francisco CA, 1979).

32. D. G. Glynn, T. A. Gulliver and M. K. Gupta, On some quaternary self-orthogonal codes, *ARS Combinatoria*, **85** (2007), 129–154.

33. M. Grassl, Bounds on minimum distances of linear codes, Available online at http://www.codetables.de.

34. M. Grassl and G. White, New codes from chains of quasi-cyclic codes, *Proc. 2005 IEEE Int. Symposium Information Theory*, (2005), pp. 2095–2099.

35. T. A. Gulliver and V. K. Bhargava, Some best rate 1/p and (p-1)/p quasi-cyclic codes over GF(3) and GF(4), *IEEE Trans. Inform. Theory* **38**, 4 (1992), 1369–1374.

36. T. A. Gulliver and V. K. Bhargava, Two new rate 2/p binary quasi-cyclic codes, *IEEE Trans. Inform. Theory* **40**, 5 (1994), 1667–1668.

37. T. A. Gulliver, New optimal ternary linear codes, *IEEE Trans. Inform. Theory* **41** (1995), 1182–1185.

38. T. A. Gulliver and V. K. Bhargava, New good rate $(m\text{-}1)/pm$ ternary and quaternary quasi-cyclic codes, *Des. Codes Cryptogr.* **7** (1996), 223–233.

39. R. W. Hamming, Error-detecting and error-correcting codes., *Bell Syst. Tech. J.* **29** (1950), 147–160.

40. A. R. Hammons Jr, P. V. Kumar, A. R. Calderbank, N. J. A. Sloane and P. Sole, The Z_4-linearity of Kerdock, Preperata, Goethals, and related codes, *IEEE Trans. Inform. Theory* **40**, 2 (1994), 301–319.

41. R. Hill and E. Kolev, A survey of recent results on optimal linear codes, eds. in Combinatorial Designs and their Applications eds. F. C. Holroyd, K. A. S. Quinn, C. Rowley and B. S. Webb (Chapman & Hall, CRC, Boca Raton, London, New York, Washington, DC, 1999).

42. W. C. Huffman and V. Pless, *Fundamentals of Error-correcting Codes*, (Cambridge University Press 2003).

43. K. Lally and P. Fitzpatrick, Algebraic structures of quasicyclic codes, *Dis. Appl. Math.* **111** (2001), 157–175.

44. I. Landjev, A. Rouseva, T. Maruta and R. Hill, On optimal codes over the field with five elements, *Des. Codes Cryptogr.* **29** (2003), 165–175.

45. D. S. Johnson, The NP-completeness column: An ongoing guide, *J. Algorithms* **3** (1982), 182–195.

46. D. S. Johnson, The NP-completeness column: An ongoing guide, *J. Algorithms* **7** (1986), 584–601.

47. D. E. Knuth, *The Art of Computer Programming, Vol. 4: Combinatorial Algorithms, Pre-Fascicle 2C*, to be published

48. R. Lidl and H. Niederreiter, *Introduction to Finite Fields and Their Applications* (Cambridge University Press, 1986).

49. R. Lidl and G. Pilz, *Applied Abstract Algebra* (Springer 1998).

50. S. Litsyn, Table of non-linear binary codes, Available online at http://www.eng.tau.ac.il/~litysn/tableand/index.html.

51. A. Lobstein and G. D. Cohen, Sur la complexité d'un probléme de codage, *Theor. Inform. Appl.* **21** (1987), 25–32.

52. H. Lüneburg, Gray-Codes, *Abh. Math. Sem. Hamburg* **52** (1982), 208–227.

53. F. J. MacWilliams and N. J. A. Sloane, *The Theory of Error Correcting Codes* (North Holland, New York, 1977).

54. B. D. McKay, AutosonA distributed batch system for Unix workstation networks (Version 1.3), Tech. Rep. TR-CS-96-03, Comput. Sci. Dept., Australian Nat. Univ., Canberra, (1996).

55. A. A. Nechaev, Kerdock code in cyclic form, *Disc. Math. (USSR)* **1** (1989) 123–139 (in Russian), English translation: *Disc. Math. Appl.* **1** (1991), pp. 364–384.

56. V. Pless and N. J. Pierce, Self dual codes over GF(q) satisfy a modified Varshamov bound, *Inform. Cont.* **23** (1973), 35–40.

57. V. Pless, J. S. Leon and J. Fields, All $\mathbb{Z}_4$ codes of Type II and length 16 are known, *J. Combin. Theory Ser. A* **78** (1997), 32–50.

58. V. Pless, Introduction to the Theory of Error-Correcting Codes, 3rd edn. (John Wiley & Sons, Inc, 1998).

59. V. S. Pless and Z. Qian, Cyclic codes and quadratic residue codes over $\mathbb{Z}_4$, *IEEE Trans. Inform. Theory* **42** (1996), 1594–1600.

60. V. Pless, On the uniqueness of the Golay codes, *J. Combinat. Theory* **5** (1968), 215–228.

61. E. M. Rains, Optimal self-dual codes over $\mathbb{Z}_4$, *Disc. Math.* **203** (1999), 215–228.

62. E. M. Rains and N. J. A. Sloane, Self-dual codes, The Handbook of Coding Theory, (eds.) V. Pless and W. C. Huffman, North-Holland, New York, (1998).

63. F. S. Roberts, *Applied Combinatorics* (Prentice-Hall, Englewood Cliffs, NJ, 1984).

64. S. Roman, *Coding and Information Theory, Graduate Texts in Mathematics* 134 (Springer-Verlag, 1992).

65. S. Roman, *Field Theory, Graduate Texts in Mathematics* 158 (Springer-Verlag, 1995).

66. C. E. Shannon, A mathematical theory of communication, *Bell Syst. Tech. J.* **27** (1948), 379–423, 623–656.

67. W. Z. Shetter, This essay is redundant. Available: http://mypage.iu.edu/~shetter/miniatures/redund.htm (2000).

68. I. Siap, N. Aydin and D. K. Ray-Chaudhuri, New ternary quasi-cyclic codes with better minimum distances, *IEEE Trans. Inform. Theory* **46** (2000), 1554–1558.

69. I. Siap, N. Aydin and D. Ray-Chaudhuri, New 1-generator quasi-twisted codes over GF(5), *Codes and Association Schemes*, Piscataway, NJ, 1999; *DIMACS Ser. Disc. Math. Theoret. Comput. Sci.* (Amer. Math. Soc., Providence, RI, 2001), pp. 265–275.

70. I. Siap, New codes over GF(8), *Ars Combin.* **71** (2004), 239–247.

71. J. Stern, Approximating the number of error locations within a constant ratio is NP-complete, *Lect. Notes Comput. Sci.* **673** (1993), 325–331.

72. T. M. Thompson, *From Error Correcting Codes through Sphere Packings to Simple Groups* (The Mathematical Association of America, 1983).

73. M. van Dijk, M. Egner, S. Greferath and M. Wassermann, On two doubly even self-dual binary codes of length 160 and minimum weight 24, *IEEE Trans. Inform. Theory* **51** (2005), 408–411.

74. J. H. van Lint, *Introduction to Coding Theory* (Springer, 1999).

75. A. Vardy, Algorithmic complexity in coding theory and the minimum distance problem, *Proc. of the Twenty-Ninth Annual ACM Symposium on Theory of Computing, STOC'97*, El Paso, Texas, (1997), pp. 92–109.

76. A. Vardy, The intractability of computing the minimum distance of a code, *IEEE Trans. Inform. Theory* **43**, 6 (1997), 1757–1766.

77. J. Walker, A new approach to the main conjecture on algebraic-geometric MDS codes, *Des. Codes Cryptogr.* **45**, 1 (1996), 115–120.

78. Z. X. Wan, Quaternary Codes, (World Scientific, Singapore, 1997).

79. G. White and M. Grassl, A new minimum weight algorithm for additive codes, *IEEE Int. Symposium on Information Theory* (2006), pp. 1119–1123.

Part III

SOURCE CODING/CHANNEL CAPACITY/ NETWORK CODING

Chapter 8

APPLICATIONS OF UNIVERSAL SOURCE CODING TO STATISTICAL ANALYSIS OF TIME SERIES

BORIS RYABKO

Siberian State University of Telecommunications and Informatics,
Institute of Computational Technology of Siberian
Branch of Russian Academy of Science,
Kirov Street 86, Novosibirsk, 630102, Russia.
boris@ryabko.net
URL: http://boris.ryabko.net/

We show how universal codes can be used for solving some of the most important statistical problems for time series. By definition, a universal code (or a universal lossless data compressor) can compress any sequence generated by a stationary and ergodic source asymptotically to the Shannon entropy, which, in turn, is the best achievable ratio for lossless data compressors.

We consider finite-alphabet and real-valued time series and the following problems: estimation of the limiting probabilities for finite-alphabet time series and estimation of the density for real-valued time series, the on-line prediction, regression, classification (or problems with side information) for both types of the time series and the following problems of hypothesis testing: goodness-of-fit testing, or identity testing, and testing of serial independence. It is important to note that all problems are considered in the framework of classical mathematical statistics and, on the other hand, everyday methods of data compression (or archivers) can be used as a tool for the estimation and testing.

It turns out, that quite often the suggested methods and tests are more powerful than known ones when they are applied in practice.

8.1. Introduction

Since Shannon published the paper "A mathematical theory of communication" [47], the ideas and results of information theory have played an important role in cryptography [26, 48], mathematical statistics [3, 8, 25], and many other fields [6, 7], which are far from telecommunications. Universal coding, which is a part of information theory, also has been efficiently applied in many fields since its discovery [13, 21]. Thus, application of results of universal coding, initiated in 1988 [35], created a new approach to prediction [1, 19, 27, 28]. Maybe the most unexpected application of data compression ideas arises in experiments that show that some ant species are capable of compressing messages and are capable of adding and subtracting small numbers [30, 43].

In this chapter, we describe a new approach to estimation, prediction, and hypothesis testing for time series, which was suggested recently [35, 38, 42]. This approach is based on ideas of universal coding (or universal data compression). We would like to emphasize that everyday methods of data compression (or archivers) can be directly used as a tool for estimation and hypothesis testing. It is important to note that the modern archivers (like *zip, arj, rar*, etc.) are based on deep theoretical results of the source coding theory [10, 20, 24, 32, 46] and have shown their high efficiency in practice because archivers can find many kinds of latent regularities and use them for compression.

It is worth noting that this approach was applied to the problem of randomness testing [42]. This problem is quite important for practice; in particular, the National Institute of Standards and Technology of USA (NIST) has suggested "A statistical test suite for random and pseudorandom number generators for cryptographic applications" [33], which consists of 16 tests. It has turned out that tests which are based on universal codes are more powerful than the tests suggested by NIST [42].

The outline of this chapter is as follows. The next section contains some necessary definitions and facts about predictors, codes, hypothesis testing, and description of one universal code. The Secs. 8.3 and 8.4 are devoted to problems of estimation and hypothesis testing, correspondingly, for the case of finite-alphabet time series. The case of infinite alphabets is considered in Sec. 8.5. All proofs are given in Appendix, but some intuitive indication are given in the body of the chapter.

8.2. Definitions and Statements of the Problems

8.2.1. *Estimation and Prediction for I.I.D. Sources*

First we consider a source with unknown statistics, which generates sequences $x_1 x_2 \cdots$ of letters from some set (or alphabet) A. It will be convenient now to describe briefly the prediction problem. Let the source generate a message $x_1 \cdots x_{t-1} x_t$, $x_i \in A$ for all i, and the following letter

x_{t+1} needs to be predicted. This problem can be traced back to Laplace [11, 29] who considered the problem of estimation of the probability that the sun will rise tomorrow, given that it has risen every day since Creation. In our notation the alphabet A contains two letters 0 (*"the sun rises"*) and 1 (*"the sun does not rise"*), t is the number of days since Creation, $x_1 \cdots x_{t-1} x_t = 00 \cdots 0$.

Laplace suggested the following predictor:

$$L_0(a|x_1 \cdots x_t) = (\nu_{x_1 \cdots x_t}(a) + 1)/(t + |A|), \tag{8.1}$$

where $\nu_{x_1 \cdots x_t}(a)$ denotes the count of letter a occurring in the word $x_1 \cdots x_{t-1} x_t$. It is important to note that the predicted probabilities cannot be equal to zero even though a certain letter did not occur in the word $x_1 \cdots x_{t-1} x_t$.

Example 8.1. Let $A = \{0, 1\}, x_1 \cdots x_5 = 01010$, then the Laplace prediction is as follows: $L_0(x_6 = 0|x_1 \cdots x_5 = 01010) = (3 + 1)/(5 + 2) = 4/7, L_0(x_6 = 1|x_1 \cdots x_5 = 01010) = (2 + 1)/(5 + 2) = 3/7$. In other words, $3/7$ and $4/7$ are estimations of the unknown probabilities $P(x_{t+1} = 0|x_1 \cdots x_t = 01010)$ and $P(x_{t+1} = 1|x_1 \cdots x_t = 01010)$. (In what follows we will use the shorter notation: $P(0|01010)$ and $P(1|01010)$.)

We can see that Laplace considered prediction as a set of estimations of unknown (conditional) probabilities. This approach to the problem of prediction was developed in 1988 [35] and now is often called on-line prediction or universal prediction [1, 19, 27, 28]. As we mentioned above, it seems natural to consider conditional probabilities to be the best prediction, because they contain all information about the future behavior of the stochastic process. Moreover, this approach is deeply connected with game-theoretical interpretation of prediction [17, 37] and, in fact, all obtained results can be easily transferred from one model to the other.

Any predictor γ defines a measure (or an estimation of probability) by the following equation:

$$\gamma(x_1 \cdots x_t) = \prod_{i=1}^{t} \gamma(x_i|x_1 \cdots x_{i-1}). \tag{8.2}$$

And, vice versa, any measure γ (or estimation of probability) defines a predictor:

$$\gamma(x_i|x_1 \cdots x_{i-1}) = \gamma(x_1 \cdots x_{i-1} x_i)/\gamma(x_1 \cdots x_{i-1}). \tag{8.3}$$

Example 8.2. Let us apply the Laplace predictor for estimation of probabilities of the sequences 01010 and 010101. From (8.2) we obtain $L_0(01010) = \frac{1}{2}\frac{1}{3}\frac{2}{4}\frac{2}{5}\frac{3}{6} = \frac{1}{60}$ and $L_0(010101) = \frac{1}{60}\frac{3}{7} = \frac{1}{140}$. Vice versa, if for some measure (or a probability estimation) χ we have $\chi(01010) = \frac{1}{60}$ and $\chi(010101) = \frac{1}{140}$, then we obtain from (8.3) the following prediction, or the estimation of the conditional probability, $\chi(1|01010) = \frac{1/140}{1/60} = \frac{3}{7}$.

Now we concretize the class of stochastic processes which will be considered. Generally speaking, we will deal with so-called stationary and ergodic time series (or sources), whose definition will be given later, but now we consider may be the simplest class of such processes, which are called i.i.d. sources. By definition, they generate independent and identically distributed random variables from some set A. In our case A will be either some alphabet or a real-valued interval.

The next natural question is how to measure the errors of prediction and estimation of probability. Mainly we will measure these errors by the Kullback–Leibler (KL) divergence which is defined by

$$D(P, Q) = \sum_{a \in A} P(a) \log \frac{P(a)}{Q(a)}, \qquad (8.4)$$

where $P(a)$ and $Q(a)$ are probability distributions over an alphabet A (here and below $\log \equiv \log_2$ and $0 \log 0 = 0$). The probability distribution $P(a)$ can be considered as unknown whereas $Q(a)$ is its estimation. It is well-known that for any distributions P and Q, the KL divergence is non-negative and equals 0 if and only if $P(a) = Q(a)$ for all a [14]. So, if the estimation Q is equal to P, the error is 0, otherwise the error is a positive number.

The KL divergence is connected with the so-called variation distance

$$\|P - Q\| = \sum_{a \in A} |P(a) - Q(a)|,$$

via the following inequality (Pinsker's inequality)

$$\sum_{a \in A} P(a) \log \frac{P(a)}{Q(a)} \geq \frac{\log e}{2} \|P - Q\|^2. \qquad (8.5)$$

Let γ be a predictor, i.e. an estimation of an unknown conditional probability and $x_1 \cdots x_t$ be a sequence of letters created by an unknown source P. The KL divergence between P and the predictor γ is equal to

$$\rho_{\gamma,P}(x_1 \cdots x_t) = \sum_{a \in A} P(a|x_1 \cdots x_t) \log \frac{P(a|x_1 \cdots x_t)}{\gamma(a|x_1 \cdots x_t)}. \tag{8.6}$$

For fixed t it is a random variable, because $x_1, x_2, \ldots, x_t$ are random variables. We define the average error at time t by

$$\rho^t(P\|\gamma) = E(\rho_{\gamma,P}(\cdot)) = \sum_{x_1 \cdots x_t \in A^t} P(x_1 \cdots x_t)\rho_{\gamma,P}(x_1 \cdots x_t), \tag{8.7}$$

$$= \sum_{x_1 \cdots x_t \in A^t} P(x_1 \cdots x_t) \sum_{a \in A} P(a|x_1 \cdots x_t) \log \frac{P(a|x_1 \cdots x_t)}{\gamma(a|x_1 \cdots x_t)}.$$

Analogously, if $\gamma()$ is an estimation of a probability distribution we define the errors *per letter* as follows:

$$\bar{\rho}_{\gamma,P}(x_1 \cdots x_t) = t^{-1}(\log(P(x_1 \cdots x_t)/\gamma(x_1 \cdots x_t))) \tag{8.8}$$

and

$$\bar{\rho}^t(P\|\gamma) = t^{-1} \sum_{x_1 \cdots x_t \in A^t} P(x_1 \cdots x_t) \log(P(x_1 \cdots x_t)/\gamma(x_1 \cdots x_t)), \tag{8.9}$$

where, as before, $\gamma(x_1 \cdots x_t) = \prod_{i=1}^{t} \gamma(x_i|x_1 \cdots x_{i-1})$. (Here and below we denote by A^t and A^* the set of all words of length t over A and the set of all finite words over A correspondingly: $A^* = \bigcup_{i=1}^{\infty} A^i$.)

Claim 8.1 (Ryabko [35]). *For any i.i.d. source P generating letters from an alphabet A and an integer t the average error (8.7) of the Laplace predictor and the average error of the Laplace estimator are upper bounded as follows:*

$$\rho^t(P\|L_0) \le ((|A| - 1) \log e)/(t + 1), \tag{8.10}$$

$$\bar{\rho}^t(P\|L_0) \le (|A| - 1) \log t/t + O(1/t), \tag{8.11}$$

where $e \simeq 2.718$ is the Euler number.

So, we can see that the average error of the Laplace predictor goes to zero for any i.i.d. source P when the length t of the sample $x_1 \cdots x_t$ tends to infinity. Such methods are called universal, because the error goes to zero for any source, or process. In this case they are universal for the set of all i.i.d. sources generating letters from the finite alphabet A, but later we consider universal estimators for the set of stationary and ergodic sources. It is worth noting that the first universal code for which the estimation (8.11) is valid, was suggested independently by Fitingof [13] and Kolmogorov [21] in 1966.

The value

$$\bar{\rho}^t(P\|\gamma) = t^{-1} \sum_{x_1\cdots x_t \in A^t} P(x_1\cdots x_t)\log(P(x_1\cdots x_t)/\gamma(x_1\cdots x_t))$$

has one more interpretation connected with data compression. Now we consider the main idea whereas the more formal definitions will be given later. First we recall the definition of the Shannon entropy $h_0(P)$ for an i.i.d. source P

$$h_0(P) = -\sum_{a\in A} P(a)\log P(a). \tag{8.12}$$

It is easy to see that $t^{-1}\sum_{x_1\cdots x_t \in A^t} P(x_1\cdots x_t)\log(P(x_1\cdots x_t)) = -h_0(P)$ for the i.i.d. source. Hence, we can represent the average error $\bar{\rho}^t(P\|\gamma)$ in (8.9) as

$$\bar{\rho}^t(P\|\gamma) = t^{-1} \sum_{x_1\cdots x_t \in A^t} P(x_1\cdots x_t)\log(1/\gamma(x_1\cdots x_t)) - h_0(P).$$

More formal and general consideration of universal codes will be given later, but here we briefly show how estimations and codes are connected. The point is that one can construct a code with codelength $\gamma_{\mathrm{code}}(a|x_1\cdots x_t) \approx -\log_2\gamma(a|x_1\cdots x_n)$ for any letter $a \in A$ (since Shannon's original research, it has been well-known that, using block codes with large block length or more modern methods of arithmetic coding [31], the approximation may be as accurate as you like). If one knows the real distribution P, one can base coding on the true distribution P and not on the prediction γ. The difference in performance measured by average code length is given by

$$\sum_{a\in A} P(a|x_1\cdots x_t)(-\log_2\gamma(a|x_1\cdots x_t)) - \sum_{a\in A} P(a|x_1\cdots x_t)$$

$$\times(-\log_2 P(a|x_1\cdots x_t)) = \sum_{a\in A} P(a|x_1\cdots x_t)\log_2\frac{P(a|x_1\cdots x_t)}{\gamma(a|x_1\cdots x_t)}.$$

Thus this excess is exactly the error defined above (8.6). Analogously, if we encode the sequence $x_1\cdots x_t$ based on a predictor γ the redundancy per letter is defined by (8.8) and (8.9). So, from mathematical point of view, the estimation of the limiting probabilities and universal coding are identical. But $-\log\gamma(x_1\cdots x_t)$ and $-\log P(x_1\cdots x_t)$ have a very natural interpretation. The first value is a code word length (in bits), if the "code" γ is applied for compressing the word $x_1\cdots x_t$ and the second one is the minimal possible codeword length. The difference is the redundancy of the code and, at the same time, the error of the predictor. It is worth noting that there are many other deep interrelations between the universal coding, prediction, and estimation [32, 35].

We can see from the claim and the Pinsker inequality (8.5) that the variation distance of the Laplace predictor and estimator goes to zero, too. Moreover, it can be easily shown that the error (8.6) (and the corresponding variation distance) goes to zero with probability 1, when t goes to infinity. (Informally, it means that the error (8.6) goes to zero for almost all sequences $x_1\cdots x_t$ according to the measure P.) Obviously, such properties are very desirable for any predictor and for larger classes of sources, like Markov and stationary ergodic (they will be briefly defined in the next subsection). However, it is proven [35] that such predictors do not exist for the class of all stationary and ergodic sources (generating letters from a given finite alphabet). More precisely, if, for example, the alphabet has two letters, then for any predictor γ and for any $\delta > 0$ there exists a source P such that with probability 1 $\rho_{\gamma,P}(x_1\cdots x_t) \geq 1/2 - \delta$ infinitely often when $t \to \infty$. In other words, the error of any predictor may not go to 0, if the predictor is applied to an arbitrary stationary and ergodic source, that is why it is difficult to use (8.6) and (8.7) to compare different predictors. On the other hand, it is shown [35] that there exists a predictor R, such that the following Cesaro average $t^{-1}\sum_{i=1}^{t}\rho_{R,P}(x_1\cdots x_i)$ goes to 0 (with probability 1) for any stationary and ergodic source P, where t goes to infinity. (This predictor will be described in the next subsection.) That is why we will focus our attention on such averages. From the definitions (8.6), (8.7) and properties of the logarithm we can see that for any probability distribution γ

$$t^{-1}\sum_{i=1}^{t}\rho_{\gamma,P}(x_1\cdots x_i) = t^{-1}(\log(P(x_1\cdots x_t)/\gamma(x_1\cdots x_t))),$$

$$t^{-1}\sum_{i=1}^{t}\rho^i(P\|\gamma) = t^{-1}\sum_{x_1\cdots x_t\in A^t} P(x_1\cdots x_t)\log(P(x_1\cdots x_t)/\gamma(x_1\cdots x_t)) .$$

Taking into account these equations, we can see from the definitions (8.8) and (8.9) that the Chesaro averages of the prediction errors (8.6) and

(8.7) are equal to the errors of estimation of limiting probabilities (8.8) and
(8.9). That is why we will use values (8.8) and (8.9) as the main measures
of the precision throughout the chapter.

A natural problem is to find a predictor and an estimator of the
limiting probabilities whose average error (8.9) is minimal for the set of
i.i.d. sources. This problem was considered and solved by Krichevsky [23,
24]. He suggested the following predictor:

$$K_0(a|x_1 \cdots x_t) = (\nu_{x_1 \cdots x_t}(a) + 1/2)/(t + |A|/2), \qquad (8.13)$$

where, as before, $\nu_{x_1 \cdots x_t}(a)$ is the number of occurrences of the letter a in
the word $x_1 \cdots x_t$. We can see that the Krychevsky predictor is quite close
to the Laplace's one (8.35).

Example 8.3. Let $A = \{0, 1\}, x_1 \cdots x_5 = 01010$. Then $K_0(x_6 = 0|01010) = (3+1/2)/(5+1) = 7/12, K_0(x_6 = 1|01010) = (2+1/2)/(5+1) = 5/12$ and $K_0(01010) = \frac{1}{2}\frac{1}{4}\frac{1}{2}\frac{3}{8}\frac{1}{2} = \frac{3}{256}$.

The Krichevsky measure K_0 can be represented as follows:

$$K_0(x_1 \cdots x_t) = \prod_{i=1}^{t} \frac{\nu_{x_1 \cdots x_{i-1}}(x_i) + 1/2}{i - 1 + |A|/2} = \frac{\prod_{a \in A}(\prod_{j=1}^{\nu_{x_1 \cdots x_t}(a)}(j - 1/2))}{\prod_{i=0}^{t-1}(i + |A|/2)}.$$

$$(8.14)$$

It is known that

$$(r + 1/2)((r + 1) + 1/2) \cdots (s - 1/2) = \frac{\Gamma(s + 1/2)}{\Gamma(r + 1/2)}, \qquad (8.15)$$

where $\Gamma()$ is the gamma function [22]. So, (8.14) can be presented as
follows:

$$K_0(x_1 \cdots x_t) = \frac{\prod_{a \in A}(\Gamma(\nu_{x_1 \cdots x_t}(a) + 1/2)/\Gamma(1/2))}{\Gamma(t + |A|/2)/\Gamma(|A|/2)}. \qquad (8.16)$$

The following claim shows that the error of the Krichevsky estimator is a
half of the Laplace's one.

Claim 8.2. *For any i.i.d. source P generating letters from a finite alphabet A the average error (8.9) of the estimator K_0 is upper bounded as follows:*

$$\bar{\rho}_t(K_0, P) \equiv t^{-1} \sum_{x_1 \cdots x_t \in A^t} P(x_1 \cdots x_t) \log(P(x_1 \cdots x_t)/K_0(x_1 \cdots x_t))$$

$$\equiv t^{-1} \sum_{x_1 \cdots x_t \in A^t} P(x_1 \cdots x_t) \log(1/K_0(x_1 \cdots x_t)) - h_0(p)$$

$$\leq ((|A| - 1) \log t + C)/(2t), \tag{8.17}$$

where C is a constant.

Moreover, in a certain sense this average error is minimal: it is shown by Krichevsky [23] that for any predictor γ there exists such a source P^* that

$$\bar{\rho}_t(\gamma, P^*) \geq ((|A| - 1) \log t + C')/(2t).$$

Hence, the bound $((|A| - 1) \log t + C)/(2t)$ cannot be reduced and the Krichevsky estimator is the best (up to $O(1/t)$) if the error is measured by the KL divergence ρ.

8.2.2. *Consistent Estimations and On-line Predictors for Markov's and Stationary Ergodic Processes*

Now we briefly describe consistent estimations of unknown probabilities and efficient on-line predictors for general stochastic processes (or sources of information).

First we give a formal definition of stationary ergodic processes. The time shift T on A^∞ is defined as $T(x_1, x_2, x_3, \cdots) = (x_2, x_3, \cdots)$. A process P is called stationary if it is T-invariant: $P(T^{-1}B) = P(B)$ for every Borel set $B \subset A^\infty$. A stationary process is called ergodic if every T-invariant set has probability 0 or 1: $P(B) = 0$ or 1 whenever $T^{-1}B = B$ [5, 14].

We denote by $M_\infty(A)$ the set of all stationary and ergodic sources and let $M_0(A) \subset M_\infty(A)$ be the set of all i.i.d. processes. We denote by $M_m(A) \subset M_\infty(A)$ the set of Markov's sources of order (or with memory, or connectivity) not larger than $m, m \geq 0$. By definition

$\mu \in M_m(A)$ if

$$\mu(x_{t+1} = a_{i_1} | x_t = a_{i_2}, x_{t-1} = a_{i_3}, \ldots, x_{t-m+1} = a_{i_{m+1}}, \cdots) \qquad (8.18)$$
$$= \mu(x_{t+1} = a_{i_1} | x_t = a_{i_2}, x_{t-1} = a_{i_3}, \ldots, x_{t-m+1} = a_{i_{m+1}}),$$

for all $t \geq m$ and $a_{i_1}, a_{i_2}, \ldots \in A$. Let $M^*(A) = \bigcup_{i=0}^{\infty} M_i(A)$ be the set of all finite-order sources.

The Laplace and Krichevsky predictors can be extended to general Markov's processes. The trick is to view a Markov's source $p \in M_m(A)$ as resulting from $|A|^m$ i.i.d. sources. We illustrate this idea by an example [44]. So assume that $A = \{O, I\}$, $m = 2$ and assume that the source $p \in M_2(A)$ has generated the sequence

$$OOIOIIOOIIIOIO.$$

We represent this sequence by the following four subsequences:

$$* * I * * * * * I * * * **,$$
$$* * *O * I * * * * I * * * O,$$
$$* * * * I * * O * * * * *I*,$$
$$* * * * * * O * * * IO * *.$$

These four subsequences contain letters which follow OO, OI, IO, and II, respectively. By definition, $p \in M_m(A)$ if $p(a|x_t \cdots x_1) = p(a|x_t \cdots x_{t-m+1})$, for all $0 < m \leq t$, all $a \in A$ and all $x_1 \cdots x_t \in A^t$. Therefore, each of the four generated subsequences may be considered to be generated by an i.i.d. source. Further, it is possible to reconstruct the original sequence if we know the four $(= |A|^m)$ subsequences and the two $(= m)$ first letters of the original sequence.

Any predictor γ for i.i.d. sources can be applied to Markov's sources. Indeed, in order to predict, it is enough to store in the memory $|A|^m$ sequences, one corresponding to each word in A^m. Thus, in the example, the letter x_3 which follows OO is predicted based on the i.i.d. method γ corresponding to the x_1x_2-subsequence $(= OO)$, then x_4 is predicted based on the i.i.d. method corresponding to x_2x_3, i.e. to the OI-subsequence, and so forth. When this scheme is applied along with either L_0 or K_0 we denote the obtained predictors as L_m and K_m, correspondingly, and define the probabilities for the first m letters as follows: $L_m(x_1) = L_m(x_2) = \cdots = L_m(x_m) = 1/|A|$, $K_m(x_1) = K_m(x_2) = \cdots = K_m(x_m) = 1/|A|$. For example, having taken into account (8.16), we can present the Krichevsky predictors for $M_m(A)$ as follows:

$$K_m(x_1 \cdots x_t) = \begin{cases} \dfrac{1}{|A|^t}, & \text{if } t \leq m, \\ \dfrac{1}{|A|^m} \displaystyle\prod_{v \in A^m} \dfrac{\prod_{a \in A}((\Gamma(\nu_x(va) + 1/2)/\Gamma(1/2))}{(\Gamma(\bar{\nu}_x(v) + |A|/2)/\Gamma(|A|/2))}, & \text{if } t > m, \end{cases}$$

$$(8.19)$$

where $\bar{\nu}_x(v) = \sum_{a \in A} \nu_x(va), x = x_1 \cdots x_t$. It is worth noting that the representation (8.14) can be more convenient for carrying out calculations if t is small.

Example 8.4. For the word $OOIOIIOOIIIOIO$ considered in the previous example, we obtain $K_2(OOIOIIOOIIIOIO) = 2^{-2} \frac{1}{2} \frac{3}{4} \frac{1}{2} \frac{1}{4} \frac{1}{2} \frac{3}{8} \frac{1}{2}$ $\frac{1}{4} \frac{1}{2} \frac{1}{2} \frac{1}{4} \frac{1}{2}$. Here groups of multipliers correspond to subsequences $II, OIIO$, IOI and OIO.

In order to estimate the error of the Krichevsky predictor K_m we need a general definition of the Shannon entropy. Let P be a stationary and ergodic source generating letters from a finite alphabet A. The m-order (conditional) Shannon entropy and the limiting Shannon entropy are defined as follows:

$$h_m(P) = \sum_{v \in A^m} P(v) \sum_{a \in A} P(a/v) \log P(a/v), \qquad h_\infty(\tau) = \lim_{m \to \infty} h_m(P).$$

$$(8.20)$$

(If $m = 0$ we obtain the definition (8.12).) It is also known that for any m [5, 14].

$$h_\infty(P) \leq h_m(P). \qquad (8.21)$$

Claim 8.3. *For any stationary and ergodic source P generating letters from a finite alphabet A the average error of the Krichevsky predictor K_m is upper bounded as follows:*

$$-t^{-1} \sum_{x_1 \cdots x_t \in A^t} P(x_1 \cdots x_t) \log(K_m(x_1 \cdots x_t)) - h_m(P)$$

$$\leq (|A|^m(|A| - 1) \log t + C)/(2t), \qquad (8.22)$$

where C is a constant.

The following so-called empirical Shannon entropy, which is an estimation of the entropy (8.20), will play a key role in the hypothesis testing. It will be convenient to consider its definition here, because this notation will be used in the proof of the next claims. Let $v = v_1 \cdots v_k$ and $x = x_1 x_2 \cdots x_t$ be words from A^*. Denote the rate of a word v occurring in the sequence $x = x_1 x_2 \cdots x_k, x_2 x_3 \cdots x_{k+1}, x_3 x_4 \cdots x_{k+2}, \ldots, x_{t-k+1} \cdots x_t$ as $\nu_x(v)$. For example, if $x = 000100$ and $v = 00$, then $\nu_x(00) = 3$. For any $0 \le k < t$ the empirical Shannon entropy of order k is defined as follows:

$$h_k^*(x) = - \sum_{v \in A^k} \frac{\bar{\nu}_x(v)}{(t-k)} \sum_{a \in A} \frac{\nu_x(va)}{\bar{\nu}_x(v)} \log \frac{\nu_x(va)}{\bar{\nu}_x(v)}, \qquad (8.23)$$

where $x = x_1 \cdots x_t$, $\bar{\nu}_x(v) = \sum_{a \in A} \nu_x(va)$. In particular, if $k = 0$, we obtain $h_0^*(x) = -t^{-1} \sum_{a \in A} \nu_x(a) \log(\nu_x(a)/t)$.

Let us define the measure R, which, in fact, is a consistent estimator of probabilities for the class of all stationary and ergodic processes with a finite alphabet. First we define a probability distribution $\{\omega = \omega_1, \omega_2, \ldots\}$ on integers $\{1, 2, \ldots\}$ by

$$\omega_1 = 1 - 1/\log 3, \ldots, \omega_i = 1/\log(i+1) - 1/\log(i+2), \ldots. \qquad (8.24)$$

(In what follows we will use this distribution, but results described below are obviously true for any distribution with non-zero probabilities.) The measure R is defined as follows:

$$R(x_1 \cdots x_t) = \sum_{i=0}^{\infty} \omega_{i+1} K_i(x_1 \cdots x_t). \qquad (8.25)$$

It is worth noting that this construction can be applied to the Laplace measure (if we use L_i instead of K_i) and any other family of measures.

Example 8.5. Let us calculate $R(00), \ldots, R(11)$. From (8.14) and (8.19) we obtain:

$$K_0(00) = K_0(11) = \frac{1/2}{1} \frac{3/2}{1+1} = 3/8,$$

$$K_0(01) = K_0(10) = \frac{1/2}{1+0} \frac{1/2}{1+1} = 1/8,$$

$$K_i(00) = K_i(01) = K_i(10) = K_i(11) = 1/4; \quad i \ge 1.$$

Having taken into account the definitions of ω_i (8.24) and the measure R (8.25), we can calculate $R(z_1 z_2)$ as follows:

$$R(00) = \omega_1 K_0(00) + \omega_2 K_1(00) + \cdots = (1 - 1/\log 3)3/8$$
$$+ (1/\log 3 - 1/\log 4)1/4 + (1/\log 4 - 1/\log 5)1/4$$
$$+ \cdots = (1 - 1/\log 3)3/8 + (1/\log 3)1/4 \approx 0.296.$$

Analogously, $R(01) = R(10) \approx 0.204$, $R(11) \approx 0.296$.

The main properties of the measure R are connected with the Shannon entropy (8.20).

Theorem 8.1 (Ryabko [35]). *For any stationary and ergodic source P the following equalities are valid:*

$$i) \qquad \lim_{t \to \infty} \frac{1}{t} \log(1/R(x_1 \cdots x_t)) = h_\infty(P)$$

with probability 1,

$$ii) \qquad \lim_{t \to \infty} \frac{1}{t} \sum_{u \in A^t} P(u) \log(1/R(u)) = h_\infty(P).$$

So, if one uses the measure R for data compression in such a way that the codeword length of the sequence $x_1 \cdots x_t$ is (approximately) equal to $\log(1/R(x_1 \cdots x_t))$ bits, he/she obtains the best achievable data compression ratio $h_\infty(P)$ per letter. On the other hand, we know that the redundancy of a universal code and the error of corresponding predictor are equal. Hence, if one uses the measure R for estimation and/or prediction, the error (per letter) will go to zero.

8.2.3. *Hypothesis Testing*

Here we briefly describe the main notions of hypothesis testing and the two particular problems considered below. A statistical test is formulated to test a specific null hypothesis (H_0). Associated with this null hypothesis is the alternative hypothesis (H_1) [33]. For example, we will consider the two following problems: goodness-of-fit testing (or identity testing) and testing of serial independence. Both problems are well-known in mathematical statistics and there is an extensive literature dealing with their non-parametric testing [2, 8, 9, 12].

The goodness-of-fit testing is described as follows: a hypothesis H_0^{id} is that the source has a particular distribution π and the alternative hypothesis H_1^{id} that the sequence is generated by a stationary and ergodic source which differs from the source under H_0^{id}. One particular case, mentioned in Sec. 8.1, is when the source alphabet A is $\{0,1\}$ and the main hypothesis H_0^{id} is that a bit sequence is generated by the Bernoulli i.i.d. source with equal probabilities of 0's and 1's. In all cases, the testing should be based on a sample $x_1 \cdots x_t$ generated by the source.

The second problem is as follows: the null hypothesis H_0^{SI} is that the source is Markovian of order not larger than $m, (m \geq 0)$, and the alternative hypothesis H_1^{SI} is that the sequence is generated by a stationary and ergodic source which differs from the source under H_0^{SI}. In particular, if $m = 0$, this is the problem of testing for independence of time series.

For each applied test, a decision is derived that accepts or rejects the null hypothesis. During the test, a test statistic value is computed on the data (the sequence being tested). This test statistic value is compared to the critical value. If the test statistic value exceeds the critical value, the null hypothesis is rejected. Otherwise, the null hypothesis is accepted. So, statistical hypothesis testing is a conclusion-generation procedure that has two possible outcomes: either accept H_0 or accept H_1.

Errors of the two following types are possible: The Type I error occurs if H_0 is true but the test accepts H_1 and, vice versa, the Type II error occurs if H_1 is true, but the test accepts H_0. The probability of Type I error is often called the level of significance of the test. This probability can be set prior to the testing and is denoted α. For a test, α is the probability that the test will say that H_0 is not true when it really is true. Common values of α are about 0.01. The probabilities of Type I and Type II errors are related to each other and to the size n of the tested sequence in such a way that if two of them are specified, the third value is automatically determined. Practitioners usually select a sample size n and a value for the probability of the Type I error — the level of significance [33].

8.2.4. Codes

We briefly describe the main definitions and properties (without proofs) of lossless codes, or methods of (lossless) data compression. A data compression method (or code) φ is defined as a set of mappings φ_n such that $\varphi_n : A^n \to \{0,1\}^*, n = 1, 2, \ldots$ and for each pair of different words $x, y \in A^n \varphi_n(x) \neq \varphi_n(y)$. It is also required that each sequence $\varphi_n(u_1)\varphi_n(u_2)\cdots\varphi_n(u_r), r \geq 1$, of encoded words from the set $A^n, n \geq 1$, could be uniquely decoded into $u_1 u_2 \cdots u_r$. Such codes are called uniquely decodable. For example, let $A = \{a, b\}$, the code $\psi_1(a) = 0, \psi_1(b) = 00$, obviously, is not uniquely decodable. In what follows we call uniquely decodable codes just "codes". It is well-known that if φ is a code then the lengths of the codewords satisfy the following inequality (Kraft's inequality) [14]: $\Sigma_{u\in A^n} 2^{-|\varphi_n(u)|} \leq 1$. It will be convenient to reformulate this property as follows:

Claim 8.4. *Let φ be a code over an alphabet A. Then for any integer n there exists a measure μ_φ on A^n such that*

$$-\log \mu_\varphi(u) \leq |\varphi(u)| \tag{8.26}$$

for any u from A^n.

(Obviously, this claim is true for the measure $\mu_\varphi(u) = 2^{-|\varphi(u)|}/\Sigma_{u\in A^n} 2^{-|\varphi(u)|}$).

It was mentioned above that, in a certain sense, the opposite claim is true, too. Namely, for any probability measure μ defined on $A^n, n \geq 1$, there exists a code φ_μ such that

$$|\varphi_\mu(u)| = -\log \mu(u). \tag{8.27}$$

(More precisely, for any $\varepsilon > 0$ one can construct such a code φ_μ^*, that $|\varphi_\mu^*(u)| < -\log \mu(u) + \varepsilon$ for any $u \in A^n$. Such a code can be constructed by applying a so-called arithmetic coding [31].) For example, for the above described measure R we can construct a code R_{code} such that

$$|R_{\text{code}}(u)| = -\log R(u). \tag{8.28}$$

As we mentioned above there exist universal codes. For their description we recall that sequences $x_1 \ldots x_t$, generated by a source P, can be "compressed" to the length $-\log P(x_1 \cdots x_t)$ bits (see (8.27)) and, on the other hand, for any source P there is no code ψ for which the average codeword length $(\Sigma_{u\in A^t} P(u)|\psi(u)|)$ is less than $-\Sigma_{u\in A^t} P(u) \log P(u)$. Universal codes can reach the lower bound $-\log P(x_1 \cdots x_t)$ asymptotically for any stationary and ergodic source P in average and with probability 1.

The formal definition is as follows: a code U is universal if for any stationary and ergodic source P the following equalities are valid:

$$\lim_{t \to \infty} |U(x_1 \ldots x_t)|/t = h_\infty(P) \qquad (8.29)$$

with probability 1, and

$$\lim_{t \to \infty} E(|U(x_1 \cdots x_t)|)/t = h_\infty(P), \qquad (8.30)$$

where $E(f)$ is the expected value of f, $h_\infty(P)$ is the Shannon entropy of P, see (8.21). So, informally speaking, a universal code estimates the probability characteristics of a source and uses them for efficient "compression".

In this chapter we mainly consider finite-alphabet and real-valued sources, but sources with countable alphabet also were considered by many authors [4, 16, 18, 39, 40]. In particular, it is shown that, for infinite alphabet, without any condition on the source distribution it is impossible to have universal source code and/or universal predictor, i.e. such a predictor whose average error goes to zero, when the length of a sequence goes to infinity. On the other hand, there are some necessary and sufficient conditions for existence of universal codes and predictors [4, 18, 39].

8.3. Finite Alphabet Processes

8.3.1. *The Estimation of (Limiting) Probabilities*

The following theorem shows how universal codes can be applied for probability estimation.

Theorem 8.2. *Let U be a universal code and*

$$\mu_U(u) = 2^{-|U(u)|}/\Sigma_{v \in A^{|u|}} 2^{-|U(v)|}. \qquad (8.31)$$

Then, for any stationary and ergodic source P the following equalities are valid:

$$i) \qquad \lim_{t \to \infty} \frac{1}{t}(-\log P(x_1 \cdots x_t) - (-\log \mu_U(x_1 \cdots x_t))) = 0$$

with probability 1,

$$ii) \qquad \lim_{t \to \infty} \frac{1}{t} \sum_{u \in A^t} P(u) \log(P(u)/\mu_U(u)) = 0.$$

The informal outline of the proof is as follows: $\frac{1}{t}(-\log P(x_1 \cdots x_t))$ and $\frac{1}{t}(-\log \mu_U(x_1 \cdots x_t))$ goes to the Shannon entropy $h_\infty(P)$, that is why the difference is 0.

So, we can see that, in a certain sense, the measure μ_U is a consistent non-parametric estimation of the (unknown) measure P.

Nowadays there are many efficient universal codes (and universal predictors connected with them), which can be applied to estimation. For example, the above described measure R is based on a universal code [34, 35] and can be applied for probability estimation. More precisely, Theorem 8.2 (and the following theorems) are true for R, if we replace μ_U by R.

It is important to note that the measure R has some additional properties, which can be useful for applications. The following theorem describes these properties (whereas all other theorems are valid for all universal codes and corresponding measures, including the measure R).

Theorem 8.3 (Ryabko [34, 35]). *For any Markov's process P with memory k*

i) the error of the probability estimator, which is based on the measure R, is upper-bounded as follows:

$$\frac{1}{t} \sum_{u \in A^t} P(u) \log(P(u)/R(u)) \leq \frac{(|A| - 1)|A|^k \log t}{2t} + O\left(\frac{1}{t}\right),$$

ii) the error of R is asymptotically minimal in the following sense: for any measure μ there exists a k-memory Markov's process p_μ such that

$$\frac{1}{t} \sum_{u \in A^t} p_\mu(u) \log(p_\mu(u)/\mu(u)) \geq \frac{(|A| - 1)|A|^k \log t}{2\,t} + O\left(\frac{1}{t}\right),$$

iii) Let Θ be a set of stationary and ergodic processes such that there exists a measure μ_Θ for which the estimation error of the probability goes to 0 uniformly:

$$\lim_{t \to \infty} \sup_{P \in \Theta} \left(\frac{1}{t} \sum_{u \in A^t} P(u) \log(P(u)/\mu_\Theta(u)) \right) = 0.$$

Then the error of the estimator which is based on the measure R, goes to 0 uniformly too:

$$\lim_{t \to \infty} \sup_{P \in \Theta} \left(\frac{1}{t} \sum_{u \in A^t} P(u) \log(P(u)/R(u)) \right) = 0.$$

8.3.2. *Prediction*

As we mentioned above, any universal code U can be applied for prediction. Namely, the measure μ_U (8.31) can be used for prediction as the following conditional probability:

$$\mu_U(x_{t+1}|x_1 \cdots x_t) = \mu_U(x_1 \cdots x_t x_{t+1})/\mu_U(x_1 \cdots x_t). \qquad (8.32)$$

The following theorem shows that such a predictor is quite reasonable. Moreover, it gives a possibility to apply practically used data compressors for prediction of real data (like EUR/USD rate) and obtain quite precise estimation [41].

Theorem 8.4. *Let U be a universal code and P be any stationary and ergodic process. Then*

$$i) \qquad \lim_{t \to \infty} \frac{1}{t} \left\{ E\left(\log \frac{P(x_1)}{\mu_U(x_1)}\right) + E\left(\log \frac{P(x_2|x_1)}{\mu_U(x_2|x_1)}\right) \right.$$

$$\left. + \cdots + E\left(\log \frac{P(x_t|x_1 \cdots x_{t-1})}{\mu_U(x_t|x_1 \cdots x_{t-1})}\right) \right\} = 0,$$

$$ii) \qquad \lim_{t \to \infty} E\left(\frac{1}{t}\sum_{i=0}^{t-1}(P(x_{i+1}|x_1 \cdots x_i) - \mu_U(x_{i+1}|x_1 \cdots x_i))^2\right) = 0,$$

and

$$iii) \qquad \lim_{t \to \infty} E\left(\frac{1}{t}\sum_{i=0}^{t-1}|P(x_{i+1}|x_1 \cdots x_i) - \mu_U(x_{i+1}|x_1 \cdots x_i)|\right) = 0.$$

An informal outline of the proof is as follows:

$$\frac{1}{t}\left\{ E\left(\log \frac{P(x_1)}{\mu_U(x_1)}\right) + E\left(\log \frac{P(x_2|x_1)}{\mu_U(x_2|x_1)}\right) \right.$$

$$\left. + \cdots + E\left(\log \frac{P(x_t|x_1 \cdots x_{t-1})}{\mu_U(x_t|x_1 \cdots x_{t-1})}\right) \right\}$$

is equal to $\frac{1}{t}E\left(\log \frac{P(x_1 \cdots x_t)}{\mu_U(x_1 \cdots x_t)}\right)$. Taking into account Theorem 8.2, we obtain the first statement of the theorem.

Comment. The measure R described above has one additional property if it is used for prediction. Namely, for any Markov's process $P(P \in M^*(A))$

the following is true:

$$\lim_{t \to \infty} \log \frac{P(x_{t+1}|x_1 \cdots x_t)}{R(x_{t+1}|x_1 \cdots x_t)} = 0$$

with probability 1, where $R(x_{t+1}|x_1 \cdots x_t) = R(x_1 \cdots x_t x_{t+1})/ R(x_1 \cdots x_t)$ [36].

Comment. It is known [45] that, in fact, the statements ii) and iii) are equivalent.

8.3.3. *Problems with Side Information*

Now we consider the so-called problems with side information, which are described as follows: there is a stationary and ergodic source whose alphabet A is presented as a product $A = X \times Y$. We are given a sequence $(x_1, y_1), \ldots, (x_{t-1}, y_{t-1})$ and side information y_t. The goal is to predict, or estimate, x_t. This problem arises in statistical decision theory, pattern recognition, and machine learning. Obviously, if someone knows the conditional probabilities $P(x_t|(x_1, y_1), \ldots, (x_{t-1}, y_{t-1}), y_t)$ for all $x_t \in X$, he has all information about x_t, available before x_t is known. That is why we will look for the best (or, at least, good) estimations for this conditional probabilities. Our solution will be based on results obtained in the previous subsection. More precisely, for any universal code U and the corresponding measure μ_U (8.31) we define the following estimate for the problem with side information:

$$\mu_U(x_t|(x_1, y_1), \ldots, (x_{t-1}, y_{t-1}), y_t)$$
$$= \frac{\mu_U((x_1, y_1), \ldots, (x_{t-1}, y_{t-1}), (x_t, y_t))}{\sum_{x_t \in X} \mu_U((x_1, y_1), \ldots, (x_{t-1}, y_{t-1}), (x_t, y_t))}.$$

The following theorem shows that this estimate is quite reasonable.

Theorem 8.5. *Let U be a universal code and let P be any stationary and ergodic process. Then*

$$i) \quad \lim_{t \to \infty} \frac{1}{t} \left\{ E \left(\log \frac{P(x_1|y_1)}{\mu_U(x_1|y_1)} \right) + E \left(\log \frac{P(x_2|(x_1, y_1), y_2)}{\mu_U(x_2|(x_1, y_1), y_2)} \right) \right.$$
$$\left. + \cdots + E \left(\log \frac{P(x_t|(x_1, y_1), \ldots, (x_{t-1}, y_{t-1}), y_t)}{\mu_U(x_t|(x_1, y_1), \ldots, (x_{t-1}, y_{t-1}), y_t)} \right) \right\} = 0,$$

$$ii) \qquad \lim_{t \to \infty} E\left(\frac{1}{t}\sum_{i=0}^{t-1}(P(x_{i+1}|(x_1,y_1),\ldots,(x_i,y_i),y_{i+1}))\right.$$

$$\left. - \mu_U(x_{i+1}|(x_1,y_1),\ldots,(x_i,y_i),y_{i+1}))^2\right) = 0,$$

and

$$iii) \qquad \lim_{t \to \infty} E\left(\frac{1}{t}\sum_{i=0}^{t-1}|P(x_{i+1}|(x_1,y_1),\ldots,(x_i,y_i),y_{i+1}))\right.$$

$$\left. - \mu_U(x_{i+1}|(x_1,y_1),\ldots,(x_i,y_i),y_{i+1})|\right) = 0.$$

The proof is very close to the proof of the previous theorem.

8.3.4. *The Case of Several Independent Samples*

In this part we consider a situation which is important for practical applications, but needs cumbersome notations. Namely, we extend our consideration to the case where the sample is presented as several independent samples $x^1 = x_1^1 \cdots x_{t_1}^1$, $x^2 = x_1^2 \cdots x_{t_2}^2, \ldots, x^r = x_1^r \cdots x_{t_r}^r$ generated by a source. More precisely, we will suppose that all sequences were independently created by one stationary and ergodic source. (The point is that it is impossible just to combine all samples into one, if the source is not i.i.d.) We denote them by $x^1 \diamond x^2 \diamond \cdots \diamond x^r$ and define $\nu_{x^1 \diamond x^2 \diamond \cdots \diamond x^r}(v) = \sum_{i=1}^{r} \nu_{x^i}(v)$. For example, if $x^1 = 0010, x^2 = 011$, then $\nu_{x^1 \diamond x^2}(00) = 1$. The definition of K_m and R can be extended to this case:

$$K_m(x^1 \diamond x^2 \diamond \cdots \diamond x^r) = \left(\prod_{i=1}^{r} |A|^{-\min\{m,t_i\}}\right)$$

$$\times \prod_{v \in A^m} \frac{\prod_{a \in A}((\Gamma(\nu_{x^1 \diamond x^2 \diamond \cdots \diamond x^r}(va) + 1/2)/\Gamma(1/2))}{(\Gamma(\bar{\nu}_{x^1 \diamond x^2 \diamond \cdots \diamond x^r}(v) + |A|/2)/\Gamma(|A|/2))}, \quad (8.33)$$

whereas the definition of R is the same (see (8.25)). (Here, as before, $\bar{\nu}_{x^1 \diamond x^2 \diamond \cdots \diamond x^r}(v) = \sum_{a \in A} \nu_{x^1 \diamond x^2 \diamond \cdots \diamond x^r}(va)$. Note, that $\bar{\nu}_{x^1 \diamond x^2 \diamond \cdots \diamond x^r}() = \sum_{i=1}^{r} t_i$ if $m = 0$.)

The following example is intended to show the difference between the case of many samples and one.

Example 8.6. Let there be two independent samples $y = y_1 \cdots y_4 = 0101$ and $x = x_1 \cdots x_3 = 101$, generated by a stationary and ergodic source with the alphabet $\{0,1\}$. One wants to estimate the (limiting) probabilities $P(z_1 z_2), z_1, z_2 \in \{0,1\}$ (here $z_1 z_2 \cdots$ can be considered as an independent sequence, generated by the source) and predict $x_4 x_5$ (i.e. estimate conditional probability $P(x_4 x_5 | x_1 \cdots x_3 = 101, y_1 \cdots y_4 = 0101)$. For solving both problems we will use the measure R (see (8.25)). First we consider the case where $P(z_1 z_2)$ is to be estimated without knowledge of sequences x and y. Those probabilities were calculated in the example 8.5 and we obtained: $R(00) \approx 0.296$, $R(01) = R(10) \approx 0.204$, $R(11) \approx 0.296$. Let us now estimate the probability $P(z_1 z_2)$ taking into account that there are two independent samples $y = y_1 \cdots y_4 = 0101$ and $x = x_1 \cdots x_3 = 101$. First of all we note that such estimates are based on the formula for conditional probabilities:

$$R(z|x \diamond y) = R(x \diamond y \diamond z)/R(x \diamond y).$$

Then we estimate the frequencies: $\nu_{0101\diamond101}(0) = 3$, $\nu_{0101\diamond101}(1) = 4$, $\nu_{0101\diamond101}(00) = \nu_{0101\diamond101}(11) = 0$, $\nu_{0101\diamond101}(01) = 3$, $\nu_{0101\diamond101}(10) = 2$, $\nu_{0101\diamond101}(010) = 1$, $\nu_{0101\diamond101}(101) = 2$, $\nu_{0101\diamond101}(0101) = 1$, whereas frequencies of all other three-letters and four-letters words are 0. Then we calculate:

$$K_0(0101 \diamond 101) = \frac{1}{2}\frac{3}{4}\frac{5}{6}\frac{7}{8}\frac{1}{10}\frac{3}{12}\frac{5}{14} \approx 0.00244,$$

$$K_1(0101 \diamond 101) = (2^{-1})^2 \frac{1}{2}\frac{3}{4}\frac{5}{6}1\frac{1}{2}\frac{3}{4}1 \approx 0.0293,$$

$$K_2(0101 \diamond 101) \approx 0.01172, \quad K_i(0101 \diamond 101) = 2^{-7}, \quad i \geq 3,$$

$$R(0101 \diamond 101) = w_1 K_0(0101 \diamond 101) + w_2 K_1(0101 \diamond 101)$$

$$+ \cdots \approx 0.369\ 0.00244 + 0.131\ 0.0293 + 0.06932\ 0.01172$$

$$+ 2^{-7}/\log 5 \approx 0.0089.$$

In order to avoid repetitions, we estimate only one probability $P(z_1 z_2 = 01)$. Carrying out similar calculations, we obtain $R(0101 \diamond 101 \diamond 01) \approx 0.00292$, $R(z_1 z_2 = 01 | y_1 \cdots y_4 = 0101, x_1 \cdots x_3 = 101) = R(0101 \diamond 101 \diamond 01)/R(0101 \diamond 101) \approx 0.32812$. If we compare this value and the estimation $R(01) \approx 0.204$, which is not based on the knowledge of samples x and y, we can see that the measure R uses additional information quite naturally (indeed, 01 is quite frequent in $y = y_1 \cdots y_4 = 0101$ and $x = x_1 \cdots x_3 = 101$).

Such generalization can be applied to many universal codes, but, generally speaking, there exist codes U for which $U(x^1 \diamond x^2)$ is not defined and, hence, the measure $\mu_U(x_1 \diamond x_2)$ is not defined. That is why we will describe properties of the universal code R, but not of universal codes in general. For the measure R all asymptotic properties are the same for the cases of one sample and several samples. More precisely, the following statement is true:

Claim 8.5. *Let* $x^1, x^2, \ldots, x^r$ *be independent sequences generated by a stationary and ergodic source and let* t *be a total length of these sequences* ($t = \sum_{i=1}^{r} |x^i|$). *Then, if* $t \to \infty$, *(and* r *is fixed) the statements of the Theorems 8.2–8.5 are valid, when applied to* $x^1 \diamond x^2 \diamond \cdots \diamond x^r$ *instead of* $x_1 \cdots x_t$. *(In Theorems 8.2–8.5* μ_U *should be changed to* R.)*

The proofs are completely analogous to the proofs of the Theorems 8.2–8.5.

Now we can extend the definition of the empirical Shannon entropy (8.23) to the case of several words $x^1 = x_1^1 \cdots x_{t_1}^1$, $x^2 = x_1^2 \cdots x_{t_2}^2, \ldots, x^r = x_1^r \cdots x_{t_r}^r$. We define $\nu_{x^1 \diamond x^2 \diamond \cdots \diamond x^r}(v) = \sum_{i=1}^{r} \nu_{x^i}(v)$. For example, if $x^1 = 0010, x^2 = 011$, then $\nu_{x^1 \diamond x^2}(00) = 1$. Analogously to (8.23),

$$
h_k^*(x^1 \diamond x^2 \diamond \cdots \diamond x^r)
$$
$$
= -\sum_{v \in A^k} \frac{\bar{\nu}_{x^1 \diamond \cdots \diamond x^r}(v)}{(t - kr)} \sum_{a \in A} \frac{\nu_{x^1 \diamond \cdots \diamond x^r}(va)}{\bar{\nu}_{x^1 \diamond \cdots \diamond x^r}(v)} \log \frac{\nu_{x^1 \diamond \cdots \diamond x^r}(va)}{\bar{\nu}_{x^1 \diamond \cdots \diamond x^r}(v)}, \quad (8.34)
$$

where $\bar{\nu}_{x^1 \diamond \cdots \diamond x^r}(v) = \sum_{a \in A} \nu_{x^1 \diamond \cdots \diamond x^r}(va)$.

For any sequence of words $x^1 = x_1^1 \cdots x_{t_1}^1$, $x^2 = x_1^2 \cdots x_{t_2}^2, \ldots, x^r = x_1^r \cdots x_{t_r}^r$ from A^* and any measure θ we define $\theta(x^1 \diamond x^2 \diamond \cdots \diamond x^r) = \prod_{i=1}^{r} \theta(x^i)$. The following lemma gives an upper bound for unknown probabilities.

Lemma 8.1. *Let* θ *be a measure from* $M_m(A), m \geq 0$, *and* $x^1, \ldots, x^r$ *be words from* A^*, *whose lengths are not less than* m. *Then*

$$
\theta(x^1 \diamond \cdots \diamond x^r) \leq 2^{-(t-rm)h_m^*(x^1 \diamond \cdots \diamond x^t)}, \quad (8.35)
$$

where $\theta(x^1 \diamond \cdots \diamond x^r) = \prod_{i=1}^{r} \theta(x^i)$.

8.4. Hypothesis Testing

8.4.1. *Goodness-of-Fit or Identity Testing*

Now we consider the problem of testing H_0^{id} against H_1^{id}. Let us recall that the hypothesis H_0^{id} is that the source has a particular distribution π and the alternative hypothesis H_1^{id} that the sequence is generated by a stationary and ergodic source which differs from the source under H_0^{id}. Let the required level of significance (or the Type I error) be $\alpha, \alpha \in (0,1)$. We describe a statistical test which can be constructed based on any code φ.

The main idea of the suggested test is quite natural: compress a sample sequence $x_1 \ldots x_t$ by a code φ. If the length of the codeword $(|\varphi(x_1 \cdots x_t)|)$ is significantly less than the value $-\log \pi(x_1 \cdots x_t)$, then H_0^{id} should be rejected. The key observation is that the probability of all rejected sequences is quite small for any φ, that is why the Type I error can be made small. The precise description of the test is as follows: *The hypothesis H_0^{id} is accepted if*

$$-\log \pi(x_1 \cdots x_t) - |\varphi(x_1 \cdots x_t)| \leq -\log \alpha. \qquad (8.36)$$

Otherwise, H_0^{id} is rejected. We denote this test by $T_\varphi^{id}(A, \alpha)$.

Theorem 8.6. *i) For each distribution $\pi, \alpha \in (0,1)$ and a code φ, the Type I error of the described test $T_\varphi^{id}(A, \alpha)$ is not larger than α and ii) if, in addition, π is a finite-order stationary and ergodic process over A^∞ (i.e. $\pi \in M^*(A))$ and φ is a universal code, then the Type II error of the test $T_\varphi^{id}(A, \alpha)$ goes to 0, when t tends to infinity.*

8.4.2. *Testing for Serial Independence*

Let us recall that the null hypothesis H_0^{SI} is that the source is Markovian of order not larger than $m, (m \geq 0)$, and the alternative hypothesis H_1^{SI} is that the sequence is generated by a stationary and ergodic source which differs from the source under H_0^{SI}. In particular, if $m = 0$, this is the problem of testing for independence of time series.

Let there be given a sample $x_1 \cdots x_t$ generated by an (unknown) source π. The main hypothesis H_0^{SI} is that the source π is Markovian whose order is not greater than $m, (m \geq 0)$, and the alternative hypothesis H_1^{SI} is that the sequence is generated by a stationary and ergodic source which differs from the source under H_0^{SI}. The described test is as follows.

Let φ be any code. By definition, the hypothesis H_0^{SI} is accepted if

$$(t - m)h_m^*(x_1 \cdots x_t) - |\varphi(x_1 \cdots x_t)| \leq \log(1/\alpha), \qquad (8.37)$$

where $\alpha \in (0,1)$. Otherwise, H_0^{SI} is rejected. We denote this test by $T_\varphi^{SI}(A, \alpha)$.

Theorem 8.7. *i) For any code φ the Type I error of the test $T_\varphi^{SI}(A, \alpha)$ is less than or equal to $\alpha, \alpha \in (0,1)$ and, ii) if, in addition, φ is a universal code, then the Type II error of the test $T_\varphi^{SI}(A, \alpha)$ goes to 0, when t tends to infinity.*

8.5. Real-Valued Time Series

8.5.1. *Density Estimation and Its Application*

Here we address the problem of non-parametric estimation of the density for time series. Let X_t be a time series and the probability distribution of X_t is unknown, but it is known that the time series is stationary and ergodic. We have seen that Shannon–MacMillan–Breiman theorem played a key role in the case of finite-alphabet processes. In this part we will use its generalization to the processes with densities, which was established by Barron [3]. First we describe considered processes with some properties needed for the generalized Shannon–MacMillan–Breiman theorem to hold. In what follows, we restrict our attention to processes that take bounded real valued. However, the main results may be extended to processes taking values in a compact subset of a separable metric space.

Let B denote the Borel subsets of R, and B^k denote the Borel subsets of R^k, where R is the set of real numbers. Let R^∞ be the set of all infinite sequences $x = x_1, x_2, \ldots$ with $x_i \in$ R, and let B^∞ denote the usual product sigma field on R^∞, generated by the finite dimensional cylinder sets $\{A_1, \ldots A_k, R, R, \ldots\}$, where $A_i \in B, i = 1, \ldots, k$. Each stochastic process $X_1, X_2, \ldots, X_i \in$ R, is defined by a probability distribution on (R^∞, B^∞). Suppose that the joint distribution P_n for $(X_1, X_2, \ldots, X_n)$ has a probability density function $p(x_1 x_2 \cdots x_n)$ with respect to a sigma-finite measure M_n. Assume that the sequence of dominating measures M_n is Markov of order $m \geq 0$ with a stationary transition measure. A familiar case for M_n is Lebesgue measure. Let $p(x_{n+1}|x_1 \cdots x_n)$ denote the conditional density given by the ratio $p(x_1 \cdots x_{n+1})/p(x_1 \cdots x_n)$ for $n > 1$. It is known that for stationary and ergodic processes there exists a so-called relative entropy rate $\tilde{h}$ defined by

$$\tilde{h} = \lim_{n \to \infty} -E(\log p(x_{n+1}|x_1 \cdots x_n)), \tag{8.38}$$

where E denotes expectation with respect to P. We will use the following generalization of the Shannon–MacMillan–Breiman theorem:

Claim 8.6 (Barron [3]). *If $\{X_n\}$ is a P-stationary ergodic process with density $p(x_1 \cdots x_n) = \mathrm{d}P_n/\mathrm{d}M_n$ and $\tilde{h}_n < \infty$ for some $n \geq m$, the sequence of relative entropy densities $-(1/n)\log p(x_1 \cdots x_n)$ convergence almost surely to the relative entropy rate, i.e.*

$$\lim_{n \to \infty} (-1/n) \log p(x_1 \cdots x_n) = \tilde{h} \tag{8.39}$$

with probability 1 (according to P).

Now we return to the estimation problems. Let $\{\Pi_n\}, n \geq 1$, be an increasing sequence of finite partitions of R that asymptotically generates the Borel sigma-field B and let $x^{[k]}$ denote the element of Π_k that contains the point x. (Informally, $x^{[k]}$ is obtained by quantizing x to k bits of precision.) For integers s and n we define the following approximation of the density

$$p^s(x_1 \cdots x_n) = P\left(x_1^{[s]} \cdots x_n^{[s]}\right) / M_n\left(x_1^{[s]} \cdots x_n^{[s]}\right). \tag{8.40}$$

We also consider

$$\tilde{h}_s = \lim_{n \to \infty} -E(\log p^s(x_{n+1}|x_1 \cdots x_n)). \tag{8.41}$$

Applying the Claim 8.2 to the density $p^s(x_1 \cdots x_t)$, we obtain that a.s.

$$\lim_{t \to \infty} -\frac{1}{t} \log p^s(x_1 \cdots x_t) = \tilde{h}_s. \tag{8.42}$$

Let U be a universal code, which is defined for any finite alphabet. In order to describe a density estimate we will use the probability distribution $\omega_i, i = 1, 2, \ldots$, see (8.24). (In what follows we will use this distribution, but results described below are obviously true for any distribution with non-zero probabilities.) Now we can define the density estimate r_U as follows:

$$r_U(x_1 \cdots x_t) = \sum_{i=0}^{\infty} \omega_i \mu_U \left(x_1^{[i]} \cdots x_t^{[i]} \right) / M_t \left(x_1^{[i]} \cdots x_t^{[i]} \right), \qquad (8.43)$$

where the measure μ_U is defined by (8.31). (It is assumed here that the code $U(x_1^{[i]} \cdots x_t^{[i]})$ is defined for the alphabet, which contains $|\Pi_i|$ letters.)

It turns out that, in a certain sense, the density $r_U(x_1 \cdots x_t)$ estimates the unknown density $p(x_1 \cdots x_t)$.

Theorem 8.8. *Let X_t be a stationary ergodic process with densities $p(x_1 \cdots x_t) = \mathrm{d}P_t/\mathrm{d}M_t$ such that*

$$\lim_{s \to \infty} \tilde{h}_s = \tilde{h} < \infty, \qquad (8.44)$$

where $\tilde{h}$ and $\tilde{h}_s$ are relative entropy rates, see (8.38) and (8.41). Then

$$\lim_{t \to \infty} \frac{1}{t} \log \frac{p(x_1 \cdots x_t)}{r_U(x_1 \cdots x_t)} = 0 \qquad (8.45)$$

with probability 1 and

$$\lim_{t \to \infty} \frac{1}{t} E \left(\log \frac{p(x_1 \ldots x_t)}{r_U(x_1 \cdots x_t)} \right) = 0. \qquad (8.46)$$

We have seen that the requirement (8.44) plays an important role in the proof. The natural question is whether there exist processes for which (8.44) is valid. The answer is positive. For example, let a process possess values in the interval $[-1, 1]$, M_n be Lebesgue measure and the considered process is Markovian with conditional density

$$p(x|y) = \begin{cases} 1/2 + \alpha \, \mathrm{sign}(y), & \text{if } x < 0 \\ 1/2 - \alpha \, \mathrm{sign}(y), & \text{if } x \geq 0, \end{cases}$$

where $\alpha \in (0, 1/2)$ is a parameter and

$$\mathrm{sign}(y) = \begin{cases} -1, & \text{if } y < 0, \\ 1, & \text{if } y \geq 0. \end{cases}$$

In words, the density depends on a sign of the previous value. If the value is positive, then the density is more than $1/2$, otherwise it is less than $1/2$. It is easy to see that (8.44) is true for any $\alpha \in (0, 1)$.

The following two theorems are devoted to the conditional probability $r_U(x|x_1 \cdots x_m) = r_U(x_1 \cdots x_m x)/r_U(x_1 \cdots x_m)$ which, in turn, is connected with the prediction problem. We will see that the conditional density $r_U(x|x_1 \cdots x_m)$ is a reasonable estimation of the unknown density $p(x|x_1 \cdots x_m)$.

Theorem 8.9. *Let $B_1, B_2, \ldots$ be a sequence of measurable sets. Then the following equalities are true:*

$$
i) \quad \lim_{t \to \infty} E\left(\frac{1}{t} \sum_{m=0}^{t-1} (P(x_{m+1} \in B_{m+1}|x_1 \cdots x_m) \right.
$$

$$
\left. - R_U(x_{m+1} \in B_{m+1}|x_1 \cdots x_m))^2 \right) = 0, \qquad (8.47)
$$

$$
ii) \quad E\left(\frac{1}{t} \sum_{m=0}^{t-1} |P(x_{m+1} \in B_{m+1}|x_1 \cdots x_m) \right.
$$

$$
\left. - R_U(x_{m+1} \in B_{m+1}|x_1 \cdots x_m)) \right| = 0,
$$

where $R_U(x_{m+1} \in B_{m+1}|x_1 \cdots x_m) = \int_{B_{m+1}} r_U(x|x_1 \cdots x_m) \mathrm{d}M_{1/m}$.

We have seen that in a certain sense the estimation r_U approximates the unknown density p. The following theorem shows that r_U can be used instead of p for estimation of average values of certain functions.

Theorem 8.10. *Let f be an integrable function, whose absolute value is bounded by a certain constant $\bar{M}$ and all conditions of Theorem 8.2 are true. Then the following equality is valid:*

$$
i) \quad \lim_{t \to \infty} \frac{1}{t} E\left(\sum_{m=0}^{t-1} \left(\int f(x)p(x|x_1 \cdots x_m)\mathrm{d}M_m \right. \right.
$$

$$
\left. \left. - \int f(x)r_U(x|x_1 \cdots x_m)\mathrm{d}M_m \right)^2 \right) = 0, \qquad (8.48)
$$

$$
ii) \quad \lim_{t \to \infty} \frac{1}{t} E\left(\sum_{m=0}^{t-1} \left| \int f(x)p(x|x_1 \cdots x_m)\mathrm{d}M_m \right. \right.
$$

$$
\left. \left. - \int f(x)r_U(x|x_1 \cdots x_m)\mathrm{d}M_m \right| \right) = 0.
$$

It is worth noting that this approach was used for prediction of real processes [41].

8.5.2. *Hypothesis Testing*

In this subsection, we consider a case where the source alphabet A is infinite, say, a part of R^n. Our strategy is to use finite partitions of A and to consider hypotheses corresponding to the partitions. This approach can be directly applied to the goodness-of-fit testing, but it cannot be applied to the serial independence testing. The point is that if someone combines letters (or states) of a Markov's chain, the chain order (or memory) can increase. For example, if the alphabet contains three letters, there exists a Markov's chain of order one, such that combining two letters into one transforms the chain into a process with infinite memory. That is why in this part we will consider the independence testing for i.i.d. processes only (i.e. processes from $M_0(A)$).

In order to avoid repetitions, we will consider a general scheme, which can be applied to both tests using notations $H_0^\aleph$, $H_1^\aleph$, and $T_\varphi^\aleph(A, \alpha)$, where $\aleph$ is an abbreviation of one of the described tests (i.e. *id* and *SI*.)

Let us give some definitions. Let $\Lambda = \lambda_1, \ldots, \lambda_s$ be a finite (measurable) partition of A and let $\Lambda(x)$ be an element of the partition Λ which contains $x \in A$. For any process π, we define a process π_Λ over a new alphabet Λ by the equation

$$\pi_\Lambda(\lambda_{i_1} \cdots \lambda_{i_k}) = \pi(x_1 \in \lambda_{i_1}, \ldots, x_k \in \lambda_{i_k}),$$

where $x_1 \cdots x_k \in A^k$.

We will consider an infinite sequence of partitions $\hat{\Lambda} = \Lambda_1, \Lambda_2, \ldots$ and say that such a sequence discriminates between a pair of hypotheses $H_0^\aleph(A), H_1^\aleph(A)$ about processes, if for each process ϱ, for which $H_1^\aleph(A)$ is true, there exists a partition Λ_j for which $H_1^\aleph(\Lambda_j)$ is true for the process ϱ_{Λ_j}.

Let $H_0^\aleph(A), H_1(A)^\aleph$ be a pair of hypotheses, $\hat{\Lambda} = \Lambda_1, \Lambda_2, \ldots$ be a sequence of partitions, α be from $(0, 1)$ and φ be a code. The scheme for both tests is as follows:

The hypothesis $H_0^\aleph(A)$ is accepted if for all $i = 1, 2, 3, \ldots$ the test $T_\varphi^\aleph(\Lambda_i, (\alpha\omega_i))$ accepts the hypothesis $H_0^\aleph(\Lambda_i)$. Otherwise, $H_0^\aleph$ is rejected. We denote this test $\mathbf{T}_{\alpha,\varphi}^\aleph(\hat{\Lambda})$.

Comment. It is important to note that one does not need to check an infinite number of inequalities when applying this test. The point is that the hypothesis $H_0^\aleph(A)$ has to be accepted if the left part in (8.36) or (8.37) is less than $- \log(\alpha \omega_i)$. Obviously, $- \log(\alpha \omega_i)$ goes to infinity if i increases. That is why there are many cases, where it is enough to check a finite number of hypotheses $H_0^\aleph(\Lambda_i)$.

Theorem 8.11. *i) For each* $\alpha \in (0,1)$, *sequence of partitions* $\hat{\Lambda}$ *and a code* φ, *the Type I error of the described test* $\mathbf{T}_{\alpha,\varphi}^\aleph(\hat{\Lambda})$ *is not larger than* α, *and ii) if, in addition,* φ *is a universal code and* $\hat{\Lambda}$ *discriminates between* $H_0^\aleph(A), H_1(A)^\aleph$, *then the Type II error of the test* $\mathbf{T}_{\alpha,\varphi}^\aleph(\hat{\Lambda})$ *goes to 0, when the sample size tends to infinity.*

8.6. Conclusion

Time series is a popular model of real stochastic processes which has a lot of applications in industry, economy, meteorology, and many other fields. Despite this, there are many practically important problems of statistical analysis of time series which are still open. Among them we can name the problem of estimation of the limiting probabilities and densities, on-line prediction, regression, classification, and some problems of hypothesis testing (goodness-of-fit testing and testing of serial independence). This chapter describes a new approach to all the problems mentioned above, which, on the one hand, gives a possibility to solve the problems in the framework of the classical mathematical statistics and, on the other hand, allows to apply methods of real data compression to solve these problems in practise. Such applications to randomness testing [42] and prediction of currency exchange rates [41] showed high efficiency, that is why the suggested methods look very promising for practical applications. Of course, problems like prediction of price of oil, gold, etc. and testing of different random number generators can be used as case studies for students.

8.7. Problems for Chapter

Problem 8.1. Suppose 010101 is a sequence generated by a source whose alphabet is $\{0,1\}$. Calculate the probabilities $L_0(01010)$ and $K_0(01010)$. Predict the next symbol by the predictors L_0 and K_0 (i.e. calculate the conditional probabilities $L_0(0|01010), L_0(1|01010)$ and $K_0(0|01010), K_0(1|01010)$).

Problem 8.2. Suppose the sequence 010101 is generated by the first-order Markov chain and the alphabet is $\{0, 1\}$ (i.e. the source belongs to $M_1(\{0, 1\})$).

i) Represent this sequence by two ones generated by i.i.d. sources.
ii) Repeat all calculations from Problem 8.1 for L_1 and K_1. Compare results of Problems 8.1 and 8.2. Explain the difference.

Problem 8.3. For the sequence 001100110011 calculate the following empirical Shannon entropies: h_0^*, h_1^*, and h_2^*.

Problem 8.4. Repeat all calculations from Problem 8.1 for the measure R. Compare obtained results with solutions of Problems 8.1 and 8.2.

Problem 8.5. Let $\varphi(a) = 000, \varphi(b) = 01, \varphi(c) = 001, \varphi(d) = 1$ be a code over the alphabet $\{a, b, c, d\}$. Calculate the corresponding measure μ_φ.

Problem 8.6 (problems with side information). Let alphabets X and Y be as follows: $A = \{0, 1\}$, $Y = \{a, b, c\}$, respectively. There is a sequence $(x_1, y_1), \ldots, (x_4, y_4) = (0, a), (1, c), (1, b), (0, a)$ and it is known that $y_5 = a$. Predict x_5 based on the measure R, i.e. estimate the following conditional probabilities:

$$R(x_5 = 0 | (x_1, y_1), \ldots, (x_4, y_4) = (0, a), (1, c), (1, b), (0, a), y_5 = a),$$

$$R(x_5 = 1 | (x_1, y_1), \ldots, (x_4, y_4) = (0, a), (1, c), (1, b), (0, a), y_5 = a).$$

Problem 8.7 (several independent samples). Let there be two independent samples $\bar{x}^1 = x_1^1 \cdots x_5^1 = 10101$ and $\bar{x}^1 = x_1^2 \cdots x_6^2 = 010101$, generated by a stationary and ergodic source with the alphabet 0,1. Based on the measure R estimate the (limiting) probability $P(x_1 x_2 x_3 = 010 | \bar{x}^1 \diamond \bar{x}^2 = 10101 \diamond 010101)$ and predict x_7^2 (i.e. calculate conditional probability $R(x_7^2 = 0 | \bar{x}^1 \diamond \bar{x}^2 = 10101 \diamond 010101)$ and $R(x_7^2 = 1 | \bar{x}^1 \diamond \bar{x}^2 = 10101 \diamond 010101)$.

Problem 8.8 (several independent samples). For the sequences 010101 and 010 calculate the following empirical Shannon entropies: $h_0^*(010101 \diamond 010)$, $h_1^*(010101 \diamond 010)$, and $h_2^*(010101 \diamond 010)$.

Problem 8.9 (hypothesis testing). Let H_0^{id} be a hypothesis that a source π is i.i.d. and generates letters from the alphabet $\{0, 1\}$ with equal probabilities, i.e. $\pi(0) = \pi(1) = 0.5$. The hypothesis H_1^{id} is that the sequence is generated by a stationary and ergodic source which differs from the source under H_0^{id} and the level of significance (α) is 0.01. There is a sample sequence 0101010101. The problem is to test H_0^{id} based on the two following codes:

i) the Laplace code $L_{0\text{code}}$ whose codeword length is given by $L_{0\text{code}}(u) = -\log L_0(u)$ and

ii) the R_{code} whose codeword length is given by $R_{\text{code}}(u) = -\log R(u)$

Problem 8.10 (hypothesis testing). Let H_0^{SI} be a hypothesis that is that the source is Markovian of order not larger than 1, and the alternative hypothesis H_1^{SI} is that the source is stationary and ergodic which differs from the source under H_0^{SI}. There is a sequence 001001001001 and let the level of significance be 0.01. The problem is to test H_0^{SI} against H_1^{SI} based on the code R_{code}.

Problem 8.11 (hypothesis testing). Use the same sequence and the same α as in the previous problem for testing H_0^{SI} that the source is Markovian of order not larger than 2, and the alternative hypothesis H_1^{SI} is that the source is stationary and ergodic which differs from the source under H_0^{SI}. Compare results of two last problems and explain the difference.

8.8. Solutions to Problems

Proof (Claim 8.1). We employ the general inequality

$$D(\mu\|\eta) \le \log e \left(-1 + \sum_{a \in A} \mu(a)^2/\eta(a) \right),$$

valid for any distributions μ and η over A (follows from the elementary inequality for natural logarithm $\ln x \le x - 1$), and find:

$$\rho^t(P\|L_0) = \sum_{x_1 \cdots x_t \in A^t} P(x_1 \cdots x_t) \sum_{a \in A} P(a|x_1 \cdots x_t) \log \frac{P(a|x_1 \cdots x_t)}{\gamma(a|x_1 \cdots x_t)},$$

$$= \log e \left(\sum_{x_1 \cdots x_t \in A^t} P(x_1 \cdots x_t) \sum_{a \in A} P(a|x_1 \cdots x_t) \ln \frac{P(a|x_1 \cdots x_t)}{\gamma(a|x_1 \cdots x_t)} \right),$$

$$\le \log e \left(-1 + \sum_{x_1 \cdots x_t \in A^t} P(x_1 \cdots x_t) \sum_{a \in A} \frac{P(a)^2(t + |A|)}{\nu_{x_1 \cdots x_t}(a) + 1} \right).$$

Applying the well-known Bernoulli formula, we obtain

$\rho^t(P\|L_0)$

$$= \log e \left(-1 + \sum_{a \in A} \sum_{i=0}^{t} \frac{P(a)^2(t + |A|)}{i + 1} \binom{t}{i} P(a)^i (1 - P(a))^{t-i} \right),$$

$$= \log e \left(-1 + \frac{t + |A|}{t + 1} \sum_{a \in A} P(a) \sum_{i=0}^{t} \binom{t + 1}{i + 1} P(a)^{i+1} (1 - P(a))^{t-i} \right),$$

$$\leq \log e \left(-1 + \frac{t + |A|}{t + 1} \sum_{a \in A} P(a) \sum_{j=0}^{t+1} \binom{t + 1}{j} P(a)^j (1 - P(a))^{t+1-j} \right).$$

Again, using the Bernoulli formula, we finish the proof

$$\rho^t(P\|L_0) = \log e \frac{|A| - 1}{t + 1}.$$

The second statement of the claim follows from the well-known asymptotic equality

$$1 + 1/2 + 1/3 + \cdots + 1/t = \ln t + O(1),$$

the obvious presentation

$$\bar{\rho}^t(P\|L_0) = t^{-1}(\rho^0(P\|L_0) + \rho^1(P\|L_0) + \cdots + \rho^{t-1}(P\|L_0))$$

and (8.10).   □

Proof (Claim 8.2). The first equality follows from the definition (8.9), whereas the second from the definition (8.12). From (8.16) we obtain:

$$- \log K_0(x_1 \cdots x_t) = - \log \left(\frac{\Gamma(|A|/2)}{\Gamma(1/2)^{|A|}} \frac{\prod_{a \in A} \Gamma(\nu^t(a) + 1/2)}{\Gamma((t + |A|/2)} \right),$$

$$= c_1 + c_2|A| + \log \Gamma(t + |A|/2) - \sum_{a \in A} \Gamma(\nu^t(a) + 1/2),$$

where c_1 and c_2 are constants. Now we use the well-known Stirling formula

$$\ln \Gamma(s) = \ln \sqrt{2\pi} + (s - 1/2) \ln s - s + \theta/12,$$

where $\theta \in (0,1)$ [22]. Using this formula, we rewrite the previous equality as

$$-\log K_0(x_1 \cdots x_t) = -\sum_{a \in A} \nu^t(a) \log(\nu^t(a)/t) + (|A|-1) \log t/2 + \bar{c}_1 + \bar{c}_2|A|,$$

where $\bar{c}_1$ and $\bar{c}_2$ are constants. Hence,

$$\sum_{x_1 \cdots x_t \in A^t} P(x_1 \cdots x_t)(-\log(K_0(x_1 \cdots x_t))) \leq t \Bigg(\sum_{x_1 \cdots x_t \in A^t} P(x_1 \cdots x_t)$$

$$\times \Bigg(-\sum_{a \in A} \nu^t(a) \log(\nu^t(a)/t) \Bigg) \Bigg) + (|A|-1) \log t/2 + c|A|.$$

Applying the well-known Jensen inequality for the concave function $-x \log x$ we obtain the following inequality:

$$\sum_{x_1 \cdots x_t \in A^t} P(x_1 \cdots x_t)(-\log(K_0(x_1 \cdots x_t)))$$

$$\leq -t \Bigg(\sum_{x_1 \cdots x_t \in A^t} P(x_1 \cdots x_t)((\nu^t(a)/t)) \Bigg)$$

$$\log \sum_{x_1 \cdots x_t \in A^t} P(x_1 \cdots x_t)(\nu^t(a)/t) + (|A|-1) \log t/2 + c|A| .$$

The source P is i.i.d., that is why the average frequency $\sum_{x_1 \cdots x_t \in A^t} P(x_1 \cdots x_t)\nu^t(a)$ is equal to $P(a)$ for any $a \in A$ and we obtain from the last two formulas the following inequality:

$$\sum_{x_1 \cdots x_t \in A^t} P(x_1 \cdots x_t)(-\log(K_0(x_1 \cdots x_t)))$$

$$\leq t \Bigg(-\sum_{a \in A} P(a) \log P(a) \Bigg) + (|A|-1) \log t/2 + c|A| \qquad (8.49)$$

On the other hand,

$$\sum_{x_1 \cdots x_t \in A^t} P(x_1 \cdots x_t)(\log P(x_1 \cdots x_t))$$

$$= \sum_{x_1 \cdots x_t \in A^t} P(x_1 \cdots x_t) \sum_{i=1}^{t} \log P(x_i)$$

$$= t \Bigg(\sum_{a \in A} P(a) \log P(a) \Bigg). \qquad (8.50)$$

From (8.49) and (8.50), we can see that

$$t^{-1} \sum_{x_1 \cdots x_t \in A^t} P(x_1 \ldots x_t) \log \frac{P(x_1 \cdots x_t)}{(K_0(x_1 \cdots x_t)} \le ((|A| - 1) \log t / 2 + c) / t.$$

$\square$

Proof (Claim 8.3). First we consider the case where $m = 0$. The proof for this case is very close to the proof of the previous claim. Namely, from (8.16) we obtain:

$$-\log K_0(x_1 \cdots x_t) = -\log \left(\frac{\Gamma(|A|/2)}{\Gamma(1/2)^{|A|}} \frac{\prod_{a \in A} \Gamma(\nu^t(a) + 1/2)}{\Gamma(t + |A|/2)} \right),$$

$$= c_1 + c_2 |A| + \log \Gamma(t + |A|/2) - \sum_{a \in A} \Gamma(\nu^t(a) + 1/2),$$

where c_1 and c_2 are constants. Now we use the well-known Stirling formula

$$\ln \Gamma(s) = \ln \sqrt{2\pi} + (s - 1/2) \ln s - s + \theta/12,$$

where $\theta \in (0, 1)$ [22]. Using this formula we rewrite the previous equality as

$$-\log K_0(x_1 \cdots x_t) = -\sum_{a \in A} \nu^t(a) \log(\nu^t(a)/t) + (|A| - 1) \log t / 2 + \bar{c}_1 + \bar{c}_2 |A|,$$

where $\bar{c}_1$ and $\bar{c}_2$ are constants. Having taken into account the definition of the empirical entropy (8.23), we obtain

$$-\log K_0(x_1 \cdots x_t) \le t h_0^*(x_1 \cdots x_t) + (|A| - 1) \log t / 2 + c|A|.$$

Hence,

$$\sum_{x_1 \cdots x_t \in A^t} P(x_1 \cdots x_t)(-\log(K_0(x_1 \cdots x_t)))$$

$$\le t \left(\sum_{x_1 \cdots x_t \in A^t} P(x_1 \cdots x_t) h_0^*(x_1 \cdots x_t) + (|A| - 1) \log t / 2 + c|A| \right).$$

Having taken into account the definition of the empirical entropy (8.23), we apply the well-known Jensen inequality for the concave function $-x \log x$

and obtain the following inequality:

$$\sum_{x_1 \cdots x_t \in A^t} P(x_1 \cdots x_t)(-\log(K_0(x_1 \cdots x_t))) \le +c|A| -$$

$$t \left(\sum_{x_1 \cdots x_t \in A^t} P(x_1 \ldots x_t)(\nu^t(a)/t) \right) \log \sum_{x_1 \cdots x_t \in A^t} P(x_1 \cdots x_t)$$

$$\times (\nu^t(a)/t) + (|A| - 1)\log t/2.$$

P is stationary and ergodic, that is why the average frequency $\sum_{x_1 \cdots x_t \in A^t} P(x_1 \cdots x_t)\nu^t(a)$ is equal to $P(a)$ for any $a \in A$ and we obtain from the last two formulas the following inequality:

$$\sum_{x_1 \cdots x_t \in A^t} P(x_1 \cdots x_t)(-\log(K_0(x_1 \cdots x_t)))$$

$$\le th_0(P) + (|A| - 1)\log t/2 + c|A|,$$

where $h_0(P)$ is the first-order Shannon entropy, see (8.12).

We have seen that any source from $M_m(A)$ can be presented as a "sum" of $|A|^m$ i.i.d. sources. From this we can easily see that the error of a predictor for the source from $M_m(A)$ can be upper bounded by the error of i.i.d. source multiplied by $|A|^m$. In particular, we obtain from the last inequality and the definition of the Shannon entropy (8.20) the upper bound (8.22). □

Proof (Theorem 8.1). We can see from Definition (8.25) of R and Claim 8.19 that the average error is upper bounded as follows:

$$-t^{-1} \sum_{x_1 \cdots x_t \in A^t} P(x_1 \cdots x_t) \log(R(x_1 \cdots x_t)) - h_k(P)$$

$$\le (|A|^k(|A| - 1)\log t + \log(1/\omega_i) + C)/(2t),$$

for any $k = 0, 1, 2, \ldots$. Taking into account that for any $P \in M_\infty(A)$ $\lim_{k \to \infty} h_k(P) = h_\infty(P)$, we can see that

$$\left(\lim_{t \to \infty} t^{-1} \sum_{x_1 \cdots x_t \in A^t} P(x_1 \cdots x_t) \log(R(x_1 \cdots x_t)) - h_\infty(P) \right) = 0.$$

The second statement of the theorem is proven. The first one can be easily derived from the ergodicity of P [5, 14]. □

Proof (Theorem 8.2). The proof is based on the Shannon–MacMillan– Breiman theorem which states that for any stationary and ergodic source P

$$\lim_{t \to \infty} -\log P(x_1 \cdots x_t)/t = h_\infty(P)$$

with probability 1 [5, 14]. From this equality and (8.29) we obtain the statement i). The second statement follows from the definition of the Shannon entropy (8.21) and (8.30). □

Proof (Theorem 8.4). i) immediately follows from the second statement of Theorem 8.2 and properties of log. The statement ii) can be proven as follows:

$$\lim_{t \to \infty} E\left(\frac{1}{t}\sum_{i=0}^{t-1}(P(x_{i+1}|x_1 \cdots x_i) - \mu_U(x_{i+1}|x_1 \cdots x_i))^2\right)$$

$$= \lim_{t \to \infty} \frac{1}{t}\sum_{i=0}^{t-1}\sum_{x_1 \cdots x_i \in A^i} P(x_1 \cdots x_i)$$

$$\times \left(\sum_{a \in A}|P(a|x_1 \cdots x_i) - \mu_U(a|x_1 \cdots x_i)|\right)^2$$

$$\leq \lim_{t \to \infty} \frac{\text{const}}{t}\sum_{i=0}^{t-1}\sum_{x_1 \cdots x_i \in A^i} P(x_1 \cdots x_i)\sum_{a \in A} P(a|x_1 \cdots x_i)$$

$$\times \log\frac{P(a|x_1 \cdots x_i)}{\mu_U(a|x_1 \ldots x_i)}$$

$$= \lim_{t \to \infty}\left(\frac{\text{const}}{t}\sum_{x_1 \cdots x_t \in A^t} P(x_1 \cdots x_t)\log(P(x_1 \cdots x_t)/\mu(x_1 \cdots x_t))\right).$$

Here the first inequality is obvious, the second follows from the Pinsker's inequality (8.5), the others from properties of expectation and log. iii) can be derived from ii) and the Jensen inequality for the function x^2. □

Proof (Theorem 8.5). The following inequality follows from the non-negativity of the KL divergency (see (8.5)), whereas the equality is obvious.

$$E\left(\log\frac{P(x_1|y_1)}{\mu_U(x_1|y_1)}\right) + E\left(\log\frac{P(x_2|(x_1,y_1),y_2)}{\mu_U(x_2|(x_1,y_1),y_2)}\right) + \cdots$$

$$\leq E\left(\log\frac{P(y_1)}{\mu_U(y_1)}\right) + E\left(\log\frac{P(x_1|y_1)}{\mu_U(x_1|y_1)}\right)$$

$$+ E\left(\log\frac{P(y_2|(x_1,y_1))}{\mu_U(y_2|(x_1,y_1))}\right) + E\left(\log\frac{P(x_2|(x_1,y_1),y_2)}{\mu_U(x_2|(x_1,y_1),y_2)}\right) + \cdots$$

$$= E\left(\log\frac{P(x_1,y_1)}{\mu_U(x_1,y_1)}\right) + E\left(\log\frac{P((x_2,y_2)|(x_1,y_1))}{\mu_U((x_2,y_2)|(x_1,y_1))}\right) + \cdots$$

Now we can apply the first statement of the previous theorem to the last sum as follows:

$$\lim_{t\to\infty}\frac{1}{t}E\left(\log\frac{P(x_1,y_1)}{\mu_U(x_1,y_1)}\right) + E\left(\log\frac{P((x_2,y_2)|(x_1,y_1))}{\mu_U((x_2,y_2)|(x_1,y_1))}\right)$$

$$+ \cdots E\left(\log\frac{P((x_t,y_t)|(x_1,y_1)\ldots(x_{t-1},y_{t-1}))}{\mu_U((x_t,y_t)|(x_1,y_1)\ldots(x_{t-1},y_{t-1}))}\right) = 0.$$

From this equality and the last inequality we obtain the proof of i). The proof of the second statement can be obtained from the similar representation for ii) and the second statement of Theorem 8.4. iii) can be derived from ii) and the Jensen inequality for the function x^2. □

Proof (Lemma 8.1). First we show that for any source $\theta^* \in M_0(A)$ and any words $x^1 = x_1^1\ldots x_{t_1}^1, \ldots, x^r = x_1^r\cdots x_{t_r}^r$,

$$\theta^*(x^1 \diamond \cdots \diamond x^r) = \prod_{a\in A}(\theta^*(a))^{\nu_{x^1\diamond\cdots\diamond x^r}(a)}$$

$$\leq \prod_{a\in A}(\nu_{x^1\diamond\cdots\diamond x^r}(a)/t)^{\nu_{x^1\diamond\cdots\diamond x^r}(a)}, \qquad (8.51)$$

where $t = \sum_{i=1}^r t_i$. Here the equality holds, because $\theta^* \in M_0(A)$. The inequality follows from Claim 1. Indeed, if $p(a) = \nu_{x^1\diamond\cdots\diamond x^r}(a)/t$ and $q(a) = \theta^*(a)$, then

$$\sum_{a\in A}\frac{\nu_{x^1\diamond\cdots\diamond x^r}(a)}{t}\log\frac{(\nu_{x^1\diamond\cdots\diamond x^r}(a)/t)}{\theta^*(a)} \geq 0.$$

From the latter inequality we obtain (8.51). Taking into account Definitions (8.34) and (8.51), we can see that the statement of lemma is true for this particular case.

For any $\theta \in M_m(A)$ and $x = x_1 \cdots x_s$, $s > m$, we present $\theta(x_1 \cdots x_s)$ as $\theta(x_1 \cdots x_s) = \theta(x_1 \cdots x_m) \prod_{u \in A^m} \prod_{a \in A} \theta(a/u)^{\nu_x(ua)}$, where $\theta(x_1 \cdots x_m)$ is the limiting probability of the word $x_1 \cdots x_m$. Hence, $\theta(x_1 \cdots x_s) \leq \prod_{u \in A^m} \prod_{a \in A} \theta(a/u)^{\nu_x(ua)}$. Taking into account the inequality (8.51), we obtain $\prod_{a \in A} \theta(a/u)^{\nu_x(ua)} \leq \prod_{a \in A} (\nu_x(ua)/\bar\nu_x(u))^{\nu_x(ua)}$ for any word u. Hence,

$$\theta(x_1 \cdots x_s) \leq \prod_{u \in A^m} \prod_{a \in A} \theta(a/u)^{\nu_x(ua)}$$

$$\leq \prod_{u \in A^m} \prod_{a \in A} (\nu_x(ua)/\bar\nu_x(u))^{\nu_x(ua)}.$$

If we apply those inequalities to $\theta(x^1 \diamond \cdots \diamond x^r)$, we immediately obtain the following inequalities

$$\theta(x^1 \diamond \cdots \diamond x^r) \leq \prod_{u \in A^m} \prod_{a \in A} \theta(a/u)^{\nu_{x^1 \diamond \cdots \diamond x^r}(ua)}$$

$$\leq \prod_{u \in A^m} \prod_{a \in A} (\nu_{x^1 \diamond \cdots \diamond x^r}(ua)/\bar\nu_{x^1 \diamond \cdots \diamond x^r}(u))^{\nu_{x^1 \diamond \cdots \diamond x^r}(ua)}.$$

Now the statement of the lemma follows from Definition (8.34). □

Proof (Theorem 8.6). Let C_α be a critical set of the test $T_\varphi^{id}(A, \alpha)$, i.e., by definition, $C_\alpha = \{u : u \in A^t \ \& - \log \pi(u) - |\varphi(u)| > -\log \alpha\}$. Let μ_φ be a measure for which Claim 8.2 is true. We define an auxiliary set $\hat{C}_\alpha = \{u : -\log \pi(u) - (-\log \mu_\varphi(u)) > -\log \alpha\}$. We have $1 \geq \sum_{u \in \hat{C}_\alpha} \mu_\varphi(u) \geq \sum_{u \in \hat{C}_\alpha} \pi(u)/\alpha = (1/\alpha)\pi(\hat{C}_\alpha)$. (Here the second inequality follows from the definition of $\hat{C}_\alpha$, whereas all others are obvious.) So, we obtain that $\pi(\hat{C}_\alpha) \leq \alpha$. From definitions of $C_\alpha, \hat{C}_\alpha$, and (8.26) we immediately obtain that $\hat{C}_\alpha \supset C_\alpha$. Thus, $\pi(C_\alpha) \leq \alpha$. By definition, $\pi(C_\alpha)$ is the value of the Type I error. The first statement of the theorem is proven.

Let us prove the second statement of the theorem. Suppose that the hypothesis $H_1^{id}(A)$ is true. That is, the sequence $x_1 \cdots x_t$ is generated by some stationary and ergodic source τ and $\tau \neq \pi$. Our strategy is to show that

$$\lim_{t \to \infty} -\log \pi(x_1 \cdots x_t) - |\varphi(x_1 \cdots x_t)| = \infty \tag{8.52}$$

with probability 1 (according to the measure τ). First we represent (8.52) as

$$-\log \pi(x_1 \ldots x_t) - |\varphi(x_1 \cdots x_t)|$$

$$= t \left(\frac{1}{t} \log \frac{\tau(x_1 \cdots x_t)}{\pi(x_1 \cdots x_t)} + \frac{1}{t}(-\log \tau(x_1 \cdots x_t) - |\varphi(x_1 \cdots x_t)|) \right).$$

From this equality and the property of a universal code (8.29) we obtain

$$-\log \pi(x_1 \cdots x_t) - |\varphi(x_1 \cdots x_t)| = t \left(\frac{1}{t} \log \frac{\tau(x_1 \cdots x_t)}{\pi(x_1 \cdots x_t)} + o(1) \right). \tag{8.53}$$

From (8.29) and (8.21) we can see that

$$\lim_{t \to \infty} -\log \tau(x_1 \ldots x_t)/t \le h_k(\tau) \tag{8.54}$$

for any $k \ge 0$ (with probability 1). It is supposed that the process π has a finite memory, i.e. belongs to $M_s(A)$ for some s. Having taken into account the definition of $M_s(A)$ (8.18), we obtain the following representation:

$$-\log \pi(x_1 \ldots x_t)/t = -t^{-1} \sum_{i=1}^{t} \log \pi(x_i/x_1 \cdots x_{i-1})$$

$$= -t^{-1} \left(\sum_{i=1}^{k} \log \pi(x_i/x_1 \cdots x_{i-1}) \right.$$

$$\left. + \sum_{i=k+1}^{t} \log \pi(x_i/x_{i-k} \cdots x_{i-1}) \right)$$

for any $k \ge s$. According to the ergodic theorem there exists a limit

$$\lim_{t \to \infty} t^{-1} \sum_{i=k+1}^{t} \log \pi(x_i/x_{i-k} \cdots x_{i-1}),$$

which is equal to $h_k(\tau)$ [5, 14]. So, from the two last equalities we can see that

$$\lim_{t \to \infty} (-\log \pi(x_1 \cdots x_t))/t = -\sum_{v \in A^k} \tau(v) \sum_{a \in A} \tau(a/v) \log \pi(a/v).$$

Taking into account this equality, (8.54) and (8.53), we can see that

$$- \log \pi(x_1 \cdots x_t) - |\varphi(x_1 \cdots x_t)|$$

$$\geq t \left(\sum_{v \in A^k} \tau(v) \sum_{a \in A} \tau(a/v) \log(\tau(a/v)/\pi(a/v)) \right) + o(t)$$

for any $k \geq s$. From this inequality and Claim 8.1 we can obtain that $- \log \pi(x_1 \cdots x_t) - |\varphi(x_1 \cdots x_t)| \geq ct + o(t)$, where c is a positive constant, $t \to \infty$. Hence, (8.52) is true and the theorem is proven. $\qquad \square$

Proof (Theorem 8.7). Let us denote the critical set of the test $T_\varphi^{SI}(A, \alpha)$ as C_α, i.e., by definition, $C_\alpha = \{x_1 \cdots x_t : (t-m)h_m^*(x_1 \cdots x_t) - |\varphi(x_1 \cdots x_t)|) > \log(1/\alpha)\}$. From Claim 8.2 we can see that there exists such a measure μ_φ that $- \log \mu_\varphi(x_1 \cdots x_t) \leq |\varphi(x_1 \cdots x_t)|$. We also define

$$\hat{C}_\alpha = \{x_1 \cdots x_t : (t-m)h_m^*(x_1 \cdots x_t)$$
$$- (- \log \mu_\varphi(x_1 \cdots x_t))) > \log(1/\alpha)\}. \qquad (8.55)$$

Obviously, $\hat{C}_\alpha \supset C_\alpha$. Let θ be any source from $M_m(A)$. The following chain of equalities and inequalities is true:

$$1 \geq \mu_\varphi(\hat{C}_\alpha) = \sum_{x_1 \cdots x_t \in \hat{C}_\alpha} \mu_\varphi(x_1 \cdots x_t)$$

$$\geq \alpha^{-1} \sum_{x_1 \cdots x_t \in \hat{C}_\alpha} 2^{(t-m)h_m^*(x_1 \cdots x_t)}$$

$$\geq \alpha^{-1} \sum_{x_1 \cdots x_t \in \hat{C}_\alpha} \theta(x_1 \cdots x_t) = \theta(\hat{C}_\alpha).$$

(Here both equalities and the first inequality are obvious, the second and the third inequalities follow from (8.55) and the Lemma, respectively.) So, we obtain that $\theta(\hat{C}_\alpha) \leq \alpha$ for any source $\theta \in M_m(A)$. Taking into account that $\hat{C}_\alpha \supset C_\alpha$, where C_α is the critical set of the test, we can see that the probability of the Type I error is not greater than α. The first statement of the theorem is proven.

The proof of the second statement will be based on some results of Information theory. We obtain from (8.29) that for any stationary and ergodic p

$$\lim_{t \to \infty} t^{-1} |\varphi(x_1 \cdots x_t)| = h_\infty(p) \qquad (8.56)$$

with probability 1. It can be seen from (8.23) that h_m^* is an estimate for the m-order Shannon entropy (8.20). Applying the ergodic theorem we obtain $\lim_{t \to \infty} h_m^*(x_1 \cdots x_t) = h_m(p)$ with probability 1 [5, 14]. It is known in Information theory that $h_m(\varrho) - h_\infty(\varrho) > 0$, if ϱ belongs to $M_\infty(A) \backslash M_m(A)$ [5, 14]. It is supposed that H_1^{SI} is true, i.e. the considered process belongs to $M_\infty(A) \backslash M_m(A)$. So, from (8.56) and the last equality we obtain that $\lim_{t \to \infty}((t - m) h_m^*(x_1 \cdots x_t) - |\varphi(x_1 \cdots x_t)|) = \infty$. This proves the second statement of the theorem. □

Proof (Theorem 8.8). First we prove that with probability 1 there exists the following limit $\lim_{t \to \infty} \frac{1}{t} \log(p(x_1 \cdots x_t)/r_U(x_1 \cdots x_t))$ and this limit is finite and non-negative. Let $A_n = \{x_1, \ldots, x_n : p(x_1, \ldots, x_n) \neq 0\}$. Define

$$z_n(x_1 \cdots x_n) = r_U(x_1 \cdots x_n)/p(x_1 \cdots x_n) \tag{8.57}$$

for $(x_1, \ldots, x_n) \in A$ and $z_n = 0$ elsewhere.
Since

$$E_P(z_n|x_1, \ldots, x_{n-1})$$

$$= E\left(\frac{r_U(x_1 \cdots x_n)}{p(x_1 \cdots x_n)} \middle| x_1, \ldots, x_{n-1}\right)$$

$$= \frac{r_U(x_1 \cdots x_{n-1})}{p(x_1 \cdots x_{n-1})} E_P\left(\frac{r_U(x_n|x_1 \cdots x_{n-1})}{p(x_n|x_1 \cdots x_{n-1})}\right)$$

$$= z_{n-1} \int_A^- \frac{r_U(x_n|x_1 \cdots x_{n-1}) dP(x_n|x_1 \cdots x_{n-1})}{dP(x_n|x_1 \cdots x_{n-1})/dM_n(x_n|x_1 \cdots x_{n-1})}$$

$$= z_{n-1} \int_A^- r_U(x_n|x_1 \cdots x_{n-1}) dM_n(x_n|x_1 \cdots x_{n-1}) \leq z_{n-1}$$

the stochastic sequence (z_n, B^n) is, by definition, a non-negative supermartingale with respect to P, with $E(z_n) \leq 1$, [49]. Hence, Doob's submartingale convergence theorem implies that the limit z_n exists and is finite with P-probability 1 (see [49] Theorem 7.4.1Shir). Since all terms are nonnegative so is the limit. Using Definition (8.57) with P-probability 1 we have

$$\lim_{n \to \infty} p(x_1 \ldots x_n)/r_U(x_1 \ldots x_n) > 0,$$

$$\lim_{n \to \infty} \log(p(x_1 \ldots x_n)/r_U(x_1 \ldots x_n)) > -\infty,$$

and

$$\lim_{n\to\infty} n^{-1} \log(p(x_1 \cdots x_n)/r_U(x_1 \cdots x_n)) \geq 0. \tag{8.58}$$

Now we note that for any integer s the following obvious equality is true: $r_U(x_1 \cdots x_t) = \omega_s \mu_U(x_1^{[s]} \cdots x_t^{[s]})/M_t(x_1^{[s]} \cdots x_t^{[s]})(1 + \delta)$ for some $\delta > 0$. From this equality, (8.31), and (8.43) we immediately obtain that a.s.

$$\lim_{t\to\infty} \frac{1}{t} \log \frac{p(x_1 \cdots x_t)}{r_U(x_1 \cdots x_t)} \leq \lim_{t\to\infty} \frac{-\log \omega_t}{t}$$

$$+ \lim_{t\to\infty} \frac{1}{t} \log \frac{p(x_1 \cdots x_t)}{\mu_U\left(x_1^{[s]} \cdots x_t^{[s]}\right)/M_t\left(x_1^{[s]} \cdots x_t^{[s]}\right)}$$

$$\leq \lim_{t\to\infty} \frac{1}{t} \log \frac{p(x_1 \cdots x_t)}{2^{-|U(x_1^{[s]} \cdots x_t^{[s]})|}/M_t\left(x_1^{[s]} \cdots x_t^{[s]}\right)}. \tag{8.59}$$

The right part can be presented as follows:

$$\lim_{t\to\infty} \frac{1}{t} \log \frac{p(x_1 \cdots x_t)}{2^{-|U(x_1^{[s]} \cdots x_t^{[s]})|}/M_t\left(x_1^{[s]} \cdots x_t^{[s]}\right)}$$

$$= \lim_{t\to\infty} \frac{1}{t} \log \frac{p^s(x_1 \cdots x_t) M_t(x_1^{[s]} \cdots x_t^{[s]})}{2^{-|U(x_1^{[s]} \cdots x_t^{[s]})|}}$$

$$+ \lim_{t\to\infty} \frac{1}{t} \log \frac{p(x_1 \cdots x_t)}{p^s(x_1 \cdots x_t)}. \tag{8.60}$$

Having taken into account that U is a universal code, (8.40) and Theorem 8.2, we can see that the first term is equal to zero. From (8.39) and (8.42) we can see that a.s. the second term is equal to $\tilde{h}_s - \tilde{h}$. This equality is valid for any integer s and, according to (8.44), the second term equals zero, too, and we obtain that

$$\lim_{t\to\infty} \frac{1}{t} \log \frac{p(x_1 \cdots x_t)}{r_U(x_1 \cdots x_t)} \leq 0.$$

Having taken into account (8.58), we can see that the first statement is proven.

From (8.59) and (8.60) we can can see that

$$E \log \frac{p(x_1 \cdots x_t)}{r_U(x_1 \cdots x_t)} \le E \log \frac{p_t^s(x_1, \ldots, x_t) M_t \left(x_1^{[s]} \cdots x_t^{[s]} \right)}{2^{-|U(x_1^{[s]} \cdots x_t^{[s]})|}}$$

$$+ E \log \frac{p(x_1 \cdots x_t)}{p^s(x_1, \ldots, x_t)}. \tag{8.61}$$

The first term is the average redundancy of the universal code for a finite-alphabet source, hence, according to Theorem 8.2, it tends to 0. The second term tends to $\tilde{h}_s - \tilde{h}$ for any s and from (8.44) we can see that it is equal to zero. The second statement is proven. $\qquad\square$

Proof (Theorem 8.9). Obviously,

$$E \left(\frac{1}{t} \sum_{m=0}^{t-1} (P(x_{m+1} \in B_{m+1}|x_1 \cdots x_m) - R_U(x_{m+1} \in B_{m+1}|x_1 \cdots x_m))^2 \right)$$

$$\le \frac{1}{t} \sum_{m=0}^{t-1} E(|P(x_{m+1} \in B_{m+1}|x_1 \cdots x_m) - R_U(x_{m+1} \in B_{m+1}|x_1 \cdots x_m)|$$

$$+ |P(x_{m+1} \in \bar{B}_{m+1}|x_1 \cdots x_m) - R_U(x_{m+1} \in \bar{B}_{m+1}|x_1 \cdots x_m)|)^2. \tag{8.62}$$

From the Pinsker inequality (8.5) and convexity of the KL divergence (8.6) we obtain the following inequalities:

$$\frac{1}{t} \sum_{m=0}^{t-1} E(|P(x_{m+1} \in B_{m+1}|x_1 \cdots x_m) - R_U(x_{m+1} \in B_{m+1}|x_1 \cdots x_m)|$$

$$+ |P(x_{m+1} \in \bar{B}_{m+1}|x_1 \cdots x_m) - R_U(x_{m+1} \in \bar{B}_{m+1}|x_1 \cdots x_m)|)^2$$

$$\le \frac{\text{const}}{t} \sum_{m=0}^{t-1} E \left(\log \frac{P(x_{m+1} \in B_{m+1}|x_1 \cdots x_m)}{R_U(x_{m+1} \in B_{m+1}|x_1 \cdots x_m)} \right.$$

$$\left. + \log \frac{P(x_{m+1} \in \bar{B}_{m+1}|x_1 \cdots x_m)}{R_U(x_{m+1} \in \bar{B}_{m+1}|x_1 \cdots x_m)} \right)$$

$$\le \frac{\text{const}}{t} \sum_{m=0}^{t-1} \left(\int p(x_1 \cdots x_m) \left(\int p(x_{m+1}|x_1 \cdots x_m) \right. \right.$$

$$\left. \left. \times \log \frac{p(x_{m+1}|x_1 \cdots x_m)}{r_U(x_{m+1}|x_1 \cdots x_m)} dM dM_m \right) \right). \tag{8.63}$$

Having taken into account that the last term is equal to $\frac{\text{const}}{t} E(\log \frac{p(x_1 \cdots x_t)}{r_U(x_1 \cdots x_t)})$, from (8.62), (8.63), and (8.46) we obtain (8.47). ii) can be derived from i) and the Jensen inequality for the function x^2. □

Proof (Theorem 8.10). The last inequality of the following chain follows from Pinsker's one, whereas all others are obvious.

$$\left(\int f(x)\, p(x|x_1 \cdots x_m)\, \mathrm{d}M_m - \int f(x)\, r_U(x|x_1 \cdots x_m)\, \mathrm{d}M_m \right)^2$$

$$= \left(\int f(x)\, (p(x|x_1 \cdots x_m) - r_U(x|x_1 \cdots x_m))\, \mathrm{d}M_m \right)^2$$

$$\leq \bar{M}^2 \left(\int (p(x|x_1 \cdots x_m) - r_U(x|x_1 \cdots x_m))\, \mathrm{d}M_m \right)^2$$

$$\leq \bar{M}^2 \left(\int |p(x|x_1 \cdots x_m) - r_U(x|x_1 \cdots x_m)|\mathrm{d}M_m \right)^2$$

$$\leq \text{const} \int p(x|x_1 \cdots x_m) \log \frac{p(x|x_1 \cdots x_m)}{r_U(x|x_1 \cdots x_m)}\mathrm{d}M_m.$$

From these inequalities we obtain:

$$E\left(\sum_{m=0}^{t-1} \left(\int f(x)\, p(x|x_1 \cdots x_m)\, \mathrm{d}M_m - \int f(x)\, r_U(x|x_1 \cdots x_m)\, \mathrm{d}M_m \right)^2 \right)$$

$$\leq \sum_{m=0}^{t-1} \text{const}\, E\left(\int p(x|x_1 \cdots x_m) \log \frac{p(x|x_1 \cdots x_m)}{r_U(x|x_1 \cdots x_m)}\mathrm{d}M_{1/m} \right).$$

$$(8.64)$$

The last term can be presented as follows:

$$\sum_{m=0}^{t-1} E\left(\int p(x|x_1 \cdots x_m) \log \frac{p(x|x_1 \cdots x_m)}{r_U(x|x_1 \cdots x_m)}\mathrm{d}M_{1/m} \right)$$

$$= \sum_{m=0}^{t-1} \int p(x_1 \cdots x_m)$$

$$\int p(x|x_1 \cdots x_m) \log \frac{p(x|x_1 \cdots x_m)}{r_U(x|x_1 \cdots x_m)}\mathrm{d}M_{1/m}\mathrm{d}M_m$$

$$= \int p(x_1 \cdots x_t) \log(p(x_1 \cdots x_t)/r_U(x_1 \cdots x_t))\mathrm{d}M_t.$$

From this equality, (8.64) and Corollary 1 we obtain (8.48). ii) can be derived from (8.64) and the Iensen inequality for the function x^2. □

Proof (Theorem 8.11). The following chain proves the first statement of the theorem:

$$P\{H_0^\aleph(A) \text{ is rejected } /H_0 \text{ is true}\}$$

$$= P\left\{\bigcup_{i=1}^{\infty}\{H_0^\aleph(\Lambda_i) \text{ is rejected } /H_0 \text{ is true}\}\right\}$$

$$\leq \sum_{i=1}^{\infty} P\{H_0^\aleph(\Lambda_i)/H_0 \text{ is true}\} \leq \sum_{i=1}^{\infty}(\alpha\omega_i) = \alpha.$$

(Here both inequalities follow from the description of the test, whereas the last equality follows from (8.24).)

The second statement also follows from the description of the test. Indeed, let a sample is created by a source ϱ, for which $H_1(A)^\aleph$ is true. It is supposed that the sequence of partitions $\hat{\Lambda}$ discriminates between $H_0^\aleph(A)$ and $H_1^\aleph(A)$. By definition, it means that there exists j for which $H_1^\aleph(\Lambda_j)$ is true for the process ϱ_{Λ_j}. It immediately follows from Theorem 8.1–8.4 that the Type II error of the test $T_\varphi^\aleph(\Lambda_j, \alpha\omega_j)$ goes to 0, when the sample size tends to infinity. □

8.9. Keywords

Empirical Shannon entropy of order k.

$$h_k^*(x) = -\sum_{v \in A^k} \frac{\bar{\nu}_x(v)}{(t-k)} \sum_{a \in A} \frac{\nu_x(va)}{\bar{\nu}_x(v)} \log \frac{\nu_x(va)}{\bar{\nu}_x(v)}$$

where $x = x_1 \cdots x_t$, $\bar{\nu}_x(v) = \sum_{a \in A} \nu_x(va)$. In particular, if $k = 0$, we obtain $h_0^*(x) = -t^{-1} \sum_{a \in A} \nu_x(a) \log(\nu_x(a)/t)$.

Goodness-of-fit testing

The hypothesis H_0^{id} is that the source has a particular distribution π and the alternative hypothesis H_1^{id} that the sequence is generated by a stationary and ergodic source which differs from the source under H_0^{id}.

i.i.d. sources

Generates independent and identically distributed random variables.

Krichevsky predictor for i.i.d. sources

$L_0(a|x_1 \cdots x_t) = (\nu_{x_1 \cdots x_t}(a) + 1/2)/(t + |A|/2)$, where $\nu_{x_1 \cdots x_t}(a)$ denotes the count of letter a occurring in the word $x_1 \ldots x_{t-1} x_t$.

Kullback–Leibler (KL) divergence

$$D(P, Q) = \sum_{a \in A} P(a) \log \frac{P(a)}{Q(a)},$$

where $P(a)$ and $Q(a)$ are probability distributions over an alphabet A (here and below $\log \equiv \log_2$ and $0 \log 0 = 0$).

Laplace predictor for i.i.d. sources

$L_0(a|x_1 \cdots x_t) = (\nu_{x_1 \cdots x_t}(a) + 1)/(t + |A|)$, where $\nu_{x_1 \cdots x_t}(a)$ denotes the count of letter a occurring in the word $x_1 \ldots x_{t-1} x_t$.

Markov sources of order (or with memory) m, m $\geq$ 0.

$$\mu(x_{t+1} = a_{i_1} | x_t = a_{i_2}, x_{t-1} = a_{i_3}, \ldots, x_{t-m+1} = a_{i_{m+1}}, \ldots)$$

$$= \mu(x_{t+1} = a_{i_1} | x_t = a_{i_2}, x_{t-1} = a_{i_3}, \ldots, x_{t-m+1} = a_{i_{m+1}})$$

for all $t \geq m$ and $a_{i_1}, a_{i_2}, \ldots \in A$.

Measure R

$$R(x_1 \cdots x_t) = \sum_{i=0}^{\infty} \omega_{i+1} K_i(x_1 \cdots x_t),$$

where K_i is the Krichevsky predictors for the set of i-memory Markov sources, $\{\omega = \omega_1, \omega_2, \ldots\}$ is the following probability distribution on integers: $\omega_1 = 1 - 1/\log 3, \ldots, \omega_i = 1/\log(i+1) - 1/\log(i+2), \ldots$.

Pinsker inequality

$$\sum_{a \in A} P(a) \log \frac{P(a)}{Q(a)} \geq \frac{\log e}{2} ||P - Q||^2,$$

where $||P - Q|| = \sum_{a \in A} |P(a) - Q(a)|$.

Serial independence test

The null hypothesis H_0^{SI} is that the source is Markovian of order not larger than $m, (m \geq 0)$, and the alternative hypothesis H_1^{SI} that the sequence is generated by a stationary and ergodic source which differs from the source under H_0^{SI}. In particular, if $m = 0$, this is the problem of testing for independence of time series.

Shannon entropy

Shannon entropy for the i.i.d. source P is

$$h_0(P) = -\sum_{a \in A} P(a) \log P(a).$$

The m-order (conditional) Shannon entropy and the limiting Shannon entropy are defined as follows:

$$h_m(P) = \sum_{v \in A^m} P(v) \sum_{a \in A} P(a/v) \log P(a/v), \quad h_\infty(\tau) = \lim_{m \to \infty} h_m(P).$$

Stationary ergodic processes

The time shift T on A^∞ is defined as $T(x_1, x_2, x_3, \dots) = (x_2, x_3, \dots)$. A process P is called stationary if it is T-invariant: $P(T^{-1}B) = P(B)$ for every Borel set $B \subset A^\infty$. A stationary process is called ergodic if every T-invariant set has probability 0 or 1: $P(B) = 0$ or 1 whenever $T^{-1}B = B$.

Universal code

U is a universal code if for any stationary and ergodic source P the following equalities are valid:

$$\lim_{t \to \infty} |U(x_1 \cdots x_t)|/t = h_\infty(P),$$

with probability 1, and

$$\lim_{t \to \infty} E(|U(x_1 \cdots x_t)|)/t = h_\infty(P),$$

where $E(f)$ is the expected value of f, $h_\infty(P)$ is the Shannon entropy of P, see (8.21).

Acknowledgment

This research was supported by Russian Foundation for Basic Research (grant no. 06-07-89025).

References

1. P. Algoet, Universal schemes for learning the best nonlinear predictor given the infinite past and side information, *IEEE Trans. Inform. Theory.* **45** (1999), 1165–1185.

2. G. J. Babu, A. Boyarsky, Y. P. Chaubey and P. Gora, New statistical method for filtering and entropy estimation of a chaotic map from noisy data, *Int. J. Bifur. Chaos* **14**, 11 (2004), 3989–3994.

3. A. R. Barron, The strong ergodic theorem for dencities: Generalized Shannon–McMillan–Breiman theorem, *Ann. Probab.* **13**, 4 (1985), 1292–1303.

4. L. Györfi, I. Páli and E. C. van der Meulen, There is no universal code for infinite alphabet, *IEEE Trans. Inform. Theory.* **40** (1994), 267–271.

5. P. Billingsley, *Ergodic Theory and Information* (John Wiley & Sons, 1965).

6. R. Cilibrasi and P. M. B. Vitanyi, Clustering by Compression, *IEEE Trans. Inform. Theory.* **51**, 4 (2005).

7. R. Cilibrasi, R. de Wolf and P. M. B. Vitanyi, Algorithmic clustering of music, *Comput. Music J.* **28**, 4 (2004), 49–67.

8. I. Csiszár and P. Shields, Notes on information theory and statistics, *Foundations and Trends in Communications and Information Theory* (2004).

9. I. Csiszár and P. Shields, The consistency of the BIC Markov order estimation, *Ann. Stat.* **6** (2000), 1601–1619.

10. M. Effros, K. Visweswariah, S. R. Kulkarni and S. Verdu, Universal lossless source coding with the Burrows Wheeler transform, *IEEE Trans. Inform. Theor.* **45** (1999), 1315–1321.

11. W. Feller, *An Introduction to Probability Theory and Its Applications*, Vol. 1. (John Wiley & Sons, New York, 1970).

12. L. Finesso, C. Liu and P. Narayan, The optimal error exponent for Markov order estimation, *IEEE Trans. Inform. Theor.* **42** (1996).

13. B. M. Fitingof, Optimal encoding for unknown and changing statistica of messages, *Prob. Inform. Transm.* **2**, 2 (1966), 3–11.

14. R. G. Gallager, *Information Theory and Reliable Communication* (John Wiley & Sons, New York, 1968).

15. E. N. Gilbert, Codes based on inaccurate source probabilities, *IEEE Trans. Inform. Theor.* **17** (1971).

16. N. Jevtic, A. Orlitsky and N. P. Santhanam. A lower bound on compression of unknown alphabets, *Theor. Comput. Sci.* **332** (2004), 293–311.

17. J. L. Kelly, A new interpretation of information rate, *Bell Syst. Tech. J.* **35** (1956), 917–926.

18. J. Kieffer, A unified approach to weak universal source coding, *IEEE Trans. Inform. Theor.* **24** (1978), 674–682.

19. J. Kieffer, Prediction and Information Theory, Preprint (available at ftp://oz.ee.umn.edu/users/kieffer/papers/prediction.pdf/), (1998).

20. J. C. Kieffer and En-Hui Yang, Grammar-based codes: A new class of universal lossless source codes, *IEEE Trans. Inform. Theor.* **46**, 3 (2000), 737–754.

21. A. N. Kolmogorov, Three approaches to the quantitative definition of information, *Probl. Inform. Transm.* **1** (1965), 3–11.

22. D. E. Knuth, *The Art of Computer Programming*, Vol. 2. (Addison Wesley, 1981).

23. R. Krichevsky, A relation between the plausibility of information about a source and encoding redundancy, *Prob. Inform. Transm.* **4**, 3 (1968), 48–57.

24. R. Krichevsky, *Universal Compression and Retrival* (Kluver Academic Publishers, 1993).

25. S. Kullback, *Information Theory and Statistics* (Wiley, New York, 1959).

26. U. Maurer, Information-theoretic cryptography, *Advances in Cryptology — CRYPTO '99, Lecture Notes in Computer Science* Vol. 1666 (Springer-Verlag, 1999) pp. 47–64.

27. D. S. Modha and E. Masry, Memory-universal prediction of stationary random processes, *IEEE Trans. Inform. Theor.* **44**, 1 (1998), 117–133.

28. A. B. Nobel, On optimal sequential prediction, *IEEE Trans. Inform. Theor.* **49**, 1 (2003), 83–98.

29. A. Orlitsky, N. P. Santhanam and J. Zhang, Always good turing: Asymptotically optimal probability estimation, *Science*, **302** (2003).

30. Zh. Reznikova, *Animal Intelligence. From Individual to Social Cognition* (CUP, 2007).

31. J. Rissanen, Generalized Kraft inequality and arithmetic coding, *IBM J. Res. Dev.* **20**, 5 (1976), 198–203.

32. J. Rissanen, Universal coding, information, prediction, and estimation, *IEEE Trans. Inform. Theor.* **30**, 4 (1984), 629–636.

33. A. Rukhin *et al.*, *A statistical test suite for random and pseudorandom number generators for cryptographic applications.* (NIST Special Publication 800-22 (with revision dated May, 15, 2001)). http://csrc.nist.gov/rng/SP800-22b.pdf.

34. B. Ya, Ryabko, Twice-universal coding *Probl. Inform. Trans.* **20**, 3 (1984), 173–177.

35. B. Ya, Ryabko, Prediction of random sequences and universal coding, *Probl. Inform. Trans.* **24**, 2 (1988), 87–96.

36. B. Ya. Ryabko, A fast adaptive coding algorithm, *Probl. Inform. Trans.* **26**, 4 (1990), 305–317.

37. B. Ya. Ryabko, The complexity and effectiveness of prediction algorithms, *J. Complex.* **10**, 3 (1994), 281–295.

38. B. Ryabko, J. Astola and A. Gammerman, Application of Kolmogorov complexity and universal codes to identity testing and nonparametric testing of serial independence for time series, *Theor. Comput. Sci.* **359** (2006), 440–448.

39. B. Ryabko, J. Astola and A. Gammerman, Adaptive coding and rediction of sources with large and infinite alphabets, *IEEE Trans. Inform. Theor.* **54**, 8 (2008).

40. B. Ryabko, J. Astola and K. Egiazarian, Fast codes for large alphabets, *Comm. Inform. Syst.* **3**, 2 (2003), 139–152.

41. B. Ryabko and V. Monarev, Experimental investigation of forecasting methods based on data compression algorithms. *Probl. Inform. Trans.* **41**, 1 (2005), 65–69.

42. B. Ryabko and V. Monarev, Using information theory approach to randomness testing, *J. Stat. Plann. Infer.* **133**, 1 (2005), 95–110.

43. B. Ryabko and Zh. Reznikova, Using Shannon entropy and Kolmogorov complexity to study the communicative system and cognitive capacities in ants, *Complexity* **2**, 2 (1996), 37–42.

44. B. Ryabko and F. Topsoe, On asymptotically optimal methods of prediction and adaptive coding for Markov sources, *J. Complex.* **18**, 1 (2002), 224–241.

45. D. Ryabko and M. Hutter, Sequence prediction for non-stationary processes, *Proc. Combinatorial and Algorithmic Foundations of Pattern and Association Discovery,* Dagstuhl Seminar, (2006), Germany, http://www.dagstuhl.de/06201/ see also http://arxiv.org/pdf/cs.LG/0606077.

46. S. A. Savari, A probabilistic approach to some asymptotics in noiseless communication, *IEEE Trans. Inform. Theor.* **46**, 4 (2000), 1246–1262.

47. C. E. Shannon, A mathematical theory of communication, *Bell Sys. Tech. J.* **27**, 379–423 (1948), 623–656.

48. C. E. Shannon, Communication theory of secrecy systems, *Bell Sys. Tech. J.* **28** (1948), 656–715.

49. A. N. Shiryaev, *Probability*, 2nd edn (Springer, 1995).

Chapter 9

INTRODUCTION TO NETWORK CODING FOR ACYCLIC AND CYCLIC NETWORKS

ÁNGELA I. BARBERO* and ØYVIND YTREHUS[†]

*Department of Applied Mathematics,
University of Valladolid, 47011 Valladolid, Spain

†Department of Informatics,
University of Bergen, N-5020 Bergen, Norway
*angbar@wmatem.eis.uva.es
†oyvind@ii.uib.no

This chapter contains two parts. The first part gives an introduction to a relatively new communication paradigm: Network coding, where network nodes, instead of just forwarding symbols or packets, also are allowed to combine symbols and packets before forwarding them. The second part focuses on deterministic encoding algorithms for multicast, in particular for networks, which may contains cycles.

9.1. Introduction

A conventional communication network operates in a *store and forward* mode: A router (a computer dedicated to forwarding information packets in the network) will accept incoming packets, store a copy of the packet in its internal memory, and forward the message toward the ultimate destination essentially without modifying it. In the network coding literature, this conventional mode of operation is sometimes referred to as *routing*. Although this term is ambiguous (and should, in particular, not be confused with the process of obtaining a route), we will use it here.

Network coding generalizes the routing model by assuming that routers are allowed to modify the packets before forwarding them. Typically, a packet that is forwarded by a network coding router is constructed by a linear combination of packets that the router has received, although other modifications are conceivable. This improved versatility of the routers can be exploited in order to give benefits in a wide range of applications. The concept is still young, however, intense research efforts have discovered advantages with respect to increased multicast efficiency, increased efficiency of combinations of multiple unicast sessions in wireless network, and improved communication robustness, to name a few application areas. It turns out that even a simple distributed communication protocol, where routers forward *random* linear combinations of incoming packets, can give significant benefits in several aspects of communication. If information about network topology and traffic patterns is available centrally, it is possible to optimize network coding operation in order to optimize, for example, the transmission capacity, combined or maximum delay, or total transmission power consumption.

This chapter gives an introduction to the theory of network coding. Section 9.2 presents the network model under consideration, as well as the notation that will be used throughout the chapter. In Sec. 9.4, we turn to general strategies for describing and obtaining the network encoding equations. Random network coding is described in Sec. 9.5, while Sec. 9.6 introduces an efficient centralized approach for acyclic networks. We will demonstrate, however, in Sec. 9.7 that many networks encountered in the wild do indeed contain cycles, and in Sec. 9.8 we proceed to adapt the encoding algorithms to this situation. Alphabet size of the encoding is considered in Sec. 9.9, and we give a version (LIFE*) of the centralized encoding algorithm that will encode in binary for any admissible network, cyclic or not. The chapter concludes with a sequence of examples of application of the LIFE* algorithm.

For readers who want to learn, or to learn more, about this fascinating subject, a number of introductory texts have emerged over the last few years. The books [23, 24, 30, 61] contain comprehensive discussions of various aspects of network coding. For readers who want very brief tutorial papers, we can also recommend [12, 25]. The purpose of the current chapter is twofold: On the one hand, we attempt to provide an introduction to the area at a level of detail between a short paper and a complete book. Second, our subjective approach to network coding concerns the deterministic multicast encoding of cyclic networks. We believe that this can be of high practical importance (Sec. 9.7) and will present this topic at a level of detail that cannot be found elsewhere in the literature.

9.2. Background

A simple communication network is often described by a graph $G = (N, E)$, where N is the set of nodes and E the set of edges. In physical terms, N is the set of routers and host computers (computers with user processes or processes that generate or consume user information) connected to the network, while E is a set of links connecting pairs of computers. It may be necessary to modify the graph to approximate the physical characteristics of certain networks, such as wireless networks, but for now we will ignore these details. We will use the terms *edge* and *link* as synonyms, and by *node* we will mean any router or host computer. For a directed link $e \in E$, we will denote by start(e) and end(e) its starting and ending node, respectively. In many cases such an edge will be denoted simply as the pair (start(e), end(e)).

Suppose that an edge represents a discrete memoryless channel. Information theory provides the means to determine the channel capacity as a function of the channel parameters, and coding theory provides the methods to use the channel at a code rate close to capacity and at low error rates. It is therefore common to assume in network coding theory that the edges are noiseless and with a specified *capacity* (but see Sec. 9.3.1.) Each link (i, j) in the graph has a label $C_{i,j} \geq 0$ representing the capacity of that link.

The starting point of network coding is computing flows from sources to sinks, and hence we will introduce some basic concepts and results needed in this chapter. They come from a classical topic in graph theory, the study of transportation networks and how to maximize the amount of goods that can be transported through them. We will introduce notation gradually as we need it.

A *transportation network* is a directed graph $G = (N, E)$ with one source $a \in N$ (a node with input index, i.e. the number of incoming edges, zero) and one sink $z \in N$ (a node with output index, i.e. number of outgoing edges, zero). The capacity of a link represents the maximum amount of the substance that can be carried on that link per time unit.

Examples of transportation networks occur in everyday life as networks of oil pipes, road networks, etc. Figure 9.1 shows an example of a transportation network with source a and sink z. The number on each link is the capacity of the link.

We will use the notations $I(i)$ and $O(i)$ for the sets of edges ending and starting at node $i \in N$ respectively, that is, $I(i) = \{e \in E \mid \text{end}(e) = i\}$,

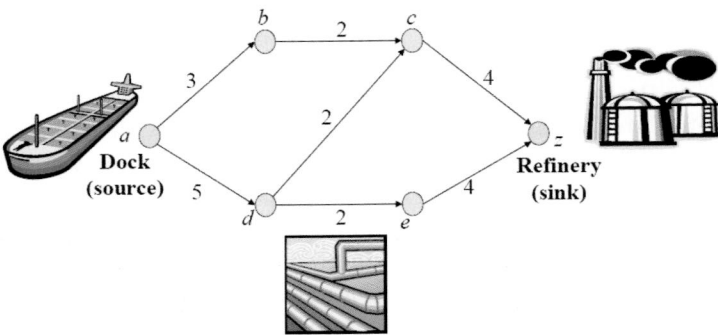

Fig. 9.1.   An example of a transportation network.

and $O(i) = \{e \in E \mid \text{start}(e) = i\}$. As we said before, for the source a and
the sink z we have $|I(a)| = |O(z)| = 0$.

A **flow** $F = (F_{i,j})_{(i,j) \in E}$ is an assignment of a value $F_{i,j}$ to each link
(i, j) such that

- $0 \le F_{i,j} \le C_{i,j}$.
- $\forall i \in V \setminus \{a, z\} \sum_{j \in O(i)} F_{i,j} = \sum_{j \in I(i)} F_{j,i}$.

The last property is known as the *flow conservation law* and states that
at the intermediate nodes nothing is created or destroyed: all that enters
through the input channels is delivered through the output channels.

The *value of flow* F is defined as $|F| = \sum_{j \in I(z)} F_{j,z}$. The flow
conservation law guarantees that the flow value can also be computed as
$|F| = \sum_{j \in O(a)} F_{a,j}$.

A *maximum flow*, or *max flow*, is a flow whose value is the maximum
possible for that transportation network. Although a maximum flow is not
unique, obviously all the maximum flows will have the same value.

Figure 9.2(a) shows a valid flow for the network in Fig. 9.1 and
Fig. 9.2(b) shows a maximum flow for the same network.

A *cut* in a transportation network is a set of edges that, when removed,
will completely disconnect the source from the sink. The *value of a cut* is
the sum of the capacities of the edges in the cut. A *minimum cut* or *min
cut* is a cut with minimum value and again, although there can be several
different minimum cuts for the same network, the value is unique.

Figure 9.3(a) shows an example of a cut and Figs. 9.3(b) and 9.3(c)
show two different min cuts for the network presented in Fig. 9.1.

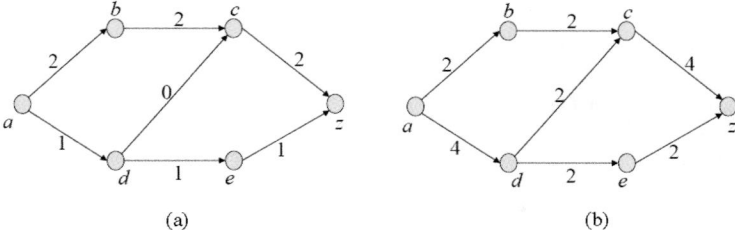

Fig. 9.2.   Flow and max flow for network in Fig. 9.1.

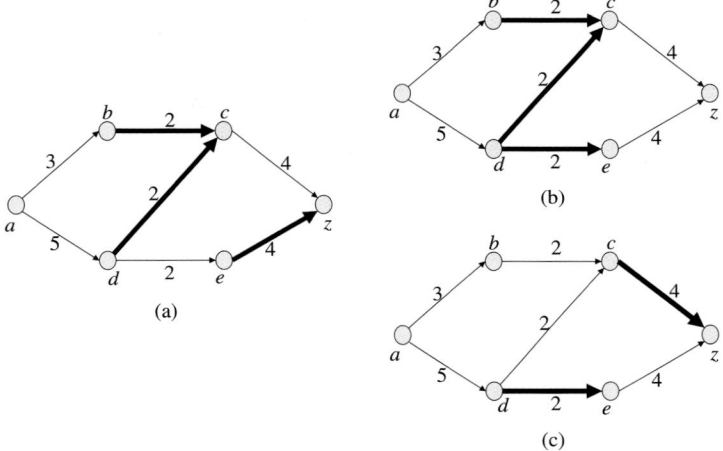

Fig. 9.3.   Cut and mincuts for network in Fig. 9.1.

A well-known theorem in graph theory establishes that the value of a maximum flow equals the value of a minimum cut. In Figs. 9.2 and 9.3 we see that this common value is 6 for the network in Fig. 9.1.

Section 9.3.4 contains a brief discussion of how to find the flow.

9.2.1. *Digital Communication Networks*

In a transportation network in which we are considering the routing of a physical substance or commodity (cars, for instance, in a road network, or oil in a network of oil pipes), the same object (the same car) or portion of substance (the same liter of oil) cannot be sent to two different sinks. However, the nature of information makes it possible for a node to copy a packet of information then forward the copies in different directions. This suggests an alternative network model for communication purposes.

As a *multicast network* we will consider a model that consists of a directed graph $G = (N, E)$ in which a certain subset of nodes $T = \{t_1, \ldots, t_r\}$, $r \geq 1$, are considered sinks, also called terminals or receivers. In principle, we still consider a single source node $S \in N$ with rate h, that is, the source node will generate h information units per time unit. Later on, we will modify the model slightly to consider h different source nodes, each with rate 1. The edges will also be called links or channels.

There are different scenarios of transmission on networks. In this chapter, we will deal mainly with multicast. In the notes at the end of this section, we will mention other scenarios.

Definition 9.1. In the *multicast* model, each sink needs to recover the h bits of information generated by the source at each time instant.

This model of communication through networks is the best-known case, for which necessary and sufficient conditions for the feasibility of the scheme were established in the paper by Ahlswede *et al.* [2], which can be considered the starting point of network coding.

Here we will reformulate the main result from that paper.

Theorem 9.1. *Let $G = (N, E)$ be a directed graph with source $S \in N$ and sinks $\{t_1, t_2, \ldots, t_r\} \subset N$. Let $C_{i,j}$ be the capacity of each link $(i, j) \in E$. Let h be the information rate of the source S, that is, S produces h information symbols per time unit. We will call the scheme $(G, C = (C_{i,j})_{(i,j) \in E}, h)$ admissible if there exists a way of sending information from the source S to each of the sinks $t_i, i = 1, 2, \ldots, r$ at rate assymptotically h.*

The scheme $(G, C = (C_{i,j})_{(i,j) \in E}, h)$ is admissible if and only if for each $i \in \{1, 2, \ldots, r\}$ the value of the max flow from S to t_i is at least h.

As we can see, the conditions for feasibility of the multicast transmission consist of the existence of a flow of value $\geq h$ when we consider the network as a transportation network with source S and sink t_i, and this must hold for each $t_i \in T$ *independently*. This means that flows to different sinks may share, and thus make a more efficient use of network links.

Figure 9.4 (from the paper by Ahlswede *et al.* [2]) shows an example of a network where full rate multicasting can be achieved with only routing (i.e. without using network coding.)

Figure 9.4(a) shows the network. The label of each link is the capacity. The source S generates information at rate five symbols, $b_1, b_2, \ldots, b_5$, per

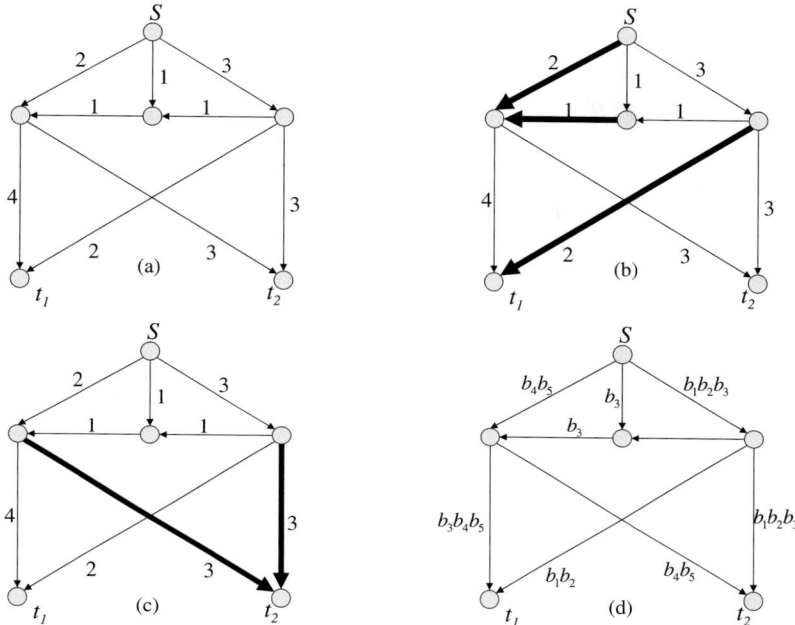

Fig. 9.4. An example of a network that can achieve multicast with only routing.

time unit. Figure 9.4(b) shows that the value of a min cut to terminal t_1 is 5, while Fig. 9.4(c) shows a min cut to terminal t_2 with value 6. The theorem states that it is possible to multicast five information symbols per time unit from the source to each of the two terminals. In fact it can be done as shown in Fig. 9.4(d), by simply rerouting the symbols arriving at each intermediate node in a suitable manner.

However, in other cases, in order to achieve the result guaranteed by the theorem, one needs to allow the intermediate nodes to perform coding. This means that an intermediate node can take the symbols that arrive at its input channels and combine them in a suitable way to compute the symbol that will be transmitted over each output channel. The encoding might, for some networks, require as well that the node has some finite memory, as we will see later in the chapter.

In Fig. 9.5(a), we present the simplest network that requires encoding at an intermediate node. It was presented in the paper by Ahlswede *et al.* [2] and it is commonly known in the literature as the *butterfly* network, having become a kind of coat of arms for the field of network coding.

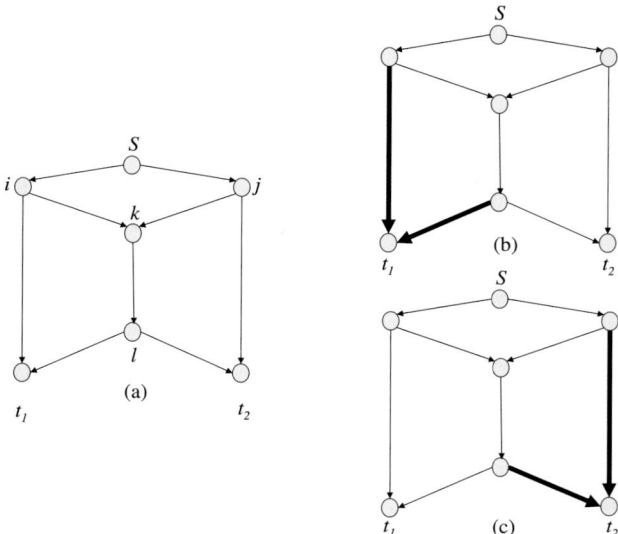

Fig. 9.5. An example of a network that requires coding to achieve multicast according to its max flow.

The source S releases information symbols at a rate of two symbols per time unit. The sinks are the nodes denoted as t_1 and t_2. All the links have unit capacity, i.e. they can transmit one information symbol per time unit. Figure 9.5(b) shows a minimum cut for terminal t_1 with value 2 and Fig. 9.5(c) shows a minimum cut for terminal t_2 with value 2. Hence, Theorem 9.1 concludes that it is possible to transfer two information symbols a and b from the source S to both terminals. However, it is easy to see that by simply copying and routing it cannot be achieved. The two bits will arrive at node k and the output link (k, l) can only accept one. If this one is a, terminal t_1 will not receive the two information symbols released by S: it will get symbol a twice but no symbol b. Analogously, if bit b is sent over (k, l), terminal t_2 will not be satisfied.

Nevertheless, the solution shown in Fig. 9.6 allows sink t_1 to receive symbols a and $a + b$, and from there very easily recover the original symbols a and b by performing a simple algebraic operation. The same goes for terminal t_2. The "trick" of allowing an intermediate node to combine input symbols and the receivers to undo the corresponding combinations allows us to increase the throughput, achieving optimality as stated in Theorem 9.1. Network coding is all about this "trick".

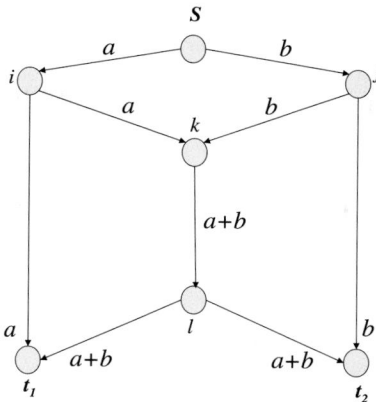

Fig. 9.6.   A network encoding for network in Fig. 9.5.

9.3. Thoughts for Practicioners

The explosion of research interest in network coding in recent years has uncovered a multitude of interesting ways to exploit the network coding paradigm. The format of a chapter does not allow us to explain all these research threads in the detail they deserve, and so we have chosen instead to discuss centralized deterministic multicast encoding in detail. However, in this subsection we attempt to give a brief presentation of some important alternative aspects of network coding. Some of the applications rely on randomized network coding, and the discussion of these will continue in Sec. 9.5.

9.3.1. *Separation of Network Coding and Channel Coding, and Network Coding on Noisy Networks*

Separation of network coding and channel coding: It has been shown [47, 54, 57] that separation of network coding and channel coding is asymptotically optimal for multi-sink networks with a single source. This justifies the network model with noiseless links, at least from a theoretical point of view.

It may still be reasonable to investigate error correcting network codes:

- Even though separation is asymptotically optimal, there may be practical advantages of a joint coding scheme. Such a joint scheme may reduce or

eliminate the need for separate encoding and decoding at each edge.
There can also be advantages in terms of complexity or delay.
- Even though channel coding can make the edges noiseless, there is still a
  real probability of errors, in particular packet erasures, in the *nodes* due
  to, for example, traffic congestion.

Several papers [34, 40, 60, 63] investigate deterministic network coding for
noisy channels. Random network coding can easily handle most erasures,
see, for example, [53].

9.3.2. *Wireless Networks and Networks with Broadcast Nodes*

A network where all or some nodes operate by broadcasting their packets
to a set of neighbors is constrained with respect to the operations its nodes
can perform.

- *Broadcast*: The broadcast nature of transmissions means that nodes
  cannot (easily) send different packets to distinct neighbors. This
  restriction is sometimes modeled by representing the network as a
  hypergraph [12], with the broadcast link being a hyperlink, as shown
  in Fig. 9.7(a). We find it convenient instead to represent the network
  with nodes as shown in Fig. 9.7(b), which is equivalent to the one in
  Fig. 9.7(a) if the network is lossless and noiseless. Here the broadcast
  node is split into two subnodes: One encoding subnode (which can send
  to the transmitting subnode only) and one transmitting subnode (which
  can only forward incoming packets, without performing any network
  encoding). One advantage of approach (b) is that we can use basically the
  same computer software as in the non-broadcast case. Another advantage

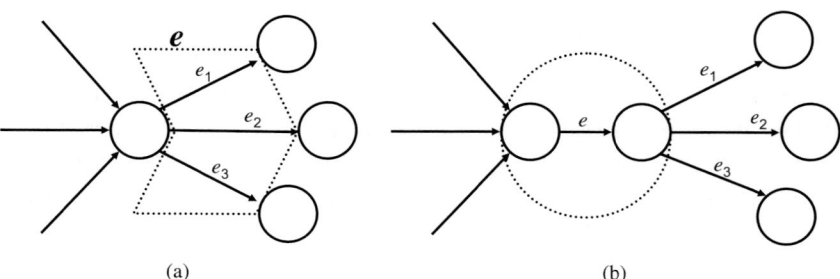

(a)   (b)

Fig. 9.7.   A broadcast edge in a network (a) modeled with a hyperlink, and (b) modeled
as a split node.

is that (we believe) this representation clarifies the equivalence between a wireless/broadcast network and a certain other non-broadcast network. We will use this representation in Sec. 9.7 to argue that it is important to consider cyclic networks.

- *Interference*: Nodes also need to cope with receiving multiple signals simultaneously. The resulting interference affects the actual capacity and throughput of the node. An approach to analyze these effects is outlined in [4].

- *Scheduling*: In order to avoid the limitations imposed by broadcast channels and interference, one might attempt to schedule the transmissions of packets to avoid simultaneous transmissions, for example, by using time division or frequency division multiplexing [9, 50].

9.3.3. *Other Communication Scenarios*

There are variations of the multicast communication model that also benefit from network coding. See, for example, [61] for a discussion of generalized multicast, broadcast, and information dispersion scenarios. In the generalized multicast model, receivers may subscribe to different subsets of the source information streams. In this model, a linear network coding solution may not exist while a non-linear one does [14, 48]. These variations also include the information dissemination or gossip models [13], which is useful in any situation where nodes need to broadcast information to all other nodes in the network. A typical example of such dissemination is the open shortest path first (OSPF), which is an Internet protocol used for determining the topology of a network, that works by broadcasting packets containing information on all local connections so that all nodes can create their own network map. Random network coding can speed up this dissemination process [20, 21, 29, 32].

Further, multiple simultaneous unicast sessions [56] through a wired or wireless network may benefit from networking coding. In particular, wireless broadcast network nodes through which equal-sized packets flow continuously in both directions will, by the observation made in Sec. 9.3.2, act as the encoding node k in the butterfly network of Fig. 9.5. Hence, using the simple encoding strategy of Fig. 9.6, these nodes will double their effective transmission throughput.

Random network coding can also help in a simple unicast scenario on a network with erasure links. See Sec. 9.5.3.

9.3.4. *Computing Flows*

Algorithms to compute max flow on transportation networks fit in two classes. The path augmenting algorithms [15, 19] and the preflow push algorithms. Several versions exist, and a distinguishing feature is to what extent the algorithms determine optimum (e.g. shortest paths, lowest power consumption) flows. We will refer the reader to the books [3, 11, 49] for more information about this subject in general, as this is an important tool in network coding.

The basic max flow algorithms considered in the references are directed at single source, single sink networks. This does not reflect accurately the multiple sink networks we consider here. A simple adaptation, which does not necessarily yield optimum solutions, is to use the single source, single sink versions for each sink in succession. In order to benefit from edge reuse, as network coding allows, edges that have already been used by a previous sink can be reassigned a cost of zero.

The problem of finding flows is related to the problem of extracting a minimum cost subnetwork from an available communication network, for use by a given multicast session. This can be formulated as an optimization problem [44, 45, 58, 59] and solved with optimization techniques. In fact, while the problem of extracting a minimum energy multicast (Steiner) tree from a random wireless network is NP-hard in the case of routing, the similar problem in the network coding case can be solved in polynomial time, even with decentralized algorithms.

9.3.5. *The Gain of Network Coding*

Deb *et al.* [12] observed 13 to 49% energy savings in random network coding over routing in random wireless multicast networks. Liu *et al.* [41] proved that the gain in terms of *throughput benefit ratio* is bounded by a constant. For theoretical models, Jaggi *et al.* [34] showed that arbitrarily large throughput benefit ratios can be obtained if the network is "properly" designed. As the reader may deduce, there are different criteria for network coding gain. We refer to the references for more details.

9.4. How to Encode

Once the main theorem is established, the next step is to find a way to design the way of combining symbols at each node. For small networks like the ones in Figs. 9.4 and 9.5, the task of determining the network

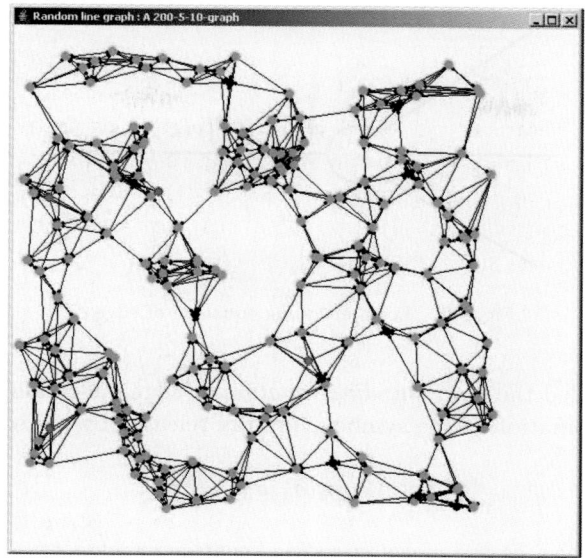

Fig. 9.8.   Another example of a network.

encoding equations may appear to be trivial. However, in larger and more complicated networks like the one shown in Fig. 9.8, it becomes less evident how to encode, and it is clear that efficient algorithms to design the encoding procedures are needed.

We need to introduce some more notation here. Let $\sigma_1, \sigma_2, \ldots, \sigma_h \in \mathbb{F}_q$ be the symbols generated by the source S at a given time instant. For the time being, we will assume that transmission on the links of the network is instantaneous and is synchronized in such a way that delays play no role in the transmission and/or in the design of the encoding. Later on we will see that in fact, delay plays an important role in the design of codes for networks that contain cycles of certain types. Appropriate use of delay in encoding can also reduce the required field size. We will return to these issues in Sec. 9.8 and beyond.

Let $v(e) \in \mathbb{F}_q$ denote the symbol that is being sent on link $e \in E$. We will distinguish two different ways of expressing that symbol:

- As a combination of the symbols that entered at node start(e),

$$v(e) = f_e(\{v(e')\}_{e' \in I(\text{start}(e))}).$$

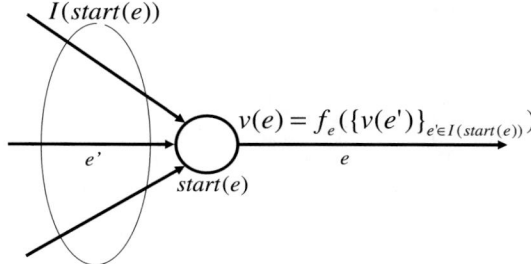

Fig. 9.9.   Local encoding equation of edge e.

This is called the *local encoding equation* at edge e (see Fig. 9.9).

• As a combination of the symbols initially released by the source

$$v(e) = g_e(\sigma_1, \sigma_2, \ldots, \sigma_h).$$

This is known as the *global encoding equation* at edge e.

We assume that the global encoding equation of edge e can be derived recursively from the local encoding equation of e and from the global encoding equations of the edges in $I(\text{start}(e))$.

Each terminal node $t_i \in T$ receives a collection of symbols $\{v(e)\}_{e \in I(t_i)}$ and should be able to retrieve the information symbols by solving the corresponding system of global equations $\{g_e(\sigma_1, \sigma_2, \ldots, \sigma_h) = v(e)\}_{e \in I(t_i)}$. The goal of network coding is to determine the functions f_e that combine the symbols for each $e \in E$ in such a way that the set of equations received by each sink $t_i \in T$ has $(\sigma_1, \sigma_2, \ldots, \sigma_h)$ as unique solution. See the paper by Koetter and Medard [35] for more details on the algebraic description of network coding.

The functions f_e for $e \in E$ might in principle be very complicated, but in 2003, Li *et al.* [36] concluded in their paper that linear encoding suffices for acyclic networks as long as the field size q is sufficiently large, i.e. that the result of the main theorem can be achieved by means of allowing the intermediate nodes to perform only linear combinations of the symbols received over their input channels.

This means that each local encoding equation is of the form

$$v(e) = \sum_{e' \in I(\text{start}(e))} m_e(e') v(e'), \tag{9.1}$$

where $m_e(e') \in \mathbb{F}_q$ is the coefficient that determines the contribution of the symbol $v(e')$ for each $e' \in I(\text{start}(e))$ in the linear combination to form $v(e)$.

When the local encoding equations are linear, the global encoding equations also become linear and will take the form

$$v(e) = \alpha(e, 1)\sigma_1 + \alpha(e, 2)\sigma_2 + \cdots + \alpha(e, h)\sigma_h, \qquad (9.2)$$

where $\alpha(e, i) \in \mathbb{F}_q$ is the coefficient that specifies the influence of symbol $\sigma_i \in \mathbb{F}_q$ on $v(e)$, and can be determined once the local encoding for each edge has been determined.

Now the condition of network multicast for each terminal node $t_i \in T$ reads as follows:

Rank condition: The system of linear equations $\{\alpha(e, 1)\sigma_1 + \alpha(e, 2)\sigma_2 + \cdots + \alpha(e, h)\sigma_h = v(e)\}_{e \in I(t_i)}$ on the set of unknowns $(\sigma_1, \sigma_2, \ldots, \sigma_h)$ has rank h.

The process of actually determining the encoding equations can be approached from different angles. We will distinguish between the two following situations:

(A) When the complete network topology is centrally available, in which case we can apply a *centralized* algorithm for network encoding.
(B) When it is not, in which case a *distributed* algorithm is required.

In case (A), we will consider *deterministic linear encoding* algorithms and in case (B), *random linear encoding*.

In the deterministic approach, the local encoding equations for each edge are determined beforehand in order to meet the rank condition. For this, it is necessary to know beforehand the whole topology of the network. Once the encoding for a certain network has been designed, each node will use the given local equation to combine the input symbols for each transmission.

The random approach, on the contrary, works in a decentralized way. Each node in the network, for each transmission, chooses the coefficients that will be used for the local encoding equation of each of its output channels. The final fulfillment of the rank condition relies on the use of a very large field, so that the probability that a random matrix has maximum rank is very high.

Both models have advantages and inconveniences and are suitable for two different kinds of situations.

Random linear encoding:

- Requires no preprocessing and no previous knowledge of the topology of the network, which makes it suitable for very large networks and specially for highly dynamic networks in which nodes are constantly entering or quitting the network. For the same reason it is resilient against failures in the network (like edges or nodes temporarily collapsing).
- Requires a certain amount of overhead on the transmission on each channel, since the coefficients used on each local encoding have to be transmitted together with the resulting information packet in order for the nodes downstream to be able to compute the global encoding equations of each outgoing edge and eventually for the terminals to be able to determine the system of linear equations that they must solve to recover the information symbols.
- The success of the scheme relies on the use of a very large field, or a large message size, in order to ensure a large probability that the final systems of global equations at the terminals are all of maximum rank.
- The terminals have to solve a new system of equations for each transmission, hence, the operational cost of solving the system for one transmission cannot be saved for the next transmission.

Deterministic linear coding:

- Requires full knowledge of the network in order to design the encoding. On the other hand, the preprocessing to design the encoding is computed once and used for every transmission. The cost of the algorithms we will present in Sec. 9.6 and beyond will be polynomial of low degree in the number of edges of the network. This makes the model suitable for reasonably stable networks of moderate size. Each time there is a change in the topology of the network that affects an edge that is being used in the present encoding scheme, the algorithm has to be run again in order to find a new network code.
- Once the encoding has been designed, the nodes will use it for every transmission, and no overhead is necessary on the transmissions. The terminals know the coefficients of the global encoding equations that they will receive.
- The code designed with the algorithms explained in this tutorial chapter can use, in most cases, the smallest possible field. In many networks, binary processing suffices.

- The terminals solve the same system of equations for each transmission, only changing the independent term. Hence, the operational cost of decoding per transmitted symbol will be low.

To summarize,

> Random linear encoding is a robust and adaptable scheme that produces suboptimal network encoding. For a known stable network, deterministic encoding in which a good encoding algorithm is employed will usually be more efficient than random coding in terms of efficiency, alphabet size, and decoding complexity.

9.5. Random Network Coding

In random network coding, the source(s) partitions the information to be transmitted into K packets, say $p_1, \ldots, p_K$. We need to assume that each packet has a fixed length of k symbols from a finite field $\mathbb{F}_Q$ (where Q in general needs to be much larger than the field size q required for deterministic network coding). The encoding model for random network coding makes it convenient to describe the encoding with respect to packets rather than to network edges. Let p be a given packet transmitted on some edge in the network. We associate with p a global encoding vector $A(p) = (\alpha(p, 1), \ldots, \alpha(p, K)) \in \mathbb{F}_Q^K$ so that

$$p = \alpha(p, 1)p_1 + \alpha(e, 2)p_2 + \cdots + \alpha(p, K)p_K.$$

(Each source and) each node n in the network maintains a memory $M(n)$ of packets received at n. Although slight variations are conceivable, the encoding rule of random network coding is that when node n is about to transmit a new packet p on a given edge, the local coding coefficients $m_{p'}(p) \in \mathbb{F}_Q$ in the local encoding equation

$$p = \sum_{p' \in M(n)} m_{p'}(p)p'$$

are all selected at random, usually according to a uniform distribution. It follows that the global encoding vector of p is

$$A(p) = \sum_{p' \in M(n)} m_{p'}(p)A(p') \in \mathbb{F}_Q^K.$$

The global encoding vector can be transmitted in a packet header. Provided the number of packets K is much less than the length of the packets k, the overhead is negligible.

Decoding commences when a receiver has received a collection of packets so that the corresponding global encoding vectors form a matrix of rank K. Then the original packets $p_1, \ldots, p_K$ can be determined by solving a set of K linear equations, for example, by the Gaussian elimination. The complexity of the decoding is therefore (normally) of order $O(K^3)$ operations over $\mathbb{F}_Q$, or of order $O(K^2)$ per decoded packet.

9.5.1. *Alphabet Size for Random Network Coding*

In pure noiseless mode, the receiver should be able to reconstruct the original set of information packets after receiving K packets. Ho *et al.* [27, 31] have shown that for an admissible multicast, on an arbitrary network with random uniformly selected network coding coefficients in finite field $\mathbb{F}_Q$, all the sinks can decode the source processes with probability at least $(1 - d/Q)^\eta$ for $Q > d$, where d is the number of sinks and η is the number of links with associated randomized coefficients. This result may give some guidelines about how large to select Q. However, note that a received packet that happens to be linearly dependent on the rest can be regarded as an erased packet. Then the receivers can simply be operated in erasure correction mode, which simply means that the receiver can simply go on receiving packets until a decoding is possible. (This requires of course that the network nodes keep transmitting packets until further notice, and it does involve a minor amount of network management.)

See Sec. 9.9 for a discussion of alphabet sizes for deterministic network coding.

9.5.2. *Decoding Complexity*

The erasure correction mode alluded to in Sec. 9.5.1 is akin to digital fountain coding [42, 52]. However, due to the distributed nature of the encoding process, random network coding cannot finely tune the encoding process so as to achieve the extremely fast decoding of LT codes [42] or raptor codes [52]. For example, in [55], packets in a video stream are encoded with the Raptor codes and re-encoded at network nodes. However, the simplicity of the Raptor decoding is lost.

9.5.3. *A Simple Erasure Network Where Network Coding Helps*

Consider a simple network consisting of a source, an intermediate node, and a sink. The source can send information to the intermediate node through a link with erasures, and the intermediate node can also send information to the sink through a link with erasures. A capacity achieving erasure correcting code applied to each link individually would achieve an overall capacity equal to the worst of the two links, at the cost of delay (the intermediate node would have to decode incoming traffic before creating a new codeword for transmission to the sink). On the other hand, end-to-end erasure correction is not optimum, since the decoder in that case would experience an erasure rate, which is worse than for either single link. Network coding can achieve the channel capacity with less delay than experienced by coding on each channel. See, for example, [25].

9.5.4. *Avalanche*

Avalanche [26] is a commercial application for file content distribution. It has been analyzed in [1, 62]. The project has recently turned into Microsoft Secure Content Distribution (MSCD).

9.6. **Deterministic linear network coding: Efficient Algorithms**

In this section, we will start presenting a family of efficient linear network encoding algorithms for multicast. They have its origin in the linear information flow (LIF) algorithm for acyclic networks, which was presented independently in [51, 33]. Later the algorithm was presented in full in the joint paper by Jaggi *et al.* [34]. Other algorithms for the same problem have been presented, including [16]. We will concentrate on the LIF family of algorithms since they are more efficient.

In this section, we will explain the LIFE algorithm presented by Barbero and Ytrehus [5], which deals with certain types of cyclic networks. Later sections will present other members of this family of algorithms, designed to deal with all kinds of cycles in the networks.

The goal of the algorithms presented in this and the following sections is to give an efficient construction of a network code when the topology of the network is known. We will modify slightly the model of communication network introduced in Sec. 9.2 in the following manner: instead of one unique source with rate h we consider a set S of h different unit rate sources

$S = \{s_1, s_2, \ldots, s_h\} \subset V$ and assume that source s_i produces symbol $\sigma_i \in \mathbb{F}_q$ for each $i = 1, \ldots, h$. Besides, from now on we will consider that each link has unit capacity and is error free. We will discuss the size of q later on. For now we just assume that it is "large enough."

From the model presented in Sec. 9.2.1, we can pass to this model by simply creating h virtual sources $s_1, \ldots, s_h$ and connect them with unit rate virtual edges to the one source S considered in the model in Sec. 9.2.1. Also, integer capacity edges can now be remodeled by considering as many parallel edges as necessary connecting two nodes.

We still consider instantaneous transmissions and will leave the matter of delay for the next section.

As usual in graph theory, a *path* from node u to node v of length l is a sequence $\{e_i \in E : i = 1, \ldots, l\}$ such that $u = \text{start}(e_1)$, $v = \text{end}(e_l)$, and $\text{start}(e_{i+1}) = \text{end}(e_i)$, for $i = 1, \ldots, l - 1$.

In this new network model, the necessary and sufficient condition for network multicast set in Theorem 9.1 can be written as follows: For each source $s \in S$ and each sink $t \in T$ there exists a *flow path* $f^{s,t}$, i.e. a path from the source s to the sink t, in such a way that for each $t \in T$ the set of flow paths $f^t = \{f^{s_i,t}, i = 1, \ldots h\}$ are all edge disjoint.

In the butterfly network of Fig. 9.10(a) we can see the two paths arriving in terminal t_1 and appreciate that they are edge disjoint. In Fig. 9.10(b) we see the flow paths arriving in t_2, again, edge disjoint.

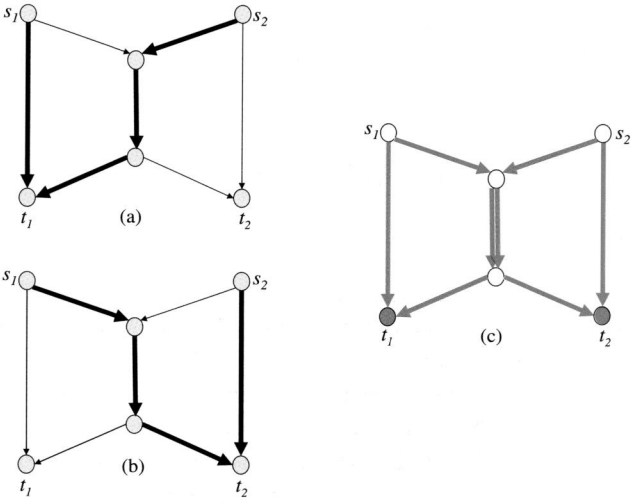

Fig. 9.10. Flow paths for network in Fig. 9.5 modified.

This guarantees that the conditions of Theorem 9.1 are satisfied and that each terminal can receive the two information symbols released by the two sources.

When putting all the paths together in Fig. 9.10(c) we can see that a path to t_1 and a path to t_2 share an edge, which is allowed.

Following the notation of the paper by Sanders *et al.* [51] we will call f^t the *flow for sink t*. The minimal subgraph of G that contains all flows (and their associated nodes) will be called the *flow path graph*, and can be determined from G by a suitable polynomial algorithm. See the remarks in Sec. 9.3.4 for details. Figure 9.10(c) shows the flow path graph of the corresponding network.

Edges that are not in any flow path will be discarded and from now on we will, for convenience, assume that the network G is actually the flow path graph of the given network.

The main contribution of the LIF algorithm is to use the flow paths to construct the encoding. In this, the LIF family of algorithms resembles other polynomial time linear network encoding algorithms, such as the one presented by Erez and Feder [16]. Besides, for each edge all the flow paths going through that edge will be considered, in such a way that no reprocessing of that edge will be needed, keeping the computational complexity not only polynomial but with a very low degree. In this it outperforms the algorithm by Erez and Feder [16], which proceeds by following the flow for each sink separately, having to recompute the encoding of those edges that participate in several flows as many times as needed.

We will introduce some further definitions.

Definition 9.2. A *link cyclic* network is a network where there exists a cyclic subset of edges, i.e. a set $\{e_1, e_2, \ldots, e_k, e_{k+1} = e_1\} \subset E$ for some positive integer k such that $\text{end}(e_i) = \text{start}(e_{i+1})$ for $1 \leq i < k$. The set of edges $\{e_1, e_2, \ldots, e_k, e_{k+1} = e_1\}$ is a *link cycle*. If no such cycle exists, the network is *link acyclic* or simply *acyclic*.

Suppose e is an edge that lies on the flow path f^{s_i, t_j}. We will denote by $f_{\leftarrow}^{t_j}(e)$ the predecessor of edge e in that path. There is no ambiguity in the notation: e can lie on several flow paths f^{s_i, t_j} for different t_j, but not for different s_i and the same t_j, since all the flow paths arriving in t_j are edge disjoint. Thus, once t_j is fixed, e can only have one predecessor in the flow to t_j, that is, $f_{\leftarrow}^{t_j}(e)$ is the predecessor of e in the only flow path arriving in t_j that contains e. In the same manner, we denote by $f_{\rightarrow}^{t_j}(e)$ the successor of e in that flow path.

Let $T(e) \subseteq T$ denote the set of sinks t that use e in some flow path in f^t, and let $P(e) = \{f^t_\leftarrow(e) \mid t \in T(e)\}$ denote the set of all predecessors of edge e. Note that, since G is actually a flow path graph, all edges will have some predecessor ($P(e) \neq \emptyset$), except those with $\text{start}(e) = s_i$ for $i \in \{1, \ldots, h\}$.

Consider the order induced in the edges by the flow paths in $f = \bigcup_{t \in T} f^t$. When two edges e_1 and e_2 lie on the same flow path and e_1 is the predecessor of e_2 in that path, we will write $e_1 \prec e_2$. We will use transitivity to define relationships among other pairs of edges that lie on the same path but are not consecutive. We observe that in each path the relation $\prec$ defines a total order in the edges that form that path, since a path that contains cycles, that is, edges that satisfy $e_1 \prec e_2 \prec \cdots \prec e_n \prec e_1$ can be simplified by avoiding taking the trip around that cycle as shown in Fig. 9.11.

Definition 9.3. A *flow acyclic* network is a network where the relation $\prec$ defines a partial ordering in E. If the relation does not define a partial ordering, the network is *flow cyclic*.

Note that a flow cyclic network is always link cyclic; however, the converse is not true.

Figure 9.12 shows networks of different types. Example 1 is acyclic, Example 2 is link cyclic but flow acyclic, Examples 3, 4, and 5 are flow cyclic. Later on, we will see that there is still a difference between the networks of Examples 3 and 4 and the network of Example 5.

The flow paths for all the networks will be shown in other figures later in the chapter, and the networks will be used as examples of the encoding algorithms that will be explained.

Similar to the LIF algorithm, the linear information flow on the edges (LIFE) algorithm proceeds by maintaining, for each iteration of the main

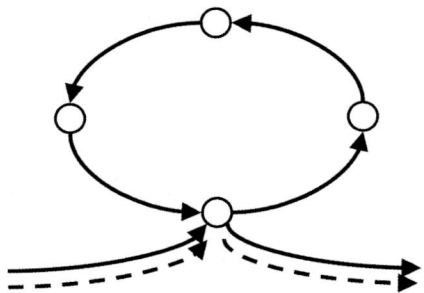

Fig. 9.11.   Cyclic flow paths can always be avoided.

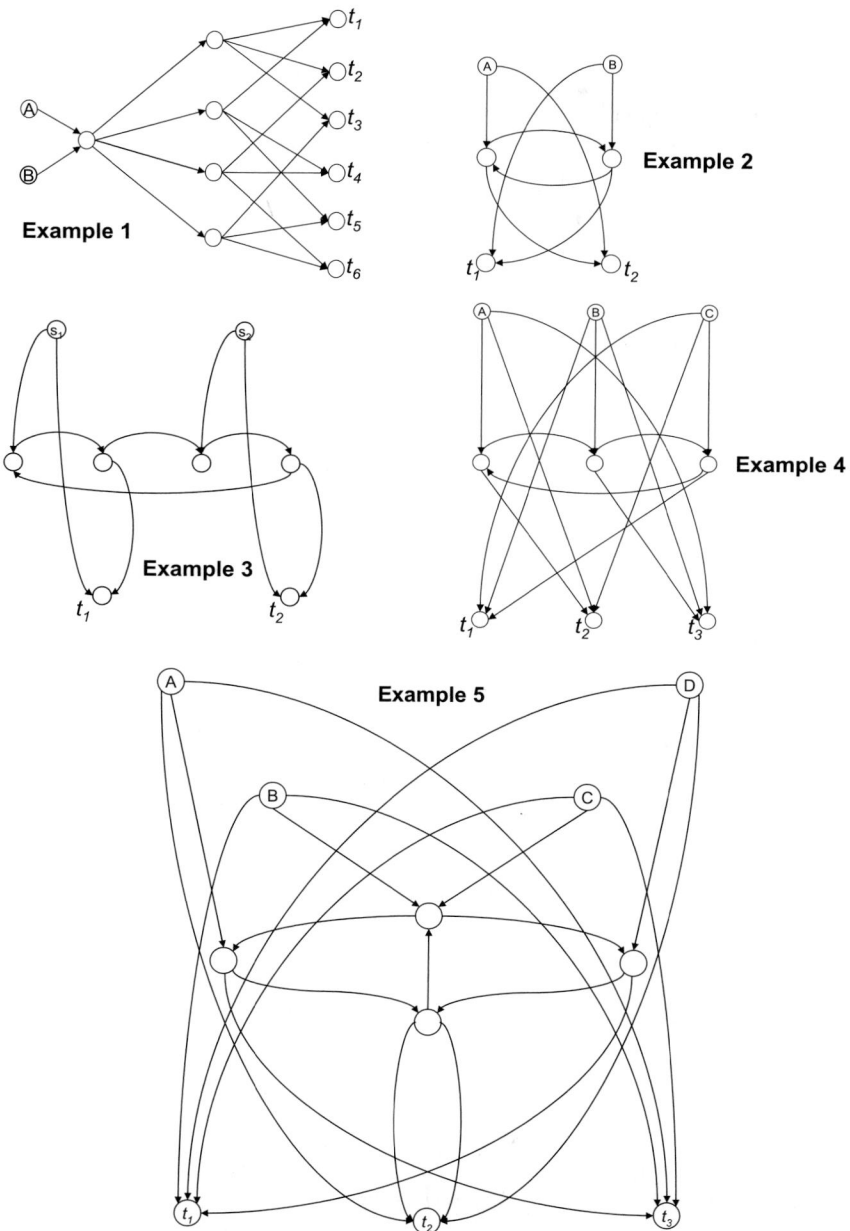

Fig. 9.12. Examples of different types of networks.

loop, the following sets for each $t \in T$:

- A set $E_t \subset E$, $|E_t| = h$, such that E_t contains *the most recently visited edge* on each flow path in f^t.
- A set of global encoding equations, $V_t = \{(v(e) = \sum_{i=1}^{h} \alpha(e, i)\sigma_i)| \ e \in E_t\}$.

We will impose the *full rank invariant: The set of equations V_t always have full rank.*

ALGORITHM LIFE (Linear Information Flow on Edges)

(1) *Input*:

A graph $G = (N, E)$ that is actually a flow path graph as explained before.

(2) *Initialization*:

$\forall t$:

- $E_t = \{e | e \in f^{s_i, t}, \text{start}(e) = s_i, i = 1, \ldots, h\}$,
- $V_t = \{v(e) = \sigma_i | e \in E_t, \text{start}(e) = s_i\}$,

that is, if $e \in f^{s_i, t}$ with $start(e) = s_i$, then $v(e) = \sigma_i$.

(3) *Main Loop*:

- Select an edge e for which the encoding equations have not yet been determined, but for which the global encoding equations of all the predecessor edges in $P(e)$ have been determined. Then update the set of current edges,

$$E_t := (E_t \setminus \{f_{\leftarrow}^{t}(e)\}) \cup \{e\}, \text{ for each } t \in T(e).$$

- Further, for each $t \in T(e)$, replace in V_t the global encoding equation for $f_{\leftarrow}^{t}(e)$ with that of the new edge e. Provided the field $\mathbb{F}_q$ is sufficiently large, it is always possible to determine a local encoding as given by Eq. (9.1) in such a way that the full rank invariant is satisfied. See [51] for details.

(4) *Output*:

- For each edge, the local encoding as given by Eq. (9.1) is produced.
- At the end of the algorithm, $E_t = \{e \mid end(e) = t\}$ and V_t is a set of h linearly independent equations from which t can recover the input.

This algorithm will produce linear encoding for all flow acyclic networks, regardless whether they are link cyclic or acyclic, for instance, it will work with the networks in Examples 1 and 2 in Fig. 9.12.

To ensure that the full rank invariant can be maintained, the field must be large enough, and it was proven in the paper by Sanders *et al.* [51] that $q \geq r = |T|$ suffices. In fact using the same argument they use, it is proven that in fact it suffices $q \geq \pi_{max}$, where π_{max} is the maximum number of flow paths that share an edge in G. Obviously $\pi_{max} \leq r$, since flow paths to the same sink must be edge disjoint, but in most cases π_{max} is much smaller than r.

By keeping the full rank invariant on each step, it is guaranteed that, at the end, the rank condition is satisfied at the terminals and hence each of them can decode by simply solving the system of the h linear equations received, whose h unknowns are $\sigma_1, \sigma_2, \ldots, \sigma_h$. As pointed out before, the main computational in solving these linear systems is performed only once at preprocessing time, since the systems that each terminal receive in all the transmissions share the matrix and only change the independent term. It follows that the decoding takes $\mathcal{O}(h^2)$ operations, or $\mathcal{O}(h)$ per decoded symbol.

The main loop is executed $|E|$ times and following the discussion on complexity in the paper by Jaggi *et al.* [34] we conclude that the LIFE, like the LIF, algorithm can be implemented in expected time $O(|E| \cdot |T| \cdot h^2)$.

The LIFE algorithm does not work with flow cycles, and in order to deal with those, further considerations concerning the actual scheduling of the transmissions on each channel must be done. We will address that in the following section.

9.6.1. *Notes*

The notation used here follows mainly the one introduced in the paper by Sanders *et al.* [51], used also in the papers by Barbero and Ytrehus and the definitions are taken from the papers by Barbero and Ytrehus [5, 6]. The examples in Fig. 9.12 are borrowed from Example 1 [34], Example 2 [5], Example 3 [5] Katriel, Example 4 [2], and Example 5 [5].

9.6.1.1. *Maintaining the full rank invariant*

In the main loop of the LIFE algorithm, in order to find the local coefficients for $v(e)$ that satisfy the full rank invariant, one can use trial and error [34], trying all vectors $(m_e(p))_{p \in P(e)} \in \mathbb{F}_q^{|P(e)|}$ until one gives a global equation for $v(e)$ that gives rank h to the set V_t. (This approach should not be confused with random network coding.)

To maintain the full rank invariant in each step is a sufficient, although not necessary condition.

9.6.1.2. *Difference between LIF and LIFE*

The original LIF algorithm [51] works by visiting nodes in topological order, and this makes it deadlock when the network is link cyclic, even if it is flow acyclic.

9.7. Many Networks are Cyclic

Many papers in the literature avoid the problem of cyclic networks by considering only networks that are directed acyclic graphs. As we have seen, the LIFE algorithm can deal with link cyclic networks. Before proceeding in Sec. 9.8 with algorithms for *flow* cyclic networks, we pause for a moment to argue why this is important.

We will assume the following practical operating conditions for a network coding protocol:

(1) A network is available for use. In a given situation, the network may already be partially in use for other purposes, but we assume that the network and its nodes and edges have a finite available capacity that may or may not be precisely known.
(2) When a specific communication need arises, for example, in the initiation phase of a multicast session, the protocol must determine if the network admits the requested multicast. If so, the protocol must proceed to determine what part of the network should be used and to allocate the necessary resources (see Sec. 9.3.4), for example, by selecting a flow path subgraph. Sometimes the allocation is easy, sometimes harder, depending on the connectivity of the underlying graph.
(3) In the case of a centralized deterministic encoding algorithm, the final step of the setup phase is to determine the encoding equations and to distribute them to the nodes and sinks involved.

The protocol faces a complicated task, and several trade-offs are possible. It is, for example, possible to attempt to avoid a cyclic flow path graph [28] in step (2). However, this may be a difficult task and may increase substantially the complexity of finding the flow path graph. Furthermore, insisting on avoiding cycles may make a certain multicast request inadmissible. In a wireless setting, the existence of acyclic flow paths may require higher graph connectivity, i.e. higher number of neighbors and edges, which requires higher average transmission power and increased interference problems. Elimination of cyclic flow paths also may require the introduction of "almost parallel" partial paths, and hence loses opportunities for path sharing by network coding. To make matters worse, the algorithm in [28] is aggressive in the sense that it disallows not only flow cyclic, but even link cyclic flow path graphs.

Simulations described in [8] indicate that random multicast networks are highly likely to be cyclic, especially when several randomly located source nodes are involved. If these networks are only *link* cyclic, an algorithm such as LIFE can still deal with them, as shown in Sec. 9.6. Still we observed in [8] that many networks are even *flow* cyclic, unless there are only very few sources and sinks or they are highly co-located.

Moreover, as noted in Sec. 9.3.2, a broadcast node as shown in Fig. 9.7(a) is equivalent to a non-broadcast node pair as shown in Fig. 9.7(b). The corresponding equivalence in terms of the symbols sent over the edges is shown in Figs. 9.13(a) and 9.13(b). If we apply this equivalence to a link cyclic network like the one in Example 2 of Fig. 9.12, but in which all the intermediate nodes are supposed to be broadcast nodes we see that the resulting equivalent network with non-broadcast nodes may be flow cyclic as in Fig. 9.14(b). Note also that this network is essentially equivalent to Example 3 in Fig. 9.12.

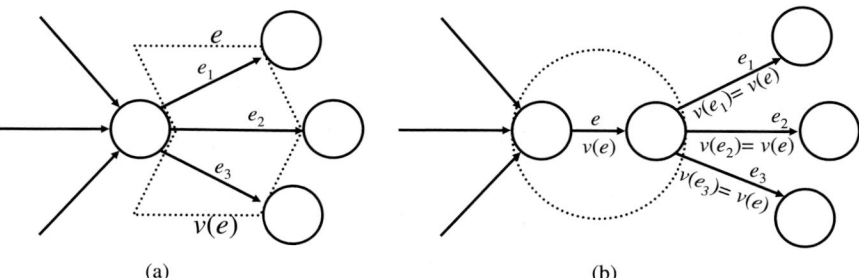

(a)   (b)

Fig. 9.13.   Symbols sent on edges exiting broadcast nodes when using the two models shown in Fig. 9.7.

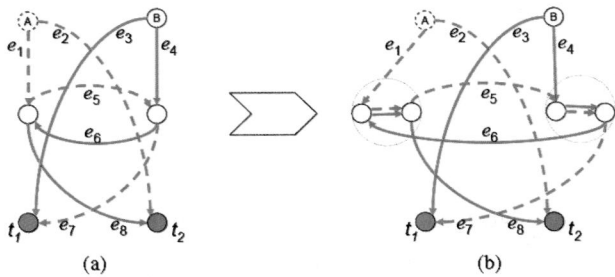

Fig. 9.14.   (a) A link cyclic network with broadcast nodes, (b) the same network with the broadcast nodes modeled as in Fig. 9.7(b).

9.8. Networks with Cycles in the Flow

Let us consider a flow cyclic network $G = (N, E)$ (actually a flow path graph).

We will distinguish two types of flow cycles, given by the following Definitions.

Definition 9.4. A *simple flow cycle* is a link cycle $\{e_1, e_2, \ldots, e_k, e_{k+1} = e_1\}$ such that for each $i = 1, \ldots, k$ there exists a flow path $f^{s_{j(i)}, t_{k(i)}}$ that implies $e_i \prec e_{i+1}$. We observe that there must be at least two distinct flow paths traversing the cycle.

Definition 9.5. A *flow knot* or simply a *knot* is formed by two or more simple flow cycles that share one or more edges.

Figure 9.15(a) shows a simple flow cycle and Fig. 9.15(b) shows a knot. Each flow traversing them is represented with a distinct color.

Examples 3, 4, and 5 in Fig. 9.12 are flow cyclic networks. Example 5 (which is a more elaborate butterfly) contains a knot. A more complicated knot is shown in Fig. 9.29.

We will first assume that the network G is flow cyclic but contains only simple cycles. In the second part of the section, we will deal with networks that contain knots.

Let $S = \{s_1, s_2, \ldots, s_h\} \subset N$ be the set of unit rate information sources and let $T = \{t_1, t_2, \ldots, t_r\} \subset N$ be the set of sinks. We will assume that each source (synchronously) generates one symbol at each (discrete) time instant. Let $x \in \mathbb{Z}$ denote the time. We will denote by $\sigma_i(x) \in \mathbb{F}_q$ the symbol produced by source s_i at time x.

The complete set of symbols $\sigma_i(x), i = 1, \ldots, h$ will be called generation x. The object of the algorithm is to find an assignment of equations in such

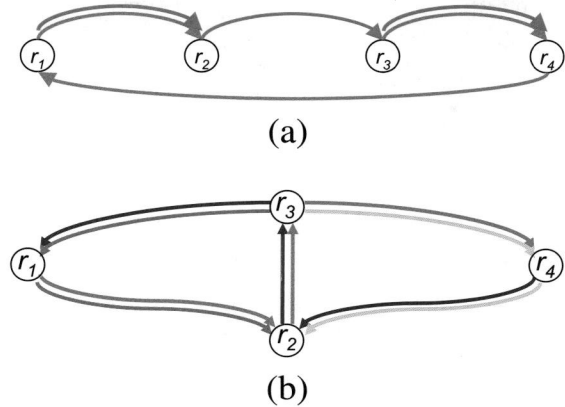

Fig. 9.15.   A simple cycle (a) and a knot (b).

a way that each sink $t \in T$ at each time $x + d_t$ can complete the decoding of
the whole generation x, where d_t is a (finite) constant for t denoting the total
delay associated with that sink. It is assumed that $\sigma_i(x) = 0, i = 1, \ldots, h$
for any $x < 0$.

It is straightforward (but messy) to modify the discussion to include
cases where edges have different or variable delay. Here, we assume for
simplicity that the transmission of a symbol on each link e has a unit delay
associated with it, that is to say, if a symbol s is being carried by edge e at
time x, that symbol, once processed at node $\text{end}(e)$, can be carried by an
edge e' at time $x + 1$, for any edge e' with $\text{start}(e') = \text{end}(e)$. In the same
manner, an edge e with $\text{start}(e) = s_i$ will carry symbol $\sigma_i(x - 1)$ at time x.

Hence the actual symbol transmitted on edge e at time x is expressed
as a global encoding of the source symbols:

$$v(e; x) = \sum_{i=1}^{h} \sum_{y=1}^{d_{\text{start}(e)}} \alpha(e, i, y) \sigma_i(x - y), \qquad (9.3)$$

where d_n is the maximum delay from any source to node n, and $\alpha(e, i, y)$
is a coefficient denoting the influence of the symbol $\sigma_i(x - y)$ in the linear
combination. For simple flow cyclic networks, there exists an encoding for
which d_n is finite $\forall n \in N$, as will be shown with the examples. The *value*
$v(e; x)$ is a *snapshot* of the information transmitted on edge e at time x.
This global encoding is still achieved by a time invariant local encoding of

each edge, like in (9.1), which in this section takes the form

$$v(e; x) = \sum_{p \in P(e)} m_e(p)v(p; x - 1) \tag{9.4}$$

for edges e that do not belong to a flow cycle and

$$v(e; x) = \sum_{p \in P(e)} m_e(p)v(p; x-1) - \sum_{p \in P_{\text{noncycle}}(e)} m_e^*(p)v(p; x-k-1) \tag{9.5}$$

for edges that lie on a flow cycle of length k. Observe that the coefficients $m_e(p)$ and $m_e^*(p)$ do not depend on the time variable x.

The explanation of the second sum in (9.5) will be given in the description of the LIFE-CYCLE algorithm.

Definition 9.6 (The LIFE-CYCLE algorithm). The *LIFE-CYCLE algorithm* is equal to the LIFE algorithm with the following exceptions:

- The encoding of (9.1) is replaced by (9.5) or (9.4) depending on whether the edge belongs to a flow cycle or not.
- The main loop is enhanced with a subalgorithm CYCLE-PROCESSING, which will be described below and which will be invoked only when the algorithm cannot find any single edge for which all the predecessors have been encoded, which means that it has encountered a flow cycle, which would cause the LIFE algorithm to deadlock.

Thus, the LIFE-CYCLE algorithm works like the LIFE algorithm by gradually extending flow paths, maintaining a full rank invariant for each sink. When a flow cycle is encountered, the CYCLE-PROCESSING algorithm will process the flow cycle as a whole, assigning local encoding functions for all edges in the cycle at once, in such a way that old symbols do not propagate indefinitely (Definition 9.7) while the *modified full rank invariant* (Definition 9.8) is satisfied.

A sufficient (but not necessary) condition for routers and sinks to be able to operate with finite memory is to avoid the coexistence of different generations of the same source symbol in the global encoding equation for any edge. A careless local encoding of a cycle edge, in contrast, may allow the effect of individual symbols to propagate indefinitely. In order to avoid this, we apply the *principle of old symbol removal* (OSR):

Definition 9.7 (Principle of OSR). Suppose there is a simple cycle of length k made up because some (say λ) of the flow paths impose a sequence of constraints $(e_1 \prec \cdots \prec e_{l_1}), (e_{l_1} \prec \cdots \prec e_{l_2}), \ldots, (e_{l_{\lambda-1}} \prec \cdots \prec e_{k+1} = e_1)$. The principle of OSR states that the following procedure has to be carried out for all edges e_j on the cycle.

There are two kinds of predecessor edges. There will always be exactly one predecessor on the cycle, namely the edge $p_{\text{cycle}}(e_j)$, which is e_{j-1} (remember that $e_1 = e_{k+1}$). Further there may be one or more predecessors not on the cycle. Those will be in the set

$$P_{\text{noncycle}}(e_j) = \{f_{\leftarrow}^t(e_j) | t \in T(e_j)\} \backslash \{p_{\text{cycle}}(e_j)\}.$$

The local encoding of edge e_j at time x will use the encoding $v(p; x-1)$ of its predecessors $p \in P_{\text{noncycle}}(e_j)$. However, k time instances ago, similar symbols were introduced into the cycle. The effect of $v(p; x-k-1)$ for any $p \in P_{\text{noncycle}}(e_j)$ should be removed from the encoding at this point, in order to limit the propagation of symbols. This can be done by subtracting from $v(p_{\text{cycle}}(e_j), x-1)$ the value $m_{e_j}^*(p) \cdot v(p; x-k-1)$ for each $p \in P_{\text{noncycle}}(e_j)$, where $m_{e_j}^*(p)$ is a coefficient that depends on the local encoding of other edges on the cycle.

Remark 9.1. The principle of OSR is a tool for containing the symbol propagation, which we anticipate will be a good idea especially in a large network, but it is not necessary for finding an encoding. We will assume that the principle is used in this section; however, there exist alternative network codes that do not use the OSR principle. This can be an appealing encoding if the network is small and if we want for some reason to avoid endowing some nodes with memory. We will not discuss this issue further in this chapter.

By the end of each iteration of the main loop of the LIFE and LIFE-CYCLE algorithms, we require that the encoding for each sink t satisfies the *modified full rank invariant*:

Definition 9.8 (Modified full rank invariant). The set of equations in V_t must be solvable, in the sense that at each time instant x the sink can recover a full set of h source symbols, one from each source, but not necessarily from the same generation.

The modified full rank invariant, together with the assumption that all symbols $\sigma_i(x) = 0$ for $x < 0$ from any source s, will allow the sink t to recover symbol $\sigma_i(x)$ (generated at time $x \geq 0$) at time $x + d_t$.

If during the main loop iteration we have processed all edges on a cycle at once (by the CYCLE-PROCESSING subalgorithm), the modified full

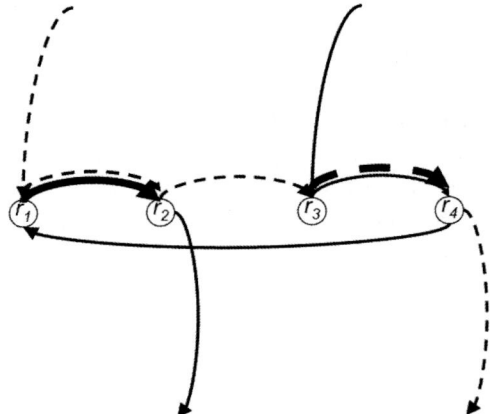

Fig. 9.16. Last edges in flow paths traversing a cycle.

rank invariant needs to be conserved with respect (only) to the last edges on each flow path that uses at least one edge on the cycle.

In Fig. 9.16 the solid flow path traversing the flow cycle needs to check the full rank invariant only with respect to the edge (r_1, r_2) marked bold in the figure, which is the last of the path contained in the cycle. In the same manner, the dashed path needs to check the full rank invariant at the edge (r_3, r_4) marked bold dashed.

We now state the CYCLE-PROCESSING algorithm:

The CYCLE-PROCESSING Algorithm

(1) *Input*:

A directed multigraph $G = (N, E)$, which is in fact a flow path graph,

A cyclic set of edges $C_E = \{e_1, \ldots, e_{k+1} = e_1\} \subseteq E$,

A set of vertices $C_N = \{\text{start}(e) \mid e \in C_E\} \subseteq N$,

Previously determined global encoding equations for all the edges entering the cycle C_E, that is $v(e; x - y), \forall y \geq 1, \forall e \in P(C) := \{P_{\text{noncycle}}(e_j), 1 \leq j \leq k\}$.

(2) *Main loop*:

    (a) /* *The modified full rank invariant concerns the set of last edges in the flow paths, which will be updated to* */

$$E_t := (E_t \setminus P^t(C_E)) \cup F^t(C_E), \quad \text{for each } t \in T(C_E),$$

---

where $T(C_E) = \{t \in T \mid f^t \cap C_E \neq \emptyset\}$ is the set of sinks that have flow paths passing through the cycle C_E, $P^t(C_E) = \{e \in E \backslash C_E \mid e \in f^t$ and $\text{end}(e) \in C_N\}$ is the set of predecessors of C_E on flow paths to t, and $F^t(C_E) = \{e \in C_E \mid e \in f^t$ and $f^t_\rightarrow(e) \notin C_E\}$ is the set of last edges on C_E on flow paths to t.

(b) /* *In order to update the corresponding set of equations V_t, we will need to consider all the encoding equations in the cycle, that is, the snapshot $v(e; x), \forall e \in C_E$, which for each e can be expressed as* */

$$v(e; x) = m_e(p_{\text{cycle}}(e))v(p_{\text{cycle}}(e); x - 1)$$
$$+ \sum_{p \in P_{\text{noncycle}}(e)} m_e(p)v(p; x - 1)$$
$$- \sum_{p \in P_{\text{noncycle}}(e)} m_e^*(p)v(p; x - k - 1). \qquad (9.6)$$

/* *See Remark 3 below* */

(3) *Output*:

Global and local encoding coefficients for the cycle.

We can make the following observations about the properties of the LIFE-CYCLE algorithm.

Remark 9.2. Comments on (9.6):

(1) $m_e(p_{\text{cycle}}(e))$ in (9.6) can be chosen to be 0 or 1. When $m_e(p_{\text{cycle}}(e))$ is non-zero, we can assume that it is equal to 1 since multiplying $v(e; x)$ with a non-zero scalar does not affect the modified full rank invariant. $m_e(p_{\text{cycle}}(e))$ can be zero only if the flow path graph is non-minimal (in a sense that we will not discuss further here), but in this case there really is no flow cycle and the encoding becomes straightforward. (In the next section, the LIFE* algorithm always discards zero coefficients, in order to keep the flow path graph already computed.)
(2) The coefficients $m_e(p)$ in the sum of (9.6) will be selected in order to maintain the modified full rank invariant, and as long as q is large enough (see below), this will always be possible.
(3) The coefficients $m_e^*(p)$ in the sum of (9.6) must be chosen to obey the principle of OSR, and these coefficients can be determined only when we know the values of the other coefficients.

Complexity is analogous to that of the LIFE algorithm and we also have analogous considerations about the field size. When $q \geq \pi_{\max}$ the algorithm succeeds finding encoding equations that satisfy the rank conditions, where now $\pi_{\max}$ is the maximum number of flow paths involved in either a single edge or in a simple cycle.

Example 9.1. We consider here encoding for the network in Example 3 of Fig. 9.12. Figure 9.17 shows the flow paths. We are using a color and pattern code: paths going to terminal t_1 are green and paths going to terminal t_2 are red. Paths exiting source s_1 are dashed while paths exiting source s_2 are solid (in other flow figures we will use similar color and pattern codes to distinguish the different paths).

The flow paths show that for this network $\pi_{\max} = 2$ and hence, as mentioned before, this network allows a binary encoding. In order to facilitate reading, we denote the source symbols by $a(x) = \sigma_1(x)$ and $b(x) = \sigma_2(x)$, respectively. Also, we will denote edges (s_i, t_j) as $s_i \to t_j$ in order to make the formulas below more easily readable. The algorithm starts by assigning encoding coefficients to edges $s_1 \to r_1, s_1 \to t_1, s_2 \to r_3$,

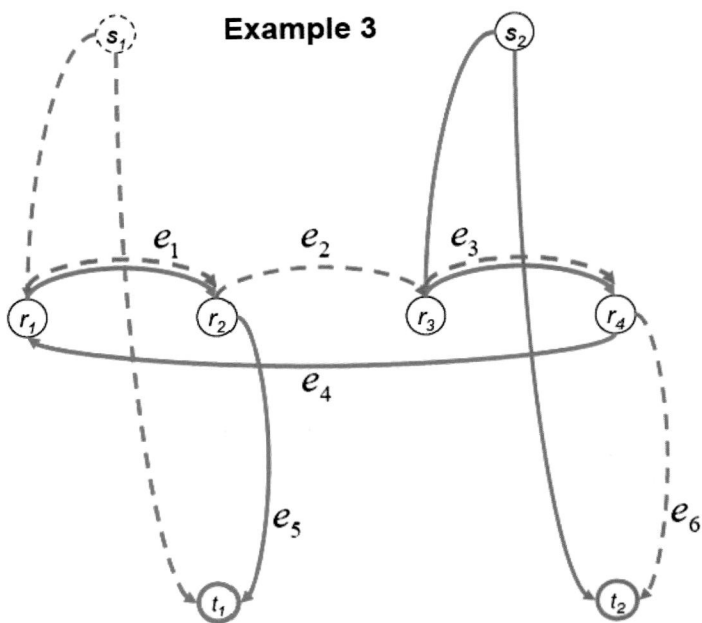

Fig. 9.17.   Flow paths for Example 3 in Fig. 9.12.

and $s_2 \to t_2$:

$$v(s_1 \to r_1; x) = v(s_1 \to t_1; x) = a(x - 1),$$
$$v(s_2 \to r_3; x) = v(s_2 \to t_2; x) = b(x - 1),$$

but then hits the cycle. The CYCLE-PROCESSING algorithm resolves the cycle as follows:

- Identify the cycle $C_E = \{e_1 = r_1 \to r_2, e_2 = r_2 \to r_3, e_3 = r_3 \to r_4, e_4 = r_4 \to r_1\}$.
- Before processing the cycle, we have that $E_{t_1} = \{s_1 \to t_1, s_2 \to r_3\}$ and $E_{t_2} = \{s_2 \to t_2, s_1 \to r_1\}$. The updates of these sets will be $E_{t_1} = \{s_1 \to t_1, e_1\}$ and $E_{t_2} = \{s_2 \to t_2, e_3\}$.
- Thus we need to determine the local encoding coefficients m_1 and m_2 so that the equation sets corresponding to E_{t_1} and E_{t_2} remain of full rank, using the encoding equations:

$$v(e_1; x) = v(e_4; x - 1) + m_1 \cdot a(x - 2) - m_1^* a(x - 6)$$
$$= \alpha_1 \cdot a(x - 2) + \beta_1 \cdot b(x - 4),$$
$$v(e_2; x) = v(e_1; x - 1) = \alpha_1 \cdot a(x - 3) + \beta_1 \cdot b(x - 5),$$
$$v(e_3; x) = v(e_2; x - 1) + m_2 \cdot b(x - 2) - m_2^* b(x - 6)$$
$$= \alpha_2 \cdot a(x - 4) + \beta_2 \cdot b(x - 2),$$
$$v(e_4; x) = v(e_3; x - 1) = \alpha_2 \cdot a(x - 5) + \beta_2 \cdot b(x - 3).$$

- These equations imply that $\alpha_1 = m_1 = m_1^* = \alpha_2$, and $\beta_1 = m_2 = m_2^* = \beta_2$, which gives

$$v(e_1; x) = m_1 \cdot a(x - 2) + m_2 \cdot b(x - 4),$$
$$v(e_2; x) = m_1 \cdot a(x - 3) + m_2 \cdot b(x - 5),$$
$$v(e_3; x) = m_1 \cdot a(x - 4) + m_2 \cdot b(x - 2),$$
$$v(e_4; x) = m_1 \cdot a(x - 5) + m_2 \cdot b(x - 3).$$

By the full rank conditions, we have

$$V_{t_1} = \{a(x - 1), m_1 \cdot a(x - 2) + m_2 \cdot b(x - 4)\}, \quad \text{so } m_2 = 1,$$
$$V_{t_2} = \{b(x - 1), m_1 \cdot a(x - 4) + m_2 \cdot b(x - 2)\}, \quad \text{so } m_1 = 1,$$

and we obtain the encoding shown in Fig. 9.18. Note that the delays are $d_{t_1} = d_{t_2} = 5$.

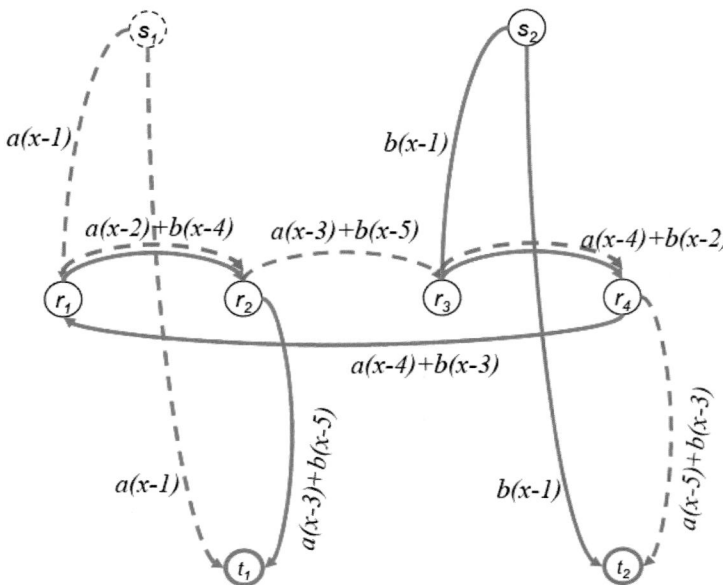

Fig. 9.18.   Encoding obtained by applying LIFE-CYCLE to Example 3.

We finish the example by showing how decoding can be done at the terminals. We will focus on terminal t_1. It will receive at time x the set of global equations

$$V_{t_1} = \{a(x-1), a(x-3) + b(x-5)\}.$$

This means that at time $x = 1$ it receives $a(0)$ on edge (s_1, t_1).

At time $x = 5$ it receives $a(2) + b(0)$ on e_5, which together with symbol $a(2)$ received on (s_1, t_1) at time $x = 3$ allows it to recover $b(0)$ and to complete the retrieval of generation 0.

From then on, it can, at each time instant, complete the recovery of a new generation in such a way that at time x it completes generation $x - 5$.

The memory required is, as we can see, finite and at most d_{t_1}.

Finally, when the network presents not only simple cycles, but also knots, as the one shown in Example 5 in Fig. 9.12, a slight modification of the cycle processing part of the algorithm will allow us to encode. That modification concerns only the way in which the expression of the old symbols circulating the knot is computed in order to be appropriately removed at the entrance nodes according to the OSR principle. The key tool here will be Mason's formula [38], used to compute the transfer functions on all the edges of the knot in order to compute the exact expression of the old symbols that must be removed at each entrance node.

We will explain this in more detail in Sec. 9.9, when presenting the LIFE* algorithm that also uses Mason's formula. Several detailed examples of the use of the formula will be provided.

9.8.1. *Notes*

In the literature, the use of delay and encoding over several generations is sometimes referred to as *convolutional network coding*. We prefer not to use this name, since we feel that tying this network coding technique to the theory of convolutional codes may confuse rather than illuminate.

9.9. On the Alphabet Size, and an Efficient Algorithm to Encode Over the Binary Field

In the previous sections, we have seen that the field size plays an important role in network coding. The linear network encoding for multicast in a network that satisfies the conditions of Theorem 9.1 can be given as long as the field is large enough. How large is "large enough" is an interesting question, and there are different approaches to the answer.

A first interesting answer in this respect is given by the results in the paper by Sanders *et al.* [51] and by Jaggi *et al.* [34]. A network with r sinks can admit a valid linear network encoding over the finite field $\mathbb{F}_q$ when $q \geq r$.

Soljanin and Fragouli [22], by applying techniques of graph coloring, gave an accurate bound for networks with $h = 2$ unit rate sources setting that a field of size

$$q \geq \lfloor \sqrt{2r - 7/4} + 1/2 \rfloor \tag{9.7}$$

is sufficient for linear multicast encoding, and the result is conjectured to be true for any value of h.

This upper bound is tight in the sense that for any r there exists a network configuration with 2 unit rate sources and r sinks that requires a field of cardinality $q \geq \lfloor \sqrt{2r - 7/4} + 1/2 \rfloor$ in order to be encoded. Nevertheless, the size dictated by the bound is not necessary in most cases.

The problem of computing the minimum field size needed by a given network for linear multicast encoding has been proven to be NP-hard. In practice most networks seem to require very small fields. Hence, we present a more practical approach.

We start by recalling that once the flow paths have been computed for a given network, a field of size $q \geq \pi_{\max}$ suffices. In most cases, this gives a bound on the field size, which is smaller than (9.7), but for many networks this bound is not tight either.

For practical purposes, it is desirable to operate with binary operations. One possibility is to adopt an opportunistic approach, trying to use the binary field until it is impossible to proceed any further.

The opportunistic version of LIFE (or of LIFE-CYCLE) works by proceeding with encoding over $\mathbb{F}_2$ until no more edges can be encoded over that field while satisfying the full rank invariant. If at any point the full rank invariant for the present edge can no longer be satisfied over $\mathbb{F}_2$, the algorithm switches to an extension field $\mathbb{F}_2(\alpha)$, where α is a primitive element, root of a primitive polynomial $x^2 + x + 1$ over $\mathbb{F}_2$. The encoding equations calculated so far do not need to be revised, since they will keep working over the new extension field. When no more edges can be encoded over the present field, a new switch to a next extension field is required. In this manner, we proceed by successively "jumping" to larger extensions only when needed.

The final result is not necessarily an encoding over the smallest field possible for the given network. However, on the other hand, all the arithmetic is kept binary, and the operational complexity of running the algorithm is exactly the same as before, since each edge is visited and encoded only once.

We have an example of the performance of this opportunistic LIFE algorithm when applying it to encoding the network of Example 1 in Fig. 9.12, which is an instance of the networks known as combination networks.

A (k, n) combination network is an acyclic network with k rate 1 sources converging in one node (or one k-rate source), a first layer of n nodes all connected to that common node (or source) and finally a second layer with $\binom{n}{k}$ terminal nodes such that each one is connected to a different subset of cardinality k of the nodes in the first layer.

Example 1 (flow paths shown in Fig. 9.19) is the (2,4) combination network, and it is well known that it requires a field larger than $\mathbb{F}_2$ to be encoded. When proceeding with the LIFE algorithm we can see that three of the edges $e_3, e_4, e_5,$ and e_6 can be encoded using $\mathbb{F}_2$ but not the fourth one. Ignoring the uncoded edges e_1 and e_2, we see that $\pi_{\max} = 3$, and the

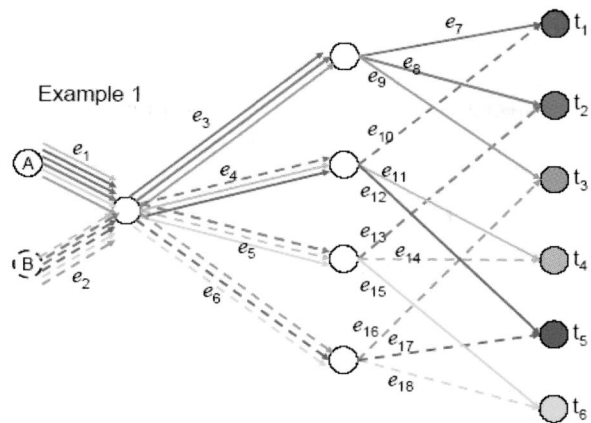

Fig. 9.19. A flow path graph for the $(2, 4)$ combination network.

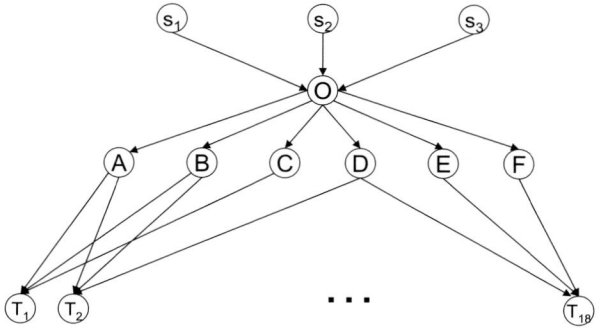

Fig. 9.20. Almost a combination network.

LIF/LIFE algorithm finds an encoding over $\mathbb{F}_3$. The opportunistic version of the algorithm will work by starting with $\mathbb{F}_2$. For the fourth edge it will jump to $\mathbb{F}_2(\alpha)$, a binary extension field of cardinality 4. In the resulting encoding, all nodes will operate in binary except four nodes that will work in $\mathbb{F}_4$.

In many cases, the opportunistic algorithm finds an encoding in a field smaller than the one suggested by the bounds.

Consider the network in Fig. 9.20 in which there are 18 receivers $T_1, \ldots, T_{18}$. Not all edges are drawn.

In the table, we describe the flow paths chosen by listing for each receiver t the last routers in the flow paths $f^{s_1,t}$, $f^{s_2,t}$, and $f^{s_3,t}$ in that order (note that this determines both the network and the flow paths chosen).

Table 9.1.   List of receivers and their neighbors in Figure 9.20.

T_1	ABC	T_7	ADE	T_{13}	FDC
T_2	ABD	T_8	ADF	T_{14}	EFC
T_3	ABE	T_9	AFE	T_{15}	EBD
T_4	ABF	T_{10}	EBC	T_{16}	FBD
T_5	ADC	T_{11}	FBC	T_{17}	EBF
T_6	AFC	T_{12}	EDC	T_{18}	FDE

The opportunistic LIFE will produce the following encoding for the six non-trivial encoding edges:

$$(O, A) \rightarrow (1, 0, 0), \quad (O, B) \rightarrow (0, 1, 0),$$
$$(O, C) \rightarrow (0, 0, 1), \quad (O, D) \rightarrow (0, 1, 1),$$
$$(O, E) \rightarrow (1, 0, 1), \quad (O, F) \rightarrow (\alpha, \alpha, 1).$$

The above example has shown a network for which the bound (9.7) takes the value 6, which means a field of size at least 7, while the opportunistic LIFE algorithm finds an encoding again over $\mathbf{F}_2(\alpha)$ (field size 4). The result obtained by using the opportunistic-LIFE algorithm not only needs a smaller field, but also has the advantage that all the routers except for F will work in $\mathbb{F}_2$ and also 8 of the 18 receivers will work in $\mathbb{F}_2$.

9.9.1. *Further Remarks*

As we pointed out, in the acyclic parts of the network the flow paths define a partial order in the edges. The LIFE algorithm follows this topological order, but it is not always unique (since it is not, in principle, a total order). The different choices to process the edges respecting the partial order can result in differences in the field size required.

We illustrate this remark by means of Example 1(b) (see Fig. 9.21), which is constructed from Example 1 by removing terminal t_6 and changing one of the connections of terminal t_3 (the flow graph for this network is essentially unique except for what we will discuss below).

If we apply the opportunistic LIFE algorithm to encode the flow path graph shown in Fig. 9.21(a) to encode edges e_3, e_4, e_5, and e_6 in that order, the encoding will require jumping to $\mathbb{F}_2(\alpha)$, while processing them in the order e_5, e_4, e_3, and e_6 will give a final encoding for the whole network using only $\mathbb{F}_2$.

This shows that the order in which the edges are visited (always respecting the topological order) matters. Unfortunately, there seems not

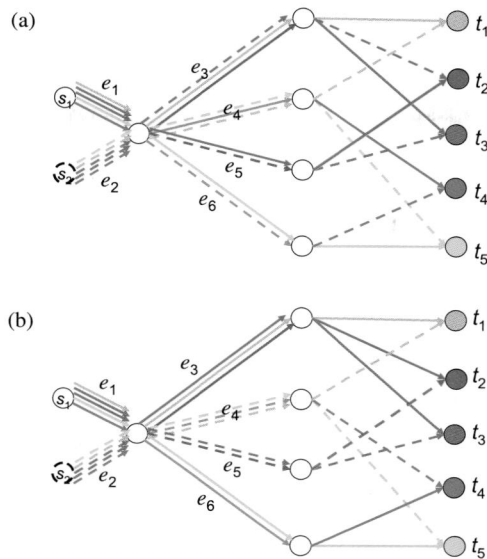

Fig. 9.21. Example (1b).

be a simple way of detecting beforehand, which order is more convenient for a general network, and checking all the possibilities might lead to non-polynomial complexity.

Another aspect that influences the required field size is the choice of flow path graphs. The size of the largest field needed when applying this version of the algorithm might depend on the flow path graph selected for the network (when this is not unique). Even such a slight variation as considering the two different options when two paths going to the same sink ($f^{s,t}$ and $f^{s',t}$) meet at an intermediate node (in the color-pattern code for representing the flow paths this means that the distribution of the patterns change, even when it does not change the distribution of the colors). In Fig. 9.21(b) we can see another flow path graph for the same network as in Fig. 9.21(a). It is almost the same flow path graph except that the assignment of patterns has been rearranged. Now, regardless the order in which we visit the edges e_3, e_4, e_5, and e_6, the encoding produced by the opportunistic version of the LIFE algorithm is always binary (in fact it becomes trivial since, in this flow path graph, each of the edges e_3, e_4, e_5, and e_6 has a unique predecessor.) Hence the choice of flow path graph for a given network can be crucial.

When different flow path graphs exist (not only in terms of patterns, but in terms of colors), in general a "dense" flow path graph is more efficient in terms of energy and edge reuse but also requires a higher field, although the exact relationship between the flow path graph and the field size is hard to determine.

9.9.2. The LIFE* Algorithm

Another approach exploits the natural delay introduced when traversing the edges of the network. We will devote the rest of the section to explain it and illustrate it with some examples. Here we will also go into detail in the use of the Masons' formula.

Again we assume that the transmission of a symbol on each link e has a unit delay associated with it, so that if a symbol s is being carried by edge e at time x, that symbol, once processed at node $end(e)$, can be carried by an edge e' at time $x + 1$, for any edge e' with $start(e') = end(e)$.

Apart from this intrinsic delay, the encoding process might assign extra delays in order to satisfy certain required conditions.

Let D denote the linear delay operator. The way in which the operator works is as follows:

$$D(\sigma(x)) = D\sigma(x) = \sigma(x - 1),$$

and it is extended by linearity.

$$D(\sigma(x) + \xi(x)) = D\sigma(x) + D\xi(x) = \sigma(x - 1) + \xi(x - 1),$$

$$D^i\sigma(x) = D(D^{i-1}\sigma(x)) = \sigma(x - i) \quad \forall\, i \in \mathbb{Z}_0^+,$$

$$(D^i + D^j)\sigma(x) = D^i\sigma(x) + D^j\sigma(x) = \sigma(x - i) + \sigma(x - j).$$

Also, abusing notation, we extend the operator in order to work with negative exponents in the following manner:

$$D^{-1}\sigma(x) = \sigma(x + 1).$$

In the whole process of encoding, we will observe two basic principles.

- *Pass on information (PI) principle* [5]: When encoding each edge $e \in E$, all the symbols carried by the incoming edges $f_{\leftarrow}^t(e), t \in T(e)$ will contribute. This is because assigning coefficient 0 to any of those predecessors means that the flow carried on that predecessor is stopped at that point, meaning in turn that the actual flow path graph used is different from the one established initially. [a]

- *OSR principle* [5]: To avoid symbols circulating endlessly on each flow cycle, they should be removed at the entrance point by the node through which they entered the cycle. As with the LIFE-CYCLE algorithm, we will explain how to do this in 9.9.2.2 and in detail by means of Examples in 9.10.

In what follows, we will consider, as usual, two ways of expressing the symbol $v_e(x) \in \mathbb{F}_2$ to be transmitted on edge e at time x.

The first one, the *local encoding equation*, that specifies the action of each node. The general local encoding equation is now written as

$$v_e(x) = \sum_{p \in P(e)} \pi(p, e)\tau(p, e)v_p(x), \qquad (9.8)$$

where $\pi(p, e)$ is a polynomial in the operator D, which denotes the encoding, i.e. they expresses the additional delay introduced for the corresponding flow path at the node start(e). The term $\tau(p, e)$ denotes a function of the operator D, which we will call the transfer function from p to e and which accounts for the natural delay that, as previously noted, is inherent to the transmission on each edge. The way to compute this function will be explained in detail for each case. The process of determining the network code consists of finding a set $\{\pi(p, e), e \in E, p \in P(e)\}$.

Remark. To simplify the encoding, we propose to use only monomial $\pi(p, e)$'s, that is, $\pi(p, e) = D^{i_e(p)}$ for some $i_e(p) \in \mathbb{Z}_0^+$. Thus, the encoding consists of simply adding the encoding vectors of all predecessor edges, each one artificially delayed for zero or more time units as necessary. The algorithm can be straightforwardly modified to allow all polynomials in $\mathbb{F}_2(D)$ to be coefficients of the encoding combination. We have chosen the proposed scheme because it is convenient, and because it exploits the flow path graph that has been computed beforehand. We do not claim that the restriction gives other benefits in terms of delay. Since there exists an encoding for any flow path graph, once a flow path graph has been

---

[a]Observe that this principle makes sense if and only if the flow path graph is "reasonably efficient". It may require some effort to calculate a reasonably efficient flow path graph, but if that work has been done it is wise, in terms of complexity, to rely on the provided flow path graph. The connection between flow path graph calculation and network coding can offer interesting complexity trade-offs; however, this discussion is beyond the scope of this tutorial chapter.

computed we want to follow it and profit from that in order to reduce the
computational complexity of finding the encoding equations for each edge.
With this simplification, the encoding takes the form

$$v_e(x) = \sum_{p \in P(e)} D^{i_e(p)} \tau(p, e) v_p(x), \qquad (9.9)$$

where $i_e(p) \in \mathbb{Z}_0^+$ are the exponents that express the additional delays
introduced at the node start(e).

The exponents $i_e(p)$ do not depend on the time variable x. We remark
that employing a time invariant encoding, as in this case, ensures that
streams of symbols from each source can be pipelined through the network.

The *global encoding equation* now takes the form

$$v_e(x) = \sum_{i=1}^{h} \sum_{y \in \mathbb{Z}_0^+} \alpha(e, i, y) \sigma_i(x - y), \qquad (9.10)$$

where $\alpha(e, i, y) \in \mathbb{F}_2$ is the coefficient that specifies the influence of each
source symbol on $v_e(x)$.

Since $\sigma_i(z) = 0$ for all $z < 0$, the sum over y in (9.10) is finite, because
the terms with $y > x$ will all be 0.

Another way of expressing the global encoding equation, more useful
for our purposes, is by means of the field of rational functions $\mathbb{F}_2(D)$, where
D is the delay operator.

$$v_e(x) = \sum_{i=1}^{h} F_{e,i}(D) \sigma_i(x), \qquad (9.11)$$

where $F_{e,i}(D) \in \mathbb{F}_2(D)$. Again, all the monomials in D^y with $y > x$ will
give null terms in the sum.

The goal of the LIFE* algorithm is to determine the local encoding
equations of each edge e in each flow path, and the global encoding equation
of the last edge in each flow path. In fact, during the course of the algorithm,
all the global encoding equations will be determined.

In particular, the global encoding should be such that each sink t can,
after a certain constant delay d_t, extract all the h information symbols
belonging to a generation from the h sources. Each sink t, at time x, receives
h symbols, namely $\{v_{e_i}(x) | e_i \in f^{s_i, t}, t = \text{end}(e_i), i = 1, \ldots, h\}$. If the
corresponding h global encoding equations are linearly independent over
the field $\mathbb{F}_2(D)$, the sink t can reconstruct the h input symbols of generation

x at time $x + d_t$, by solving (in a simple manner) the corresponding set of equations.

As in the LIFE algorithm, in the case of a flow acyclic network, all the edges in E can be visited according to the partial order $\prec$ induced by the flow paths, starting by the edges with no predecessor ($e \in E \mid \text{start}(e) = s_i$ for some i), and proceeding in such a way that an edge e will not be visited until all its predecessors (all $p \in P(e)$) have already been visited and their global encoding equations computed. If the global encoding equations of all predecessors of e are substituted into (9.9), we get the global encoding equation for e.

When the network is flow cyclic the algorithm comes to a point at which no more edges can be visited in topological order, that is, the algorithm hits the flow cycle. Once the edges taking part in that cycle have been identified, the encoding of the whole cycle has to be treated as a whole. Here the OSR principle is useful. An appropriate version of Mason's formula (see [38]) for the transfer function on a circuit will be the main tool to deal with it. We will explain this in detail in the second part of this section.

Similar to the LIF and the LIFE algorithms, the new LIFE* algorithm proceeds by maintaining, through the iterations of the main loop, the following sets for each $t \in T$:

- A set $E_t \subset E$, $|E_t| = h$, such that E_t contains *the most recently visited edge* on each flow path in f^t.
- An $h \times h$ matrix M_t, in which the element i, j will be the coefficient of $\sigma_i(x)$ in the global encoding equation of the jth element of the subset E_t:

$$
M_t = \begin{pmatrix} F_{e_{t,1},1}(D) & \cdots & F_{e_{t,h},1}(D) \\ \vdots & & \vdots \\ F_{e_{t,1},h}(D) & \cdots & F_{e_{t,h},h}(D) \end{pmatrix},
$$

where $e_{t,1}, \ldots, e_{t,h}$ are the edges in E_t.

Through each step of the algorithm we will impose the *full rank condition: The matrix M_t must have rank h, for all $t \in T$.* This condition is sufficient, but not necessary, for obtaining a valid network code. At the final step, for each $t \in T$ the set E_t will be that of the h edges, which arrive in t. If these edges carry the symbols $r_{t,1}(x), \ldots, r_{t,h}(x)$ at time x, we have

$$
(\sigma_1(x), \ldots, \sigma_h(x))M_t = (r_{t,1}(x), \ldots, r_{t,h}(x)).
$$

By the full rank condition the matrix M_t is invertible, and the symbols $(\sigma_1(x), \ldots, \sigma_h(x))$ can be recovered from the received ones as

$$(\sigma_1(x), \ldots, \sigma_h(x)) = (r_{t,1}(x), \ldots, r_{t,h}(x))M_t^{-1}.$$

We will now explain how the algorithm will proceed.

9.9.2.1. *The flow acyclic parts of the network*

Whenever there is an edge e that can be visited in topological order (that is to say, one for which all the predecessors have been already visited), the algorithm will proceed to update the sets E_t and V_t in the following manner:

$$E_t := \{E_t \setminus \{f_\leftarrow^t(e)\}\} \cup \{e\} \quad \text{for each } t \in T(e).$$

We consider the general local encoding equation for edge e

$$v_e(x) = \sum_{p \in P(e)} D^{i_e(p)} \tau(p, e) v_p(x),$$

where $i_e(p) \in \mathbb{Z}_0^+$ are the unknowns and represent the extra delay added at edge e to maintain the full rank invariant.

In the acyclic case with unit link delay, $\tau(p, e) = D$, $\forall p \in P(e)$. If the natural link delay is anything else, or not necessarily the same for all links, it will be straightforward to adapt the corresponding equations. The local encoding formula (9.9) in the acyclic case with unit link delay takes the form

$$v_e(x) = \sum_{p \in P(e)} D^{i_e(p)} D v_p(x) = \sum_{p \in P(e)} D^{i_e(p)+1} v_p(x). \tag{9.12}$$

When the global encoding equations for all the $v_p(x)$, already determined in the previous steps of the algorithm, are substituted in the above expression, we will have the global encoding equation for e.

Further, for each $t \in T(e)$, replace in each matrix M_t the column corresponding to the encoding equation of $f_\leftarrow^t(e)$ with the new one $v_e(x)$.

The unknowns will be chosen to have values that satisfy the full rank invariant for all the matrices $M_t, t \in T(e)$, which have been updated following the update of the corresponding set E_t.

Conjecture 9.1. *There exists a finite value I, that depends on the graph and in particular on the set $T(e)$, so that some set $\{i_e(p) : p \in P(e)$ and*

$i_e(p) < I\}$, when applied to (9.9) and (9.12), will satisfy the full rank condition.

This conjecture is in a way similar to Lemma 6 in [51]. There the field size required to guarantee the full rank invariant is proven to be $|T(e)|$, that is to say, each coefficient of the linear combination can be chosen among $|T(e)|$ possibilities. If we assume $I = |T(e)|$, the coefficients of the combinations in (9.9) can be chosen also among $|T(e)|$ different possibilities, namely $D^0 = 1, D, D^2, \ldots, D^{|T(e)|-1}$. Nevertheless, the proof cannot use the Linear Algebra arguments used there because the set with which we are working is not a vector space.

The conjecture is further supported by Theorems 9.1 and 9.3 in [39], and by software simulations for random networks that we have carried out. We omit the details of these simulations.

For networks with random structure, most encodings need no extra delay at all, and in the few remaining cases a delay of one unit in one of the incoming paths is enough to solve the problem for most of them. This observation also tells us that in order to find the encoding for each edge, which means finding the delay exponents $i_e(p)$, an efficient approach will be to use a greedy algorithm (see [34]) that starts by considering $i_e(p) = 0$, $\forall p \in P(e)$, checks if the full rank condition is satisfied, and in case it is not, proceeds to increment the delay exponents one by one until a solution is found. According to our simulations, the average number of tries needed to find the solution for each edge will be very low.

9.9.2.2. *Dealing with flow cycles*

When the algorithm encounters a flow cycle, the set of edges forming part of the cycle has to be computed. Let us recall some notation already used for the LIFE-CYCLE algorithm. We denote by C_E the set of edges and C_N the set of nodes that are ends of those edges in the cycle.

Let us call $P(C)$ the set of predecessor edges of the cycle,

$$P(C) = \{e \in E \backslash C_E \mid e \in P(e') \text{ for some } e' \in C_E\}.$$

We assume that the encoding equations for all the edges $e \in P(C)$ have been already determined in previous steps of the algorithm. The goal of this step of the algorithm will be similar to the CYCLE-PROCESSING part of the LIFE-CYCLE algorithm: to find the encoding equation of all the edges in C_E simultaneously. In a sense it is as if the cycle as a whole is being treated as a kind of "superedge". All the individual edges in it will have the same structure of equation, which will be a combination of the equations of the edges in $P(C)$, that is,

$$v_C(x) = \sum_{p \in P(C)} D^{ic(p)} \tau(p, C) v_p(x). \tag{9.13}$$

Here $\tau(p, C)$ is a notation with which we simply mean a transfer function that will have to be computed separately for each particular edge in C_E. Thus, each edge in C_E will have a slightly different version of that basic structure due to the fact that they lay in different parts of the cycle and will consequently observe the incoming equations with different delay. We will distinguish two cases, namely, when the flow cycle is simple, or when it is a knot. The distinction will be made just for the clarity of explanation, since the simple case is just a particular case of the knot case.

Once the exact encoding equation has been determined for all the edges in C_E, the full rank invariant has to be checked only for the last edges in the cycle for each flow path, that is to say, for each t with $f^t \cap C_E \neq \emptyset$ the corresponding full rank condition must be satisfied by the edges $e \in f^t \cap C_E$ such that $f^t_\rightarrow(e) \notin C_E$.

9.9.2.3. *The simple flow cycle case*

Suppose $C_E = \{e_1, e_2, \ldots, e_k\}$ with $\text{end}(e_i) = \text{start}(e_{i-1})$ for $i = 1, \ldots, k-1$ and $\text{end}(e_k) = \text{start}(e_1)$.

Let us consider the local encoding equation of the cycle (9.13). We show now how to use the OSR principle.

We will focus on a certain edge in C_E, for instance e_1. Suppose the local (and global) encoding equations of e_k, the predecessor of e_1 in the cycle, have been determined exactly:

$$v_{e_k}(x) = \sum_{p \in P(C)} D^{ie_k(p)} \tau(p, e_k) v_p(x).$$

If e_1 had no predecessor outside the cycle, that is to say $P(e_1) = \{e_k\}$, then the encoding equation of e_1 would simply be $v_{e_1}(x) = D v_{e_k}(x) = \sum_{p \in P(C)} D^{ie_k(p)} \tau(p, e_k) D v_p(x)$, which means $\tau(p, e_1) = D\tau(p, e_k) \ \forall p \in P(C)$. In the same manner, it is clear in general that if $p \in P(C) \backslash P(e_1)$, then $\tau(p, e_1) = D\tau(p, e_k)$.

Now suppose that there is one particular edge $p_1 \in P(C)$, which is the unique predecessor of e_1 not in C_E, this implies $\tau(p_1, e_1) = D$. Not removing at that point the old contribution of that predecessor would mean that the new contribution would mix with the old ones on each loop of the cycle and would keep circulating forever. In order to avoid this we want to remove the old contribution that came on edge p_1 k time instants ago (where k is obviously the length of the cycle) and contribute to the circulation in the cycle with only the newest symbol coming on p_1. This is done as follows:

$$
\begin{aligned}
v_{e_1}(x) &= D \cdot v_{e_k}(x) + D^{i_C(p_1)} \tau(p_1, e_1)[-v_{p_1}(x - k) + v_{p_1}(x)] \\
&= D \cdot v_{e_k}(x) + D^{i_C(p_1)} D[-v_{p_1}(x - k) + v_{p_1}(x)].
\end{aligned}
$$

Here we have used the minus operator $(-)$, although all the operations are always on the binary field, in order to emphasize which symbols are being removed from the circulation.

In general, if $e \in C_E$ has several predecessors not lying in the cycle, the local encoding equation of e in terms of the predecessors of e takes the form

$$
\begin{aligned}
v_e(x) &= D \cdot v_{p_C(e)}(x) + \sum_{p \in P(e) \cap P(C)} D^{i_C(p)} \tau(p, e)[-v_p(x - k) + v_p(x)] \\
&= D \cdot v_{p_C(e)}(x) + \sum_{p \in P(e) \cap P(C)} D^{i_C(p)} D[-v_p(x - k) + v_p(x)],
\end{aligned}
$$

where $p_C(e)$ is the single edge in $P(e) \cap C_E$.

The result of doing this at the entrance in the cycle of each predecessor $p \in P(C)$ is that only one "instance" of the symbols carried by each $p \in P(C)$ will be circulating on each edge $e \in C_E$. It is easy to see that the transfer function from each predecessor of the cycle to each edge in the cycle will be $\tau(p, e) = d(p, e) \; \forall p \in P(C), e \in C_E$, where $d(p, e)$ is the "distance" measured in number of edges in the cycle that lay between $\text{end}(p)$ and $\text{end}(e)$.

To summarize, the local encoding equation of each edge $e \in C_E$ in terms of the predecessors of the cycle is

$$
v_e(x) = \sum_{p \in P(C)} D^{i_C(p) + d(p, e)} v_p(x),
$$

where again, the only unknowns are the values $i_C(p) \in \mathbb{Z}_0^+$.

An example of the use of this procedure can be found in Example 4 in Sec. 9.10.

9.9.2.4. *The knot case*

Suppose the set C_E is not just a simple flow cycle but forms a knot.

Again the local encoding equation for the whole knot will share a common structure

$$v_C(x) = \sum_{p \in P(C)} D^{ic\,(p)} \tau(p, C) v_p(x).$$

Once more the idea is to treat the whole knot as a kind of "superedge".

Here the main idea is the same as before: the old symbols must be removed from the circulation. In order to do it one needs to know how those arrive at each edge of the knot, and for this we need as a tool Mason's formula (see [38, 46]) for the computation of the transfer function on a cyclic circuit.

We apply Mason's formula to the directed line graph associated with C_E in the following manner: Two edges e' and e in C_E are considered adjacent (and an arc starting in e' and ending in e will be drawn in the line graph) if and only if there exists a flow f^t for some $t \in T$ such that $f^t_{\leftarrow}(e) = e'$.

For each symbol entering the knot, we have to compute the corresponding transfer function over all the edges in C_E. For this we will consider the entrance point of the symbol (the edge at which that symbol enters) and the exit point (that is, the edge at which we want the transfer function of that symbol), and will apply Mason's formula between these points in the line graph mentioned above, and use the delay operator D as the branch gain (see [38]) of each edge. Again, each branch gain can be taken to be whatever function of D models best the actual behavior of the transmission on that edge and the corresponding equations can be adapted consequently.

We show in detail how to compute the transfer functions by means of Example 6 in in Sec. 9.10.

The local encoding equation of edge $e \in C_E$ in terms of its predecessors is

$$v_e(x) = \sum_{p \in P(e) \cap C_E} D v_p(x) D + \left[\sum_{p \in P(e) \cap P(C)} D^{ic\,(p)} v_p(x) \right.$$
$$\left. - \sum_{e' \in P(e) \cap C_E} \sum_{p' \in P'(e) \cap P(C)} D^{ic\,(p')} \tau(p', e') v_{p'}(x) \right], \quad (9.14)$$

where $P'(e) = \{p \in E \mid \text{end}(p) = \text{start}(e)\}$ (for a network that is a flow path graph). Note that $P(e) \subseteq P'(e)$ but the converse is not true in general. For instance, in Example 5 $P'(e_{13}) = \{e_5, e_8, e_{17}\}$ while $P(e_{13}) = \{e_5, e_{17}\}$ (see Fig. 9.24). (When the network we are working with is already a flow path graph, $P'(e) = I(e)$.)

The second sum in the formula brings the updated versions of the symbols that enter the knot at that point, while the double sum in the third term of the formula takes care of removing the old symbols.

This results in the following local encoding equation of each edge in the cycle in terms of the predecessors of the cycle:

$$v_e(x) = \sum_{p \in P(C)} D^{ic(p)} \tau(p, e) v_p(x). \tag{9.15}$$

One can see that the previous case is just a particular case of this one, since the line graph that will be associated with a simple flow cycle will always contain a simple cycle itself, and the corresponding transfer function between each $p \in P(C)$ and each edge e in the cycle will be $\tau(p, e) = d(p, e)$ as was shown in Sec. 9.9.2.3.

A final observation at this point is that when the network is flow acyclic or contains only simple flow cycles, the global encoding equations will only contain polynomials on D, and not rational functions. Rational functions will be the result of using Mason's formula on knots.

Details of *decoding* will be considered in Sec. 9.10.

We conclude the current section by summarizing the complete LIFE* algorithm.

Algorithm Life*

Input: A flow path graph $G = (N, E)$.

Initializing: $\forall t$:

- $E_t = \{e | e \in f^{s_i, t}, \text{start}(e) = s_i, i = 1, \ldots, h\}$,
- $\{v_e(x) = D\sigma_i(x) = \sigma_i(x - 1) | \ e \in E_t, \text{start}(e) = s_i\}$ or, equivalently, $M_t = DI_h, \ \forall t \in T$.

Main loop: Select an edge e for which the encoding equations have not yet been determined, but for which the global encoding equations of all the predecessor edges in $P(e)$ have been determined. Then proceed with the update of the set of current edges and current encoding equations as described in Sec. 9.9.2.1.

> If selecting such and edge is not possible, then a flow cycle has been encountered. Follow the procedure explained in Sec. 9.9.2.2.
>
> *Output:* For each edge, the local encoding as given by (9.9) is produced. At the end of the algorithm, $E_t = \{e \,|\, \text{end}(e) = t\}$ and M_t is still a matrix of rank h from which t can recover the input.

Examples of application of the algorithm can be found in Sec. 9.10.

9.9.3. *Notes*

The algorithm LIFE* will execute the main loop at most $|E|$ times. The exact complexity depends on details of the algorithm not discussed here. However, the complexity of the LIF and LIFE algorithms are similar. For discussions on the complexity of the algorithm, we also refer the reader to [51].

The encoding presented here follows a flow path graph given for a network. This flow path graph is not necessarily unique, and the choices made when computing the flow path graph determine much of the possible encodings that can be achieved. Which flow path graph is the best choice remains an open problem. First one should consider in which way the solution wants to be optimal (minimal delay, minimal number of links used, minimal number of encoding nodes, etc.). Some notions of minimality in the flow path graph can be considered that we will not discuss here. Also we will not discuss the different strategies that can be used in order to compute a flow path graph.

Once the flow path graph for the network has been computed, the algorithm proceeds by following a topological order of the edges whenever that is possible (until a flow cycle or knot is found). However, this topological order is not unique. In many cases there is a certain choice to be made at each step about which edge will be encoded next of the several that follow in the order. This choice might in certain cases influence the total amount of delay necessary for the encoding. Examples can be given in which different choices of order lead to different final amounts of delay. Which ordering is most convenient for each flow path graph is also an open problem.

Another consideration to take into account is that the presence of added delay means that the nodes at which the delay has to be introduced must have memory elements to store the symbols that have to be "artificially" delayed. In most cases the sinks will need to use memory in order to be able to solve the equations. In any case the maximum delay used is finite. In case no extra delay needs to be added, the maximum delay needed on each path will correspond to the total length of that path from source to sink.

9.9.3.1. *Network precoding*

The inverse matrix of the encoding equation system may contain rational functions with denominators not on the form of D^i, for some constant i. If so, the encoding is "catastrophic" in the sense that an error occurring in one of the transmissions can result in an infinite sequence of errors at the decoding sink. In order to avoid that, once the encoding has been computed using the LIFE* algorithm, one can compute the polynomial which is the least common multiple of all denominators of the rational functions resulting in the encoding process and introduce a precoding of the symbols generated by the sources, multiplying them by that least common multiple before they are introduced in the network. Alternatively, we can carry out this precoding locally in the nodes where a path enters a knot. We omit the details.

After this precoding is introduced, the network code as viewed from the perspective of the sink is polynomial, and any error that might occur will cause only a limited error propagation that can be handled by a suitable error correcting or erasure restoring code.

9.9.3.2. *Complexity*

The LIFE* algorithm is able to encode any given network with low-polynomial complexity and over the binary field. The addition of delay at some nodes is not a major drawback. In fact, any network encoding will in practice have intrinsic delay associated with it, and the delay will differ over the various paths. Thus in most cases, LIFE* does not need to introduce extra delay. In the few cases in which we actually need to introduce extra delay, this extra delay is what allows us to get the encoding on the binary field, which would have been impossible otherwise.

For networks where the LIF/LIFE algorithms work, LIFE* will perform with essentially the same complexity as the others, i.e. there is no known more efficient algorithm in these cases. If the network contains knots, which many practical networks do, no other known algorithm works, but for LIFE* the complexity may become dominated by the calculation of Mason's formula. The greedy approach to finding the coding coefficients for each edge performs essentially as in the acyclic cases, also for knots.

9.9.3.3. *Further remarks*

As already noted when discussing the opportunistic version of the LIFE algorithm, the choice of flow path graph and even the order in which the edges are visited (always according to the topological order imposed by the chosen flow path graph) will affect the resulting encoding for the given network in terms of maximum added delay needed.

In fact, the two ideas discussed in this section, using an extension of degree 2 of the given field, or adding one delay unit, are equivalent.

9.10. Examples

We will show here how the LIFE* algorithm will find encodings for different types of networks shown in Fig. 9.12.

9.10.1. *Example 1*

As mentioned before, the $(2, 4)$ combination network presented in Fig. 9.12 is known ([5, 22, 51]) to be a network which, when extra delay is not used, requires a finite field larger than $\mathbb{F}_2$. We show here how the LIFE* algorithm will work on the flow graph given in Fig. 9.19.

In order to better follow the progress of the algorithm, we have assigned labels $e_1, \ldots, e_{18}$ to the edges in the network following a topological order. Also, for simplicity in the notation we have called the two sources A and B, and $a(x)$ and $b(x)$ are the binary symbols released by the sources at time x.

The flow paths are represented with a code of colors and patterns in order to make visually easy to follow. Each sink has a color assigned and each of the two sources has a pattern assigned (solid for A, dash for B). The flow path from a source to a sink will be drawn in the color of the sink and with the pattern of the source.

The LIFE* algorithm starts by setting the following initial values:

$$E_{t_1} = E_{t_2} = \cdots = E_{t_6} = \{e_1, e_2\},$$
$$v_{e_1}(x) = Da(x) = a(x - 1),$$
$$v_{e_2}(x) = Db(x) = b(x - 1).$$

Hence

$$M_{t_i} = \begin{pmatrix} D & 0 \\ 0 & D \end{pmatrix}, \quad i = 1, \ldots, 6.$$

Now the algorithm enters the main loop:

- For encoding e_3 we can observe that the only predecessor is e_1, and thus

$$v_{e_3}(x) = D^{i_{e_3}(e_1)+1}v_{e_1}(x) = D^{i_{e_3}(e_1)+2}a(x).$$

Edge e_3 is in the flows to sinks t_1, t_2, and t_3; hence, we update the corresponding sets of edges and matrices

$$E_{t_1} = E_{t_2} = E_{t_3} = \{e_3, e_2\},$$

$$M_{t_1} = M_{t_2} = M_{t_3} = \begin{pmatrix} D^{i_{e_3}(e_1)+2} & 0 \\ 0 & D \end{pmatrix}.$$

Clearly the choice $i_{e_3}(e_1) = 0$ makes all the matrices non-singular. Thus, the encoding of e_3 is

$$v_{e_3}(x) = Dv_{e_1}(x) = D^2 a(x) = a(x-2).$$

- e_4 has two predecessors, e_1 and e_2:

$$v_{e_4}(x) = D^{i_{e_4}(e_1)+1}v_{e_1}(x) + D^{i_{e_4}(e_2)+1}v_{e_2}(x)$$
$$= D^{i_{e_4}(e_1)+2}a(x) + D^{i_{e_4}(e_2)+2}b(x).$$

Edge e_4 is in the flows to sinks t_1, t_4, and t_5; hence, we update the corresponding sets of edges and matrices:

$$E_{t_1} = \{e_3, e_4\}, \quad E_{t_4} = \{e_4, e_2\}, \quad E_{t_5} = \{e_4, e_2\},$$

$$M_{t_1} = \begin{pmatrix} D^2 & D^{i_{e_4}(e_1)+2} \\ 0 & D^{i_{e_4}(e_2)+2} \end{pmatrix}, \quad M_{t_4} = M_{t_5} = \begin{pmatrix} D^{i_{e_4}(e_1)+2} & 0 \\ D^{i_{e_4}(e_2)+2} & D \end{pmatrix}.$$

Again one can see that the choice $i_{e_4}(e_1) = i_{e_4}(e_2) = 0$ makes all three matrices non-singular. Hence,

$$v_{e_4}(x) = Dv_{e_1}(x) + Dv_{e_2}(x) = D^2 a(x) + D^2 b(x) = a(x-2) + b(x-2).$$

- Edge e_5 has e_1 and e_2 as predecessors, and the form of the encoding is

$$v_{e_5}(x) = D^{i_{e_5}(e_1)+1}v_{e_1}(x) + D^{i_{e_5}(e_2)+1}v_{e_2}(x)$$
$$= D^{i_{e_5}(e_1)+2}a(x) + D^{i_{e_5}(e_2)+2}b(x).$$

Edge e_5 takes part in the flows to sinks t_2, t_4, and t_6, and the corresponding updating of edge sets and matrices is as follows:

$$E_{t_2} = \{e_3, e_5\}, \quad E_{t_4} = \{e_4, e_5\}, \quad E_{t_6} = \{e_5, e_2\},$$

$$M_{t_2} = \begin{pmatrix} D^2 & D^{i_{e_5}(e_1)+2} \\ 0 & D^{i_{e_5}(e_2)+2} \end{pmatrix}, \quad M_{t_4} \begin{pmatrix} D^2 & D^{i_{e_5}(e_1)+2} \\ D^2 & D^{i_{e_5}(e_2)+2} \end{pmatrix},$$

$$M_{t_6} \begin{pmatrix} D^{i_{e_5}(e_1)+2} & 0 \\ D^{i_{e_5}(e_2)+2} & D \end{pmatrix}.$$

Now clearly any value of $i_{e_5}(e_1)$ and $i_{e_5}(e_2)$ will make matrices M_{t_2} and M_{t_6} non-singular, but in order to get M_{t_4} non-singular we need those two values to be different, hence setting both equal to 0 does not work in this case. A possible choice would be $i_{e_5}(e_1) = 1$, $i_{e_5}(e_2) = 0$, which gives us the next encoding:

$$v_{e_5}(x) = D^2v_{e_1}(x) + Dv_{e_2}(x) = D^3a(x) + D^2b(x) = a(x-3) + b(x-2).$$

- In the same manner, we work with edge e_6, which has only one predecessor, namely e_2 and following the same procedure as before we can see that setting the only unknown exponent to 0 will give a correct encoding:

$$v_{e_6}(x) = Dv_{e_2}(x) = D^2b(x) = b(x-2).$$

Remark. The particular case we have seen in the encodings of edges e_3 and e_6, that is to say, an edge with only one predecessor is always solved in the same manner, copying the symbol carried by the predecessor and adding the natural delay unit. This corresponds to

$$v_e(x) = Dv_p(x),$$

when p is the only predecessor of e, that is to say, $P(e) = \{p\}$. (This means that the exponent $i_e(p)$ has been chosen to be 0.)

The updated matrices will keep full rank since for each $t \in T(e)$ the corresponding updated matrix will be the result of multiplying by D the elements of one of the columns of the old matrix, which does not alter the rank of the matrix.

The rest of the encoding steps in this example are trivial in that sense, since all the rest of the edges have an only predecessor.

To complete the example we will illustrate how sinks can decode, this will also show what the delay means at the receiver end.

Let us focus on sink t_6. According to the encoding just computed this sink will receive at time x the symbols $a(x-4) + b(x-3)$ and $b(x-3)$.

Since we are assuming $a(x) = b(x) = 0$ for all negative x, sink t_6 will receive zeros on both channels until time $x = 3$ in which it receives $0 + b(0)$ and $b(0)$. This obviously gives him the knowledge only of symbol $b(0)$. However, at time $x = 4$ it receives $a(0) + b(1)$ and $b(1)$. The knowledge of $b(1)$ allows it to recover $a(0)$, which completes the recovery of the symbols of generation 0. Proceeding in the same way it will complete the recovering of the symbols of generation x at time $x + 4$. The total delay observed by sink t_6 is 4, which in this case coincides with the maximum power of D used in the encoding equations arriving in t_6.

This can also be interpreted in terms of matrices.

$$M_{t_6} \begin{pmatrix} D^4 & 0 \\ D^3 & D^3 \end{pmatrix}.$$

If we create a vector with the symbols that arrive at t_6 at time x and denote it as $[r_{t_6,1}(x), r_{t_6,2}(x)]$, the encoding process can be described as

$$[a(x), b(x)]M_{t_6} = [r_{t_6,1}(x), r_{t_6,2}(x)]$$

(which is equivalent to saying that $r_{t_6,1}(x) = a(x-4) + b(x-3)$, $r_{t_6,2}(x) = b(x-3)$).

Now the decoding process can be described as

$$[a(x), b(x)] = [r_{t_6,1}(x), r_{t_6,2}(x)]M_{t_6}^{-1} = [r_{t_6,1}(x), r_{t_6,2}(x)]\frac{1}{D^4}\begin{pmatrix} 1 & 0 \\ 1 & D \end{pmatrix},$$

that is to say

$$a(x) = r_{t_6,1}(x+4) + r_{t_6,2}(x+4),$$
$$b(x) = r_{t_6,2}(x+3),$$

which again shows how, to recover the symbols in generation x, sink t_6 has to wait until receiving symbols at time $x + 4$.

Remark. In general, the delay experienced by each sink is lower bounded by the maximum length of the flow paths arriving at it from the h different sources and upper bounded by the maximum power of the delay operator D used in the global encoding equations of the edges arriving at that sink.

The upper bound is not always tight. To illustrate this, let us consider the decoding that sink t_4 in the example has to do. Despite the maximum power of D for that sink is 4, it is easy to see that t_4 will complete the recovery of generation x at time $x+3$. (However, in addition it is absolutely necessary for t_4 to keep one memory element in order to be able to decode.)

9.10.2. *Example 2*

This example shows a network that is link cyclic but flow acyclic. The encoding process will work analogous to what was shown in the previous example.

The edges in Fig. 9.22 have been numbered according to a topological order and one can observe that each edge has in fact only one predecessor; hence, the encoding becomes trivial. We simply show here the result obtained:

$$v_{e_1}(x) = v_{e_2}(x) = a(x-1),$$
$$v_{e_3}(x) = v_{e_4}(x) = b(x-1),$$
$$v_{e_5}(x) = a(x-2),$$
$$v_{e_6}(x) = b(x-2),$$
$$v_{e_7}(x) = a(x-3),$$
$$v_{e_8}(x) = b(x-3).$$

Example 2

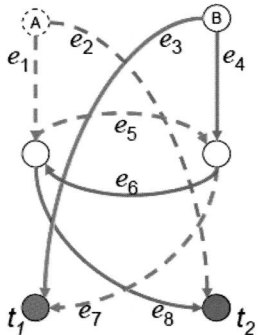

Fig. 9.22.   The unique flow path graph for Example 2.

As we can see, link cyclic but flow acyclic networks do not present any additional problem for encoding, they behave exactly as the acyclic networks did.

9.10.3. *Example 4*

Here we deal with a flow cyclic network that contains a simple flow cycle. Figure 9.23 shows the unique flow path graph for this network.

The initialization will give us the encoding of the first nine edges:

$$v_{e_1}(x) = v_{e_2}(x) = v_{e_3}(x) = a(x-1),$$
$$v_{e_4}(x) = v_{e_5}(x) = v_{e_6}(x) = b(x-1),$$
$$v_{e_7}(x) = v_{e_8}(x) = v_{e_9}(x) = c(x-1).$$

Now no more edges can be visited following a topological order. The cycle has set of edges $C_E = \{e_{10}, e_{11}, e_{12}\}$ and set of predecessors $P(C) = \{e_1, e_5, e_9\}$.

The structure of the local encoding equation in the cycle will be

$$v_C(x) = D^{ic(e_1)}\tau(e_1, C)v_{e_1}(x) + D^{ic(e_5)}\tau(e_5, C)v_{e_5}(x)$$
$$+ D^{ic(e_9)}\tau(e_9, C)v_{e_9}(x).$$

Example 4

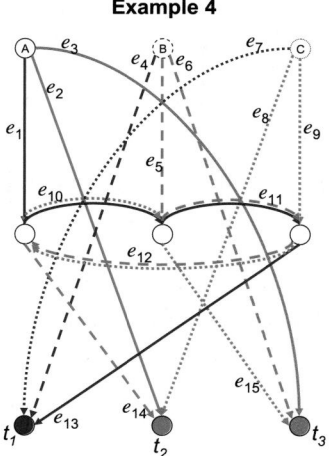

Fig. 9.23.   The unique flow path graph for Example 4.

Since we are in the case of a simple flow cycle, we can follow the formula given in Sec. 9.9.2.3 for the local encoding equation of each of the three edges in C_E.

$$v_{e_j}(x) = D^{ic(e_1)+d(e_1,e_j)}v_{e_1}(x) + D^{ic(e_5)+d(e_5,e_j)}v_{e_5}(x)$$
$$+ D^{ic(e_9)+d(e_9,e_j)}v_{e_9}(x), \quad j = 10, 11, 12.$$

Inspection of the graph shows that

$$d(e_1,e_{10}) = 1, \quad d(e_5,e_{10}) = 3, \quad d(e_9,e_{10}) = 2,$$
$$d(e_1,e_{11}) = 2, \quad d(e_5,e_{11}) = 1, \quad d(e_9,e_{11}) = 3,$$
$$d(e_1,e_{12}) = 3, \quad d(e_5,e_{12}) = 2, \quad d(e_9,e_{12}) = 1,$$

and the substitution in the above expression gives

$$v_{e_{10}}(x) = D^{ic(e_1)+1}a(x-1) + D^{ic(e_5)+3}b(x-1) + D^{ic(e_9)+2}c(x-1),$$
$$v_{e_{11}}(x) = D^{ic(e_1)+2}a(x-1) + D^{ic(e_5)+1}b(x-1) + D^{ic(e_9)+3}c(x-1),$$
$$v_{e_{12}}(x) = D^{ic(e_1)+3}a(x-1) + D^{ic(e_5)+2}b(x-1) + D^{ic(e_9)+1}c(x-1).$$

The full rank invariant condition must be checked for edge e_{10} in the flow to sink t_3, for edge e_{11} in the flow to sink t_1, and for edge e_{12} in the flow to sink t_2. This gives us the following matrices:

$$M_{t_1} = \begin{pmatrix} D^{ic(e_1)+3} & 0 & 0 \\ D^{ic(e_5)+2} & D & 0 \\ D^{ic(e_9)+4} & 0 & D \end{pmatrix}, \quad M_{t_2} = \begin{pmatrix} D & D^{ic(e_1)+4} & 0 \\ 0 & D^{ic(e_5)+3} & 0 \\ 0 & D^{ic(e_9)+2} & D \end{pmatrix},$$

$$M_{t_3} = \begin{pmatrix} D & 0 & D^{ic(e_1)+2} \\ 0 & D & D^{ic(e_5)+4} \\ 0 & 0 & D^{ic(e_9)+3} \end{pmatrix}.$$

Clearly any value of $i_C(e_1), i_C(e_5)$ and $i_C(e_9)$ satisfies the full rank invariant and we choose the simplest one setting the three unknowns to be 0. The global encoding equations of the edges in the cycle are as follows:

$$v_{e_{10}}(x) = a(x-2) + b(x-4) + c(x-3),$$
$$v_{e_{11}}(x) = a(x-3) + b(x-2) + c(x-4),$$
$$v_{e_{12}}(x) = a(x-4) + b(x-3) + c(x-2).$$

The encoding of edges e_{13}, e_{14}, and e_{15} is now trivial since each of them has only one predecessor:

$$v_{e_{13}}(x) = a(x-4) + b(x-3) + c(x-5),$$
$$v_{e_{14}}(x) = a(x-5) + b(x-4) + c(x-3),$$
$$v_{e_{15}}(x) = a(x-3) + b(x-5) + c(x-4).$$

The delay at the final receivers is 4, even when the maximum exponent of D in the equations received by the sinks is 5. Besides there is some extra memory needed in order to decode. For instance, recovering the element $a(1)$ and hence completing the generation 1 can be done by sink t_1 at time $x = 5$, provided it kept in memory the element $c(0)$.

A slight modification could be done for the encoding of the edges whose predecessors lie in the cycle, in such a way that they get the last updated values, for instance, edge e_{13} can benefit from the fact that edge e_9 enters in the same node from which e_{13} exits and hence get an updated version of the symbol carried by e_9, then the encoding of e_{13} would be

$$v_{e_{13}}(x) = a(x-4) + b(x-3) + c(x-2),$$

which is actually the same encoding that the edge e_{12} has, and results in smaller memory needed at the receiver t_1.

Remark. In general, using this last observation, the local encoding equation of an edge $e \notin C_E$ with a predecessor $p_C \in P'(e) \cap P(C)$ would be

$$v_e(x) = D^{i_e(s_C)} v_{s_C}(x) + \sum_{p \in P(e) \setminus \{p_C\}} D^{i_e(p)} D v_p(x),$$

where s_C is the successor of p_C that lies in the cycle, that is to say, the edge in C_E with $\text{start}(s_C) = \text{start}(e) = \text{end}(p_C)$.

We finally remark that in a simple flow cycle, the element s_C is unique, even if there are several elements p_C in $P'(e) \cap P(C)$.

9.10.4. *Example 5*

In this example, we show how to work with a knot. Figure 9.24 shows the essentially unique flow path graph for the network given.

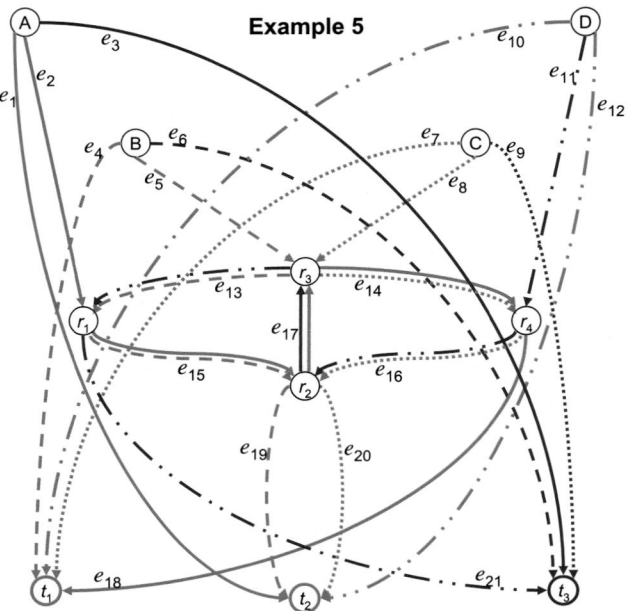

Fig. 9.24.   A flow path graph for the network in Example 5.

The initialization values are

$$v_{e_1}(x) = v_{e_2}(x) = v_{e_3}(x) = a(x-1),$$
$$v_{e_4}(x) = v_{e_5}(x) = v_{e_6}(x) = b(x-1),$$
$$v_{e_7}(x) = v_{e_8}(x) = v_{e_9}(x) = c(x-1),$$
$$v_{e_{10}}(x) = v_{e_{11}}(x) = v_{e_{12}}(x) = d(x-1).$$

No more edges can be visited in topological order because $C_E = \{e_{13}, e_{14}, e_{15}, e_{16}, e_{17}\}$ form a flow cycle. In fact it is a non-simple cycle, since it contains two flow cycles $e_{13} \prec e_{15} \prec e_{17} \prec e_{13}$ and $e_{14} \prec e_{16} \prec e_{17} \prec e_{14}$, both sharing the edge e_{17}. Hence we are in presence of a flow knot. (See Fig. 9.25(a))

The predecessors of the knot are $P_C = \{e_2, e_5, e_8, e_{11}\}$.

The general structure of the local encoding equation for the knot is

$$v_C(x) = D^{ic(e_2)}\tau(e_2, C)v_{e_2}(x) + D^{ic(e_5)}\tau(e_5, C)v_{e_5}(x)$$
$$+ D^{ic(e_8)}\tau(e_8, C)v_{e_8}(x) + D^{ic(e_{11})}\tau(e_{11}, C)v_{e_{11}}(x). \qquad (9.16)$$

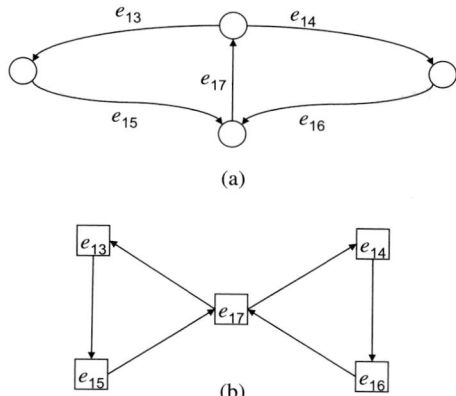

Fig. 9.25.   The knot in Example 5 and its line graph.

In order to find the local encoding of each edge in the knot, we have to use Mason's formula on a line graph to compute the transfer function for each of the symbols carried by the predecessors of the knot.

The line graph is shown in Fig. 9.25(b). The nodes correspond to the edges in C_E. The edge connecting e_{13} to e_{15} is drawn because of the flow path from source B to sink t_2, the edge connecting e_{15} to e_{17} is determined by the flow from source A to sink t_1. In the same manner, we draw the other connections in the line graph following the flow paths.

The branch gain of each connection is D. Mason's formula is as follows:

$$\tau(e_j, e_k) = \frac{\sum F_i(e_j, e_k) \Delta_i(e_j, e_k)}{\Delta},$$

where $\Delta = 1 + \sum c_i - \sum c_i c_j + \sum c_i c_j c_k + \cdots$. We assume that c_i is the gain of cycle i. The sum $\sum c_i$ sums the gain over all cycles, $\sum c_i c_j$ sums over all pairs of *non-touching* cycles, $\sum c_i c_j c_k$ sums over all triples of *non-touching* cycles, and so on. Moreover, $F_i(e_j, e_k)$ is the function corresponding to the ith forward path form e_j to e_k and $\Delta_i(e_j, e_k)$ is defined as Δ but counting only the cycles in the circuit that are disjoint with respect to the ith forward path. Here, we are using the notations in [38], and we refer the reader there for a more detailed explanation of Mason's formula.

- We will now focus on the symbol that enters through edge e_2 into e_{15}. Its itinerary through the knot is shown in Fig. 9.26(a).

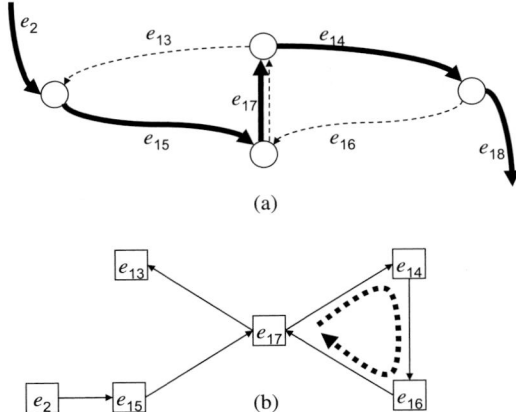

Fig. 9.26.   The itinerary of symbols carried by e_2 in the knot of Example 5.

The edges used by the flow path that carries that symbol are represented as bold lines; however, we also observe that when the symbol arrives at node $\text{end}(e_{17})$ this node will distribute it not only to edge e_{14}, but since there is a flow path connecting edge e_{17} with edge e_{13}, the symbol in question will travel also on edge e_{13} and in the same manner, we can see that it will travel also on edge e_{16}. That is represented in dashed lines in Fig. 9.26(a). On the other hand, using memory, the node $\text{end}(e_2) = \text{end}(e_{13}) = \text{start}(e_{15})$ can remove the contribution of the old symbol that arrives back at it through e_{13}, so we can then remove the connection between edges e_{13} and e_{15} from the line graph. Hence, the actual line graph followed by the symbols that enter the knot through edge e_2 is shown in Fig. 9.26(b).

In that graph, there is only one cycle, which has length 3, namely $\{e_{17}, e_{14}, e_{16}, e_{17}\}$. All the transfer functions will have as denominator the function $\Delta = 1 + D^3$.

For the transfer function corresponding to edge e_{15}, Fig. 9.27(a) shows that the only forward path is node disjoint with the only cycle of the graph; hence, $F_1(e_2, e_{15}) = D$ and $\Delta_1(e_{15}, e_{15}) = 1 + D^3$. Finally $\tau(e_2, e_{15}) = \frac{D \cdot (1+D^3)}{1+D^3} = D$, which is the expectable result.

Now we compute the transfer function corresponding to edge e_{17}. Figure 9.27(b) shows the only forward path, which is not node disjoint with the only cycle of the graph; hence, $F_1(e_2, e_{17}) = D^2$ and $\Delta_1(e_2, e_{17}) = 1$. Finally $\tau(e_2, e_{17}) = \frac{D^2 \cdot 1}{1+D^3}$.

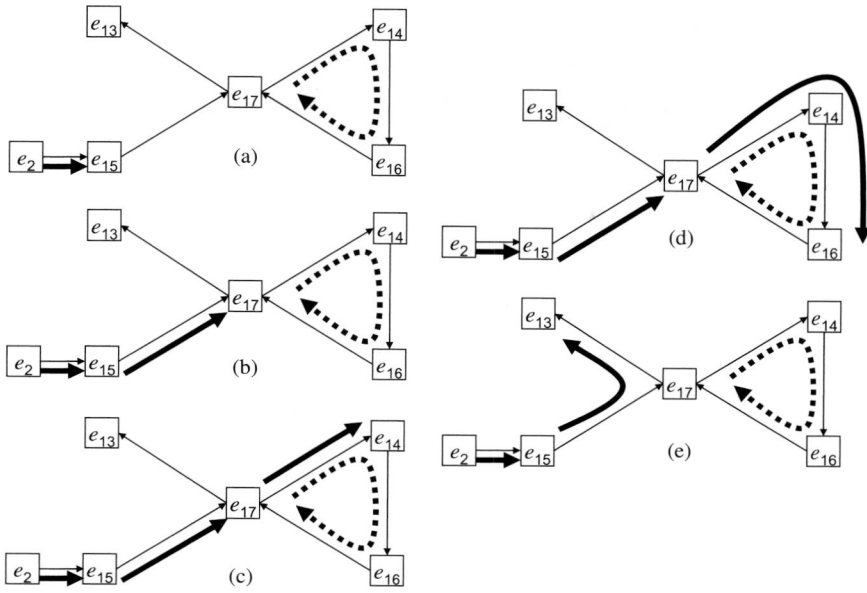

Fig. 9.27.   Computation of Mason's formula from edge e_2.

In the same manner, all the other transfer functions can be computed (drawings of the corresponding forward paths can be seen in Figs. 9.27(c)–9.27(e)).

$$\tau(e_2, e_{14}) = \frac{D^3 \cdot 1}{1 + D^3}, \quad \tau(e_2, e_{16}) = \frac{D^4 \cdot 1}{1 + D^3}, \quad \tau(e_2, e_{13}) = \frac{D^3 \cdot 1}{1 + D^3}.$$

Of special interest is the transfer function corresponding to edge e_{13} since it gives the function of the old symbol that has to be removed when passing again through node $\text{end}(e_2) = \text{end}(e_{13}) = \text{start}(e_{15})$.

- If we now focus on the circulation in the knot of the symbol carried by edge e_5 we observe (Fig. 9.28) that the flow path will transport it through edges e_{13} and e_{15} (bold line in the figure) and node $\text{end}(e_{15})$ will send it back to node $\text{end}(e_5)$ through edge e_{17} (dashedline in the figure), but node $\text{end}(e_5)$ will not send that symbol on edge e_{14} and hence the symbol does not travel along the two cycles in the knot, but only on the cycle $\{e_{13}, e_{15}, e_{17}, e_{13}\}$. The computation of the transfer functions is

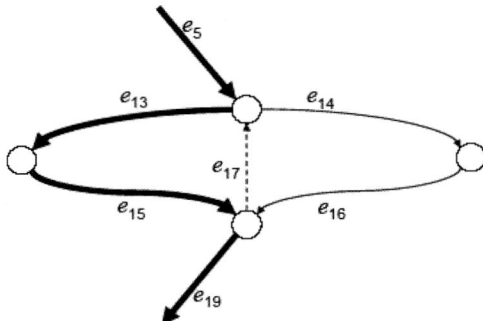

Fig. 9.28.   Itinerary of symbols carried by e_5 in the knot of Example 5.

then straightforward:

$$\tau(e_5, e_{13}) = D, \quad \tau(e_5, e_{15}) = D^2, \quad \tau(e_5, e_{17}) = D^3, \quad \tau(e_5, e_{14}) = 0,$$
$$\tau(e_5, e_{16}) = 0.$$

- The symbol entering the knot via edge e_8 follows a similar trajectory to that of the one entering via edge e_5:

$$\tau(e_8, e_{14}) = D, \quad \tau(e_8, e_{16}) = D^2, \quad \tau(e_8, e_{17}) = D^3, \quad \tau(e_8, e_{13}) = 0,$$
$$\tau(e_8, e_{15}) = 0.$$

- Finally, the symbol entering via edge e_{11} follows an itinerary identical (considering symmetry) to that entering via e_2 already studied:

$$\tau(e_{11}, e_{16}) = \frac{D \cdot (1 + D^3)}{1 + D^3} = D, \quad \tau(e_{11}, e_{17}) = \frac{D^2 \cdot 1}{1 + D^3},$$
$$\tau(e_{11}, e_{13}) = \frac{D^3 \cdot 1}{1 + D^3}, \quad \tau(e_{11}, e_{15}) = \frac{D^4 \cdot 1}{1 + D^3}, \quad \tau(e_{11}, e_{14}) = \frac{D^3 \cdot 1}{1 + D^3}.$$

Now that all the transfer functions have been computed one can use formula in Eq. (9.15) to compute the encoding of each edge in the knot.

We will go in detail with the computation of the encoding of edge e_{13}:

$$v_{e_{13}}(x) = D^{ic(e_2)}\tau(e_2, e_{13})v_2(x) + D^{ic(e_5)}\tau(e_5, e_{13})v_5(x)$$
$$+ D^{ic(e_8)}\tau(e_8, e_{13})v_8(x) + D^{ic(e_{11})}\tau(e_{11}, e_{13})v_{11}(x)$$
$$= D^{ic(e_2)}\frac{D^3}{1 + D^3}a(x - 1) + D^{ic(e_5)}Db(x - 1) + D^{ic(e_8)}0c(x - 1)$$
$$+ D^{ic(e_{11})}\frac{D^3}{1 + D^3}d(x - 1).$$

In the same manner, the encoding of any other edge in the knot can be computed using Eq. 9.15.

$$v_{e_{14}}(x) = D^{ic(e_2)}\frac{D^3}{1+D^3}a(x-1) + D^{ic(e_5)}0b(x-1) + D^{ic(e_8)}Dc(x-1)$$
$$+ D^{ic(e_{11})}\frac{D^3}{1+D^3}d(x-1),$$

$$v_{e_{15}}(x) = D^{ic(e_2)}Da(x-1) + D^{ic(e_5)}D^2b(x-1) + D^{ic(e_8)}0c(x-1)$$
$$+ D^{ic(e_{11})}\frac{D^4}{1+D^3}d(x-1),$$

$$v_{e_{16}}(x) = D^{ic(e_2)}\frac{D^4}{1+D^3}a(x-1) + D^{ic(e_5)}0b(x-1) + D^{ic(e_8)}D^2c(x-1)$$
$$+ D^{ic(e_{11})}Dd(x-1),$$

$$v_{e_{17}}(x) = D^{ic(e_2)}\frac{D^2}{1+D^3}a(x-1) + D^{ic(e_5)}D^3b(x-1)$$
$$+ D^{ic(e_8)}D^3c(x-1) + D^{ic(e_{11})}\frac{D^2}{1+D^3}d(x-1).$$

Now we will show how each edge can compute its encoding based on its predecessors and using formula in Eq. (9.14). Again we will go in detail with edge e_{13}:

$$P(e_{13}) = \{e_5, e_{17}\}, \quad P(e_{13}) \cap P(C) = \{e_5\}, \quad P(e_{13}) \cap C_E = \{e_{17}\},$$
$$P'(e_{13}) = \{e_5, e_8, e_{17}\}, \quad P'(e_{13}) \cap P(C) = \{e_5, e_8\}.$$

The direct application of formula in Eq. (9.14) to this case gives the following:

$$v_{e_{13}}(x) = Dv_{e_{17}}(x) + DD^{ic(e_5)}v_{e_5}(x) - D(D^{ic(e_5)}\tau(e_5, e_{17})v_{e_5}(x)$$
$$+ D^{ic(e_8)}\tau(e_8, e_{17})v_{e_8}(x))$$
$$= D\left(D^{ic(e_2)}\frac{D^2}{1+D^3}a(x-1) + D^{ic(e_5)}D^3b(x-1) \right.$$
$$\left. + D^{ic(e_8)}D^3c(x-1) + D^{ic(e_{11})}\frac{D^2}{1+D^3}d(x-1) \right)$$
$$+ D^{ic(e_5)}Db(x-1) - D^{ic(e_5)}DD^3b(x-1) - D^{ic(e_8)}DD^3c(x-1)$$

$$= D^{ic(e_2)}\frac{D^3}{1+D^3}a(x-1) + D^{ic(e_5)}Db(x-1) + D^{ic(e_8)}0c(x-1)$$

$$+ D^{ic(e_{11})}\frac{D^3}{1+D^3}d(x-1).$$

In a similar way, all the other edges can get their encoding using those of its predecessors and formula in Eq. (9.14).

The next point is determining the unknowns $i_C(e)$ for each edge in C by checking the full rank conditions. To be precise, the full rank condition must be checked for edge e_{13} in the flow to sink t_1, for edge e_{14} in the flow to sink t_3 and for edges e_{15} and e_{16} in the flow to sink t_2. A careful exam of the corresponding matrices will show that $i_C(e_2) = i_C(e_5) = i_C(e_8) = i_C(e_{11}) = 0$ is a valid choice.

Finally, a similar remark to the one made in Example 3 gives us $v_{e_{18}}(x) = v_{e_{16}}(x)$ and $v_{e_{21}}(x) = v_{e_{15}}(x)$.

Also we have $v_{e_{19}}(x) = Dv_{e_{15}}(x), v_{e_{20}}(x) = Dv_{e_{16}}(x)$.

The final encoding matrices at the sinks are

$$M_{t_1} = \begin{pmatrix} \dfrac{D^5}{1+D^3} & 0 & 0 & 0 \\ 0 & D & 0 & 0 \\ D^3 & 0 & D & 0 \\ D^2 & 0 & 0 & D \end{pmatrix}, \quad M_{t_2} = \begin{pmatrix} D & D^3 & \dfrac{D^6}{1+D^3} & 0 \\ 0 & D^4 & 0 & 0 \\ 0 & 0 & D^4 & 0 \\ 0 & \dfrac{D^6}{1+D^3} & D^3 & D \end{pmatrix},$$

$$M_{t_3} = \begin{pmatrix} D & 0 & 0 & D^2 \\ 0 & D & 0 & D^3 \\ 0 & 0 & D & 0 \\ 0 & 0 & 0 & \dfrac{D^5}{1+D^3} \end{pmatrix}.$$

As can be seen the elements in the matrices are now rational functions (typical case after traversing a knot). This, however, represents no extra difficulty at the sink. For instance, if we focus on sink t_2, the received symbols at that sink are $r_{t_2,1}(x) = v_1(x)$, $r_{t_2,2}(x) = v_{19}(x)$, $r_{t_2,3}(x) = v_{20}(x)$, and $r_{t_2,4}(x) = v_{12}(x)$. Their expressions in terms of the source symbols are given in each column of the matrix M_{t_2}. In order to retrieve the source symbols t_2 will multiply the received ones by the inverse of matrix M_{t_2}:

$$[a(x), b(x), c(x), d(x)] = [r_{t_2,1}(x), r_{t_2,2}(x), r_{t_2,3}(x), r_{t_2,4}(x)]M_{t_2}^{-1}$$

$$= \left[r_{t_2,1}(x+1), r_{t_2,1}(x+2) + r_{t_2,2}(x+4) \right.$$

$$+ \frac{D}{1+D^3} r_{t_2,4}(x), \frac{D}{1+D^3} r_{t_2,1}(x) + r_{t_2,3}(x+4)$$

$$\left. + r_{t_2,4}(x+2), r_{t_2,4}(x+1) \right].$$

The way to deal with expressions like $\frac{D}{1+D^3} r_{t_2,1}(x)$ at t_1 is to keep in memory three local variables that we will call $r_{t_2,1,i}(x)$. They will all be initialized as 0, and at time x one of them will be updated and the other two remain the same as follows:

$$r_{t_2,1,i_x}(x) = r_{t_2,1,i_x}(x-1) + r_{t_2,1}(x),$$
$$r_{t_2,1,i}(x) = r_{t_2,1,i}(x-1) \quad \text{for } i \in \{0,1,2\} \backslash \{i_x\},$$

where $x = 3q_x + i_x$ with $i_x \in \{0,1,2\}$, that is to say, i_x is the remainder of the integer division of x by 3.

In this way, $\frac{D}{1+D^3} r_{t_2,1}(x) = r_{t_2,1,i_{(x-1)}}(x-1)$, and the receiver t_2 does not need to keep an infinite memory, despite the aspect of the equations received.

9.10.5. *Example 6*

We will briefly show here one more example in which we deal with a more complicated network.

The network is shown in Fig. 9.29. It has six unit rate sources, labeled $A, B, \ldots, F$, and three sinks, labeled t_1, t_2, t_3.

The other nodes have been labeled as $1, \ldots, 7$.

The flow path graph for such a network is (essentially) unique. The flow paths starting in sources A and B are shown in Fig. 9.30 in solid bold and dashed bold lines, respectively. The flow paths from the other sources are the same but with a rotation of $120°$ to the right or to the left.

This implies that the flow path graph of that network contains a knot formed by six simple cycles, each of length 3. Figure 9.31(a) shows the knot

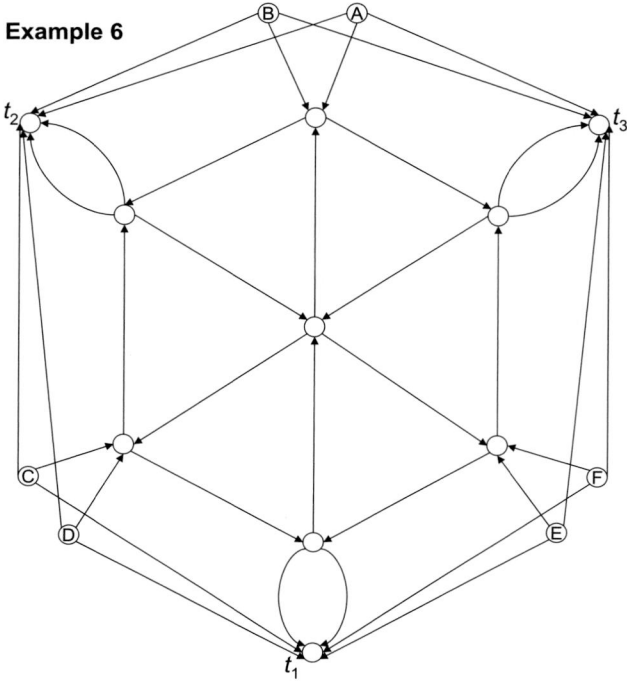

Fig. 9.29.   The network of Example 6.

while Fig. 9.31(b) shows the corresponding line graph constructed following the flow paths through the knot.

We will focus on the way the symbols are carried by the edge that connects source A with node 1 travel through the node. Mason's formula will be used to compute the transfer functions from that edge (which is one of the predecessors of the knot) to any other edge in the knot. Let us call that edge α, that is, $\alpha = (A, 1) \in P(C)$. In Fig. 9.32(a) we show the itinerary of the symbol from α. The bold solid lines are the actual flow path and the dashed lines are the edges not belonging to the flow path of the symbol but that will nevertheless carry instances of that symbol due to the connections in the knot. We can see that the only edge in which that symbol does not travel is e_{11}. Figure 9.32(b) shows the corresponding modified line graph that allows us to compute the transfer function $\tau(\alpha, 12)$. The bold dashed arrows show the position of the four simple cycles in that graph and the two bold lines (one black and the other gray) show the trajectories of

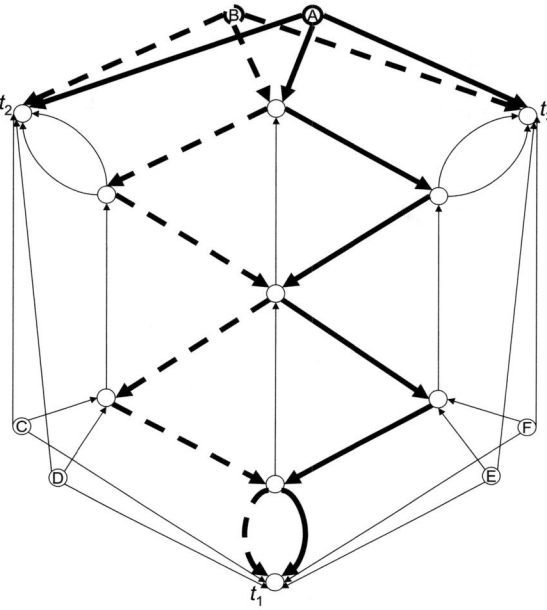

Fig. 9.30.   Flow paths in the network of Example 6.

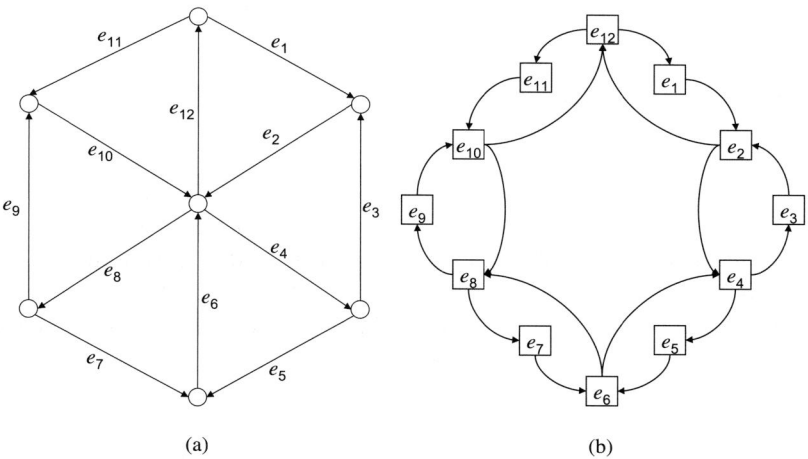

(a)   (b)

Fig. 9.31.   The knot and its line graph in the network of Example 6.

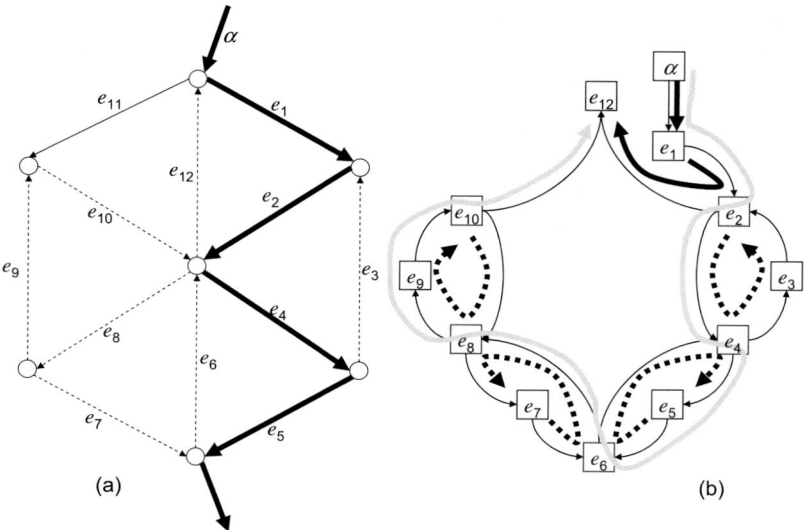

Fig. 9.32.   Itinerary of the symbol from edge α through the knot in the network of Example 6.

the two different forward paths from α to e_{12}. Direct application of Mason's formula gives

$$\tau(\alpha, e_{12}) = \frac{D^3(1 + 3D^3 + D^6) + D^9 \mathrm{i}}{1 + 4D^3 + 3D^6} = \frac{D^3}{1 + D^3}.$$

The rest of the transfer functions from α to edges in C are

$$\tau(\alpha, e_1) = D, \quad \tau(\alpha, e_2) = D^2 + \frac{D^5}{1 + D^6}, \quad \tau(\alpha, e_3) = \frac{D^4}{1 + D^6},$$

$$\tau(\alpha, e_4) = \frac{D^3}{1 + D^6}, \quad \tau(\alpha, e_5) = \frac{D^4}{1 + D^6}, \quad \tau(\alpha, e_6) = \frac{D^5}{1 + D^3},$$

$$\tau(\alpha, e_7) = \frac{D^7}{1 + D^6}, \quad \tau(\alpha, e_8) = \frac{D^6}{1 + D^6}, \quad \tau(\alpha, e_9) = \frac{D^7}{1 + D^6},$$

$$\tau(\alpha, e_{10}) = \frac{D^8}{1 + D^6}, \quad \tau(\alpha, e_{11}) = 0,$$

The transfer functions from the other predecessors of the knot to the edges in the knot are analogous. In fact, they can be derived from the ones already computed by simply taking into account the multiple symmetries that this network presents.

If we denote by β the edge that connects source B with node 1, γ the one that connects C with 5, δ the one that connects D with 5, and finally ϵ and ϕ the edges connecting sources E and F, respectively, to node 3, following the procedure of LIFE* we obtain the following global encoding equation for edge e_5:

$$v_{e_5}(x) = D^{i_C(\alpha)}\frac{D^5}{1+D^6}a(x) + D^{i_C(\beta)}\frac{D^8}{1+D^6}b(x) + D^{i_C(\gamma)}\frac{D^8}{1+D^6}c(x)$$

$$+ D^{i_C(\delta)}\frac{D^5}{1+D^6}d(x) + D^{i_C(\epsilon)}D^2e(x). \tag{9.17}$$

Encoding for other edges in the network will be analogous.

The full rank condition will be satisfied when choosing $i_C(\alpha) = i_C(\beta) = i_C(\gamma) = i_C(\delta) = i_C(\epsilon) = i_C(\phi) = 0$.

The local encoding equation of edge e_5 in terms of its predecessors is as follows:

$$v_{e_5}(x) = Dv_{e_4}(x) + DD^{i_C(\epsilon)}v_\epsilon(x)) - D(D^{i_C(\epsilon)}\tau(\epsilon, e_4)v_\epsilon(x)$$

$$+ D^{i_C(\phi)}\tau(\phi, e_4)v_\phi(x)).$$

The decoding matrix for sink t_1 has the form

$$M_{t_1}^{-1} = \frac{1}{D^6}\begin{pmatrix} 1 & D^3 & 0 & 0 & 0 & 0 \\ D^3 & 1 & 0 & 0 & 0 & 0 \\ \dfrac{D^8}{1+D^6} & \dfrac{D^{11}}{1+D^6} & D^5 & 0 & 0 & 0 \\ \dfrac{D^{11}}{1+D^6} & \dfrac{D^2}{1+D^6} & 0 & D^5 & 0 & 0 \\ D^2 + \dfrac{D^{11}}{1+D^6} & D^5 + \dfrac{D^8}{1+D^6} & 0 & 0 & D^5 & 0 \\ \dfrac{D^8}{1+D^6} & \dfrac{D^5}{1+D^6} & 0 & 0 & 0 & D^5 \end{pmatrix}.$$

9.11. Keywords and Terminology

Broadcast

A communication scenario where one or more nodes send information to all nodes in the network, with the intent or acceptance that everyone will be able to reconstruct all the information.

Capacity
The maximum rate at which a communication system can transmit information. For a single communication link, this can be determined by Shannon's theorems. For a network where each link's capacity can be determined independently and where node processing delay is zero, the capacity can also be determined by flow algorithms. However, for other networks, such as wireless broadcast networks, it is in general not known how to determine the capacity.

Deterministic network coding
A network encoding scheme where each node encodes according to a predetermined rule.

Flow
A **flow** is an assignment of values $F_{i,j}$ to each link (i,j) in the network such that

- $F_{i,j}$ is a non-negative number upper bounded by the capacity of link (i,j).
- For each node i that is not a source or a sink, it holds that $\sum_{j \in O(i)} F_{i,j} = \sum_{j \in I(i)} F_{j,i}$, where $O(i)$ and $I(i)$ are the set of outgoing links to i and the set of incoming links to i, respectively.

Flow for a specified sink t
The term *flow* is also often used in the network coding literature, and also in this chapter, to describe, for a network of unit capacity links, a set of edge-disjoint flow paths that collectively carry all the source information to a specified sink.

Flow cycle
A link cycle in a flow graph, with the property that for each edge e on the flow cycle there is a flow for which the flow predecessor of e coincides with the cycle predecessor of e. This means that there is no well-defined partial ordering of the edges; hence, the LIFE algorithm cannot be applied.

Flow graph
A graph that contains the flow for each sink, and that does not contain edges or nodes that do not appear in a flow for any sink.

Global encoding equation
The linear system that determines the relationship between a symbol on a link and the original set of source symbols.

Knot
A collection of flow cycles that share edges.

Link cycle
Any cycle in a flow graph.

Local encoding equation
The linear system that determines the relationship between a symbol on a link and the symbols that are received at the encoding node, which is the start of that link.

Mason's formula
A procedure for tracing signal flow in a network.

Multicast
A communication scenario where one or more nodes send information to a specified subset of nodes in the network, with the intent or acceptance that each node in the subset will be able to reconstruct all the information.

Random network coding
A network encoding scheme where each node encode each symbol or packet by selecting a new local encoding equation at random.

Routing
A way to find directions in a graph. To a certain degree routing can be replaced by network coding.

Unicast
A communication scenario where a single node sends information to another single node in the network.

9.12. Review Questions and Answers

9.12.1. *Review Questions*

(1) List some areas in which network coding is beneficial.
(2) In which way is a digital communication network different from a transportation network?
(3) Compute the channel capacity of the network in Fig. 9.33, given that each link is a binary erasure channel with erasure probability ε.

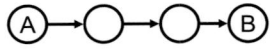

Fig. 9.33. A network for Exercise 3.

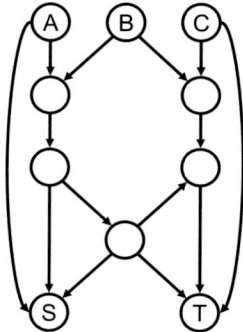

Fig. 9.34. A network for Exercise 5.

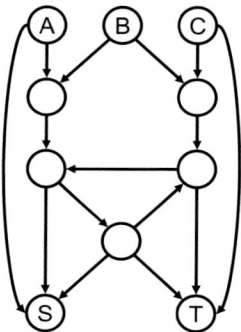

Fig. 9.35. A network for Exercise 6.

(4) Describe the difference between random network coding and deterministic network coding.

(5) In Fig. 9.34, A, B, and C are sources that want to multicast information to the sinks S and T. Is this possible? If so, determine a network encoding for the network in Fig. 9.34.

(6) Repeat Exercise 5 for the network in Fig. 9.35.

(7) Suppose that nodes A and B are having a two-way conversation with the help of *one* intermediary node using wireless broadcast communication. Assume that both A and B have continuous streams of information symbols to send. Draw the network using the representation as in Fig. 9.7 and derive the corresponding flow path diagrams, show that this is an example of multicast communication, and derive an efficient encoding scheme.

(8) Repeat Exercise 7, but now with *two* intermediary nodes.
(9) Repeat Exercise 8, but now with *three* intermediary nodes.
(10) Make a new simple example, non-isomorphic to the ones in this chapter, of a network which is flow cyclic.

9.12.2. *Review Answers*

(1) Network coding was first introduced for multicast, but in recent years a number of advantages have been demonstrated, many of which have been recounted in this chapter. It is likely that future research will uncover further advantages.

(2) In a digital communication network, a node can generate copies of incoming symbols and send the copy on several or all outgoing links. This does not apply to a transportation network sending physical objects or physical matter. The communication network nodes can also mix incoming symbols before resending, similar to a transportation network for, say, oil, while a transportation network for railroad cars should refrain from a similar practice.

(3) The A → B channel capacity is equal to the capacity of each individual erasure channel, namely $1 - \varepsilon$. An overall erasure correcting code will not achieve this, since the overall channel has an erasure probability of about 3ε. A capacity achieving code for each individual channel, but requires complete decoding at each step and hence increased delay. However, network coding can achieve the capacity with a shorter delay [25].

(4) In deterministic network coding, each node combines incoming packets into new packets that will be sent on outgoing lines, according to a predetermined rule. In random network encoding the way to combine the packets is selected at random for each packet; and the combination must be attached to the packet header. See Sec. 9.4.

(5) Yes, this is possible. Figure 9.36 shows a solution.

(6) By selecting the flow paths as shown in Fig. 9.37 we see that the flow path graph is not cyclic and that there is an encoding solution.

(7) Figure 9.38(a) shows a solution. It can be seen that this network is special case of the simple butterfly network in Fig. 9.5, and the coding can be carried out accordingly.

(8) Figure 9.38(b) shows a solution. It can be seen that this network is special case of the flow cyclic network in Example 3 of Fig. 9.12, and the coding can be carried out accordingly.

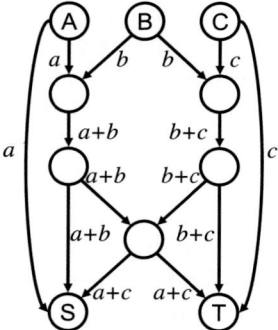

Fig. 9.36.   A solution for Exercise 5.

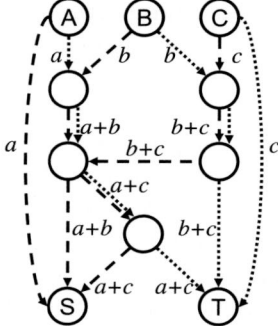

Fig. 9.37.   A solution for Exercise 6.

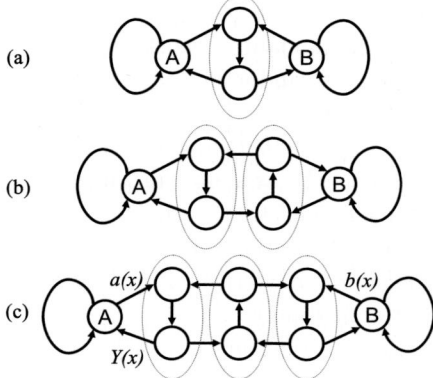

Fig. 9.38.   Solution for (a) Exercise 7, (b) Exercise 8, and (c) Exercise 9.

(9) Figure 9.38(c) shows a solution. It can be seen that this network contains a knot.

Suppose, for convenience, that there is a delay of one time unit per link in the graph and that at a certain time (x), nodes A and B send symbols $a(x)$ and $b(x)$, respectively. The symbol $Y(x)$ received by node A at the same time depends on the local encoding equations of the (encoding) nodes in the network. If the LIFE* algorithm and Mason's formula are used, one solution is $Y(x) = a(x-2) + \sum_{i=0}^{\infty} b(x-6-4i)$. Does this mean that A needs to keep an infinite memory of the received b sequence? Other solutions exist, and we challenge the reader to find at least one.

Acknowledgments

The authors wish to thank the anonymous reviewers for useful comments. This work is supported by MEC (grant PR2007-0175) and NFR (OWL project).

References

1. S. Acedański, S. Deb, M. Médard and R. Koetter, *How Good is Random Linear Coding Based Distributed Networked Storage?*, NetCod, Italy, (April 2005).
2. R. Ahlswede, N. Cai, S.-Y. R. Li and R. W. Yeung, Network information flow, *IEEE Trans. Inform. Theory* **46** (2000), 1204–1216.
3. R. K. Ahuja, T. L. Magnanti and J. B. Orlin, *Network Flows: Theory, Algorithms, and Applications* (Prentice-Hall, 1993).
4. A. S. Avestimehr, S. N. Diggavi and D. N. C. Tse, A deterministic model for wireless relay networks and its capacity, *Proc. IEEE Information Theory Workshop on Information Theory for Wireless Networks*, Bergen, (1–6, July 2007), pp. 6–11.
5. Á. Barbero and Ø. Ytrehus, Cycle-logical treatment of "cyclopathic" networks, *IEEE Trans. Inform. Theory* **52** (2006), 2795–2805.
6. Á. Barbero and Ø. Ytrehus, Knotworking, *Proc. ITA 2006,* San Diego, (February 2006) (electronic publication).
7. Á. Barbero and Ø. Ytrehus, Heuristic algorithms for small field multicast encoding, *Proc. ITW'06 Chengdu*, (October 2006), pp. 428–432.
8. Á. Barbero and Ø. Ytrehus, Facts of LIFE, *Proc. 16th Int. Conf. Computer Communications and Networks, (ICCCN),* Turtle Bay, Hawaii, (August 2007), pp. 832–837.
9. P. Chaporkar and A. Proutiere, Adaptive network coding and scheduling for maximizing throughput in wireless networks, *Proc. 13th Annual ACM Int. Conf. Mobile Computing and Networking Table of Contents*, Montrïeal, Quïebec, Canada, (2007), pp. 135–146.

10. P. A. Chou, Y. Wu and K. Jain, Practical network coding, *Proc. 41st Annual Allerton Conf. Communication, Control and Computing*, (October 2003).

11. T. H. Cormen, C. L. Leiserson and R. L. Rivest, *Introduction to Algorithms, 2nd edn.* (MIT Press, and McGraw-Hill, 2001).

12. S. Deb, M. Effros, T. Ho, D. R. Karger, R. Koetter, D. S. Lun, M. Médard and N. Ratnakar, Network coding for wireless applications: A brief tutorial, *Proc. IWWAN*, London, UK, (May 2005).

13. S. Deb, M. Médard and C. Choute, Algebraic gossip: A network coding approach to optimal multiple rumor mongering, *IEEE Trans. Inform. Theory* **52** (2006), 2486–2507.

14. R. Dougherty, C. Freiling and K. Zeger, Insufficiency of linear coding in network information flow, *IEEE Trans. Inform. Theory* **51** (2005), 2745–2759.

15. P. Elias, A. Feinstein and C. E. Shannon, Note on maximum flow through a network, *IRE Trans. Inform. Theory IT-2* (1956), 117–119.

16. E. Erez and M. Feder, On codes for network multicast, *Proc. 41st Annual Allerton Conf. Communication Control and Computing*, (October 2003).

17. E. Erez and M. Feder, Convolutional network coding, *Proc. ISIT'04, Chicago*, (June 2004), p. 146.

18. E. Erez and M. Feder, Convolutional network codes for cyclic networks, *Proc. Netcod 2005*, Italy, (April 2005).

19. L. R. Ford and D. R. Fulkerson, *Flows in Networks* (Princeton University Press, 1962).

20. C. Fragouli and A. Markopoulou, A network coding approach to network monitoring, *Proc. 43rd Annual Allerton Conf. Communication, Control and Computing*, (September 2005).

21. C. Fragouli and A. Markopoulou, Active topology inference using network coding, *Proc. 44th Annual Allerton Conf. Communication, Control and Computing*, (September 2006).

22. C. Fragouli and E. Soljanin, Information flow decomposition for network coding, *IEEE Trans. Inform. Theory* **52** (2006), 829–848.

23. C. Fragouli and E. Soljanin, *Network Coding Fundamentals* (Now Publishers, 2007).

24. C. Fragouli and E. Soljanin, *Network Coding Applications* (Now Publishers, 2008).

25. C. Fragouli, J.-Y. Le Boudec and J. Widmer, Network coding: An instant primer, *ACM SIGCOMM*, (2008).

26. C. Gkantsidis and P. R. Rodriguez, Network coding for large scale content distribution, *IEEE Proc., INFOCOM 2005, 24th Annual Joint Conf. IEEE Computer and Communications Societies*, (2005), pp. 2235–2245.

27. T. Ho, R. Koetter, M. Médard, D. R. Karger and M. Effros, The benefits of coding over routing in a randomized setting, *Proc. IEEE Int. Symp. Information Theory (ISIT)*, Yokohama, Japan, (June/July 2003), p. 442.

28. T. Ho, B. Leong, R. Koetter and M. Médard, Distributed asynchronous algorithms for multicast network coding, *1st Workshop on Network Coding*, WiOpt (2005).

29. T. Ho, B. Leong, Y.-H. Chang, Y. Wen and R. Koetter, Network monitoring in multicast networks using network coding, *Proc. IEEE Int. Symp. Information Theory (ISIT)*, Adelaide, (September 2005), pp. 1977–1981.

30. T. Ho and D. Lun, *Network Coding — An Introduction* (Cambridge University Press, 2008).

31. T. Ho, M. Médard, R. Koetter, D. R. Karger, M. Effros, J. Shi and B. Leong, Toward a random operation of networks, *IEEE Trans. Inform. Theory* (2006), submitted for publication.

32. M. Jafarisiavoshani, C. Fragouli and S. Diggavi, On subspace properties for randomized network coding, *Proc. IEEE Information Theory Workshop on Information Theory for Wireless Networks*, Bergen, (1–6, July 2007), pp. 17–21.

33. S. Jaggi, P. A. Chou and K. Jain, Low complexity algebraic multicast network codes, *Proc. IEEE Int. Symp. Information Theory (ISIT)*, Yokohama, Japan, (June/July 2003), p. 368.

34. S. Jaggi, P. Sanders, P. A. Chou, M. Effros, S. Egner, K. Jain and L. M. G. M. Tolhuizen, Polynomial time algorithms for multicast network code construction, *IEEE Trans. Inform. Theory* **51** (2005) 1973–1982.

35. R. Koetter and M. Médard, An algebraic approach to network coding, *IEEE/ACM Trans. Network* **11**, 5 (2003), 782–795.

36. S.-Y. R. Li, R. W. Yeung and N. Cai, Linear Network Coding, *IEEE Trans. Inform. Theory* **49** (2003), 371–381.

37. S.-Y. R. Li and R. W. Yeung, On convolutional network coding, *Proc. ISIT'06 Seattle,* (June 2006), pp. 1743–1747.

38. S. Lin and D. Costello, *Error Control Coding* (Prentice-Hall 2004).

39. H. (Francis) Lu, Binary linear network codes, *Proc. IEEE Information Theory Workshop on Information Theory for Wireless Networks*, Bergen, (1–6, July 2007), pp. 224–227.

40. R. Matsumoto, Construction algorithm for network error-correcting codes attaining the singleton bound, *IEICE Trans. Fundam Electron. Commun. Comput. Sci.* **E90-A**, 9 (2007), 1729–1735.

41. J. Liu, D. Goeckel and D. Towsley, Bounds on the gain of network coding and broadcasting in wireless networks, *Proc. 26th IEEE Int. Conf. Computer Communications (INFOCOM)*, (2007), pp. 724–732.

42. M. Luby, LT codes, *Proc. 43rd Annual IEEE Symposium on Foundations of Computer Science*, (16–19, November 2002), pp. 271–280.

43. D. S. Lun, M. Médard and M. Effros, On coding for reliable communication over packet networks, *Proc. 42nd Annual Allerton Conf. Communication, Control, and Computing*, (September–October 2004).

44. D. S. Lun, M. Médard, T. Ho and R. Koetter, On network coding with a cost criterion, *Proc. Int. Symp. Inform. Theory and its Applications (ISITA)*, (2004).

45. D. S. Lun, N. Ratnakar, M. Médard, R. Koetter, D. R. Karger, T. Ho, E. Ahmed and F. Zhao, Minimum-cost multicast over coded packet networks, *IEEE Trans. Inform. Theory* **52**, 6 (2006), 2608–2623.

46. M. C. Zhou, C.-H. Wang and X. Zhao, Automating Mason's rule and its application to analysis of stochastic Petri nets, *IEEE Trans. Control Syst. Technol.* **3**, 2 (1995), 238–244.

47. N. Ratnakar and G. Kramer, On the separation of channel and network coding in Aref networks, *Proc. IEEE Int. Symp. Information Theory (ISIT)*, Adelaide, (September 2005), pp. 1716–1719.

48. S. Riis, Linear versus nonlinear Boolean functions in network flow, *Proc. 2004 Conf. Information Sciences and Systems (CISS 2004), 2004 Conf. Information Sciences and Systems* (CISS 2004), 2004.

49. K. H. Rosen (ed.), *Handbook of Discrete and Combinatorial Mathematics* (CRC Press, 1999).

50. Y. E. Sagduyu and A. Ephremides, Joint scheduling and wireless network coding, *Proc. First Workshop on Network Coding*, Riva del Garda, Italy, (April 2005).

51. P. Sanders, S. Egner and L. Tolhuizen, Polynomial time algorithms for network information flow, *Proc. SPAA'03,* San Diego, 7–9, (June 2003), pp. 286–294.

52. A. Shokrollahi, Raptor Codes, *IEEE Trans. Inform. Theory* **52**, 6 (2006), 2551–2567.

53. D. Silva, F. R. Kschischang and R. Koetter, A rank-metric approach to error control in random network coding, *Proc. IEEE Inform. Theory Workshop on Information Theory for Wireless Networks*, Bergen, (1–6, July 2007), pp. 228–232.

54. L. Song, R. W. Yeung and N. Cai, A separation theorem for single-source network coding, *IEEE Trans. Inform. Theory* **52**, 5 (2006), 1861–1871.

55. N. Thomos and P. Frossard, Raptor network video coding, *Proc. Int. Workshop on Mobile Video*, Augsburg, (28–28, September 2007), pp. 19–24.

56. D. Traskov, N. Ratnakar, D. S. Lun, R. Koetter and M. Médard, Network coding for multiple unicasts: An approach based on linear optimization, *Proc. IEEE Int. Symp. Information Theory (ISIT)*, Seattle, (July 2006), pp. 1758–1762.

57. D. Tuninetti and C. Fragouli, On the throughput improvement due to limited complexity processing at relay nodes, *Proc. IEEE Int. Symp. Information Theory (ISIT)*, Adelaide, (September 2005), pp. 1081–1085.

58. Y. Xi and E. M. Yeh, Distributed algorithms for minimum cost multicast with network coding, *Proc. 43rd Annual Allerton Conf. Communication, Control and Computing,* (September 2005).

59. Y. Xi and E. M. Yeh, Distributed algorithms for minimum cost multicast with network coding in wireless networks, *Proc. 4th Int. Symp. Modeling and Optimization in Mobile, Ad Hoc and Wireless Networks (WiOpt '06),* Boston, MA, (April 2006).

60. S. Yang and R. W. Yeung, Refined coding bounds for network error correction, *Proc. IEEE Information Theory Workshop on Information Theory for Wireless Networks*, Bergen, (1–6 July 2007), pp. 126–130.

61. R. W. Yeung, S.-Y. Robert Li, N. Cai and Z. Zhang, *Network Coding Theory* (World Scientific, 2006).
62. R. W. Yeung, Avalanche: A network coding analysis, *Communications in Information and Systems* **7**, 4 (2007), 353–358.
63. Z. Zhang, X. Yan and H. Balli, Some key problems in network error correction coding theory, *Proc. IEEE Information Theory Workshop on Information Theory for Wireless Networks*, Bergen, (1–6, July 2007), pp. 131–135.

Chapter 10

DISTRIBUTED JOINT SOURCE–CHANNEL CODING ON A MULTIPLE ACCESS CHANNEL

VINOD SHARMA* and R RAJESH[†]

*Department of Electrical Communication Engineering,
Indian Institute of Science, Bangalore 560012, India*
* vinod@ece.iisc.ernet.in
[†] rajesh@pal.ece.iisc.ernet.in

We consider the problem of transmission of several distributed sources over a multiple access channel (MAC) with side information at the sources and the decoder. Source-channel separation does not hold for this channel. Sufficient conditions are provided for transmission of sources with a given distortion. The source and/or the channel could have continuous alphabets (thus Gaussian sources and Gaussian MACs are special cases). Various previous results are obtained as special cases. We also provide several good joint source-channel coding schemes for a discrete/continuous source and discrete/continuous alphabet channel. Channels with feedback and fading are also considered.

Keywords: Multiple access channel; side information; lossy joint source-channel coding; channels with feedback; fading channels.

10.1. Introduction

In this chapter, we consider the transmission of various correlated sources over a multiple access channel (MAC). We survey the results available when the system may have side information at the sources and/or at the decoder. We also consider an MAC with feedback or when the channel experiences time varying fading.

This system does not satisfy source–channel separation [1]. Thus for optimum transmission one needs to consider joint source–channel coding. Thus we will provide several good joint source–channel coding schemes. Although this topic has been studied for last several decades, one recent motivation is the problem of estimating a random field via sensor networks.

Sensor nodes have limited computational and storage capabilities and very limited energy [2]. These sensor nodes need to transmit their observations to a fusion center, which uses this data to estimate the sensed random field. Since transmission is very energy intensive, it is important to minimize it.

The proximity of the sensing nodes to each other induces high correlations between the observations of adjacent sensors. One can exploit these correlations to compress the transmitted data significantly. Furthermore, some of the nodes can be more powerful and can act as cluster heads [3]. Neighboring nodes can first transmit their data to a cluster head, which can further compress information before transmission to the fusion center. The transmission of data from sensor nodes to their cluster-head is usually through an MAC. At the fusion center the underlying physical process is estimated. The main trade-off possible is between the rates at which the sensors send their observations and the distortion incurred in the estimation at the fusion center. The availability of side information at the encoders and/or the decoder can reduce the rate of transmission [4, 5].

The above considerations open up new interesting problems in multi-user information theory and the quest for finding the optimal performance for various models of sources, channels, and side information have made this an active area of research. The optimal solution is unknown except in a few simple cases. In this chapter a joint source-channel coding approach is discussed under various assumptions on side information and distortion criteria. Sufficient conditions for transmission of discrete/continuous alphabet sources over a discrete/continuous alphabet MAC are given. These results generalize the previous results available on this problem.

The chapter is organized as follows. Section 10.2 provides the background and surveys the related literature. Transmission of distributed sources over an MAC with side information is considered in Sec. 10.3. The sources and the channel alphabets can be continuous or discrete. Several previous results are recovered as special cases in Sec. 10.4. Section 10.5 considers the important case of transmission of discrete correlated sources over a Gaussian MAC (GMAC) and presents a new coding scheme. Section 10.6 discusses several joint source–channel coding schemes for transmission of Gaussian sources over a GMAC and compares their performance. It also suggests coding schemes for general continuous sources over a GMAC. Transmission of correlated sources over orthogonal channels is considered in Sec. 10.7. Section 10.8 discusses an MAC with feedback. An MAC with multi-path fading is addressed in Sec. 10.9. Section 10.10 provides practical schemes for joint source–channel coding. Section 10.11

gives the directions for future research and Sec. 10.12 concludes the chapter.

10.2. Background

In the following we survey the related literature. Ahlswede [6] and Liao [7] obtained the capacity region of a discrete memoryless MAC with independent inputs. Cover *et al.* [1] made further significant progress by providing sufficient conditions for losslessly transmitting correlated observations over an MAC. They proposed a "correlation preserving" scheme for transmitting the sources. This mapping is extended to a more general system with several principle sources and several side-information sources subject to cross-observations at the encoders in [8]. However single letter characterization of the capacity region is still unknown. Indeed Duek [9] proved that the conditions given in [1] are only sufficient and may not be necessary. In [10] a finite letter upper bound for the problem is obtained. It is also shown in [1] that the source–channel separation does not hold in this case. The authors in [11] obtain a condition for separation to hold in an MAC.

The capacity region for distributed lossless source coding problem is given in the classic paper by Slepian and Wolf [12]. Cover [13] extended Slepian–Wolf's (S–W) results to an arbitrary number of discrete, ergodic sources using a technique called "random binning". Other related papers on this problem are [8, 14].

Inspired by S–W results, Wyner and Ziv [4] obtained the rate distortion function for source coding with side information at the decoder. Unlike for the lossless case, it is shown that the knowledge of the side information at the encoders in addition to the decoder, permits the transmission at a lower rate. The latter result, when encoder and decoder have side information, was first obtained by Gray and is known as conditional rate distortion function [15]. Related work on side-information coding is provided in [16–18]. The lossy version of S–W problem is called multi-terminal source coding problem and despite numerous attempts (e.g. [19, 20]) the exact rate region is not known except for a few special cases. The first major advancement was by Berger and Tung [15], that is, an inner and an outer bound on the rate distortion region was obtained. Lossy coding of continuous sources at the high resolution limit is given in [21], where an explicit single-letter bound is obtained. Gastpar [22] derived an inner and an outer bound with side information and proved the tightness of his bounds when the sources are conditionally independent given the side information. The authors in [23] obtained inner and outer bounds on the rate region with side information at the encoders and the decoder. [24, 25] extend the result in [23] by requiring the encoders to communicate over an MAC, i.e. they obtain sufficient conditions for transmission of correlated sources over an MAC with given distortion constraints. In [26] achievable rate region for an MAC with correlated sources and feedback is given.

The distributed Gaussian source coding problem is discussed in [20, 27]. Exact rate region is provided in [27]. The capacity of a GMAC with feedback is given in [28]. In [29] a necessary and two sufficient conditions for transmitting a jointly Gaussian source over a GMAC are provided. It is shown that the amplify and forward (AF) scheme is optimal below a certain SNR determined by source correlations. The performance comparison of the schemes given in [29] with a separation based scheme is given in [30]. GMAC under received power constraints is studied in [31] and it is shown that the source–channel separation holds in this case.

In [32], the authors discuss a joint source-channel coding coding scheme over an MAC and show the scaling behavior for the Gaussian channel. A Gaussian sensor network in distributed and collaborative setting is studied in [33]. The authors show that it is better to compress the local estimates than to compress the raw data. The scaling laws for a many-to-one data-gathering channel are discussed in [34]. It is shown that the transport capacity of the network scales as $\mathcal{O}(\log N)$ when the number of sensors N grows to infinity and the total average power remains fixed. The scaling laws for the problem without side information are discussed in [35] and it is shown that separating source coding from channel coding may require exponential growth, as a function of number of sensors, in communication bandwidth. A lower bound on best achievable distortion as a function of the number of sensors, total transmit power, the degrees of freedom of the underlying process, and the spatio-temporal communication bandwidth is given.

The joint source–channel coding problem also bears relationship to the CEO problem [36]. In this problem, multiple encoders observe different, noisy versions of a single information source and communicate it to a single decoder called the CEO which is required to reconstruct the source within a certain distortion. The Gaussian version of the CEO problem is studied in [37].

When time division multiple access (TDMA), code division multiple access (CDMA) or frequency division multiple access (FDMA) is used then an MAC becomes a system of orthogonal channels. These protocols, although suboptimal are frequently used in practice and hence have been extensively studied [38, 39]. Lossless transmission of correlated sources over orthogonal channels is addressed in [40]. The authors prove that the source–channel separation holds for this system. They also obtain the exact rate region. [41] extends these results to the lossy case and shows that separation holds for the lossy case too. Distributed scalar quantizers were designed for correlated Gaussian sources and independent Gaussian channels in [42].

The information-theoretic and communication features of a fading MAC are given in an excellent survey paper [43]. A survey of practical schemes for distributed source coding for sensor networks is given in [44]. Practical schemes for distributed source coding are also provided in [45, 46].

10.3. Transmission of Correlated Sources Over an MAC

In this section, we consider the transmission of memoryless dependent sources, through a memoryless MAC (Fig. 10.1). The sources and/or the channel input/output alphabets can be discrete or continuous. Furthermore, side information may be available at the encoders and the decoder. Thus our system is very general and covers many systems studied over the years as special cases.

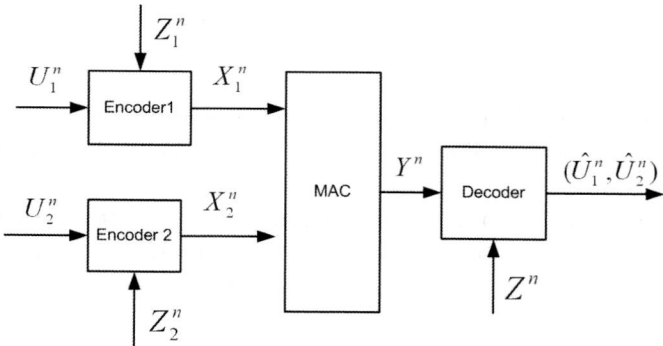

Fig. 10.1.   Transmission of correlated sources over an MAC with side information.

We consider two sources (U_1, U_2) and side information random variables Z_1, Z_2, and Z with a known joint distribution $F(u_1, u_2, z_1, z_2, z)$. Side information Z_i is available to encoder i, $i \in \{1, 2\}$ and the decoder has side information Z. The random vector sequence $\{(U_{1n}, U_{2n}, Z_{1n}, Z_{2n}, Z_n), n \geq 1\}$ formed from the source outputs and the side information with distribution F is independent identically distributed (*iid*) in time. The sources transmit their codewords X_i's to a single decoder through a memoryless MAC. The channel output Y has distribution $p(y \,|\, x_1, x_2)$ if x_1 and x_2 are transmitted at that time. The decoder receives Y and also has access to the side information Z. The encoders at the two users do not communicate with each other except via the side information. It uses Y and Z to estimate the sensor observations U_i as $\hat{U}_i$, $i \in \{1, 2\}$. It is of interest to find encoders and a decoder such that $\{U_{1n}, U_{2n}, n \geq 1\}$ can be transmitted over the given MAC with $E[d_1(U_1, \hat{U}_1)] \leq D_1$ and $E[d_2(U_2, \hat{U}_2)] \leq D_2$ where d_i are non-negative distortion measures and D_i are the given distortion constraints. If the distortion measures are unbounded we assume that y_i^*, $i \in \{1, 2\}$ exist such that $E[d_i(U_i, y_i^*)] < \infty$, $i \in \{1, 2\}$. Source channel separation does not hold in this case.

For discrete sources a common distortion measure is the Hamming distance

$$d(x, x') = 1 \quad \text{if } x \neq x',$$

$$d(x, x') = 0 \quad \text{if } x = x'.$$

For continuous alphabet sources the most common distortion measure is $d(x, x') = (x - x')^2$.

We will denote $\{U_{ij}, j = 1, 2, \ldots, n\}$ by $U_i^n, i = 1, 2$.

Definition 10.1. The source (U_1^n, U_2^n) can be transmitted over the MAC with distortions $\mathbf{D} \triangleq (D_1, D_2)$, if for any $\epsilon > 0$ there is an n_0 such that for all $n > n_0$ there exist encoders $f_{E,i}^n : \mathcal{U}_i^n \times \mathcal{Z}_i^n \to \mathcal{X}_i^n$, $i \in \{1, 2\}$ and a decoder $f_D^n : \mathcal{Y}^n \times \mathcal{Z}^n \to (\hat{\mathcal{U}}_1^n, \hat{\mathcal{U}}_2^n)$ such that $\frac{1}{n} E[\sum_{j=1}^n d(U_{ij}, \hat{U}_{ij})] \leq D_i + \epsilon$, $i \in \{1, 2\}$ where $(\hat{U}_1^n, \hat{U}_2^n) = f_D(Y^n, Z^n)$, $\mathcal{U}_i$, $\mathcal{Z}_i$, $\mathcal{Z}$, $\mathcal{X}_i$, $\mathcal{Y}$, and $\hat{\mathcal{U}}_i$ are the sets in which U_i, Z_i, Z, X_i, Y, and $\hat{U}_i$ take values.

We denote the joint distribution of (U_1, U_2) by $p(u_1, u_2)$ and let $p(y \,|\, x_1, x_2)$ be the transition probabilities of the MAC. Since the MAC is memoryless, $p(y^n \,|\, x_1^n, x_2^n) = \prod_{j=1}^n p(y_j \,|\, x_{1j}, x_{2j})$. $X \leftrightarrow Y \leftrightarrow Z$ will indicate that $\{X, Y, Z\}$ form a Markov chain.

Now we state the main theorem.

Theorem 10.1. *A source can be transmitted over the multiple access channel with distortions (D_1, D_2) if there exist random variables (W_1, W_2, X_1, X_2) such that*

1. $p(u_1, u_2, z_1, z_2, z, w_1, w_2, x_1, x_2, y) \quad = \quad p(u_1, u_2, z_1, z_2, z)p(w_1 \mid u_1, z_1)$
   $p(w_2 \mid u_2, z_2)p(x_1 \mid w_1)p(x_2 \mid w_2)p(y \mid x_1, x_2)$
   and
2. *there exists a function $f_D : \mathcal{W}_1 \times \mathcal{W}_2 \times \mathcal{Z} \to (\hat{\mathcal{U}}_1 \times \hat{\mathcal{U}}_2)$ such that $E[d(U_i, \hat{U}_i)] \leq D_i$, $i = 1, 2$, where $(\hat{U}_1, \hat{U}_2) = f_D(W_1, W_2, Z)$ and the constraints*

$$I(U_1, Z_1; W_1 \mid W_2, Z) < I(X_1; Y \mid X_2, W_2, Z),$$

$$I(U_2, Z_2; W_2 \mid W_1, Z) < I(X_2; Y \mid X_1, W_1, Z), \qquad (10.1)$$

$$I(U_1, U_2, Z_1, Z_2; W_1, W_2 \mid Z) < I(X_1, X_2; Y \mid Z),$$

are satisfied where $\mathcal{W}_i$ are the sets in which W_i take values.

In Theorem 10.1 the encoding scheme involves distributed quantization (W_1, W_2) of the sources (U_1, U_2) and the side information Z_1, Z_2 followed by correlation preserving mapping to the channel codewords (X_1, X_2). The decoding approach involves first decoding (W_1, W_2) and then obtaining estimate $(\hat{U}_1, \hat{U}_2)$ as a function of (W_1, W_2) and the decoder side information Z. The proof of the theorem is given in [25, 47] and is an extension of the proof in [24].

If the channel alphabets are continuous (e.g. GMAC) then in addition to the conditions in Theorem 10.1 certain power constraints $E[X_i^2] \leq P_i$, $i = 1, 2$ are also needed.

For the discrete sources to recover the results with lossless transmission one can use the Hamming distance as the distortion measure.

If the source–channel separation holds then one can talk about the capacity region of the channel. For example, when there is no side information Z_1, Z_2, Z and the sources are independent then we obtain the rate region

$$R_1 \leq I(X_1; Y \mid X_2), \quad R_2 \leq I(X_2; Y \mid X_1),$$

$$R_1 + R_2 \leq I(X_1, X_2; Y). \qquad (10.2)$$

This is the well known rate region of an MAC [38]. To obtain Eq. (10.2) from Eq. (10.1), take (Z_1, Z_2, Z) independent of (U_1, U_2). Also, take U_1, U_2

discrete and $W_i = U_i$, $i = 1, 2$. Other special cases will be provided in Sec. 10.4.

In Theorem 10.1 it is possible to include other distortion constraints. For example, in addition to the bounds on $E[d(U_i, \hat{U}_i)]$ one may want a bound on the joint distortion $E[d((U_1, U_2), (\hat{U}_1, \hat{U}_2))]$. Then the only modification needed in the statement of the above theorem is to include this also as a condition in defining f_D.

If we only want to estimate a function $g(U_1, U_2)$ at the decoder and not (U_1, U_2) themselves, then again one can use the techniques in proof of Theorem 10.1 to obtain sufficient conditions. Depending upon g, the conditions needed may be weaker than those needed in Eq. (10.1).

In our problem setup the side conditions Z_i can be included with source U_i and then consider this problem as one with no side information at the encoders. However, the above formulation has the advantage that our conditions (Eq. (10.1)) are explicit in Z_i.

The main problem in using Theorem 10.1 is in obtaining good source–channel coding schemes providing (W_1, W_2, X_1, X_2), which satisfy the conditions in the theorem for a given source (U_1, U_2) and channel. A substantial part of this chapter will be devoted to this problem.

10.3.1. *Extension to Multiple Sources*

The above results can be generalized to the multiple (≥ 2) source case. Let $\mathcal{S} = 1, 2, \ldots, M$ be the set of sources with joint distribution $p(u_1, u_2, \ldots, u_M)$.

Theorem 10.2. *Sources $(U_i^n, i \in \mathcal{S})$ can be communicated in a distributed fashion over the memoryless multiple access channel $p(y \mid x_i, i \in \mathcal{S})$ with distortions $(D_i, i \in \mathcal{S})$ if there exist auxiliary random variables $(W_i, X_i, i \in \mathcal{S})$ satisfying*

1. $p(u_i, z_i, z, w_i, x_i, y, i \in \mathcal{S}) = p(u_i, z_i, z, i \in \mathcal{S})p(y \mid x_i, i \in \mathcal{S})\prod_{j \in \mathcal{S}} p(w_j \mid u_j, z_j)p(x_j \mid w_j)$
2. *There exists a function $f_D : \prod_{j \in \mathcal{S}} \mathcal{W}_j \times \mathcal{Z} \to (\hat{\mathcal{U}}_i, i \in \mathcal{S})$ such that $E[d(U_i, \hat{U}_i)] \leq D_i$, $i \in \mathcal{S}$ and the constraints*

$$I(U_A, Z_A; W_A \mid W_{A^c}, Z) < I(X_A; Y \mid X_{A^c}, W_{A^c}, Z) \quad \text{for all } A \subset \mathcal{S} \tag{10.3}$$

are satisfied (in case of continuous channel alphabets we also need the power constraints $E[X_i^2] \leq P_i$, $i = 1, \ldots, M$).

10.3.2. *Example*

We provide an example to show the reduction possible in transmission rates by exploiting the correlation between the sources, the side information and the permissible distortions.

Consider (U_1, U_2) with the joint distribution: $P(U_1 = 0; U_2 = 0) = P(U_1 = 1; U_2 = 1) = 1/3$; $P(U_1 = 1; U_2 = 0) = P(U_1 = 0; U_2 = 1) = 1/6$. If we use independent encoders which do not exploit the correlation among the sources then we need $R_1 \geq H(U_1) = 1$ bits and $R_2 \geq H(U_2) = 1$ bits for lossless coding of the sources. If we use the coding scheme in [12], then $R_1 \geq H(U_1 \mid U_2) = 0.918$ bits, $R_2 \geq H(U_2 \mid U_1) = 0.918$ bits and $R_1 + R_2 \geq H(U_1, U_2) = 1.918$ bits suffice.

Next consider an MAC such that $Y = X_1 + X_2$ where X_1 and X_2 take values from the alphabet $\{0, 1\}$ and Y takes values from the alphabet $\{0, 1, 2\}$. This does not satisfy the separation conditions [11]. The sum capacity C of such a channel with independent X_1 and X_2 is 1.5 bits and if we use source–channel separation, the given sources cannot be transmitted losslessly because $H(U_1, U_2) > C$. Now we use a joint source–channel code to improve the capacity of the channel. Take $X_1 = U_1$ and $X_2 = U_2$. Then the capacity of the channel is improved to $I(X_1, X_2; Y) = 1.585$ bits. This is still not enough to transmit the sources over the given MAC. Next we exploit the side information.

The side-information random variables are generated as follows. Z_1 is generated from U_2 by using a binary symmetric channel (BSC) with crossover probability $p = 0.3$. Similarly Z_2 is generated from U_1 by using the same BSC. Let $Z = (Z_1, Z_2, V)$, where $V = U_1 \cdot U_2 \cdot N$, N is a binary random variable with $P(N = 0) = P(N = 1) = 0.5$ independent of U_1 and U_2 and "·" denotes the logical AND operation. This denotes the case when the decoder can observe the encoder side information and also has some extra side information. Then from Eq. (10.1) if we use just the side information Z_1 the sum rate for the sources needs to be 1.8 bits. By symmetry the same holds if we only have Z_2. If we use Z_1 and Z_2 then we can use the sum rate 1.683 bits. If only V is used then the sum rate needed is 1.606 bits. So far we can still not transmit (U_1, U_2) losslessly if we use the coding $U_i = X_i$, $i = 1, 2$. If all the information in Z_1, Z_2, and V are used then we need $R_1 + R_2 \geq 1.4120$ bits. Thus with the aid of Z_1, Z_2, and Z we can transmit (U_1, U_2) losslessly over the MAC even with independent X_1 and X_2.

Next we consider the distortion criterion to be the Hamming distance and the allowable distortion as 4%. Then for compressing the individual sources without side information we need $R_i \geq H(p) - H(d) = 0.758$ bits,

$i = 1, 2$, where $H(x) = -x \log_2(x) - (1-x) \log_2(1-x)$. Thus we still cannot transmit (U_1, U_2) with this distortion when (X_1, X_2) are independent. If U_1 and U_2 are encoded, exploiting their correlations, (X_1, X_2) can be correlated. Next assume the side information $Z = (Z_1, Z_2)$ to be available at the decoder only. Then we need $R_1 \geq I(U_1; W_1) - I(Z_1; W_1)$, where W_1 is an auxiliary random variable generated from U_1. W_1 and Z_1 are related by a cascade of a BSC with cross-over probability 0.3 with a BSC with cross-over probability 0.04. This implies that $R_1 \geq 0.6577$ bits and $R_2 \geq 0.6577$ bits.

10.4. Special Cases

In the following, we provide several systems studied in literature as special cases. The practically important special cases of GMAC and orthogonal channels will be studied in detail in later sections. There, we will discuss several specific joint source–channel coding schemes for these and compare their performance.

10.4.1. *Lossless MAC with Correlated Sources*

Take $(Z_1, Z_2, Z) \perp (U_1, U_2)$ ($X \perp Y$ denotes that r.v. X is independent of r.v. Y) and $W_1 = U_1$ and $W_2 = U_2$ where U_1 and U_2 are discrete sources. Then the constraints of Eq. (10.1) reduce to

$$H(U_1 \mid U_2) < I(X_1; Y \mid X_2, U_2),$$
$$H(U_2 \mid U_1) < I(X_2; Y \mid X_1, U_1), \tag{10.4}$$
$$H(U_1, U_2) < I(X_1, X_2; Y),$$

where X_1 and X_2 are the channel inputs, Y is the channel output, and $X_1 \leftrightarrow U_1 \leftrightarrow U_2 \leftrightarrow X_2$ is satisfied. These are the conditions obtained in [1].

10.4.2. *Lossy MAC*

Take $(Z_1, Z_2, Z) \perp (U_1, U_2, W_1, W_2)$. In this case the constraints in Eq. (10.1) reduce to

$$I(U_1; W_1 \mid W_2) < I(X_1; Y \mid X_2, W_2),$$
$$I(U_2; W_2 \mid W_1) < I(X_2; Y \mid X_1, W_1), \tag{10.5}$$
$$I(U_1, U_2; W_1, W_2) < I(X_1, X_2; Y).$$

This is an immediate generalization of [1] to the lossy case.

10.4.3. *Lossless Multiple Access Communication with Common Information*

Consider $U_1 = (U_1', U_0')$, $U_2 = (U_2', U_0')$ where U_0', U_1', U_2' are independent of each other. U_0' is interpreted as the common information at the two encoders. Then, taking $(Z_1, Z_2, Z) \perp (U_1, U_2)$, $W_1 = U_1$ and $W_2 = U_2$ we obtain sufficient conditions for lossless transmission as

$$H(U_1') < I(X_1; Y \mid X_2, U_0'),$$
$$H(U_2') < I(X_2; Y \mid X_1, U_0'), \qquad (10.6)$$
$$H(U_1') + H(U_2') + H(U_0') < I(X_1, X_2; Y).$$

This provides the capacity region of the MAC with common information available in [48].

Our results generalize this system with lossy transmission also.

10.4.4. *Lossy Distributed Source Coding with Side Information*

The multiple access channel is taken as a dummy channel which reproduces its inputs. In this case we obtain that the sources can be coded with rates R_1 and R_2 to obtain the specified distortions at the decoder if

$$R_1 > I(U_1, Z_1; W_1 \mid W_2, Z),$$
$$R_2 > I(U_2, Z_2; W_2 \mid W_1, Z), \qquad (10.7)$$
$$R_1 + R_2 > I(U_1, U_2, Z_1, Z_2; W_1, W_2 \mid Z).$$

This recovers the result in [23], and generalizes the results in [4, 5, 12].

10.4.5. *Correlated Sources with Lossless Transmission Over Multi-User Channels with Receiver Side Information*

If we consider $(Z_1, Z_2) \perp (U_1, U_2)$, $W_1 = U_1$ and $W_2 = U_2$ then we recover the conditions

$$H(U_1 \mid U_2, Z) < I(X_1; Y \mid X_2, U_2, Z),$$
$$H(U_2 \mid U_1, Z) < I(X_2; Y \mid X_1, U_1, Z), \qquad (10.8)$$
$$H(U_1, U_2 \mid Z) < I(X_1, X_2; Y \mid Z),$$

in Theorem 2.1 in [49].

10.4.6. *Mixed Side Information*

The aim is to determine the rate distortion function for transmitting a source X with the aid of side information (Y, Z) (system in Fig. 10.1(c) of [50]). The encoder is provided with Y and the decoder has access to both Y and Z. This represents the mixed side information (MSI) system, which combines the conditional rate distortion system and the Wyner–Ziv system. This has the system in Figs. 10.1(a) and 10.1(b) of [50] as special cases. The results of Fig. 10.1(c) can be recovered from our theorem if we take X, Y, Z, and W in [50] as $U_1 = X, Z = (Z, Y), Z_1 = Y, W_1 = W$. U_2 and Z_2 are taken to be constants. The acceptable rate region is given by $R > I(X, W | Y, Z)$, where W is a random variable with the property $W \leftrightarrow (X, Y) \leftrightarrow Z$ and for which there exists a decoder function such that the distortion constraints are met.

10.5. Discrete Alphabet Sources Over GMAC

This system is practically very useful. For example, in a sensor network, the observations sensed by the sensor nodes are discretized and then transmitted over a GMAC. The physical proximity of the sensor nodes makes their observations correlated. This correlation can be exploited to compress the transmitted data. We present a distributed "correlation preserving" joint source–channel coding scheme yielding jointly Gaussian channel codewords which will be shown to compress the data efficiently. This coding scheme was developed in [51]. The proofs of Lemmas 10.1–10.4 below are in [51, 47].

Sufficient conditions for lossless transmission of two discrete sources (U_1, U_2) (generating *iid* sequences in time) over a general MAC with no side information are obtained in Eq. (10.4) and reproduced below for convenience

$$H(U_1 \,|\, U_2) < I(X_1; Y \,|\, X_2, U_2),$$
$$H(U_2 \,|\, U_1) < I(X_2; Y \,|\, X_1, U_1), \tag{10.9}$$
$$H(U_1, U_2) < I(X_1, X_2; Y),$$

where $X_1 \leftrightarrow U_1 \leftrightarrow U_2 \leftrightarrow X_2$ is satisfied.

In this section, we further specialize the above results for lossless transmission of discrete correlated sources over an additive memoryless GMAC: $Y = X_1 + X_2 + N$ where N is a Gaussian random variable independent of X_1 and X_2. The noise N satisfies $E[N] = 0$ and

$\text{Var}(N) = \sigma_N^2$. We will also have the transmit power constraints: $E[X_i^2] \leq P_i$, $i = 1, 2$. Since source–channel separation does not hold for this system, a joint source–channel coding scheme is needed for optimal performance.

The dependence of R.H.S. of Eq. (10.9) on input alphabets prevents us from getting a closed form expression for the admissibility criterion. Therefore we relax the conditions by taking away the dependence on the input alphabets. This will allow us to obtain good joint source–channel codes.

Lemma 10.1. *Under our assumptions,* $I(X_1; Y \mid X_2, U_2) \leq I(X_1; Y \mid X_2)$.

Thus from Eq. (10.9),

$$H(U_1 \mid U_2) < I(X_1; Y \mid X_2, U_2) \leq I(X_1; Y \mid X_2), \qquad (10.10)$$

$$H(U_2 \mid U_1) < I(X_2; Y \mid X_1, U_1) \leq I(X_2; Y \mid X_1), \qquad (10.11)$$

$$H(U_1, U_2) < I(X_1, X_2; Y). \qquad (10.12)$$

The relaxation of the upper bounds is only in Eqs. (10.10) and (10.11) and not in Eq. (10.12).

We show that the relaxed upper bounds are maximized if (X_1, X_2) is jointly Gaussian and the correlation ρ between X_1 and X_2 is high (the highest possible ρ may not give the largest upper bound in the three inequalities in Eqs. (10.10)–(10.12).

Lemma 10.2. *A jointly Gaussian distribution for* (X_1, X_2) *maximizes* $I(X_1; Y \mid X_2)$, $I(X_2; Y \mid X_1)$ *and* $I(X_1, X_2; Y)$ *simultaneously.*

The difference between the bounds in Eq. (10.10) is

$$I(X_1, Y \mid X_2) - I(X_1, Y \mid X_2, U_2) = I(X_1 + N; U_2 \mid X_2). \qquad (10.13)$$

This difference is small if correlation between (U_1, U_2) is small. In that case $H(U_1 \mid U_2)$ and $H(U_2 \mid U_1)$ will be large and Eqs. (10.10) and (10.11) can be active constraints. If correlation between (U_1, U_2) is large, $H(U_1 \mid U_2)$ and $H(U_2 \mid U_1)$ will be small and Eq. (10.12) will be the only active constraint. In this case the difference between the two bounds in Eqs. (10.10) and (10.11) is large but not important. Thus, the outer bounds in Eqs. (10.10) and (10.11) are close to the inner bounds whenever the constraints Eqs. (10.10) and (10.11) are active. Often Eq. (10.12) will be the only active constraint.

An advantage of outer bounds in Eqs. (10.10) and (10.11) is that we will be able to obtain a good source–channel coding scheme. Once

(X_1, X_2) are obtained we can check for sufficient conditions Eq. (10.9). If these are not satisfied for the (X_1, X_2) obtained, we will increase the correlation ρ between (X_1, X_2) if possible (see details below). Increasing the correlation in (X_1, X_2) will decrease the difference in (10.13) and increase the possibility of satisfying Eq. (10.9) when the outer bounds in Eqs. (10.10) and (10.11) are satisfied.

We evaluate the (relaxed) rate region Eqs. (10.10)–(10.12) for the Gaussian MAC with jointly Gaussian channel inputs (X_1, X_2) with the transmit power constraints. For maximization of this region we need mean vector $[0\ 0]$ and covariance matrix $K_{X_1, X_2} = \begin{pmatrix} P_1 & \rho\sqrt{P_1 P_2} \\ \rho\sqrt{P_1 P_2} & P_2 \end{pmatrix}$ where ρ is the correlation between X_1 and X_2. Then Eqs. (10.10)–(10.12) provide the relaxed constraints

$$H(U_1 \mid U_2) < 0.5 \log \left[1 + \frac{P_1 (1 - \rho^2)}{\sigma_N{}^2} \right], \tag{10.14}$$

$$H(U_2 \mid U_1) < 0.5 \log \left[1 + \frac{P_2 (1 - \rho^2)}{\sigma_N{}^2} \right], \tag{10.15}$$

$$H(U_1, U_2) < 0.5 \log \left[1 + \frac{P_1 + P_2 + 2\rho\sqrt{P_1 P_2}}{\sigma_N{}^2} \right]. \tag{10.16}$$

The upper bounds in the first two inequalities in Eqs. (10.14) and (10.15) decrease as ρ increases. But the third upper bound Eq. (10.16) increases with ρ and often the third constraint is the limiting constraint.

This motivates us to consider the GMAC with correlated jointly Gaussian inputs. The next lemma provides an upper bound on the correlation between (X_1, X_2) in terms of the distribution of (U_1, U_2).

Lemma 10.3. *Let (U_1, U_2) be the correlated sources and $X_1 \leftrightarrow U_1 \leftrightarrow U_2 \leftrightarrow X_2$ where X_1 and X_2 are jointly Gaussian. Then the correlation between (X_1, X_2) satisfies $\rho^2 \leq 1 - 2^{-2I(U_1, U_2)}$.*

10.5.1. *A Coding Scheme*

In this section, we develop a coding scheme for mapping the discrete alphabets into jointly Gaussian correlated code words, which also satisfy the Markov condition. The heart of the scheme is to approximate a jointly Gaussian distribution with the sum of product of Gaussian marginals. Although this is stated in the following lemma for two dimensional vectors (X_1, X_2), the results hold for any finite dimensional vectors.

Lemma 10.4. *Any jointly Gaussian two dimensional density can be uniformly arbitrarily closely approximated by a weighted sum of product of marginal Gaussian densities:*

$$\sum_{i=1}^{N} \frac{p_i}{\sqrt{2\pi c_{1i}}} e^{\frac{-1}{2c_{1i}}(x_1-a_{1i})^2} \frac{q_i}{\sqrt{2\pi c_{2i}}} e^{\frac{-1}{2c_{2i}}(x_2-a_{2i})^2}. \tag{10.17}$$

From Lemma 10.4 a joint Gaussian density with any correlation can be expressed by a linear combination of marginal Gaussian densities. But the coefficients p_i and q_i in Eq. (10.17) may be positive or negative. To realize our coding scheme, we would like to have the p_i's and q_i's to be non-negative. This introduces constraints on the realizable Gaussian densities in our coding scheme. For example, from Lemma 10.3, the correlation ρ between X_1 and X_2 cannot exceed $\sqrt{1 - 2^{-2I(U_1;U_2)}}$. Also there is still the question of getting a good linear combination of marginal densities to obtain the joint density for a given N in Eq. (10.17).

This motivates us to consider an optimization procedure for finding $p_i, q_i, a_{1i}, a_{2i}, c_{1i}$, and c_{2i} in Eq. (10.17) that provides the best approximation to a given joint Gaussian density. We illustrate this with an example. Consider U_1, U_2 to be binary. Let $P(U_1 = 0; U_2 = 0) = p_{00}$; $P(U_1 = 0; U_2 = 1) = p_{01}$; $P(U_1 = 1; U_2 = 0) = p_{10}$; and $P(U_1 = 1; U_2 = 1) = p_{11}$. We can consider

$$f(X_1 = . \,|\, U_1 = 0) = p_{101}\mathcal{N}(a_{101}, c_{101}) + p_{102}\mathcal{N}(a_{102}, c_{102})$$
$$+ \cdots + p_{10r_1}\mathcal{N}(a_{10r_1}, c_{10r_1}), \tag{10.18}$$

$$f(X_1 = . \,|\, U_1 = 1) = p_{111}\mathcal{N}(a_{111}, c_{111}) + p_{112}\mathcal{N}(a_{112}, c_{112})$$
$$+ \cdots + p_{11r_2}\mathcal{N}(a_{11r_2}, c_{11r_2}), \tag{10.19}$$

$$f(X_2 = . \,|\, U_2 = 0) = p_{201}\mathcal{N}(a_{201}, c_{201}) + p_{202}\mathcal{N}(a_{202}, c_{202})$$
$$+ \cdots + p_{20r_3}\mathcal{N}(a_{20r_3}, c_{20r_3}), \tag{10.20}$$

$$f(X_2 = . \,|\, U_2 = 1) = p_{211}\mathcal{N}(a_{211}, c_{211}) + p_{212}\mathcal{N}(a_{212}, c_{212})$$
$$+ \cdots + p_{21r_4}\mathcal{N}(a_{21r_4}, c_{21r_4}). \tag{10.21}$$

where $\mathcal{N}(a, b)$ denotes Gaussian density with mean a and variance b. Let $\underline{p}$ be the vector with components $p_{101}, \ldots, p_{10r_1}, p_{111}, \ldots, p_{11r_2}, p_{201}, \ldots, p_{20r_3}, p_{211}, \ldots, p_{21r_4}$. Similarly we denote by $\underline{a}$ and $\underline{c}$ the vectors with components $a_{101}, \ldots, a_{10r_1}, a_{111}, \ldots, a_{11r_2}, a_{201}, \ldots, a_{20r_3}, a_{211}, \ldots, a_{21r_4}$ and $c_{101}, \ldots, c_{10r_1}, c_{111}, \ldots, c_{11r_2}, c_{201}, \ldots, c_{20r_3}, c_{211}, \ldots, c_{21r_4}$.

Let $f_\rho(x_1, x_2)$ be the jointly Gaussian density that we want to approximate. Let it has zero mean and covariance matrix $K_{X_1, X_2} = \begin{pmatrix} 1 & \rho \\ \rho & 1 \end{pmatrix}$. Let $g_{\underline{p}, \underline{a}, \underline{c}}$ be the sum of marginal densities with parameters $\underline{p}, \underline{a}$, and $\underline{c}$ approximating f_ρ. The best g is obtained by solving the following minimization problem:

$$\min_{\underline{p}, \underline{a}, \underline{c}} \int [g_{\underline{p}, \underline{a}, \underline{c}}(x_1, x_2) - f_\rho(x_1, x_2)]^2 dx_1 dx_2, \qquad (10.22)$$

subject to

$$(p_{00} + p_{01}) \sum_{i=1}^{r_1} p_{10i} a_{10i} + (p_{10} + p_{11}) \sum_{i=1}^{r_2} p_{11i} a_{11i} = 0,$$

$$(p_{00} + p_{10}) \sum_{i=1}^{r_3} p_{20i} a_{20i} + (p_{01} + p_{11}) \sum_{i=1}^{r_4} p_{21i} a_{21i} = 0,$$

$$(p_{00} + p_{01}) \sum_{i=1}^{r_1} p_{10i}(c_{10i} + a_{10i}^2) + (p_{10} + p_{11}) \sum_{i=1}^{r_2} p_{11i}(c_{11i} + a_{11i}^2) = 1,$$

$$(p_{00} + p_{10}) \sum_{i=1}^{r_3} p_{20i}(c_{20i} + a_{20i}^2) + (p_{01} + p_{11}) \sum_{i=1}^{r_4} p_{21i}(c_{21i} + a_{21i}^2) = 1,$$

$$\sum_{i=1}^{r_1} p_{10i} = 1, \quad \sum_{i=1}^{r_2} p_{11i} = 1, \quad \sum_{i=1}^{r_3} p_{20i} = 1, \quad \sum_{i=1}^{r_4} p_{21i} = 1,$$

$$p_{10i} \geq 0, \quad c_{10i} \geq 0 \quad \text{for } i \in \{1, 2, \ldots, r_1\},$$
$$p_{11i} \geq 0, \quad c_{11i} \geq 0 \quad \text{for } i \in \{1, 2, \ldots, r_2\},$$
$$p_{20i} \geq 0, \quad c_{20i} \geq 0 \quad \text{for } i \in \{1, 2, \ldots, r_3\},$$
$$p_{21i} \geq 0, \quad c_{21i} \geq 0 \quad \text{for } i \in \{1, 2, \ldots, r_4\}.$$

The above constraints are such that the resulting distribution g for (X_1, X_2) will satisfy $E[X_i] = 0$ and $E[X_i^2] = 1$, $i = 1, 2$.

The above coding scheme will be used to obtain a codebook as follows. If user 1 produces $U_1 = 0$, then with probability p_{10i} the encoder 1 obtains codeword X_1 from the distribution $\mathcal{N}(a_{10i}, c_{10i})$. Similarly we obtain the codewords for $U_1 = 1$ and for user 2. Once we have found the encoder maps the encoding and decoding are as described in the proof of Theorem 1 in [25] (also in [24]). The decoding is done by joint typicality of the received Y^n with (U_1^n, U_2^n).

This coding scheme can be extended to any discrete alphabet case. We give an example below to illustrate the coding scheme.

10.5.2. *Example*

Consider (U_1, U_2) with the joint distribution: $P(U_1 = 0; U_2 = 0) = P(U_1 = 1; U_2 = 1) = P(U_1 = 0; U_2 = 1) = 1/3; P(U_1 = 1; U_2 = 0) = 0$ and power constraints $P_1 = 3; P_2 = 4$. Also consider a Gaussian multiple access channel with $\sigma_N^2 = 1$. If the sources are mapped into independent channel code words, then the sum rate condition in Eq. (10.16) with $\rho = 0$ should hold. The LHS evaluates to 1.585 bits whereas the RHS is 1.5 bits. Thus condition Eq. (10.16) is violated and hence the sufficient conditions in Eq. (10.9) are also violated.

In the following, we explore the possibility of using correlated (X_1, X_2) to see if we can transmit this source on the given MAC. The inputs (U_1, U_2) can be distributedly mapped to jointly Gaussian channel code words (X_1, X_2) by the technique mentioned above. The maximum ρ which satisfies Eqs. (10.14) and (10.15) are 0.7024 and 0.7874, respectively, and the minimum ρ which satisfies Eq. (10.16) is 0.144. Thus, we can pick a ρ which satisfies Eqs. (10.14)–(10.16). From Lemma 10.3, ρ is upper bounded by 0.546. Therefore we want to obtain jointly Gaussian (X_1, X_2) satisfying $X_1 \leftrightarrow U_1 \leftrightarrow U_2 \leftrightarrow X_2$ with correlation $\rho \in [0.144, 0.546]$. If we pick a ρ that satisfies the original bounds, then we will be able to transmit the sources (U_1, U_2) reliably on this MAC. Without loss of generality the jointly Gaussian channel inputs required are chosen with mean vector $[0 \ 0]$ and covariance matrix $K_{X_1, X_2} = \begin{pmatrix} 1 & \rho \\ \rho & 1 \end{pmatrix}$. The ρ chosen is 0.3 and hence is such that it meets all the conditions Eqs. (10.14)–(10.16). Also, we choose $r_1 = r_2 = r_3 = r_4 = 2$. We solve the optimization problem Eq. (10.22) via MATLAB to get the function g. The normalized minimum distortion, defined as $\int [g_{p,a,c}(x_1, x_2) - f_\rho(x_1, x_2)]^2 dx_1 dx_2 / \int f_\rho^2(x_1, x_2) dx_1 dx_2$ is 0.137% when the marginals are chosen as:

$$f(X_1 \,|\, U_1 = 0) = \mathcal{N}(0.0002, 0.9108), \quad f(X_1 \,|\, U_1 = 1) = \mathcal{N}(0.0001, 1.0446),$$

$$f(X_2 \,|\, U_2 = 0) = \mathcal{N}(0.0021, 1.1358), \quad f(X_2 \,|\, U_2 = 1) = \mathcal{N}(0.0042, 0.7283).$$

The approximation (a cross-section of the two dimensional densities) is shown in Fig. 10.2(a).

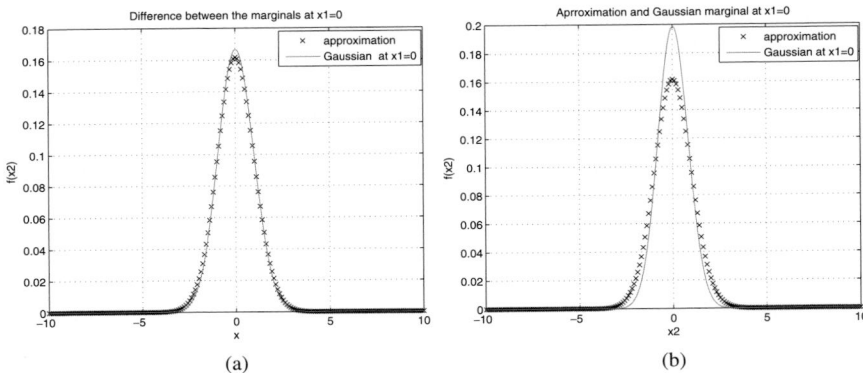

Fig. 10.2. Cross-section of the approximation of the joint Gaussian (a) $\rho = 0.3$ and (b) $\rho = 0.6$.

If we take $\rho = 0.6$ which violates Lemma 10.3 then the approximation is shown in Fig. 10.2(b). We can see from Fig. 10.2(b) that the error in this case is more. Now the normalized marginal distortion is 10.5%.

The original upper bound in Eqs. (10.10) and (10.11) for this example with $\rho = 0.3$ is $I(X_1; Y \mid X_2, U_2) = 0.792$, $I(X_2; Y \mid X_1, U_1) = 0.996$. Also, $I(X_1; Y \mid X_2) = 0.949$, $I(X_2; Y \mid X_1) = 1.107$. $H(U_1 \mid U_2) = H(U_2 \mid U_1) = 0.66$ and we conclude that the original bounds too are satisfied by the choice of $\rho = 0.3$.

10.6. Source-Channel Coding for Gaussian Sources Over GMAC

In this section, we consider transmission of correlated Gaussian sources over a GMAC. This is an important example for transmitting continuous alphabet sources over a GMAC. For example, one comes across it if a sensor network is sampling a Gaussian random field. Also, in the application of detection of change [52] by a sensor network, it is often the detection of change in the mean of the sensor observations with the sensor observation noise being Gaussian.

We will assume that (U_{1n}, U_{2n}) is jointly Gaussian with mean zero, variances σ_i^2, $i = 1, 2$ and correlation ρ. The distortion measure will be mean square error (MSE). The (relaxed) sufficient conditions from Eq. (10.6) for transmission of the sources over the channel are given by

(these continue to hold because Lemmas 10.1–10.3 are still valid)

$$I(U_1; W_1 \mid W_2) < 0.5 \log \left[1 + \frac{P_1(1 - \tilde{\rho}^2)}{\sigma_N{}^2} \right],$$

$$I(U_2; W_2 \mid W_1) < 0.5 \log \left[1 + \frac{P_2(1 - \tilde{\rho}^2)}{\sigma_N{}^2} \right], \tag{10.23}$$

$$I(U_1, U_2; W_1, W_2) < 0.5 \log \left[1 + \frac{P_1 + P_2 + 2\tilde{\rho}\sqrt{P_1 P_2}}{\sigma_N{}^2} \right].$$

We consider three specific coding schemes to obtain W_1, W_2, X_1, and X_2 where (W_1, W_2) satisfy the distortion constraints and (X_1, X_2) are jointly Gaussian with an appropriate $\tilde{\rho}$ such that Eq. (10.23) is satisfied. These coding schemes have been widely used. We compare their performance also.

10.6.1. *Amplify and Forward Scheme*

In the AF scheme the channel codes X_i are just scaled source symbols U_i. Since (U_1, U_2) are themselves jointly Gaussian, (X_1, X_2) will be jointly Gaussian and retain the dependence of inputs (U_1, U_2). The scaling is done to ensure $E[X_i{}^2] = P_i, i = 1, 2$. For a single user case this coding is optimal [38].

At the decoder inputs U_1 and U_2 are directly estimated from Y as $\hat{U}_i = E[U_i|Y]$, $i = 1, 2$. Because U_i and Y are jointly Gaussian this estimate is linear and also satisfies the minimum mean square error (MMSE) and maximum likelihood (ML) criteria.

The MMSE distortion for this encoding–decoding scheme is

$$\overline{D_1} = \frac{\sigma_1{}^2 \left[P_2(1 - \rho^2) + \sigma_N{}^2 \right]}{P_1 + P_2 + 2\rho\sqrt{P_1 P_2} + \sigma_N{}^2}, \quad \overline{D_2} = \frac{\sigma_2{}^2 \left[P_1(1 - \rho^2) + \sigma_N{}^2 \right]}{P_1 + P_2 + 2\rho\sqrt{P_1 P_2} + \sigma_N{}^2}. \tag{10.24}$$

Since encoding and decoding require minimum processing and delay in this scheme, if it satisfies the required distortion bounds D_i, it should be the scheme to implement. This scheme has been studied in [29] and found to be optimal below a certain SNR for two-user symmetric case ($P_1 = P_2, \sigma_1 = \sigma_2, D_1 = D_2$). However unlike for single user case, in this case user 1 acts as interference for user 2 (and vice versa). Thus one should not expect this scheme to be optimal under high SNR case. That this is indeed true was shown in [30]. It was also shown there, that at high SNR, for $P_1 \neq P_2$, it may indeed be better in AF to use less power than P_1, P_2. This can also be

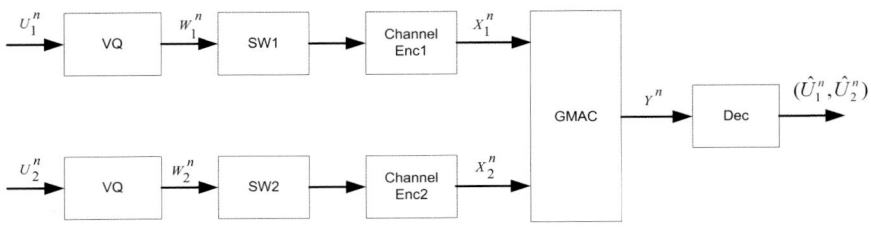

Fig. 10.3. Separation based scheme.

interpreted as using AF on $U_1 - \alpha_1 E[U_2 \,|\, U_1]$ and $U_2 - \alpha_2 E[U_1 \,|\, U_2]$ at the two encoders at high SNR, which will reduce the correlations between the transmitted symbols.

10.6.2. *Separation Based Scheme*

In separation based (SB) approach (Fig. 10.3) the jointly Gaussian sources are vector quantized to W_1^n and W_2^n. The quantized outputs are S–W encoded [12]. This produces code words, which are (asymptotically) independent. These independent code words are encoded to capacity achieving Gaussian channel codes (X_1^n, X_2^n) with correlation $\tilde{\rho} = 0$. This is a very natural scheme and has been considered by various authors [1, 38, 11].

Since source–channel separation does not hold for this system, this scheme is not expected to be optimal. But because this scheme decouples source coding from channel coding, it is preferable to a joint source–channel coding scheme with comparable performance.

10.6.3. *Lapidoth–Tinguely Scheme*

In this scheme, obtained in [29], (U_1^n, U_2^n) are vector quantized to 2^{nR_1} and 2^{nR_2} $(\tilde{U}_1^n, \tilde{U}_2^n)$ vectors where R_1 and R_2 will be specified below. Also, W_1^n and W_2^n are 2^{nR_1} and 2^{nR_2}, n length code words obtained independently with distributions $\mathcal{N}(0,1)$. For each $\tilde{u}_i^n$, we pick the codeword w_i^n that is closest to it. This way we obtain Gaussian codewords W_1^n and W_2^n, which retain the correlations of (U_1^n, U_2^n). X_1^n and X_2^n are obtained by scaling W_1^n and W_2^n to satisfy the transmit power constraints. We will call this LT scheme. (U_1, U_2, W_1, W_2) are (approximately) jointly Gaussian with covariance matrix.

$$
\begin{pmatrix}
\sigma_1^2 & \rho\sigma_1\sigma_2 & \sigma_1^2(1 - 2^{-2R_1}) & \rho\sigma_1\sigma_2(1 - 2^{-2R_2}) \\
\rho\sigma_1\sigma_2 & \sigma_2^2 & \rho\sigma_1\sigma_2(1 - 2^{-2R_1}) & \sigma_2^2(1 - 2^{-2R_2}) \\
\sigma_1^2(1 - 2^{-2R_1}) & \rho\sigma_1\sigma_2(1 - 2^{-2R_1}) & \sigma_1^2(1 - 2^{-2R_1}) & \dfrac{\tilde{\rho}^2\sigma_1\sigma_2}{\rho} \\
\rho\sigma_1\sigma_2(1 - 2^{-2R_2}) & \sigma_2^2(1 - 2^{-2R_2}) & \dfrac{\tilde{\rho}^2\sigma_1\sigma_2}{\rho} & \sigma_2^2(1 - 2^{-2R_2})
\end{pmatrix}.
$$

$$(10.25)$$

In Eq. (10.25) $\tilde{\rho} = \rho\sqrt{(1 - 2^{-2R_1})(1 - 2^{-2R_2})}$.
We obtain the (R_1, R_2) above from Eq. (10.23). From

$$
I(U_1; W_1 \mid W_2) = H(W_1 \mid W_2) - H(W_1 \mid W_2, U_1),
$$

and the fact that the Markov chain condition $W_1 \leftrightarrow U_1 \leftrightarrow U_2 \leftrightarrow W_2$ holds,

$$
H(W_1 \mid W_2, U_1) = H(W_1 \mid U_1)
$$

and

$$
I(U_1; W_1 \mid W_2) = 0.5 \log[(1 - \tilde{\rho}^2)2^{2R_1}].
$$

Thus from Eq. (10.23) we need R_1 and R_2 which satisfy

$$
R_1 \leq 0.5 \log\left[\frac{P_1}{\sigma_N{}^2} + \frac{1}{(1 - \tilde{\rho}^2)}\right]. \tag{10.26}
$$

Similarly, we also need

$$
R_2 \leq 0.5 \log\left[\frac{P_2}{\sigma_N{}^2} + \frac{1}{(1 - \tilde{\rho}^2)}\right], \tag{10.27}
$$

$$
R_1 + R_2 \leq 0.5 \log\left[\frac{\sigma_N{}^2 + P_1 + P_2 + 2\tilde{\rho}\sqrt{P_1 P_2}}{(1 - \tilde{\rho}^2)\sigma_N{}^2}\right]. \tag{10.28}
$$

The inequalities Eqs. (10.26)–(10.28) are the same as in [29]. Thus we recover the conditions in [29] from our general result (Eq. (10.1)). Taking $\hat{U}_i = E[U_i \mid W_1, W_2]$, $i = 1, 2$, we obtain the distortions

$$
D_1 = \text{var}(U_1 \mid W_1, W_2) = \frac{\sigma_1{}^2 2^{-2R_1}\left[1 - \rho^2\left(1 - 2^{-2R_2}\right)\right]}{(1 - \tilde{\rho}^2)}, \tag{10.29}
$$

$$
D_2 = \text{var}(U_2 \mid W_1, W_2) = \frac{\sigma_2{}^2 2^{-2R_2}\left[1 - \rho^2\left(1 - 2^{-2R_1}\right)\right]}{(1 - \tilde{\rho}^2)}. \tag{10.30}
$$

The minimum distortion is obtained when $\tilde{\rho}$ is such that the sum rate is met with equality in Eq. (10.28). For the symmetric case at the minimum distortion, $R_1 = R_2$.

10.6.4. Asymptotic Performance of the Three Schemes

We compare the performance of the three schemes. These results are from [30]. For simplicity we consider the symmetric case: $P_1 = P_2 = P$, $\sigma_1 = \sigma_2 = \sigma$, $D_1 = D_2 = D$. We will denote the SNR $P/\sigma_N{}^2$ by S.

Consider the AF scheme. From Eq. (10.24)

$$D(S) = \frac{\sigma^2 \left[S\left(1 - \rho^2\right) + 1\right]}{2S(1 + \rho) + 1}.$$
(10.31)

Thus $D(S)$ decreases to $\sigma^2(1 - \rho)/2$ strictly monotonically at rate $O(1)$ as $S \to \infty$.

Also,

$$\lim_{S \to 0} \left| \frac{D(S) - \sigma^2}{S} \right| = \sigma^2(1 + \rho)^2.$$
(10.32)

Thus, $D(S) \to \sigma^2$ at rate $O(S)$ as $S \to 0$.

Consider the SB scheme at High SNR. From [27], if each source is encoded with rate R then it can be decoded at the decoder with distortion

$$D^2 = 2^{-4R}(1 - \rho^2) + \rho^2 2^{-8R}.$$
(10.33)

At high SNR, from the capacity result for independent inputs, we have $R < 0.25 \log S$ [38]. Then from Eq. (10.33) we obtain

$$D \geq \sqrt{\frac{\sigma^4(1 - \rho^2)}{S} + \frac{\sigma^4 \rho^2}{S^2}}$$
(10.34)

and this lower bound is achievable. As $S \to \infty$, this lower bound approaches zero at rate $O(\sqrt{S})$. Thus this scheme outperforms AF at high SNR.

At low SNR, $R \approx \frac{S}{2}$ and hence from Eq. (10.33)

$$D \geq \rho^2 \sigma^4 2^{-4S} + \sigma^2(1 - \rho^2)2^{-2S}.$$
(10.35)

Thus $D \to \sigma^2$ at rate $O(S^2)$ as $S \to 0$ at high ρ and at rate $O(S)$ at small ρ. Therefore, we expect that at low SNR, at high ρ this scheme will be worse than AF but at low ρ it will be comparable.

Consider the LT scheme. In the high SNR region we assume that $\tilde\rho = \rho$ since $R = R_1 = R_2$ are sufficiently large. Then from Eq. (10.28) $R \approx 0.25 \log[2S/(1-\rho)]$ and the distortion can be approximated by

$$D \approx \sigma^2 \sqrt{(1-\rho)/2S}. \qquad (10.36)$$

Therefore, $D \to 0$ as $S \to \infty$ at rate $O(\sqrt{S})$. This rate of convergence is same as for SB. However, the R.H.S. in (10.34) is greater than that of (10.36) and at low ρ the two are close. Thus at high SNR LT always outperforms SB but the improvement is small for low ρ.

At low SNR

$$R \approx \frac{S(1+\tilde\rho)}{2} - \frac{\log(1-\tilde\rho^2)}{4}$$

and evaluating D from Eq. (10.29) we get

$$D = \frac{\sigma^2 2^{-\overline{S}}(1 - \rho^2(1 - \sqrt{1-\tilde\rho^2}2^{-\overline{S}}))}{\sqrt{1-\tilde\rho^2}}, \qquad (10.37)$$

where $\overline{S} = S(1+\tilde\rho)$. Therefore, $D \to \sigma^2$ as $S \to 0$ at rate $O(S^2)$ at high ρ and at rate $O(S)$ at low ρ. These rates are same as that for SB. In fact, dividing the expression for D at low SNR for SB by that for LT, we can show that the two distortions tend to σ^2 at the same rate for all ρ.

The necessary conditions (NC) to be able to transmit on the GMAC with distortion (D, D) for the symmetric case are [29]

$$D \geq \begin{cases} \dfrac{\sigma^2[S(1-\rho^2)+1]}{2S(1+\rho)+1}, & \text{for } S \leq \dfrac{\rho}{1-\rho^2}, \\[4mm] \sigma^2 \sqrt{\dfrac{(1-\rho^2)}{2S(1+\rho)+1}}, & \text{for } S > \dfrac{\rho}{1-\rho^2}. \end{cases} \qquad (10.38)$$

The above three schemes along with Eq. (10.38) are compared below using exact computations. Figure 10.4 shows the distortion as a function of SNR for unit variance jointly Gaussian sources with correlations $\rho = 0.1$ and 0.75.

From these plots we confirm our theoretical conclusions provided above.

10.6.5. *Continuous Sources Over a GMAC*

For general continuous alphabet sources (U_1, U_2) we vector quantize U_1^n and U_2^n into $\tilde U_1^n$ and $\tilde U_2^n$. Then to obtain correlated Gaussian codewords

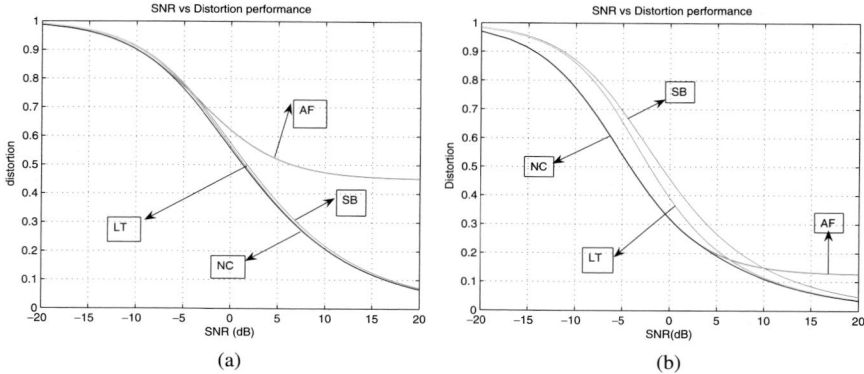

Fig. 10.4.   SNR vs distortion performance (a) $\rho = 0.1$ and (b) $\rho = 0.75$.

(X_1^n, X_2^n) we can use two schemes adapted from the cases studied above. In the first scheme, we use the scheme developed in Sec. 10.5.1. In the second scheme we use LT scheme explained in Sec. 10.6.3.

10.7. Correlated Sources Over Orthogonal MAC

One standard way to use the MAC is via TDMA, FDMA, CDMA or orthogonal frequency division multiple access (OFDMA) [38–40]. These protocols although suboptimal are used due to practical considerations. These protocols make an MAC a set of parallel orthogonal channels (for CDMA, it happens if we use orthogonal codes). We study transmission of correlated sources through such a system.

10.7.1. *Transmission of Correlated Sources Over Orthogonal Channels*

Consider the setup in Fig. 10.1 when $Y = (Y_1, Y_2)$ and $p(y \mid x_1, x_2) = p(y_1, y_2 \mid x_1, x_2) = p(y_1 \mid x_1)p(y_2 \mid x_2)$. Then the conditions in Eq. (10.1) become

$$I(U_1, Z_1; W_1 \mid W_2, Z) < I(X_1; Y_1 \mid W_2, Z) \leq I(X_1; Y_1), \qquad (10.39)$$

$$I(U_2, Z_2; W_2 \mid W_1, Z) < I(X_2; Y_2 \mid W_1, Z) \leq I(X_2; Y_2), \qquad (10.40)$$

$$I(U_1, U_2, Z_1, Z_2; W_1, W_2 \mid Z) < I(X_1, X_2; Y_1, Y_2 \mid Z)$$

$$\leq I(X_1; Y_1) + I(X_2; Y_2). \qquad (10.41)$$

The outer bounds in Eqs. (10.39)–(10.41) are attained if the channel codewords (X_1, X_2) are independent of each other. Also, the distribution of (X_1, X_2) maximizing these bounds are not dependent on the distribution of (U_1, U_2). This implies that source–channel separation holds for this system for lossless transmission even with side information (Z_1, Z_2, Z) (for the sufficient conditions Eq. (10.1)). Thus by choosing (X_1, X_2) which maximize the outer bounds in Eqs. (10.39)–(10.41) we obtain capacity region for this system which is independent of the side conditions. Also, for a GMAC this is obtained by independent Gaussian r.v.s X_1 and X_2 with distributions $\mathcal{N}(0, P_i)$, $i = 1, 2$, where P_i are the power constraints. Furthermore, the L.H.S. of the inequalities are simultaneously minimized when W_1 and W_2 are independent. Thus, the source coding (W_1, W_2) on (U_1, Z_1) and (U_2, Z_2) can be done as in S–W coding (by first vector quantizing in case of continuous valued U_1, U_2) but also taking into account the fact that the side information Z is available at the decoder. In this section this coding scheme will be called SB.

If we take $W_1 = U_1$ and $W_2 = U_2$ and the side information $(Z_1, Z_2, Z) \perp (U_1, U_2)$, we can recover the conditions in [40].

10.7.2. *Gaussian Sources and Orthogonal Gaussian Channels*

Now we consider the transmission of jointly Gaussian sources over orthogonal Gaussian channels. Initially it will also be assumed that there is no side information (Z_1, Z_2, Z).

Now (U_1, U_2) are zero mean jointly Gaussian random variables with variances σ_1^2 and σ_2^2, respectively, and correlation ρ. Then $Y_i = X_i + N_i$, $i = 1, 2$ where N_i is Gaussian with zero mean and $\sigma_{N_i}^2$ variance. Also N_1 and N_2 are independent of each other and also of (U_1, U_2).

In this scenario, the R.H.S. of the inequalities in Eqs. (10.39)–(10.41) are maximized by taking $X_i \sim \mathcal{N}(0, P_i)$, $i = 1, 2$ independent of each other where P_i is the average transmit power constraint on user i. Then $I(X_i, Y_i) = 0.5 \log(1 + P_i/\sigma_{N_i}^2)$, $i = 1, 2$.

Based on the comments at the end of Sec. 10.7.1, for two users, using the results from [27] we obtain the necessary and sufficient conditions for transmission on an orthogonal GMAC with given distortions D_1 and D_2.

We can specialize the above results to a TDMA, FDMA or CDMA based transmission scheme. The specialization to TDMA is given here. Suppose source 1 uses the channel α fraction of time and user 2, $1 - \alpha$

fraction of time. In this case we can use average power P_1/α for the first user and $P_2/(1-\alpha)$ for the second user whenever they transmit. The conditions Eqs. (10.39)–(10.41) for the optimal scheme become

$$I(U_1; W_1|W_2) < 0.5\alpha \log\left[1 + \frac{P_1}{\alpha\sigma_{N_1}{}^2}\right], \qquad (10.42)$$

$$I(U_2; W_2|W_1) < 0.5(1-\alpha) \log\left[1 + \frac{P_2}{(1-\alpha)\sigma_{N_2}{}^2}\right], \qquad (10.43)$$

$$I(U_1, U_2; W_1, W_2) < 0.5\alpha \log\left[1 + \frac{P_1}{\alpha\sigma_{N_1}{}^2}\right]$$
$$+ 0.5(1-\alpha) \log\left[1 + \frac{P_2}{(1-\alpha)\sigma_{N_2}{}^2}\right]. \qquad (10.44)$$

In the following, we compare the performance of the AF scheme (explained in Sec. 10.6.1) with the SB scheme. Unlike for the GMAC there is no interference between the two users when orthogonal channels are used. Therefore, in this case we expect AF to perform quite well.

For AF, the minimum distortions (D_1, D_2) are

$$D_1 = \frac{(\sigma_1\sigma_{N_1})^2\left[P_2(1-\rho^2) + \sigma_{N_2}^2\right]}{P_1 P_2(1-\rho^2) + \sigma_{N_2}^2 P_1 + \sigma_{N_1}^2 P_2 + \sigma_{N_1}^2\sigma_{N_2}^2}, \qquad (10.45)$$

$$D_2 = \frac{(\sigma_2\sigma_{N_2})^2\left[P_1(1-\rho^2) + \sigma_{N_1}^2\right]}{P_1 P_2(1-\rho^2) + \sigma_{N_2}^2 P_1 + \sigma_{N_1}^2 P_2 + \sigma_{N_1}^2\sigma_{N_2}^2}. \qquad (10.46)$$

Thus, as $P_1, P_2 \to \infty$, D_1, D_2 tend to zero. We also see that D_1 and D_2 are minimum when the average powers used are P_1 and P_2. These conclusions are in contrast to the case of a GMAC where the distortion for the AF does not approach zero as $P_1, P_2 \to \infty$ and the optimal powers needed may not be the maximum average allowed P_1 and P_2 [30].

We compare the performance of AF with SB for the symmetric case where $P_1 = P_2 = P$, $\sigma_1^2 = \sigma_2^2 = \sigma^2$, $D_1 = D_2 = D$, and $\sigma_{N_1} = \sigma_{N_2} = \sigma_N^2$. These results are from [53].

We denote the minimum distortions achieved in SB and AF by $D(SB)$ and $D(AF)$, respectively. σ^2 is taken to be unity without loss of generality. We denote P/σ_N^2 by S. Then

$$D(SB) = \sqrt{\frac{1-\rho^2}{(1+S)^2} + \frac{\rho^2}{(1+S)^4}}, \quad D(AF) = \frac{S(1-\rho^2) + 1}{1 + 2S + S^2(1-\rho^2)}.$$
$$(10.47)$$

We see from the above equations that when $\rho = 0$, $D(SB) = D(AF) = 1/(1 + S)$. At high S, $D(AF) \approx 1/S$ and $D(SB) \approx \sqrt{1 - \rho^2}/S$. Eventually both $D(SB)$ and $D(AF)$ tend to zero as $S \to \infty$. When $S \to 0$ both $D(SB)$ and $D(AF)$ go to σ^2.

By squaring Eq. (10.47) we can show that $D(AF) \geq D(SB)$ for all S. But in [53] we have shown that $D(AF) - D(SB)$ is small when S is small or large or whenever ρ is small.

$D(AF)$ and $D(SB)$ are plotted for $\rho = 0.3$ and 0.7 using exact computations in Figs. 10.5(a) and 10.5(b).

The above results can be easily extended to the multiple source case. For SB, for the source coding part, the rate region for multiple user case (under a symmetry assumption) is given in [54]. This can be combined with the capacity achieving Gaussian channel codes over each independent channel to obtain the necessary and sufficient conditions for transmission.

Let N be the number of sources which are jointly Gaussian with zero mean and covariance matrix K_U. Let P be the symmetric power constraint. Let K_U have the same structure as given in [54]. Let $C_{UY} = \sqrt{P}[1\rho \cdots \rho]$ be a $1 \times N$ vector. The minimum distortion achieved by the AF scheme is given as $D(AF) = 1 - C_{UY}(PK_U + \sigma_N^2 I)^{-1}C'_{UY}$.

10.7.3. *Side Information*

Let us consider the case when side information Z_i is available at encoder i, $i = 1, 2$ and Z is available at the decoder. One use of the side information Z_i at the encoders is to increase the correlation between the sources. This can be optimally done [55], if we take appropriate linear combination of

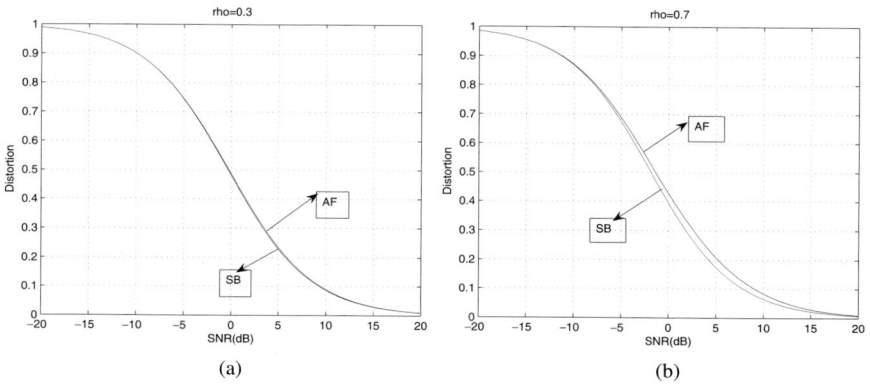

Fig. 10.5.   SNR vs distortion performance (a) $\rho = 0.3$ and (b) $\rho = 0.7$.

(U_i, Z_i) at encoder i. The following results are from [53]. We are not aware of any other result on performance of joint source–channel schemes with side information. We are currently working on obtaining similar results for the general MAC.

10.7.3.1. AF with Side Information

Side information at encoder only: A linear combination of the source outputs and side information $L_i = a_i U_i + b_i Z_i$, $i = 1, 2$ is amplified and sent over the channel. We find the linear combinations, which minimize the sum of distortions. For this we consider the following optimization problem:

Minimize

$$D(a_1, b_1, a_2, b_2) = E[(U_1 - \hat{U}_1)^2] + E[(U_2 - \hat{U}_2)^2] \tag{10.48}$$

subject to

$$E[X_1^2] \le P_1, \quad E[X_2^2] \le P_2,$$

where

$$\hat{U}_1 = E[U_1 \mid Y_1, Y_2], \quad \hat{U}_2 = E[U_2 \mid Y_1, Y_2].$$

Side information at decoder only: In this case the decoder side information Z is used in estimating (U_1, U_2) from (Y_1, Y_2). The optimal estimation rule is

$$\hat{U}_1 = E[U_1 \mid Y_1, Y_2, Z], \quad \hat{U}_2 = E[U_2 \mid Y_1, Y_2, Z]. \tag{10.49}$$

Side information at both encoder and decoder: Linear combinations of the sources are amplified as above and sent over the channel. To find the optimal linear combination, solve an optimization problem similar to Eq. (10.48) with $(\hat{U}_1, \hat{U}_2)$ as given in Eq. (10.49).

10.7.3.2. SB with Side Information

For a given (L_1, L_2) we use the source–channel coding scheme explained at the end of Sec. 10.7.1. The side information Z at the decoder reduces the source rate region. This is also used at the decoder in estimating $(\hat{U}_1, \hat{U}_2)$. The linear combinations L_1 and L_2 are obtained which minimize Eq. (10.48) through this coding–decoding scheme.

10.7.3.3. *Comparison of AF and SB with Side Information*

We provide the comparison of AF with SB for $U_1, U_2 \sim \mathcal{N}(0,1)$. Also we take the side information with a specific structure which seems natural in this set up. Let $Z_1 = s_1 U_2 + V_1$ and $Z_2 = s_2 U_1 + V_2$, where $V_1, V_2 \sim \mathcal{N}(0,1)$ are independent of each other and independent of the sources, and s_1 and s_2 are constants that can be interpreted as the side channel SNR. We also take $Z = (Z_1, Z_2)$.

We have compared AF and SB with different ρ and s_1, s_2 by explicitly computing the minimum $(D_1 + D_2)/2$ achievable. We take $P_1 = P_2$. For $s_1 = s_2 = 0.5$ and $\rho = 0.4$, we provide the results in Fig. 10.6. From the figure one sees that without side information, the performance of AF and SB is very close for different SNRs. The difference in their performance increases with side information for moderate values of SNR because the effect of the side information is to effectively increase the correlation between the sources. Even for these cases at low and high SNRs the performance of AF is close to that of SB. These observations are in conformity with our conclusions in the previous section.

Our other conclusions, based on computations not provided here are the following. For the symmetric case, for SB, encoder-only side information reduces the distortion marginally. This happens because a distortion is incurred for (U_1, U_2) while making the linear combinations (L_1, L_2). For the AF we actually see no improvement and the optimal linear combination has $b_1 = b_2 = 0$. For decoder-only side information the performance is improved

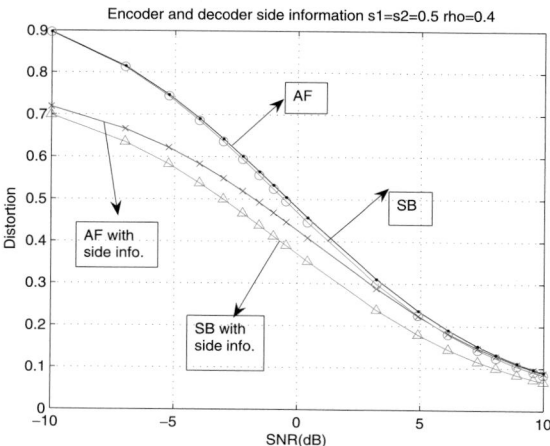

Fig. 10.6. AF and SB with both encoder and decoder side information.

for both AF and SB as the side information can be used to obtain better estimates of (U_1, U_2). Adding encoder side information further improves the performance only marginally for SB; the AF performance is not improved.

In the asymmetric case some of these conclusions may not be valid.

10.8. MAC with Feedback

In this section, we consider a memoryless MAC with feedback. The channel output Y_{k-1} is available to the encoders at time k.

Gaarder and Wolf [56] showed that, unlike in the point to point case, feedback increases the capacity region of a discrete memoryless MAC. In [57] an achievable region

$$R_1 < I(X_1; Y \mid X_2, U), \quad R_2 < I(X_2; Y \mid X_1, U),$$
$$R_1 + R_2 < I(X_1, X_2; Y) \tag{10.50}$$

where $p(u, x_1, x_2, y) = p(u)p(x_1 \mid u)p(x_2 \mid u)p(y \mid x_1, x_2)$. It was demonstrated in [58] that the same rate region is achievable, if there is a feedback link to only one of the transmitters. This achievable region was improved in [59].

The achievable region for an MAC, where each node receives possibly different channel feedback, is derived in [60]. The feedback signal in their set-up is correlated but not identical to the signal observed by the receiver. A simpler and larger rate region for the same set-up was obtained in [61].

Kramer [62] used the notion of "directed information" to derive an expression for the capacity region of the MAC with feedback. However, no single letter expressions were obtained.

If the users generate independent sequences, then the capacity region $\mathcal{C}_{fb}$ of the white Gaussian MAC is [28]

$$\mathcal{C}_{fb} = \bigcup_{0 \le \rho \le 1} \left\{ (R_1, R_2) : R_1 \le 0.5 \log \left[1 + \frac{P_1(1 - \rho^2)}{\sigma_N{}^2} \right], \right.$$

$$R_2 \le 0.5 \log \left[1 + \frac{P_2(1 - \rho^2)}{\sigma_N{}^2} \right],$$

$$\left. R_1 + R_2 \le 0.5 \log \left[1 + \frac{P_1 + P_2 + 2\rho\sqrt{P_1 P_2}}{\sigma_N{}^2} \right] \right\}. \tag{10.51}$$

The capacity region for a given ρ in Eq. (10.51) is same as in Eqs. (10.14)–(10.16) for a channel without feedback but with correlation ρ between channel inputs (X_1, X_2). Thus the effect of feedback is to allow arbitrary correlation in (X_1, X_2).

An achievable region for an GMAC with noisy feedback is provided in [63]. Gaussian MAC with different feedback to different nodes is considered in [64]. An achievable region based on cooperation among the sources is also given.

[65] obtains an achievable region when non-causal state information is available at both encoders. The authors also provide the capacity region for a Gaussian MAC with additive interference and feedback. It is found that feedback of the output enhances the capacity of the MAC with state information. Interference when causally known at the transmitters can be exactly cancelled and hence has no impact on the capacity region of a two user MAC. Thus the capacity region is the same as given in Eq. (10.51).

In [66], it is shown that feedback does not increase the capacity of the Gelfand–Pinsker channel [67] and feedforward does not improve the achievable rate-distortion performance in the Wyner–Ziv system [4].

MAC with feedback and correlated sources (MACFCS) is studied in [26, 68]. This has an MAC with correlated sources and a MAC with feedback as special cases. Gaussian MACFCS with a total average power constraint is considered in [68]. Different achievable rate regions and a capacity outer bound are given for the MACFCS in [26]. For the first achievable region a decode and forward based strategy is used where the sources first exchange their data, and then cooperate to send the full information to the destination. For two other achievable regions, S–W coding is performed first to remove the correlations among the source data and it is followed by the coding for the MAC with feedback or MAC disregarding the feedback. The authors also show that different coding strategies perform better under different source correlation structures.

The transmission of bivariate Gaussian sources over a Gaussian MAC with feedback is analyzed in [69]. The authors show that for the symmetric case, for SNR less than a threshold which is determined by the source correlation, feedback is useless and minimum distortion is achieved by uncoded transmission.

10.9. MAC with Fading

A Gaussian MAC with a finite number of fading states is considered. We provide results when there are M independent sources. (Recently we have extended our Theorem 10.1 to the case of fading channels. We describe this result at the end of this section.) The channel state information (CSI) may be available at the receiver and/or the transmitters. Consider the channel [43]

$$Y_k = \sum_{l=1}^{M} h_{lk} X_{lk} + N_k, \qquad (10.52)$$

where X_{lk} is the channel input and h_{lk} is the fading value at time k for user l. The fading processes $\{h_{lk}, \ k \geq 1\}$ of all users are jointly stationary and ergodic and the stationary distribution has a continuous bounded density. The fading process for the different users are independent. $\{N_k\}$ is the additive white Gaussian noise. All the users are power constrained to P, i.e. $E[X_l^2] \leq P$ for all l.

Since the source–channel separation holds, we provide the capacity region of this channel.

10.9.1. *CSI at Receiver Only*

When the channel fading process $\{h_{lk}\}$ is available at the receiver only, the achievable rate region is the set of rates $(R_1, R_2, \ldots, R_M)$ satisfying

$$\sum_{l \in \mathcal{S}} R_l \leq E\left[\log\left(1 + \frac{\sum_{l \in \mathcal{S}} \nu_l P}{\sigma^2}\right)\right] \qquad (10.53)$$

for all subsets $\mathcal{S}$ of $\{1, 2, \ldots, M\}$, $\nu_l = |h_l|^2$, and $\sigma^2 = E[N^2]$. The expectation is over all fading powers $\{\nu_l\}$, $l \in \mathcal{S}$. One of the performance measures is normalized sum rate per user

$$R = \frac{1}{M} \sum_{l=1}^{M} R_l = \frac{1}{M} E\left[\log\left(1 + \frac{MP\frac{1}{M}\sum_{l=1}^{M}\nu_l}{\sigma^2}\right)\right],$$

$$\leq \frac{1}{M} \log\left(1 + \frac{MP\frac{1}{M}\sum_{l=1}^{M}E[\nu_l]}{\sigma^2}\right). \qquad (10.54)$$

If $E[\nu_l] = 1$ for each l, then the upper bound equals the capacity of the AWGN channel $\frac{1}{M}\log\left[1 + \frac{MP}{\sigma^2}\right]$. Also, as M increases, if $\{\nu_l\}$ are *iid*, by law of large numbers (LLN), R will be close to this upper bound. Thus averaging over many users mitigates the effect of fading. This is in contrast to the time/frequency/space averaging.

The capacity achieving distribution is iid Gaussian for each user and the code for one user is independent of the code for another user (in other words AF is optimal in this case).

10.9.2. *CSI at Both Transmitter and Receiver*

The additional element that is introduced when CSI is provided to the transmitters in addition to the receiver is dynamic power control, which can be done in response to the changing channel state.

Given a joint fading power $\nu = (\nu_1, \ldots, \nu_M)$, $P_i(\nu)$ denotes the transmit power allocated to user i. Let P_i be the average power constraint for user i. For a given power control policy $\mathcal{P}$

$$
\mathcal{C}_f(\mathcal{P}) = \left\{ \mathbf{R} : \mathbf{R}(\mathcal{S}) \le E\left[\frac{1}{2}\log\left(1 + \frac{\sum_{i \in \mathcal{S}} \nu_i P_i(\nu)}{\sigma^2}\right)\right] \right.
$$

$$
\left. \text{for all } \mathcal{S} \subset \{1, 2, \ldots, M\} \right\} \qquad (10.55)
$$

denotes the rate region achievable. The capacity region is

$$
\mathcal{C}(\mathbf{P}) = \bigcup_{\mathcal{P} \in \mathcal{F}} \mathcal{C}_f(\mathcal{P}), \qquad (10.56)
$$

where $\mathcal{F}$ is the set of feasible power control policies,

$$
\mathcal{F} \equiv \{\mathcal{P} : E[P_i(\nu)] \le P_i, \text{ for } i = 1, \ldots, M\}. \qquad (10.57)
$$

Since the capacity region is convex, the above characterization implies that time sharing is not required.

The explicit characterization of the capacity region exploiting its polymatroid structure is given in [70]. For $P_i = P$ for each i and each h_i having the same distribution, the optimal power control is that only the user with the best channel transmits at a time. The instantaneous power assigned to the ith user, observing the realization of the fading powers $\nu_1, \nu_2, \ldots, \nu_M$ is

$$
P_l(\nu_j, j = 1, \ldots, M) = \begin{cases} \dfrac{1}{\lambda} - \dfrac{1}{\nu_l}, & \nu_l > \lambda, \ \nu_l > \nu_j, j \ne l, \\ 0 & \text{otherwise} \end{cases} \qquad (10.58)
$$

where λ is chosen such that the average power constraint is satisfied. This function is actually the well known water filling function [71] optimal for a single user. This strategy does not depend on the fading statistics but for the constant λ. The capacity achieving distribution is Gaussian (thus AF for each user in its assigned slot is optimal).

Unlike in the single user case the optimal power control may yield substantial gain in capacity for the multi-user case. This happens because if M is large, with high probability at least one of the iid fading powers will be large providing a good channel for the respective user at that time instant.

The optimal strategy is also valid for non-equal average powers. The only change being that the fading values are normalized by the Lagrange's coefficients [72]. The extension of this strategy to frequency selective channels is given in [73].

An explicit characterization of the ergodic capacity region and a simple encoding–decoding scheme for a fading GMAC with common data is given in [74]. Optimum power allocation schemes are also provided.

Recently we have extended [75] our Theorem 10.1 to a system with correlated sources and a fading MAC where the channel output at a time has distribution $p(y|x_1, x_2, h_1, h_2)$ where x_1, x_2 are the channel inputs and h_1, h_2 are the fading states. The fading states form iid sequences and is known at the encoders and the decoder. There is no side information (Z_1, Z_2, Z). Then the sufficient conditions for transmission are stated in Theorem 10.3.

Theorem 10.3. *A source can be transmitted over a fading multiple access channel with distortions (D_1, D_2) if there exist random variables (W_1, W_2, X_1, X_2) such that*

1. $p(u_1, u_2, w_1, w_2, x_1, x_2, y, h_1, h_2) = p(u_1, u_2)p(w_1 \,|\, u_1)p(w_2 \,|\, u_2)p(x_1|w_1, h_1, h_2)p(x_2 \,|\, w_2, h_1, h_2)p(y \,|\, x_1, x_2, h_1, h_2)$.
2. *There exists a function $f_D\colon \mathcal{W}_1 \times \mathcal{W}_2 \times \mathcal{H}_1 \times \mathcal{H}_2 \to (\hat{\mathcal{U}}_1 \times \hat{\mathcal{U}}_2)$ such that $E[d(U_i, \hat{U}_i)] \leq D_i$, $i = 1, 2$ and the constraints*

$$I(U_1; W_1 \,|\, W_2) < I(X_1; Y \,|\, X_2, W_2, H_1, H_2),$$

$$I(U_2; W_2 \,|\, W_1) < I(X_2; Y \,|\, X_1, W_1, H_1, H_2), \qquad (10.59)$$

$$I(U_1, U_2; W_1, W_2) < I(X_1, X_2; Y \,|\, H_1, H_2),$$

are satisfied where $\mathcal{W}_i$ is the set in which W_i take values.

It is also shown that now for the GMAC case the policy in Eq. (10.58) is no longer optimal.

More recently we have extended this result to the case where there is only partial CSI at the transmitters and the receiver.

10.10. Thoughts for Practitioners

Practical schemes for distributed source coding, channel coding, and joint source–channel coding for MAC are of interest. The achievability proofs assume infinite length code words and ignore delay and complexity which make them of limited interest in practical scenarios.

[76] reviews a panorama of practical joint source–channel coding methods for single user systems. The techniques given are hierarchical protection, channel optimized vector quantizers (COVQ), self organizing hypercube (SOH), modulation organized vector quantizer, and hierarchical modulation.

For lossless distributed source coding, Slepian and Wolf [12] provide the rate region. The underlying idea for construction of practical codes for this system is to exploit the duality between the source and channel coding. The approach is to partition the space of all possible source outcomes into disjoint bins that are cosets of a good linear channel code. Such constructions lead to constructive and non-asymptotic schemes.

Wyner was the first to suggest such a scheme in [77]. Inspired by Wyner's scheme, Turbo/LDPC based practical code design is given in [78] for correlated binary sources. The correlation between the sources were modeled by a "virtual" binary symmetric channel (BSC) with crossover probability p. The performance of this scheme is very close to the S–W limit $H(p)$. S–W code designs using powerful turbo and LDPC codes for other correlation models and more than two sources is given in [79].

LDPC based codes were also proposed in [46] where a general iterative S–W decoding algorithm that incorporates the graphical structure of all the encoders and operates in a "Turbo like" fashion is proposed. The authors in [80] propose LDPC codes for binary S–W coding problem with maximum likelihood (ML) decoding. This gives an upper bound on performance with iterative decoding. They also show that a linear code for S–W source coding can be used to construct a channel code for an MAC with correlated additive white noise.

In distributed source coding using syndromes (DISCUS) [45] Trellis coded modulation (TCM), Hamming codes, and Reed–Solomon (RS) codes are used for S–W coding. For the Gaussian version of DISCUS, the source is first quantized and then discrete DISCUS is used at both encoder and decoder.

Source coding with fidelity criterion subject to the availability of side information is addressed in [4]. First the source is quantized to the extent allowed by the fidelity requirement. Then S–W coding is used to remove the information at the decoder due to the side information. Since S–W coding is based on channel codes, Wyner–Ziv coding can be interpreted as a source–channel coding problem. The coding incurres a quantization loss due to source coding and binning loss due to channel coding. To achieve

Wyner–Ziv limit powerful codes need to be employed for both source coding and channel coding.

It was shown in [81] that nested lattice codes can achieve the Wyner–Ziv limit asymptotically, for large dimensions. A practical nested lattice code implementation is provided in [82]. For the BSC correlation model, linear binary block codes are used for lossy Wyner–Ziv coding in [81, 83].

Lattice codes and Trellis based codes [84] have been used for both source and channel coding for the correlated Gaussian sources. A nested lattice construction based on similar sublattices for high correlation is proposed in [82]. Another approach to practical code constructions is based on Slepian–Wolf coded nested quantization (SWC-NQ), which is a nested scheme followed by binning. Asymptotic performance bounds of SWC-NQ are established in [85]. A combination of a scalar quantizer and a powerful S–W code is also used for nested Wyner–Ziv coding. Wyner–Ziv coding based on TCQ and LDPC are provided in [86]. A comparison of different approaches for both Wyner–Ziv coding and classical source coding are provided in [44].

Low density generator matrix (LDGM) codes are proposed for joint source–channel coding of correlated sources in [87].

Practical code construction for the special case of the CEO problem are provided in [88, 89].

10.11. Directions for Future Research

In this chapter we have provided sufficient conditions for transmission of correlated sources over an MAC with specified distortions. It is of interest to find a single letter characterization of the necessary conditions and to establish the tightness of the sufficient conditions. It is also of interest to extend the above results and coding schemes to sources correlated in time and an MAC with memory. The error exponents are also of interest.

Most of the achievability results in this chapter use random codes, which are inefficient because of large codeword length. It is desirable to obtain power efficient practical codes for side information aware compression that performs very close to the optimal scheme.

For the fading channels, fairness of the rates provided to different users, the delay experienced by the messages of different users and channel tracking are issues worth pondering. It is also desirable to find the performance of these schemes in terms of scaling behavior in a network scenario. The combination of joint source–channel coding and network coding is also a new area of research. Another emerging area is the use of joint source–channel codes in MIMO systems and co-operative communication.

10.12. Conclusions

In this chapter, sufficient conditions are provided for transmission of correlated sources over an MAC. Various previous results on this problem are obtained as special cases. Suitable examples are given to emphasis the superiority of joint source–channel coding schemes. Important special cases: Correlated discrete sources over a GMAC and Gaussian sources over a GMAC are discussed in more detail. In particular a new joint source–channel coding scheme is presented for discrete sources over a GMAC. Performance of specific joint source–channel coding schemes for Gaussian sources are also compared. Practical schemes like TDMA, FDMA, and CDMA are brought into this framework. We also consider an MAC with feedback and a fading MAC. Various practical schemes motivated by joint source–channel coding are also presented.

10.13. Problems

(1) Consider a GMAC $Y_k = X_{1k} + X_{2k} + W_k$ where $\{W_k\}$ is iid $\mathcal{N}(0, \sigma^2)$. Let P_1 and P_2 be the average powers with which we can transmit at the channel inputs 1 and 2.

(a) Plot the rate region of the MAC.

(b) Let $\{U_{1k}\}$ be an iid source with distribution $\mathcal{N}(0, \sigma^2)$. Let $\{U_{2k}\}$ be another continuous source with mean μ and variance σ_2^2. $\{U_{1k}\}$ is independent of $\{U_{2k}\}$. Under what conditions we can transmit the two sources over the MAC with mean square distortions $\sigma_1^2/2$ and $\sigma_2^2/2$.

(c) What is the minimum mean square distortion we can get for user 1. Find the minimum mean square distortion that user 2 can get simultaneously.

(d) Now assume that the two sources in (b) are jointly Gaussian with correlation ρ. These are iid in time. What is the minimum sum of mean square distortions we can get when transmitting these sources on the above MAC.

(e) Let there be feedback of Y_{k-1} to the encoders at time k. Now repeat (a), (b), and (c).

(2) Consider a channel with inputs X_1 and X_2 and output $Y = X_1 + \text{sgn}(X_2)$ where

$$\text{sgn}(x) = \begin{cases} 1 & \text{if } x \geq 0, \\ -1 & \text{if } x < 0. \end{cases}$$

The inputs can transmit at powers $E[X_1^2] \leq P_1$, $E[X_2^2] \leq P_2$.

(a) Find the capacity region of this channel.
(b) Find the coding schemes that achieve the capacity region.

(3) Let Z_1, Z_2, and Z_3 be independent r.v.s with $P(Z_i = 1) = p = 1 - P(Z_i = 0)$, $i = 1, 2, 3$. Define $X_1 = Z_1, X_2 = Z_1 + Z_2$, and $X_3 = Z_1 + Z_2 + Z_3$. Find the S–W region for the sources X_1, X_2, and X_3.

(4) Find the S–W region for (X, Y) where $Y = f(X)$ and f is a deterministic function.

(5) Given two discrete r.v.s X and Y. Define conditional rate distortion function $R_{X|Y}(D) = \inf_{p(\hat{x}|x,y)} I(X; \hat{X}|Y)$ where inf is over the conditional distribution such that $E[d(X, \hat{X})] \leq D$. Prove that $R_X(D) \geq R_{X|Y}(D) \geq R_X(D) - I(X, Y)$.

(6) Consider an MAC with two senders X_1, X_2 and one receiver Y. Let $X_1, X_2 \in \{0, 1\}$ and $Y \in \{0, 1, e\}$. The transition matrix for the channel is $P(Y = 0 | X_1 = 0, X_2 = 0) = 1$ and $P(Y = 1 | X_1 = 1, X_2 = 1) = 1$, $Y = e$ otherwise. Find the capacity of the channel.

(7) Consider an GMAC with fading when the fading processes of the two users are stationary, ergodic sequences, and independent of each other. Each encoder and the decoder knows the channel state causally. Then the optimal multiple access strategy for achieving sum capacity of the uplink fading channel, when the users have identical channel statistics and power constraints is to allow only the user with best channel to transmit and use water filling in time on each user. Find the optimal strategy in the general case when the channel statistics and power constraints of the users are arbitrary.

(8) Consider the AF scheme with $P_1 = 10$ and $P_2 = 50$.

(a) What is the distortion $D_1 + D_2$ for transmitting unit variance Gaussian sources with correlation $\rho = 0.5$.
(b) Reduce P_2 to 20. Find $D_1 + D_2$.
(c) Find the power allocation (P_1^*, P_2^*) that minimizes $D_1 + D_2$ such that $P_i^* \leq P_i$, $i = 1, 2$.

(9) Consider the LT scheme. Show by formulating an optimization scheme, that, unlike for AF, under power constraints (P_1, P_2) the minimum $D_1 + D_2$ is achieved at (P_1, P_2).

(10) Derive necessary conditions for transmitting jointly Gaussian sources over a GMAC by considering a co-operative encoder, which has both the sources as inputs. The decoder needs to estimate the source outputs with distortions D_1 and D_2.

10.14. Keywords

Wireless sensor network (WSN)
It is a wireless network consisting of spatially distributed inexpensive sensors with limited battery power.

Multiple access channel (MAC)
A channel shared by many distributed sources to send information to a single destination.

Slepian–Wolf coding
A distributed lossless source coding scheme which codes at rates equal to the joint entropy of the sources. It uses a new idea called "random binning".

Gaussian multiple access channel (GMAC)
A special practically important MAC where the output is the sum of the inputs and a white Gaussian noise.

Joint source–channel coding
A coding paradigm in which the source and the channel characteristics are jointly considered during encoding and decoding.

Side information
An information source correlated with the main information source and helps in decoding and/or encoding the main information.

Amplify and forward (AF)
A scheme where the inputs are scaled to meet the channel input cost constraints and then forwarded through the channel. It is a symbol by symbol encoding/decoding scheme.

Fading multiple access channel
A MAC with time varying channel gain due to multi-path fading in wireless channels.

Multi-terminal source coding
A coding technique to code multiple sources distributed in space where the encoders may have information only about their own sources. The decoder needs to decode the coded words to meet certain fidelity criteria.

Orthogonal multiple access channel
Special case of an MAC where the MAC can be considered as comprising of independent point to point channels.

Channel state information (CSI)
It represents the fading state of the channel. It may be made available causally/non-causally at the decoder and/or the encoders.

References

1. T. M. Cover, A. E. Gamal and M. Salehi, Multiple access channels with arbitrarily correlated sources, *IEEE Trans. Inform. Theor.* **26**, 6 (1980), 648–657.
2. I. F. Akylidiz, W. Su, Y. Sankarasubramaniam and E. Cayirici, A survey on sensor networks, *IEEE Comm. Mag.* (2002), 1–13.
3. S. J. Baek, G. Veciana and X. Su, Minimizing energy consumption in large-scale sensor networks through distributed data compression and hierarchical aggregation, *IEEE JSAC.* **22**, 6 (2004), 1130–1140.
4. A. Wyner and J. Ziv, The rate distortion function for source coding with side information at the receiver, *IEEE Trans. Inform. Theor.* **IT-22**, (1976), 1–11.
5. M. Gastpar, Wyner-Ziv problem with multiple sources, *IEEE Trans. Inform. Theory.* **50**, 11 (2004), 2762–2768.
6. R. Ahlswede, Multiway communication channels, *Proc. Second Int. Symp. Inform. Transmission,* Armenia, USSR (Hungarian Press, 1971).
7. H. Liao, Multiple access channels, Ph.D Dissertion, Dept. Elec. Engg., Univ of Hawaii, Honolulu, (1972).
8. R. Ahlswede and T. Han, On source coding with side information via a multiple access channel and related problems in information theory, *IEEE Trans. Inform. Theor.* **29**, 3 (1983), 396–411.
9. G. Dueck, A note on the multiple access channel with correlated sources, *IEEE Trans. Inform. Theor.* **27**, 2 (1981), 232–235.
10. W. Kang and S. Ulukus, An outer bound for mac with correlated sources, *Proc. 40th Annual Conf. Information Sciences and Systems,* (March, 2006), pp. 240–244.

11. S. Ray, M. Medard, M. Effros and R. Kotter, On separation for multiple access channels, *Proc. IEEE Inform. Theory Workshop*, (2006).

12. D. Slepian and J. K. Wolf, Noiseless coding of correlated information sources, *IEEE Trans. Inform. Theor.* **19**, 4 (1973), 471–480.

13. T. M. Cover, A proof of the data compression theorem of Slepian and Wolf for ergodic sources, *IEEE Trans. Inform. Theor.* **21**, 2 (1975), 226–228.

14. J. Barros and S. D. Servetto, Reachback capacity with non-interfering nodes, *Proc. ISIT*, (2003), pp. 356–361.

15. T. Berger, Multiterminal source coding, in Information Theory Approach to Communication, (ed.) G. Longo (Springer-Verlag, N.Y., 1977).

16. R. J. Barron, B. Chen and G. W. Wornell, The duality between information embedding and source coding with side information and some applications, *IEEE Trans. Inform. Theor.* **49**, 5 (2003), 1159–1180.

17. S. S. Pradhan, J. Chou and K. Ramachandran, Duality between source coding and channel coding and its extension to the side information case, *IEEE Trans. Inform. Theor.* **49**, 5 (2003), 1181–1203.

18. S. C. Draper and G. W. Wornell, Side information aware coding stategies for sensor networks, *IEEE J. Sel. Area. Comm.* **22**, (2004), 1–11.

19. T. Berger and R. W. Yeung, Multiterminal source coding with one distortion criterion, *IEEE Trans. Inform. Theor.* **35**, 2 (1989), 228–236.

20. Y. Oohama, Gaussian multiterminal source coding, *IEEE Trans. Inform. Theor.* **43**, 6 (1977), 1912–1923.

21. R. Zamir and T. Berger, Multiterminal source coding with high resolution, *IEEE Trans. Inform. Theor.* **45**, 1 (1999), 106–117.

22. M. Gastpar, Wyner-Ziv problem with multiple sources, *IEEE Trans. Inform. Theor.* **50**, 11 (2004), 2762–2768.

23. V. K. Varsheneya and V. Sharma, Lossy distributed source coding with side information, *Proc. National Conf. Communication (NCC)*, New Delhi, (January, 2006).

24. V. K. Varsheneya and V. Sharma, Distributed coding for multiple access communication with side information, *Proc. IEEE Wireless Communication and Networking Conference (WCNC)*, (April, 2006).

25. R. Rajesh, V. K. Varsheneya and V. Sharma, Distributed joint source–channel coding on a multiple access channel with side information, *Proc. Int. Symposium on Information Theory (ISIT)*, Toronto, Canada, (2008).

26. L. Ong and M. Motani, Coding stategies for multiple–access channels with feedback and correlated sources, *IEEE Trans. Inform. Theor.* **53**, 10 (2007), 3476–3497.

27. A. B. Wagner, S. Tavildar and P. Viswanath, The rate region of the quadratic Gaussian two terminal source coding problem, *IEEE Trans. Inform. Theor.* **54**, 5 (2008), 1938–1961.

28. L. H. Ozarow, The capacity of the white Gaussian multiple access channel with feedback, *IEEE Trans. Inform. Theor.* **30**, 4 (1984), 623–629.

29. A. Lapidoth and S. Tinguely, Sending a bi-variate Gaussian source over a Gaussian MAC, *IEEE ISIT 06*, (2006).

30. R. Rajesh and V. Sharma, Source channel coding for Gaussian sources over a Gaussian multiple access channel, *Proc. 45 Allerton Conf. Computing Control and Communication, Monticello, IL*, (2007).
31. M. Gastpar, Multiple access channels under received-power constraints, *Proc. IEEE Inform. Theory Workshop*, (2004), pp. 452–457.
32. M. Gastpar and M. Vetterli, Source–channel communication in sensor networks, *Proc. IPSN'03*, (2003), pp. 162–177.
33. P. Ishwar, R. Puri, K. Ramchandran and S. S. Pradhan, On rate constrained distributed estimation in unreliable sensor networks, *IEEE JSAC*, (2005), pp. 765–775.
34. H. E. Gamal, On scaling laws of dense wireless sensor networks: The data gathering channel, *IEEE Trans. Inform. Theor.* **51**, 3 (2005), 1229–1234.
35. M. Gastpar and M. Vetterli, Power spatio-temporal bandwidth and distortion in large sensor networks, *IEEE JSAC.* **23**, 4 (2005), 745–754.
36. T. Berger, Z. Zhang and H. Viswanathan, The CEO problem, *IEEE Trans. Inform. Theor.* **42**, 3 (1996), 887–902.
37. Y. Oohama, The rate distortion function for quadratic Gaussian CEO problem, *IEEE Trans. Inform. Theor.* **44**, 3 (1998), 1057–1070.
38. T. M. Cover and J. A. Thomas, *Elements of Information Theory*, Series in Telecommunication (Wiley, N.Y., 2004).
39. J. G. Proakis, *Digital Communication*, International edition (McGraw-Hill, 2001).
40. J. Barros and S. D. Servetto, Network information flow with correlated sources, *IEEE Trans. Inform. Theor.* **52**, 1 (2006) 155–170.
41. J. J. Xiao and Z. Q. Luo, Multiterminal source channel communication over an orthogonal multiple access channel, *IEEE Trans. Inform. Theor.* **53**, 9 (2007), 3255–3264.
42. N. Wernersson, J. Karlsson and M. Skoglund, Distributed scalar quantisers for Gaussian channels, ISIT, Nice, France, (June, 2007).
43. E. Biglieri, J. Proakis and S. Shamai, Fading channels: Information theoretic and communication aspects, *IEEE Trans. Inform. Theor.* **44**, 6 (1998), 2619–2692.
44. Z. Xiong, A. Liveris and S. Cheng, Distributed source coding for sensor networks, *IEEE Sig. Proces. Mag.* (2004), 80–94.
45. S. S. Pradhan and K. Ramchandran, Distributed source coding using syndromes DISCUS: Design and Construction, *IEEE Trans. Inform. Theor.* **49**, 3 (2003), 626–643.
46. T. P. Coleman, A. H. Lee, M. Medard and M. Effros, Low-complexity approaches to Slepian–Wolf near-lossless distributed data compression, *IEEE Trans. Inform. Theor.* **52**, 8 (2006), 3546–3561.
47. R. Rajesh and V. Sharma, Distributed joint source–channel coding on a multiple access channel with side information, Technical Report, Available: Arxiv, http://arxiv.org/PS_cache/arxiv/pdf/0803/0803.1445v1.pdf. (2008).
48. D. Slepian and J. K. Wolf, A coding theorem for multiple access channels with correlated sources, *Bell Syst. Tech. J.* **52**, 7 (1973), 1037–1076.

49. D. Gunduz and E. Erkip, Transmission of correlated sources over multiuser channels with receiver side information, *UCSD ITA Workshop*, San Diego, CA, (January, 2007).

50. M. Fleming and M. Effros, On rate distortion with mixed types of side information, *IEEE Trans. Inform. Theor.* **52**, 4 (2006), 1698–1705.

51. R. Rajesh and V. Sharma, A joint source-channel coding scheme for transmission of correlated discrete sources over a Gaussian multiple access channel, *Procs. Intl. Symp. Information Theory and Applications (ISITA)* (Auckland, New Zealand, 2008).

52. V. V. Veeravalli, Decentralized quickest change detection, *IEEE Trans. Inform. Theor.* **47**, 4 (2001), 1657–1665.

53. R. Rajesh and V. Sharma, Correlated Gaussian sources over orthogonal Gaussian channels, submitted.

54. A. B. Wagner, S. Tavildar and P. Viswanath, The rate region of the quadratic Gaussian two terminal source coding problem, Preprint.

55. L. Brieman and H. Friedman, Estimating optimal transformations for multiple regression and correlation, *J. Am. Stat. Assoc.* **80**, 391 (1983), 580–598.

56. N. T. Gaarder and J. K. Wolf, The capacity region of a multiple access discrete memoryless channel can increase with feedback, *IEEE Trans. Inform. Theor.* **IT-21**, (1975).

57. T. M. Cover and C. S. K. Leung, An achievable rate region for the multiple-access channel with feedback, *IEEE Trans. Inform. Theor.* **27**, 3 (1981), 292–298.

58. F. M. J. Willems and E. V. Meulan, Partial feedback for the discrete memoryless multiple access channel, *IEEE Trans. Inform. Theor.* **29**, 2 (1983), 287–290.

59. S. Bross and A. Lapidoth, An improved achievable region for discrete memoryless two-user multiple-access channel with noiseless feedback, *IEEE Trans. Inform. Theor.* **51**, 3 (2005), 811–833.

60. A. B. Carleial, Multiple access channels with different generalized feedback signals, *IEEE Trans. Inform. Theor.* **IT-28**, 6 (1982), 841–850.

61. F. M. J. Willems, E. V. Meulan and J. P. M. Schalkwijk, Achievable rate region for the multiple access channel with generalized feedback, *Proc. Allerton Conf.*, Monticello, IL, (1983).

62. G. Kramer, Capacity results for discrete memoryless network, *IEEE Trans. Inform. Theor.* **49**, (2003), 4–21.

63. A. Lapidoth and M. A. Wigger, On Gaussian MAC with imperfect feedback, *24th IEEE Convention of Electrical and Electronics Engineers in Israel (IEEEI 06), Eilat*, (November, 2006).

64. A. Sendonaris, E. Erkip and B. Aazhang, User cooperation diversity-part I, *IEEE Trans. Commun.* **51**, 11 (2003), 1927–1938.

65. W. Wu, S. Vishwanath and A. Arapostatis, On the capacity of multiple access channels with side information and feedback, *Proc. Int. Symp. Information Theory*, (July, 2006).

66. N. Merhav and T. Weissman, Coding for the feedback Gel'fand–Pinsker channel and feed forward Wyner–Ziv source, *IEEE Trans. Inform. Theor.* **52**, (2006), 4207–4211.

67. S. Gel'fand and M. Pinsker, Coding for channels with random parameters, *Probl. Control Inform. Theor.* **9**, 1 (1980), 19–31.

68. A. D. Murugan, P. K. Gopala and H. El-Gamal, Correlated sources over wireless channels: Cooperative source-channel coding, *IEEE J. Sel. Area. Comm.* **22**, 6 (2004), 988–998.

69. A. Lapidoth and S. Tinguely, Sending a bivariate Gaussian source over a Gaussian MAC with feedback, *IEEE ISIT*, Nice, France, (June, 2007).

70. D. Tse and S. V. Hanly, Multiaccess fading channels-part i: Polymatroid structure, optimal resource allocation and throughput capacities, *IEEE Trans. Inform. Theor.* **44**, 7 (1998), 2796–2815.

71. A. J. Goldsmith and P. P. Varaiya, Capacity of fading channels with channel side information, *IEEE Trans. Inform. Theor.* **43**, 6 (1997), 1986–1992.

72. R. Knopp and P. A. Humblet, Information capacity and power control in single-cell multiuser communication, *Proc. Int. Conf. Communication, ICC'95*, Seattle, WA, (June, 1995), pp. 331–335.

73. R. Knopp and P. A. Humblet, Multiple-accessing over frequency-selective fading channels channels, *6th IEEE Int. Symp. on Personal Indoor and Mobile Radio Communication, PIMRC'95*, Toronto, Canada, (September, 1995), pp. 1326–1331.

74. N. Liu and S. Ulukus, Capacity region and optimum power control stategies for fading Gaussian multiple access channels with common data, *IEEE Trans. Comm.* **54**, 10 (2006), 1815–1826.

75. R. Rajesh and V. Sharma, Transmission of correlated sources over a fading multiple access channel, submitted.

76. S. B. Z. Azami, P. Duhamel and O. Rioul, Combined source–channel coding: Panorama of methods, *CNES Workshop on Data Compression*, Toulouse, France, (November 1996).

77. A. Wyner, Recent results in Shannon theory, *IEEE Trans. Inform. Theor.* **20**, 1 (1974), 2–10.

78. A. Liveris, Z. Xiong and C. Georghiades, Compression of binary sources with side information at the decoder using LDPC codes, *IEEE Comm. Lett.* **6**, 10 (2002), 440–442.

79. C. Lan, K. N. A. Liveris, Z. Xiong and C. Georghiades, Slepain–Wolf coding of multiple m-ary sources using LDPC codes, *Proc. DCC'04, Snowbird, UT*, (2004), p. 549.

80. J. Muramastu, T. Uyematsu and T. Wadayama, Low density parity check matrices for coding of correlated sources, *IEEE Trans. Inform. Theor.* **51**, 10 (2005), 3645–3654.

81. R. Zamir, S. Shamai and U. Erez, Nested linear/lattice codes for structured multi-terminal binning, *IEEE Trans. Inform. Theor.* **48**, 6 (2002), 1250–1276.

82. S. Servetto, Lattice quantization with side information: Codes asymptotics and applications in sensor networks, *IEEE Trans. Inform. Theor.* **53**, 2 (2007), 714–731.

83. S. Shamai, S. Verdu and R. Zamir, Systematic lossy source/channel coding, *IEEE Trans. Inform. Theor.* **44**, 2 (1998), 564–578.

84. M. V. Eyuboglu and G. D. Forney, Lattice and trellis quantization with lattice and trellis-bounded codebooks-high rate theory for memoryless sources, *IEEE Trans. Inform. Theor.* **39**, 1 (1993), 46–59.

85. Z. Liu, S. Cheng, A. Liveris and Z. Xiong, Slepian–Wolf coded nested quantization (swc-nq) for Wyner–Ziv coding: Performance analysis and code design, *IEEE Trans. Inform. Theor.* **52**, (2006), 4358–4379.

86. Y. Yang, S. Cheng, Z. Xiong and W. Zhao, Wyner–Ziv coding based on TCQ and LDPC codes, *Proc. Asilomer Conf. Signals, Systems and Computers*, Pacific Grove, CA, (2003), pp. 825–829.

87. W. Zong and J. G. Frias, LDGM codes for channel coding and joint source–channel coding of correlated sources, *EURASIP J. Appl. Sig. Process.* (2005), 942–953.

88. S. Pradhan and K. Ramchandran, Distributed source coding: Symmetric rates and application to sensor networks, *Proc. DCC'00*, Snowbird, UT, (2000), pp. 302–311.

89. Y. Yang, V. Stankovic, Z. Xiong and W. Zhao, Distributed source coding: Symmetric rates and application to sensor networks, *Proc. DCC'04*, Snowbird, UT, (2004), p. 572.

Part IV

OTHER SELECTED TOPICS IN INFORMATION
AND CODING THEORY

Chapter 11

LOW-DENSITY PARITY-CHECK CODES AND THE RELATED PERFORMANCE ANALYSIS METHODS

XUDONG MA

FlexRealm Silicon Inc., Virginia, USA
xma@ieee.org

Low-density parity-check (LDPC) codes are a class of error-correction codes, which attract much attention recently. The LDPC codes can approach within one-tenth dB of the Shannon limits. The codes have already found many applications in practice. This chapter presents a tutorial exposition of LDPC codes and the related performance analysis methods.

11.1. Introduction

11.1.1. *Background*

Low-density parity-check (LDPC) codes are a class of error correction codes, which attract much attention recently. Along with Turbo codes, LDPC codes are one of the two known classes of error correction codes, which can approach the Shannon capacity limit (the theoretical limit of maximal transmission rates over communication channels with power and bandwidth constraints). These capacity-approaching codes have already found applications in deep-space communications, satellite communications, and wireless communications. Compared with Turbo codes, LDPC codes have the advantages of having lower decoding complexity and delay.

11.1.2. *Overview of this Chapter*

In this chapter, we will present a tutorial exposition of LDPC codes and the related performance analysis methods. The chapter is organized as follows.

We begin with an introduction to Tanner graphs in Sec. 11.2. In Sec. 11.3, we discuss LDPC code construction schemes. In Sec. 11.4, we discuss the factor graph approach for signal detection and estimation. The de facto decoding algorithm for LDPC codes — the message-passing algorithm — is discussed from the viewpoint of factor graph approaches in Sec. 11.5. We discuss an important asymptotic performance analysis method for the message-passing algorithm — density evolution — in Sec. 11.6. Another important performance analysis method — the finite-length analysis — based on pseudo-codewords is discussed in Sec. 11.7. Some thoughts for practitioners are presented in Sec. 11.10. Potential future research directions are discussed in Sec. 11.11. The conclusions are presented in Sec. 11.12. Examples and exercises are also provided in order to illustrate the various studied concepts and techniques.

11.2. Tanner Graph

LDPC codes are linear block codes with sparse parity-check matrices. A linear block code $\mathcal{C}$ is a set of binary vectors $\boldsymbol{x}$, such that $\boldsymbol{x} \in \mathcal{C}$, if and only if $\boldsymbol{Hx} = \boldsymbol{0}$, where $\boldsymbol{H}$ is a given matrix, $\boldsymbol{0}$ is the all zero vector and all elements of $\boldsymbol{x}$ and $\boldsymbol{H}$ are in $GF(2)$ (the Galois field of order two). A LDPC code is a linear block code, where each row and column of the matrix $\boldsymbol{H}$ contains only a small number of non-zero elements.

Tanner graphs, which were invented by Tanner in 1980s [21], are widely adopted graphical representations for LDPC codes. A Tanner graph is a bipartite graph with one partition consisting of all variable nodes and the other partition consisting of all check nodes. Each variable node represents one codeword bit and each check node represents one parity-check constraint. There exists an edge between a variable node and a check node if and only if the corresponding bit is involved in the parity-check constraint.

Example 11.1. A Tanner graph is shown in Fig. 11.1. The corresponding parity-check matrix is shown as follows:

$$\boldsymbol{H} = \begin{bmatrix} 1 & 1 & 0 & 1 & 0 & 0 \\ 0 & 0 & 1 & 1 & 1 & 0 \\ 0 & 1 & 0 & 0 & 1 & 1 \end{bmatrix}. \tag{11.1}$$

In Tanner graph representation, the variable nodes are usually drawn as circles, and the check nodes are usually drawn as squares.

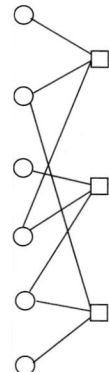

Fig. 11.1. The Tanner graph representation.

11.3. Constructing LDPC Codes

Various approaches for constructing LDPC codes have been proposed. They can be classified into two categories, algebraic constructions and random constructions. In this chapter, we focus more on random constructions. Because for this class of codes, there exists a rich theory for performance analysis. However, an example of algebraic constructions is also provided. We also refer the readers, who are interested in algebraic constructions, to the recent works, [10, 11, 23], etc.

11.3.1. *Algebraic Construction*

A simple algebraic code construction scheme is based on cyclically shifted identity matrices. A cyclically shifted identity matrix is a shifted version of an identity matrix. That is, the matrix is obtained from an identity matrix by shifting each row of the identity matrix by k times, where k is a given integer. The parity-check matrix by the construction is of the form,

$$\boldsymbol{H} = \begin{bmatrix} \boldsymbol{H}_{11} & \boldsymbol{H}_{12} & \boldsymbol{H}_{13} & \cdots & \boldsymbol{H}_{1m} \\ \boldsymbol{H}_{21} & \boldsymbol{H}_{22} & \boldsymbol{H}_{23} & \cdots & \boldsymbol{H}_{2m} \\ \cdots & \cdots & \cdots & \cdots & \cdots \\ \vdots & \vdots & \vdots & \vdots & \vdots \\ \cdots & \cdots & \cdots & \cdots & \cdots \\ \boldsymbol{H}_{n1} & \boldsymbol{H}_{n2} & \boldsymbol{H}_{n3} & \cdots & \boldsymbol{H}_{nm} \end{bmatrix}, \tag{11.2}$$

where each submatrix $\boldsymbol{H}_{ij}$ is a cyclically shifted identity matrix.

Example 11.2. The one-time cyclically right-shifted identity matrix I_1 of size 5×5 and three-times cyclically right-shifted identity matrix I_3 of size 5×5 are shown as follows:

$$I_1 = \begin{bmatrix} 0 & 1 & 0 & 0 & 0 \\ 0 & 0 & 1 & 0 & 0 \\ 0 & 0 & 0 & 1 & 0 \\ 0 & 0 & 0 & 0 & 1 \\ 1 & 0 & 0 & 0 & 0 \end{bmatrix}, \tag{11.3}$$

$$I_3 = \begin{bmatrix} 0 & 0 & 0 & 1 & 0 \\ 0 & 0 & 0 & 0 & 1 \\ 1 & 0 & 0 & 0 & 0 \\ 0 & 1 & 0 & 0 & 0 \\ 0 & 0 & 1 & 0 & 0 \end{bmatrix}. \tag{11.4}$$

The zero-time cyclically shifted identity matrix I_0 of size 5×5 is the identity matrix of size 5×5.

$$I_0 = \begin{bmatrix} 1 & 0 & 0 & 0 & 0 \\ 0 & 1 & 0 & 0 & 0 \\ 0 & 0 & 1 & 0 & 0 \\ 0 & 0 & 0 & 1 & 0 \\ 0 & 0 & 0 & 0 & 1 \end{bmatrix} \tag{11.5}$$

Example 11.3. The parity-check matrix of a well-known code [27] based on cyclically shifted identity matrices is shown as follows:

$$H = \begin{bmatrix} J_1 & J_2 & J_4 & J_8 \\ J_5 & J_{10} & J_{20} & J_9 \\ J_{25} & J_{19} & J_7 & J_{14} \end{bmatrix},$$

where J_k represents the k-times cyclically right-shifted identity matrix of size 31×31. The code has blocklength 124, rate 0.27, dimension 33, and minimal distance 24.

11.3.2. *Standard Random Construction*

In the standard random construction, LDPC codes are randomly chosen from code ensembles, where a code ensemble is a set of codes with certain properties. A standard LDPC code ensemble $\mathcal{C}(\lambda, \rho, n)$ is specified by three parameters: A left degree distribution $\lambda(x)$, a right degree distribution

$\rho(x)$, and a blocklength n [18]. The left and right degree distributions are polynomials,

$$\lambda(x) = \sum_{i=2}^{d_v} \lambda_i x^{i-1}, \quad \rho(x) = \sum_{j=2}^{d_c} \rho_j x^{j-1}. \tag{11.6}$$

An LDPC code is in the code ensemble $\mathcal{C}(\lambda, \rho, n)$, if and only if,

- the blocklength of the code is n;
- in the corresponding Tanner graph, the fraction of edges that are connected to variable nodes with degree i is λ_i; and
- in the corresponding Tanner graph, the fraction of edges that are connected to check nodes with degree j is ρ_j.

In the construction scheme, an LDPC code is constructed from a code ensemble $\mathcal{C}(\lambda, \rho, n)$ by first generating half-edges from the variable nodes and check nodes. The half-edges from the variable nodes are first permutated and then connected to half-edges from the check nodes one by one. The construction scheme can be visualized in Fig. 11.2.

11.3.3. *Random Constructions Based on Graph Lifts*

Several code construction schemes based on random graph lifts have been proposed [16, 22]. These approaches start from a small Tanner graph called *protograph* and then use *graph lifts* to generate a larger Tanner graph from the small protograph. Various advantages of these construction schemes have been demonstrated.

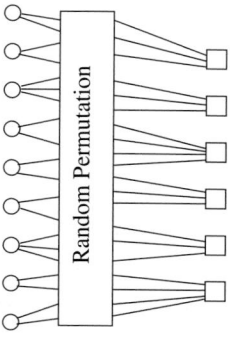

Fig. 11.2.   The standard random construction.

11.4. Signal Estimation and Detection Using Factor Graphs

11.4.1. *Factor Graphs*

Factor graphs represent global product functions and illustrate how the global functions can be factored into local functions. In signal estimation and detection, factor graphs provide a convenient technical language, especially for describing an important class of algorithms, the message-passing algorithms [12]. Many important signal estimation and detection algorithms can be considered as the special cases of message passing algorithms, for example, Kalman filtering and LDPC code decoding algorithms.

A factor graph is a bipartite graph with one partition consisting of function nodes and the other partition consisting of variable nodes. Each function node represents one local function. Each variable node represents one variable. The global function is the product of all the local functions. Edges indicate the involvement between functions and variables; that is, there is an edge between one variable node and one function node if and only if the variable is an argument of the local function.

Example 11.4. A factor graph is shown in Fig. 11.3. We use circles to represent variable nodes and squares to represent function nodes. The factor graph represents the probability distribution function of a binary sequence $X_1 X_2 X_3$ with length three. The random variables X_i, $1 \leq i \leq 3$, are independent and identically distributed with a probability distribution $p(x)$. The global function is

$$f_{X_1 X_2 X_3}(x_1, x_2, x_3) = f_{X_1}(x_1) f_{X_2}(x_2) f_{X_3}(x_3), \qquad (11.7)$$

where $f_{X_i}(\cdot) = p(\cdot)$, for $i = 1, 2, 3$, are the local functions.

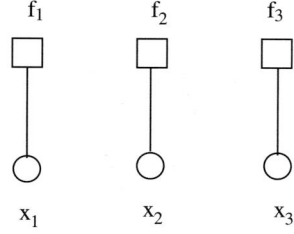

Fig. 11.3.   The factor graph.

Example 11.5. Factor graphs can be used to represent probability distributions in the decoding of LDPC codes. Let $\boldsymbol{x}_1^n = \{x_1, \ldots, x_n\}$ denote the transmitted codeword. Let $\boldsymbol{y}_1^n = \{y_1, \ldots, y_n\}$ be the received channel outputs. Assume the channel is memoryless. Define functions f_j as

$$f_j(\boldsymbol{x}_1^n) = \begin{cases} 1, & \text{if } j\text{th parity check is satisfied} \\ 0, & \text{otherwise} \end{cases}. \qquad (11.8)$$

Define functions $g_i(x_i, y_i)$ as the conditional probability that y_i is the channel output, given that x_i is transmitted. That is,

$$g(x_i, y_i) = \mathbb{P}(y_i \text{ received}|x_i \text{ transmitted}). \qquad (11.9)$$

The probability density function $p(\boldsymbol{x}_1^n, \boldsymbol{y}_1^n)$ can be factored as

$$p(\boldsymbol{x}_1^n, \boldsymbol{y}_1^n) = \alpha \left(\prod_{j=1}^m f_j(\boldsymbol{x}_1^n) \right) \left(\prod_{i=1}^n g(x_i, y_i) \right), \qquad (11.10)$$

where α is a normalization constant, and m is the number of parity-checks. An example of such a factor graph for LDPC decoding is shown in Fig. 11.4.

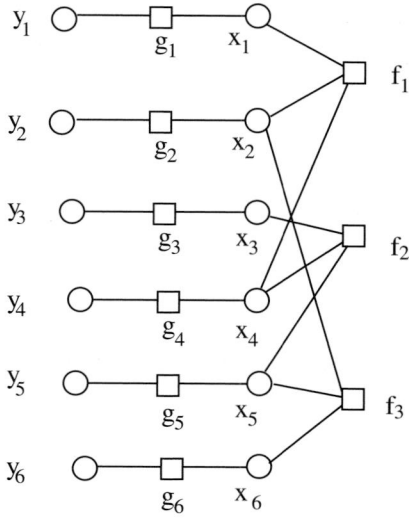

Fig. 11.4. An example of factor graph representations of probability distributions in LDPC decoding.

11.4.2. Message-Passing Algorithm for Tree-Structured Factor Graphs

The message-passing algorithm is a parallel algorithm for marginal distribution calculation. In this section, we discuss the message-passing algorithms for the special case that the factor graphs do not contain cycles. We refer such factor graphs as tree-structured factor graphs. The more general cases will be discussed in the next section.

Consider a factor graph with n variable nodes and m function nodes. The global probability distribution function f can be factored as

$$f(x_1, \ldots, x_n) = \prod_{j=1}^{m} f_j(\mathcal{X}_j), \tag{11.11}$$

where $\mathcal{X}_j$ denotes the subset of variables, which are arguments of the jth function $f_j(\cdot)$. We wish to calculate all the marginal distributions $f_{X_i}(\cdot)$, where $f_{X_i}(\cdot)$ is the marginal distributions with respect to a variable X_i,

$$f_{X_i}(s) = \sum_{x_1} \sum_{x_2} \cdots \sum_{x_{i-1}} \sum_{x_{i+1}} \cdots \sum_{x_n} f(x_1, \ldots, x_{i-1}, s, x_{i+1}, \ldots, x_n). \tag{11.12}$$

In the message-passing algorithm, we define messages on all edges. The messages are partial calculation results for marginal distribution calculations. Consider a function node f_0 and a neighboring variable node x_0. If we remove the edge between the nodes f_0 and x_0, the factor graph will become a graph consisting of two unconnected smaller graphs. Consider the graph which contains the function node f_0. Denote the set of function nodes and variable nodes in the graph except the function node f_0 by A. Similarly, denote the set of function nodes and variable nodes in the other graph except the variable node x_0 by B. Denote the product of all local functions in the set A and the function f_0 as $f_A(\cdot)$, while denote the product of all local functions in the set B as $f_B(\cdot)$.

The message on the edge between f_0 and x_0, with a direction from the function node f_0 to the variable node x_0, is defined as

$$m_{f_0 \to x_0}(s) = \sum_{x_1} \sum_{x_2} \cdots \sum_{x_k} f_A|_{x_0=s}(x_0, x_1, \ldots, x_k), \tag{11.13}$$

where $\{x_1, \ldots, x_k\}$ denotes the set of variable nodes in the set A. The message on the edge between f_0 and x_0, with a direction from the variable

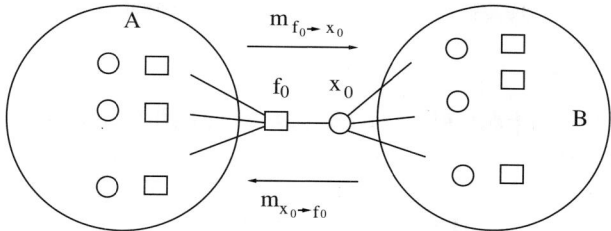

Fig. 11.5. The interpretation of messages.

node x_0 to the function node f_0, is defined as,

$$m_{x_0 \to f_0}(s) = \sum_{y_1} \sum_{y_2} \cdots \sum_{y_l} f_B|_{x_0=s}(x_0, y_1, \ldots, y_l), \qquad (11.14)$$

where $\{y_1, \ldots, y_l\}$ denotes the set of variable nodes in the set B. Each message is a function with a domain identical to the alphabet of the variable x_0. The definition of messages is illustrated in Fig. 11.5.

The message-passing algorithm computes messages in a recursive manner. For a message with a direction from a variable node x_0 to a function node f_0, the message $m_{x_0 \to f_0}(s)$ can be recursively calculated from the messages $m_{f_i \to x_0}(s)$, $i = 1, \ldots, k$, where $f_0, f_1, \ldots, f_k$ are all the neighboring function nodes of the variable node x_0. It can be checked that

$$m_{x_0 \to f_0}(s) = \prod_{i=1}^{k} m_{f_i \to x_0}(s). \qquad (11.15)$$

The message update rule is depicted in Fig. 11.6. Similarly, for a message with a direction from a function node f_0 to a variable node x_0, the message $m_{f_0 \to x_0}(s)$ can be recursively calculated from the messages $m_{x_i \to f_0}$, $i = 1, \ldots, k$, where $x_0, x_1, \ldots, x_k$ are all the neighboring variable nodes of the function node f_0. It can be checked that

$$m_{f_0 \to x_0}(s) = \sum_{x_1, \ldots, x_k} \left(f_0|_{x_0=s}(x_0, x_1, \ldots, x_k) \prod_{i=1}^{k} m_{x_i \to f_0}(x_i) \right). \qquad (11.16)$$

The message update rule is depicted in Fig. 11.7. The calculating of all messages can be decomposed into calculating messages with a direction from a leaf of the factor graph in the above recursive manner. The latter ones can be easily calculated directly.

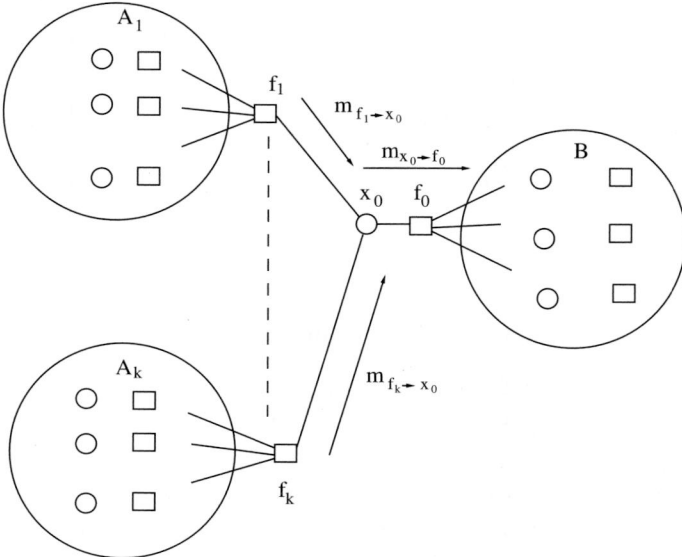

Fig. 11.6.   The messages update rule for a message with a direction from a variable node to a function node.

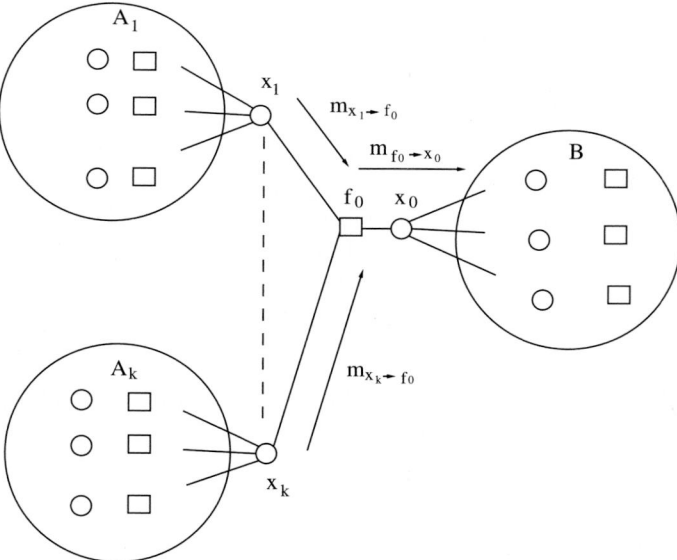

Fig. 11.7.   The messages update rule for a message with a direction from a function node to a variable node.

The marginal distributions of each variable node can be determined from messages. Let x_0 be a variable node. Assume that there are k neighboring function nodes of x_0. It can be checked that

$$f_{X_i}(s) = \sum_{x_1} \sum_{x_2} \cdots \sum_{x_{i-1}} \sum_{x_{i+1}} \cdots \sum_{x_n} f(x_1, \ldots, x_{i-1}, s, x_{i+1}, \ldots, x_n),$$

$$= \prod_{i=1}^{k} m_{f_i \to x_0}(s). \tag{11.17}$$

The marginal distribution calculation from messages is depicted in Fig. 11.8.

11.4.3. *Loopy Propagation*

The marginal distributions can be exactly calculated by the message-passing algorithm in polynomial time, if the global probability distribution functions can be represented by tree-structured factor graphs. However, if factor graphs contains cycles, the calculation of exact marginal probability distributions is usually NP-hard. In such cases, a heuristic version of the message-passing algorithm called *loopy propagation message-passing*

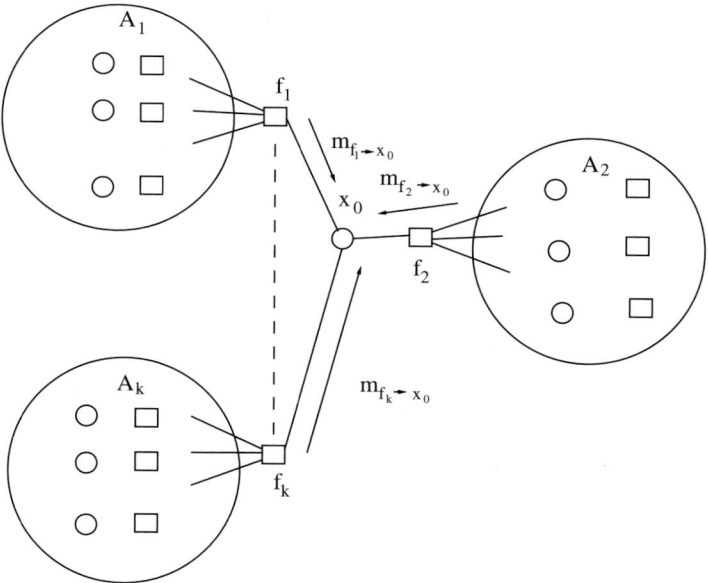

Fig. 11.8.   Calculating marginal distributions from messages.

algorithm is usually adopted. The loopy propagation message-passing algorithm calculates the messages using the same set of rules of the message-passing algorithms on tree-structured factor graphs. The calculated results are only approximate values of marginal distributions.

11.5. LDPC Decoding by Loopy Propagation

The message-passing algorithm for LDPC code decoding is an instance of loopy propagation. Consider factor graphs for LDPC code decoding, such as the one shown in Fig. 11.4. The marginal distribution for each codeword bit can be approximately calculated by the loopy-propagation algorithm. For a variable x_0, the bit-wise MAP (maximum a posterior) decoding algorithm outputs 1, if the marginal distribution $\mathbb{P}(x_0 = 1|\boldsymbol{y}_1^n) \geq \mathbb{P}(x_0 = 0|\boldsymbol{y}_1^n)$, and 0 otherwise. More details are explained as follows.

Assume that the LDPC code has a blocklength n, and m parity-checks. Denote the variable nodes corresponding to codeword bits by x_i, $i = 1, \ldots, n$. Denote the function nodes corresponding to parity-check constraints by f_j, $j = 1, \ldots, m$. Denote the function nodes corresponding to channel conditional probabilities by g_i, $i = 1, \ldots, n$. The message-passing decoding algorithm for the LDPC code usually follows the following flooding schedule for message updating. At each iteration, the algorithm first updates all messages directed from variable nodes x_i to function nodes f_j, and then updates all messages directed from function nodes f_j to variable nodes x_i. The messages directed from function nodes g_i to variable nodes x_i are fixed during all iterations. The messages directed from variable nodes x_i to function nodes g_i are not calculated, because the decoding decisions are independent of these messages. The initial values of messages directed from function nodes f_j to variable nodes x_i are all set to be one. That is, for a message directed from a function node f_j to a variable node x_i,

$$m_{f_j \to x_i}(0) = m_{f_j \to x_i}(1) = 1. \tag{11.18}$$

The initial values of messages directed from function nodes g_i to variable nodes x_i are calculated according to the channel conditional probability distribution. That is, for a message directed from a function node g_i to a variable node x_i,

$$m_{g_i \to x_i}(0) = \mathbb{P}\left(y_i \text{ is received } |0 \text{ is transmitted}\right), \tag{11.19}$$

$$m_{g_i \to x_i}(1) = \mathbb{P}\left(y_i \text{ is received } |1 \text{ is transmitted}\right). \tag{11.20}$$

During each iteration, the message updating rule for a message directed from a variable node x_0 to a function node f_0 is as follows:

$$m_{x_0 \to f_0}(s) = m_{g_i \to x_0}(s) \prod_{i=1}^{k} m_{f_i \to x_0}(s), \qquad (11.21)$$

where $f_0, f_1, \ldots, f_k$ are all neighboring function nodes of x_0 that correspond to parity checks. Similarly, the message updating rule for a message directed from a function node f_0 to a variable node x_0 is as follows:

$$m_{f_0 \to x_0}(s) = \sum_{x_1} \sum_{x_2} \cdots \sum_{x_k} \left(f_0(x_0 = s, x_1, \ldots, x_k) \prod_{i=1}^{k} m_{x_i \to f_0}(x_i) \right),$$
$$(11.22)$$

where $x_0, x_1, \ldots, x_k$ are all neighboring variable nodes of f_0.

For the special case of binary LDPC decoding, there exist more simple formula for the message updating rules in Eqs. (11.21) and (11.22). We define the log-likelihood ratios for messages as follows,

$$\Lambda_{f_0 \to x_0} = \ln \left(\frac{m_{f_0 \to x_0}(0)}{m_{f_0 \to x_0}(1)} \right) = \log_e \left(\frac{m_{f_0 \to x_0}(0)}{m_{f_0 \to x_0}(1)} \right), \qquad (11.23)$$

$$\Lambda_{x_0 \to f_0} = \ln \left(\frac{m_{x_0 \to f_0}(0)}{m_{x_0 \to f_0}(1)} \right) = \log_e \left(\frac{m_{x_0 \to f_0}(0)}{m_{x_0 \to f_0}(1)} \right), \qquad (11.24)$$

$$\Lambda_{g_0 \to x_0} = \ln \left(\frac{m_{g_0 \to x_0}(0)}{m_{g_0 \to x_0}(1)} \right) = \log_e \left(\frac{m_{g_0 \to x_0}(0)}{m_{g_0 \to x_0}(1)} \right), \qquad (11.25)$$

where $m_{f_0 \to x_0}$, $m_{x_0 \to f_0}$, and $m_{g_0 \to x_0}$ are messages. The hyperbolic trigonometric functions are defined as,

$$\sinh(x) = \frac{e^x - e^{-x}}{2}, \qquad (11.26)$$

$$\cosh(x) = \frac{e^x + e^{-x}}{2}, \qquad (11.27)$$

$$\tanh(x) = \frac{\sinh(x)}{\cosh(x)} = \frac{e^x - e^{-x}}{e^x + e^{-x}} = \frac{e^{2x} - 1}{e^{2x} + 1}. \qquad (11.28)$$

The message updating rules can be simplified into the following well known *sum law* and *tanh law* (see, for example, [26]). Sketch proofs are provided in order to make this chapter self-contained.

Proposition 11.1 (sum law). *For a message directed from a variable node x_0 to a function node f_0, the message update rule can be written as*

$$\Lambda_{x_0 \to f_0} = \Lambda_{g_0 \to x_0} + \sum_{j=1}^{k} \Lambda_{f_i \to x_0}, \qquad (11.29)$$

where g_0 is the neighboring function node that corresponds to the channel conditional probability function, $f_0, f_1, \ldots, f_k$ are all the neighboring function nodes of x_0 that correspond to parity checks.

Proof. By the definition of log-likelihood ratios. □

Proposition 11.2 (tanh law). *For a message directed from a function node f_0 to a variable node x_0, the message update rule can be written as*

$$\tanh\left(\frac{\Lambda_{f_0 \to x_0}}{2}\right) = \prod_{i=1}^{k} \tanh\left(\frac{\Lambda_{x_i \to f_0}}{2}\right), \qquad (11.30)$$

where $x_0, x_1, \ldots, x_k$ are all the neighboring variable nodes of f_0.

Proof. By the definition of the function f_0, we have

$$m_{f_0 \to x_0}(0) = \sum_{x_1, \ldots, x_k : x_1 + \cdots + x_k = 0} \left(\prod_{i=1}^{k} m_{x_i \to f_0}(x_i)\right), \qquad (11.31)$$

$$m_{f_0 \to x_0}(1) = \sum_{x_1, \ldots, x_k : x_1 + \cdots + x_k = 1} \left(\prod_{i=1}^{k} m_{x_i \to f_0}(x_i)\right). \qquad (11.32)$$

Therefore,

$$m_{f_0 \to x_0}(0) - m_{f_0 \to x_0}(1) = \prod_{i=1}^{k} \left(m_{x_i \to f_0}(0) - m_{x_i \to f_0}(1)\right), \qquad (11.33)$$

$$m_{f_0 \to x_0}(0) + m_{f_0 \to x_0}(1) = \prod_{i=1}^{k} \left(m_{x_i \to f_0}(0) + m_{x_i \to f_0}(1)\right). \qquad (11.34)$$

By the definition of the tanh function,

$$\tanh\left(\frac{\Lambda_{f_0\to x_0}}{2}\right) = \frac{\frac{m_{f_0\to x_0}(0)}{m_{f_0\to x_0}(1)} - 1}{\frac{m_{f_0\to x_0}(0)}{m_{f_0\to x_0}(1)} + 1} = \frac{m_{f_0\to x_0}(0) - m_{f_0\to x_0}(1)}{m_{f_0\to x_0}(0) + m_{f_0\to x_0}(1)},$$

$$= \prod_{i=1}^{k} \frac{m_{x_i\to f_0}(0) - m_{x_i\to f_0}(1)}{m_{x_i\to f_0}(0) + m_{x_i\to f_0}(1)},$$

$$= \prod_{i=1}^{k} \tanh\left(\frac{\Lambda_{x_i\to f_0}}{2}\right). \qquad (11.35)$$

The proposition is proven. $\qquad\square$

Example 11.6. Consider a binary erasure channel (BEC) as shown in Fig. 11.9. The channel takes binary inputs and outputs 0, 1, or e, where e stands for a missing bit. The channel parameter ξ is the probability that a transmitted bit is lost. The channel outputs a transmitted bit correctly with probability $1-\xi$, and outputs e with probability ξ. We call the missing bits erasures. It can be checked that the capacity is $C = 1 - \xi$.

In the case of BECs, the log-likelihood ratios for messages can only take three values: $+\infty$, $-\infty$, and 0. The message update rules in Eqs. (11.58) and (11.30) can be simplified as follows:

$$\Lambda_{x_0\to f_0} = \begin{cases} +\infty, & \text{one of } \Lambda_{f_i\to x_0} \text{ and } \Lambda_{g_0\to x_0} \text{ is equal to } +\infty \\ -\infty, & \text{one of } \Lambda_{f_i\to x_0} \text{ and } \Lambda_{g_0\to x_0} \text{ is equal to } -\infty, \\ 0, & \text{all of } \Lambda_{f_i\to x_0} \text{ and } \Lambda_{g_0\to x_0} \text{ are equal to } 0 \end{cases}$$

$$(11.36)$$

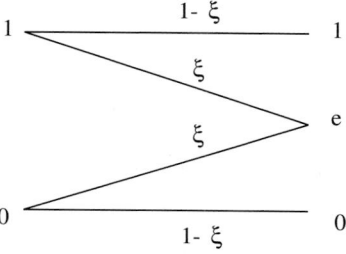

Fig. 11.9. Binary erasure channel.

$$\Lambda_{f_0 \to x_0} = \begin{cases} +\infty, & \text{all } \Lambda_{x_i \to f_0} \neq 0, \text{ number of } -\infty \text{ in } \Lambda_{x_i \to f_0} \text{ is even} \\ -\infty, & \text{all } \Lambda_{x_i \to f_0} \neq 0, \text{ number of } -\infty \text{in } \Lambda_{x_i \to f_0} \text{ is odd} \\ 0, & \text{one of } \Lambda_{x_i \to f_0} = 0 \end{cases},$$

(11.37)

for each log-likelihood ratios $\Lambda_{x_0 \to f_0}$ and $\Lambda_{f_0 \to x_0}$ during one iteration.

11.6. Density Evolution

11.6.1. *Concentration Theorem*

Density evolution is an asymptotic method for calculating the limiting bit error probabilities, after certain decoding iterations, as the blocklengths go to infinity [18]. The theoretical foundations of density evolution are concentration theorems [18]. The concentration theorems show that, after a given number of decoding iterations, the bit error probabilities converge to their limiting values with probabilities approaching one, as the blocklengths go to infinity, where the code is uniformly chosen from a given code ensemble at random.

The concentration theorem also shows that the limiting values of bit error probabilities are exactly the bit error probabilities of infinitely long codes. Therefore, the bit error probabilities of long LDPC codes can be approximately calculated by calculating the limiting bit error probabilities, which can be efficiently recursively calculated, because we can assume that the incoming messages at each variable node and check nodes are independent and identically distributed. This approach of approximately calculating the decoding bit error probabilities is called *density evolution*.

For many classes of channels, there exist ordering of the channels, such that the channel with order α_1 can be considered as a physically degraded channel of the channel with order α_2, if $\alpha_1 < \alpha_2$. Examples include binary input additive white Gaussian noise (BIAWGN) channels, binary symmetric channels (BSC), and BECs. It has been shown in [18], that for many classes of channels with such ordering, there exists a critical value $\alpha(\lambda, \rho)$ for each fixed pair of degree distributions $\lambda(x)$, $\rho(x)$, such that

- the ensemble average bit error probabilities of $\mathcal{C}(\lambda, \rho, n)$ approach zero for all channels with order $\alpha > \alpha(\lambda, \rho)$ as the blocklength n go to infinity, and
- the ensemble average bit error probabilities of $\mathcal{C}(\lambda, \rho, n)$ are bounded away from zero for all channel with order $\alpha \leq \alpha(\lambda, \rho)$.

This critical value $\alpha(\lambda, \rho)$ is a very important parameter for an LDPC code ensemble. It is called a *threshold* for the pair of degree distributions $\lambda(x), \rho(x)$ and can be calculated by density evolution.

11.6.2. *Density Evolution for BIAWGN Channels*

A BIAWGN channel is shown in Fig. 11.10. The input signal X takes values 1 or -1. The noise Z is white with Gaussian distribution $N(0, \sigma_n^2)$. The received signal is $Y = X + Z$.

We will use the following notation.

- Let g_i, f_j, x_i, $\Lambda_{g_i \to x_i}$, $\Lambda_{x_i \to f_j}$, and $\Lambda_{f_j \to x_i}$ be defined as in Sec. 11.5.
- Let $h^l_{g_i \to x_i}$, $h^l_{x_i \to f_j}$, and $h^l_{f_j \to x_i}$ denote the probability density functions of the log-likelihood ratios $\Lambda_{g_i \to x_i}$, $\Lambda_{x_i \to f_j}$, and $\Lambda_{f_j \to x_i}$, after l decoding iterations.
- Let $h^l_{g \to x}$, $h^l_{x \to f}$, and $h^l_{f \to x}$ denote the probability distributions of messages averaged over all variable nodes and check nodes, after l decoding iterations. It is clear that $h^l_{g \to x}$ corresponds to messages from g_i to x_i, $h^l_{x \to f}$ corresponds to messages from x_i to f_j, and $h^l_{f \to x}$ corresponds to messages from f_j to x_i.

Because BIAWGN channels are symmetric and the LDPC codes are linear block codes, the error probabilities are independent of the transmitted codewords. Hence, in order to simplify the discussions, we can assume that the transmitted codeword is the all zero codeword. We call a message $m_{f_j \to x_i}$ *incorrect*, if $\Lambda_{f_j \to x_i} < 0$. The fraction of incorrect messages, after l decoding iterations, is

$$\int_{-\infty}^{0} h^l_{f \to x}(x) \mathrm{d}x. \tag{11.38}$$

The performance of the code can be determined if we can calculate the distribution $h^l_{f \to x}(x)$. However, calculating the distribution $h^l_{f \to x}$ or even

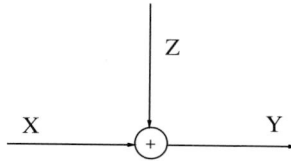

Fig. 11.10.   Binary input additive white Gaussian noise channels.

the asymptotic limit is quite difficult, because $h^l_{f\to x}$ is a function and the computation is in an infinite dimensional space.

Density Evolution with Gaussian approximation [2] is a reduced-complexity scheme on approximately calculating the probability density functions (other approximations have also been proposed, see example [1]). In such schemes, the distributions $h^l_{f\to x}$ and $h^l_{x\to f}$ are approximated by Gaussian distributions. Since a Gaussian distribution can be specified by its two parameters, the recursive calculation of distributions in infinite dimensional spaces can be reduced to the recursive calculation of the distribution parameters in finite dimensional spaces.

A first thought on this problem may suggest that two parameters need to be recursively calculated for each distribution. A careful examination shows that the calculation on one parameter for each distribution is sufficient. This is because the distributions $h^l_{g\to x}$, $h^l_{x\to f}$, and $h^l_{f\to x}$ satisfy a so called *symmetric condition* [18]. If a Gaussian distribution $h(x) = N(m, \sigma^2)$ satisfies the symmetric condition, then $\sigma^2 = 2m$. The distribution can be uniquely determined by its mean value.

Let us denote the mean of the log-likelihood ratios from check nodes to variables nodes at l-th iteration by $\mu^l_{c\to v}$. For an irregular LDPC code with the left degree sequence $\lambda(x) = \sum_i \lambda_i x^{i-1}$ and right degree sequence $\rho(x) = \sum_j \rho_j x^{i-1}$. The mean $\mu^l_{c\to v}$ can be calculated recursively as follows [2],

$$\mu^{l+1}_{c\to v} = \sum_{j=2}^{d_c} \rho_j \phi^{-1} \left(1 - \left[1 - \sum_{i=2}^{d_c} \lambda_i \phi \left(\mu_{g\to v} + (i-1)\mu^l_{c\to v} \right) \right]^{j-1} \right),$$

$$(11.39)$$

where d_v and d_c are the maximal left and right degrees, $\mu_{g\to v}$ is the mean of the log-likelihood ratios of messages from channel output nodes to variable nodes, and the function $\phi(x)$ is defined as

$$\phi(x) = \begin{cases} 1 - \dfrac{1}{\sqrt{2\pi x}} \displaystyle\int_{-\infty}^{\infty} \tanh\left(\dfrac{u}{2}\right) e^{-\frac{(u-x)^2}{4x}} \, du, & x > 0 \\ 1, & x = 0 \end{cases} . \quad (11.40)$$

11.6.3. *Density Evolution for BECs*

In the message-passing decoding for BECs, the log-likelihood ratios $\Lambda_{c\to v}$ and $\Lambda_{v\to c}$ can only take values ∞, $-\infty$ or 0. A message is said to be an erasure, if the corresponding log-likelihood ratio is 0. Let us consider the LDPC codes with the left degree sequence $\lambda(x) = \sum_i \lambda_i x^{i-1}$ and right degree sequence $\rho(x) = \sum_j \rho_j x^{j-1}$. Let $p_{e,c \to v}^{(l)}$ $(p_{e,v \to c}^{(l)})$ denote the probability that a message from a check node to a variable node (a variable node to a check node) is erasure along a randomly chosen edge during the l-th iteration. The erasure probability $p_{e,c \to v}^{(l)}$ can be calculated as,

$$p_{e,c \to v}^{(l)} = \sum_{j=2}^{d_c} \mathbb{P}(c \text{ has degree } j)\mathbb{P}\left(m_{c \to v}^{(l)} \text{ is erasure} \mid \text{the degree of } c \text{ is } j\right),$$

$$= \sum_{j=2}^{d_c} \rho_j \left\{1 - [1 - p_{e,v \to c}^{(l)}]^{j-1}\right\} = 1 - \sum_{j=2}^{d_c} \rho_j \left[1 - p_{e,v \to c}^{(l)}\right]^{j-1}.$$

$$(11.41)$$

The erasure probability $p_{e,v \to c}^{(l)}$ can be calculated as,

$$p_{e,v \to c}^{(l)} = \sum_{i=2}^{d_v} \mathbb{P}(v \text{ has degree } i)\mathbb{P}(m_{v \to c} \text{ is erasure} \mid \text{the degree of } v \text{ is } i),$$

$$= \xi \sum_{i=2}^{d_v} \lambda_i \left[p_{e,c \to v}^{(l-1)}\right]^{i-1},$$

$$(11.42)$$

where ξ is the channel parameter.

In summary,

$$p_{e,v \to c}^{(l)} = \xi \sum_{i=2}^{d_v} \lambda_i \left\{1 - \sum_{j=2}^{d_c} \rho_j \left[1 - p_{e,v \to c}^{(l-1)}\right]^{j-1}\right\}^{i-1}$$

$$= \xi \lambda \left(1 - \rho \left(1 - p_{e,v \to c}^{i-1}\right)\right).$$

$$(11.43)$$

The code can be successfully decoded with high probability if and only if,

$$\xi \lambda (1 - \rho (1 - x)) < x,$$

$$(11.44)$$

for all $0 < x \leq \xi$.

11.7. Finite-Length Analysis

Unlike asymptotic analysis methods (such as density evolution), a finite-length analysis evaluates the performance of finite-length codes. In the modern finite-length analysis of LDPC codes, the concept of *pseudo-codewords* plays a center role. The idea can be illustrated by looking at the so-called *computational trees*.

Given a variable node x_0 and a number of decoding iteration L, the computational tree for the node x_0 and iteration L shows how the computation of the decoding decision and messages can be traced back in the first L iterations. The computational tree is a rooted tree of variable nodes and check nodes where,

- the root is the variable node x_0;
- the nodes at the depth $2d, d = 0, \ldots, L - 1$ are variable nodes;
- the nodes at the depth $2d + 1, d = 0, \ldots, L - 1$ are check nodes;
- a variable node x at the depth $2d$ corresponds to the computational of the message from the variable node x to its parent node f (in the computational tree) during the $L - d$ iterations; and
- a check node c at the depth $2d + 1$ corresponds to the computational of the message from the check node c to its parent node x during the $L - d$ iterations.

Example 11.7. Consider an LDPC code shown in Fig. 11.11. The code has three variable nodes $1, 2, 3$ and three check nodes a, b, c. The computational tree for the variable node 1 and two decoding iterations is shown in Fig. 11.12. The corresponding nodes in the Tanner graph are labeled besides the nodes in the computational tree.

It is easy to check that if the local subgraph of the variable node x_0 does not contain a cycle, then the computational tree is exactly the local subgraph. In this case, there is a one-to-one correspondence between the nodes in the local subgraph and the nodes in the computational tree. The decoding decision at the variable node x_0 after L decoding iteration is exactly the result of bit-wise MAP decoding based on the channel observations at the local subgraph.

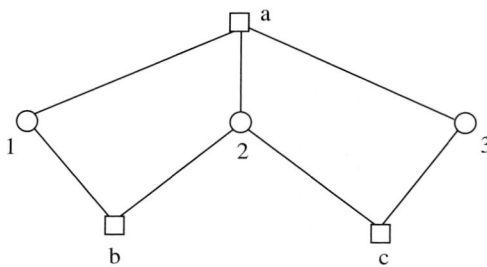

Fig. 11.11.   A simple LDPC code.

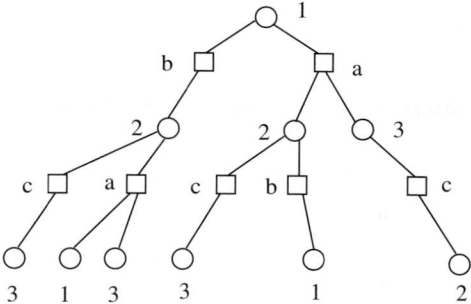

Fig. 11.12.   The computational tree at node 1.

On the other hand, if the local subgraph contains cycles, then the computational tree is an unwrapped version of the local subgraph. In this case, a node in the local subgraph may correspond to multiple nodes in the computational tree. It is clear that each valid configuration (codeword) on the local subgraph induces a valid configuration on the computational tree. A key observation is that some valid configurations on the computational tree do not correspond to any valid configuration on the local subgraph.

Let us revisit Example 11.7. It can be checked that the rate of the code is zero. The only codeword (valid configuration on the Tanner graph) is the all zero codeword. However, there exist non-zero valid configurations on the computational tree. For example, a valid configuration on the computational tree is shown in Fig. 11.13, where variable nodes taking the value one are drawn as filled circles and the variable nodes taking the value zero are drawn as empty circles. Because the message-passing decoding

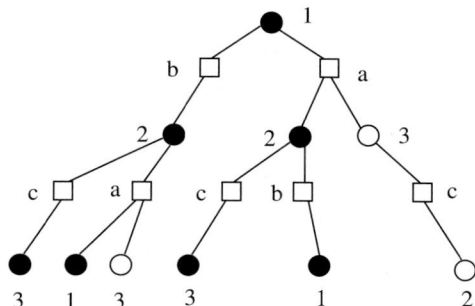

Fig. 11.13.   A computational tree and a pseudo-codeword.

is the bit-wise MAP decoder on the computational tree, the decoder may favor a non-zero valid configuration on the computational tree under certain conditions. In these cases, the decoder decodes x_0 to 1.

A valid configuration on the computational tree is called a *pseudo-codeword* in this chapter (our definition here is different from the one in [24], however, the two definitions are fundamentally related). It can be seen that the performance of message-passing decoding is determined by the distributions of pseudo-codewords.

Due to the difficulties in enumerating and detecting pseudo-codewords, previously proposed finite-length analysis methods are generally based on subgraph structures, which are related to pseudo-codewords. For BEC, the finite-length analysis methods are generally based on stopping sets. For AWGN channels, the finite-length analysis methods are usually based on trapping sets.

Definition 11.1. A stopping set S is a set of variable nodes, such that, for each neighboring check node c of S, there exist at least two edges with one end being c and the other end being in S. We say that the stopping set S has weight w, if the cardinality of the stopping set is w.

Definition 11.2. An (a, b) trapping set S is a set of variable nodes, such that,

• the cardinality of the set S is a,
• there exist exactly b check nodes c, such that c is connected to S by an odd number of edges.

A rule of thumb in LDPC code design is that stopping sets with low weights and trapping sets with small a and b should be avoided, because such subgraph structures introduce pseudo-codewords, which degrade the error probability performance significantly.

The performance analysis of LDPC codes based on pseudo-codeword analysis has become a very active research area recently. For further reading, we refer the interested readers to recent works, such as [4–8, 24]

11.8. Problems

Problem 11.1. Implement the message-passing decoding algorithm using loopy propagation. Determine the bit error probabilities of the code in Example 11.3 under AWGN channels by simulation.

Problem 11.2. Find a standard random LDPC ensemble with rate 0.5, such that the randomly constructed codes from the ensemble can be successfully decoded with high probability under a BEC with channel parameter 0.46. Construct one such code from the code ensemble. Determine the bit error probabilities of the code by simulation.

Problem 11.3. Consider the message-passing rules in Sec. 11.5 — the *sum law* and *tanh law*. Show that these message updating rules can be approximated by the following *min–sum rules*, if the signal-to-noise ratio is high.

$$\Lambda_{x_0 \to f_0} = \Lambda_{g_0 \to x_0} + \sum_{j=1}^{k} \Lambda_{f_i \to x_0}, \tag{11.45}$$

$$\Lambda_{f_0 \to x_0} = \left(\prod_{i=1}^{k} \text{Sign} \left(\Lambda_{x_i \to f_0} \right) \right) \min \left\{ |\Lambda_{x_1 \to f_0}|, \ldots, |\Lambda_{x_k \to f_0}| \right\}, \tag{11.46}$$

where $\text{sign}(\Lambda_{x_i \to f_0})$ denotes the sign of $\Lambda_{x_i \to f_0}$, $|\cdot|$ denotes the absolute values.

Problem 11.4. Make a comparison between the *sum–tanh rules* and the *min–sum rules*. What are their advantages and disadvantages?

Problem 11.5. Implement the message-passing decoding using the min–sum rules. Compare the error probability performance of the decoding algorithm with min–sum rules and the decoding algorithm with the sum and tanh rules.

Problem 11.6. Consider a scenario that an LDPC code has been constructed. Suppose that the code has low error probabilities performance under the Maximal-Likelihood (ML) decoding and MAP decoding. Now, if we observe that the code has very poor error probability performance under the message-passing decoding, can you explain the potential reasons for that? (Tips: look at Example 11.7).

Problem 11.7. Give an example of computational trees and pseudo-codewords.

Problem 11.8. Plot the factor graph for decoding a parallel-concatenated Turbo code. Suppose that the channel is memoryless.

Problem 11.9. Consider a discrete time channel with real-valued inputs $X_1, X_2, \ldots, X_i, \ldots, X_n$, where X_i is the channel input at the ith time slot. The channel output Y_i at the ith time slot is

$$Y_i = \alpha_i X_i + Z_i, \qquad (11.47)$$

where $Z_i \sim \mathcal{N}(0,1)$, is Gaussian noise with zero mean and variance one, α_i is a real-valued channel gain. Assume that the receiver has a way to estimate α_i, so that α_i is known perfectly at the receiver. A message is transmitted over this channel using binary LDPC codes and the BPSK (binary phase shift keying) modulation. How should we design the message-passing decoding algorithm for this channel?

Problem 11.10. Consider the following channel with state. The state of the channel at the ith time slot is denoted by $S_i \in \{g, b\}$. The real-valued input and output of the channel at the ith time slot are denoted by X_i and Y_i, respectively,

$$Y_i = X_i + Z_i, \qquad (11.48)$$

where Z_i are independent zero-mean Gaussian noise. The variance of Z_i is dependent on the channel state S_i,

$$Z_i \sim \mathcal{N}(0, 10), \quad \text{if } S_i = b, \qquad (11.49)$$

$$Z_i \sim \mathcal{N}(0, 1), \quad \text{if } S_i = g. \qquad (11.50)$$

The channel states S_i are independent of X_i, Y_i and form a Markov chain, such that

$$\mathbb{P}(S_{i+1} = g | S_i = b) = \mathbb{P}(S_{i+1} = b | S_i = g) = 0.1, \qquad (11.51)$$

$$\mathbb{P}(S_{i+1} = g | S_i = g) = \mathbb{P}(S_{i+1} = b | S_i = b) = 0.9. \qquad (11.52)$$

Assume that $S_0 = g$ with probability 1 and S_i are unknown at both the transmitter and receiver. A message is transmitted over this channel using binary LDPC codes and the BPSK modulation. Plot the factor graph for LDPC decoding. How should we design the decoding algorithm? Implement your proposed decoding algorithm.

11.9. Problem Solution

Solution (to Problem 11.3). If the signal-to-noise ratio is high,

$$\left| \tanh\left(\frac{\Lambda_{x_i \to f_0}}{2}\right) \right| \approx 1, \quad \text{for } i = 1, \ldots, k. \tag{11.53}$$

Therefore the product of the absolute values is approximately equal to the value of the smallest absolute value. The claim follows.

Solution (to Problem 11.4). One main advantage of the min–sum decoding is that it is easy to implement in firmware. However, the bit error probability is slightly higher, if the min–sum law is used.

Solution (to Problem 11.6). The most possible reason is that there exist some pseudo-codewords, which degrade the error probability performance significantly. For example, in Example 11.7, the bit-wise MAP decoder always decodes each bit to zero, because all-zero codeword is the only codeword in the code-book. On the other hand, a message-passing decoder may decode some bits to one.

Solution (to Problem 11.8). There are many ways to decompose the joint probability distribution and draw the factor graph. An example of my version is shown in Fig. 11.14. The variable nodes x_i represent the codeword symbols. The variable nodes y_i represent the received channel outputs. The variable nodes s_i represent the states of the trellises. The function nodes e_i represent the equal functions. The function e_i takes the value one if the two

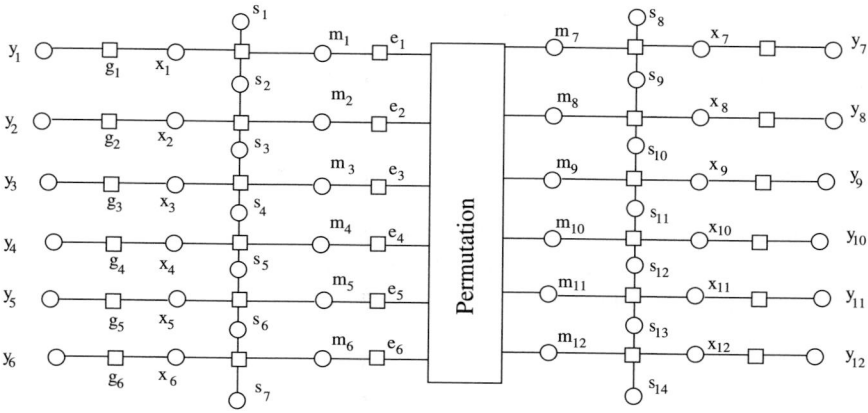

Fig. 11.14. Factor graph for Turbo code.

neighboring variable nodes take the same value, the function takes the value zero otherwise. The variable nodes $m_1, \ldots, m_6$ represent the transmitted messages.

Solution (to Problem 11.9). The messages $m_{g_i \to x_i}$ are calculated as follows.

$$m_{g_i \to x_i}(0) = \frac{1}{\sqrt{2\pi}} \exp\left\{-\frac{(y_i + \alpha_i)^2}{2}\right\}, \tag{11.54}$$

$$m_{g_i \to x_i}(1) = \frac{1}{\sqrt{2\pi}} \exp\left\{-\frac{(y_i - \alpha_i)^2}{2}\right\}. \tag{11.55}$$

The computation of other messages follows the sum and tanh law as discussed in Sec. 11.5.

Solution (to Problem 11.10). The joint probability distribution can be decomposed as follows.

$$\mathbb{P}(x_1, \ldots, x_n, y_1, \ldots, y_n, s_1, \ldots, s_n) = \tag{11.56}$$

$$\left(\prod_{i=1}^{n-1} \mathbb{P}(s_{i+1}|s_i)\right) \left(\prod_{i=1}^{n} \mathbb{P}(y_i|x_i, s_i)\right)$$

$$\mathbb{P}(x_1, \ldots, x_n, \text{ is transmitted}). \tag{11.57}$$

An example of such factor graphs is shown in Fig. 11.15, where

- the variable nodes x_i represent the transmitted signal bits,
- the variable nodes y_i represent the channel outputs,
- the variable nodes s_i represent the channel states,
- the function nodes f_i represent the parity-check constraints,
- the function nodes g_i represent the channel conditional distributions $\mathbb{P}(y_i|x_i, s_i)$, and
- the function nodes h_i represent the transition probabilities of the states $\mathbb{P}(s_{i+1}|s_i)$.

The message-passing rules between the nodes f_i and x_i are the sum and tanh law as in Sec. 11.5. The message passing rules between other nodes are summarized as follows:

- The message-passing rule from the nodes x_i to g_i is,

$$\Lambda_{x_i \to g_i} = \sum_{j=1}^{k} \Lambda_{f_{l(j)} \to x_i}, \tag{11.58}$$

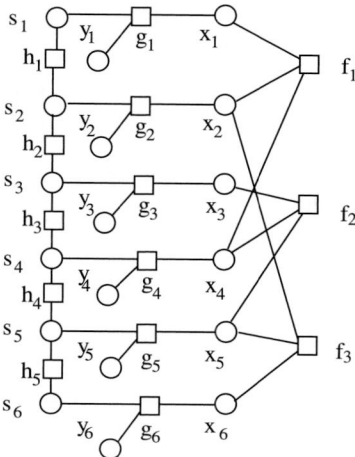

Fig. 11.15. The factor graph for a channel with state.

where $f_{l(j)}$ are all the k neighboring function nodes that correspond to check constraints.

• The message-passing rule from the nodes g_i to x_i is

$$
m_{g_i \to x_i}(0) = m_{s_i \to g_i}(b)\frac{1}{\sqrt{20\pi}}\exp\left\{-\frac{(y_i+1)^2}{20}\right\}
$$
$$
+ m_{s_i \to g_i}(g)\frac{1}{\sqrt{2\pi}}\exp\left\{-\frac{(y_i+1)^2}{2}\right\}, \qquad (11.59)
$$

$$
m_{g_i \to x_i}(1) = m_{s_i \to g_i}(b)\frac{1}{\sqrt{20\pi}}\exp\left\{-\frac{(y_i-1)^2}{20}\right\}
$$
$$
+ m_{s_i \to g_i}(g)\frac{1}{\sqrt{2\pi}}\exp\left\{-\frac{(y_i-1)^2}{2}\right\}. \qquad (11.60)
$$

• The message-passing rule from the nodes g_i to s_i is

$$
m_{g_i \to x_i}(b) = m_{x_i \to g_i}(0)\frac{1}{\sqrt{20\pi}}\exp\left\{-\frac{(y_i+1)^2}{20}\right\}
$$
$$
+ m_{x_i \to g_i}(1)\frac{1}{\sqrt{20\pi}}\exp\left\{-\frac{(y_i-1)^2}{20}\right\}, \qquad (11.61)
$$

$$m_{g_i \to x_i}(g) = m_{x_i \to g_i}(0) \frac{1}{\sqrt{2\pi}} \exp\left\{-\frac{(y_i+1)^2}{2}\right\}$$

$$+ m_{x_i \to g_i}(1) \frac{1}{\sqrt{2\pi}} \exp\left\{-\frac{(y_i-1)^2}{2}\right\}. \quad (11.62)$$

- The message-passing rule from the nodes s_i to g_i is

$$m_{s_i \to g_i}(b) = m_{h_{i-1} \to s_i}(b) m_{h_i \to s_i}(b), \quad (11.63)$$

$$m_{s_i \to g_i}(g) = m_{h_{i-1} \to s_i}(g) m_{h_i \to s_i}(g). \quad (11.64)$$

- The message-passing rule from the nodes s_i to h_i is

$$m_{s_i \to h_i}(b) = m_{h_{i-1} \to s_i}(b) m_{g_i \to s_i}(b), \quad (11.65)$$

$$m_{s_i \to h_i}(g) = m_{h_{i-1} \to s_i}(g) m_{g_i \to s_i}(g), \quad (11.66)$$

$$m_{s_i \to h_{i-1}}(b) = m_{h_i \to s_i}(b) m_{g_i \to s_i}(b), \quad (11.67)$$

$$m_{s_i \to h_{i-1}}(g) = m_{h_i \to s_i}(g) m_{g_i \to s_i}(g). \quad (11.68)$$

- The message-passing rule from the nodes h_i to s_i is

$$m_{h_i \to s_i}(b) = 0.1 m_{s_{i+1} \to h_i}(g) + 0.9 m_{s_{i+1} \to h_i}(b), \quad (11.69)$$

$$m_{h_i \to s_i}(g) = 0.1 m_{s_{i+1} \to h_i}(b) + 0.9 m_{s_{i+1} \to h_i}(g), \quad (11.70)$$

$$m_{h_i \to s_{i+1}}(b) = 0.1 m_{s_i \to h_i}(g) + 0.9 m_{s_i \to h_i}(b), \quad (11.71)$$

$$m_{h_i \to s_{i+1}}(g) = 0.1 m_{s_i \to h_i}(b) + 0.9 m_{s_i \to h_i}(g). \quad (11.72)$$

11.10. Thoughts for Practitioners

In the design and implementation of LDPC codes in practice, two issues need to be considered. The first issue is the convergence of the iterative decoding algorithm. For sufficiently long randomly constructed codes, the convergence of the decoding algorithm can be accurately predicted by density evolution. The second issue is that the low weight pseudo-codewords (stopping sets, trapping sets, etc.) should be avoided. Both the two issues have been discussed in this chapter. We hope that this chapter can serve as a useful information resource for practice.

There are two categories of LDPC code construction schemes, the random constructions, and the algebraic constructions. In this chapter, we focus largely on randomly constructed codes. For sufficiently long blocklengths (e.g. several thousands bits), the state-of-the-art randomly constructed codes have excellent capacity-approaching and error probability performance, and should be seriously considered for error-correction-coding practice. For short blocklengths (e.g. several hundred bits), the randomly constructed codes have inferior performance compared with algebraically constructed codes. Therefore, in these cases, algebraic LDPC codes are usually considered in practice.

11.11. Directions for Future Research

As we have seen in this chapter, several code construction schemes and performance analysis methods have been proposed. However, the code construction and performance analysis methods are still active research topics. There are still many unsolved problems in these areas. For example, there does not exist a method, which can generally predict the error probability performance accurately for finite-length codes for a wide range of channel models.

Another active research topic is the VLSI implementation of LDPC encoding and decoding. An interesting research direction is to design LDPC codes that facilitate efficient VLSI implementations.

11.12. Conclusion

We present a tutorial exposition of LDPC codes and the related performance analysis methods. We have reviewed two classes of LDPC code construction schemes, the random constructions and the algebraic constructions. We have discussed the factor graph approach for signal detection and estimation and the message-passing decoding from the viewpoint of factor graphs. We have reviewed two classes of performance analysis methods, density evolution and finite-length analysis based on pseudo-codewords. Finally, thoughts for practitioners are also provided.

11.13. Keywords

Low-density parity-check codes

Low-density parity-check codes are linear block codes with sparse parity-check matrices.

Factor graph

Factor graphs are graphical representation of global functions and how the global functions can be decomposed into the product of local functions. A factor graph is a bipartite graph with one partition consisting of function nodes and the other partition consisting of variable nodes. Each function node represents one local function. Each variable node represents one variable. The global function is the product of all the local functions. Edges indicate the involvement between functions and variables; that is, there is an edge between one variable node and one function node if and only if the variable is an argument of the local function.

Message-passing algorithm

The message-passing algorithm is a parallel algorithm for marginal distribution calculation.

Density evolution

Density evolution is an asymptotic method for calculating the limiting bit error probabilities, after certain decoding iterations, as the blocklengths go to infinity.

Pseudo-codewords

A pseudo-codeword is a valid configuration on a computational tree.

Stopping sets

A stopping set $\mathcal{S}$ is a set of variable nodes, such that, for each neighboring check node c of $\mathcal{S}$, there exist at least two edges with one end being c and the other end being in $\mathcal{S}$.

Trapping sets

An (a, b) trapping set $\mathcal{S}$ is a set of variable nodes, such that, the cardinality of the set $\mathcal{S}$ is a; and there exist exactly b check nodes c, such that c is connected to $\mathcal{S}$ by an odd number of edges.

Tanner graph

Tanner graphs are graphical representations for LDPC codes. A Tanner graph is a bipartite graph with one partition consisting of all variable nodes and the other partition consisting of all check nodes. Each variable node represents one codeword bit and each check node represents one parity-check constraint. There exists an edge between a variable node and a check node if and only if the corresponding bit is involved in the parity-check constraint.

Loopy propagation

Loopy propagation is a heuristic version of the message-passing algorithm. The loopy propagation message-passing algorithm calculates the messages using the same set of rules of the message-passing algorithms on tree-structured factor graphs. The calculated results are only approximate values of marginal distributions.

Concentration theorem

The concentration theorem shows that, as the blocklengths go to infinity, the limiting values of bit error probabilities after a fixed number of decoding iterations go to the bit error probabilities of infinitely long codes without cycles.

Binary erasure channel

Binary erasure channel is a channel model with binary inputs and outputs 0, 1, or e, where e stands for a missing bit. The channel parameter ξ is the probability that a transmitted bit is lost. The channel outputs a transmitted bit correctly with probability $1 - \xi$, and outputs e with probability ξ.

Standard random construction

The standard random construction is a LDPC code construction scheme, where codes are randomly chosen from standard code ensembles.

References

1. M. Ardakani and F. R. Kschischang, A more accurate one-dimensional analysis and design of irregular LDPC codes, *IEEE Trans. Commun.*, **52**, 12, (December 2004), 2106–2114.
2. S. Y. Chung, T. Richardson and R. Urbanke, Analysis of sum–product decoding of low-density parity-check codes using a Gaussian approximation, *IEEE Trans. Inform. Theor.* **47**, 2, (February 2001), 657–670.
3. S. Chung, G. Forney Jr., T. J. Richardson and R. Urbanke, On the design of low-density parity-check codes within 0.0045 dB of the Shannon limit, *IEEE Communication Letters* **5**, 2, (February 2001), 58–60.
4. C. Di, D. Proietti, I. Telatar, T. J. Richardson and R. Urbanke, Finite-length analysis of low-density parity-check codes on the binary erasure channel, *IEEE Trans. on Inform. Theory* **48**, 6, (June 2002), 1570–1579.
5. J. Feldman, Decoding error-correcting codes via linear-programming, Ph.D. Thesis, Massachusetts Institute of Technology, Cambridge, MA, (2003).
6. J. Feldman, M. J. Wainwright and D. R. Karger, Using linear programming to decode binary linear codes, *IEEE Trans. Inform. Theor.* **51**, 3, (2005), 954–972.

7. B. J. Frey, R. Koetter and A. Vardy, Signal space characterization of iterative decoding, *IEEE Trans. Inform. Theor.*, (February 2001).

8. G. D. Forney, R. Koetter, F. R. Kschischang and A. Reznik, On the effective weights of pseudocodewords for codes defined on graphs with cycles, *Codes, Syst. Graph. Model.* (Springer-Verlag, New York, 2001), pp. 102–112,

9. R. G. Gallager, *Low-density parity-check codes* (MIT Press, 1963.)

10. J.-L. Kim, U. N. Peled, I. Perepelitsa, V. Pless and S. Friedland, Explicit construction of families of LDPC codes with no 4-cycles, *IEEE Trans. Inform. Theor.* **50**, (October 2004), 2378–2388.

11. Y. Kou, S. Lin and M. P. C. Fossorier, Low-density parity-check codes based on finite geometries, a rediscovery and new results, *IEEE Trans. Inform. Theor.* **47**, 7, (November 2001), 2711–2736.

12. F. R. Kschischang, B. J. Frey, and H. Loeliger, Factor graphs and the sum–product algorithm", *IEEE Trans. Inform. Theor.* **47**, 2, (February 2001), 498–519.

13. M. Luby, M. Mitzenmacher, M. Shokrollahi and D. Spielman, Improved low-density parity-check codes using irregular graphs and belief propagation, *Proceedings on IEEE Int. Symp. Inform. Theory.*, (1998).

14. M. Luby, M. Mitzenmacher, A. Shokrollahi and D. Spielman, Efficient erasure correction codes, *IEEE Trans. Inform. Theor.* **47**, 2, (February 2001), 569–584.

15. M. Luby, M. Mitzenmacher, M. Shokrollahi and D. Spielman, Improved low-density parity-check codes using irregular graphs, *IEEE Trans. Inform. Theor.* **47**, 2, (February 2001), 585–598.

16. X. Ma and E.-H. Yang, Constructing LDPC codes by 2-lifts, in *Proc. 2007 IEEE Int. Symposium on Information Theory* (ISIT 2007), Nice, France, 24–29, (June 2007).

17. D. MacKay and M. Postol, Weakness of Margulis and Ramanujan–Margulis low-density parity check codes, *Electron. Notes Theor. Comput. Sci.* **74** (2003).

18. T. Richardson and R. Urbanke, The capacity of low-density parity-check codes under message-passing decoding, *IEEE Trans. Inform. Theor.* **47**, 2, (February 2001), 599–618.

19. T. Richardson, M. Shokrollahi and R. Urbanke, Design of capacity-approaching irregular low-density parity-check codes, *IEEE Trans. Inform. Theor.* **47**, 2, (February 2001), 619–637.

20. T. Richardson, Error floors of LDPC codes, in *Proc. 41st Annual Allerton Conf. Comm., Contr. and Computing*, Monticello, IL, (2003).

21. R. M. Tanner, A recursive approach to low-complexity codes, *IEEE Trans. Inform. Theor.* **27**, 5, (September 1981), 533–547.

22. J. Thorpe, K. Andrews and S. Dolinar, Methodologies for designing LDPC codes using protographs and circulants, *Proc. IEEE Symp. on Inform. Theory*, (2004), p. 238.

23. B. Vasic and O. Milenkovic, Combinatorial construction of low-density parity-check codes, *IEEE Trans. Inform. Theor.* **50**, 6, (June 2004), 1156–1176.

24. P. O. Vontobel and R. Koetter, Graph-cover decoding and finite-length analysis of message-passing iterative decoding of LDPC codes, manuscript accepted for *IEEE Transactions on Information Theory*, (2007).
25. N. Wiberg, Ph.D. Thesis, Linkoping University, (1996).
26. S. Wicker and S. Kim, *Fundamentals of Codes, Graphs, and Iterative Decoding* (Kluwer Academic Publishers), (2003).
27. R. Smarandache and P.O. Vontobel, On regular quasi-cyclic LDPC Codes from binomials, *Proc. IEEE Intl. Symp. Information Theory*, (Chicago, 2004), p. 274.

Chapter 12

VARIABLE-LENGTH CODES AND
FINITE AUTOMATA

MARIE-PIERRE BÉAL*, JEAN BERSTEL*,
BRIAN H. MARCUS†, DOMINIQUE PERRIN*,
CHRISTOPHE REUTENAUER‡
and PAUL H. SIEGEL¶

*Institut Gaspard-Monge (IGM), Université Paris-Est, France

†University of British Columbia, Vancouver, Canada

‡LaCIM, Université du Québec à Montréal, Canada

¶Department of Electrical and Computer Engineering, UCSD, USA

The aim of this chapter is to present, in appropriate perspective, some selected topics in the theory of variable-length codes. One of the domains of applications is lossless data compression. The main aspects covered include optimal prefix codes and finite automata and transducers. These are a basic tool for encoding and decoding variable-length codes. Connections with codes for constrained channels and sources are developed in some detail. Generating series are used systematically for computing the parameters of encodings such as length and probability distributions. The chapter contains numerous examples and exercises with solutions.

12.1. Introduction

Variable-length codes occur frequently in the domain of data compression. Historically, they appeared at the beginning of modern information theory with the seminal work of Shannon. One of the first algorithmic results is the construction, by Huffman, of an optimal variable-length code for a given weight distribution. Although their role in data communication has been limited by their weak tolerance to faults, they are nonetheless commonly used in contexts where error handling is less critical or is treated by other methods.

Variable-length codes are strongly related to automata, and one of the aims of this chapter is to highlight connections between these domains. Automata are labeled graphs, and their use goes beyond the field of coding. Automata can be used to implement encoders and decoders, such as for compression codes, modulation codes, and convolutional error correcting codes.

The use of variable-length codes in data compression is widespread. Huffman's algorithm is still frequently used in various contexts, and under various forms, including in its adaptive version. In particular, Huffman codes are frequently used in the compression of motion pictures. In another setting, search trees are strongly related to ordered prefix codes, and optimal ordered prefix codes, as constructed later, correspond to optimal binary search trees.

Coding for constrained channels is required in the context of magnetic or optical recording. The constraints that occur can be represented by finite automata, and a coding method makes use of finite transducers. In this context, the constraints are defined by an automaton, which in turn is used to define a state-dependent encoder. Even if this encoder operates at fixed rate, it can also be considered as a memoryless encoder based on two variable-length codes with the same length distribution, that is, with the same number of words for each length.

Although convolutional error correcting codes are fixed-length codes, their analysis involves use of finite automata because encoders and decoders are described in terms of labeled graphs.

Specific properties of automata correspond to properties of variable-length codes. Typically, unambiguity in automata corresponds to unique decipherability.

Variable-length codes are also used for the representation of natural languages. They play a role, for instance, in phonetic transcription of languages and in the transformation from text to speech, or in speech recognition. We will mention examples in Sec. 12.10.

The mathematical relationship between codes and automata is very deep, as shown in early pioneering investigations by Schützenberger. He discovered and developed a new branch of algebra relating unique decipherability of codes with the theory of semigroups. There are still difficult open problems in the theory of variable-length codes. One of them is the commutative equivalence conjecture. It has practical significance in relation with optimal coding. We will discuss this and other open problems in Sec. 12.13.

The material covered by the chapter is as follows. We start with a few definitions and examples, and then address the problem of constructing optimal codes under various types of constraints. Typically, we consider alphabetic coding under cost conditions.

We study in detail prefix codes used for representing integers, such as Elias and Golomb codes.

Automata and transducers are introduced insofar as coding and decoding operations are concerned. These are applied to two special domains, namely coding with constraints on channels and constraints on sources. They are important in applications, which are described here.

Reversible codes, also called bifix codes, have both practical significance and deep mathematical properties, which we only sketch here.

The final section is concerned with synchronization. This is important in the context of error recovery, and we present very recent theoretical results such as the road coloring theorem.

The chapter is written at a level accessible to non-specialists. There are few formal proofs, but key algorithms are described in considerable detail and many illustrative examples are given. Exercises that reinforce and extend the material are given at the end of most sections, and sketches of solutions are provided. Some elementary questions and answers are also included.

12.2. Background

The topic of this chapter is an introduction to some syntactic and algorithmic problems of data transmission. In this sense, it is connected with three main fields:

(1) coding and information theory;
(2) automata and formal language theory;
(3) algorithms.

In this section, we describe these connections, their historical roots and the notions to which they relate.

The relationship between codes and automata can be traced back to Shannon, who used labeled graphs to represent information sources. Later, the notion of *information lossless* machine was introduced as a model for reversible encoding [44]. These are the unambiguous transducers defined below. The term lossless has remained in common use with the notion of *lossless methods* for data compression. The main motivation for studying variable-length codes is in data compression. In many coding systems, encoding by variable-length codes is used in connection with other coding components for error correction. The search for efficient data compression methods leads to algorithmic problems such as the design of optimal prefix codes under various criteria.

One of the main tools for encoding and decoding as presented here is finite automata and transducers. In this context, finite automata are particular cases of more general models of machines, such as pushdown automata that can use an auxiliary stack and, more generally, Turing machines that are used to define the notion of computability. The theory of automata has also developed independently of coding theory with motivations in algorithms on strings of symbols, in the theory of computation and also as a model for discrete processes.

The basic result of automata theory is *Kleene's theorem* asserting the equivalence between finite automata and regular expressions and providing algorithms to convert from automata to regular expressions and conversely. This conversion is actually used in the context of convolutional codes to compute the *path weight enumerator* (also called the *transfer function*) of a convolutional code [48].

Inside the family of finite automata, several subclasses have been studied, which correspond to various types of restrictions. An important one is the class of *aperiodic* automata, which contains the classes of local automata frequently used in connection with coding, in particular with sliding block encoders and decoders.

One of the possible extensions of automata theory is the use of multiplicities, which can be integers, real numbers, or elements of other semirings (see [12, 18]). The ordinary case corresponds to the Boolean semiring with just two elements $0, 1$. This leads to a theory in which sets of strings are replaced by functions from strings to a semiring. This point of view has in particular the advantage of allowing the handling of *generating series* and gives a method, due to Schützenberger, to compute them. We will often use this method to compute generating series.

The well-known Huffman algorithm, described below, is the ancestor of a family of algorithms used in the field of information searching. Indeed, it can be used to build search trees as well as optimal prefix codes for source compression. The design and analysis of search algorithms is part of an important body of knowledge encompassing many different ideas and methods, including, for example, the theory of hashing functions (see [43]). Text processing algorithms find application in a variety of domains, ranging from bioinformatics to the processing of large data sets such as those maintained by Internet search engines (see [46] for an introduction).

The topic of coding with constraints is related to symbolic dynamics, which is a field in its own right. Its aim is to describe dynamical systems and mappings between them. Codes for constrained channels are a particular case of these mappings (see [45]).

12.3. Thoughts for Practitioners

In this chapter, a wide variety of variable-length coding techniques are presented. We consider source and channel models that take into account the statistical properties of the source and the costs associated with transmission of channel symbols. Given these models, we define a measure of code optimality by which to evaluate the code design algorithms. The resulting families of variable-length codes are intended for use in a range of practical applications, including image and video coding, speech compression, magnetic and optical recording, data transmission, natural language representation, and tree search algorithms.

In practice, however, there are often system-related issues that are not explicitly reflected in the idealized source and channel models, and therefore are not taken into account by the code design algorithms. For example, there are often tradeoffs between the efficiency of the code and the complexity of its implementation in software or hardware. Encoding and decoding operations may be subject to latency constraints or a requirement for synchronous input–output processing. There is often a need for resilience against errors introduced by the channel, as well as robustness in the presence of variability in the source and channel characteristics. There may also be a system interface that dictates the incorporation of specific code properties or even a specific code. Finally, intellectual property concerns, such as the existence of patents on certain codes or coding methods, can play a role in practical code design. Such realities provide a challenge to the coding practitioner in applying the various design methods, as well as a stimulus for further research and innovation.

To illustrate some of these points, we examine two applications where variable-length coding has found pervasive use — data compression and digital magnetic recording.

Codes for data compression. In Sec. 12.6, we present the classical Huffman algorithm for designing an optimal prefix code for a memoryless source with specified symbol statistics and equal channel symbol costs. Codes produced by this algorithm and its variants have been extensively employed in data compression applications. Practical system issues are often addressed during the code design process, or by using a modification of the standard design approach.

As will be described later in the chapter, for a binary channel alphabet, the Huffman algorithm builds a binary code tree from the bottom up by combining two nodes with the smallest probabilities. If there are multiple possibilities for selecting such pairs of nodes, all choices lead to codes with the same average codeword length. However, this is not true of the variance of the codeword lengths, and a large variance could have an impact on the implementation complexity if the code is incorporated into a data transmission system that calls for a constant transmission rate. This problem can be mitigated if the practitioner follows a simple rule for judiciously combining nodes during the generation of the code tree, resulting in a Huffman code with the smallest possible variance. There are also variants of the Huffman coding that help to control the maximum length of a codeword, a parameter that also may influence implementation complexity of the code.

Another practical consideration in applying Huffman codes may be the ease of representing the code tree. The class of *canonical* Huffman codes has a particularly succinct description: the codewords can be generated directly from a suitably ordered list of codeword lengths. Canonical codes are also very amenable to fast decoding and are of particular interest when the source alphabet is large. Fortunately, a code designed using the standard Huffman algorithm can be directly converted into a canonical code.

The practitioner may also encounter situations where the source statistics are not known in advance. In order to deal with this situation, one can use *adaptive* Huffman coding techniques. Application of adaptive coding, though, requires careful attention to a number of implementation related issues.

Huffman codes are generally not resilient to channel symbol errors. Nevertheless they have some inherent synchronization capabilities and, for certain length distributions, one can design synchronized Huffman codes. Typically, though, in order to ensure recovery within a reasonable time after a channel error, substantial modifications to the coding scheme are necessary.

Despite the availability of more efficient data compression methods, such as arithmetic coding, Huffman coding and its variants continue to play a role in many text and multimedia compression systems. They are relatively effective, simple to implement, and, as just discussed, they offer some flexibility in coping with a variety of practical system issues. For further details, see, for example, [58].

Codes for digital magnetic recording. In Sec. 12.9, we consider the problem of encoding binary source sequences into binary channel sequences in which there are at least d and at most k 0's between consecutive 1's, for specified $0 \leq d < k$. This $[d, k]$-*constraint* is used as a channel model in magnetic recording.

One simple approach to encoding a binary source sequence into the $[d, k]$-constraint uses the idea of a *bit-stuffing* encoder. The bit-stuffing encoder generates a code sequence by inserting extra bits into the source sequence to prevent violations of the $[d, k]$-constraint. It uses a counter to keep track of the number of consecutive 0's in the generated sequence. When the number reaches k, the encoder inserts a 1 followed by d 0's. Whenever the source bit is a 1, the encoder inserts d 0's. The decoder is correspondingly simple. It also keeps track of the number of consecutive 0's. Whenever the number reaches k, it removes the following $d + 1$ bits (the 1 and d 0's that had been inserted by the encoder). When the decoder encounters a 1, it removes the following d bits (the d 0's that had been inserted by the encoder). The bit-stuffing encoder can also be recast as a variable-length code, as shown for the special case $[d, k] = [2, 7]$ in Fig. 12.1. A source sequence can be uniquely parsed into a sequence of the source words shown, possibly followed by a prefix of a word. Each source word is then encoded into a corresponding channel word according to the table. Although the encoding and decoding operations are extremely simple conceptually, and the decoder resynchronizes with only a moderate delay following an erroneous channel bit, this code is not suitable for use in a disk drive because it does not have fixed encoding and decoding rate.

source	channel
1	001
01	0001
001	00001
0001	000001
00001	0000001
00000	00000001

Fig. 12.1.   Variable-length [2,7] bit-stuffing code.

source	channel
10	0100
11	1000
000	000100
010	100100
011	001000
0010	00100100
0011	00001000

Fig. 12.2.   The $[2, 7]$ Franaszek code.

In Sec. 12.9, we present another variable-length $[2, 7]$ code, the Franaszek code, whose encoder mapping is shown in Fig. 12.2. This encoder has a fixed encoding rate of $1/2$, since each source word is mapped to a codeword with twice its length. In fact, as shown in Sec. 12.9, this code can be implemented by a rate 1:2, 6-state encoder that converts each source bit synchronously into two channel bits according to simple state-dependent rules. The decoder is also synchronous, and it decodes a pair of channel bits into a single source bit based upon the contents of an 8-bit window containing the channel bit pair along with the preceding pair and the next two upcoming pairs. This sliding-window decoding limits to four bits the propagation of decoding errors caused by an erroneous channel bit.

This code is also efficient, in the sense that the theoretical upper bound on the rate of a code for the $[2, 7]$ constraint is approximately 0.5172. Moreover, among $[2, k]$-constraints, the $[2, 7]$-constraint has the smallest k constraint that can support a rate $1/2$ code. This code was patented by IBM and was extensively used in its commercial disk drive products.

The construction of the Franaszek code involved certain choices along the way. For example, a different choice of the assignment of source words to channel codewords would potentially affect the implementation complexity as well as the worst-case decoder error propagation. Virtually all code design methodologies share this characteristic, and the practitioner has to exercise considerable technical insight in order to construct the best code for the particular situation.

For example, using the state-splitting algorithm discussed later in the chapter, one can construct another rate 1:2 code for the $[2, 7]$-constraint that requires only five encoder states; however, the resulting maximum error propagation of the decoder is increased to five bits. It is also possible to construct fixed-rate encoders for the [2,7]-constraint that have a higher code rate than the Franaszek code; however, the encoder and decoder implementation will very likely be more complex.

While some of the earliest $[d, k]$-constrained codes designed for disk drives were standardized by the interface between the drive and the host computer drive controller, the encoding and decoding functions eventually migrated into the drive itself. This allowed coding practitioners to exercise their creativity and invent new codes to meet their particular system requirements. For further details about practical constrained code design, see, for example, [39, 47].

12.4. Definitions and Notation

We start with some definitions and notation on words and languages.

Some notation. Given a finite set A called the *alphabet*, each element of A is a *letter*, and a finite sequence of letters is called a *word*. The *length* of a word is the number of its letters. The length of a word w is denoted by $|w|$. The *empty word*, usually denoted by ε, is the unique word of length 0. The set of all words over the alphabet A is denoted by A^*. We denote by juxtaposition the *concatenation* of words. If w, x, y, z are words, and if $w = xy$, then x is a *prefix*, and y is a *suffix* of w. If $w = xyz$, then y is called a *factor* (or also a subword or a block) of w.

Given sets X and Y of words over some alphabet A, we denote by XY the set of all words xy, for x in X and y in Y. We write X^n for the n-fold product of X, with $X^0 = \{\varepsilon\}$. We denote by X^* the set of all words that are products of words in X, formally

$$X^* = \{\varepsilon\} \cup X \cup X^2 \cup \cdots \cup X^n \cup \cdots .$$

If $X = \{x\}$, we write x^* for $\{x\}^*$. Thus, $x^* = \{x^n \mid n \geq 0\} = \{\varepsilon, x, x^2, x^3, \ldots, \}$. The operations of union, set product, and star (*) are used to describe sets of words by so-called *regular expressions*.

Generating series. Given a set of words X, the *generating series* of the lengths of the words of X is the series in the variable z defined by

$$f_X(z) = \sum_{x \in X} z^{|x|} = \sum_{n \geq 0} a_n z^n,$$

where a_n is the number of words in X of length n. It is easily checked that

$$f_{X \cup Y}(z) = f_X(z) + f_Y(z), \tag{12.1}$$

whenever X and Y are disjoint, and

$$f_{XY}(z) = f_X(z)f_Y(z), \tag{12.2}$$

when the product XY is *unambiguous*, that is whenever $xy = x'y'$ with $x, x' \in X$, $y, y' \in Y$ we have $x = x'$, $y = y'$. We will also make use of an extension (introduced later) of these series to the case where words are equipped with a cost.

Encoding. We start with a *source* alphabet B and a *channel* alphabet A. Consider a mapping γ that associates with each symbol b in B a non-empty word over the alphabet A. This mapping is extended to words over B by $\gamma(s_1 \cdots s_n) = \gamma(s_1) \cdots \gamma(s_n)$. We say that γ is an *encoding* if it is *uniquely decipherable* (UD) in the sense that

$$\gamma(w) = \gamma(w') \Rightarrow w = w'$$

for each pair of words w, w'. In this case, each $\gamma(b)$ for b in B is a *codeword*, and the set of all codewords is called a *variable-length code* or VLC for short. We will call this a *code* for short instead of the commonly used term UD code.

Every property of an encoding has a natural formulation in terms of a property of the associated code, and *vice versa*. We will generally not distinguish between codes and encodings.

Let C be a code. Since C^n and C^m are disjoint for $n \neq m$ and since the products C^n are unambiguous, we obtain from (12.1) and (12.2) the following fundamental equation:

$$f_{C^*}(z) = \sum_{n \geq 0}(f_C(z))^n = \frac{1}{1 - f_C(z)}. \tag{12.3}$$

Alphabetic encoding. Suppose now that both B and A are ordered. The *order* on the alphabet is extended to words lexicographically, that is, $u < v$ if either u is a proper prefix of v or $u = zaw$ and $v = zbw'$ for some words z, w, w' and letters a, b with $a < b$.

An encoding γ is said to be *ordered* or *alphabetic* if $b < b'$ implies $\gamma(b) < \gamma(b')$.

A set C of non-empty words over an alphabet A is a *prefix code* (*suffix code*) if no element of C is a proper prefix (suffix) of another one. An encoding γ of B is called a *prefix encoding* if the set $\gamma(B)$ is a prefix code.

Prefix codes are especially interesting because they are instantaneously decipherable in a left to right parsing.

Table 12.1. A binary ordered encoding of
the five most frequent English words.

b	$\gamma(b)$
A	000
AND	001
OF	01
THE	10
TO	11

Example 12.1. The set B is composed of five elements in bijection with the five most common words in English, which are A, AND, OF, THE, and TO. An ordered prefix encoding γ of these words over the binary alphabet $\{0, 1\}$ is given in Table 12.1.

A common way to represent an encoding — one that is especially enlightening for prefix encodings — is by a rooted planar tree labeled in an appropriate way.

Assume the channel alphabet A has q symbols. The tree considered has nodes, which all have at most q children. The edge from a node to a child is labeled with one symbol of the channel alphabet A. If this alphabet is ordered, then the children are ordered accordingly from left to right. Some of the children may be missing.

Each path from the root to a node in the tree corresponds to a word over the channel alphabet, obtained by concatenating the labels on its edges. In this manner, a set of words is associated with a set of nodes in a tree, and conversely. If the set of words is a prefix code, then the set of nodes is the set of leaves of the tree.

Thus, a prefix encoding γ from a source alphabet B into words over a channel alphabet A is represented by a tree, and each leaf of the tree may in addition be labeled with the symbol b corresponding to the codeword $\gamma(b)$, which labels the path to this leaf. Figure 12.3 represents the tree associated with the ordered encoding γ of Table 12.1.

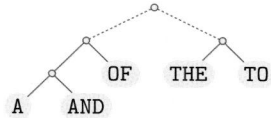

Fig. 12.3. The tree of the binary ordered encoding of the five most frequent English words.

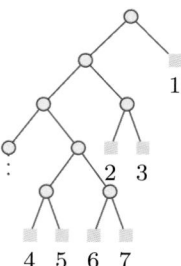

Fig. 12.4. The Elias code.

Example 12.2. The *Morse code* associates with each alphanumeric character a sequence of dots and dashes. For instance, A is encoded by ". –" and J is encoded by ". – – –." Provided each codeword is terminated with an additional symbol (usually a space, called a "pause"), the Morse code becomes a prefix code.

Example 12.3. There are many representations of integers. The unary representation of an integer n is composed of a sequence of n symbols 1. The usual binary representation of positive integers is exponentially more succinct than the unary representation, and thus is preferable for efficiency. However, it is not adapted to representation of sequences of integers, since it is not UD: for instance, 11010 may represent the number 26, or the sequence $6, 2$, or the sequence $1, 2, 2$. The *Elias code* [19] maps a positive integer into a word composed of its binary representation preceded by a number of zeros equal to the length of this representation minus one. For instance, the Elias encoding of 26 is 000011010. It is easily seen that the set of Elias encodings of positive integers is a prefix code. The corresponding tree is given in Fig. 12.4.

Example 12.4. There exist codes, even quite simple ones, which are neither prefix nor suffix. This is the case of the encoding of a, b, and c by 00, 10, and 100. To see that this is indeed UD, one considers the occurrences of 1. If the number of 0's following a 1 is even, this block is decoded as $ca \cdots a$; otherwise, as $ba \cdots a$.

12.5. Basic Properties of Codes

We start with stating a basic numerical inequality on codes. It gives a restriction on the distribution of lengths of codewords.

Kraft–McMillan inequality. For any code C over an alphabet A with k letters, one has the inequality, called the *Kraft–McMillan inequality* (see, for instance, [5])

$$\sum_{c \in C} k^{-|c|} \leq 1. \tag{12.4}$$

We prove the inequality below. Before that, we note that (12.4) is easy to prove for a finite prefix code. Indeed, the inequality above can also be written as

$$\sum_i a_i k^{-i} \leq 1,$$

where a_i is the number of code words in C of length i. Multiply both sides by k^n, where n is the maximal length of codewords. One gets $\sum_i a_i k^{n-i} \leq k^n$. The left-hand side counts the number of words of length n that have a prefix in C, and the inequality expresses the fact that each word of length n has at most one prefix in C.

For general codes, the inequality (12.4) can be proved as follows. Consider the *generating series* of the lengths of the words of the code C

$$f_C(z) = \sum_{c \in C} z^{|c|} = \sum_{n \geq 0} a_n z^n,$$

where a_n is the number of codewords of length n. Then, since C is a code, we have using (12.3):

$$f_{C^*}(z) = \frac{1}{1 - f_C(z)}.$$

Set $f_{C^*}(z) = \sum_{n \geq 0} b_n z^n$. Since C^* is a subset of A^*, one has $b_n \leq k^n$. Thus, the radius of convergence of $f_{C^*}(z)$ is at least equal to $1/k$. Since $f_C(r)$ is increasing for real positive r and $f_C(0) = 0$, the radius of convergence of $f_{C^*}(z)$ is precisely the positive real number r such that $f_C(r) = 1$. Thus, $f_C(1/k) \leq 1$. This proves the Kraft–McMillan inequality.

There is a converse statement for the Kraft–McMillan inequality: For any sequence $\ell_1, \ldots, \ell_n$ of positive integers such that $\sum_i k^{-\ell_i} \leq 1$, there exists a prefix code $C = \{c_1, \ldots, c_n\}$ over A such that $|c_i| = \ell_i$.

This can be proved by induction on n as follows. It is clearly true for $n = 1$. Suppose that $n > 1$ and consider $\ell_1, \ldots, \ell_n$ satisfying $\sum_{i=1}^n k^{-\ell_i} \leq 1$. Since also $\sum_{i=1}^{n-1} k^{-\ell_i} \leq 1$, there exists by induction hypothesis a prefix code

$C = \{c_1, \ldots, c_{n-1}\}$ such that $|c_i| = \ell_i$ for $1 \le i \le n-1$. We multiply both sides of the inequality $\sum_{i=1}^{n-1} k^{-\ell_i} \le 1$ by k^{ℓ_n}, and we obtain

$$\sum_{i=1}^{n-1} k^{\ell_n - \ell_i} \le k^{\ell_n} - 1. \tag{12.5}$$

Each of the terms $k^{\ell_n - \ell_i}$ of the left-hand side of (12.5) counts the number of words of length $\ell_n - \ell_i$, and can be viewed as counting the number of words of length ℓ_n with fixed prefix c_i of length ℓ_i. Since the code C is prefix, the sets of words of length ℓ_n with fixed prefix c_i are pairwise disjoint; hence, the left-hand side of (12.5) is the number of words of length ℓ_n on the alphabet A, which have a prefix in C. Thus, (12.5) implies that there exists a word c_n of length ℓ_n over the alphabet A, which does not have a prefix in C. The set $\{c_1, \ldots, c_n\}$ is then a prefix code. This proves the property.

Entropy.   Consider a source alphabet B. We associate with each symbol $b \in B$ a weight which we denote by $\text{weight}(b)$. For now, we assume that B is finite. The symbol weights are often *normalized* to sum to 1, in which case they can be interpreted as probabilities. The *entropy* of the source $B = \{b_1, \ldots, b_n\}$ with probabilities $p_i = \text{weight}(b_i)$ is the number

$$H = -\sum_{i=1}^{n} p_i \log p_i,$$

where log is the logarithm to base 2. Actually, this expression defines what is called the entropy of order 1. The same expression defines the entropy of order k, when the p_i's are the probabilities of the blocks of length k and n is replaced by n^k.

Channel.   In the context of encoding the symbols of a source alphabet B, we consider a channel alphabet A with a *cost*, denoted $\text{cost}(a)$, associated with each channel symbol $a \in A$. The cost of a symbol is a positive integer. The symbol costs allow us to consider the case where the channel symbols have non-uniform lengths, and the cost of each symbol can be interpreted as the time required to send the symbol. A classic example is the alphabet composed of two symbols $\{., -\}$, referred to as *dot* and *dash*, with costs 1 and 2, respectively. This alphabet is sometimes referred to as the *telegraph channel.*

The *channel capacity* is $\log 1/\rho$ where ρ is the real positive root of

$$\sum_{a \in A} z^{\text{cost}(a)} = 1.$$

In the case of an alphabet with k symbols, each having cost equal to 1, this reduces to $\rho = 1/k$. In the case of the telegraph channel, the capacity is the positive root of $\rho + \rho^2 = 1$, which is $\rho \approx 0.618$.

The cost of a word w over the alphabet A is denoted by $\text{cost}(w)$. It is by definition the sum of the costs of the letters composing it. Thus,

$$\text{cost}(a_1 \cdots a_n) = \text{cost}(a_1) + \cdots + \text{cost}(a_n).$$

We extend the notation of generating series as follows. For a set of words, X, denote by

$$f_X(z) = \sum_{x \in X} z^{\text{cost}(x)}$$

the *generating series of the costs* of the elements of X. For convenience, the cost function is omitted in the notation. Note that if the cost function assigns to each word its length, the generating series of costs reduces to the generating series of lengths considered earlier. Equations (12.1)–(12.3) hold for general cost functions.

For a code C over the alphabet A, the following inequality holds, which is a generalization of the Kraft–McMillan inequality (12.4):

$$\sum_{c \in C} \rho^{\text{cost}(c)} \leq 1. \tag{12.6}$$

The proof follows the same argument as for the Kraft–McMillan inequality using the generating series of the costs of the words of C

$$f_C(z) = \sum_{c \in C} z^{\text{cost}(c)} = \sum_{n \geq 0} a_n z^n,$$

where a_n is the number of words in C of cost equal to n. Similarly,

$$f_{A^*}(z) = \sum_{w \in A^*} z^{\text{cost}(w)} = \sum_{n \geq 0} b_n z^n,$$

where b_n is the number of words in A^* of cost equal to n. Now

$$f_{A^*}(z) = \frac{1}{1 - f_A(z)},$$

and the radius of convergence of $f_{A^*}(z)$ is the number ρ, which is the real positive root of $\sum_{a \in A} z^{\text{cost}(a)} = 1$.

Optimal encoding. Consider an encoding γ, which associates with each symbol b in B a word $\gamma(b)$ over the alphabet A. The *weighted cost* is

$$W(\gamma) = \sum_{b \in B} \text{weight}(b)\text{cost}(\gamma(b)).$$

When the weights are probabilities, Shannon's fundamental theorem on discrete noiseless channels [62] implies a lower bound on the weighted cost,

$$W(\gamma) \geq \frac{H}{\log 1/\rho}.$$

To show this, we set $B = \{b_1, \ldots, b_n\}$, $p_i = \text{weight}(b_i)$, and $q_i = \rho^{\text{cost}(\gamma(b_i))}$. We can then write $(\log \rho)W(\gamma) = \sum p_i \log q_i$. Invoking the well-known inequality $\ln x \leq x-1$, where $\ln$ denotes the natural logarithm, and applying (12.6), we find

$$(\log \rho)W(\gamma) - \sum p_i \log p_i = \sum p_i \log \frac{q_i}{p_i} = (\log e) \sum p_i \ln \frac{q_i}{p_i}$$

$$\leq (\log e) \sum p_i \left(\frac{q_i}{p_i} - 1\right)$$

$$\leq (\log e) \left(\sum q_i - 1\right) \leq 0,$$

from which the bound follows.

The *optimal encoding problem* is the problem of finding, for given sets B and A with associated weight and cost functions, an encoding γ such that $W(\gamma)$ is minimal.

The *optimal prefix encoding problem* is the problem of finding an optimal prefix encoding. Most research on optimal coding has been devoted to this second problem, both because of its practical interest and because of the conjecture that an optimal encoding can always be chosen to be prefix (see [41] and also the discussion below).

There is another situation that will be considered below, where the alphabets A and B are ordered and the encoding is required to be ordered.

In the case of *equal letter costs*, that is, where all $\text{cost}(a)$ are equal, the cost of a letter may be assumed to be 1. The cost of a codeword $\gamma(b)$ is then merely the length $|\gamma(b)|$. In this case, the weighted cost is called the *average length* of codewords. An optimal encoding can always be chosen to be prefix in view of the Kraft–McMillan inequality, as mentioned above.

Commutative equivalence. In the case of *unequal letter costs*, the answer to the problem of finding a prefix encoding, which has the same weighted cost as an optimal encoding, is not known. This is related to an important conjecture, which we now formulate. Two codes C and D are *commutatively equivalent* if there is a one-to-one correspondence between C and D such that any two words in correspondence have the same number of occurrences of each letter (that is, they are anagrams). Observe that the encodings corresponding to commutatively equivalent codes have the same weight, and therefore one is optimal if the other is.

Example 12.5. The code $C = \{00, 10, 100\}$ seen in Example 12.4 is neither prefix nor suffix. It is commutatively equivalent to the prefix code $D = \{00, 01, 100\}$ and to the suffix code $D' = \{00, 10, 001\}$.

It is conjectured that any finite *maximal* code (that is, a code that is not strictly contained in another code) is commutatively equivalent to a prefix code. This would imply that, in the case of maximal codes, the optimal encoding can be obtained with a prefix code. For a discussion, see [10, 13]. The conjecture is known to be false if the code is not maximal. A counter-example has been given by Shor [63]. Note that if equality holds in the Kraft–McMillan inequality (12.4), then the code must be maximal. Conversely, it can be shown (see, [11]) that for a finite maximal code, (12.4) is an equality.

12.6. Optimal Prefix Codes

In this section, we describe methods used to obtain optimal prefix codes under various constraints, such as equal or unequal letter costs, as well as equal or unequal letter weights, and finally encodings which are alphabetic or not. The different cases are summarized in Fig. 12.5, where the vertices are associated with the inventors of some corresponding algorithm. For instance, vertex 3 denotes the problem of finding an optimal alphabetic tree in the case of unequal weights and unequal letter costs, and it is solved by Itai's algorithm described later. All these algorithms, except Karp's algorithm, have polynomial running time. We consider first the two cases (vertices 4 and 1 in Fig. 12.5) of unequal weights without the constraint on the encoding to be alphabetic.

Unequal letter costs. The computational complexity of the optimal prefix encoding problem in the case of unequal letter costs (vertex 4 in Fig. 12.5) is

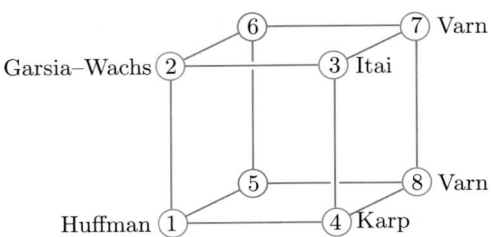

Fig. 12.5.   Various hypotheses. Front plane $(1, 2, 3, 4)$: unequal weights. Right plane $(3, 4, 7, 8)$: unequal letter costs. Top plane $(2, 3, 6, 7)$: alphabetic encodings. The two unnamed vertices $5, 6$ are easy special cases.

still unknown, in the sense that neither polynomial time algorithm is known for this, nor is it known whether the corresponding recognition problem (is there a code of cost $\leq m$?) is NP-complete. It has been shown to be reducible to an integer programming problem by Karp [41].

We explain how an optimal prefix code can be found by solving an integer programming problem. Let ρ be the positive real number such that $f_A(\rho) = 1$. Thus, $\log 1/\rho$ is the channel capacity. Recall that, for any code C, one has

$$f_C(\rho) \leq 1. \tag{12.7}$$

However, given a series $f(z)$, the inequality $f(\rho) \leq 1$ is not sufficient to imply the existence of a prefix code C such that $f(z) = f_C(z)$. For example, if the alphabet A has a single letter of cost 2, then $f_A(z) = z^2$, and so $\rho = 1$. The polynomial $f(z) = z$ satisfies $f(\rho) = 1$; however, there can be no codeword of cost 1.

Despite this fact, the existence of a code C with prescribed generating series of costs can be formulated in terms of solutions for a system of linear equations, as we describe now.

Let C be a prefix code over the channel alphabet A, with source alphabet B equipped with weights denoted $\text{weight}(b)$ for $b \in B$, and let P be the set of prefixes of words in C which do not belong to C. Set

$$f_A(z) = \sum_{i \geq 1} a_i z^i, \quad f_C(z) = \sum_{i \geq 1} c_i z^i, \quad f_P(z) = \sum_{i \geq 1} p_i z^i.$$

Here a_i is the number of channel symbols of cost i, c_i is the number of codewords of cost i, and p_i is the number of words in P of cost i. The following equality holds between the sets C, P, and A (see Exercise 12.6.2).

$$PA \cup \{\varepsilon\} = P \cup C.$$

Since the unions are disjoint, it follows that

$$c_1 + p_1 = a_1,$$
$$c_2 + p_2 = p_1 a_1 + a_2,$$
$$\vdots$$
$$c_n + p_n = p_{n-1} a_1 + \cdots p_1 a_{n-1} + a_n,$$
$$\vdots$$

(12.8)

Conversely, if c_i, a_i are non-negative integers satisfying these equations, there is a prefix code C such that $f_C(z) = \sum_{i \geq 1} c_i z^i$.

Thus, an optimal prefix code can be found by solving the problem of finding non-negative integers u_b for $b \in B$, which will be the costs of the codewords, and integers c_i, p_i for $i \geq 1$, which minimize the linear form $\sum_b u_b \text{weight}(b)$ such that Eq. (12.8) hold and with c_i equal to the number of b such that $u_b = i$.

There have been many approaches to partial solutions of the optimal prefix encoding problem [25, 49]. The most recent one is a polynomial time approximation scheme, which has been given in [29]. This means that, given $\epsilon > 0$, there exists a polynomial time algorithm computing a solution with weighted cost $(1 + \epsilon)W$, where W is the optimal weighted cost.

Equal letter costs. The case of equal letter costs (vertex 1 in Fig. 12.5) is solved by the well-known *Huffman algorithm* [37]. The principle of this algorithm in the binary case is as follows. Select two symbols b_1, b_2 in B with lowest weights, replace them by a fresh symbol b with weight $\text{weight}(b) = \text{weight}(b_1) + \text{weight}(b_2)$, and associate with b a node with children labeled b_1 and b_2. Then iterate the process. The result is a binary tree corresponding to an optimal prefix code. The complexity of the algorithm is $O(n \log n)$, or $O(n)$ if the weights are available in increasing order. The case where all weights are equal (vertex 5 in Fig. 12.5) is an easy special case.

Example 12.6. Consider the alphabets $B = \{a, b, c, d, e, f\}$ and $A = \{0, 1\}$, and the weights given in the following table:

	a	b	c	d	e	f
weight	2	2	3	3	3	5

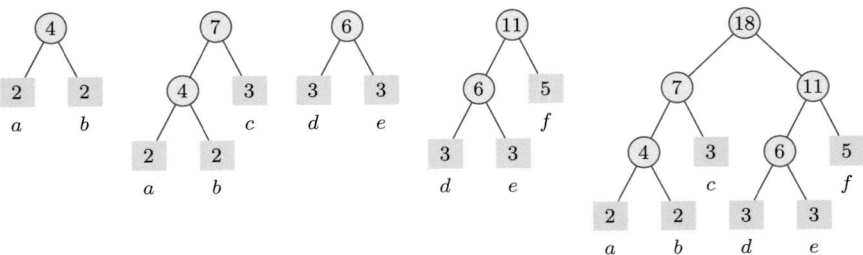

Fig. 12.6.   Computing an optimal Huffman encoding by combining trees.

The steps of the algorithm are presented below:

a b c d e f	ab c d e f	ab c de f	(ab)c de f	(ab)c (de)f	((ab)c)((de)f)
2 2 3 3 3 5	4 3 3 3 5	4 3 6 5	7 6 5	7 11	18

The corresponding trees are given in Fig. 12.6.

Alphabetic coding.   We suppose that both the source and the channel alphabets are ordered (these are vertices $2, 3, 6, 7$ in Fig. 12.5). Recall that an encoding γ is said to be *ordered* or *alphabetic* if $b < b' \Rightarrow \gamma(b) < \gamma(b')$. The *optimal alphabetic prefix encoding problem* is the problem of finding an optimal ordered prefix encoding.

Alphabetic encoding is motivated by searching problems. Indeed, a prefix code can be used as a searching procedure to retrieve an element of an ordered set. Each node of the associated tree corresponds to a query, and the answer to this query determines the subtree in which to continue the search.

Example 12.7. In the binary tree of Figure 12.3, one looks for an occurrence of an English word. The query associated with the root can be the comparison of the first letter of the word to the letter T.

In contrast to the non-alphabetic case, there exist polynomial-time solutions to the optimal alphabetic prefix encoding problem. It has been considered mainly in the case where the channel alphabet is binary. Again, there is a distinction between equal letter costs and unequal letter costs.

Example 12.8. Consider the alphabet $B = \{a, b, c\}$, with weight(a) = weight(c) = 1 and weight(b) = 4. Figure 12.7(a) shows an optimum tree

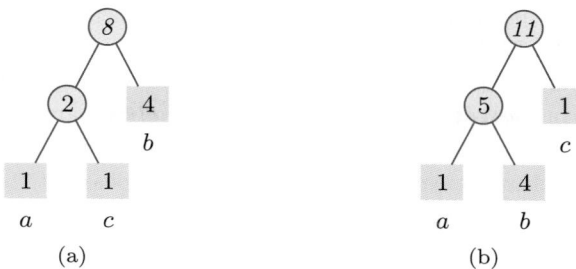

Fig. 12.7. Two trees for the given weights. Optimum coding tree (a) has weighted cost 8, it is optimal but not ordered. Optimum ordered tree (b) is ordered and has weighted cost 11.

for these weights, and Fig. 12.7(b) an optimum ordered tree. This example shows that Huffman's algorithm does not give the optimal ordered tree.

We consider now the case of equal letter costs, represented by vertex 2 in Fig. 12.5. Let $B = \{b_1, \ldots, b_n\}$ be an ordered alphabet with n letters, and let p_i be the weight of letter b_i. There is a simple algorithm for computing an optimal ordered tree based on dynamic programming. It runs in time $O(n^3)$ and can be improved to run in time $O(n^2)$ (see [43]).

We present a more sophisticated algorithm due to Garsia and Wachs [23]. The intuitive idea of the algorithm is to use a variant of Huffman's algorithm by grouping together pairs of elements with minimal weight, which are consecutive in the ordering. The algorithm can be implemented to run in time $O(n \log n)$.

Example 12.9. Consider the following weights for an alphabet of five letters:

	a	b	c	d	e
weight	25	20	12	10	14

The algorithm is composed of three parts. In the first part, called the *combination* part, one starts with the sequence of weights

$$p = (p_1, \ldots, p_n)$$

and constructs an optimal binary tree T' for a permutation $b_{\sigma(1)}, \ldots, b_{\sigma(n)}$ of the alphabet. The leaves, from left to right, have weights

$$p_{\sigma(1)}, \ldots, p_{\sigma(n)}.$$

In general, this permutation is not the identity; hence, the tree is not ordered, see Fig. 12.8(a).

The second part, the *level assignment*, consists of computing the levels of the leaves. In the last part, called the *recombination* part, one constructs a tree T, which has the weights $p_1, \ldots, p_n$ associated with its leaves from left to right, and where each leaf with weight p_i appears at the same level as in the previous tree T'. This tree is ordered, see Fig. 12.8(b).

Since the leaves have the same level in T and in T', the corresponding codewords have the same length, and therefore the trees T and T' have the same weighted cost. Thus, T is an optimal ordered tree.

We now give the details of the algorithm and illustrate it with this specific example. For ease of description, it is convenient to introduce some terminology. A sequence $(p_1, \ldots, p_k)$ of numbers is 2-*descending* if $p_i > p_{i+2}$ for $1 \le i \le k - 2$. Clearly a sequence is 2-descending if and only if the sequence of "two-sums" $(p_1 + p_2, \ldots, p_{k-1} + p_k)$ is strictly decreasing.

Let $p = (p_1, \ldots, p_n)$ be a sequence of (positive) weights. The *left minimal pair* or simply *minimal pair* of p is the pair (p_{k-1}, p_k), where $(p_1, \ldots, p_k)$ is the longest 2-descending chain that is a prefix of p. The index k is the *position* of the pair. In other words, k is the integer such that

$$p_{i-1} > p_{i+1} \ (1 < i < k) \quad \text{and} \quad p_{k-1} \le p_{k+1}$$

with the convention that $p_0 = p_{n+1} = \infty$. The *target* is the index j with $1 \le j < k$ such that

$$p_{j-1} \ge p_{k-1} + p_k > p_j, \ldots, p_k.$$

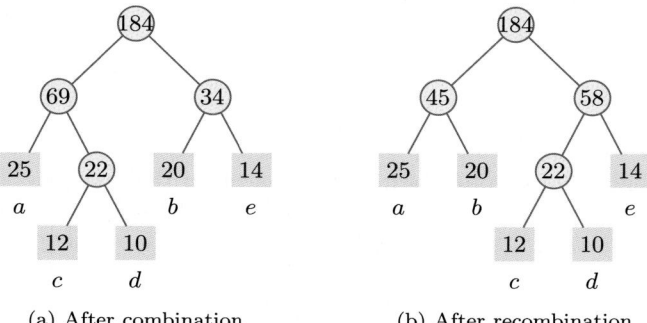

(a) After combination   (b) After recombination

Fig. 12.8.   The two steps of the algorithm: (a) the unordered tree obtained in the combination phase, and (b) the final ordered tree, obtained by recombination. Both have weighted cost 184.

Example 12.10. For $(14, 15, 10, 11, 12, 6, 8, 4)$, the minimal pair is $(10, 11)$, the target is 1, whereas for the sequence $(28, 8, 15, 20, 7, 5)$, the minimal pair is $(8, 15)$ and the target is 2.

The three phases of the algorithm work as follows. Let $(p_1, \ldots, p_n)$ be a sequence of weights.

Combination. Associate with each weight a tree composed of a single leaf. Repeat the following steps as long as the sequence of weights has more than one element:

(i) Compute the *left minimal pair* (p_{k-1}, p_k).
(ii) Compute the *target j*.
(iii) Remove the weights p_{k-1} and p_k.
(iv) Insert $p_{k-1} + p_k$ between p_{j-1} and p_j.
(v) Associate with $p_{k-1} + p_k$ a new tree with weight $p_{k-1} + p_k$, and which has a left (right) subtree the corresponding tree for p_{k-1} (for p_k).

Level assignment. Compute, for each letter b in B, the level of its leaf in the tree T'.

Recombination. Construct an ordered tree T in which the leaves of the letters have the levels computed by the level assignment.

Example 12.11. Consider again the following weights for an alphabet of five letters:

	a	b	c	d	e
weight	25	20	12	10	14

The initial sequence of trees is given in Fig. 12.9. The left minimal pair is 12, 10; its target is 2; hence, the leaves for c and d are combined into a tree which is inserted just to the right of the first tree, as shown in the left-hand side of Fig. 12.10. Now the minimal pair is $(20, 14)$ (there is an infinite weight at the right end); hence, the leaves for letters b and e are combined, and inserted at the beginning. The resulting sequence of trees is shown in the right-hand side of Fig. 12.10.

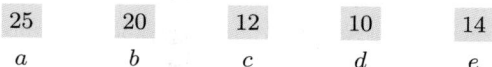

Fig. 12.9. The initial sequence of trees.

Fig. 12.10.   The next two steps.

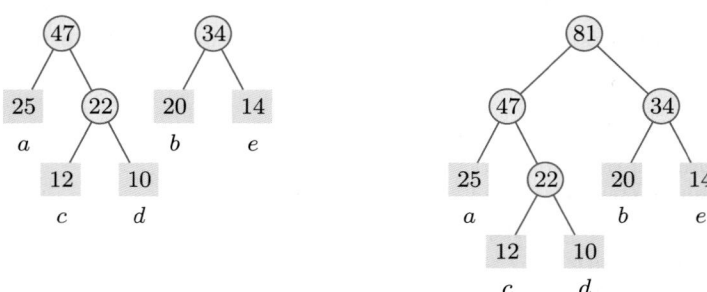

Fig. 12.11.   The two last steps of the combination part.

Next, the last two trees are combined and inserted at the beginning as shown in the left-hand side of Fig. 12.11, and finally, the two remaining trees are combined, yielding the tree shown in the right-hand side of the figure.

The tree T' obtained at the end of the first phase is not ordered. The prescribed levels for the letters of the example are as follows:

	a	b	c	d	e
level	2	2	3	3	2

The optimal ordered tree with these levels is given by recombination. It is the tree given in Fig. 12.8(b).

For a correctness proof, see [42, 43]. The time bound is given in [43]. The Garsia–Wachs algorithm is simpler than a previous algorithm due to Hu and Tucker [35], which was also described in the first edition of Knuth's book, preceding [43]. For a proof and a detailed description of the Hu–Tucker algorithm and variations, see [34, 36].

Alphabetic coding with unequal costs.   This is the most general case for alphabetic encoding (vertex 3 in Fig. 12.5). There is a dynamic programming algorithm due to Itai [40], which computes an optimal solution in polynomial time.

Given a source alphabet $B = \{1, \ldots, n\}$ with n symbols, and weights $\text{weight}(1), \ldots, \text{weight}(n)$, one looks for an optimal alphabetic encoding γ on the ordered channel alphabet A with costs $\text{cost}(a)$, for a in A. The weighted cost is $\sum_{i=1}^{n} \text{weight}(i)\text{cost}(\gamma(i))$. For convenience, the first (resp. the last) letter in A is denoted by α (resp. by ω). We also write $a + 1$ for the letter following a in the order on A.

Define $W_{a,b}[i, j]$ as the minimal weight of an alphabetic encoding for the symbols k with $i \leq k \leq j$, using codewords for which the initial symbol x satisfies $a \leq x \leq b$.

The following equations provide a method to compute the minimal weight $W_{\alpha,\omega}[1, n]$. First, for $a < b$, $i < j$,

$$W_{a,b}[i, j] = \min\{W_{a+1,b}[i, j], V_{a,b}[i, j], W_{a,a}[i, j]\}, \qquad (12.9)$$

where

$$V_{a,b}[i, j] = \min_{i \leq k < j} (W_{a,a}[i, k] + W_{a+1,b}[k + 1, j]).$$

This formula expresses the fact that either the first codeword does not start with the letter a, or it does, and the set of codewords starting with a encodes the interval $[i, k]$ for some $k < j$, or finally all codewords start with a. Next, for $i < j$,

$$W_{a,a}[i, j] = \text{cost}(a) \left(\sum_{k=i}^{j} \text{weight}(k) \right)$$
$$+ \min_{\substack{i \leq k < j \\ \alpha \leq x < \omega}} \{W_{x,x}[i, k] + W_{x+1,\omega}[k + 1, j]\}. \qquad (12.10)$$

In this case, all codewords start with the letter a. Moreover, the second letter cannot be the same for all codewords (otherwise this letter can be removed and this improves the solution). Finally, for $a \leq b$, the boundary conditions are:

$$W_{a,b}[i, i] = \min_{a \leq x \leq b}\{W_{x,x}[i, i]\}, \quad W_{x,x}[i, i] = \text{cost}(x)\text{weight}(i). \qquad (12.11)$$

The appropriate way to compute the W's is by increasing values of the difference $j - i$, starting with (12.11) and, for a fixed value of $j - i$, by increasing lengths of the source alphabet intervals, starting with (12.10), followed by (12.9).

This method gives an algorithm running in time $O(q^2 n^3)$ where q is the size of the channel alphabet. Indeed, each evaluation of (12.10) requires time $O((j - i)q)$, and each evaluation of (12.9) is done in time $O(j - i)$. Itai [40] has given an improvement of the algorithm that leads to a better bound of $O(q^2 n^2)$.

Example 12.12. Consider a five symbol source alphabet and a three letter channel alphabet $\{a, b, c\}$ with weights and costs as follows:

i	1	2	3	4	5
weight(i)	5	8	2	10	4

x	a	b	c
cost(x)	1	3	2

The algorithm computes the following tables:

$$
W_{a,a} = \begin{pmatrix} 5 & 34 & 48 & 85 & 115 \\ - & 8 & 22 & 54 & 84 \\ - & - & 2 & 34 & 56 \\ - & - & - & 10 & 32 \\ - & - & - & - & 4 \end{pmatrix}, \quad
W_{b,b} = \begin{pmatrix} 15 & 60 & 78 & 135 & 173 \\ - & 24 & 42 & 94 & 132 \\ - & - & 6 & 58 & 88 \\ - & - & - & 30 & 60 \\ - & - & - & - & 12 \end{pmatrix},
$$

$$
W_{c,c} = \begin{pmatrix} 10 & 47 & 63 & 110 & 144 \\ - & 16 & 32 & 74 & 108 \\ - & - & 4 & 46 & 72 \\ - & - & - & 20 & 46 \\ - & - & - & - & 8 \end{pmatrix}, \quad
W_{a,b} = \begin{pmatrix} 5 & 29 & 40 & 78 & 97 \\ - & 8 & 14 & 52 & 66 \\ - & - & 2 & 32 & 46 \\ - & - & - & 10 & 22 \\ - & - & - & - & 4 \end{pmatrix},
$$

$$
W_{b,c} = \begin{pmatrix} 10 & 31 & 47 & 89 & 123 \\ - & 16 & 28 & 62 & 88 \\ - & - & 4 & 26 & 52 \\ - & - & - & 20 & 38 \\ - & - & - & - & 8 \end{pmatrix}, \quad
W_{a,c} = \begin{pmatrix} 5 & 21 & 33 & 60 & 86 \\ - & 8 & 12 & 34 & 60 \\ - & - & 2 & 22 & 40 \\ - & - & - & 10 & 18 \\ - & - & - & - & 4 \end{pmatrix}.
$$

Hence, the minimal weight of an encoding is $W_{a,c}[1,5] = 86$. Since $W_{a,a}[1,2] + W_{b,c}[3,5] = W_{a,a}[1,3] + W_{b,c}[4,5] = 86$, there are, by (12.9), two optimal trees. Inspection of the matrices yields the trees given in Fig. 12.12.

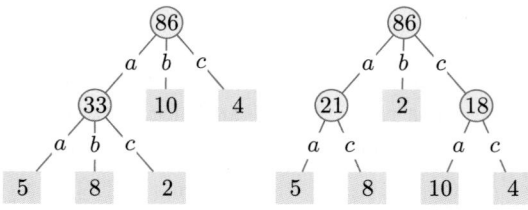

Fig. 12.12.   Trees built with Itai's algorithm.

Optimal encodings with equal weights. In the case where all source symbols have the same weight (vertices 5–8 in Fig. 12.5), this weight can be assumed to be 1. The weighted cost becomes simply

$$W(\gamma) = \sum_{b \in B} \text{cost}(\gamma(b)).$$

The prefix coding problem in this case is known as the *Varn coding problem*. It has an amazingly simple $O(n \log n)$ time solution [67].

We assume that A is a k-letter alphabet and that $n = q(k-1) + 1$ for some integer q. Hence, the prefix code (or the tree) obtained is complete with q internal nodes and n leaves. Varn's algorithm starts with a tree composed solely of its root, and iteratively replaces a leaf of minimal cost by an internal node, which has k leaves, one for each letter. The number of leaves increases by $k - 1$; hence, in q steps, one gets a tree with n leaves. Note that this solves also the cases numbered $5, 6$ in Fig. 12.5.

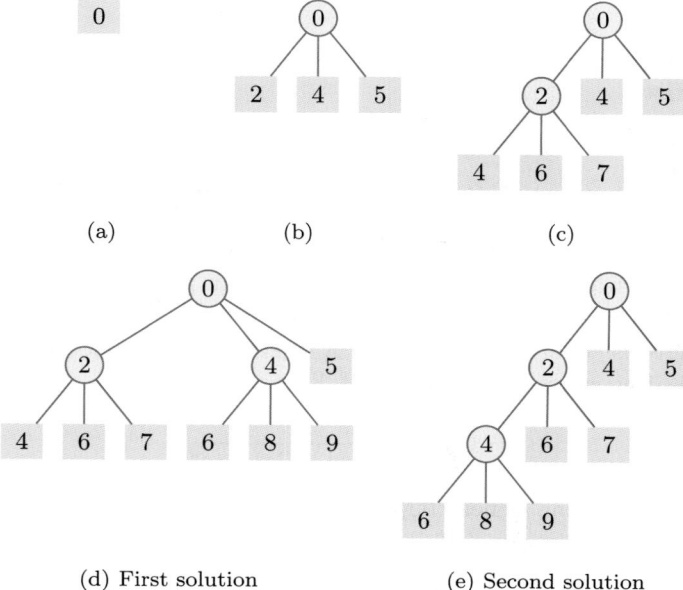

(a) (b) (c)

(d) First solution   (e) Second solution

Fig. 12.13. Varn's algorithm for seven words over a 3-letter alphabet. At each step, a leaf of minimal cost is replaced by a node with all possible leaves. There are two choices for the last step. Both give a minimal tree.

Example 12.13. Assume we are looking for a code with seven words over the ternary alphabet $\{a, b, c\}$, and that the cost for letter a is 2, for letter b is 4, and for letter c is 5. The algorithm starts with a tree composed of a single leaf, and then builds the tree by iteration. There are two solutions, both of cost 45, given in Fig. 12.13. Tree 12.13(d) defines the prefix code $\{aa, ab, ac, ba, bb, bc, c\}$, and tree 12.13(e) gives the code $\{aaa, aab, aac, ab, ac, b, c\}$.

12.6.1. *Exercises*

Exercise 12.6.1. Show that for any distribution p_i of probabilities, the sequence $\ell_i = \lceil \log 1/p_i \rceil$ satisfies the inequality $\sum 2^{-\ell_i} \le 1$. Conclude that for any source with equal letter costs, there is a prefix code with weighted cost $W \le H + 1$ where H is the entropy of the source with probabilities p_i.

Exercise 12.6.2. Let A be a k-letter alphabet. A k-ary tree is *complete* if each of its nodes has 0 or k children. A prefix code is complete if its tree is complete. Let C be a finite complete prefix code and let P be the set of prefixes of the words of C, which are not in C. Show that

$$PA \cup \{\varepsilon\} = P \cup C.$$

Deduce that

$$\mathrm{Card}(C) - 1 = \mathrm{Card}(P)(k - 1),$$

where, for a finite set S, $\mathrm{Card}(S)$ denotes the cardinality of S.

Exercise 12.6.3. For a non-empty binary word s of length p, denote by Q the set of words w of length strictly less than p such that sw has s as a suffix. Let X be the set of binary words which have s as a suffix but no other factor is equal to s. Show that X is a maximal prefix code and that the generating series of the lengths of X is

$$f_X(z) = \frac{z^p}{z^p + (1 - 2z)f_Q(z)}, \tag{12.12}$$

where $f_Q(z)$ is the generating series of the lengths of Q (called the *autocorrelation polynomial* of s).

Exercise 12.6.4. Show that, for $s = 101$, one has

$$f_X(z) = \frac{z^3}{1 - 2z + z^2 - z^3}.$$

Table 12.2. The cardinalities of the codes $C_{n,s}$.

	11	10	111	110	101	1111	1110	1010	1001
3	1	2	0	1	1				
4	1	3	1	2	1	0	1	0	1
5	2	4	1	4	2	1	2	2	2
6	3	5	2	7	4	1	4	3	3
7	5	6	4	12	7	2	8	4	6
8	8	7	7	20	12	4	15	9	11
9	13	8	13	33	21	8	28	18	21
10	21	9	24	54	37	15	52	32	39
11	34	10	44	88	65	29	96	60	73
12	55	11	81	143	114	56	177	115	136

Exercise 12.6.5. A set of binary words is said to be a *prefix synchronized code* if all words have the same length n and share a common prefix s which does not appear elsewhere in the sequences of codewords. For each s and n, there is a maximal set $C_{n,s}$, which satisfies this condition. For example, $C_{5,101} = \{10100, 10111\}$. Table 12.2 shows the cardinalities of some of the sets $C_{n,s}$.

Let U be the set of binary words u such that s is a proper prefix of u and us does not have s as a factor except as a prefix or a suffix. Show that

$$f_U(z) = \frac{2z - 1}{z^p + (1 - 2z)f_Q(z)} + 1,$$

where p is the length of s and $f_Q(z)$ is the autocorrelation polynomial of the word s. (Note: Use the fact that $s \cup \{0, 1\}X = X \cup Us$ where X is as in Exercise 12.6.3.) Show that for each $n \geq p$, $C_{n,s}$ is the set of words of length n in U.

Exercise 12.6.6. Let π be a Bernoulli distribution on the source alphabet B. A *Tunstall code* of order n is a maximal prefix code with n codewords over the alphabet B, which has maximal average length with respect to π. Such a code is used to encode the words of the source alphabet by binary blocks of length k for $n \leq 2^k$. For example, if $B = \{a, b\}$ and $\pi(a) = 0.8$, $\pi(b) = 0.2$, the code $C = \{aaa, aab, ab, b\}$ is a Tunstall code of order 4. Its average length is 2.44 and thus coding each word of C by a binary block of length 2 realizes a compression with rate $2/2.44 \approx 0.82$.

Show how Varn's algorithm can be used to build a Tunstall code. (Tunstall codes were introduced in [66], see also [58].)

12.7. Prefix Codes for Integers

Some particular codes are used for compression purposes to encode numerical data subject to a known probability distribution. They appear in particular in the context of digital audio and video coding. The data encoded are integers and thus these codes are infinite. We will consider several families of these codes, beginning with the Golomb codes introduced in [30]. We have already seen the Elias code, which belongs to one of these families.

Golomb codes. The *Golomb code* of order $m \geq 1$, denoted by G_m, is the maximal infinite prefix code:

$$G_m = 1^*0R_m.$$

Thus, words in G_m are composed of a (possibly empty) block of 1's, followed by 0, and followed by a word in R_m, where the prefix codes R_m are defined as follows. If $m = 2^k$ is a power of 2, then R_m is the set of all binary words of length k. In particular, R_1 is composed of the empty word only. For other values of m, the description is more involved. Set $m = 2^k + \ell$, with $0 < \ell < 2^k$. Setting $n = 2^{k-1}$,

$$R_m = \begin{cases} 0R_\ell \cup 1R_{2n} & \text{if } \ell \geq n, \\ 0R_n \cup 1R_{n+\ell} & \text{otherwise.} \end{cases}$$

The codes R_m for $m = 1$ to 7 are represented in Fig. 12.14. Note that, in particular, the lengths of the codewords differ at most by one.

The Golomb codes of order $1, 2, 3$ are represented in Fig. 12.15. The encoding of the integers is alphabetic. Note that, except possibly for the first level, there are exactly m words of each length. One way to define directly the encoding of an integer is as follows. Set $r = \lceil \log m \rceil$. Define the *adjusted binary representation* of an integer $n < m$ as its representation on $r-1$ bits if $n < 2^r - m$ and on r bits otherwise (adding 0's on the left if necessary).

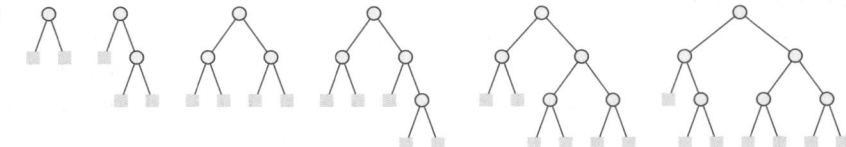

Fig. 12.14.   The sets R_1 to R_7.

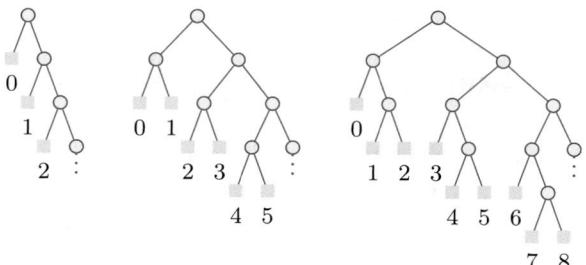

Fig. 12.15. The Golomb codes of orders $1, 2, 3$.

The encoding of the integer n in G_m is formed of n/m 1's followed by 0, followed by the adjusted binary representation of n modulo m.

A *geometric distribution* on the set of integers is given by

$$\pi(n) = p^n q, \tag{12.13}$$

for positive real numbers p, q with $p + q = 1$. Such a distribution may arise from *run-length encoding* where a sequence $0^n 1$ is encoded by n. If the source produces 0 and 1's independently with probability p and q, the probability of $0^n 1$ is precisely $\pi(n)$. This is of practical interest if p is large since then long runs of 0 are expected and the run-length encoding realizes a logarithmic compression.

We will show that for a source of integers with the geometric distribution corresponding to a given p, there is an integer $m \geq 1$ such that the Golomb code G_m is an optimal prefix code.

For this, consider, following [22], the integer m such that

$$p^m + p^{m+1} \leq 1 < p^m + p^{m-1}. \tag{12.14}$$

For each p with $0 < p < 1$, there is a unique integer m satisfying (12.14). Indeed, (12.14) is equivalent to

$$p^m(1 + p) \leq 1 < p^{m-1}(1 + p)$$

or equivalently

$$m \geq -\frac{\log(1 + p)}{\log p} > m - 1.$$

Note that when m is large, (12.14) implies $p^m \sim 1/2$, and that (12.14) holds for $p^m = 1/2$. Let us show that the application of the Huffman algorithm to a geometric distribution given by (12.13) can produce the Golomb code of order m where m is defined by (12.14). This shows the optimality of the Golomb code. Actually, we will operate on a truncated, but growing source since Huffman's algorithm works only on finite alphabets.

Set $Q = 1 - p^m$. By the choice of m, one has $p^{-1-m} \geq 1/Q > p^{1-m}$. We consider, for $k \geq -1$, the bounded alphabet

$$B_k = \{0, \dots, k+m\}.$$

In particular, $B_{-1} = \{0, \dots, m-1\}$. We consider on B_k the distribution

$$\pi(i) = \begin{cases} p^i q & \text{for } 0 \leq i \leq k, \\ p^i q/Q & \text{for } k < i \leq k+m. \end{cases}$$

Clearly $\pi(i) > \pi(k)$ for $i < k$ and $\pi(k+i) > \pi(k+m)$ for $1 < i < m$. Observe that also $\pi(i) > \pi(k+m)$ for $i < k$ since $\pi(k+m) = p^{k+m}q/Q \leq p^{k+m}q/p^{m+1} = \pi(k-1)$. Also $\pi(k+i) > \pi(k)$ for $1 < i < m$ since indeed $\pi(k+i) > \pi(k+m-1) = p^{k+m-1}q/Q > p^k q = \pi(k)$. As a consequence, the symbols k and $k+m$ are those of minimal weight. Huffman's algorithm replaces them with a new symbol, say k', which is the root node of a tree with, say, left child k and right child $k+m$. The weight of k' is

$$\pi(k') = \pi(k) + \pi(k+m) = p^k q(1 + p^m/Q) = p^k q/Q.$$

Thus, we may identify $B_k \backslash \{k, k+m\} \cup \{k'\}$ with B_{k-1} by assigning to k the new value $\pi(k) = p^k q/Q$. We get for B_{k-1} the same properties as for B_k and we may iterate.

After m iterations, we have replaced B_k by B_{k-m}, and each of the symbols $k-m+1, \dots, k$ now is the root of a tree with two children. Assume now that $k = (h+1)m - 1$ for some h. Then after hm steps, one gets the alphabet $B_{-1} = \{0, \dots, m-1\}$, and each of the symbols i in B_{-1} is the root of a binary tree of height h composed of a unique right path of length h, and at each level one left child $i+m$, $i+2m, \dots, i+(h-1)m$. This corresponds to the code $P_h = \{0, 10, \dots, 1^{h-1}0, 1^h\}$. The weights of the symbols in B_{-1} are decreasing, and moreover $\pi(m-2) + \pi(m-1) > \pi(0)$ because $p^{m-2} + p^{m-1} > 1$. It follows from Exercise 12.7.2 below that the code R_m is optimal for this probability distribution.

Thus we have shown that the application of Huffman's algorithm to the truncated source produces the code $R_m P_k$. When h tends to infinity, the sequence of codes converges to $R_m 1^* 0$. Since each of the codes in the sequence is optimal, the code $R_m 1^* 0$ is an optimal prefix code for the geometric distribution. The Golomb code $G_m = 1^*0R_m$ has the same length distribution and so is also optimal.

Golomb–Rice codes. The *Golomb–Rice code* of order k, denoted GR_k, is the particular case of the Golomb code for $m = 2^k$. It was introduced in [54]. Its structure is especially simple and allows an easy explicit description. The encoding assigns to an integer $n \geq 0$ the concatenation of two binary words, the *base* and the *offset*. The base is the unary expansion (over the alphabet $\{1\}$) of $\lfloor n/2^k \rfloor$ followed by a 0. The offset is the remainder of the division written in binary on k bits. Thus, for $k = 2$, the integer $n = 9$ is coded by $110|01$. The binary trees representing the Golomb–Rice code of orders $0, 1, 2$ are represented in Fig. 12.16.

Another description of the Golomb–Rice code of order k is given by the regular expression

$$GR_k = 1^*0\{0, 1\}^k. \tag{12.15}$$

This indicates that the binary words forming the code are composed of a base of the form 1^i0 for some $i \geq 0$ and an offset, which is an arbitrary binary sequence of length k.

It follows from (12.15) that the generating series of the lengths of words of the Golomb–Rice code of order k is

$$f_{GR_k}(z) = \frac{2^k z^{k+1}}{1 - z} = \sum_{i \geq k+1} 2^k z^i.$$

The *weighted generating series* of a code C with probability distribution π is

$$p_C(z) = \sum_{x \in C} \pi(x) z^{|x|}.$$

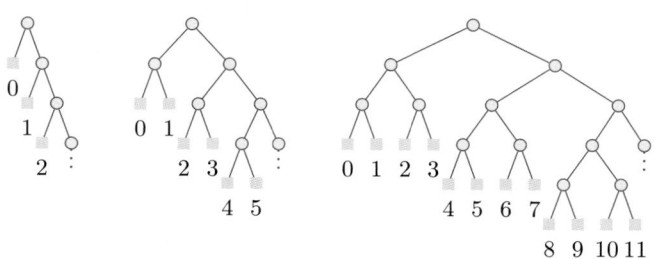

Fig. 12.16. The Golomb–Rice codes of orders $0, 1, 2$.

The average length of C is then

$$\lambda_C(z) = \sum_{x \in C} |x| \pi(x) = p'_C(1).$$

For a uniform Bernoulli distribution on the channel symbols, the weighted generating series for the resulting probabilities of the Golomb–Rice codes GR_k and the corresponding average length λ_{GR_k} are

$$p_{GR_k}(z) = f_{GR_k}\left(\frac{z}{2}\right) = \frac{z^{k+1}}{2-z},$$

$$\lambda_{GR_k} = p'_{GR_k}(1) = k + 2. \qquad (12.16)$$

Observe that in the case where $p^m = 1/2$, the series $p_{GR_k}(z)$, and thus also the average length λ_{GR_k}, happens to be the same for the probability distribution on the code induced by the geometric distribution on the source.

Indeed, the sum of the probabilities of the codewords for rm to $(r+1)m - 1$ is $p^{rm}(1 - p^m)$, and since $p^m = 1/2$, this is equal to 2^{-r-1}. The sum of the probabilities of m words of length $r + k + 1$ with respect to a uniform Bernoulli distribution is $m2^{-r-k-1} = 2^{-r-1}$ (recall that $m = 2^k$). Thus, we get the same value in both cases, as claimed.

Exponential Golomb codes. The *exponential Golomb codes*, denoted EG_k for $k \geq 0$, form a family of codes whose length distributions make them better suited than the Golomb–Rice codes for encoding the integers endowed with certain probability distributions. They are introduced in [64]. The case $k = 0$ is the *Elias code* already mentioned and introduced in [19]. Exponential Golomb codes are used in practice in digital transmissions. In particular, they are a part of the video compression standard technically known as H.264/MPEG-4 Advanced Video Coding (AVC) [55].

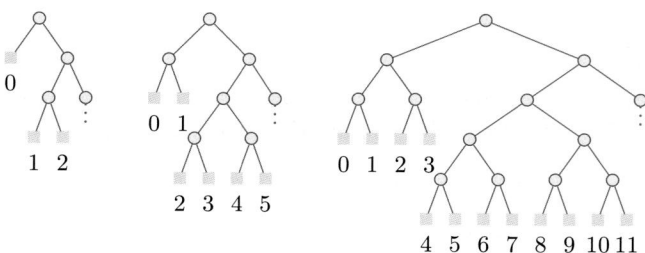

Fig. 12.17.   The exponential Golomb codes of orders $0, 1, 2$.

The codeword representing an integer n tends to be shorter for large integers. The base of the codeword for an integer n is obtained as follows. Let x be the binary expansion of $1 + \lfloor n/2^k \rfloor$ and let i be its length. The base is made of the unary expansion of $i - 1$ followed by x with its initial 1 replaced by a 0. The offset is, as before, the binary expansion of the remainder of the division of n by 2^k, written on k bits. Thus, for $k = 1$, the codeword for 9 is $11001|1$. Figure 12.17 represents the binary trees of the exponential Golomb codes of orders $0, 1, 2$. An expression describing the exponential Golomb code of order k is

$$EG_k = \bigcup_{i \geq 0} 1^i 0 \{0, 1\}^{i+k},$$

and we have the simple relation

$$EG_k = EG_0 \{0, 1\}^k.$$

The generating series of the lengths of words in EG_k is

$$f_{EG_k}(z) = \frac{2^k z^{k+1}}{1 - 2z^2}.$$

The weighted generating series for the probabilities of codewords corresponding to a uniform Bernoulli distribution and the average length are

$$p_{EG_k}(z) = \frac{z^{k+1}}{2 - z^2},$$

$$\lambda_{EG_k} = k + 3.$$

For handling signed integers, there is a simple method, which consists of adding a bit to the words of one of the previous codes. More sophisticated methods, adapted to a two-sided geometric distribution, have been developed in [50].

12.7.1. *Exercises*

Exercise 12.7.1. Show that the entropy $H = -\sum \pi(n) \log \pi(n)$ of the source emitting integers with a geometric distribution (12.13) is

$$H = -(p \log p + q \log q)/q.$$

Verify directly that, for $p^{2^k} = 1/2$, the average length of the Golomb–Rice code GR_k satisfies $\lambda_{GR_k} \leq H + 1$.

Exercise 12.7.2. Let $m \geq 3$. A non-increasing sequence $w_1 \geq w_2 \geq \cdots \geq w_m$ of integers is said to be *quasi-uniform* if it satisfies $w_{m-1} + w_m \geq w_1$. Let T be an optimal binary tree corresponding to a quasi-uniform sequence of weights. Show that the heights of the leaves of T differ at most by one.

12.8. Encoders and Decoders

In this section, we present the basic notions of automata theory, as far as encoding and decoding processes are concerned. The notion of finite automata allows us to define a state-dependent process, which can be used for both encoding and decoding. We give the definition of two closely related notions, namely automata and transducers, which are both labeled directed graphs.

A *finite automaton* $\mathcal{A}$ over some (finite) alphabet A is composed of a finite set Q of *states*, together with two distinguished subsets I and T called the sets of *initial* and *terminal* states, and a set E of *edges*, which are triples (p, a, q) where p and q are states and a, is a symbol. An edge is also denoted by $p \xrightarrow{a} q$. It starts in p, ends in q, and has label a.

Similarly, a *finite transducer* $\mathcal{T}$ uses an input alphabet A and an output alphabet B. It is composed of a finite set Q of *states*, together with two distinguished subsets I and T called the sets of *initial* and *terminal* states, and a set E of *edges*, which are quadruples (p, u, v, q) where p and q are states and u is a word over A and v is a word over B. An edge is also denoted by $p \xrightarrow{u|v} q$. The main purpose for transducers is decoding. In this case, A is the channel alphabet and B is the source alphabet.

A path in an automaton or in a transducer is a finite sequence of consecutive edges. The label of the path is obtained by concatenating the labels of the edges (in the case of a transducer, one concatenates separately the input and the output labels). We write $p \xrightarrow{w} q$ for a path in an automaton labeled with w starting in state p and ending in state q. Similarly, we write $p \xrightarrow{x|y} q$ for a path in a transducer. A path is *successful* if it starts in an initial state and ends in a terminal state.

An automaton $\mathcal{A}$ *recognizes* a set of words, which is the set of labels of its successful paths. The sets recognized by finite automata are called *regular sets*.

A transducer $\mathcal{T}$ defines a binary relation between words on the two alphabets as follows. A pair (u, v) is in the relation if it is the label of a successful path. This is called the relation *realized* by $\mathcal{T}$. This relation can be viewed as a multi-valued mapping from the input words into the output words, and also as a multi-valued mapping from the output words into the input words. For practical purposes, this definition is too general and will be specialized. We consider transducers called *literal*, which by definition means that each input label is a single letter. For example, an encoding γ, as defined at the beginning of the chapter, can be realized by a one-state literal transducer, with the set of labels of edges being simply the pairs $(b, \gamma(b))$ for b in B.

Example 12.14. Consider the encoding defined by $\gamma(a) = 00$, $\gamma(b) = 1$, and $\gamma(c) = 01$. The corresponding encoding transducer is given in Fig. 12.18.

Transducers for decoding are more interesting. For the purpose of coding and decoding, we are concerned with transducers which define single-valued mappings in both directions. We need two additional notions.

An automaton is called *deterministic* if it has a unique initial state and if, for each state p and each letter a, there is at most one edge starting in p and labeled with a. This implies that, for each state p and each word w, there exists at most one path starting in p and labeled with w.

Consider a finite deterministic automaton with a unique terminal state, which is equal to the initial state i. The closed paths from i to i such that no initial segment ends in i are called *first return paths*. The set of labels of these paths is a regular prefix code C (that is a prefix code, which is a regular set), and the set recognized by the automaton is the set C^*. Conversely, any regular prefix code is obtained in this way.

For example, the Golomb codes are regular, whereas exponential Golomb codes are not.

More generally, an automaton is called *unambiguous* if, for all states p, q, and all words w, there is at most one path from p to q labeled with w. Clearly, a deterministic automaton is unambiguous.

Fig. 12.18. A simple encoder. The only state is both initial and terminal.

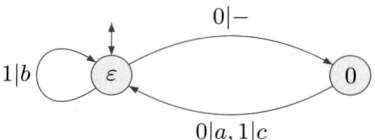

Fig. 12.19.   A deterministic decoder. A dash represents the empty word. An incoming (outgoing) arrow indicates the initial (terminal) state.

A literal transducer defines naturally an automaton over its input alphabet, called its *input automaton*. For simplicity, we discard the possibility of multiple edges in the resulting automaton. A literal transducer is called *deterministic* (resp. *unambiguous*) if its associated input automaton is deterministic (resp. unambiguous). Clearly, the relation realized by a deterministic transducer is a function.

An important result is that for any encoding (with finite source and channel alphabets), there exists a literal unambiguous transducer, which realizes the associated decoding. When the code is prefix, the transducer is actually deterministic.

The construction is as follows. Let γ be an encoding. Define a transducer $\mathcal{T}$ by taking a state for each proper prefix of some codeword. The state corresponding to the empty word ε is the initial and terminal state. There is an edge $p \xrightarrow{a|-} pa$, where $-$ represents the empty word, for each prefix p and letter a such that pa is a prefix, and an edge $p \xrightarrow{a|b} \varepsilon$ for each p and letter a with $pa = \gamma(b)$. When the code is prefix, the decoder is deterministic. In the general case, the property of unique decipherability is reflected by the fact that the transducer is unambiguous.

Example 12.15. The decoder corresponding to the prefix code of Example 12.14 is represented in Fig. 12.19.

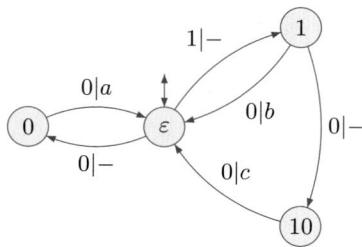

Fig. 12.20.   An unambiguous decoder for a code, which is not prefix.

Example 12.16. Consider the code $C = \{00, 10, 100\}$ of Example 12.4. The decoder given by the construction is represented in Fig. 12.20.

As a consequence of this construction, it can be shown that decoding can always be realized in linear time with respect to the length of the encoded string (considering the number of states as a constant). Indeed, given a word $w = a_1 \cdots a_n$ of length n to be decoded, one computes the sequence of sets S_i of states accessible from the initial state for each prefix $a_1 \cdots a_i$ of length i of w, with the convention $S_0 = \{\varepsilon\}$. Of course, the terminal state ε is in S_n. Working backwards, we set $q_n = \varepsilon$ and we identify in each set S_i the unique state q_i such that there is an edge $q_i \xrightarrow{a_i} q_{i+1}$ in the input automaton. The uniqueness comes from the unambiguity of the transducer. The corresponding sequence of output labels gives the decoding. This construction is based on the *Schützenberger covering* of an unambiguous automaton, see [57].

Example 12.17. Consider again the code $C = \{00, 10, 100\}$. The decoding of the sequence 10001010000 is represented on Fig. 12.21. Working from left to right produces the tree of possible paths in the decoder of Fig. 12.20. Working backwards from the state ε in the last column produces the successful path indicated in boldface.

The notion of deterministic transducer is too constrained for the purpose of coding and decoding because it does not allow a lookahead on the input or equivalently a delay on the output. The notion of sequential transducer to be introduced now fills this gap.

A *sequential transducer* is composed of a deterministic transducer and an output function. This function maps the terminal states of the transducer into words on the output alphabet. The function realized by a sequential transducer is obtained by appending, to the value of the deterministic transducer, the image of the output function on the arrival state. Formally, the value on the input word x is

$$f(x) = g(x)\sigma(i \cdot x),$$

Fig. 12.21.   The decoding of 10001010000.

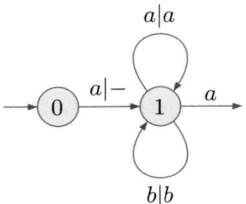

Fig. 12.22. A sequential transducer realizing a cyclic shift on words starting with the letter a, with $\sigma(0) = \varepsilon$ and $\sigma(1) = a$.

where $g(x)$ is the value of the deterministic transducer on the input word x, where $i \cdot x$ is the state reached from the input state i by the word x, and where σ is the output function. This is defined only if the state $i \cdot x$ is a terminal state.

Example 12.18. The sequential transducer given in Fig. 12.22 realizes the partial function $aw \mapsto wa$, for each word w. The output function σ is given by $\sigma(0) = \varepsilon$ and $\sigma(1) = a$.

It is a well-known property of finite automata that any finite automaton is equivalent to a finite deterministic automaton. The process realizing this transformation is known as the *determinization algorithm* . This remarkable property does not hold in general for transducers.

Nonetheless, there is an effective procedure to compute a sequential transducer $\mathcal{S}$ that is equivalent to a given literal transducer $\mathcal{T}$, whenever such a transducer exists, see [46]. The algorithm goes as follows.

The states of $\mathcal{S}$ are sets of pairs (u, p). Each pair (u, p) is composed of an output word u and a state p of $\mathcal{T}$. For a state s of $\mathcal{S}$ and an input letter a, one first computes the set $\bar{s}$ of pairs (uv, q) such that there is a pair (u, p) in s and an edge $p \xrightarrow{a|v} q$ in $\mathcal{T}$. In a second step, one chooses some common prefix z of all words uv, and one defines the set $t = \{(w, q) \mid (zw, q) \in \bar{s}\}$. There is a transition from state s to a state t labeled with (a, z). The initial state is (ε, i), where i is the initial state of $\mathcal{T}$. The terminal states are the sets t containing a pair (u, q) with q terminal in $\mathcal{T}$. The output function σ is defined on state t of $\mathcal{S}$ by $\sigma(t) = u$. If there are several pairs (u, q) in t with distinct u for the same terminal state q, then the given transducer does not compute a function and thus it is not equivalent to a sequential one.

The process of building new states of $\mathcal{S}$ may not halt if the lengths of the words, which are the components of the pairs, are not bounded.

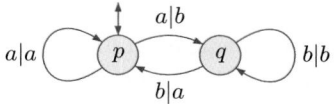

Fig. 12.23. Another transducer realizing a cyclic shift on words starting with the letter a.

There exist *a priori* bounds for the maximal length of the words appearing whenever the determinization is possible, provided that, at each step, the longest common prefix is chosen. This makes the procedure effective.

Example 12.19. Consider the transducer given in Fig. 12.23. The result of the determinization algorithm is the transducer of Fig. 12.22. State 0 is composed of the pair (ε, p), and state 1 is formed of the pairs (a, p) and (b, q).

Example 12.20. Consider the code $C = \{00, 10, 100\}$ of Example 12.4. Its decoder is represented in Fig. 12.20. The determinization algorithm applied to this transducer produces, for the input word 10^{2n}, the state consisting of $(ba^{n-1}, 0)$ and (ca^{n-1}, ε). Thus, the algorithm does not terminate, and there is no equivalent sequential transducer.

12.8.1. *Exercises*

Exercise 12.8.1. A set of words C is said to be *weakly prefix* if there is an integer $d \geq 0$ such that the following condition holds for any elements c, c' of C and any words w, w' of C^*. If the prefix of length $|c| + d$ of cw is a prefix of $c'w'$, then $c = c'$. The least integer d such that this property holds is called the *deciphering delay* and a weakly prefix code is also said to have *finite deciphering delay* .
Show that a weakly prefix set is UD.

Exercise 12.8.2. Which of the following sets C and C' is a weakly prefix code? $C = \{00, 10, 100\}$ and $C' = \{0, 001\}$.

Exercise 12.8.3. Show that a finite code is weakly prefix if and only if it can be decoded by a sequential transducer.

12.9. Codes for Constrained Channels

The problem considered in this section arises in connection with the use of communication channels which impose input constraints on the sequences that can be transmitted. User messages are encoded to satisfy the constraint; the encoded messages are transmitted across the channel and then decoded by an inverse to the encoder.

In this context, we will use a more general notion of encoding. Instead of a memoryless substitution of source symbols by codewords, we will consider finite transducers: the codeword associated with a source symbol depends not only on this symbol but also on a state depending on the past.

We require, in order to be able to decode, that the encoder is unambiguous on its output in the sense that for each pair of states and each word w, there is at most one path between these states with output label w. For ease of use, we will also assume that the input automaton of the encoder is deterministic.

Example 12.21. The encoder in Fig. 12.26 is deterministic. The source alphabet is $\{0, 1\}$ and the channel alphabet is $\{a, b, c\}$. The sequence 0011 is encoded by $acbb$ if the encoding starts in state 1. A more complicated example of a deterministic encoder that will be described later is depicted in Fig. 12.31.

A stronger condition for a decoder than unambiguity is that of having *finite look-ahead*. This means that there exists an integer $D \geq 0$ such that, for all states q and all channel sequences w of length $D + 1$, all paths that begin at state q and have channel label w share the same first edge (and therefore the first source symbol). In other words, decoding is a function of the current state, the current channel symbol, and the upcoming string of D channel symbols [45]. These decoders correspond to sequential transducers as introduced in Sec. 12.8.

However, even this condition is heavily state-dependent, and so a channel error that causes the decoder to lose track of the current state may propagate errors forever. For this reason, the following stronger condition is usually employed.

A *sliding block decoder* operates on words of channel symbols with a window of fixed size. The decoder uses m symbols before the current one and a symbols after it (m is for memory and a for anticipation). According to the value of the symbols between time $n - m$ and time $n + a$, the value of the nth source symbol is determined. Figure 12.24 depicts a schematic view of a sliding block decoder. It is not hard to show that sliding-block decodability implies the weaker finite lookahead property mentioned above. Note that a sliding block decoder avoids possible problems with error propagation because any channel error can affect the decoding only while it is in the decoder window and thus can corrupt at most $m + a + 1$ source symbols.

Transducers realizing sliding block decoders are of a special kind. An automaton is called *local* if the knowledge of a finite number of symbols in the past and in the future determines the current state. More precisely, for given integers $m, a \geq 0$ (m stands for memory and a for anticipation), an automaton is said to be (m, a)-*local* if for words u and v of length m and a, respectively, $p \xrightarrow{u} q \xrightarrow{v} r$ and $p' \xrightarrow{u} q' \xrightarrow{v} r'$ imply that $q = q'$. A transducer is said to be *local* if its input automaton is local.

A sliding block decoder can be realized by a local transducer with the same parameters. Conversely, a transducer whose input automaton is (m, a)-local, and such that the input label and the output label of each edge have the same length, is a sliding block decoder.

While input constraints have been imposed in some communication channels, they arise more commonly in data recording channels, such as those found in magnetic and optical disk drives. Constraints on inputs are of various types and have changed over the course of the 50-year history of magnetic disk drives. For illustration, we focus on the $[d, k]$-*constraint*, where $0 \leq d \leq k$. A binary sequence is said to satisfy the $[d, k]$-constraint if the number of contiguous symbols 0 between consecutive symbols 1 is at least d and at most k.

These constraints arise for the following reasons. An electrical current in the write head, situated over the spinning disk, creates a magnetic field, which is reversed when the current polarity is reversed. These write field reversals, in turn, induce reversals in the orientation of the magnetization along the recorded track on the disk. During the data recovery process, the read head senses the recorded pattern of magnetization along the track. Each transition in magnetization direction produces a correspondingly oriented pulse in the readback voltage.

Two problems should be avoided. The first is called *intersymbol interference*. If polarity changes are too close together, the induced magnetic

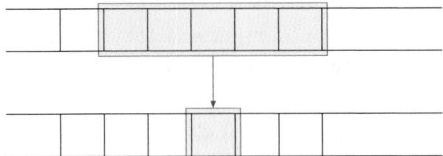

Fig. 12.24.   A sliding block decoder.

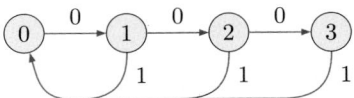

Fig. 12.25.   The $[1, 3]$-constraint.

fields tend to interfere and the pulses in the readback signal are harder to detect. The second problem is called *clock drift*. This problem arises when the pulses are separated by intervals of time which are too large. When the read head senses a pulse it sends information through a phase lock loop which keeps the bit clock running accurately; if the separation between pulses is too large, the clock can lose synchronization and skip through a complete bit period.

Several values of the parameters d, k are of practical importance. The simplest case is the constraint $[1, 3]$. This means that the blocks 11 and 0000 are forbidden. The binary sequences satisfying this constraint are those which label paths in the graph of Fig. 12.25.

We consider an encoding of all binary sequences by sequences satisfying the $[1, 3]$-constraint. This encoding is not realizable by a sequential encoder without modifying the output alphabet, because there are more binary source sequences of length n than admissible binary channel sequences of length n. However, it is possible to operate at rate 1:2, by encoding a source bit by one of the 2-bit symbols $a = 00$, $b = 01$, or $c = 10$. A particular way of doing this is the modified frequency modulation (MFM, see [47]) code of Fig. 12.26, which was used on floppy disks for many years.

Observe that the second bit of the output is always equal to the input bit, and so the decoder can operate symbol by symbol, producing a 1-bit input from a 2-bit output. The first bit of the output is chosen in such a way that there are no consecutive 1's and no block of four 0's.

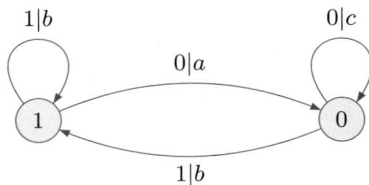

Fig. 12.26.   The MFM code. Here the state names reflect the input bit.

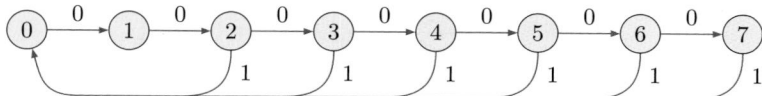

Fig. 12.27. The $[2,7]$-constraint.

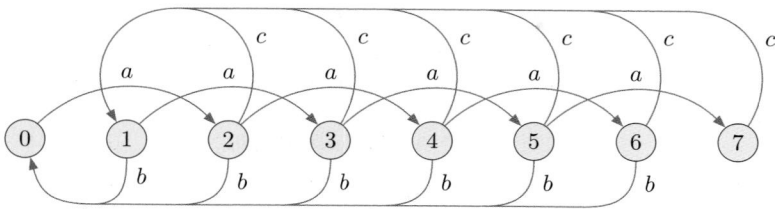

Fig. 12.28. The squared $[2,7]$-constraint.

A more complex example is the $[2,7]$-constraint illustrated in Fig. 12.27. Again, the sequences satisfying the constraint are the labels of paths in the graph.

For the purpose of coding arbitrary binary sequences by sequences satisfying the $[2,7]$-constraint, we again consider a representation obtained by changing the alphabet to $a = 00$, $b = 01$, and $c = 10$, as shown in Fig. 12.28.

The result of the development that follows will be the sequential transducer for encoding known as the Franaszek encoder depicted in Fig. 12.31. It must be checked that this encoder satisfies the $[2,7]$-constraint and that, for practical applications, it admits a sliding block decoder. A direct verification is possible but complicated.

The encoder design process we now describe is the one historically followed by Franaszek [21]. It starts with the graph in Fig. 12.28, which represents the $[2,7]$-constraint of Fig. 12.27 in terms of the alphabet a, b, c. More precisely, Fig. 12.28 represents the set of $[2,7]$-constrained sequences

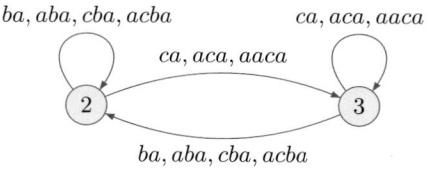

Fig. 12.29. The poles.

C	P
ba	10
ca	11
aba	000
cba	010
aca	011
$acba$	0010
$aaca$	0011

Fig. 12.30.   The Franaszek code.

obtained by writing each such sequence as a string of non-overlapping 2-bit pairs.

Next, choose the two vertices 2 and 3 (called the *poles* or the *principal states*). The paths of first return from 2 or 3 to 2 or 3 are represented in Fig. 12.29.

Observe that the set C of labels of first return paths is independent of the starting vertex. Thus all concatenations of words in C are produced by paths in the graph of Fig. 12.29 and thus admissible for the [2,7]-constraint. The set C is a prefix code called the *Franaszek code*, shown in the first column of Fig. 12.30.

It happens that the set C has the same length distribution as the maximal binary prefix code P of words appearing in the right column.

The pair of prefix codes of Fig. 12.30 is used as follows to encode a binary word at rate 1:2. A source message is parsed as a sequence of codewords in P, possibly followed by a prefix of such a word. For example, the word

$$011011100101110111100010011\cdots$$

is parsed as

$$011 \mid 011 \mid 10 \mid 010 \mid 11 \mid 10 \mid 11 \mid 11 \mid 000 \mid 10 \mid 011\cdots.$$

Next, this is encoded row-by-row using the correspondence between P and C as follows:

$$aca \mid aca \mid ba \mid cba \mid ca \mid ba \mid ca \mid ca \mid aba \mid ba \mid aca.$$

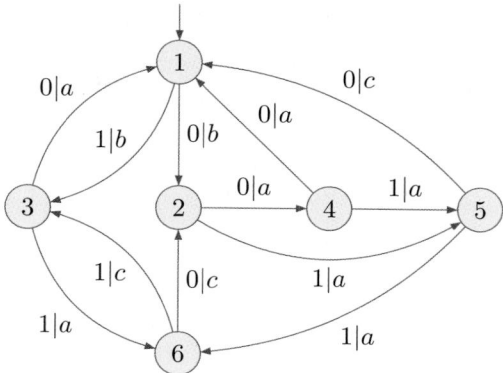

Fig. 12.31. The Franaszek encoder.

This stands for the channel encoding

$$001000 \mid 001000 \mid 0100 \mid 100100 \mid 1000 \mid 0100 \mid 1000 \mid$$

$$1000 \mid 000100 \mid 0100 \mid 001000.$$

Figure 12.31 represents an implementation of this transformation, up to some shift. The encoder is a deterministic transducer which outputs one symbol for each input symbol. State 1 is the initial state, and all states are terminal states. All inputs of at least 2 symbols produce an output sequence that begins with ba, followed by the sequence generated by the encoder described above, up to the last two symbols. For example, for the input word 011011, the corresponding path in the deterministic encoder is

$$1 \xrightarrow{0|b} 2 \xrightarrow{1|a} 5 \xrightarrow{1|a} 6 \xrightarrow{0|c} 2 \xrightarrow{1|a} 5 \xrightarrow{1|a} 6 \,.$$

Hence, the output is $ba|aca|a$, which, following the initial ba, corresponds to one codeword aca followed by the beginning of a second occurrence of aca.

This encoder can be obtained as follows from the pair (C, P) of prefix codes. Consider first the transducer of Fig. 12.32. This is obtained by composing a decoder for P with an encoder for C, and merging common prefixes. We omit the details.

We build a deterministic transducer by determinization of the transducer of Fig. 12.32, by the algorithm presented in Sec. 12.8. In this case, we are able to maintain a constant rate of output by keeping always in the pairs of the sequential transducer an output word of length 2 (this

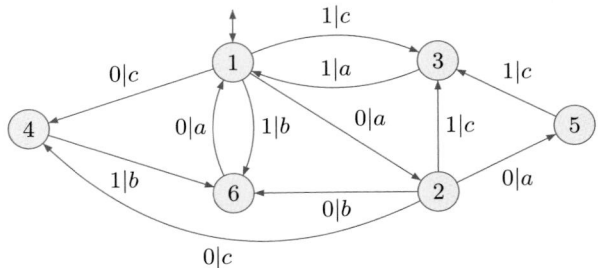

Fig. 12.32.   The construction of the Franaszek encoder.

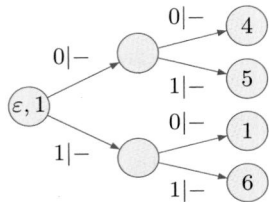

Fig. 12.33.   Initial part of the Franaszek encoder.

represents the two symbols to be output later). Thus, the output is delayed with two symbols.

The states of the result are given in Table 12.3. In this table, and also in Fig. 12.31, we only report the states which have an output delay of exactly two symbols. They correspond to the strongly connected part of the encoder. In particular, the initial state $(\varepsilon, 1)$ is not represented. Figure 12.33 gives the states reached by the first two symbols.

Figure 12.34 represents the decoder, which can be realized with a sliding window of size 4. Indeed, the diagrams of Fig. 12.34 show that for any content $xyzt$ of the window, we can define the output r corresponding to the first symbol. An inspection of Figure 12.31 shows that if $y = b$, then $r = 0$. In the same manner, if $y = c$, then $r = 1$. If $y = a$, then four cases arise. If $z = a$, then $r = 0$. If $z = c$, then $r = 0$ or 1 according to $t = b$ or a.

Table 12.3.   The states of the Franaszek encoder.

state	1	2	3	4	5	6
content	ba,1	ac,4	ab,6	ac,4	cb,6	ca,1
		aa,2	ac,3	ab,6	ac,3	
				aa,5		
output	ba					ca

Finally, if $z = b$, then $r = 0$ if $z = b$ and $r = 1$ otherwise (the x in the last frame stands for b or c).

There are many approaches to the design of encoder–decoder pairs for input constraints. The approach illustrated above was extended to a technique called the *method of poles*, developed by Béal [6–8].

Another approach is the *state-splitting algorithm*, also known as the *ACH algorithm* [1], which transforms a finite-state graph representation of a constraint into a sliding-block decodable finite-state encoder. The rough idea is as follows.

One starts with a finite-state graph representation of the constraint and then chooses a feasible encoding rate $p{:}q$; here, feasibility means that p/q does not exceed the capacity of the constraint, which is a measure of the "size" or "complexity" of the constraint and generalizes the notion of capacity of a variable-length code given in Sec. 12.5. The capacity can be computed explicitly, and for a feasible code rate, one can compute a vector, called an approximate eigenvector, which effectively assigns non-negative integer weights to the states of the graph [45]. Next, one replaces the original representation with a new one whose symbols are q-bit strings, just as in the transformation of Figs. 12.27 and 12.28. Then by an iterative sequence of state splittings, guided by an approximate eigenvector, one is guaranteed to arrive at a final representation upon which a sequential finite-state encoder can be built. The splittings are graph transformations in which states are split according to partitions of outgoing edges. In the final representation, each state has at least 2^p outgoing edges, enabling an assignment of p-bit input labels and therefore a sequential finite-state encoder at rate $p{:}q$. It is also guaranteed, under mild conditions, that such an encoder is sliding-block decodable.

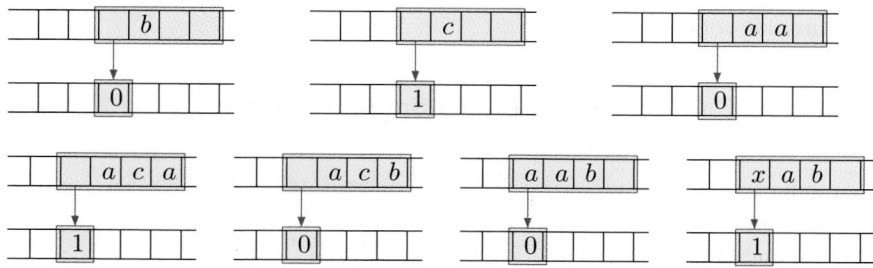

Fig. 12.34. The associated sliding block decoder.

The state splitting algorithm has been modified in many ways. For instance, one could consider finite-state variable-length representations of input constraints or transform fixed-length representations into more compact variable-length representations. There is a variable-length version of the state splitting algorithm, which again iteratively splits states. In this setting, instead of aiming for a graph with sufficient out-degree at each state, the sequence of state splittings results in a final representation in which the lengths of outgoing edges at each state satisfy a reverse Kraft inequality. This technique was introduced in [2] and further developed in [33]. In the latter reference, the method was illustrated with two examples, one of which is an alternative derivation of the Franaszek code described above.

The theory and practice of coding for constrained channels is very well developed. For more information, the reader may consult Béal [7], Immink [39], Lind–Marcus [45], and Marcus–Roth–Siegel [47] (or the latest version at http://www.math.ubc.ca/~marcus/Handbook/index.html).

12.9.1. *Exercises*

Exercise 12.9.1. A code C over the alphabet A is a *circular code* if, for any words u, v on the alphabet A, the cyclically shifted words uv and vu can be in C^* only when u and v are. Let C be a circular code and let x be a word of C. Show that a set D of words of the form $x^i y$, for $i \geq 0$ and y in $C \backslash x$, is a circular code.

Note: The terminology *circular code* stems from the fact that the unique decipherability property holds for words written on a circle (see [10]).

Exercise 12.9.2. Use Exercise 12.9.1 to show that the set $C = \{ba, ca, aba, cba, aca, acba, aaca\}$ appearing in the first column of Fig. 12.30 is a circular code.

12.10. Codes for Constrained Sources

In the same way as there exist codes for constrained channels, there exist codes for constrained sources. As for constraints on channels, the constraints on sources can be expressed by means of finite automata. This leads us to use encodings with memory.

We limit the presentation to two examples, the first drawn from linguistics, and the second reflecting a more general point of view.

The first example gives an interesting encoding from a constrained source to an unconstrained channel [31]. It starts with a simple substitution, which is not UD, but it appears to be UD by taking into account the constraints on the source. The example is taken from the field of natural language processing where codings and automata are heavily used (see, for instance, [56]).

Example 12.22. We start with a source alphabet composed of a sample set of six syllable types in the Turkish language. The concatenation of syllables is subject to constraints that will be explained below. The types of syllables are denoted by A to F as follows (with the syllables written in italics in the example):

Symbol	Structure	Example
A	0	*a*çik (open)
B	10	*ba*ba (father)
C	01	*ek*mek (bread)
D	101	al*tin* (gold)
E	011	*erk* (power)
F	1011	*türk* (turkish)

The structure of a syllable is the binary sequence obtained by coding 0 for a vowel and 1 for a consonant. We consider the structure of a syllable as its encoding. The decoding is then to recover the sequence of syllables from the encoding. Thus, the source alphabet is $\{A, \ldots, F\}$, the channel alphabet is $\{0, 1\}$, and the encoding is $A \mapsto 0, \ldots, F \mapsto 1011$.

There are linguistic constraints on the possible concatenations of syllable types, which come from the fact that a syllable ending with a consonant cannot be followed by one which begins with a vowel. These constraints are summarized by the following matrix, where a 0 in entry x, y means that x cannot be followed by y:

$$M = \begin{array}{c@{}c} & \begin{array}{cccccc} A & B & C & D & E & F \end{array} \\ \begin{array}{c} A \\ B \\ C \\ D \\ E \\ F \end{array} & \left(\begin{array}{cccccc} 1 & 1 & 1 & 1 & 1 & 1 \\ 1 & 1 & 1 & 1 & 1 & 1 \\ 0 & 1 & 0 & 1 & 0 & 1 \\ 0 & 1 & 0 & 1 & 0 & 1 \\ 0 & 1 & 0 & 1 & 0 & 1 \\ 0 & 1 & 0 & 1 & 0 & 1 \end{array} \right) \end{array}.$$

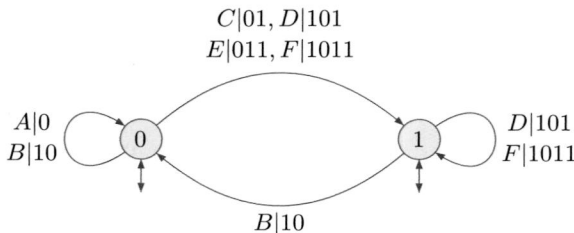

Fig. 12.35.   A 2-state encoder reflecting the legal concatenations.

The encoding can be realized in a straightforward manner with a 2-state transducer given in Fig. 12.35, which reflects the legal concatenations of source symbols.

The decoder is built by applying the methods of Sec. 12.8 to this transducer. We start by exchanging input and output labels in Fig. 12.35. In a second step, we introduce additional states in order to get literal output. An edge is broken into smaller edges. For instance, the edge $0 \xrightarrow{011|E} 1$ is replaced by the path $0 \xrightarrow{0|-} E_1 \xrightarrow{1|-} E_2 \xrightarrow{1|E} 1$. Each state x_i stands for the prefix of length i of the encoding of x. The state x_i can be used for several edges, because the arrival state depends only on x. The transducer is given in Fig. 12.36.

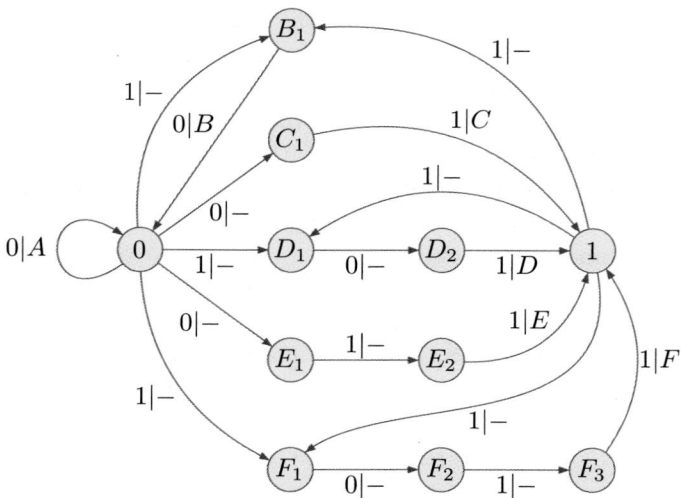

Fig. 12.36.   Literal decoder for phonetization.

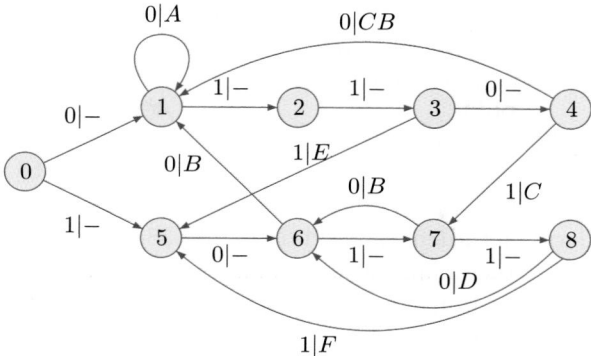

Fig. 12.37.   A sequential transducer for the phonetization.

Table 12.4.   The states of the transducer for the phonetization.

state	0	1	2	3	4	5	6	7	8
content	$\varepsilon, 0$	$A, 0$	A, B_1	C, B_1	$CB, 0$	ε, B_1	$B, 0$	B, B_1	$D, D_1,$
		ε, C_1	$C, 1$	C, D_1	C, D_2	ε, D_1	ε, D_2	$B, D_1,$	D, F_1
		ε, E_1	ε, E_2	C, F_1	C, F_2	ε, F_1	ε, F_2	B, F_1	D, B_1
			A, D_1	$E, 1$				$D, 1$	$F, 1$
			A, F_1					ε, F_3	
output		A	C	E	CB		B	D	F

It happens that the resulting transducer can be transformed into a sequential one. The result is shown in Fig. 12.37. The correspondence is given in Table 12.4. The value of the output function is given in the node.

The decoder is deterministic in its input and it outputs the decoded sequence with finite delay (this means that it uses a finite lookahead on the input). On termination of a correct input sequence, the last symbol to be produced is indicated by the corresponding state. This coding is used in [31] to build a syllabification algorithm, which produces for each word a parsing in syllables.

This kind of problem appears more generally in the framework of hyphenation in text processing software (see [16], for example), and also in "text-to-speech" synthesis [17].

The next example of codes for constrained sources is from [14]. It illustrates again how an ambiguous substitution can be converted into a UD encoding.

The three examples below have decoders, which are local (that is realizable by a sliding block decoder) for the first one, sequential for the second one, and simply unambiguous for the third one.

Example 12.23. Consider the constrained source on the symbols A, B, C with the constraint that B cannot follow A. We consider the substitution that maps A, B, and C to 0, 1, and 01, respectively. The mapping is realized by the transducer on the left-hand side of Fig. 12.38. It is easily seen that the constraint implies the unique decipherability property. Actually, a decoder can be computed by using the determinization algorithm. This yields the sequential decoder represented on the right-hand side of Fig. 12.38. It is even $(1, 0)$-local since the last letter determines the state.

We now give a second, slightly more involved example. The source has four symbols A, B, C, D with the constraints of concatenation given by the matrix

$$
M_2 = \begin{array}{c} \\ A \\ B \\ C \\ D \end{array}
\begin{array}{cccc} A & B & C & D \\ \end{array}
\left(\begin{array}{cccc}
1 & 0 & 1 & 0 \\
0 & 1 & 0 & 1 \\
1 & 1 & 1 & 1 \\
1 & 1 & 1 & 1
\end{array}\right).
$$

We consider the encoding assigning $0, 1, 01$, and 10 to A, B, C, and D, respectively. The decoder is given in Fig. 12.39. This transducer is sequential with output function indicated in the states. However, it is not local in input since there are two cycles labeled by 01.

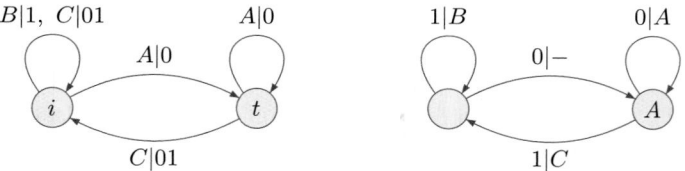

Fig. 12.38.   An encoder and a local decoder.

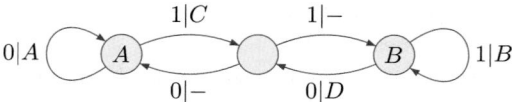

Fig. 12.39.   A sequential decoder.

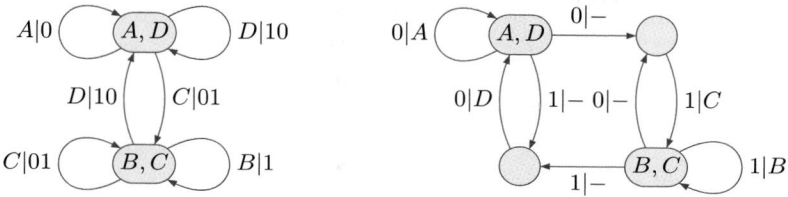

Fig. 12.40. An encoder and an unambiguous decoder.

The third example uses the same substitution but a different set of constraints, given by the matrix

$$M_3 = \begin{array}{c} \\ A \\ B \\ C \\ D \end{array} \begin{array}{c} A \quad B \quad C \quad D \\ \begin{pmatrix} 1 & 0 & 1 & 1 \\ 0 & 1 & 1 & 1 \\ 0 & 1 & 1 & 1 \\ 1 & 0 & 1 & 1 \end{pmatrix} \end{array}.$$

The substitution is realized by the transducer shown in the left-hand side of Fig. 12.40. The transducer shown in the right-hand side is a decoder. It can be checked to be unambiguous. However, it is not equivalent to a sequential transducer because a sequence $0101 \cdots$ has two potential decodings as $ADDD \cdots$ and as $CCCC \cdots$.

These examples are variable-length, and in fact variable-rate, codes. In applications where error propagation may be a problem, one can use fixed-rate block codes [38] or fixed-rate sliding-block codes [20]. The construction of the latter is dual to the state splitting method mentioned in the previous section.

The connection between variable-length codes, unambiguous automata, and local constraints has been further developed in [9, 52, 53].

12.11. Bifix Codes

Recall that a set of words C is a *suffix* code if no element of C is a proper suffix of another one. A set of words is called a *bifix code* if it is at the same time a prefix code and a suffix code. Bifix codes are also known as *reversible variable-length codes* (RVLC). The idea to study bifix codes goes back to [26, 59]. These papers already contain significant results. The first systematic study is in [60, 61]. The development of codes for video recording has renewed the interest in bifix codes [27, 69, 70].

One example is given by prefix codes which are equal to their reversal. The *reversal* of a word $w = a_1 \cdots a_n$, where $a_1, \ldots, a_n$ are symbols, is the word $\tilde{w} = a_n \cdots a_1$ obtained by reading w from right to left. The reversal of a set C, denoted $\tilde{C}$, is the set of reversals of its elements. For example, 01^*0 is a bifix code since it is prefix and equal to its reversal.

The use of bifix codes for transmission is linked with the possibility of limiting the consequences of errors occurring in the transmission using a bidirectional decoding scheme as follows. Assume that we use a binary bifix code to transmit data and that for the transmission, messages are grouped into blocks of N source symbols, encoded as N codewords. The block of N codewords is first decoded by using an ordinary left-to-right sequential decoding (Fig. 12.41). Suppose that the codewords x_1 up to x_{i-1} are correctly decoded, but that an error has occurred during transmission that makes it impossible to identify the next codeword in the rest of the message. Then a new decoding process is started, this time from right to left. If at most one error has occurred, then again the codewords from x_N down to x_{i+1} are decoded correctly. Thus, in a block of N encoded source symbols, at most one codeword will be read incorrectly.

These codes are used for the compression of moving pictures. Indeed, there are reversible codes with the same length distribution as the Golomb–Rice codes, as shown in [68]. The AVC standard mentioned previously recommends the use of these codes instead of the ordinary Golomb–Rice codes to obtain an error resilient coding (see [55]). The difference from the ordinary codes is that, in the base, the word 1^i0 is replaced by $10^{i-1}1$ for $i \geq 1$. Since the set of bases forms a bifix code, the set of all codewords is also a bifix code. Figure 12.42 represents the reversible Golomb–Rice codes of orders $0, 1, 2$.

There is also a reversible version of the exponential Golomb codes, denoted by REG_k, which are bifix codes with the same length distribution. The code REG_0 is given by

$$REG_0 = \{0\} \cup 1\{00, 10\}^*\{0, 1\}1 \, .$$

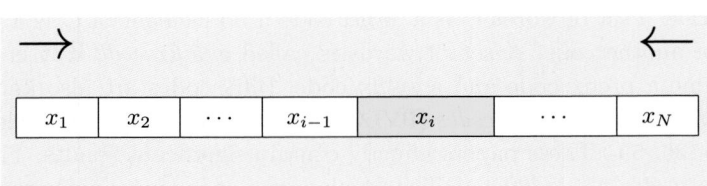

Fig. 12.41.   The transmission of a block of codewords.

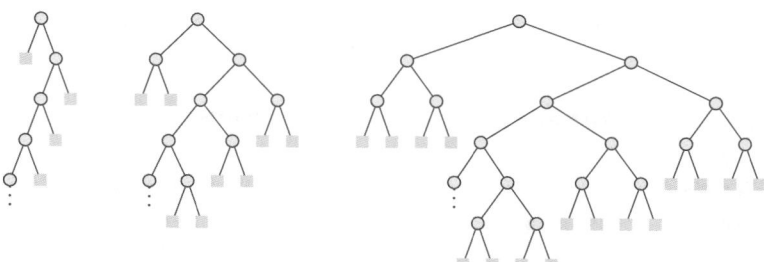

Fig. 12.42. The reversible Golomb–Rice codes of orders $0, 1, 2$.

It is a bifix code because it is equal to its reversal. This comes from the fact that the set $\{00, 10\}^*\{0, 1\}$ is equal to its reversal because it is the set of words of odd length which have a 0 at each even position, starting at position 1.

The code of order k is

$$REG_k = REG_0\{0, 1\}^k.$$

The codes REG_k are represented for $k = 0$ and 2 in Fig. 12.43.

We now consider the length distribution of bifix codes. In contrast to the case of codes or of prefix codes, it is not true that any sequence $(u_n)_{n \geq 1}$ of non-negative integers such that $\sum_{n \geq 1} u_n k^{-n} \leq 1$ is the length distribution of a bifix code on k letters. For instance, there is no bifix code on the alphabet $\{a, b\}$, which has the same distribution as the prefix code $\{a, ba, bb\}$. Indeed, such a code must contain a letter, say a, and then the only possible word of length 2 is bb. On the other hand, the following result provides a sufficient condition for a length distribution to be realizable by a bifix code [4].

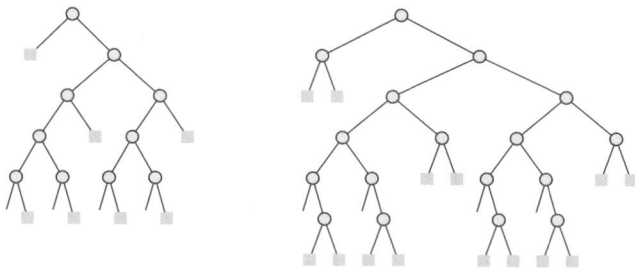

Fig. 12.43. The reversible exponential Golomb codes of orders 0 and 1.

Proposition 12.1. *For any sequence $(u_n)_{n \geq 1}$ of non-negative integers such that*

$$\sum_{n \geq 1} u_n k^{-n} \leq \frac{1}{2} \tag{12.17}$$

there exists a bifix code on an alphabet of k letters with length distribution $(u_n)_{n \geq 1}$.

Proof. We show by induction on $n \geq 1$ that there exists a bifix code X_n of length distribution $(u_i)_{1 \leq i \leq n}$ on an alphabet A of k symbols. It is true for $n = 1$ since $u_1 k^{-1} \leq 1/2$ and thus $u_1 < k$. Assume that the property is true for n. We have by Inequality (12.17)

$$\sum_{i=1}^{n+1} u_i k^{-i} \leq \frac{1}{2}$$

or equivalently, multiplying both sides by $2k^{n+1}$,

$$2(u_1 k^n + \cdots + u_n k + u_{n+1}) \leq k^{n+1}$$

whence

$$u_{n+1} \leq 2u_{n+1} \leq k^{n+1} - 2(u_1 k^n + \cdots + u_n k). \tag{12.18}$$

Since X_n is bifix by the induction hypothesis, we have

$$\mathrm{Card}(X_n A^* \cap A^{n+1}) = \mathrm{Card}(A^* X_n \cap A^{n+1}) = u_1 k^n + \cdots + u_n k.$$

Thus, we have

$$\mathrm{Card}((X_n A^* \cup A^* X_n) \cap A^{n+1})$$
$$\leq \mathrm{Card}(X_n A^* \cap A^{n+1}) + \mathrm{Card}(A^* X_n \cap A^{n+1})$$
$$\leq 2(u_1 k^n + \cdots + u_n k).$$

It follows from Inequality (12.18) that

$$u_{n+1} \leq k^{n+1} - 2(u_1 k^n + \cdots + u_n k)$$
$$\leq \mathrm{Card}(A^{n+1}) - \mathrm{Card}((X_n A^* \cup A^* X_n) \cap A^{n+1})$$
$$= \mathrm{Card}(A^{n+1} - (X_n A^* \cup A^* X_n)).$$

This shows that we can choose a set Y of u_{n+1} words of length $n + 1$ on the alphabet A, which do not have a prefix or a suffix in X_n. Then $X_{n+1} = Y \cup X_n$ is bifix, which ends the proof. $\qquad \square$

Table 12.5. Maximal 2-realizable length distributions of length $N = 2, 3,$ and 4.

N	2			3				4				
	u_1	u_2	$u(1/2)$	u_1	u_2	u_3	$u(1/2)$	u_1	u_2	u_3	u_4	$u(1/2)$
	2	0	1.0000	2	0	0	1.0000	2	0	0	0	1.0000
	1	1	0.7500	1	1	1	0.8750	1	1	1	1	0.9375
				1	0	2	0.7500	1	0	2	1	0.8125
								1	0	1	3	0.8125
								1	0	0	4	0.7500
	0	4	1.0000	0	4	0	1.0000	0	4	0	0	1.0000
				0	3	1	0.8750	0	3	1	0	0.8750
								0	3	0	1	0.8125
				0	2	2	0.7500	0	2	2	2	0.8750
								0	2	1	3	0.8125
								0	2	0	4	0.7500
				0	1	5	0.8750	0	1	5	1	0.9375
								0	*1*	*4*	*4*	*1.0000*
								0	1	3	5	0.9375
								0	1	2	6	0.8750
								0	1	1	7	0.8125
								0	1	0	9	0.8125
				0	0	8	1.0000	0	0	8	0	1.0000
								0	0	7	1	0.9375
								0	0	6	2	0.8750
								0	0	5	4	0.8750
								0	0	4	6	0.8750
								0	0	3	8	0.8750
								0	0	2	10	0.8750
								0	0	1	13	0.9375
								0	0	0	16	1.0000

The bound $1/2$ in the statement of Proposition 12.1 is not the best possible. It is conjectured in [4] that the statement holds with $3/4$ instead of $1/2$. Some attempts to prove the conjecture have led to improvements over Proposition 12.1. For example, it is proved in [71] that $1/2$ can be replaced by $5/8$. Another approach to the conjecture is presented in [15].

For convenience, we call a sequence (u_n) of integers k-*realizable* if there is a bifix code on k symbols with this length distribution.

We fix $N \geq 1$ and we order sequences $(u_n)_{1 \leq n \leq N}$ of integers by setting $(u_n) \leq (v_n)$ if and only if $u_n \leq v_n$ for $1 \leq n \leq N$. If $(u_n) \leq (v_n)$ and (v_n) is k-realizable then so is (u_n). We give in Table 12.5 the values of the maximal 2-realizable sequences for $N \leq 4$, with respect to this order. Set $u(z) = \sum_{n \geq 1} u_n z^n$. For each value of N, we list in decreasing lexicographic order the maximal realizable sequence with the corresponding value of the sum $u(1/2) = \sum u_n 2^{-n}$. The distributions with value 1 correspond to

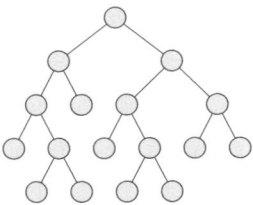

Fig. 12.44. A maximal bifix code of degree 3.

maximal bifix codes. For example, the distribution $(0, 1, 4, 4)$ highlighted
in Table 12.5 corresponds to the maximal bifix code of Fig. 12.44.

It can be checked in this table that the minimal value of the sums $u(1/2)$
is $3/4$. Since the distributions listed are maximal for componentwise order,
this shows that for any sequence $(u_n)_{1 \leq n \leq N}$ with $N \leq 4$ such that $u(1/2) \leq 3/4$, there exists a binary bifix code C such that $u(z) = \sum_{n \geq 1} u_n z^n$ is the
generating series of lengths of C.

For a maximal bifix code C which is a regular set, the generating
series of the lengths of the words of C satisfies $f_C(1/k) = 1$, where
k is the size of the alphabet. The average length of C with respect
to the uniform Bernoulli distribution is $(1/k)f'_C(1/k)$. Indeed, setting
$f_C(z) = \sum_{n \geq 1} u_n z^n$, one gets $f'_C(z) = \sum_{n \geq 1} n u_n z^{n-1}$ and thus
$(1/k)f'_C(1/k) = \sum_{n \geq 1} n u_n k^{-n}$. It is known that the average length
of a regular maximal bifix code is an integer, called the $degree$ of the
code [26, 61].

For example, the maximal bifix code C represented in Fig. 12.44 has
degree 3. One has

$$f_C(z) = z^2 + 4z^3 + 4z^4,$$

$$f'_C(z) = 2z + 12z^2 + 16z^3,$$

and thus $f_C(1/2) = 1$ and $(1/2)f'_C(1/2) = 3$.

Table 12.6 lists the length distributions of finite maximal bifix codes
of degree $d \leq 4$ over $\{a, b\}$. For each degree, the last column contains the
number of bifix codes with this distribution, with a total number of 73 of
degree 4. Note that for the highlighted distribution $(0, 1, 4, 4)$, there are two
distinct bifix codes. One is the code of Fig. 12.44, and the other is obtained
by exchanging 0 and 1.

We have seen (Eq. (12.16)) that the Golomb–Rice code of order k has
average length $k + 2$ for the uniform Bernoulli distribution on the alphabet.

Table 12.6. The length distributions of binary finite maximal bifix codes of degree at most 4.

d	1		2			3			4		
	2	1	0 4	1		0 0 8	1		0 0 0 16		1
									0 0 1 12 4		6
									0 0 2 8 8		6
									0 0 2 9 4 4		8
									0 0 3 5 8 4		6
									0 0 3 6 4 8		4
									0 0 3 6 5 4 4		4
									0 0 4 3 5 8 4		4
						0 1 4 4	2		0 1 0 5 12 4		2
									0 1 0 6 8 8		2
									0 1 0 6 9 4 4		4
									0 1 0 7 5 8 4		4
									0 1 0 7 6 5 4 4		2
									0 1 0 8 2 9 4 4		2
									0 1 1 3 9 8 4		4
									0 1 1 4 6 8 8		4
									0 1 1 4 6 9 4 4		4
									0 1 1 5 3 9 8 4		4
									0 1 2 2 4 9 12 4		2
		1		1			3				73

The same holds of course for the reversible one. The fact that the average length is an integer is a necessary condition for the existence of a reversible version of any regular prefix code, as we have already mentioned before. The average length of the Golomb code G_3 is easily seen to be $7/2$ for the uniform Bernoulli distribution. Since this is not an integer, there is no regular bifix code with the same length distribution $(0, 1, 3, 3, \ldots)$. Actually, one may verify that there is not even a binary bifix code with length distribution $(0, 1, 3, 3, 2)$.

12.11.1. *Exercises*

Exercise 12.11.1. The aim of this exercise is to describe a method, due to Girod [28] (see also [58]), which allows a decoding in both directions for any finite binary prefix code. Let C be a finite binary prefix code and let L be the maximal length of the words of C. Consider a concatenation $c_1 c_2 \cdots c_n$ of codewords. Let

$$w = c_1 c_2 \cdots c_n 0^L \oplus 0^L \tilde{c}_1 \tilde{c}_2 \cdots \tilde{c}_n, \tag{12.19}$$

where $\tilde{c}$ is the reverse of the word c and where $\oplus$ denotes addition mod 2. Show that w can be decoded in both directions.

12.12. Synchronizing Words

A word v is said to be *synchronizing* for a prefix code C if for any words u, w, one has uvw in C^* only if uv and w are in C^*. Thus, the decoding of a message where v appears has to break at the end of v. A prefix code is said to be *synchronized* if there exists a synchronizing word. An occurrence of a synchronizing word limits the propagation of errors that have occurred during the transmission, as shown in the following example.

Example 12.24. Consider the prefix code $C = \{01, 10, 110, 111\}$. The word 110 is not synchronizing. Indeed, it appears in a non-trivial way in the parsing of **111|01**. On the contrary $v = 0110$ is synchronizing. Indeed, the only possible parsings of v in some context are $\cdots 0|110|\cdots$ and $\cdots |01|10|\cdots$. In both cases, the parsing has a cut point at the end of v. To see how an occurrence of v avoids error propagation, consider the message in the first row below and the corrupted version below it produced when an error has changed the third bit of the message from 0 to 1.

$$0\ 1|0\ 1|1\ 0|1\ 1\ 1|\mathbf{0}\ \mathbf{1}|\mathbf{1}\ \mathbf{0}|0\ 1|0\ 1,$$
$$0\ 1|\mathit{1}\ 1\ 1|0\ 1|1\ 1\ \mathbf{0}|\mathbf{1}\ \mathbf{1}\ \mathbf{0}|0\ 1|0\ 1.$$

After the occurrence of 0110, the parsings on both rows become identical again. Thus, the occurrence of the word v has stopped the propagation of the error. Note that it can also happen that the resulting message does not have a decoding anymore. This happens, for example, if the second bit of the message is changed from 1 to 0. In this case, the error is detected and some appropriate action may be taken. In the previous case, the error is not detectable and may propagate for an arbitrary number of bits.

Example 12.25. The word 010 is synchronizing for the prefix code C used in the Franaszek code (see Fig. 12.30).

A synchronizing word can also be defined for a deterministic automaton. Let $\mathcal{A}$ be a deterministic automaton. A word w is said to be *synchronizing* for $\mathcal{A}$ if all paths labeled w end at the same state. Thus the state reached after reading a synchronizing word is independent of the starting state. A deterministic automaton is said to be *synchronized* if there exists a synchronizing word for the automaton.

An automaton is *strongly connected* if any pair of states is connected by some path. Let i be a state of a strongly connected deterministic automaton $\mathcal{A}$. Let C be the prefix code of first returns from i to i. Then C has a synchronizing word if and only if $\mathcal{A}$ is synchronized. First, let v be a synchronizing word for C. Then any path labeled v ends at state i. Indeed, if $p \xrightarrow{v} q$ let u, w be such that $i \xrightarrow{u} p$ and $q \xrightarrow{w} i$. Then uvw is in C^*, which implies that uv is in C^*. This implies that $q = i$ since the automaton is deterministic. Thus, v is synchronizing for the automaton $\mathcal{A}$. Conversely, if v is synchronizing for $\mathcal{A}$, then since $\mathcal{A}$ is strongly connected, there is a word w such that all paths labeled vw end at state i. Then, vw is a synchronizing word for C.

Example 12.26. Let $\mathcal{A}$ be the automaton represented in Fig. 12.45. Each word w defines an *action* on the set of states, which is the partial function that maps a state p to the state reached from p by the path labeled by the input word w. The set of first returns to state 1 is the maximal prefix code $C = \{00, 01, 110, 111\}$ of Example 12.24. In Table 12.7 the actions of words of length at most 4 are listed. The words are ordered by their length, and within words of the same length, by lexicographic order. Each column is reported only the first time it occurs. We stop at the first synchronizing word, which is 0110.

The road coloring theorem. There are prefix codes which are not synchronized. For example, if the lengths of the codewords of a non-empty prefix code are all multiples of some integer $p \geq 2$, then the code is not synchronized. The same observation holds for a strongly connected automaton. If the period of the underlying graph (i.e. the greatest common

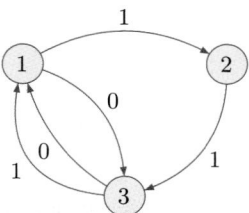

Fig. 12.45. A synchronized deterministic automaton.

Table 12.7. The action of words on the states of the automaton $\mathcal{A}$.

	0	1	00	01	10	11	001	011	100	101	110	111	0011	0110
1	3	2	1	1	—	3	2	2	—	—	1	1	3	1
2	—	3	—	—	1	1	—	—	3	2	3	2	—	—
3	1	1	3	2	3	2	1	3	1	1	—	3	2	1

divisor of the lengths of its cycles) is not 1, then the automaton is not synchronized. The *road coloring theorem* asserts the following.

Theorem 12.1. *Let G be a strongly connected graph with period 1 such that each vertex has two outgoing edges. Then there exists a binary labeling of the edges of G, which turns it into a synchronized deterministic automaton.*

The road coloring theorem was proposed as a conjecture by Adler *et al.* [3]. It was proved by Trahtman [65].

The name of this theorem comes from the following interpretation of a synchronizing word: if one assigns a color to each letter of the alphabet, the labeling of the edges of an automaton can be viewed as a coloring of the edges of the underlying graph. One may further identify the vertices of the graph with cities and the edges with roads connecting these cities. A synchronizing word then corresponds to a sequence of colors, which leads to a fixed city regardless of the starting point.

Example 12.27. Consider the automata represented on Fig. 12.46. These automata have the same underlying graph and differ only by the labeling. The automaton on the left is not synchronized. Indeed, the action of the letters on the subsets $\{1, 3\}$ and $\{2, 4\}$ exchanges these sets as shown in Fig. 12.47.

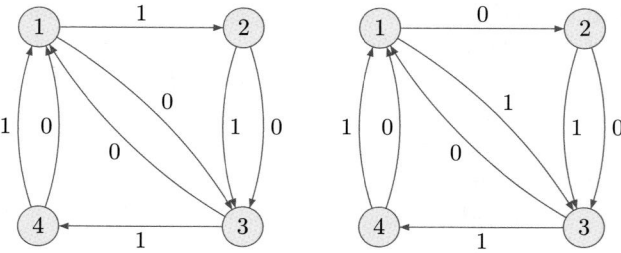

Fig. 12.46.   Two different labelings: a non-synchronized and a synchronized automaton.

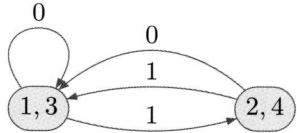

Fig. 12.47.   The action on the sets $\{1, 3\}$ and $\{2, 4\}$.

On the other hand, the automaton on the right is synchronized. Indeed, 101 is a synchronizing word.

It is important to note that for equal letter costs, an optimal prefix code can always be chosen to be synchronized (provided the greatest common divisor of the lengths is 1). Indeed, the relabeling of an automaton accepting such a code does not change its length distribution. The Franaszek code of Sec. 12.9 is actually chosen in such a way that the code on the right-hand side of Fig. 12.30 is synchronized (010 is a synchronizing word).

For unequal letter costs, it has been proved that the result holds for finite maximal codes. Indeed, it is proved in [51] (see also [11]) that any finite maximal prefix code is commutatively equivalent to a synchronized prefix code.

12.13. Directions for Future Research

The field of variable-length codes and automata has a number of challenging mathematical open problems. They are often formulated as conjectures, some of which have been open for many years. Their solution would increase our understanding of these objects and give rise to new algorithms. Let us mention the following ones of particular importance.

The structure of finite maximal codes. It is not known whether it is decidable whether a finite code is or is not contained in a finite maximal code. In contrast, it is known that any finite code is contained in a regular maximal code (see [11]).

Another open question is the status of the conjecture concerning the commutative equivalence of any finite maximal code to a prefix code. This conjecture would itself be solved if one could prove the factorization conjecture asserting that for any finite maximal code on the alphabet A, there exist two finite sets of words P, Q on the alphabet A such that any word on the alphabet A has a unique expression of the form $pc_1c_2 \cdots c_n q$ with $p \in P$, $c_i \in C$, and $q \in Q$.

Optimal prefix codes. There is still some work to be done to derive efficient methods for building an optimal prefix code corresponding to a source with an infinite number of elements.

Synchronized automata. It is conjectured that for any synchronized deterministic automaton with n states, there exists a synchronizing word of length at most $(n-1)^2$. This conjecture is known as Černy's conjecture. The length of the shortest synchronizing word has practical significance since an occurrence of a synchronizing word has an error-correcting effect. The best known upper bound is cubic. The conjecture is known to be true in several particular cases.

Constrained channels. For a given constrained channel and allowable code rate, the problem of designing codes that achieve the minimum possible number of encoder states or the minimum sliding block decoder window size remains open.

Bifix codes. No method similar to the Huffman algorithm is known for building an optimal bifix code given a source with weights. It is perhaps related to the fact that the length distributions of bifix codes are not well understood. In particular, the question whether any sequence $(u_n)_{n \geq 1}$ such that $\sum_{n \geq 1} u_n k^{-n} \leq 3/4$ is the length distribution of a bifix code over a k-letter alphabet is still open.

12.14. Conclusion

We have described basic properties of variable-length codes, and some of their uses in the design of optimal codes under various constraints. We have shown some of the relations between codes, finite state automata and finite transducers, both devices for encoding and decoding.

Optimal codes. Among variable-length codes, the most frequently used are prefix codes. These codes are instantaneously decipherable codes. The Elias, Golomb, and Golomb–Rice codes are examples of prefix codes that are infinite, and that encode non-negative integers by an algorithm that is easy to implement. Other prefix codes that encode integers have been designed. For a systematic description, see [58]. Optimal prefix codes for various constraints have been given in this chapter. It is interesting to note that the general problem is still open, and the research is still going on (see [29] for a recent contribution).

General codes. Prefix codes are special codes. Other families of codes have been studied, and a general theory of variable-length codes addresses the properties of these codes. In this context, it is natural to associate with any code C an automaton that recognizes the set C^* of all sequences of codewords. Many properties or parameters of codes are reflected by features of these automata.

It appears that properties of codes are combinatorial in nature; however, they can also be described in an algebraic way based upon the structural properties of the associated automaton or the algebraic properties of the transition monoid of the automaton, also called the syntactic monoid. See [11].

Constrained channels. We have also briefly investigated constraints on the channel. The theory of sliding block codes is much broader in scope than we could convey here. Research is motivated by applications, as we have illustrated by the example of the Franaszek code. For more information along these lines, see [7, 45].

12.15. Solutions to Exercises

Solution to 12.6.1. One has $l_i \geq \log 1/p_i$ and thus

$$\sum_i 2^{-\ell_i} \leq \sum_i 2^{-\log \frac{1}{p_i}} = \sum_i 2^{\log p_i} = \sum_i p_i = 1.$$

Let C be a prefix code with length distribution ℓ_i. Since $\ell_i \leq \log 1/p_i + 1$, its average length W satisfies

$$W = \sum p_i \ell_i \leq \sum p_i \left(\log \frac{1}{p_i} + 1 \right) = p_i \log \frac{1}{p_i} + 1 = H + 1.$$

Solution to 12.6.2. Each word of P followed by a letter is either in P or in C, but not in both. Moreover, any non-empty word in C or P is obtained in this way. This shows that $PA \cup \{\epsilon\} = P \cup C$. Counting the elements on both sides gives $\mathrm{Card}(P)\,\mathrm{Card}(A) + 1 = \mathrm{Card}(P) + \mathrm{Card}(C)$.

Solution to 12.6.3. Let P be the set of words on the alphabet A which do not have s as a factor. Then P is also the set of proper prefixes of the words of the maximal prefix code X. Thus, we have $P\{0, 1\} \cup \{\epsilon\} = X \cup P$, whence $f_P(z)(1 - 2z) = 1 - f_X(z)$. On the other hand, we have $Ps = XQ$ and thus $z^p f_P(z) = f_X(z) f_Q(z)$. Combining these relations and solving for f_X gives the desired solution.

Solution to 12.6.4. One has $Q = \{\epsilon, 01\}$ and thus $f_Q(z) = 1 + z^2$.

Solution to 12.6.5. Since $z^p + 2z f_X(z) = f_X(z) + f_U(z)$, the result follows from (12.12). Prefix synchronized codes were introduced by Gilbert [24], who conjectured that for any $n \geq 1$, the maximal cardinality is obtained for an *unbordered word* such as $11 \cdots 10$ (a word is called unbordered if no non-empty prefix is also a suffix). This conjecture was solved positively by Guibas and Odlyzko [32] who also showed that the generalization to alphabets with k symbols is true for $k \leq 4$ but false for $k \geq 5$.

Solution to 12.6.6. Consider the cost on B defined by $\text{cost}(b) = -\log \pi(b)$. Then $\text{cost}(c) = -\log \pi(c)$ for each codeword and thus a code with minimal cost (with equal weights) will have maximal average length with respect to π.

Solution to 12.7.1. One has, using the fact that $\sum_{n \geq 0} np^n q = p/q$,

$$H = -\sum_{n \geq 0} p^n q \log(p^n q) = -\sum_{n \geq 0} p^n q \log p^n - \sum_{n \geq 0} p^n q \log q$$

$$= -\sum_{n \geq 0} np^n q \log p - \log q = -p/q \log p - \log q,$$

whence the result. If $p^{2^k} = 1/2$ then $p = 2^{-2^{-k}}$. For $x \geq 0$, we have $2^{-x} \geq 1 - x \ln 2 \geq 1 - x$. Thus

$$q = 1 - p = 1 - 2^{-2^{-k}} \leq 2^{-k}.$$

Taking the logarithm of both sides gives $\log(1/q) \geq k$. On the other hand, since $\log x \geq x - 1$ for $1 \leq x \leq 2$, we have $p/q \log(1/p) \geq 1$. Combining these inequalities, we obtain $H \geq k + 1$, whence $H + 1 \geq k + 2 = \lambda_{GR_k}$.

Solution to 12.7.2. We use an induction on m. The property clearly holds for $m = 3$. Let $m \geq 4$ and set $w_0 = w_{m-1} + w_m$. The sequence $w_0 \geq w_1 \geq \cdots \geq w_{m-2}$ is quasi-uniform since $w_{m-3} + w_{m-2} \geq w_{m-1} + w_m$. By the induction hypothesis, an optimal binary tree for the sequence $w_0 \geq w_1 \geq \cdots \geq w_{m-2}$ has leaves of heights k and possibly $k+1$ for some $k \geq 1$. The leaf corresponding to w_0 has to be at height k since it has maximal weight. We replace it by two leaves of level $k+1$ with weights w_{m-1} and w_m. The result is clearly an optimal binary tree with leaves at height k and $k+1$. This argument appears in [22].

Solution to 12.8.1. Suppose that $c_1 c_2 \cdots c_n = c_1' c_2' \cdots c_m'$. We may suppose that the length of the word $c_2 \cdots c_n$ is larger than d, padding both sides on the right by an appropriate number of words c from C. Then, the condition implies that $c_1 = c_2$, and so on.

Solution to 12.8.2. The set C is the code of Example 12.4. It is not weakly prefix since for any n, $(10)(00)^n$ is a prefix of $(100)(00)^n$.

The code C' is weakly prefix. Indeed, the decoding of a word beginning with $001 \cdots$ has to start with 001.

Solution to 12.8.3. Suppose first that C has a sequential decoder. Let d be the maximal length of the values of the output function. Then C is weakly prefix with delay dL, where L is the maximal length of the codewords.

Conversely, if C is a weakly prefix code, the determinization algorithm gives a sequential decoder.

Solution to 12.9.1. First, observe that D, viewed as a code over the alphabet C, is a prefix code. Thus D is a code. Assume now that u, v are such that uv and vu are in D^*. Then, since C is circular, u and v are in C^*. Set $u = x_1 \cdots x_n$ and $v = x_{n+1} \cdots x_{n+m}$. Since uv is in D^*, we have $x_{n+m} \neq x$ and thus v is in D^*. Similarly, since vu is in D^*, we have $x_n \neq x$ and thus u is in D^*. Thus D is circular.

Solution to 12.9.2. We use repeatedly Exercise 12.9.1 to form a sequence $C_0, C_1, \ldots$ of circular codes. To form C_{i+1}, we choose one element x_i in C_i and select words of the form $x_i^k y$ with $y \neq x_i$. We start with $C_0 = \{a, b, c\}$, which is obviously circular, and we choose $x_0 = c$. Then $C_1 = \{a, b, ca, cb\}$ is a circular code. Choosing next $x_1 = cb$, we build $C_2 = \{a, b, ca, cba\}$. Then, we choose $x_2 = b$ to obtain $C_3 = \{a, ba, ca, cba\}$. A final choice of $x_3 = a$ shows that C is of the required form.

Note: this construction is used in a more general setting in the study of circular codes [10].

Solution to 12.11.1. The definition of w being symmetrical, it is enough to show that w can be decoded from left to right. By construction, c_1 is a prefix of w and the first codeword can therefore be decoded. However, this also identifies the prefix of length $L + |c_1|$ of the second term of the right-hand side of (12.19). Adding this prefix to the corresponding prefix of w gives a word beginning with $c_1 c_2$ and thus identifies c_2, and so on.

12.16. Questions and Answers

Question 12.1. Compute the entropy of the source consisting of the five most frequent words in English with the normalized weights given in Table 12.8. Compare the result with the average length of the prefix code of Example 12.1.

Answer. One obtains $H = 2.197$. The average length is $\lambda = 2.29$. Thus, the average length is slightly larger than the entropy.

Table 12.8. Normalized weights of the five most frequent English words.

A	AND	OF	THE	TO
0.116	0.174	0.223	0.356	0.131

Question 12.2. What is the result of the Huffman algorithm applied to the five most frequent words in English with the frequencies given in Question 12.1? What is the average length?

Answer. The result is represented in Fig. 12.48. The average length in this case is $\lambda = 2.247$ (which is smaller than for the code of Example 12.1).

Question 12.3. What is the result of the Garsia–Wachs algorithm applied to the coding of the five most frequent words in English with the distribution given in Question 12.1?

Answer. The result of the combination step is the tree of Fig. 12.49. The recombination step gives the tree of Example 12.1.

Question 12.4. Consider the source having six symbols of costs $3, 4, \ldots, 8$ respectively. Show that the capacity C of this source satisfies $C > 1/2$.

Answer. One has $C = \log(1/\rho)$ where ρ is the positive root of $f(z) = 1$ with $f(z) = z^3 + z^4 + z^5 + z^6 + z^7 + z^8$. Since $f(z) = z^2(z+z^2)(1+z^2+z^4)$, we have $f(\sqrt{2}/2) > 0.5 \times 1.2 \times 1.75 > 1$. This shows that $\rho < \sqrt{2}/2$ and thus $C > 1/2$.

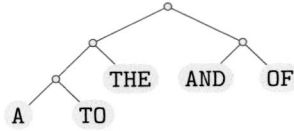

Fig. 12.48.   The result of Huffman's algorithm.

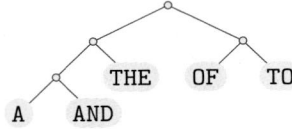

Fig. 12.49.   The result of the Garsia–Wachs algorithm.

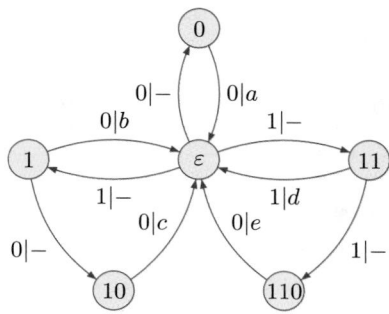

Fig. 12.50. An unambiguous decoder.

Question 12.5. What is the value of m corresponding to (12.14) for the geometric distribution such that $p = 0.95$? What values of p correspond to $m = 8$?

Answer. Since $-(\log(1+p)/\log p) = 13.35$, we obtain $m = 14$. For $m = 8$, we have $9.11 < p < 9.22$.

Question 12.6. Write the encoding of the first 10 integers for the Golomb code G_4.

Answer. The following table is obtained:

n	codeword
0	000
1	001
2	010
3	011
4	100
5	101
6	110
7	111
8	1000
9	1001

Question 12.7. Is the encoding of a, b, c, d, e by $00, 10, 100, 11, 110$ UD? Compute a decoder for this encoding.

Answer. There are only two possibilities to obtain two factorizations starting at the same point. Both alternate indefinitely and never come to

Table 12.9.

name		output
1	(ε, e)	ε
2	$(a, e), (\varepsilon, 0)$	a
3	$(aa, e), (a, 0), (\varepsilon, 00)$	aa

coincide. Thus, the code is UD.

$$\tfrac{1}{2}10^1 0_2 0^1 0_2 0^1, \qquad \tfrac{1}{2}11^1 0_2 0^1 0_2 0^1.$$

The method described in Sec. 12.8 gives the unambiguous transducer of Fig. 12.50.

Question 12.8. Design a sequential transducer decoding the weakly prefix code $C = \{0, 001\}$.

Answer. The general method gives the unambiguous transducer of Fig. 12.51 with $a = 0$ and $b = 001$. The procedure described in Sec. 12.8 applied to the transducer of Fig. 12.51 gives the sequential transducer of Fig. 12.52. We obtain three states given in Table 12.9 with the values of the output function in the last column.

Question 12.9. What is the labeled graph corresponding to the $[1,7]$-constraint?

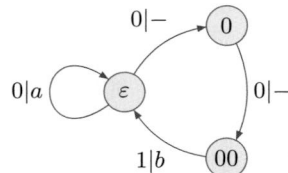

Fig. 12.51.   An unambiguous decoder.

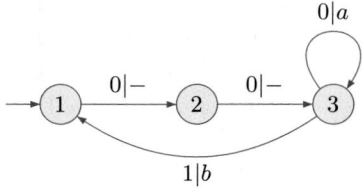

Fig. 12.52.   A sequential transducer.

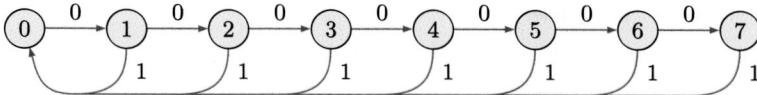

Fig. 12.53.   The [1,7] constraint.

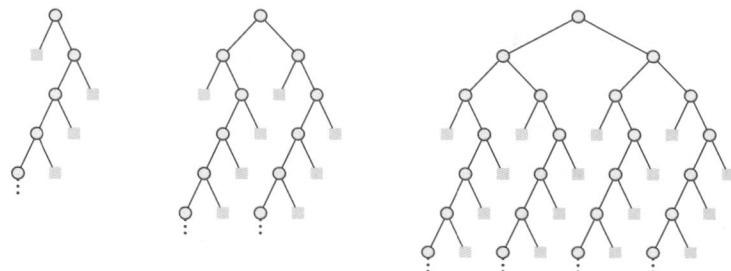

Fig. 12.54.   The reversals of the codes RGR_k, for $k = 0, 1, 2$.

Answer. The graph is the same as for the $[2, 7]$-constraint with one more arrow to allow the block 101. See Fig. 12.53.

Question 12.10. What is the reversal of the reversible Golomb–Rice code of order 1?

Answer. The expression of the reversal of RGR_k is $\widetilde{RGR}_k = RGR_0\{0, 1\}^k$ since RGR_0 is its own reversal. The corresponding trees are represented in Fig. 12.54.

12.17. Keywords

The following list of keywords is a kind of dictionary of terms used in this chapter with an explanation of each entry.

Finite automaton

A finite set of states, together with two distinguished subsets called the sets of initial and terminal states, and a set of edges which are triples consisting of a pair of states and a symbol in some (finite) alphabet A.

Unambiguous automaton

A finite automaton such that, for every pair of states and every word over the alphabet, there is at most one path (i.e. sequence of edges) from the first state of the pair to the second that is labeled with the word.

Deterministic automaton

A finite automaton with a unique initial state such that, for each state and each symbol in an alphabet, there is at most one edge starting in that state and labeled with that symbol.

Determinization algorithm

An algorithm that transforms any finite automaton into an equivalent deterministic finite automaton.

Regular set

A set of words generated by paths in a finite automaton, starting in an initial state and ending in a terminal state.

Transducer

A finite state automaton with output realizing a relation between words on an input alphabet A and words on an output alphabet B. It is similar to an automaton, but edges are quadruples consisting of a pair of states, a word over A, and a word over B. The main purpose for transducers is decoding. In this case, A is the channel alphabet and B is the source alphabet.

Literal transducer

A transducer such that each input label is a single letter.

Unambiguous transducer

A literal transducer whose associated input automaton is unambiguous.

Deterministic transducer

A literal transducer whose associated input automaton is deterministic.

Sequential transducer

A deterministic transducer and an output function. This function maps the terminal states of the transducer into words on the output alphabet. The function realized by a sequential transducer is obtained by appending to the value of the deterministic transducer the image of the output function on the terminal state.

Alphabetic coding

An order preserving encoding. The source and the channel alphabets are ordered, and the codewords are ordered lexicographically.

Optimal source coding

An encoding that minimizes the weighted cost. The weights are on the source alphabet and the costs on the channel alphabet.

Code

A set of non-empty words over an alphabet A such that every concatenation of codewords is uniquely decipherable.

Prefix (suffix) code

A set of non-empty words over an alphabet A such that no word in the set is a proper prefix (suffix) of another one.

Reversible code

A code which is prefix and which is also prefix when the words are read from right to left. Such a code is also called bifix.

Maximal code

A code (the image of an encoding) that is not strictly contained in another code.

Constrained channel

A communication channel that imposes constraints on the channel symbol sequences that can be transmitted.

Constrained source

A source that imposes constraints on the sequences of source alphabet symbols that it generates.

Symbolic dynamics

The study of symbolic representations of dynamical systems and code mappings between such representations.

Sliding block decoder

A decoder that operates on strings of channel symbols with a window of fixed size. The decoder uses m symbols before the current one and a symbols after it (m is for memory and a for anticipation). According to the value of the symbols between time $n - m$ and time $n + a$, the value of the nth source symbol is determined.

State splitting algorithm

An algorithm that transforms a finite-state graph representation of a constraint into a sliding-block decodable finite-state encoder.

Commutative equivalence

A relation between two codes in which there is a one-to-one correspondence between codewords such that corresponding pairs of words have the same number of occurrences of each alphabet symbol (that is, they are anagrams).

Channel capacity

The maximum amount of information that can be transmitted over the channel per unit cost. In many applications, the cost of a symbol corresponds to the number of channel uses or the time required to transmit it. The capacity is then a measure of the amount of information that can be transmitted over the channel per channel use.

Binary source entropy

A measure of the average number of bits per symbol required to represent a source with weighted source symbols.

Generating series

The power series whose coefficients are the number of words or the probabilities of words of each length.

Synchronizing word

A word for a finite automaton such that all paths labeled by this word end in the same state.

Geometric distribution

A probability distribution π on the set of non-negative integers such that $\pi(n+1)/\pi(n)$ is constant.

Acknowledgments

The authors would like to thank Frédérique Bassino, Julien Clément, Éric Incerti, Claire Kenyon, Éric Laporte, and Olivier Vénard for their help in the preparation of this text, and the anonymous referees for their helpful comments.

References

1. R. L. Adler, D. Coppersmith and M. Hassner, Algorithms for sliding block codes, *IEEE Trans. Inform. Theory* **29**, 1 (1983), 5–22.
2. R. L. Adler, J. Friedman, B. Kitchens and B. H. Marcus, State splitting for variable-length graphs, *IEEE Trans. Inform. Theory* **32**, 1 (1986), 108–113.
3. R. L. Adler, L. W. Goodwyn and B. Weiss, Equivalence of topological Markov shifts, *Israel J. Math.* **27**, 1 (1977), 48–63.
4. R. Ahlswede, B. Balkenhol and L. H. Khachatrian, Some properties of fix-free codes, *Proc. 1st Int. Seminar on Coding Theory and Combinatorics, Thahkadzor, Armenia* (1996), pp. 20–33.
5. R. B. Ash, *Information Theory*, Dover Publications Inc., New York (1990), Corrected reprint of the 1965 original.
6. M.-P. Béal, The method of poles: A coding method for constrained channels, *IEEE Trans. Inform. Theory* **36**, 4 (1990), 763–772.
7. M.-P. Béal, *Codage symbolique*, Masson (1993).
8. M.-P. Béal, Extensions of the method of poles for code construction, *IEEE Trans. Inform. Theory* **49**, 6 (2003), 1516–1523.
9. M.-P. Béal and D. Perrin, Codes, unambiguous automata and sofic systems, *Theor. Comput. Sci.* **356**, 1–2 (2006), 6–13.
10. J. Berstel and D. Perrin, *Theory of Codes*, Academic Press (1985).
11. J. Berstel, D. Perrin and C. Reutenauer, *Codes and Automata*, Cambridge University Press (2009).
12. J. Berstel and C. Reutenauer, *Rational Series and their Languages*, Springer-Verlag (1988).
13. V. Bruyère and M. Latteux, Variable-length maximal codes, *Automata, Languages and Programming*, (Paderborn, 1996), *Lecture Notes in Computer Science*, Vol. 1099, Springer-Verlag (1996) pp. 24–47.
14. M. Dalai and R. Leonardi, Non prefix-free codes for constrained sequences, *IEEE Int. Symposium on Information Theory* (2005), pp. 1534–1538.
15. C. Deppe and H. Schnettler, On q-ary fix-free codes and directed deBrujin graphs, *IEEE Int. Symposium on Information Theory* 2(006), pp. 1482–1485.
16. J. Désarménien, La division par ordinateur des mots français: Application à TEX, *Tech. Sci. Informa.* **5**, 4 (1986), 251–265.
17. T. Dutoit, *An Introduction to Text-To-Speech Synthesis*, Kluwer (1997).
18. S. Eilenberg, *Automata, Languages and Machines*, Vol. A, Academic Press (1974).
19. P. Elias, Universal codeword sets and representations of the integers, *IEEE Trans. Inform. Theory* **21**, 2 (1975), 194–203.
20. J. L. Fan, B. H. Marcus and R. M. Roth, Lossless sliding-block compression of constrained systems, *IEEE Trans. Inform. Theory* **46**, 2 (2000), 624–633.
21. P. A. Franaszek, Run-length-limited variable length coding with error propagation limitation, US Patent 3,689,899, 1972.
22. R. G. Gallager and D. C. van Voorhis, Optimal source codes for geometrically distributed integer alphabets, *IEEE Trans. Inform. Theory* **21**, 2 (1975), 228–230.

23. A. M. Garsia and M. L. Wachs, A new algorithm for minimum cost binary trees, *SIAM J. Comput.* **6**, 4 (1977), 622–642.

24. E. N. Gilbert, Synchronization of binary messages, *IRE Trans. Inform. Theory* **6** (1960), 470–477.

25. E. N. Gilbert, Coding with digits of unequal cost, *IEEE Trans. Inform. Theory* **41**, 2 (1995), 596–600.

26. E. N. Gilbert and E. F. Moore, Variable length binary encodings, *Bell Sys. Tech. J.* **38** (1959), 933–967.

27. D. Gillman and R. Rivest, Complete variable length fix-free codes, *Design. Codes Cryptogr.* **5** (1995), 109–114.

28. B. Girod, Bidirectionally decodable streams of prefix code words, *IEEE Commun. Lett.* **3**, 8 (1999), 245–247.

29. M. J. Golin, C. Kenyon and N. E. Young, Huffman coding with unequal letter costs, *ACM Symposium Theory on Computing* (2002), pp. 785–791.

30. S. W. Golomb, Run-length encodings, *IEEE Trans. Inform. Theory*, **12**, 3 (1966), 399–401.

31. G. Gönenç, Unique decipherability of codes with constraints with application to syllabification of Turkish words, *COLING 1973: Computational and Mathematical Linguistics: Proceedings of the International Conference on Computational Linguistics*, Vol. 1, Firenze, Italy (1973), pp. 183–193.

32. L. J. Guibas and A. M. Odlyzko, Maximal prefix-synchronized codes, *SIAM J. Appl. Math.* **35**, 2 (1978), 401–418.

33. C. D. Heegard, B. H. Marcus and P. H. Siegel, Variable-length state splitting with applications to average runlength-constrained (ARC) codes, *IEEE Trans. Inform. Theory* **37**, 3, Part 2 (1991), 759–777.

34. T. C. Hu and M.-T. Shing, *Combinatorial Algorithms*, 2nd edn. Dover Publications Inc., Mineola, NY, (2002).

35. T. C. Hu and A. C. Tucker, Optimal computer search trees and variable-length alphabetical codes, *SIAM J. Appl. Math.* **21** (1971), 514–532.

36. T. C. Hu and P. A. Tucker. Optimal alphabetic trees for binary search, *Inform. Process. Lett.* **67**, 3 (1998), 137–140.

37. D. A. Huffman, A method for the construction of minimum redundancy codes, *Proceedings of the Institute of Electronics and Radio Engineers*, Vol. 40, No. 10, September 1952, pp. 1098–1101.

38. K. A. S. Immink, A practical method for approaching the channel capacity of constrained channels, *IEEE Trans. Inform. Theory* **43**, 5 (1997), 1389–1399.

39. K. A. S. Immink, *Codes for Mass Data Storage Systems*, 2nd edn. Shannon Foundation Publishers (2004).

40. A. Itai, Optimal alphabetic trees, *SIAM J. Comput.* **5**, 1 (1976), 9–18.

41. R. M. Karp, Minimum redundancy codes for the discrete noiseless channel, *IRE Trans. Inform. Theory* **7** (1961), 27–38.

42. J. H. Kingston, A new proof of the Garsia-Wachs algorithm, *J. Algorithms* **9**, 1 (1988), 129–136.

43. D. E. Knuth, *The Art of Computer Programming, Volume III: Sorting and Searching*, 2nd edn. Addison-Wesley (1998).

44. Z. Kohavi, *Switching and Automata Theory*, 2nd edn. McGraw-Hill (1978).
45. D. Lind and B. Marcus, *An Introduction to Symbolic Dynamics and Coding* Cambridge University Press, Cambridge (1995).
46. M. Lothaire, *Applied Combinatorics on Words*, Encyclopedia of Mathematics and its Applications, Vol. 105, Cambridge University Press, Cambridge (2005).
47. B. H. Marcus, R. M. Roth and P. H. Siegel, Constrained systems and coding for recording channels, *Handbook of Coding Theory*, eds. V. S. Pless and W. C. Huffmann, Elsevier (1998).
48. R. J. McEliece, *The Theory of Information and Coding*, student edn. Encyclopedia of Mathematics and its Applications, Vol. 86, Cambridge University Press, Cambridge (2004), with a foreword by Mark Kac.
49. K. Mehlhorn, An efficient algorithm for constructing nearly optimal prefix codes, *IEEE Trans. Inform. Theory* **26**, 5 (1980), 513–517.
50. N. Merhav, G. Seroussi and M. J. Weinberger, Optimal prefix codes for sources with two-sided geometric distributions, *IEEE Trans. Inform. Theory* **46**, 1 (2000), 121–135.
51. D. Perrin and M.-P. Schützenberger, Synchronizing prefix codes and automata and the road coloring problem, *Symbolic Dynamics and its Applications* (New Haven, CT, 1991), Contemporary Mathematics, Vol. 135, American Mathematical Society Providence, RI (1992), pp. 295–318.
52. A. Restivo, Codes and local constraints, *Theoret. Comput. Sci.* **72**, 1 (1990), 55–64.
53. C. Reutenauer, Ensembles libres de chemins dans un graphe, *Bull. Soc. Math. France* **114**, 2 (1986), 135–152.
54. R. F. Rice, Some practical universal noiseless coding techniques, Technical Report, Jet Propulsion Laboratory (1979).
55. I. Richardson, *H.264 and MPEG-4 Video Compression: Video Coding for Next-generation Multimedia*, Wiley (2003).
56. E. Roche and Y. Schabes, eds., *Finite-State Language Processing*, MIT Press (1997).
57. J. Sakarovitch, *Elements of Automata Theory*, Cambridge University Press (2009).
58. D. Salomon, *Variable-Length Codes for Data Compression*, Springer-Verlag (2007).
59. M.-P. Schützenberger, On an application of semigroup methods to some problems in coding, *IEE Trans. Inform. Theory* **2** (1956), 47–60.
60. M.-P. Schützenberger, On a family of submonoids, *Publ. Math. Inst. Hungar. Acad. Sci. Ser. A* **VI** (1961), 381–391.
61. M.-P. Schützenberger, On a special class of recurrent events, *Ann. Math. Statist.* **32** (1961), 1201–1213.
62. C. E. Shannon, A mathematical theory of communication, *Bell Sys. Tech. J.* **27** (1948), 379–423, 623–656.
63. P. W. Shor, A counterexample to the triangle conjecture, *J. Combin. Theory Ser. A* **38** (1983), 110–112.

64. J. Teuhola, A compression method for clustered bit-vectors, *Inf. Process. Lett.* **7**, 6 (1978), 308–311.
65. A. N. Trahtman, The road coloring problem, *Israel J. Math.* (2008), to appear.
66. B. P. Tunstall, *Synthesis of noiseless compression codes.* PhD Thesis, Georgia Institute of Technology (1967).
67. B. Varn, Optimal variable length codes (arbitrary symbol cost and equal code word probability), *Inform. Contr.* **19** (1971), 289–301.
68. J. Wen and J. Villasenor, Reversible variable length codes for efficient and robust image and video coding, *IEEE Data Compression Conference* (1998), pp. 471–480.
69. M. W. Y. Takishima and H. Murakami, Reversible variable length codes, *IEEE Trans. Commun.* **43**, 2/3/4 (1995), 158–162.
70. C. Ye and R. W. Yeung, Some basic properties of fix-free codes, *IEEE Trans. Inform. Theory* **47**, 1 (2001), 72–87.
71. S. Yekhanin, Improved upper bound for the redundancy of fix-free codes, *IEEE Trans. Inform. Theory* **50**, 11 (2004), 2815–2818.

Chapter 13

DECODING AND FINDING THE MINIMUM DISTANCE WITH GRÖBNER BASES: HISTORY AND NEW INSIGHTS

STANISLAV BULYGIN* and RUUD PELLIKAAN†

*Department of Mathematics, University of Kaiserslautern,
P.O. Box 3049, 67653 Kaiserslautern, Germany

†Department of Mathematics and Computing Science,
Eindhoven University of Technology,
P.O. Box 513, NL-5600 MB, Eindhoven, The Netherlands
*bulygin@mathematik.uni-kl.de
†g.r.pellikaan@tue.nl

In this chapter, we discuss decoding techniques and finding the minimum distance of linear codes with the use of Gröbner bases. First, we give a historical overview of decoding cyclic codes via solving systems of polynomial equations over finite fields. In particular, we mention papers of Cooper, Reed, Chen, Helleseth, Truong, Augot, Mora, Sala, and others. Some structural theorems that use Gröbner bases in this context are presented. After that we shift to the general situation of arbitrary linear codes. We give an overview of approaches of Fitzgerald and Lax. Then we introduce our method of decoding linear codes that reduces this problem to solving a system of quadratic equations. We discuss open problems and future research possibilities.

13.1. Introduction

The chapter is devoted to decoding and finding the minimum distance of arbitrary linear codes with the use of Gröbner bases. In recent years, a lot of attention was paid to this question for cyclic codes, which form a particular subclass of linear codes. We give a survey on decoding cyclic codes with Gröbner bases and consider two approaches that exist for arbitrary linear codes. We also present a new method based on reducing the problems of decoding and finding the minimum distance to solving a system of

585

quadratic equations. We give a very brief introduction to Gröbner bases theory. Introduction material can be taken for instance from [1, 2].

Quite a lot of methods exist for decoding cyclic codes and the literature on this topic is vast. We just mention [3–7]. But all these methods do not correct up to the true error-correcting capacity. The theory of Gröbner bases is used to remedy this problem. These methods are roughly divided into the following categories:

- Newton identities method [8–14].
- Power sums method or Cooper's philosophy [14–19].

The term Cooper's philosophy first was used during the talk presented in [20]. In Sec. 13.2, necessary background on Gröbner bases is given, as well as the notation we are going to use throughout the chapter. In Sec. 13.3, we give an overview of the methods based on power sums and Newton identities together with examples. Section 13.4 is devoted to the case of arbitrary linear codes. Namely, we look at the method of Fitzgerald and Lax, and the method based on solving a quadratic system of equations. We should mention that there exist other Gröbner bases-based methods for arbitrary linear codes, e.g. generalizations of the Cooper's philosophy [21, 22], applications of Padé approximation [23, 24], FGLM-like techniques [25, 26], and key equation [27]. These methods are out of scope of this chapter. We end the chapter with the thoughts for practitioners and directions for further research, as well as conclusions, terminology list and the list of sample questions and answers to the material presented in the chapter. We made an extensive bibliography, so that the reader is able to look at numerous sources that exist in the area.

13.2. Background

13.2.1. *Gröbner Bases in Polynomial System Solving*

The theory of Gröbner basis is about solving systems of polynomial equations in several variables and can be viewed as a common generalization of Gaussian elimination in linear algebra that deals with linear systems of equations in several variables and the Euclidean algorithm that is about polynomial equations of arbitrary degree in one variable. The polynomial equations are linearized by treating the monomials as new variables. In this way the number of variables grows exponentially in the degree of the polynomials. The complexity of computing a Gröbner basis is doubly exponential in general, and exponential in our case of a finite set of solutions. In this subsection, we give a brief overview of monomial orders, Gröbner bases and their use in polynomial system solving. This subsection is only intended to refresh these notions; for a thorough exposition of the material the reader can use, e.g. [1, 28].

Let $\mathbb{F}$ be a field and let $\mathbb{F}[X_1, \ldots, X_n] = \mathbb{F}[X]$ be the polynomial ring in n variables over $\mathbb{F}$. In commutative algebra objects like polynomials, ideals, and quotients are intensively studied. If we want to do computations with these objects we must somehow impose an order on them, so that we know which way a computation will go. Let $\text{Mon}(X)$ be the set of all monomials in the variables $X = (X_1, \ldots, X_n)$.

Definition 13.1. A *monomial order* on $\mathbb{F}[X]$ is any relation $>$ on $\text{Mon}(X)$ such that

(1) $>$ is a total order on $\text{Mon}(X)$.
(2) $>$ is multiplicative, i.e. $X^\alpha > X^\beta$ implies $X^\alpha \cdot X^\gamma > X^\beta \cdot X^\gamma$ for all vectors γ with non-negative integer entries; here $X^\alpha = X_1^{\alpha_1} \cdots X_n^{\alpha_n}$.
(3) $>$ is a well-order, i.e. every non-empty subset of $\text{Mon}(X)$ has a minimal element.

Example 13.2. Here are some orders that will be used in this chapter.

- *Lexicographic order* induced by $X_1 > \cdots > X_n : X^\alpha >_{\text{LP}} X^\beta$ if and only if there exists an s such that $\alpha_1 = \beta_1, \ldots, \alpha_{s-1} = \beta_{s-1}, \alpha_s > \beta_s$.
- *Degree reverse lexicographic order* induced by $X_1 > \cdots > X_n : X^\alpha >_{\text{DP}} X^\beta$ if and only if $|\alpha| := \alpha_1 + \cdots + \alpha_n > \beta_1 + \cdots + \beta_n =: |\beta|$ or if $|\alpha| = |\beta|$ and there exists an s such that $\alpha_n = \beta_n, \ldots, \alpha_{n-s+1} = \beta_{n-s+1}, \alpha_{n-s} < \beta_{n-s}$.
- *Block order* or *product order*. Let X and Y be two ordered sets of variables, $>_1$ a monomial order on $\mathbb{F}[X]$ and $>_2$ a monomial order on $\mathbb{F}[Y]$. The block order on $\mathbb{F}[X, Y]$ is the following: $X^{\alpha_1} Y^{\beta_1} > X^{\alpha_2} Y^{\beta_2}$ if and only if $X^{\alpha_1} >_1 X^{\alpha_2}$ or if $X^{\alpha_1} =_1 X^{\alpha_2}$ and $Y^{\beta_1} >_2 Y^{\beta_2}$.

Definition 13.3. Let $>$ be a monomial order on $\mathbb{F}[X]$. Let $f = \sum_\alpha c_\alpha X^\alpha$ be a non-zero polynomial from $\mathbb{F}[X]$. Let α_0 be such that $c_{\alpha_0} \neq 0$ and $X^{\alpha_0} > X^\alpha$ for all $\alpha \neq \alpha_0$ with $c_\alpha \neq 0$. Then $\text{lc}(f) := c_{\alpha_0}$ is called the *leading coefficient* of f, $\text{lm}(f) := X^{\alpha_0}$ is called the *leading monomial* of f, $\text{lt}(f) := c_{\alpha_0} X^{\alpha_0}$ is called the *leading term* of f.

Having these notions we are ready to define the notion of a Gröbner basis.

Definition 13.4. Let I be an ideal in $\mathbb{F}[X]$. The *leading ideal* of I with respect to $>$ is defined as $L_>(I) := \langle \mathrm{lt}(f) \mid f \in I, f \neq 0 \rangle$. Then $L_>(I)$ is sometimes abbreviated by $L(I)$. A finite subset $G = \{g_1, \ldots, g_m\}$ of I is called a *Gröbner basis* for I with respect to $>$ if $L_>(I) = \langle \mathrm{lt}(g_1), \ldots, \mathrm{lt}(g_m) \rangle$.

Example 13.5. Consider two polynomials $f = X^3, g = Y^4 - X^2Y$ from $\mathbb{F}[X, Y]$, where $\mathbb{F}$ is any field. We claim that f and g constitute a Gröbner basis of an ideal $I = \langle f, g \rangle$ with respect to the degree reverse lexicographic order $>_{\mathrm{DP}}$ with $X > Y$. For this we need to show that $L(I) = \langle \mathrm{lt}(f), \mathrm{lt}(g) \rangle$. We have $\mathrm{lt}(f) = X^3$ and $\mathrm{lt}(g) = Y^4$. Thus we have to show that $\mathrm{lt}(h)$ is divisible either by X^3 or by Y^4, for any $h \in I$. A polynomial h can be written as $h = af + bg = aX^3 + b(Y^4 - X^2Y)$. If $\deg(a) > 1 + \deg(b)$, then $\mathrm{lm}(h) = \mathrm{lm}(a)X^3$. If $\deg(a) < 1 + \deg(b)$, then $\mathrm{lm}(h)$ is divisible by Y^4. If $\deg(a) = 1 + \deg(b)$ and $\mathrm{lm}(a)X^3 \neq \mathrm{lm}(b)Y^4$, then $\mathrm{lm}(h) = \mathrm{lm}(a)X^3$. If $\deg(a) = 1 + \deg(b)$ and $\mathrm{lm}(a)X^3 = \mathrm{lm}(b)Y^4$, then $\mathrm{lm}(h)$ is divisible by X^3.

Every ideal has a Gröbner basis. By doing some additional operations on the elements of a Gröbner basis, one can construct a *reduced* Gröbner basis. For the definition we refer to the literature. The reduced Gröbner basis of an ideal with respect to a given monomial order is unique.

There are several algorithms for computing Gröbner bases. Historically the first is Buchberger's algorithm [29] and its numerous improvements and optimizations are implemented in several computer algebra systems like for example SINGULAR [30], MAGMA [31], and CoCoA [32]. Also there are algorithms F4 and F5 [33, 34]. The algorithm F4 is implemented, e.g. in MAGMA and FGB [35].

For solving systems of polynomial equations with the use of Gröbner bases we need the so-called elimination orders.

Definition 13.6. Let S be some subset of variables in X. A monomial order $>$ on $\mathbb{F}[X]$ is called an *elimination order* with respect to S if for all $f \in \mathbb{F}[X]$ from the fact that $\mathrm{lm}(f) \in \mathbb{F}[X \backslash S]$ follows that $f \in \mathbb{F}[X \backslash S]$.

For example, let $>$ be the block order $(>_1, >_2)$ on $\mathbb{F}[S, T]$ ($S \subset X$ and $T = X \backslash S$), where $>_1$ is defined on $\mathbb{F}[S]$ and $>_2$ is defined on $\mathbb{F}[T]$, is an elimination order with respect to S. In particular, lexicographic order is an elimination order with respect to any subset S of X. Due to this property of the lexicographic order, we have the following theorem that can be obtained from the elimination theorem (p. 114 in [1]) and the theorem about finiteness (p. 232 in [1], also p. 83 [28]).

Theorem 13.7. *Let $f_1(X) = \cdots = f_m(X) = 0$ be a system of polynomial equations defined over $\mathbb{F}[X]$ with $X = (X_1, \ldots, X_n)$, such that it has finitely many solutions in $\bar{\mathbb{F}}^n$, where $\bar{\mathbb{F}}$ is the algebraic closure of $\mathbb{F}$. Let $I = \langle f_1, \ldots, f_m \rangle$ be an ideal defined by the polynomials in the system and let G be a Gröbner basis for I with respect to $>_{\mathrm{LP}}$. Then there are elements $g_1, \ldots, g_n \in G$ such that*

$$\begin{cases} g_n \in \mathbb{F}[X_n], & \mathrm{lt}(g_n) = c_n X_n^{m_n}, \\ g_{n-1} \in \mathbb{F}[X_{n-1}, X_n], & \mathrm{lt}(g_{n-1}) = c_{n-1} X_{n-1}^{m_{n-1}}, \\ \cdots \\ g_1 \in \mathbb{F}[X_1, \ldots, X_n], & \mathrm{lt}(g_1) = c_1 X_1^{m_1}. \end{cases}$$

It is clear how to solve the system I now. After computing G, first solve a univariate equation $g_n(X_n) = 0$. Let $a_1^{(n)}, \ldots, a_{l_n}^{(n)}$ be the roots. For every $a_i^{(n)}$ then solve $g_{n-1}(X_{n-1}, a_i^{(n)}) = 0$ to find possible values for X_{n-1}. Repeat this process until all the coordinates of all the solutions are found. Since the number of solutions is finite it is always possible.

Remark 13.8. Usually from the practical point of view finding a Gröbner basis with respect to an elimination order is harder than with respect to some degree-refining order, like the degree reverse lexicographic order. Therefore, a conversion technique like FGLM [36] comes in hand here. It enables one to convert a basis with respect to one order, for instance some degree-refining order, to another one, such as the lexicographic order. For solving, we actually need an elimination order, but sometimes it is possible to obtain a result with a degree order. More on that in Sec. 13.4.2, Theorem 13.34.

13.2.2. Notation

Let C be a linear code over the field $\mathbb{F}_q$ with q elements of *length n*, *dimension k*, and *minimum distance d*. The parameters of C are denoted by $[n, k, d]$ and its *redundancy* by $r = n - k$. The (true) *error-correcting capacity* $\lfloor (d-1)/2 \rfloor$ of the code is denoted by e. The code C can be constructed via its *generator* matrix G, which is any matrix composed of a basis of vectors in C. Alternatively, one can see C as a null-space of a *parity-check* matrix H, so $c \in C$ if and only if $Hc^T = 0$. The code C is *cyclic*, if for every codeword $\mathbf{c} = (c_0, \dots, c_{n-1})$ in C its cyclic shift $(c_{n-1}, c_0, \dots, c_{n-2})$ is again a codeword in C. When working with cyclic codes, vectors are usually presented as polynomials. So $\mathbf{c}$ is represented by the polynomial $c(x) = \sum_{i=0}^{n-1} c_i x^i$ with $x^n = 1$, more precisely $c(x)$ is an element of the factor ring $\mathbb{F}_q[X]/\langle X^n - 1 \rangle$. Cyclic codes over $\mathbb{F}_q$ of length n correspond one-to-one to ideals in this factor ring. We assume for cyclic codes that $(q, n) = 1$. Let $\mathbb{F} = \mathbb{F}_{q^m}$ be the splitting field of $X^n - 1$ over $\mathbb{F}_q$. Then $\mathbb{F}$ has a *primitive n-th root of unity* which will be denoted by a. A cyclic code is uniquely given by a *defining set* S_C which is a subset of $\mathbb{Z}_n$ such that

$$c(x) \in C \quad \text{if } c(a^i) = 0 \text{ for all } i \in S_C.$$

A code has several defining sets. The *complete defining set* of C is the set of all $i \in \mathbb{Z}_n$ such that $c(a^i) = 0$ for all $c(x) \in C$. If $c(a^i) = 0$, then $c(a^{qi}) = (c(a^i))^q = 0$. Hence a defining set is complete if and only it is invariant under multiplication by q. A *cyclotomic* set of a number $j \in \mathbb{Z}_n$ is a subset $Cl(j) := \{jq^i \bmod n \,|\, i \in \mathbb{N}\}$. A defining set is complete if and only if it is a disjoint union of some cyclotomic sets. The size of the complete defining set is equal to the redundancy $r = n - k$.

13.3. Decoding and Finding Minimum Distance of Cyclic Codes

13.3.1. *Cooper's Philosophy and Its Development*

In this subsection, we give an overview of the so-called Cooper's philosophy or the power sums method (see Sec. 13.1). The idea here is basically to write parity check equations with unknowns for error positions and error values and then try to solve with respect to these unknowns by adding some natural restrictions on them.

If i is in the defining set of C, then

$$(1, a^i, \dots, a^{(n-1)i})\mathbf{c}^T = c_0 + c_1 a^i + \cdots + c_{n-1} a^{(n-1)i} = c(a^i) = 0.$$

Hence $(1, a^i, \dots, a^{(n-1)i})$ is a parity check of C. Let $\{i_1, \dots, i_r\}$ be a defining set of C. Then a parity check matrix H of C can be represented as a matrix with entries in $\mathbb{F}$:

$$H = \begin{pmatrix} 1 & a^{i_1} & a^{2i_1} & \dots & a^{(n-1)i_1} \\ 1 & a^{i_2} & a^{2i_2} & \dots & a^{(n-1)i_2} \\ \vdots & \vdots & \vdots & \ddots & \vdots \\ 1 & a^{i_r} & a^{2i_r} & \dots & a^{(n-1)i_r} \end{pmatrix}.$$

Let $\mathbf{c}$, $\mathbf{r}$, and $\mathbf{e}$ be the transmitted codeword, the received word, and the error vector, respectively. Then $\mathbf{r} = \mathbf{c} + \mathbf{e}$. Denote the corresponding polynomials by $c(x)$, $r(x)$, and $e(x)$, respectively. If we apply the parity check matrix to $\mathbf{r}$, we obtain

$$\mathbf{s}^T := H\mathbf{r}^T = H(\mathbf{c}^T + \mathbf{e}^T) = H\mathbf{c}^T + H\mathbf{e}^T = H\mathbf{e}^T,$$

since $H\mathbf{c}^T = 0$, where $\mathbf{s}$ is the so-called *syndrome vector*. Define $s_i = r(a^i)$ for all $i = 1, \ldots, n$. Then $s_i = e(a^i)$ for all i in the complete defining set, and these s_i are called the *known syndromes*. The remaining s_i are called the *unknown syndromes*. We have that the vector $\mathbf{s}$ above has entries $\mathbf{s} = (s_{i_1}, \ldots, s_{i_r})$. Let t be the number of errors that occurred while transmitting $\mathbf{c}$ over a noisy channel. If the error vector is of weight t, then it is of the form

$$\mathbf{e} = (0, \ldots, 0, e_{j_1}, 0, \ldots, 0, e_{j_l}, 0, \ldots, 0, e_{j_t}, 0, \ldots, 0),$$

more precisely there are t indices j_l with $1 \le j_1 < \cdots < j_t \le n$ such that $e_{j_l} \ne 0$ for all $l = 1, \ldots, t$ and $e_j = 0$ for all j not in $\{j_1, \ldots, j_t\}$. We obtain

$$s_{i_u} = r(a^{i_u}) = e(a^{i_u}) = \sum_{l=1}^{t} e_{j_l}(a^{i_u})^{j_l}, \quad 1 \le u \le r. \tag{13.1}$$

The $a^{j_1}, \ldots, a^{j_t}$ but also the $j_1, \ldots, j_t$ are called the *error locations*, and the $e_{j_1}, \ldots, e_{j_t}$ are called the *error values*. Define $z_l = a^{j_l}$ and $y_l = e_{j_l}$. Then $z_1, \ldots, z_t$ are the error locations and $y_1, \ldots, y_t$ are the error values and the syndromes in Eq. (13.1) become *generalized power sum functions*

$$s_{i_u} = \sum_{l=1}^{t} y_l z_l^{i_u}, \quad 1 \le u \le r. \tag{13.2}$$

In the binary case the error values are $y_i = l$, and the syndromes are the ordinary power sums.

Now we give a description of *Cooper's philosophy* [18]. As the receiver does not know how many errors occurred, the upper bound t is replaced by the error-correcting capacity e and some z_l's are allowed to be zero, while assuming that the number of errors is at most e. The following variables are introduced: $X_1, \ldots, X_r, Z_1, \ldots, Z_e$, and $Y_1, \ldots, Y_e$, where X_u stands for the syndrome $s_{i_u}, 1 \le u \le r$; Z_l stands for the error location z_l for $1 \le l \le t$, and 0 for $t < l \le e$; and finally Y_l stands for the error value y_l for $1 \le l \le t$, and any element of $\mathbb{F}_q \setminus \{0\}$ for $t < l \le e$. The syndrome equation (13.1) is rewritten in terms of these variables as power sums:

$$f_u := \sum_{l=1}^{e} Y_l Z_l^{i_u} - X_u = 0, \quad 1 \le u \le r.$$

We also add some other equations in order to specify the range of values that can be achieved by our variables, namely:

$$\epsilon_u := X_u^{q^m} - X_u = 0, \quad 1 \le u \le r,$$

since $s_j \in \mathbb{F}$;

$$\eta_l := Z_l^{n+1} - Z_l = 0, \quad 1 \le l \le e,$$

since a^{j_l} are either n-th roots of unity or zero; and

$$\lambda_l := Y_l^{q-1} - 1 = 0, \quad 1 \le l \le e,$$

since $y_l \in \mathbb{F}_q \setminus \{0\}$. We obtain the following set of polynomials in the variables $X = (X_1, \ldots, X_r)$, $Z = (Z_1, \ldots, Z_e)$, and $Y = (Y_1, \ldots, Y_e)$:

$$F_C = \{f_u, \epsilon_u, \eta_l, \quad \lambda_l : 1 \le u \le r, 1 \le l \le e\} \subset \mathbb{F}_q[X, Z, Y]. \qquad (13.3)$$

The zero-dimensional ideal I_C generated by F_C is called the *CRHT-syndrome ideal* associated to the code C, and the variety $V(F_C)$ defined by F_C is called the *CRHT-syndrome variety* , after Chen, Reed, Helleseth, and Truong, see [14, 16, 17]. We have $V(F_C) = V(I_C)$.

Initially decoding of cyclic codes was essentially brought to finding the reduced Gröbner basis of the CRHT-ideal . It turned out that adding more polynomials to this ideal gives better results [19]. By adding polynomials

$$\chi_{l,m} := Z_l Z_m p(n, Z_l, Z_m) = 0, \quad 1 \le l < m \le e$$

to F_C, where

$$p(n, X, Y) = \frac{X^n - Y^n}{X - Y} = \sum_{i=0}^{n-1} X^i Y^{n-1-i}, \qquad (13.4)$$

we ensure that for all l and m either Z_l and Z_m are distinct or at least one of them is zero. The resulting set of polynomials:

$$F_C' := \{f_u, \epsilon_u, \eta_i, \lambda_i, \chi_{l,m} : 1 \le u \le r, 1 \le i \le e, 1 \le l < m \le e\}$$
$$\subset \mathbb{F}_q[X, Z, Y]. \qquad (13.5)$$

The ideal generated by F'_C is denoted by I'_C. By investigating the structure of I'_C and its reduced Gröbner basis with respect to lexicographic order induced by $X_1 < \cdots < X_r < Z_e < \cdots < Z_1 < Y_1 < \cdots < Y_e$, the following result is proved, see [19](Theorems 6.8 and 6.9).

Theorem 13.9. *Every cyclic code* C *possesses a general error-locator polynomial* L_C. *This means that there exists a unique polynomial* L_C *from* $\mathbb{F}_q[X_1, \ldots, X_r, Z]$ *that satisfies the following two properties*:

- $L_C = Z^e + a_{t-1} Z^{e-1} + \cdots + a_0$ *with* $a_j \in \mathbb{F}_q[X_1, \ldots, X_r]$, $0 \leq j \leq e - 1$,
- *given a syndrome* $\mathbf{s} = (s_{i_1}, \ldots, s_{i_r}) \in \mathbb{F}^r$ *corresponding to an error of weight* $t \leq e$ *and error locations* $\{k_1, \ldots, k_t\}$, *if we evaluate the* $X_u = s_{i_u}$ *for all* $1 \leq u \leq r$, *then the roots of* $L_C(\mathbf{s}, Z)$ *are exactly* $a^{k_1}, \ldots, a^{k_t}$ *and* 0 *of multiplicity* $e - t$, *in other words*

$$L_C(\mathbf{s}, Z) = Z^{e-t} \prod_{i=1}^{t} (Z - a^{k_i}).$$

Such an error-locator polynomial actually is an element of the reduced Gröbner basis of I'_C. Having this polynomial, decoding of the cyclic code C reduces to univariate factorization. The main effort here is finding the reduced Gröbner basis of I'_C. In general this is infeasible already for moderate size codes, but for small codes, though, it is possible to apply this technique successfully [37].

Example 13.10. As an example we consider finding the general error-locator polynomial for a binary cyclic BCH code C with parameters $[15, 7, 5]$ that correct 2 errors. This code has $\{1, 3\}$ as a defining set. So here $q = 2$, $m = 4$, and $n = 15$. The field $\mathbb{F}_{16}$ is the splitting field of $X^{15} - 1$ over $\mathbb{F}_2$. During this example, we show how the idea of the Cooper's philosophy is applied. For rigorous justification of the steps below, see [14, 16, 17, 19, 37]. In the above description, we have to write equations for all syndromes that correspond to elements in the complete defining set. Note that we may write the equations only for the elements from the defining set $\{1, 3\}$ as all the others are just consequences of those. Following the description above we write generators F'_C of the ideal I'_C in the ring $\mathbb{F}_{16}[X_1, X_2, Z_1, Z_2]$:

$$\begin{cases} Z_1 + Z_2 - X_1, & Z_1^3 + Z_2^3 - X_2, \\ X_1^{16} - X_1, & X_2^{16} - X_2, \\ Z_1^{16} - Z_1, & Z_2^{16} - Z_2, \\ Z_1 Z_2 p(15, Z_1, Z_2). \end{cases}$$

We suppress the equations λ_1 and λ_2 as error values are over $\mathbb{F}_2$. In order to find the general error-locator polynomial we compute the reduced Gröbner basis G of the ideal I_C' with respect to the lexicographical order induced by $X_1 < X_2 < Z_2 < Z_1$. The elements of G are:

$$\begin{cases} X_1^{16} + X_1, \\ X_2 X_1^{15} + X_2, \\ X_2^8 + X_2^4 X_1^{12} + X_2^2 X_1^3 + X_2 X_1^6, \\ Z_2 X_1^{15} + Z_2, \\ Z_2^2 + Z_2 X_1 + X_2 X_1^{14} + X_1^2, \\ Z_1 + Z_2 + X_1. \end{cases}$$

According to Theorem 6.8 (cf. [19]) the general error-correcting polynomial L_C is then a unique element of G of degree 2 with respect to Z_2. So $L_C \in \mathbb{F}_2[X_1, X_2, Z]$ is

$$L_C(X_1, X_2, Z) = Z^2 + Z X_1 + X_2 X_1^{14} + X_1^2.$$

Let us see how decoding using L_C works. Let $\mathbf{r} = (0, 1, 1, 0, 1, 0, 1, 0, 1, 0, 0, 1, 1, 0, 1)$ be a received word with at most 2 errors. In the field $\mathbb{F}_{16}$ with a primitive element a, such that $a^4 + a + 1 = 0$, a is also a 15-th root of unity. Then the syndromes are $\mathbf{s}_1 = a^2, \mathbf{s}_3 = a$. Plug them into L_C in place of X_1 and X_2 and obtain:

$$L_C(Z) = Z^2 + a^2 Z + a(a^2)^{14} + (a^2)^2 = Z^2 + a^2 Z + a^9.$$

Factorizing yields $L_C = (Z + a^3)(Z + a^6)$. According to Theorem 13.9, exponents 3 and 6 show exactly the error locations minus 1. So that errors occurred on positions 4 and 7.

Consider another example. Let $\mathbf{r} = (0, 0, 0, 1, 0, 0, 0, 1, 1, 1, 1, 1, 0, 0, 0)$ be a received word with at most 2 errors. The syndromes are now $\mathbf{s}_1 = a^8$, $\mathbf{s}_3 = a^9$. Plug them into L_C in place of X_1 and X_2 and obtain:

$$L_C(Z) = Z^2 + a^8 Z + a^9(a^8)^{14} + (a^8)^2 = Z^2 + a^8 Z.$$

Factorizing yields $L_C = Z(Z + a^8)$. Thus 1 error occurred according to Theorem 13.9, namely on position 8+1=9.

This method can be adapted to correct erasures [19], and to find the minimum distance of a code [38]. The basic approach is as follows. We are working again with the cyclic code C with parameters $[n, k, d]$ over $\mathbb{F}_q$. Let

$w \leq d$. Denote by $J_C(w)$ the set of equations in Eq. (13.5) for $t = w$, variables X_i assigned to zero, and the equations $Z_i^{n+1} - Z_i = 0$ replaced by $Z_i^n - 1 = 0$. In the binary case we have the following result that can be deduced from Theorem 3.3 and Corollary 3.4 [38]:

Theorem 13.11. *Let C be a binary $[n, k, d]$ cyclic code with $S_C = \{i_1, \ldots, i_v\}$ as defining set. Let $1 \leq w \leq n$ and let $J_C(w)$ denote the system:*

$$\begin{cases} Z_1^{i_1} + \cdots + Z_w^{i_1} = 0, \\ \qquad \vdots \\ Z_1^{i_v} + \cdots + Z_w^{i_v} = 0, \\ Z_1^n - 1 = 0, \\ \qquad \vdots \\ Z_w^n - 1 = 0 \\ p(n, Z_i, Z_j) = 0, \qquad 1 \leq i < j \leq w \end{cases}$$

Then the number of solutions of $J_C(w)$ is equal to $w!$ times the number of codewords of weight w. And for $1 \leq w \leq d$:

- *either $J_C(w)$ has no solution, which is equivalent to $w < d$,*
- *or $J_C(w)$ has a solution, which is equivalent to $w = d$.*

So, the method of finding the minimum distance is based on replacing syndrome variables by zeros and then searching for solutions of corresponding parameterized systems. In the previous theorem $J_C(w)$ is parameterized by w. We also mention the notion of *accelerator polynomials*. The idea is as follows. Since, when trying to find the minimum distance of a code we are only interested in a question whether the corresponding system $J_C(w)$ has solutions or not, we may add some polynomials $A_C(w)$ with the property that if we enlarge our system $J_C(w)$ with these polynomials and the system $J_C(w)$ had some solutions, then the new system $A_C(w) \cup J_C(w)$ also has some solutions. So not all solutions are lost. In [39] it is shown, how to choose such polynomials $A_C(w)$, so that solving the system $A_C(w) \cup J_C(w)$ takes less time, than solving $J_C(w)$.

It is possible to adapt the method to find codewords of certain weights, and thus the weight enumerator of a given code.

Example 13.12. As an example application of Theorem 13.11 we show how to determine the minimum distance of a cyclic code C from Example 13.10. This binary cyclic code C has parameters $[15, 7]$ and has a defining set $\{1,3\}$, so the assumptions of Theorem 13.11 are satisfied. We have to look at all systems $J_C(w)$ starting from $w = 1$, until we encounter a system, which has some solutions. The system $J_C(w)$ is

$$
\begin{cases}
Z_1 + \cdots + Z_w = 0, \\
Z_1^3 + \cdots + Z_w^3 = 0, \\
Z_1^{15} - 1 = 0, \\
\quad \vdots \\
Z_w^{15} - 1 = 0, \\
p(15, Z_i, Z_j) = 0, \quad 1 \le i < j \le w.
\end{cases}
$$

For $w = 1, \ldots, 4$ the reduced Göbner basis of $J_C(w)$ is $\{1\}$, so there are no solutions. For $J_C(5)$ the reduced Gröbner basis with respect to the lexicographic order is

$$
\begin{cases}
Z_5^{15} + 1, \\
Z_4^{12} + Z_4^9 Z_5^3 + Z_4^6 Z_5^6 + Z_4^3 Z_5^9 + Z_5^{12}, \\
Z_3^6 + Z_3^4 Z_4 Z_5 + Z_3^2 Z_4^2 Z_5^2 + Z_3 Z_4^4 Z_5 + Z_3 Z_4 Z_5^4 + Z_4^6 + Z_5^6, \\
g_2(Z_2, Z_3, Z_4, Z_5), \\
Z_1 + Z_2 + Z_3 + Z_4 + Z_5.
\end{cases}
$$

Here $g_2(Z_2, Z_3, Z_4, Z_5)$ is equal to

$$
\begin{cases}
Z_2^2 + Z_2 Z_3 + Z_2 Z_4 + Z_2 Z_5 + Z_3^5 Z_4^{10} Z_5^2 + Z_3^5 Z_4^9 Z_5^3 \\
\quad + Z_3^5 Z_4^8 Z_5^4 + Z_3^5 Z_4^4 Z_5^8 + Z_3^5 Z_3^3 Z_5^9 + Z_3^5 Z_4^2 Z_5^{10} \\
\quad + Z_3^4 Z_4^{11} Z_5^2 + Z_3^4 Z_4^8 Z_5^5 + Z_3^4 Z_4^5 Z_5^8 + Z_3^4 Z_4^2 Z_5^{11} \\
\quad + Z_3^3 Z_4^{10} Z_5^4 + Z_3^3 Z_4^9 Z_5^5 + Z_3^3 Z_4^8 Z_5^6 + Z_3^3 Z_4^4 Z_5^{10} \\
\quad + Z_3^3 Z_4^3 Z_5^{11} + Z_3^3 Z_4^2 Z_5^{12} + Z_3^3 Z_5^{14} + Z_3^2 Z_4^{11} Z_5^4 \\
\quad + Z_3^2 Z_4^8 Z_5^7 + Z_3^2 Z_4^5 Z_5^{10} + Z_3^2 Z_4^2 Z_5^{13} + Z_3^2 Z_4 Z_5^{14} \\
\quad + Z_3^2 + Z_3 Z_4^{10} Z_5^6 + Z_3 Z_4^9 Z_5^7 + Z_3 Z_4^8 Z_5^8 + Z_3 Z_4^4 Z_5^{12} \\
\quad + Z_3 Z_4^3 Z_5^{13} + Z_3 Z_4 + Z_4^{11} Z_5^6 + Z_4^8 Z_5^9 + Z_4^5 Z_5^{12} + Z_4^3 Z_5^{14} + Z_4^2.
\end{cases}
$$

Already the fact that the GB of $J_C(5)$ is not equal to 1 shows that there is a solution. Theorem 13.7 gives all solutions explicitly. We show how to obtain one solution here. Namely, we know already that $a^{15} + 1 = 0$, where a is a primitive element of $\mathbb{F}_{16}$, so set $Z_5 = a$ and the first equation is satisfied. Substituting $Z_5 = a$ to the second equation, we have $Z_4^{12} + a^3 Z_4^9 + a^6 Z_4^6 + a^9 Z_4^3 + a^{12} = 0$. Factorizing yields that $Z_4 = 1$ is one of the roots. Substituting $Z_5 = a$, and $Z_4 = 1$ to the third equation, we have $Z_3^6 + a Z_3^4 + a^2 Z_3^2 + Z_3 + a^{13} = 0$. Factorizing yields that $Z_3 = a^2$ is one of the roots. Substituting $Z_5 = a, Z_4 = 1$, and $Z_3 = a^2$ to the third equation, we have $Z_2^2 + a^{10} Z_2 + a^7 = 0$. Here $Z_2 = a^9$ is one of the roots. Finally, substituting $Z_5 = a, Z_4 = 1, Z_3 = a^2$, and $Z_2 = a^9$ to the last equation, we obtain that $Z_1 = a^{13}$. Thus we have proved that the system $J_C(5)$ has a solution and thus the minimum distance of C is 5, which coincides with what we had in Example 13.10. Note that the BCH bound yields $d(C) \geq 5$, so in fact it was necessary to consider only $J_C(5)$. Here it is possible to count the number of roots. Due to the equations $Z_1^{15} - 1 = 0, \ldots, Z_5^{15} - 1 = 0$ and the fact that $\mathbb{F}_{16}$ is the splitting field of $X^{15} - 1$, we have that the number of solutions is just the product of leading terms degrees of the elements in the Gröbner basis above. This number is $15 \cdot 12 \cdot 6 \cdot 2 \cdot 1 = 2160$. Dividing this number by 5! yields the number of minimum weight codewords: 18.

We mention that the first use of Gröbner bases in finding minimum distance appears in [40].

13.3.2. *Newton Identities Based Method*

The *error-locator polynomial* is defined by

$$\sigma(Z) = \prod_{l=1}^{t} (Z - z_l).$$

If this product is expanded

$$\sigma(Z) = Z^t + \sigma_1 Z^{t-1} + \cdots + \sigma_{t-1} Z + \sigma_t,$$

then the coefficients σ_i are the *elementary symmetric functions* in the error locations $z_1, \ldots, z_t$.

$$\sigma_i = (-1)^i \sum_{1 \leq j_1 < j_2 < \cdots < j_i \leq t} z_{j_1} z_{j_2} \ldots z_{j_i}, \quad 1 \leq i \leq t.$$

Techniques for decoding of cyclic codes up to half the BCH distance can be found in [6] for the binary case and in [41] for arbitrary q and independently in [3], and goes as follows. The syndromes s_i and the coefficients σ_i satisfy the following *generalized Newton identities* [6].

Theorem 13.13.

$$s_i + \sum_{j=1}^{t} \sigma_j s_{i-j} = 0, \quad \text{for all } i \in \mathbb{Z}_n. \tag{13.6}$$

Now suppose that the complete defining set of the cyclic code contains the $2t$ consecutive elements $b, \ldots, b+2t-1$ for some b. Then $d \geq 2t+1$ by the *BCH bound*. Furthermore, the set of equations in Eq. (13.6) for $i = b+t, \ldots, b+2t-1$ is a system of t linear equations in the unknowns $\sigma_1, \ldots, \sigma_t$ with the known syndromes $s_b, \ldots, s_{b+2t-1}$ as coefficients. Gaussian elimination solves the system of equations with complexity $\mathcal{O}(t^3)$. In this way we have obtained the *APGZ decoding* algorithm, after Arimoto, Peterson, Gorenstein, and Zierler.

Example 13.14. We consider the same example that was considered in [10], namely the binary 3-error-correcting cyclic code of length 31 and dimension 16 with defining set $\{1, 5, 7\}$. This code is actually a quadratic residue code and has parameters $[31, 16, 7]$. The splitting field of $X^{31} - 1$ over $\mathbb{F}_2$ is $\mathbb{F}_{32}$ with a primitive 31st root of unity a, such that $a^5 + a^2 + 1 = 0$. Note that $\mathbb{Z}_{31}$ is a disjoint union of cyclotomic classes of 1, 3, 5, 7, 11, and 15. That is to say if i is in a defining set, then $2i$ is in the complete defining set. The cyclotomic class of 1 is $\{1, 2, 4, 8, 16\}$, of 5 is $\{5, 10, 20, 9, 18\}$, and of 7 is $\{7, 14, 28, 25, 19\}$. Hence the complete defining set of C is $\{1, 2, 4, 5, 7, 8, 9, 10, 14, 16, 18, 19, 20, 25, 28\}$. It has $7, 8, 9$, and 10 as four consecutive elements. Hence the BCH bound is 5 and with the APGZ algorithm we are able to correct two errors.

Let

$$\mathbf{r} = (0, 0, 0, 0, 0, 1, 0, 0, 1, 0, 1, 1, 1, 1, 1, 0, 0, 1, 0, 0, 0, 1, 1, 1, 0, 1, 1, 0, 0, 0, 1)$$

be a received word with at most two errors. So the known syndromes from the defining set are $s_1 = a^{13}, s_5 = a^{23}, s_7 = a^{16}$. From this we can compute $s_8 = s_1^8 = a^{11}, s_9 = s_5^8 = a^{29}, s_{10} = s_5^2 = a^{15}$. The corresponding APGZ linear system is then:

$$\begin{cases} a^{29} + a^{11}\sigma_1 + a^{16}\sigma_2 = 0, \\ a^{15} + a^{29}\sigma_1 + a^{11}\sigma_2 = 0. \end{cases}$$

This system has a unique solution $\sigma_1 = a^{13}$, and $\sigma_2 = a^{10}$. The corresponding error-locator polynomial is $\sigma(Z) = Z^2 + a^{13}Z + a^{10}$, which has the roots a^3 and a^7, so the error positions are 4 and 8. So

$$\mathbf{c} = (0,0,0,1,0,1,0,1,1,0,1,1,1,1,1,0,0,1,0,0,0,1,1,1,0,1,1,0,0,0,1)$$

is the nearest codeword.

Suppose that $\{1, \ldots, 2t\} \subseteq S_C$. Define the *syndrome polynomial* $S(Z)$ by

$$S(Z) = \sum_{i=1}^{2t} s_i Z^{i-1}.$$

The Newton identities can be reformulated as the *key equation*

$$\sigma(Z)S(Z) \equiv \omega(Z) \mod Z^{2t}, \tag{13.7}$$

for some polynomial $\omega(Z)$ such that $\deg(\omega(Z)) < \deg(\sigma(Z))$. The key equation is solved by the *algorithm of Berlekamp–Massey* [4, 5] and a variant of *Euclidean algorithm* due to Sugiyama et al. [7]. Here $\omega(Z)$ is called the *error-evaluator polynomial* and is used to calculate the error values by *Forney's* formula [42], Theorem 25, p. 246. Both these algorithms are of prime importance in applications. They are more efficient than solving the system of linear equations, and are basically equivalent [43], although one might prefer one over the other depending on the application and actual implementation.

All these algorithms decode up to the BCH error-correcting capacity, which is often strictly smaller than the true capacity. A general method was outlined by Berlekamp [4, pp. 231–240], Tzeng et al. [44], and Stevens [45], where the *unknown syndromes* were treated as variables. We have

$$s_{i+n} = s_i, \quad \text{for all } i \in \mathbb{Z}_n,$$

since $s_{i+n} = r(a^{i+n}) = r(a^i)$. Furthermore

$$s_i^q = (e(a^i))^q = e(a^{iq}) = s_{qi}, \quad \text{for all } i \in \mathbb{Z}_n,$$

and

$$\sigma_i^{q^m} = \sigma_i, \quad \text{for all } 1 \leq i \leq t.$$

So the zeros of the following set of polynomials $Newton_t$ in the variables $S_1, \ldots, S_n$ and $\sigma_1, \ldots, \sigma_t$ is considered, see [8, 9].

$$Newton_t \begin{cases} \sigma_i^{q^m} - \sigma_i, & \text{for all} \quad 1 \leq i \leq t, \\ S_{i+n} - S_i, & \text{for all} \quad i \in \mathbb{Z}_n, \\ S_i^q - S_{qi}, & \text{for all} \quad i \in \mathbb{Z}_n, \\ S_i + \sum_{j=1}^t \sigma_j S_{i-j}, & \text{for all} \quad i \in \mathbb{Z}_n. \end{cases} \tag{13.8}$$

It is this method of treating the unknown syndromes as variables that we generalize to arbitrary linear codes in Sec. 13.4.2.

Solutions of $Newton_t$ are called *generic*, *formal* or *one-step* and this is considered as a preprocessing phase which has to be performed only one time. For the actual decoder for every received word $\mathbf{r}$ the variables S_i are specialized to the actual value $s_i(\mathbf{r})$ for $i \in S_C$. Alternatively one can solve $Newton_t$ together with the polynomials $S_i - s_i(\mathbf{r})$ for $i \in S_C$. This is called *online* decoding. Note that obtaining general error-locator polynomial as in the previous subsection is an example of formal decoding: this polynomial has to be found only once.

Example 13.15. Let us consider an example of decoding using Newton identities and such that the APGZ algorithm is not applicable. We consider the same 3-error-correcting cyclic code of length 31 with a defining set $\{1, 5, 7\}$ as in Example 13.14. This time we are aiming at correcting three errors. Let us write the corresponding ideal:

$$\begin{cases} \sigma_1 S_{31} + \sigma_2 S_{30} + \sigma_3 S_{29} + S_1, \\ \sigma_1 S_1 + \sigma_2 S_{31} + \sigma_3 S_{30} + S_2, \\ \sigma_1 S_2 + \sigma_2 S_1 + \sigma_3 S_{31} + S_3, \\ \sigma_1 S_{i-1} + \sigma_2 S_{i-2} + \sigma_3 S_{i-3} + S_i, & 4 \leq i \leq 31, \\ \sigma_i^{32} + \sigma_i, & i = 1, 2, 3, \\ S_{i+31} + S_i, & \text{for all } i \in \mathbb{Z}_{31}, \\ S_i^2 + S_{2i}, & \text{for all } i \in \mathbb{Z}_{31}. \end{cases}$$

Note that the equations $S_{i+31} = S_i$ and $S_i^2 = S_{2i}$ imply,

$$\begin{cases} S_1^2 + S_2, S_1^4 + S_4, S_1^8 + S_8, S_1^{16} + S_{16}, \\ S_3^2 + S_6, S_3^4 + S_{12}, S_3^8 + S_{24}, S_3^{16} + S_{17}, \\ S_5^2 + S_{10}, S_5^4 + S_{20}, S_5^8 + S_9, S_5^{16} + S_{18}, \\ S_7^2 + S_{14}, S_7^4 + S_{28}, S_7^8 + S_{25}, S_7^{16} + S_{19}, \\ S_3^2 + S_6, S_3^4 + S_{12}, S_3^8 + S_{24}, S_3^{16} + S_{17}, \\ S_{11}^2 + S_{22}, S_{11}^4 + S_{13}, S_{11}^8 + S_{26}, S_{11}^{16} + S_{21}, \\ S_{15}^2 + S_{30}, S_{15}^4 + S_{29}, S_{15}^8 + S_{27}, S_{15}^{16} + S_{23}, \\ S_{31}^2 + S_{31}. \end{cases}$$

Our intent is to write σ_1, σ_2, and σ_3 in terms of known syndromes S_1, S_5, and S_7. The next step would be to compute the reduced Gröbner basis of this system with respect to some elimination order induced by $S_{31} > \cdots > S_8 > S_6 > S_4 > \cdots > S_2 > \sigma_1 > \sigma_2 > \sigma_3 > S_7 > S_5 > S_1$. Unfortunately the computation is quite time consuming and the result is too huge to illustrate the idea. Rather, we do online decoding, i.e. compute syndromes S_1, S_5, and S_7, plug the values into the system and then find σ's. Let

$$\mathbf{r} = (0,0,0,0,0,1,0,0,1,0,1,1,1,1,1,0,0,1,0,0,0,1,1,1,0,0,1,0,0,0,1),$$

be a received word with at most three errors. So the known syndromes we need are $s_1 = a^5, s_5 = a^8$, and $s_7 = a^{26}$. Substitute these values into the system above and compute the reduced Gröbner basis of the system. The reduced Gröbner basis with respect to the degree reverse lexicographic order (here it is possible to go without an elimination order, see Remark 13.8) restricted to the variables σ_1, σ_2, and σ_3 is

$$\begin{cases} \sigma_3 + a^4, \\ \sigma_2 + a^5, \\ \sigma_1 + a^5. \end{cases}$$

Corresponding values for σ's gives rise to the error-locator polynomial:

$$\sigma(Z) = Z^3 + a^5 Z^2 + a^5 Z + a^4.$$

Factoring this polynomial yields three roots: a^3, a^7, and a^{25}, which indicate error positions.

Note also that we could have worked only with the equations for $S_1, S_5, S_7, S_3, S_{11}, S_{15}$, and S_{31}, but the Gröbner basis computation is harder then: on our computer it took 8 times more time.

Another way of finding the error-locator polynomial $\sigma(Z)$ in the binary case is described in [10]. In this case the error values are 1 and the $S_i, i \in S_C$ are power sums of the error positions and therefore symmetric under all possible permutations of these positions. Hence S_i is equal to a polynomial $w_i(\sigma_1, \dots, \sigma_t)$. These w_i's are known as *Waring functions*. By considering the ideal generated by the following polynomials

$$S_i - w_i(\sigma_1, \dots, \sigma_t), \quad i \in S_C,$$

Augot *et al.* were able to prove the uniqueness theorem for the solution $(\sigma_1^*, \dots, \sigma_t^*)$, when S_i's are assigned the concrete values of syndromes. Here the authors prefer online decoding, rather than formal one. This approach demonstrates pretty good performance in practice, but it lacks some theoretical explanations of several tricks the authors used. Further treatment of this approach is in [46].

13.4. Decoding and Finding the Minimum Distance of Arbitrary Linear Codes

13.4.1. *Decoding Affine Variety Codes*

The method proposed by Fitzgerald and Lax [47, 48] generalizes Cooper's philosophy to arbitrary linear codes. In this approach the main notion is the *affine variety code*. Let $I = \langle g_1, \dots, g_m \rangle \subseteq \mathbb{F}_q[X_1, \dots, X_s]$ be an ideal. Define $I_q := I + \langle X_1^q - X_1, \dots, X_s^q - X_s \rangle$. So I_q is a 0-dimensional ideal. Define also $V(I_q) =: \{P_1, \dots, P_n\}$. The claim [48] is that every q-ary linear code C with parameters $[n, k]$ can be seen as an *affine variety code* $C(I, L)$, that is the image of a vector space L of the *evaluation map*

$$\begin{cases} \phi : R \to \mathbb{F}_q^n, \\ \bar{f} \mapsto (f(P_1), \dots, f(P_n)), \end{cases}$$

where $R := \mathbb{F}_q[U_1, \dots, U_s]/I_q$, L is a vector subspace of R and $\bar{f}$ the coset of f in $\mathbb{F}_q[U_1, \dots, U_s]$ modulo I_q. In order to obtain this description we do the following. Given a q-ary $[n, k]$ code C with a generator matrix

$G = (g_{ij})$, we choose s, such that $q^s \geq n$, and construct s distinct points $P_1, \ldots, P_s$ in $\mathbb{F}_q^s$. Then there is an algorithm [49] that produces a Gröbner basis $\{g_1, \ldots, g_m\}$ for an ideal I of polynomials from $\mathbb{F}_q[X_1, \ldots, X_s]$ that vanish at the points $P_1, \ldots, P_s$. Then, denote by $\xi_i \in \mathbb{F}_q[X_1, \ldots, X_s]$ a polynomial that assumes the values 1 at P_i and 0 at all other P_j. The linear combinations $f_i = \sum_{i=1}^n g_{ij}\xi_j$ span the space L, so that $g_{ij} = f_i(P_j)$. In this way we obtain that the code C is the image of the evaluation above, so $C = C(I, L)$. In the same way by considering a parity check matrix instead of a generator matrix we have that the dual code is also an affine variety code.

The method of decoding is analogous to the one of CRHT with the generalization that along with the polynomials of type Eq. (13.3) one needs to add polynomials $(g_l(X_{k1}, \ldots, X_{ks}))_{l=1,\ldots,m;k=1,\ldots,t}$ for every error position. We also assume that field equations on X_{ij}'s are included among the polynomials above. Let C be a q-ary $[n, k]$ linear code such that its dual is written as an affine variety code of the form $C^{\perp} = C(I, L)$, where

$$\begin{cases} I = \langle g_1, \ldots, g_m \rangle \subseteq \mathbb{F}_q[X_1, \ldots, X_s], \\ L = \{\bar{f}_1, \ldots, \bar{f}_{n-k}\}, \\ V(I_q) = \{P_1, \ldots, P_s\}. \end{cases}$$

Let $\mathbf{r} = (r_1, \ldots, r_n)$ be a received word with error vector $\mathbf{e}$, so that $\mathbf{r} = \mathbf{c} + (e_1, \ldots, e_n)$ with t errors and $t \leq e$. Then the syndromes are computed by

$$s_i = \sum_{j=1}^n r_j f_i(P_j) = \sum_{j=1}^n e_j f_i(P_j) \quad \text{for } i = 1, \ldots, n - k.$$

Now consider the ring $\mathbb{F}_q[X_{11}, \ldots, X_{1s}, \ldots, X_{t1}, \ldots, X_{ts}, E_1, \ldots, E_t]$, where $(X_{i1}, \ldots, X_{is})$ correspond to the i-th error position and E_i to the i-th error value. Consider the ideal $\mathcal{I}_C$ generated by

$$\begin{cases} \sum_{j=1}^t E_j f_i(X_{j1}, \ldots, X_{js}) - s_i, & 1 \leq i \leq n - k, \\ g_l(X_{j1}, \ldots, X_{js}), & 1 \leq l \leq m, \\ E_k^{q-1} - 1. \end{cases}$$

The order $<$ is defined as follows. It is the block order $(<_1, <_2)$, where $<_1$ is the lexicographic order induced by $X_{11} < \cdots < X_{1s} < E_1$

and $<_2$ is any (e.g. degree reverse lexicographic) order on the variables $X_{21}, \ldots, X_{2s}, E_2, \ldots, X_{t1}, \ldots, X_{ts}, E_t$. We only impose lexicographic order on the first error variables, as there is a symmetry group acting on the solutions, so we are interested only in the first coordinate solutions. They will in turn give solutions for all the coordinates by symmetry. Then Theorem 2.2 from [48] states

Theorem 13.16. *Let G be the reduced Gröbner basis for $\mathcal{I}_C$ with respect to the order $<$. Then we may solve for the error locations and values by applying elimination theory to the polynomials in G.*

In general, finding I and L is quite technical and it turns out that for random codes this method is quite poor, because of the complicated structure of $\mathcal{I}_C$. As in Sec. 13.3.1, it is possible to replace syndromes with variables, but the ideal becomes then even more complicated. We consider an application of the method to Hermitian codes as is done in [48].

Example 13.17. Consider the Hermitian function field defined by $Y^2 + Y + X^3 = 0$ over $\mathbb{F}_4$ with a primitive element a, such that $a^2 + a + 1 = 0$. Let C be a quaternary $[8, 3, 5]$ Hermitian code. It is orthogonal to the $[8, 5, 3]$ Hermitian code defined by $L = \langle 1, X, Y, X^2, XY \rangle$. Choose $s = 2$, so that $4^2 > 8$, and the points

$$P_1 = (0, 0), P_2 = (0, 1), P_3 = (1, a), P_4 = (1, a^2),$$
$$P_5 = (a, a), P_6 = (a, a^2), P_7 = (a^2, a), P_8 = (a^2, a^2).$$

Denote $I = \langle Y^2 + Y + X^3 \rangle$. Then $C^\perp = C(I, L)$. The parity check matrix for C is

$$H = \begin{pmatrix} 1 & 1 & 1 & 1 & 1 & 1 & 1 & 1 \\ 0 & 0 & 1 & 1 & a & a & a^2 & a^2 \\ 0 & 1 & a & a^2 & a & a^2 & a & a^2 \\ 0 & 0 & 1 & 1 & a^2 & a^2 & a & a \\ 0 & 0 & a & a^2 & a^2 & 1 & 1 & a \end{pmatrix}.$$

Let $\mathbf{r} = (1, 1, 0, a, a^2, a^2, a, 1)$ be a received word with at most two errors. The corresponding syndrome is $\mathbf{s} = (1, 0, a^2, a^2, 0)$. So the ideal $\mathcal{I}_C$ in

$\mathbb{F}_4[X_1, Y_1, E_1, X_2, Y_2, E_2]$ is generated by

$$
\begin{cases}
X_1^4 + X_1, & X_2^4 + X_2, \\
Y_1^4 + Y_1, & Y_2^4 + Y_2, \\
E_1^3 + 1, & E_2^3 + 1, \\
Y_1^2 + Y_1 + X_1^3, & Y_2^2 + Y_2 + X_2^3, \\
E_1 + E_2 + 1, \\
E_1 X_1 + E_2 X_2, \\
E_1 Y_1 + E_2 Y_2 + a^2, \\
E_1 X_1^2 + E_2 X_2^2 + a^2, \\
E_1 X_1 Y_1 + E_2 X_2 Y_2.
\end{cases}
$$

We are working with a block order $(<_1, <_2)$ induced by $X_1 < Y_1 < E_1 < X_2 < Y_2 < E_2$, where $<_1$ is the lexicographic order induced by $X_1 < Y_1 < E_1$, and $<_2$ is the degree reverse lexicographic order induced by $X_2 < Y_2 < E_2$. The reduced Gröbner basis G of $\mathcal{I}_C$ with respect to this order is

$$
\begin{cases}
X_1^2 + aX_1 + a^2, \\
Y_1 + a^2, \\
E_1 + a^2 X_1 + 1, \\
X_2 + X_1 + a, \\
Y_2 + a^2, \\
E_2 + a^2 X_1.
\end{cases}
$$

Solving the first equation $X_1^2 + aX_1 + a^2 = 0$ actually gives the X-coordinates of the two error positions. They are 1 and a^2. We substitute further and obtain the error positions $(1, a^2)$ and (a^2, a^2), that is positions 4 and 8 in our numeration, and the corresponding error values a and a^2, respectively. Hence $\mathbf{c} = (1, 1, 0, 0, a^2, a^2, a, a)$ is the codeword sent.

We mention that there are generalizations of the approach of Fitzgerald and Lax, which follow the same idea as the generalizations for the CRHT-ideal. Namely, one adds the polynomials that ensure that the error locations are different. For more details, see [22]. There it is also proven that affine variety codes possess the so-called *multi-dimensional general error-locator polynomial,* which is a generalization of the general error-locator polynomial from Sec. 13.3.1.

13.4.2. *The Method of Quadratic Equations*

In this subsection, we propose a new method of decoding and finding the minimum distance of arbitrary linear codes that is based on solving a system of quadratic equations. In a sense we generalize the ideas from Sec. 13.3.2 by trying to find unknown syndromes, although the meaning of this term is different in our setting. Most of the results with proofs can be found in [50].

Definition 13.18. Let $\mathbf{b}_1, \ldots, \mathbf{b}_n$ be a basis of $\mathbb{F}_q^n$. Let B be the $n \times n$ matrix with $\mathbf{b}_1, \ldots, \mathbf{b}_n$ as rows. The *(unknown) syndrome* $\mathbf{u}(B, \mathbf{e})$ of a word $\mathbf{e}$ with respect to B is the column vector $\mathbf{u}(B, \mathbf{e}) = B\mathbf{e}^T$. It has entries $u_i(B, \mathbf{e}) = \mathbf{b}_i \cdot \mathbf{e}$ for $i = 1, \ldots, n$. The following abbreviations $\mathbf{u}(\mathbf{e})$ and $u_i(\mathbf{e})$ are used for $\mathbf{u}(B, \mathbf{e})$ and $u_i(B, \mathbf{e})$, respectively.

Remark 13.19. The matrix B is invertible, since its rank is n. The syndrome $\mathbf{u}(B, \mathbf{e})$ determines the vector $\mathbf{e}$ uniquely, since

$$B^{-1}\mathbf{u}(B, \mathbf{e}) = B^{-1}B\mathbf{e}^T = \mathbf{e}^T.$$

Our idea is going to be: having an error vector $\mathbf{e}$ to find a vector of unknown syndromes with respect to some specific basis with the matrix B. Then finding $\mathbf{e}$ itself is straightforward.

Definition 13.20. Define the coordinatewise *star product* of two vectors $\mathbf{x}, \mathbf{y} \in \mathbb{F}_q^n$ by $\mathbf{x} * \mathbf{y} = (x_1 y_1, \ldots, x_n y_n)$. Then $\mathbf{b}_i * \mathbf{b}_j$ is a linear combination of the basis vectors $\mathbf{b}_1, \ldots, \mathbf{b}_n$, so there are constants $\mu_l^{ij} \in \mathbb{F}_q$ such that

$$\mathbf{b}_i * \mathbf{b}_j = \sum_{l=1}^{n} \mu_l^{ij} \mathbf{b}_l.$$

The elements $\mu_l^{ij} \in \mathbb{F}_q$ are called the *structure constants* of the basis $\mathbf{b}_1, \ldots, \mathbf{b}_n$.

Definition 13.21. Define the $n \times n$ *matrix of (unknown) syndromes* $\mathcal{U}(\mathbf{e}) = (u_{ij}(\mathbf{e}))_{1 \le i,j \le n}$ of a word $\mathbf{e}$ by $u_{ij}(\mathbf{e}) = (\mathbf{b}_i * \mathbf{b}_j) \cdot \mathbf{e}$.

Remark 13.22. The relation between the entries of the matrix $\mathcal{U}(\mathbf{e})$ and the vector $\mathbf{u}(\mathbf{e})$ of unknown syndromes is given by

$$u_{ij}(\mathbf{e}) = \sum_{l=1}^{n} \mu_l^{ij} u_l(\mathbf{e}).$$

Definition 13.23. Let $\mathbf{b}_1, \ldots, \mathbf{b}_n$ be a basis of $\mathbb{F}_q^n$. Let B_s be the $s \times n$ matrix with $\mathbf{b}_1, \ldots, \mathbf{b}_s$ as rows, then $B = B_n$. We say that $\mathbf{b}_1, \ldots, \mathbf{b}_n$ is an *ordered MDS basis* and B an *MDS matrix* if all the $s \times s$ submatrices of B_s have rank s for all $s = 1, \ldots, n$.

Remark 13.24. Let B be an MDS matrix. Let C_s be the code with B_s as parity check matrix. Then C_s is an $[n, n-s, s+1]$ code, that is an MDS code for all $1 \le s \le n$. This motivates the name in the definition above.

Definition 13.25. Suppose $n \le q$. Let $\mathbf{x} = (x_1, \ldots, x_n)$ be an n-tuple of mutually distinct elements in $\mathbb{F}_q$. Define

$$\mathbf{b}_i = (x_1^{i-1}, \ldots, x_n^{i-1}).$$

Then $\mathbf{b}_1, \ldots, \mathbf{b}_n$ is called an *ordered Vandermonde basis* and the corresponding matrix is denoted by $B(\mathbf{x})$ and called a *Vandermonde matrix*.

It can be shown that if we have a linear $[n, k, d]$ code C over the field $\mathbb{F}_q$, then the code $C' = C\mathbb{F}_{q^m}$ has the same parameters. So without loss of generality we may assume that after a finite extension of the finite field $\mathbb{F}_q$ we have that $n \le q$. Let $\mathbf{b}_1, \ldots, \mathbf{b}_n$ be a basis of $\mathbb{F}_q^n$. From now on we assume that the corresponding matrix B is an MDS matrix.

The next proposition gives a relation between the rank of the matrix of unknown syndromes $\mathcal{U}(\mathbf{e})$ and the weight of $\mathbf{e}$.

Proposition 13.26. *Let $D(\mathbf{e})$ be the diagonal matrix with $\mathbf{e}$ on its diagonal. Then*

$$\mathcal{U}(\mathbf{e}) = BD(\mathbf{e})B^T,$$

and the rank of $\mathcal{U}(\mathbf{e})$ is equal to the weight of $\mathbf{e}$.

Proof. See also [51, Lemma 4.7]. We have that

$$u_{ij}(\mathbf{e}) = (\mathbf{b}_i * \mathbf{b}_j) \cdot \mathbf{e} = \sum_{l=1}^{n} b_{il} e_l b_{jl}.$$

Hence $\mathcal{U}(\mathbf{e}) = BD(\mathbf{e})B^T$. Now B and B^T are invertible. Hence the rank of $\mathcal{U}(\mathbf{e})$ is equal to the rank of $D(\mathbf{e})$ which is equal to the weight of $\mathbf{e}$. $\square$

We have done all the necessary preparations. Next, we will see which role the notion of unknown syndromes plays in decoding. Let C be an $\mathbb{F}_q$-linear code with parameters $[n, k, d]$. Choose a parity check matrix H of C. Let $\mathbf{h}_1, \dots, \mathbf{h}_r$ be the rows of H. The row $\mathbf{h}_i$ is a linear combination of the basis $\mathbf{b}_1, \dots, \mathbf{b}_n$, that is there are constants $a_{ij} \in \mathbb{F}_q$ such that

$$\mathbf{h}_i = \sum_{j=1}^{n} a_{ij}\mathbf{b}_j.$$

In other words $H = AB$ where A is the $r \times n$ matrix with entries a_{ij}.

Remark 13.27. Let $\mathbf{r} = \mathbf{c} + \mathbf{e}$ be a received word with $\mathbf{c} \in C$ a codeword and $\mathbf{e}$ an error vector. The syndromes of $\mathbf{r}$ and $\mathbf{e}$ with respect to H are equal and known: $s_i(\mathbf{r}) := \mathbf{h}_i \cdot \mathbf{r} = \mathbf{h}_i \cdot \mathbf{e} = s_i(\mathbf{e})$ and they can be expressed in the unknown syndromes of $\mathbf{e}$ with respect to B:

$$s_i(\mathbf{r}) = \sum_{j=1}^{n} a_{ij}u_j(\mathbf{e}),$$

since $\mathbf{h}_i = \sum_{j=1}^{n} a_{ij}\mathbf{b}_j$ and $\mathbf{b}_j \cdot \mathbf{e} = u_j(\mathbf{e})$.

Definition 13.28. Let $1 \le u, v \le n$. Then we define $\mathcal{U}_{uv}(\mathbf{e})$ as a submatrix of $\mathcal{U}(\mathbf{e})$ with the entries $(u_{ij}(\mathbf{e}))_{1 \le i \le u, 1 \le j \le v}$.

Proposition 13.29. *Let $w = wt(\mathbf{e})$. If $u \ge w$, then*

$$\mathrm{rank}(\mathcal{U}_{uv}(\mathbf{e})) = \min\{v, w\}.$$

Hence $\mathcal{U}_{nv}(\mathbf{e})$ has rank v if $v \le w$, and its rank is w if $v > w$. This also means that in the matrix $\mathcal{U}_{n,w+1}$ the $w+1$ columns are linearly dependent, which implies the existence of elements $v_1(\mathbf{e}), \dots, v_w(\mathbf{e})$, such that

$$\sum_{j=1}^{w} u_{ij}(\mathbf{e})v_j(\mathbf{e}) - u_{i,w+1}(\mathbf{e}) \quad \text{for } i = 1, \dots, n.$$

Recall, that at the beginning of this subsection we claimed our goal to be finding the unknown syndromes. Thus, replace them everywhere by variables and try to solve the corresponding system.

Definition 13.30. Let B be an MDS matrix with structure constants μ_l^{ij}. Define the linear functions U_{ij} in the variables $U_1, \ldots, U_n$ by

$$U_{ij} = \sum_{l=1}^{n} \mu_l^{ij} U_l.$$

Let $\mathcal{U}$ be the $n \times n$ matrix with entries U_{ij}. Let $\mathcal{U}_{u,v}$ be the $u \times v$ matrix with entries U_{ij} with $1 \le i \le u$ and $1 \le j \le v$.

Definition 13.31. The ideal $J(\mathbf{r})$ in the ring $\mathbb{F}_q[U_1, \ldots, U_n]$ is generated by the elements

$$\sum_{l=1}^{n} a_{jl} U_l - s_j(\mathbf{r}) \quad \text{for } j = 1, \ldots, r.$$

The ideal $I(t, \mathcal{U}, V)$ in the ring $\mathbb{F}[U_1, \ldots, U_n, V_1, \ldots, V_t]$ is generated by the elements

$$\sum_{j=1}^{t} U_{ij} V_j - U_{i,t+1} \quad \text{for } i = 1, \ldots, n.$$

Let $J(t, \mathbf{r})$ be the ideal in $\mathbb{F}_q[U_1, \ldots, U_n, V_1, \ldots, V_t]$ generated by $J(\mathbf{r})$ and $I(t, \mathcal{U}, V)$.

Remark 13.32. The ideal $J(t, \mathbf{r})$ is generated by $n - k$ linear functions and n quadratic polynomials.

Now we are ready to state the main results of this subsection.

Theorem 13.33. *Let B be an MDS matrix with structure constants μ_l^{ij} and linear functions U_{ij}. Let H be a parity check matrix of the code C such that $H = AB$ as above. Let $\mathbf{r} = \mathbf{c} + \mathbf{e}$ be a received word with $\mathbf{c}$ in C the codeword sent and $\mathbf{e}$ the error vector. Suppose that the weight of $\mathbf{e}$ is not zero and at most $\lfloor (d(C) - 1)/2 \rfloor$. Let t be the smallest positive integer such that $J(t, \mathbf{r})$ has a solution $(\mathbf{u}, \mathbf{v})$ over $\bar{\mathbb{F}}_q$. Then $wt(\mathbf{e}) = t$ and the solution is unique satisfying $\mathbf{u} = \mathbf{u}(\mathbf{e})$.*

Theorem 13.34. *With the notation as above, let t be the smallest positive integer such that $J(t, \mathbf{r})$ has a solution. Then the reduced Gröbner basis G for the ideal $J(t, \mathbf{r})$ with respect to any monomial ordering* *will be*

$$\begin{cases} U_i - u_i(\boldsymbol{e}), & i = 1, \ldots, n, \\ V_j - v_j, & j = 1, \ldots, t, \end{cases}$$

where $(\mathbf{u}(\mathbf{e}), \mathbf{v})$ is the unique solution of multiplicity one.

We would like to note that the fact that we can obtain a unique solution of multiplicity one without adding field equations is quite remarkable and really makes computations easier. We also refer to [46] for the result on uniqueness and multiplicity one for cyclic codes. The proofs, especially of the last statement, are quite involved. We demonstrate how the above works for a small example.

Example 13.35. Consider the ternary Golay code with parameters $[11, 6, 5]$. The code is 2-error correcting. We have $q = 3, n = 11$, and $t = 2$. So $m = 3$ is the smallest number, such that $3^m \geq n$. We have not taken the conventional choice $m = 5$, which is the smallest degree of an extension such that $\mathbb{F}_{3^m}^*$ has an element of order 11. A parity check matrix for this code is given by $H = (P \mid I_5)$, where

$$P = \begin{pmatrix} 1 & -1 & -1 & -1 & 1 & 0 \\ 0 & 1 & -1 & -1 & -1 & 1 \\ -1 & 1 & -1 & 0 & 1 & -1 \\ 1 & 1 & 0 & 1 & 1 & 1 \\ -1 & -1 & -1 & 1 & 0 & 1 \end{pmatrix}.$$

Let a be a primitive element of $\mathbb{F}_{27}$ with $a^3 - a + 1 = 0$. Then $B = (a^{ij})_{0 \leq i, j \leq 10}$ is an MDS matrix. Now $H = AB$, where

$$A = \begin{pmatrix} a^{12} & a^2 & a^3 & a^8 & a^{19} & a^{12} & a^2 & a^{22} & a^{11} & a^{11} & a^{23} \\ a^{23} & a^{22} & a^{13} & a^7 & a^{12} & a^0 & a^5 & a^0 & a^5 & a^5 & a^8 \\ a^6 & a^8 & a^{24} & a^{23} & a^2 & a^3 & a^{14} & a^{10} & a^{23} & a^{13} & a^{10} \\ a^{14} & a^{14} & a^{16} & a^{18} & a^6 & a^{17} & a^8 & 0 & a^{21} & a^{20} & a \\ a^{14} & a^7 & a^{13} & a^{10} & a^9 & a^{25} & a^{17} & a^{20} & a^{25} & a^5 & a^0 \end{pmatrix}.$$

Let $\mathbf{r} = (1, 1, 0, -1, 0, 1, -1, -1, -1, 1, 0)$ be a received word with two errors. Then the syndrome vector $\mathbf{s}(\mathbf{r})$ is given by $(0, -1, 1, 0, 1)^T$. We are working in the ring $\mathbb{F}_{27}[U_1, \ldots, U_{11}, V_1, V_2]$. Let us choose an order to be degree reverse lexicographic with $U_{11} > \cdots > U_6 > V_1 > V_2 > U_5 > \cdots > U_1$, cf. Theorem 13.34. The ideal $J(\mathbf{r})$ is generated by the entries of the vector AU^T, where $U = (U_1, \ldots, U_{11})$. The matrix $\mathcal{U}$ has entries U_{ij}, where

$U_{ij} = U_{i+j-1}$ for all i, j with $1 \leq i + j \leq 11$. The ideal $J(t, \mathbf{r})$ is generated by $J(\mathbf{r})$ and $J(t, \mathcal{U}, V)$, and the generators of $J(2, \mathcal{U}, V)$ are listed by

$$
\begin{cases}
V_1 U_1 & + V_2 U_2 & - U_3, \\
V_1 U_2 & + V_2 U_3 & - U_4, \\
V_1 U_3 & + V_2 U_4 & - U_5, \\
V_1 U_4 & + V_2 U_5 & - U_6, \\
V_1 U_5 & + V_2 U_6 & - U_7, \\
V_1 U_6 & + V_2 U_7 & - U_8, \\
V_1 U_7 & + V_2 U_8 & - U_9, \\
V_1 U_8 & + V_2 U_9 & - U_{10}, \\
V_1 U_9 & + V_2 U_{10} & - U_{11}, \\
V_1 U_{10} & + V_2 U_{11} & - U_{10,3}, \\
V_1 U_{11} & + V_2 U_{11,2} & - U_{11,3},
\end{cases}
$$

where

$$
\begin{aligned}
-U_{10,3} = & \; a^{16} U_1 + a^{22} U_2 + a^{25} U_3 + a^{22} U_4 + a^{20} U_5 + a^{25} U_6 + a^7 U_7 + a^{18} U_8 \\
& + a^{10} U_9 + + a^3 U_{10} + a^{16} U_{11}, \\
U_{11,2} = & \; a^3 U_1 + a^9 U_2 + a^{12} U_3 + a^9 U_4 + a^7 U_5 + a^{12} U_6 + a^{20} U_7 + a^5 U_8 \\
& + a^{23} U_9 + + a^{16} U_{10} + a^3 U_{11}, \\
-U_{11,3} = & \; a^{19} U_1 + a^{19} U_2 + a^7 U_3 + a^{12} U_4 + a^5 U_5 + a^9 U_6 + a^9 U_7 + a^{23} U_8 \\
& + a^4 U_9 + + a^{24} U_{10} + a^{25} U_{11}.
\end{aligned}
$$

The reduced Gröbner basis for the ideal $J(2, \mathbf{r})$ is

$$
\begin{cases}
U_1, U_2 - 1, U_3 + a^9, U_4 - 1, \\
U_5 + a^3, V_2 + a^9, V_1 + a^4, U_6 + a^{16}, \\
U_7 + a, U_8 + a^3, U_9 + a^{22}, U_{10} - 1, \\
U_{11} + a.
\end{cases}
$$

Let us check that this unique solution indeed gives rise to the error vector $\mathbf{e}$. Indeed, from the above we obtain that the vector $\mathbf{u}(B, \mathbf{e})$ of unknown syndromes is $(0, 1, -a^9, 1, -a^3) = (0, 1, a^{22}, 1, a^{16})$. By Remark 13.19 we can then find $\mathbf{e}$ as

$$
\mathbf{e}^T = B^{-1} \mathbf{u}(B, \mathbf{e}) = (0, 1, 0, -1, 0, 0, 0, 0, 0, 0, 0)^T.
$$

We also note that the corresponding system for $t = 1$ has the reduced Gröbner basis $\{1\}$ and therefore it has no solutions.

Example 13.36. Let us revise Example 13.15. So we are working again with the 3-error-correcting binary cyclic code of length 31 with a defining set $\{1, 5, 7\}$. So now we have $q = 2, n = 31$, and $t = 3$. So $m = 5$ and $q^m = 32$. Choose as matrix B a Vandermonde matrix with $x_i = a^{i-1}$, where a is a primitive 31st root of unity of $\mathbb{F}_{32}$ with $a^5 + a^2 + 1 = 0$. A parity check matrix of this code over $\mathbb{F}_2$ in the row echelon form $H = (I_{15} \,|\, P)$, where

$$
P = \begin{pmatrix}
1 & 1 & 0 & 1 & 1 & 0 & 0 & 0 & 1 & 0 & 1 & 0 & 0 & 1 & 0 & 0 \\
0 & 1 & 1 & 0 & 1 & 1 & 0 & 0 & 0 & 1 & 0 & 1 & 0 & 0 & 1 & 0 \\
0 & 0 & 1 & 1 & 0 & 1 & 1 & 0 & 0 & 0 & 1 & 0 & 1 & 0 & 0 & 1 \\
1 & 1 & 0 & 0 & 0 & 0 & 1 & 1 & 1 & 0 & 1 & 1 & 0 & 0 & 0 & 0 \\
0 & 1 & 1 & 0 & 0 & 0 & 0 & 1 & 1 & 1 & 0 & 1 & 1 & 0 & 0 & 0 \\
0 & 0 & 1 & 1 & 0 & 0 & 0 & 0 & 1 & 1 & 1 & 0 & 1 & 1 & 0 & 0 \\
0 & 0 & 0 & 1 & 1 & 0 & 0 & 0 & 0 & 1 & 1 & 1 & 0 & 1 & 1 & 0 \\
0 & 0 & 0 & 0 & 1 & 1 & 0 & 0 & 0 & 0 & 1 & 1 & 1 & 0 & 1 & 1 \\
1 & 1 & 0 & 1 & 1 & 1 & 1 & 0 & 1 & 0 & 1 & 1 & 1 & 0 & 0 & 1 \\
1 & 0 & 1 & 1 & 0 & 1 & 1 & 1 & 1 & 1 & 1 & 1 & 1 & 0 & 0 & 0 \\
0 & 1 & 0 & 1 & 1 & 0 & 1 & 1 & 1 & 1 & 1 & 1 & 1 & 1 & 0 & 0 \\
0 & 0 & 1 & 0 & 1 & 1 & 0 & 1 & 1 & 1 & 1 & 1 & 1 & 1 & 1 & 0 \\
0 & 0 & 0 & 1 & 0 & 1 & 1 & 0 & 1 & 1 & 1 & 1 & 1 & 1 & 1 & 1 \\
1 & 1 & 0 & 1 & 0 & 0 & 1 & 1 & 1 & 1 & 0 & 1 & 1 & 0 & 1 & 1 \\
1 & 0 & 1 & 1 & 0 & 0 & 0 & 1 & 0 & 1 & 0 & 0 & 1 & 0 & 0 & 1
\end{pmatrix}.
$$

Let as in Example 13.15

$$\mathbf{r} = (0,0,0,0,0,1,0,0,1,0,1,1,1,1,1,0,0,1,0,0,0,1,1,1,0,0,1,0,0,0,1)$$

be the received word with errors on positions 4, 8, and 26. The corresponding system is

$$
\begin{cases}
J(\mathbf{r}), \\
U_{29}V_1 + U_{30}V_2 + U_{31}V_3 + U_1, \\
U_{30}V_1 + U_{31}V_2 + U_1 V_3 + U_2, \\
U_{31}V_1 + U_1 V_2 + U_2 V_3 + U_3, \\
U_{i-3}V_1 + U_{i-2}V_2 + U_{i-1}V_3 + U_i \quad \text{for } 4 \leq i \leq 31.
\end{cases}
$$

From this system we obtain a unique solution, which gives a way to find an error vector via multiplication of the vector of unknown syndromes by B^{-1} on the left. Note that from the way we compute a syndrome in Eq. (13.1) and the way we have chosen the matrix B it follows that actually $U_i = S_{i-1}, V_j = \sigma_{t-j+1}$ for $1 \le i \le n$ and $1 \le j \le t$, where S_i and σ_j are the variables from Example 13.15. So we see that it is actually possible to decode without adding the field equations. This example also shows how we generalize the ideas from Sec. 13.3.2.

Much more serious examples can be attacked by this approach. We only mention that correcting, e.g. 5–20 errors in a binary code of length 120 and dimension 10–30 is feasible. For more details, see [50].

Next we mention how the above technique can be adapted for finding minimum distance of a code. The following is more or less a special case of Theorem 13.33 on decoding up to half the minimum distance.

Theorem 13.37. *Let B be an MDS matrix with structure constants μ_l^{ij} and linear functions U_{ij}. Let H be a parity check matrix of the code C such that $H = AB$. Let t be the smallest integer such that $J(t,0)$ has a solution $(\mathbf{u}, \mathbf{v})$ with $\mathbf{u} \ne 0$. Then t is the minimum distance of C.*

Again we supply the statement with an example.

Example 13.38. Let us find minimum distance of a cyclic $[15, 7]$ binary code with a check matrix:

$$
H = \begin{pmatrix}
1 & 1 & 0 & 1 & 0 & 0 & 0 & 1 & 0 & 0 & 0 & 0 & 0 & 0 & 0 \\
0 & 1 & 1 & 0 & 1 & 0 & 0 & 0 & 1 & 0 & 0 & 0 & 0 & 0 & 0 \\
0 & 0 & 1 & 1 & 0 & 1 & 0 & 0 & 0 & 1 & 0 & 0 & 0 & 0 & 0 \\
0 & 0 & 0 & 1 & 1 & 0 & 1 & 0 & 0 & 0 & 1 & 0 & 0 & 0 & 0 \\
0 & 0 & 0 & 0 & 1 & 1 & 0 & 1 & 0 & 0 & 0 & 1 & 0 & 0 & 0 \\
0 & 0 & 0 & 0 & 0 & 1 & 1 & 0 & 1 & 0 & 0 & 0 & 1 & 0 & 0 \\
0 & 0 & 0 & 0 & 0 & 0 & 1 & 1 & 0 & 1 & 0 & 0 & 0 & 1 & 0 \\
0 & 0 & 0 & 0 & 0 & 0 & 0 & 1 & 1 & 0 & 1 & 0 & 0 & 0 & 1
\end{pmatrix}.
$$

By computing the reduced Gröbner basis of $J(t,0)$ for $t = 1, \ldots, 4$ we see that it always consists of the elements $U_1, \ldots, U_{15}$, so there is no solution $(\mathbf{u}, \mathbf{v})$ with $\mathbf{u} \ne 0$. For $t = 5$ the reduced Gröbner basis (with respect to the degree reverse lexicographic order) is listed in Appendix A. It can be seen that, e.g. $(\mathbf{u}, \mathbf{v})$ with $\mathbf{u} = (1, 0, 0, 0, 0, 1, 0, 1, 0, 0, 1, 1, 0, 1, 1)$ and $\mathbf{v} = (1, 1, 0, 1, 0)$ is a solution of the system $J(5, 0)$. So we obtained a desired solution, thus minimum distance is 5. It can also be seen that $\mathbf{u}$ corresponds to a codeword of weight 5, namely $\mathbf{c} = (1, 0, 0, 0, 0, 0, 0, 1, 0, 0, 0, 1, 0, 1, 1)$.

13.5. Thoughts for Practitioners

From the practical point of view, we can outline three major possible directions of further study for the method of quadratic system solving

- Try to adapt an MDS matrix to a given code in order to obtain decoding schemes similar to the ones that exist for cyclic codes, e.g. generalized Newton identities. This potentially may yield polynomial-time decoding algorithms up to some designed correcting capacity for codes, for which such algorithms are not available at the moment [52, 53].
- As the quadratic system we are working with has a quite specific reduced Gröbner basis, it might be possible to find some specialized algorithms that would find such a basis, and perform much faster.

The last item applies also to the other methods based on Gröbner bases: to find an adapted algorithm that would solve a particular system faster, than a "generic" one.

13.6. Directions for Future Research

On a more theoretical side probably one of the most important questions is the question of estimating the complexity of algorithms presented in this chapter. There were some attempts to apply the theory of semi-regular sequences, but so far they were not successful. There are some rough upper bounds for complexity available, though. Definitely more research is needed here, as this question is one of the milestones in comprehending the Gröbner bases-based methods. Also the question of formal decoding is interesting for particular classes of codes like cyclic codes and some AG-codes. It would be interesting to prove the existence of sparse general error-locator polynomials in these cases. This would shed more light on the old problem of whether decoding cyclic codes is NP-hard.

For the method of quadratic equations it is possible to prove the results like the ones presented here, but for the nearest codeword problem. So the question of list decoding should be studied here further.

13.7. Conclusions

In this chapter, we gave an overview of some of the existing methods for decoding and finding the minimum distance of linear codes and cyclic codes in particular. We tried to give an exposition in a manner as the material appeared historically with the following improvements. We concentrated more on giving examples that would facilitate understanding of the methods, rather than giving some real-life comparisons, although such are also available in [50]. A SINGULAR library for generating different systems for decoding is available [54]. The original method based on solving quadratic system was presented. It turned out that the corresponding reduced Gröbner basis for our system has a very simple form, which is not true for many other methods. We hope that more research in this area may reveal quite fast and robust algorithms that are able to correct up to error-correcting capacity.

13.8. Questions

(1) In Example 13.5, we have seen that $f = X^3$, $g = Y^4 - X^2Y$ from $\mathbb{F}[X, Y]$ form a Gröbner basis of an ideal $I = \langle f, g \rangle$ with respect to the degree reverse lexicographic order $>_{\mathrm{DP}}$. Do they form a Gröbner basis with respect to the lexicographic order $>_{\mathrm{LP}}$?

(2) For constructing a binary cyclic code of length 41, we need a primitive 41-th root of unity. What is the smallest field extension of $\mathbb{F}_2$ that has an element of order 41?

(3) In Eq. (13.4), the polynomial $p(n, X, Y) = (X^n - Y^n)/(X - Y) = \sum_{i=0}^{n-1} X^i Y^{n-1-i}$ is defined. Suppose that $n \geq 2$ and $(n, q) = 1$. Let x, y be elements in an extension of $\mathbb{F}_q$. Show that x and y are distinct if at least one of them is non-zero and $p(n, x, y) = 0$.

(4) In Theorem 13.9, the general error-correcting polynomial is defined. Does its degree depend on the number of errors occurred?

(5) In Example 13.12, can we leave out the polynomials $Z_i Z_j p(15, Z_i, Z_j) = 0, 1 \leq i < j \leq w$?

(6) Let $P_i = (a_{i1}, \ldots, a_{is}), 1 \leq i \leq n$ be n distinct points in $\mathbb{F}_q^s$ ($q^s \geq n$). In Sec. 13.4.1, we needed functions $\xi_i, 1 \leq i \leq n$, such that $\xi_i(P_i) = 1, \xi_i(P_j) = 0, j \neq i$. Give a representation of ξ_i as a polynomial from $\mathbb{F}_q[X_1, \ldots, X_s]$.

(7) In Definition 13.23, we are talking about an *ordered* MDS matrix. Is the order of $\mathbf{b}_1, \ldots, \mathbf{b}_n$ really important for the MDS property to hold?

(8) In Definition 13.25, we introduced a Vandermonde matrix $B = B(\mathbf{x})$. Why is this matrix MDS?

(9) Let C be an $[n, k, d]$ code over $\mathbb{F}_q$. Show that the code $C' = C\mathbb{F}_{q^m}$ has the same parameters $[n, k, d]$ over $\mathbb{F}_{q^m}$.

(10) For those who are familiar with the notion of a reduced Gröbner basis: Consider an ideal $I \subset \mathbb{F}_{37}[X, Y, Z]$. It is known that a Gröbner basis of I with respect to the lexicographic order induced by $X > Y > Z$ is $X + 16, Y + 6$, and $Z + 1$. What is the reduced Gröbner basis of I with respect to the degree reverse lexicographic order?

13.9. Answers

(1) No. We have $\mathrm{lm}(f) = X^3$, $\mathrm{lm}(g) = X^2Y$. Consider $h = Y \cdot f + X \cdot g = XY^4 \in I$, but $\mathrm{lm}(h) = XY^4$ is divisible neither by X^3 nor by X^2Y.

(2) $\mathbb{F}_{2^{20}}$. In order to find it we need to find the smallest n such that $41 \mid (2^n - 1)$, then the extension is going to be $\mathbb{F}_{2^n}$. In our case $n = 20$.

(3) If $x = y$, then $p(n, x, y) = p(n, x, x) = nx^{n-1}$. Since we assumed $n \geq 2$ and $(n, q) = 1$, it follows that $x = y = 0$.

(4) No. The degree is always e, the error-correcting capacity. If $t < e$ errors occurred, then the specialized polynomial $L_C(\mathbf{s}, Z)$ has the root zero with multiplicity $e - t$.

(5) No. Since the system $J_C(2) : Z_1 + Z_2, Z_1^3 + Z_2^3, Z_1^{15} + 1, Z_2^{15} + 1$, would have a Gröbner basis $Z_2^{15} + 1, Z_1 + Z_2$. Thus we obtain the solutions of the form $(a, a), a \in \mathbb{F}_{16}$.

(6) $\xi_i(X_1, \ldots, X_s) = \prod_{j=1}^{s}(1 - (X_j - a_{ij})^{q-1})$. We have that if $X_j \neq a_{ij}$ for some j then $X_j - a_{ij}$ is a non-zero element in $\mathbb{F}_q$, and thus $(X_j - a_{ij})^{q-1} = 1$. On the other hand, if $X_j = a_{ij}$, then $1 - (X_j - a_{ij})^{q-1} = 1$.

(7) Yes. Consider a matrix M over $\mathbb{F}_2$ with rows $(1, 1)$ and $(0, 1)$. It is obvious that M is an MDS matrix. On the other hand, if we exchange the rows, then the obtained matrix N does not have an MDS property, because it has a zero in the first row, and thus not all the 1×1 submatrices of N_1 have full rank 1.

(8) First assume that all x_i are non-zero. Consider an $s \times s$ submatrix S of B_s with the columns indexed by $j_1, \ldots, j_s$, so that the rows of S are $(1, \ldots, 1), (x_{j_1}, \ldots, x_{j_s}), \ldots, (x_{j_1}^{s-1}, \ldots, x_{j_s}^{s-1})$. The matrix S is invertible, since its determinant can be computed to be $\prod_{i>j, i,j \in \{j_1, \ldots, j_s\}}(x_i - x_j)$. If we allow some x_i to be zero, w.l.o.g $x_1 = 0$ and consider again a matrix S, wherein $j_1 = 1$, then we may consider an $(s - 1) \times (s - 1)$ submatrix S' of S, obtained by leaving out the

first row and the first column. The rank of S' is $s-1$ by the argument above. Then, passing to the matrix S we see that its rank is s, because the first column of S is $(1,0,\ldots,0)^T$.

(9) Obviously the length of C' is n. Then, we claim that the vectors $\mathbf{b}_1,\ldots,\mathbf{b}_l$ from $\mathbb{F}_q^n$ are linearly dependent over $\mathbb{F}_q$ if and only if they are linearly dependent over $\mathbb{F}_{q^m}$. Indeed, if $\mathbf{b}_1,\ldots,\mathbf{b}_l$ are linearly dependent over $\mathbb{F}_q$, they are linearly dependent over $\mathbb{F}_{q^m}$. Other way, let there exist a non-trivial linear combination $\alpha_1\mathbf{b}_1 + \cdots + \alpha_l\mathbf{b}_l = 0$ with $\alpha_i \in \mathbb{F}_{q^m}, 1 \leq i \leq l$. Write α's as vectors over $\mathbb{F}_q$ of length m: $\alpha_i = (\alpha_{i1},\ldots,\alpha_{im}), 1 \leq i \leq l$. Since the vectors $\mathbf{b}_1,\ldots,\mathbf{b}_l$ are defined over $\mathbb{F}_q$, we have $\alpha_{1j}\mathbf{b}_1 + \cdots + \alpha_{lj}\mathbf{b}_l = 0, 1 \leq j \leq m$. As the initial linear combination was non-trivial, we obtain at least one non-trivial linear combination for $\mathbf{b}_1,\ldots,\mathbf{b}_l$ over $\mathbb{F}_q$, and thus they are linearly dependent over $\mathbb{F}_q$. Therefore, the dimension of C' is also k. Now the minimum distance can be found as a minimal number of linearly dependent columns of a parity check matrix of a code. Using the argument above, we see that the minimum distance of C' is d, as a parity check matrix H for C is also a parity check matrix of C'.

(10) The same. Denote this basis with G. It is obvious that $X + 16, Y + 6, Z + 1 \in G$ and that I is generated by these polynomials. For any other polynomial $f \in I$, it follows $\mathrm{lm}(f)$ is divisible by either of X, Y, or Z. So f does not belong to the reduced Gröbner basis.

13.10. Keywords

Gröbner basis

A finite subset $G = \{g_1,\ldots,g_m\}$ of an ideal $I \subseteq \mathbb{F}[X_1,\ldots,X_n]$ is called a *Gröbner basis* for I with respect to a monomial order $>$ if $L_>(I) = \langle \mathrm{lt}(g_1),\ldots,\mathrm{lt}(g_m)\rangle$, where $L > (I)$ is the leading ideal of I with respect to $>$; Definition 13.4.

Generalized power sum function

It is a sum of the form $\sum_{l=1}^t y_l z_l^{i_m}$. In our context we use it to compute the syndrome via error locations and error values in Eq. (13.2).

CRHT-ideal

It is an ideal constructed following Cooper's philosophy and its variety contains the information needed for decoding a cyclic code. The generators of this ideal are given in Eq. (13.3).

General error-locator polynomial

A polynomial $L_C = Z^e + a_{t-1}Z^{e-1} + \cdots + a_0$ with $a_j \in \mathbb{F}_q[X_1, \ldots, X_r]$, $0 \le j \le e-1$. When the X-variables are assigned the syndromes, then the roots of $L_C(\mathbf{s}, Z)$ give error positions; Theorem 13.9.

Newton identities

Equation (13.6) shows how syndromes are connected with the elementary symmetric functions. One can use known syndromes to find symmetric functions, and then the error positions via the *error-locator polynomial*, see below.

Error-locator polynomial

It is a polynomial $\sigma(Z) = \prod_{l=1}^{t}(Z - z_l) = Z^t + \sigma_1 Z^{t-1} + \cdots + \sigma_{t-1}Z + \sigma_t$, where σ's are elementary symmetric functions and z's are the error positions. The knowledge of the symmetric functions yields the error positions as the roots of $\sigma(Z)$.

Formal and online decoding

The former term means that in a Gröbner bases-based method one needs to compute a Gröbner basis only once and then decode every time using this precomputed Gröbner basis. The latter term means that one performs the Gröbner basis computation every time one wants to decode. Online decoding involves systems with less variables and thus is much easier to handle; on the other hand performing the Gröbner basis computation every time can be too time consuming.

Affine variety code

$C(I, L)$ is an image of a vector subspace L of the map $\phi : R \to \mathbb{F}_q^n, \bar{f} \mapsto (f(P_1), \ldots, f(P_n))$, where $R := \mathbb{F}_q[U_1, \ldots, U_s]/I_q$ and $f \in L$ is any pre-image of $\bar{f}$ under a canonical homomorphism, L is a vector subspace of R.

Unknown syndromes

Let $\mathbf{b}_1, \ldots, \mathbf{b}_n$ be a basis of $\mathbb{F}_q^n$ and let B be the $n \times n$ matrix with $\mathbf{b}_1, \ldots, \mathbf{b}_n$ as rows. The (unknown) syndrome $\mathbf{u}(B, \mathbf{e})$ of a word $\mathbf{e}$ with respect to B is the column vector $\mathbf{u}(B, \mathbf{e}) = B\mathbf{e}^T$. It has entries $u_i(B, \mathbf{e}) = \mathbf{b}_i \cdot \mathbf{e}$ for $i = 1, \ldots, n$; Definition 13.18.

MDS basis/matrix

Let $\mathbf{b}_1, \ldots, \mathbf{b}_n$ be a basis of $\mathbb{F}_q^n$. Let B_s be the $s \times n$ matrix with $\mathbf{b}_1, \ldots, \mathbf{b}_s$ as rows, then $B = B_n$. We say that $\mathbf{b}_1, \ldots, \mathbf{b}_n$ is an *ordered MDS basis* and B an *MDS matrix* if all the $s \times s$ submatrices of B_s have rank s for all $s = 1, \ldots, n$; Definition 13.23.

Acknowledgments

The first author would like to thank "DASMOD: Cluster of Excellence in Rhineland-Palatinate" for funding his research, and also personally his Ph.D. supervisor Prof. Dr. Gert-Martin Greuel and his second supervisor Prof. Dr. Gerhard Pfister for continuous support. The work of the first author has been partially inspired by the Special Semester on Gröbner Bases, 1 February–31 July, 2006, organized by RICAM, Austrian Academy of Sciences, and RISC, Johannes Kepler University, Linz, Austria.

Appendix A. For Example 13.38

$U_2,$	$U_3,$	$U_4,$	$U_5,$
$U_7,$	$U_9,$	$U_{10},$	$U_{13},$
$V_5U_1,$	$V_3U_1,$	$V_1U_1 + U_6,$	$U_6^2 + U_{11}U_1,$
$U_{11}U_6 + U_1^2,$	$V_5U_6,$	$V_4U_6 + U_8,$	$V_3U_6,$
$V_1U_6 + U_{11},$	$U_8^2 + U_{15}U_1,$	$U_{12}U_8 + U_{14}U_6,$	$V_5U_8,$
$V_4U_8 + V_2U_6,$	$V_3U_8,$	$V_2U_8 + U_{12},$	$U_{11}^2 + U_6U_1,$
$U_{12}U_{11} + U_{15}U_8,$	$V_5U_{11},$	$V_4U_{11} + V_1U_8,$	$V_3U_{11},$
$V_2U_{11} + U_{15},$	$V_1U_{11} + U_1,$	$U_{12}^2 + U_8U_1,$	$U_{14}U_{12} + U_{15}U_{11},$
$V_5U_{12},$	$V_4U_{12} + U_{14},$	$V_3U_{12},$	$U_{14}^2 + U_{12}U_1,$
$U_{15}U_{14} + U_8U_6,$	$V_5U_{14},$	$V_4U_{14} + V_2U_{12},$	$V_3U_{14},$
$V_2U_{14} + V_4U_1,$	$U_{15}^2 + U_{14}U_1,$	$V_5U_{15},$	$V_4U_{15} + V_1U_{12},$
$V_3U_{15},$	$V_2U_{15} + V_1U_{14},$	$V_1U_{15} + V_2U_1,$	$V_4U_1^2 + U_{11}U_8,$
$V_2U_1^2 + U_{15}U_6,$	$V_2U_6U_1 + U_{15}U_{11},$	$V_4^2U_1 + V_2U_1,$	$U_{14}U_8U_6 + U_{15}U_{12}U_1,$
$U_{15}U_8U_6 + U_{12}U_1^2,$	$V_2U_{12}U_6 + U_{14}U_8,$	$V_2^2U_6 + U_{14},$	$U_{14}U_{11}U_8 + U_{15}U_{12}U_6,$
$U_{15}U_{11}U_8 + U_{12}U_6U_1,$	$V_1U_{14}U_8 + U_{15}U_{12},$	$V_1^2U_8 + V_4U_1,$	$V_2^2U_{12} + V_2U_1,$
$V_1^2U_{12} + V_2V_4U_1,$	$V_1^2U_{14} + V_2^2U_1,$	$V_2^3U_1 + V_1U_8$	

References

1. D. Cox, J. Little and D. O'Shea, *Ideals, Varieties, and Algorithms*, 2nd edn. (Springer-Verlag, 1997).
2. G.-M.Greuel and G. Pfister, *A SINGULAR Introduction to Commutative Algebra* (Springer-Verlag, 2002).
3. S. Arimoto, Encoding and decoding of p-ary group codes and the correction system (in Japanese) *Inform. Process. Jpn.* **2** (1961), 320–325.
4. E. R. Berlekamp, *Algebraic Coding Theory* (Mc Graw Hill, New York, 1968).
5. J. L. Massey, Shift-register synthesis and BCH decoding, *IEEE Trans. Inform. Theor.* **IT-15** (1969), 122–127.

6. W. W. Peterson and E. J. Weldon, *Error-Correcting Codes* (MIT Pres, Cambridge 1977).

7. Y. Sugiyama, M. Kasahara, S. Hirasawa and T. Namekawa, A method for solving the key equation for decoding Goppa codes, *Inform. Cont.* **27** (1975), 87–99.

8. D. Augot, P. Charpin and N. Sendrier, The minimum distance of some binary codes via the Newton's Identities, in *Eurocodes'90*, Vol. 514(LNCS, 1990), pp. 65–73.

9. D. Augot, P. Charpin and N. Sendrier, Studying the locator polynomial of minimum weight codewords of BCH codes, *IEEE Trans. Inform. Theor.* **IT-38** (1992), 960–973.

10. D. Augot, M. Bardet and J.-C. Faugère, Efficient decoding of (binary) cyclic codes beyond the correction capacity of the code using Gröbner bases, *INRIA Report*, No. 4652, (November 2002).

11. D. Augot, M. Bardet and J.C. Faugère, On formulas for decoding binary cyclic codes, *Proc. IEEE Int. Symp. Information Theory*, (2007).

12. M. A. de Boer and R. Pellikaan, Gröbner bases for codes, in *Some Tapas of Computer Algebra*, (eds.) A. M. Cohen, H. Cuypers and H. Sterk (Springer-Verlag, Berlin, 1999), Chap. 10, pp. 237–259.

13. M. A. de Boer and R. Pellikaan, Gröbner bases for decoding, in *Some Tapas of Computer Algebra* (eds.) A. M. Cohen, H. Cuypers and H. Sterk (Springer-Verlag, Berlin, 1999), Chap. 11, pp. 260–275.

14. X. Chen, I. S. Reed, T. Helleseth and T. K. Truong, Use of Gröbner bases to decode binary cyclic codes up to the true minimum distance, *IEEE Trans. Inform. Theor.* **IT-40** (1994), 1654–1661.

15. M. Caboara and T. Mora, The Chen–Reed–Helleseth–Truong decoding algorithm and the Gianni–Kalkbrenner Gröbner shape theorem, *Appl. Algeb. Eng. Comm. Comput.* **13** (2002), 209–232.

16. X. Chen, I. S. Reed, T. Helleseth and T. K. Truong, Algebraic decoding of cyclic codes: A polynomial point of view, *Contemp. Math.* **168**, (1994), 15–22.

17. X. Chen, I. S. Reed, T. Helleseth and T. K. Truong, General principles for the algebraic decoding of cyclic codes, *IEEE Trans. Inform. Theor.* **IT-40** (1994), 1661–1663.

18. A. B. Cooper, Toward a new method of decoding algebraic codes using Gröbner bases, *Trans. 10th Army Conf. Appl. Math. Comp*, (1993), 1–11.

19. E. Orsini and M. Sala, Correcting errors and erasures via the syndrome variety, *J. Pure Appl. Algebra* **200** (2005), 191–226.

20. T. Mora and E. Orsini, Decoding cyclic codes: The Cooper philosophy, *A Talk at the Special Semester on Gröbner Bases*, (May, 2006).

21. M. Giorgetti and M. Sala, A commutative algebra approach to linear codes, *BCRI* preprint No. 58, www.bcri.ucc.ie, (2006).

22. E. Orsini and M. Sala, Improved decoding of affine–variety codes, BCRI preprint No. 68, www.bcri.ucc.ie, (2007).

23. J. B. Farr and S. Gao, Gröbner bases and generalized Padé approximation, *Math. Comp.* **75** (2005), 461–473.

24. P. Fitzpatrick and J. Flynn, A Gröbner basis technique for Padé approximation, *J. Symbo. Comput.* **24**, 5 (1992), 133–138.
25. M. Borges-Quintana, M. A. Borges-Trenard and E. Martínez-Moro, A general framework for applying FGLM techniques to linear codes, *AAECC 2006, Lecture Notes in Computer Science* (2006), 76–86.
26. P. Loustaunau and E. V. York, On the decoding of cyclic codes using Gröbner bases, *AAECC* **8**, 6 (1997), 469–483.
27. P. Fitzpatrick, On the key equation, *IEEE Trans. Inform. Theor.* **41**, 5 (1995), 1290–1302.
28. C. Lossen and A. Frühbis-Krüger, Introduction to Computer Algebra (Solving Systems of Polynomial Equations) http://www.mathematik.uni-kl.de/~lossen/ SKRIPTEN/COMPALG/compalg.ps.gz, (2005).
29. B. Buchberger, Ein Algorithmus zum Auffinden der Basiselemente des Restklassenrings, *Ph.D. Thesis*, Innsbruck, (1965).
30. G.-M. Greuel, G. Pfister and H. Schönemann. SINGULAR 3.0. A computer algebra system for polynomial computations. Centre for Computer Algebra, University of Kaiserslautern, 2007. http://www.singular.uni-kl.de.
31. Magma V2.14-4, Computational Algebra Group, School of Mathematics and Statistics, University of Sydney, Website: http://magma.maths.usyd.edu.au, (2007).
32. CoCoA: A system for doing Computations in Commutative Algebra, http:// cocoa.dima.unige.it, (2007).
33. J. C. Faugère, A new efficient algorithm for computing Gröbner bases (F4), *J. Pure Appl. Algebra* **139**, 1–3, (1999), 61–88.
34. J. C. Faugère, A new efficient algorithm for computing Gröbner bases without reduction to zero F5, (ed.) T. Mora, *Proc. 2002 Int. Symp. Symbolic and Algebraic Computation, ISSAC*, (2002), pp. 75–83.
35. FGb, http://fgbrs.lip6.fr/jcf/Software/FGb/index.html, (2007).
36. J.-C. Faugère, P. Gianni, D. Lazard and T. Mora, Efficient computation of zero-dimensional Gröbner bases by change of ordering, *J. Symb. Comput.* **16** (1993), 329–344.
37. E. Orsini and M. Sala, General error locator polynomials for binary cyclic codes with $t \le 2$ and $n < 63$, *IEEE Trans. Inform. Theor.* **53**, 3 (2007), 1095–1107.
38. M. Sala, Gröbner basis techniques to compute weight distributions of shortened cyclic codes, *J. Algebra Appl.* **6**, 3 (2007), 403–414.
39. T. Mora and M. Sala, On the Gröbner bases for some symmetric systems and their application to coding theory, *J. Symb. Comp.* **35**, 2 (2003), 177–194.
40. D. Augot, Description of minimum weight codewords of cyclic codes by algebraic system, *Finite Fields Appl.* **2**, 2 (1996), 138–152.
41. D. C. Gorenstein and N. Zierler, A class of error-correcting codes in p^m symbols, *J. SIAM* **9** (1961), 207–214.
42. F. J. MacWilliams and N. J. A. Sloane, *The Theory of Error-Correcting Codes* (Amsterdam–New York–Oxford: North Holland, 1977).
43. A. E. Heydtmann and J. M. Jensen, On the equivalence of the Berlekamp-Massey and the Euclidean algorithms for decoding, *IEEE Trans. Inform. Theor.* **46**, (2000), 2614–2624.

44. K. K. Tzeng, C. R. P. Hartmann and R. T. Chien, Some notes on iterative decoding, *Proc. 9th Allerton Conf. Circuit and Systems Theory*, (1971).

45. P. Stevens, Extensions of the BCH decoding algorithm to decode binary cyclic codes up to their maximum error correction capacities, *IEEE Trans. Inform. Theor.* **IT-34** (1988), 1332–1340.

46. D. Augot, M. Bardet and J. C. Faugère, On the decoding of cyclic codes with Newton identities, to appear in *Special Issue "Gröbner Bases Techniques in Cryptography and Coding Theory" of Journ. Symbolic Comp.*, (2008).

47. J. Fitzgerald, Applications of Gröbner bases to Linear Codes, Ph.D. Thesis, Louisiana State University, (1996).

48. J. Fitzgerald and R. F. Lax, Decoding affine variety codes using Gröbner bases, *Des. Codes Cryptogr.* **13** (1998), 147–158.

49. M. G. Marinari, H. M. Möller and T. Mora, Gröbner basis of ideals defined by functional with an application to ideals of projective points, *Appl. Algebra Eng. Comm. Comput.* **4**, 2 (1993), 103–145.

50. S. Bulygin and R. Pellikaan, Bounded distance decoding of linear error-correcting codes with Gröbner bases, to appear in *Special Issue "Gröbner Bases Techniques in Cryptography and Coding Theory" of J. Symbolic Comp.*, (2009).

51. T. Høholdt, J. H. van Lint and R. Pellikaan, Algebraic geometry codes, in *Handbook of Coding Theory*, (eds.) V. S. Pless and W. C. Huffman, Vol. 1, pp. 871–961 (Elsevier, Amsterdam, 1998).

52. R. J. McEliece, A public-key cryptosystem based on algebraic coding theory, *DSN Progress Report* **42–44** (1978), 114–116.

53. H. Niederreiter, Knapsack-type crypto systems and algebraic coding theory, *Prob. Contr. Inform. Theor.* **15**, 2 (1986), 159–166.

54. S. Bulygin, decodegb.lib, http://www.mathematik.uni-kl.de/~bulygin/files/decodegb.lib, (2009).

Chapter 14

COOPERATIVE DIVERSITY SYSTEMS FOR WIRELESS COMMUNICATION

MURAT UYSAL* and MUHAMMAD MEHBOOB FAREED†

Department of Electrical and Computer Engineering,
University of Waterloo, Waterloo, Ontario, N2L3G1, Canada
*muysal@ece.uwaterloo.ca
†mmfareed@ece.uwaterloo.ca

Cooperative diversity represents a new class of wireless communication techniques in which network nodes help each other in relaying information to realize spatial diversity advantages. This new transmission paradigm promises significant performance gains in terms of link reliability, spectral efficiency, system capacity, and transmission range. Cooperative diversity has spurred tremendous excitement within the academia and industry circles since its introduction and has been extensively studied over the last few years. This chapter presents an overview of cooperative diversity conveying the underlying basic principles and further discusses the latest advances and open issues in this rapidly evolving field.

14.1. Introduction

The increasing demand for wireless multi-media and interactive internet services is fueling intensive research efforts on higher-speed data transmission and improved power efficiency compared to current wireless communication systems. The characteristics of wireless channels impose fundamental limitations on the performance of these systems. The degrading effects in a wireless channel can be classified as *large-scale (long-term)* impairments including path loss, shadowing and *small-scale (short-term)* impairment which is commonly referred as fading [1, 2]. The former impairment is used to predict the average signal power at the receiver side and the transmission coverage area. The latter is due to the multi-path propagation which causes random fluctuations in the received signal level and affects the instantaneous signal-to-noise ratio (SNR).

For a typical mobile wireless channel in urban areas where there is no line-of-sight propagation and the number of scatters is considerably large, the application of central limit theory indicates that the complex fading channel coefficient has two quadrature components which are zero-mean Gaussian distributed. As a result, the amplitude of the fading envelope follows a Rayleigh distribution. In terms of error rate performance, Rayleigh fading converts the exponential dependency of the bit error rate (BER) on the SNR for the classical additive white Gaussian noise (AWGN) channel into an approximately inverse linear one, resulting in a large SNR penalty.

A common approach to mitigate the degrading effects of fading is the use of diversity techniques such as frequency, time, polarization, and spatial diversity [2]. Diversity improves transmission performance by making use of more than one independently faded version of the transmitted signal. If several replicas of the signals are transmitted over multiple channels that exhibit independent fading with comparable strengths, the probability that all the independently faded signals simultaneously experience deep fading is significantly reduced. A widely adopted diversity method is spatial diversity which involves the use of multiple antennas at the transmitter and/or receiver side. Multiple-input multiple-output (MIMO) antenna techniques have been extensively studied within the last decade [1] and already incorporated in the wireless standards. Although MIMO communications offer distinct advantages, there are various scenarios where the deployment of multiple antennas is not practical due to the size, power limitations, and hardware complexity of the terminals. Examples include wireless sensor networks and (WSNs) and *ad hoc* networks which are gaining popularity in recent years.

Cooperative diversity (also known as *cooperative communications* or *user cooperation*) [3–8] has emerged as a powerful alternative to reap the benefits of MIMO communications in a wireless scenario with single-antenna terminals. Cooperative communication takes advantage of the broadcast nature of wireless transmission and creates a *virtual antenna array* through cooperating nodes. For a preliminary explanation of the basic idea behind cooperative diversity, Fig. 14.1 depicts a wireless setting where a single-antenna source terminal (node A) would like to communicate with the base station. Since it is equipped with only one antenna, node A is not able to extract spatial diversity on its own. However, thanks to the broadcasting nature of the wireless transmission, the information transmitted by this node is *overheard* by other nodes in the vicinity

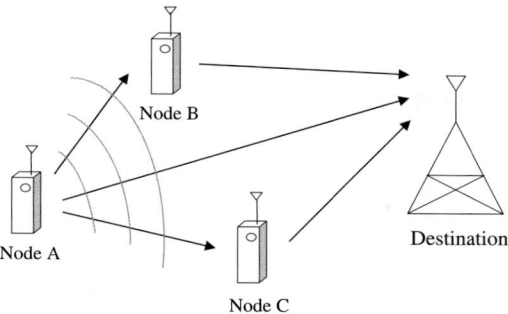

Fig. 14.1.   Cooperative communication mimics a virtual antenna array and extracts spatial diversity advantages.

(nodes B and C). Assuming that these nodes are willing to share their resources with node A, they can forward the overheard information along their own data to the destination and act as *relays* for node A. Since the fading paths from these three nodes are statistically independent, this generates a spatial diversity effectively creating a virtual array of three antennas.

Cooperative communication can be applied to both *infrastructure-based networks* such as cellular systems, WLANs (wireless local area networks), WMANs (wireless metropolitan area networks) and infrastructure-less networks such as MANETs (mobile ad-hoc networks), VANETs (vehicular area ad-hoc networks), and WSNs. Based on the envisioned application and physical features of cooperating nodes, cooperative communication can take different forms [9]: In terminal (client) cooperation, user terminals cooperate to create a virtual antenna array as depicted in Fig. 14.1. Such peer-to-peer networking is particularly attractive for applications such as WSNs where space and power limitations of terminals exclude the use of multi-antenna deployment. On the other hand, the concept of cooperative communication can be applied to form an infrastructure mesh where fixed network elements (e.g. base stations, access points) can be deployed for relaying in a multi-hop fashion to enhance the throughput and/or the coverage area. In Fig. 14.2, a cellular network is depicted where the coverage area of a base station is extended through the assistance of relaying stations with relatively lower power (notice that the range of relaying stations is smaller than that of main base station). Since transmit power requirements are significantly reduced for relays, less expensive power amplifiers can be deployed in those stations.

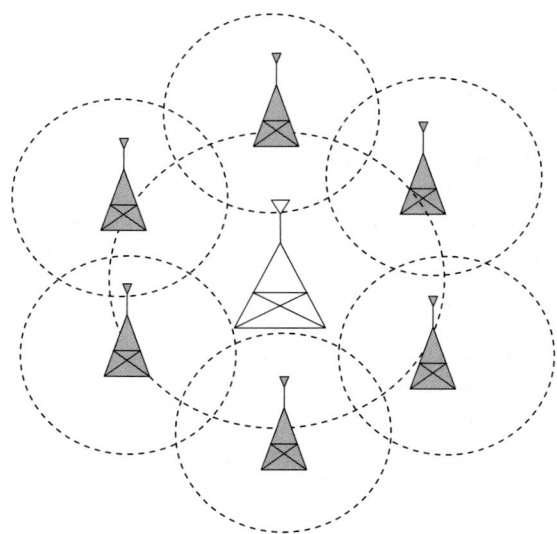

Fig. 14.2.   Extension of the coverage area of a base station through relaying stations.

14.2. Background

The recent surge of interest in cooperative communication was subsequent to the works of Sendonaris *et al.* [6, 7] and Laneman *et al.* [3–5]. However, the basic ideas behind cooperative communication can be traced back to Meulen's early work on the relay channel [10] in early 1970's. The first rigorous information theoretical analysis of the relay channel was exposed in [11] by Cover and Gamal for AWGN channels. Extending the work of authors in [11] for wireless fading channels, Sendonaris *et al.* [6, 7] have investigated the achievable rate region for relay-assisted transmission and coined the term *user cooperation.* In an independent work by Laneman *et al.* [4, 5], it is demonstrated that full spatial diversity can be achieved through user cooperation. Their proposed cooperation protocol is built upon a two-phase transmission scheme. In the first phase (i.e. broadcasting phase), the source broadcasts to the destination and relay terminals. In the second phase (i.e. relaying phase), the relays transmit processed version of their received signals to the destination using either orthogonal subchannels (i.e. repetition-based cooperative diversity), or the same subchannel (i.e. space-time coded cooperative diversity). The latter relies on the implementation of conventional orthogonal space-time block codes (STBCs) [12] in a distributed fashion among the relay nodes.

After Laneman *et al*'s work, there has a been a growing attention on the topic of cooperative diversity [13–19]. Jing and Hassibi [13] extended the idea of Linear Dispersion (LD) codes from multi-antenna systems to cooperative systems. In their two-phase protocol, the source node broadcasts in the first phase, while in second phase relay nodes use LD codes in a distributed manner to send information to the destination. In [14], Yang and Belfiore highlighted the fact that the performance of cooperative schemes suffers in terms of the diversity-multiplexing tradeoff (DMT). They proposed a new class of cooperation scheme to improve the multiplexing gain. On the other hand, to address the practical implementation of distributed space-time coding, Li and Xia [15] introduced a family of space-time trellis codes based on the stack construction which is able to achieve the full cooperative diversity order without the idealistic synchronization assumption. Motivated by non-coherent differential space-time coding, Oggier and Hassibi [16] presented a coding strategy for cooperative networks which does not require channel knowledge. Kiran and Rajan [17] further proposed a cooperative scheme which assumes partial channel information, i.e. only destination has knowledge of relay-to-destination channel.

Two main relaying techniques are studied in [4]: *Amplify-and-Forward (AaF)* and *Decode-and-Forward (DaF)*. In DaF relaying, the relay node fully decodes, re-encodes (possibly with a different codebook), and retransmits the source node's message. In AaF relaying, the relay retransmits a scaled version of the received signal without any attempt to decode it.

Another classification for relaying is also proposed in [4]: In *fixed relaying*, the relay always forwards the message that it receives from the source. The performance of fixed DaF relaying is mainly limited by the quality of source-to-relay channel. An alternative to fixed relaying is *selection relaying* which is, in nature, adaptive to the channel conditions. In this type of relaying, the source reverts to non-cooperation mode at times when the measured instantaneous SNR falls below a certain threshold and continues its own direct transmission to the destination. The work in [20] can be considered as a systematic realization of such adaptive relaying through powerful channel coding techniques. In so-called *coded cooperation* of [20], Hunter *et al.* realize the concept of user cooperation through the distributed implementation of existing channel coding methods such as convolutional and turbo codes. The basic idea is that each user tries to transmit incremental redundancy for its partner.

Whenever that is not possible, the users automatically revert to a non-cooperative mode.

The cooperation protocol proposed by Laneman *et al.* in [4] effectively implements receive diversity in a distributed manner. In [21], Nabar *et al.* establish a unified framework of cooperation protocols for single-relay wireless networks. They quantify achievable performance gains for distributed schemes in an analogy to conventional co-located multi-antenna configurations. Specifically, three TDMA-based protocols are considered in [21] which are named as Protocol I, Protocol II, and Protocol III (Table 14.1). In an independent work by Ochiai *et al.* [22], they are referred as *Transmit Diversity (TD) Protocol, Receive Diversity (RD) Protocol,* and *Simplified Transmit Diversity (STD) Protocol.*

In TD protocol, the source terminal communicates with the relay and destination during the first phase (i.e. broadcasting phase). During the second phase (i.e. relaying phase), both the relay and source terminals communicate with the destination terminal. This protocol realizes maximum degrees of broadcasting and receive collision. In [23], Azarian *et al.* have demonstrated that this protocol is optimum in terms of DMT for AaF relaying. In RD protocol, the source terminal communicates with the relay and destination terminals in the first phase. In the second phase, only the relay terminal communicates with the destination. This protocol realizes a maximum degree of broadcasting and exhibits no receive collision. This is the same cooperation protocol proposed by Laneman *et al.* in [4]. STD protocol is essentially similar to TD protocol except that the destination terminal does not receive from the source during the first phase for reasons which are possibly imposed from the upper-layer networking protocols (e.g. the destination terminal may be engaged in data transmission to another terminal during the first time slot). This protocol does not implement broadcasting but realizes receive collision.

Table 14.1.   Cooperation protocols for single-relay networks [21, 22].

| Terminal | Protocol | | | | | |
| | TD | | STD | | RD | |
	Phase 1	Phase 2	Phase 1	Phase 2	Phase 1	Phase 2
Source	●	●	●	●	●	—
Relay	○	●	○	●	○	●
Destination	○	○	—	○	○	○

●: Transmitting, ○: Receiving, — : Idle.

The generalization of these protocols for multiple relays can be found in [23, 24].

It can be noticed from the descriptions of the above protocols that the signal transmitted to both relay and destination terminals is the same over the two phases in RD protocol. Therefore, classical space-time code construction does not apply to this protocol. On the other hand, TD and STD protocols can transmit different signals to the relay and destination terminals. Hence, conventional space-time codes can be easily applied to these protocols in a distributed fashion. It should be noted that the use of STBC has been also proposed by Laneman *et al.* in [5] for RD protocol. Their proposed use of STBC, however, implements coding across the relay nodes assuming a scenario with more than one relay and differs from the distributed STBC (DSTBC) setup in [21] proposed for TD and STD protocols which involve the source terminal in a single-relay scenario.

In the rest of this chapter, we consider the generic cooperative transmission scheme depicted in Fig. 14.3 where the transmission from source node to the destination is carried out with the assistance of N relays. All the nodes are equipped with single transmit/receive antennas, unless otherwise stated. They operate in half-duplex mode (i.e. a node can either transmit or receive but not both) which is motivated by the practical implementation problem resulting from the large difference between transmit and receive powers in many applications. They are assumed to be located in a two-dimensional plane where d_{SD}, d_{SR_i}, and $d_{\mathrm{R}_i\mathrm{D}}$, $i = 1, 2, \ldots, N$ denote the distances of source-to-destination $(\mathrm{S} \to \mathrm{D})$, source-to-relay $(\mathrm{S} \to \mathrm{R}_i)$, and relay-to-destination $(\mathrm{R}_i \to \mathrm{D})$ links, respectively. To incorporate the effect of relay location into this model, we

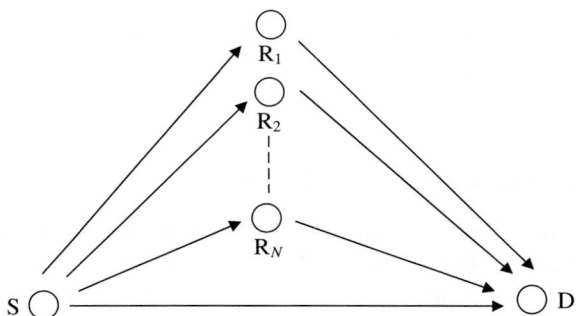

Fig. 14.3.   Cooperative transmission model.

consider an aggregated channel model which takes into account both long-term path loss and short-term Rayleigh fading. The path loss is proportional to $d^{-\alpha}$ where d is the propagation distance and α is the path loss coefficient. Typical values of α for various wireless environments can be found in [25]. Normalizing the path loss in $S \to D$ to be unity, the relative geometrical gains of $S \to R_i$ and $R_i \to D$ links are defined as $G_{SR_i} = (d_{SD}/d_{SR_i})^\alpha$ and $G_{R_iD} = (d_{SD}/d_{R_iD})^\alpha$. They can be further related to each other by law of cosines, i.e. $G_{SR_i}^{-2/\alpha} + G_{R_iD}^{-2/\alpha} - 2G_{SR_i}^{-1/\alpha} G_{R_iD}^{-1/\alpha} \cos\theta_i = 1$ where θ_i is the angle between lines $S \to R_i$ and $R_i \to D$ [22]. The fading coefficients for $S \to D$, $S \to R_i$, and $R_i \to D$ links are denoted by h_{SD}, h_{SR_i}, and h_{R_iD}, respectively. They are modeled as zero-mean complex Gaussian random variables with unit variance leading to a Rayleigh fading channel model.

We assume the deployment of RD protocol. Let x denote the transmitted M-QAM (quadrature amplitude modulation) or M-PSK (phase shift keying) modulation signal. In this protocol, the received signals at the destination and relay nodes during the broadcasting phase are given by

$$y_{D1} = h_{SD}x + n_{D1}, \tag{14.1}$$

$$y_{R_i} = \sqrt{G_{SR_i}} h_{SR_i} x + n_{R_i}, \quad i = 1, 2, \ldots, N. \tag{14.2}$$

In the relaying phase, the source node remains silent. Relay nodes forward the received signals to the destination. The relay can work in either regenerative (DaF relaying) or non-regenerative (AaF relaying) mode. The signal received from the ith relay at the destination during this phase is given by

$$y_{D2} = \sqrt{G_{R_iD}} h_{R_iD} g_{R_i}(x) + n_{D2}, \tag{14.3}$$

where $g_{R_i}(x)$ is the signal transmitted by ith relays and is a function of the transmitted signal by the source node. The particular choice of $g(x)$ depends on the type of relaying which will be elaborated in the following sections.

14.3. Amplify-and-Forward Relaying

In AaF relaying, each of the relay needs to scale its respective received signal before transmission. The scaling factor can take different forms based on the availability of channel state information (CSI) at the relay terminal. In *CSI-assisted AaF* scheme [4], the relay uses instantaneous CSI of the source-to-relay link to scale its received noisy signal before re-transmission.

This ensures that the same output power is maintained for each realization. On the other hand, *blind AaF* scheme does not have access to CSI and employs fixed power constraint. This ensures that an average output power is maintained, but allows for the instantaneous output power to be much larger than the average. Although blind AaF is not expected to perform as good as CSI-assisted AaF relaying, the elimination of channel estimation at the relay terminal promises low complexity and makes it attractive from a practical point of view. The scaling factors for blind and CSI-assisted AaF schemes are referred as *average power scaling (APS)* and *instantaneous power scaling (IPS)* constraints in [26] which are, respectively, given by

$$\beta_{\text{APS}} = \cfrac{1}{\sqrt{\underset{n,h_{\text{SR}}}{\text{E}}\left[|y_{R_i}|^2\right]}} = \frac{1}{\sqrt{G_{\text{SR}_i}E + N_0}}, \tag{14.4a}$$

$$\beta_{\text{IPS}} = \cfrac{1}{\sqrt{\underset{n}{\text{E}}[|y_{R_i}|^2]}} = \frac{1}{\sqrt{G_{\text{SR}_i}E|h_{\text{SR}_i}|^2 + N_0}}, \tag{14.4b}$$

where $\text{E}[.]$ denotes the expectation operation, E is the average energy of the signal constellation and $N_0 = E[|n_{R_i}|^2]$. The function $g_{R_i}(x)$ in (14.3), therefore, takes the form of $g_{R_i}(x) = \beta_{\text{APS}}y_{R_i}$ and $g_{R_i}(x) = \beta_{\text{IPS}}y_{R_i}$ for blind and CSI-assisted AaF schemes, respectively. After this scaling operation, the relay terminals forward the received signals to the destination. The signal received from the ith relay at the destination terminal is

$$y_{D2} = \sqrt{\frac{G_{\text{SR}_i}G_{R_iD}}{G_{\text{SR}_i}E + N_0}}h_{\text{SR}_i}h_{R_iD}x + \sqrt{\frac{G_{R_iD}}{G_{\text{SR}_i}E + N_0}}h_{R_iD}n_{R_i} + n_{D2},$$

$$\tag{14.5}$$

for blind AaF scheme and

$$y_{D2} = \sqrt{\frac{G_{\text{SR}_i}G_{R_iD}}{G_{\text{SR}_i}E|h_{\text{SR}_i}|^2 + N_0}}h_{\text{SR}_i}h_{R_iD}x$$

$$+ \sqrt{\frac{G_{R_iD}}{G_{\text{SR}_i}E|h_{\text{SR}_i}|^2 + N_0}}h_{R_iD}n_{R_i} + n_{D2}, \tag{14.6}$$

for CSI-assisted AaF scheme. The received signals during two phases are then fed to a maximum likelihood (ML) detector to yield an estimate of x.

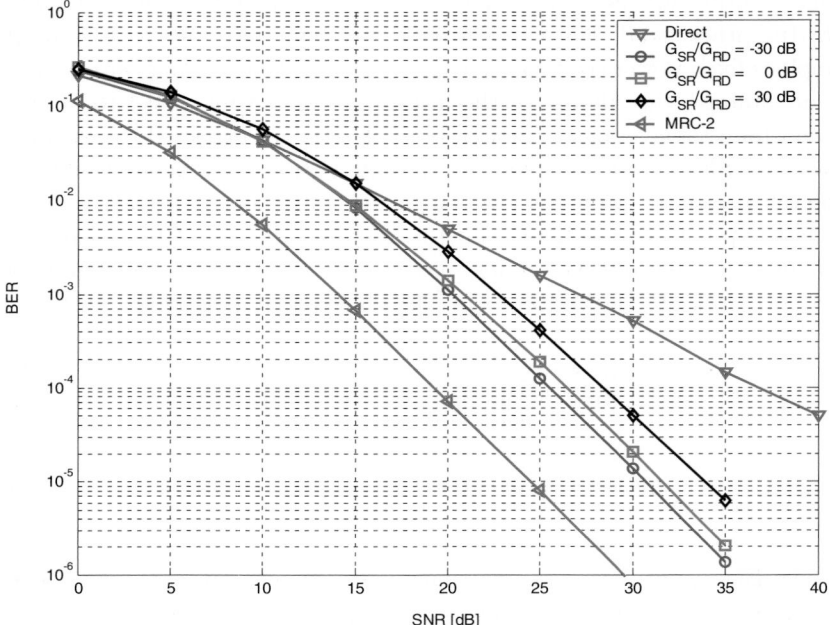

Fig. 14.4. BER performance of RD protocol with AaF relaying for various relay locations.

Next, we investigate the performance of AaF relaying through Monte-Carlo simulations. In Fig. 14.4, we present the performance of blind AaF scheme for a single-relay scenario assuming various relay locations and $\theta = \pi$. As benchmarks, the performance of direct (non-cooperative) transmission and maximal ratio combining (MRC) with two co-located receive antennas are further included. We assume 16-QAM for the cooperative scheme while 4-QAM is employed for the non-cooperative and MRC schemes to maintain the same throughput rate. The ratio of relative geometrical gains G_{SR}/G_{RD} (in dB) is used to indicate the effect of relay location. More negative this ratio is, more closely the relay is placed to destination terminal. On the other hand, positive values of this ratio indicate that the relay is more close to source terminal. Specifically, we consider $G_{SR}/G_{RD} = -30\,\text{dB}$, $0\,\text{dB}$, $30\,\text{dB}$. Particular case of $G_{SR}/G_{RD} = 0\,\text{dB}$ corresponds to a scenario where the relay is equidistant from source and destination terminals. Figure 14.4 clearly demonstrates the advantage of cooperation where a performance improvement of 8 dB is observed at a target BER of 10^{-4} for $G_{SR}/G_{RD} = 30\,\text{dB}$. The performance improvement climbs up to 11dB for $G_{SR}/G_{RD} = -30$ dB where the relay is located close to the destination. This is rather expected considering that the deployed RD protocol effectively realizes receive diversity in a distributed fashion. Although cooperative transmission significantly improves the performance, the performance of this virtual antenna array is still about 6 dB away from that of its co-located antenna counterpart (c.f. the performance of MRC scheme with two co-located antennas).

In Fig. 14.5, we compare the performance of blind and CSI-assisted AaF schemes. We assume 4-QAM, $\theta = \pi$, and $G_{SR}/G_{RD} = 30$ dB in simulations. For the single-relay scenario under consideration, both schemes are able to extract a diversity order[a] of two. Although both of them achieve the full diversity for this scenario, CSI-assisted AaF scheme outperforms its blind counterpart as expected. Specifically, we observe a performance degradation of 3.2 dB at BER $= 10^{-4}$.

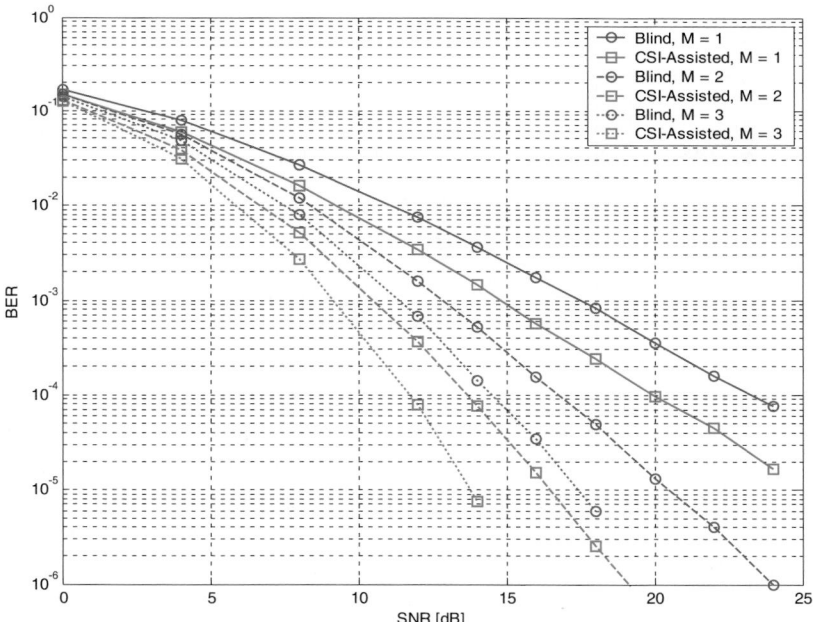

Fig. 14.5.   Performance comparison of CSI-assisted and blind AaF schemes with one, two, and three receive antennas at destination.

---

[a]Diversity order is defined as the negative asymptotic slope of error probability versus SNR on a log–log scale [1].

We further compare the performance of two competing schemes assuming multi-antenna deployment at the destination terminal. Interestingly, the diversity orders of two competing schemes do not coincide anymore. CSI-assisted scheme achieves diversity orders of 4 and 6 when the destination terminal is equipped with $M = 2$ and 3 antennas, respectively. On the other hand, the diversity order of blind scheme is limited to 3 and 4 for the same scenarios. Indeed, it has been analytically shown in [26] that CSI-assisted and blind schemes are able to achieve diversity orders of $2M$ and $M + 1$. This is due to the fact that the diversity order of blind scheme over the relaying path is governed by the link which has smaller diversity order.

In Fig. 14.6, we return our attention to a multi-relay scenario. A main design issue for such a scenario is how the relay nodes participate in the cooperation. We consider three methods:

- *All-Participant (AP)* method in which each of the relay nodes forwards its information in orthogonal time slots [27].

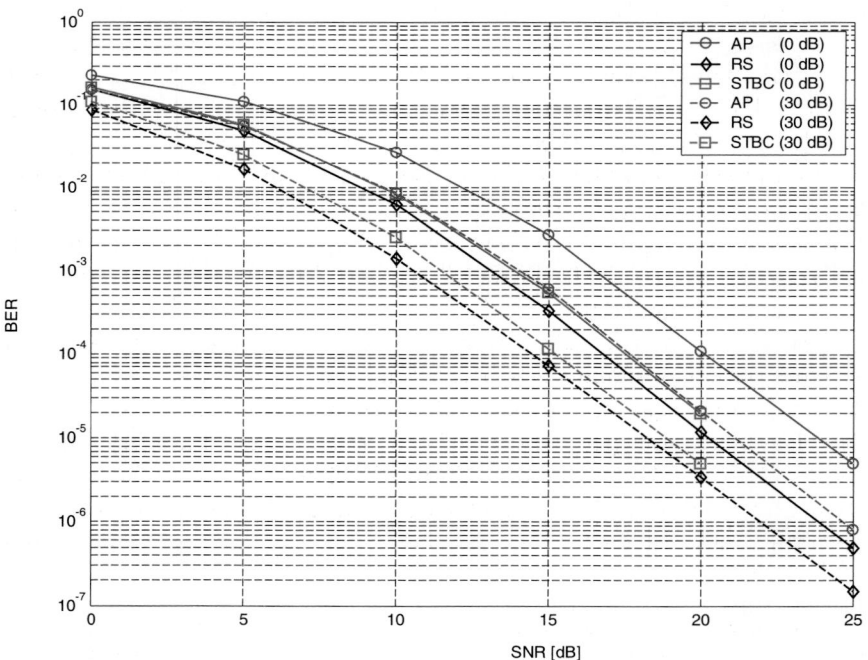

Fig. 14.6.   Performance comparison of AP, RS, and DSTBC schemes in a two-relay scenario.

- *Relay Selection (RS)* method in which one relay with best end-to-end SNR is selected at the end of broadcasting phase and participates in next phase [27, 28].
- *DSTBC* method in which conventional orthogonal STBCs of [12, 29] are applied among relay nodes.

Figure 14.6 illustrates the performance of two-relay cooperative transmission deploying the aforementioned relay participation methods. We assume $\theta_1 = \theta_2 = \pi$ in the simulations. To make a fair comparison, 8-QAM is used for AP while 4-QAM is employed for RS and DSTBC. It is observed that RS scheme yields the best performance while AP scheme yields the worst. The performance of DSTBC is close to that of RS. It operates merely 0.8 dB away from the RS scheme assuming $G_{SR}/G_{RD} = 30$ dB. We should, however, note that the performance gap between RS and DSTBC schemes would increase for more than two relays. This is as a result of the lack of full-rate STBCs for more than two antennas.[b]

In terms of practical implementation, DSTBC presents some advantages over the RS scheme. The implementation of RS scheme depends on the availability of full CSI (i.e. instantaneous channel information of all underlying links). Using this information, the destination node selects the "best" relay and sends the selection decision to all relays through feedback to inform which relay is chosen. On the other hand, DSTBCs can operate without any feedback. DSTBCs are also robust in the case of node failure. Since these codes have orthogonal structure, node failure corresponds to deletion of a column in the code matrix, but the other columns remain orthogonal. This lets the DSTBCs still exploit the residual diversity benefits from the remaining nodes.

As earlier noted, the performance of cooperative communication schemes (although significantly better than the direct transmission) is still far away from that of the physical antenna arrays. Some recent work has shown that the performance of cooperative communication schemes can be substantially improved by optimally distributing the power among cooperating nodes. In [31], Host-Madsen and Zhang have derived bounds on ergodic capacity for fading relay channels and studied power allocation

---

[b]The STBC designed for two transmit antennas, i.e. Alamouti's scheme [12, 29] is able to achieve the full diversity and full transmission rate for both real and complex signal constellations. The other STBCs proposed in [12] for more antennas enjoy the full diversity at full transmission rate only for real signal constellations. It is further proven in [30] that the rates cannot be greater than 3/4 for more than two antennas with complex signal constellation under the orthogonality assumption.

problem to maximize channel capacity. Their proposed power allocation scheme requires the feedback of CSI of all communication channels to the source for each channel realization. In [32], Ahmed and Aazhang have proposed a power allocation method relying on partial feedback information. Close-loop power allocation schemes such as in [31, 32] require the availability of CSI at the transmitter side and their implementation might be problematic in some practical applications. In [33], Hasna and Alouini have investigated the optimal power allocation problem for an open-loop transmission scheme (i.e. CSI information available only at the receiver side) to minimize the outage probability assuming both AaF and DaF relaying. Their results for AaF relaying are, however, restricted to multi-hop systems without diversity advantages. In [34], we have studied the power allocation problem for RD protocol to minimize BER as an objective function. In Fig. 14.7, we illustrate the power savings through optimum power allocation (OPA) at a fixed BER for a range of relay locations. We observe performance improvements up to ∼2.6 dB in comparison to equal power allocation (EPA) for negative values of G_{SR}/G_{RD}. For positive values, it is observed that OPA and EPA performance curves converge to each other. This indicates that OPA is more rewarding in scenarios where relay is close to destination.

Figure 14.8 compares the performance of EPA and OPA schemes with that of co-located receive/transmit antenna diversity schemes in

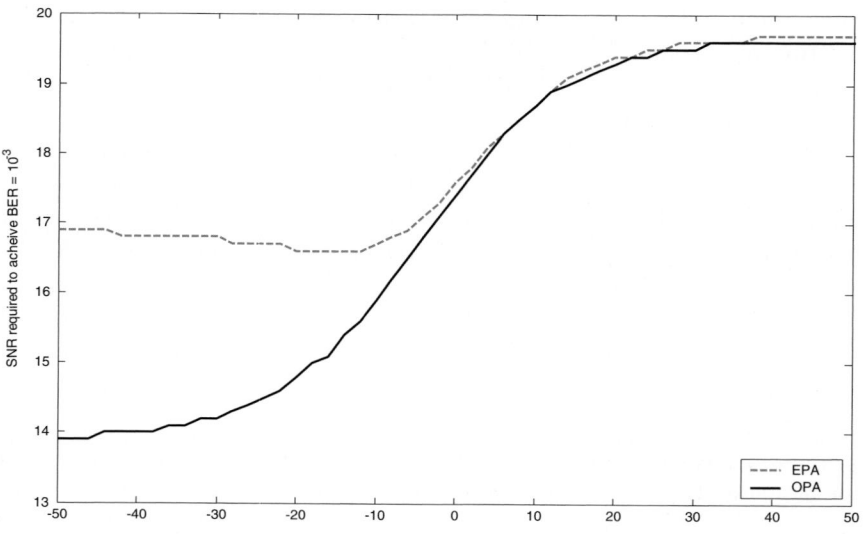

Fig. 14.7.   Optimum versus equal power allocation.

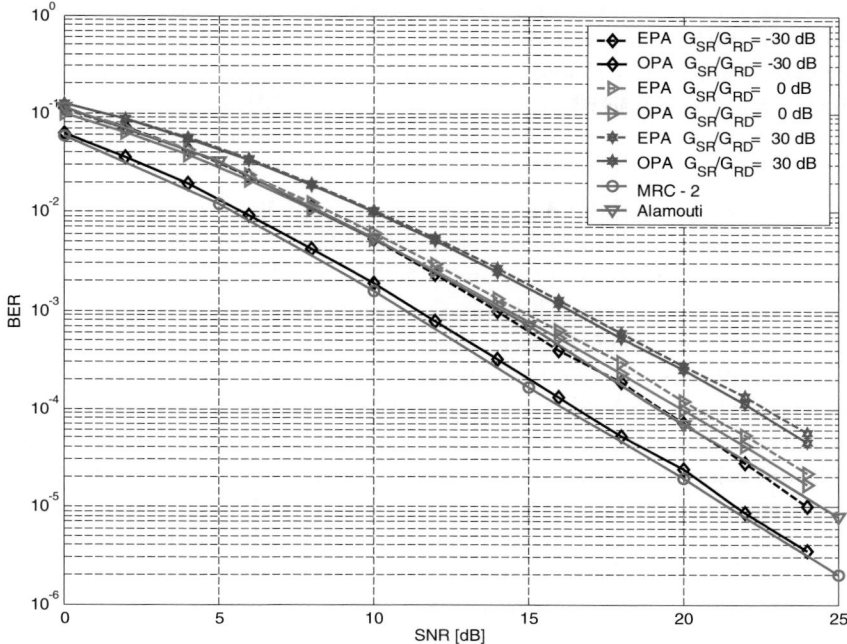

Fig. 14.8. Performance comparison of OPA, EPA, MRC, and Alamouti schemes.

a single-relay scenario. Specifically, we consider MRC with two receive antennas and Alamouti scheme with two transmit antennas. We assume 4-QAM modulation for the cooperative schemes. To make a fair comparison between cooperative and non-cooperative schemes which achieve rates of 1/2 and 1, respectively, direct transmission and co-located antenna scenarios are simulated with BPSK. Our results demonstrate that EPA operates within 3.7 dB of the MRC scheme for $G_{SR}/G_{RD} = 0$ dB. The gap between two schemes further decreases for $G_{SR}/G_{RD} = -30$ dB for which OPA is merely 0.35 dB away from the MRC scheme.

14.4. Decode-and-Forward Relaying

It is evident that AaF relaying provides soft information to the destination. DaF relaying, on the other hand, demodulates/decodes the received signal, re-encodes possibly adding some redundancy before modulating and transmitting it towards the destination. Due to demodulation process, the output signal at the relay carries no information about the degree of uncertainty in the relay's choice of the optimal demodulated symbol. Under the assumption that no coding is performed at the source node, the signals received at the relay and destination during the broadcasting phase are already given by (14.1) and (14.2). The received signal at the destination during relaying phase is given by

$$y_{D2} = \sqrt{G_{R_iD}} h_{R_iD} \hat{x} + n_{D2}, \tag{14.7}$$

where $\hat{x}$ is the modulation symbol transmitted by the relay. It can be possibly chosen from a signal constellation different than the one employed at the source. Based on (14.1) and (14.7), the destination needs to obtain an estimate of x. In the following, we discuss some of the receiver options which provide tradeoffs between complexity and performance.

A simple decoder can be designed by ignoring the possibility of errors at relay. In such a receiver, conventional MRC can be used to combine the two observations in (14.1) and (14.7). This decoder requires CSI of direct link and relay-to-destination link at destination. Based on the derivation of outage behavior, Laneman *et al.* [4] have shown that the performance of such a decoder is limited by transmission performance between source and relay terminal and can result in no diversity if the source-to-relay link has poor SNR. This is clearly demonstrated in Fig. 14.9 where the performance of DaF relaying with MRC is simulated for different relay locations in a single-relay scenario. We assume BPSK and $\theta = \pi/3$. When relay is close to source (i.e $G_{SR}/G_{RD} = 30\,\mathrm{dB}$), a diversity order of two is achieved extracting the full diversity for the single-relay scenario under consideration. As relay moves away from source, the performance degrades. Specifically, we observe a performance degradation of $10\,\mathrm{dB}$ at $G_{SR}/G_{RD} = 0\,\mathrm{dB}$ in comparison to $G_{SR}/G_{RD} = 30\,\mathrm{dB}$ to achieve a BER of 10^{-3}. When the relay is close to destination, the overall performance is dominated by the poor SNR experienced in source-to-relay link and diversity order reduces to one.

An alternative to MRC is ML decoder [7] which yields full diversity regardless of channel conditions. To construct the decoding metric, this decoder requires information about source-to-relay channel at destination in addition to CSI required by the conventional MRC method. Assuming BPSK modulation, the ML decoder for DaF relaying in a single-relay scenario is given by

$$\hat{x}_D^{ML} = \arg\max_x \left\{ \frac{1 - P_{SR}(\gamma_{SR})}{2\pi N_0} \exp\left(-\frac{|y_{SD} - h_{SD}x|^2 + |y_{RD} - h_{RD}x|^2}{2N_0} \right) \right.$$
$$\left. + \frac{P_{SR}(\gamma_{SR})}{2\pi N_0} \exp\left(-\frac{|y_{SD} - h_{SD}x|^2 + |y_{RD} + h_{RD}x|^2}{2N_0} \right) \right\}, \tag{14.8}$$

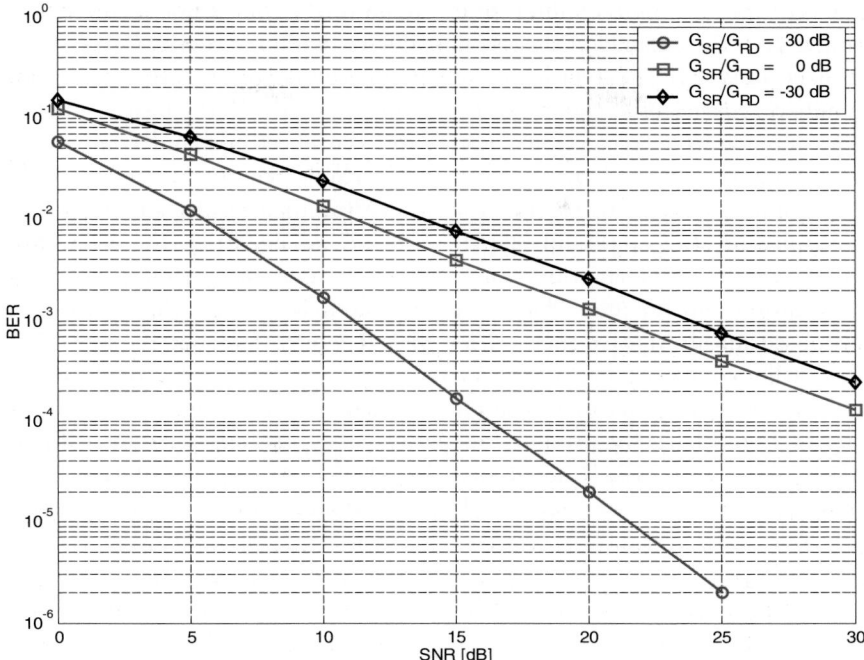

Fig. 14.9.   Performance of DaF relaying with MRC receiver for various relay locations.

where γ_{SR} and $P_{SR}(\gamma_{SR})$ are, respectively, SNR and instantaneous error probability of source-to-relay link. The complexity of this decoder increases with higher-order modulations.

To address the complexity issue, so-called λ-MRC has been proposed in [7]. In this method, destination pre-multiplies the signal coming from relay by an arbitrary weight λ before sending it to conventional MRC. It is assumed that λ takes values within the range of $[0, 1]$ and particular value depends on the quality of source-to-relay link. The performance of λ-MRC lies between those of MRC and ML decoders. Another low-complexity detector can be found in [35]. This scheme is named as cooperative MRC (C-MRC). It is similar to λ-MRC where the signal arriving at destination from relay is pre-multiplied by a weighting factor before it is fed to the conventional MRC.

The failure of fixed relaying to achieve full diversity without using a complex decoder can be rectified by use of *selection relaying* (SEL-R). In SEL-R, the relay forwards information only if it has correctly decoded the symbol. A common and practical approach for error detection in source-to-relay channel is the use of cyclic redundancy check (CRC). CRC is a technique for detecting errors in digital data. In CRC, a certain number of parity bits, called *checksum*, are appended to the message before transmission. The receiver calculates its own checksum and compares it with that transmitted by source and determine whether or not the two checksums are same, to ascertain with a certain degree of probability that transmission was successful [36]. If an error has occurred, the relay does not forward the information. The use of CRC can ensure full diversity at destination, but it suffers from throughput loss. In [37], Ibrahim *et al.* have proposed an alternative method for SEL-R which does not require CRC, but rather compares SNR of the source-to-relay link with that of the direct link to decide whether or not to participate in the relaying.

In Fig. 14.10, we present the performance of DaF relaying with C-MRC, ML, and SEL-R methods. We assume BPSK modulation, $\theta = \pi/3$ and $G_{\mathrm{SR}}/G_{\mathrm{RD}} = -30\,\mathrm{dB}$. The performance of conventional MRC

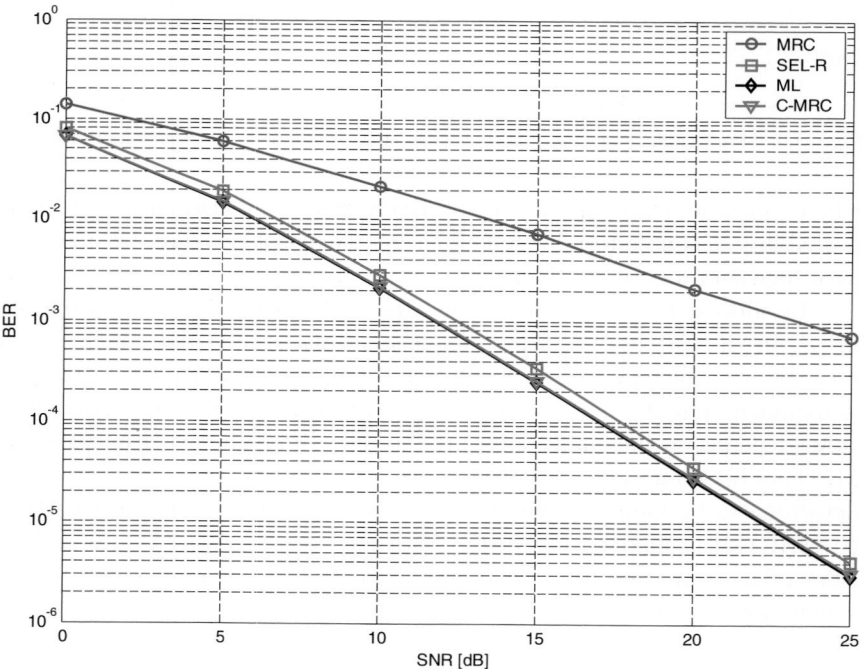

Fig. 14.10.   Performance of DaF relaying with various decoders.

is also included as a benchmark. As earlier observed in Fig. 14.9, conventional MRC is not able to extract diversity for the relay location under consideration. On the other hand, C-MRC, ML, and SEL-R achieve diversity orders of two and significantly outperform the conventional MRC.

In Fig. 14.11, we return our attention to a multi-relay scenario. We consider the following methods:

- *AP* method in which all relays forward whether or not they have decoded correctly (i.e. no error detection is employed).
- *RS* method in which one relay with best end-to-end SNR is selected at the end of broadcasting phase and participates in the next phase. This is the same as the one adopted in AaF relaying (c.f. Sec. 3).
- *RS with CRC (RS-CRC)* method in which the selected relay forwards only if it has been able to correctly decode the message. In our simulations, we assume a 16-bit CRC is inserted by the source to the transmission frame of 512 bits to facilitate the error detection process.

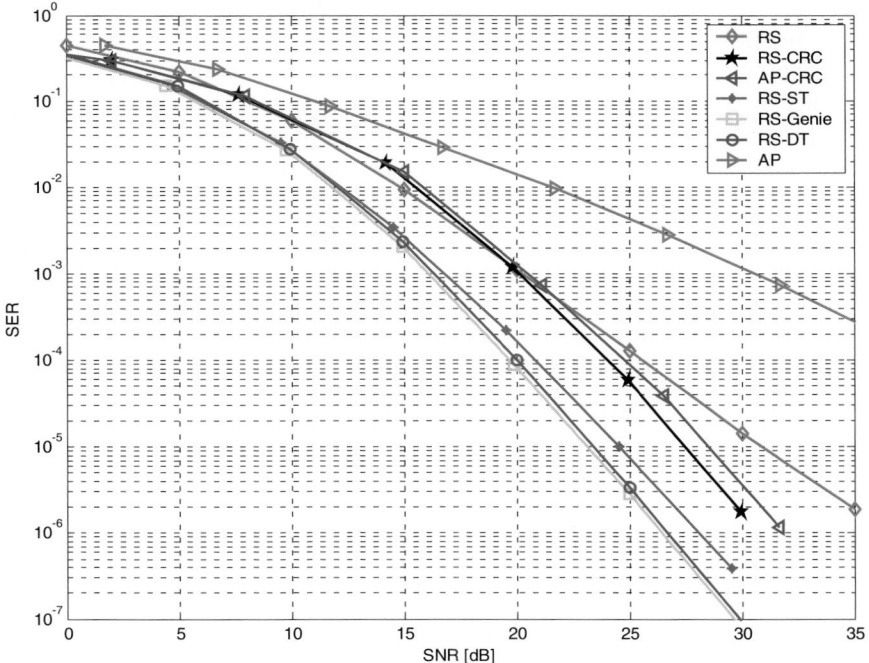

Fig. 14.11. Comparison of different DaF schemes in a multi-relay scenario.

- *AP with CRC (AP-CRC)* method in which the relays which have been able to correctly decode forward to the destination. CRC and transmission frame sizes are the same as in RS-CRC method.
- *RS with Static Threshold (RS-ST)* method in which the selected relay forwards only if the SNR of the source-to-relay link is greater than a predetermined ST.
- *RS with Genie error detection (RS-Genie)* method in which the selected relay is able to detect error on a symbol-by-symbol basis. The performance of this method can be considered as an ideal bound for that of *RS-CRC* with frame length equal to one symbol.
- *RS with Dynamic Threshold (RS-DT)* method in which the selected relay forwards only if SNR of source-to-relay link is greater than SNR of direct link [38].

It is obvious from Fig. 14.11 that performance of AP method without any error detection mechanism at the relay is worst. Due to the deployed MRC at the destination, it is not able to provide diversity in this particular scenario because of poor source-to-relay SNR. On the other hand, RS-Genie achieves full diversity order and yields the best performance which is an upper bound for the practical RS schemes under consideration. RS scheme (in which one relay with best end-to-end SNR is selected) outperforms AP by 10 dB for the target BER $= 10^{-3}$. The use of CRC improves the performance of both AP and RS schemes, but their performance is still away from the RS-Genie bound. It should be also noted that RS-CRC needs maximum two time slots, while AP-CRC needs a number of time slots depending on how many relays have been determined as "error-free." RS-ST scheme which avoids use of CRC and takes decision on a symbol-by-symbol basis performs better than CRC-based schemes. The performance can be further improved by incorporating DT. Particularly, RS-DT scheme performs within 0.3 dB of the genie bound.

14.5. Thoughts for Practitioners

Despite the abundance of research results which have appeared on theoretical aspects of cooperative diversity within the last few years, the practical implementation and related issues seem to be an open problem. Majority of the research works consider a highly-idealized cooperative communication system which consists of a set of *perfectly coordinated* links where each of the relay nodes exactly knows *if* and *when* it should transmit. Such ideal assumptions typically arise from the rather misleading analogy (at least in practical sense) to MIMO antenna links. A virtual antenna array formed through cooperative diversity inherently differs from a physical antenna array with co-located antennas. First of all, information is only known at one of the antenna elements of virtual array (i.e. source node) and should be communicated to other antenna elements (i.e. relay nodes) through wireless links which are subject to path loss, fading, and noise. Second, the number of antenna elements in a virtual array is in general random. This depends on how many nodes will participate in the transmission as relays and how many of those nodes are indeed helpful in relaying the source's information. This obviously changes as a function of time based on channel conditions. Cooperative systems, therefore, need to acquire time-dependent information (i.e. number of available nodes and their CSI) at the routing and physical layers and periodically update them. Last, but not least, virtual arrays require a significant amount of coordination (therefore overhead) which involve the design of specialized distributed algorithms based on the synchronous/asynchronous nature of relay nodes. Taking into account such considerations, one needs to devise methods which bridge the theory and practice for cooperative communication systems.

A first attempt for the practical implementation of cooperative diversity has been reported by Bletsas and Lippman in [39]. They consider a three-relay scheme where the source node retrieves weather information from the internet and sends it to the destination through the assistance of relay nodes. The received information about the weather conditions and information about which relays have participated are shown at the destination display. The demonstration in [39] is built on commercial, off-the-shelf radio hardware. Since the non-linearity of deployed low-complexity radio module front-end precludes the use of in-band simultaneous transmission, distributed space-time coding among relay nodes is not possible. Instead, RS (named as "opportunistic relaying" in [39]) is adopted. The implementation involves the deployment of timers at relay nodes. The relay nodes overhear pilot symbols transmitted from the source and destination, i.e. ready-to-send (RTS) and clear-to-send (CTS). They use RTS and CTS pilot signals to estimate the source-to-relay and destination-to-relay[c] links and construct the *channel quality metric* which will be discussed in the following. Each relay node starts a timer with an

---

[c]This is equivalent to relay-to-destination link based on reciprocity theorem.

initial value which is inversely proportional to the quality of its own end-to-end channel. The timer of relay with best end-to-end quality expires first and that node informs the destination as well as other relay nodes about its participation in relaying. As soon as one node announces itself as the "best" node, the other relay nodes back off and reset their timers for the next signaling interval. In case the relay nodes are "hidden" from each other (i.e. all relays can listen to source and destination, but they cannot listen each other), the destination notifies the relays about the selected relay with a brief broadcast message. The selected relay is used for a specific period of time which is smaller than the coherence time of the channel.

A critical design issue is the choice of channel quality metric. In [39], two different metrics are used which are given by $\gamma_i = \min(\gamma_{\mathrm{SR}_i}, \gamma_{\mathrm{R}_i\mathrm{D}})$ and $\gamma_i = 2(\gamma_{\mathrm{SR}_i}\gamma_{\mathrm{R}_i\mathrm{D}})/(\gamma_{\mathrm{SR}_i} + \gamma_{\mathrm{R}_i\mathrm{D}})$. In the former, the minimum of instantaneous SNRs in $\mathrm{S} \to \mathrm{R}_i$ and $\mathrm{R}_i \to \mathrm{D}$ links is selected. Since such a metric selects the bottleneck of the two paths, it is more appropriate for DaF relaying. In the latter, the harmonic mean of two SNRs is calculated which yields a smoother function and, therefore, is more appropriate for AaF relaying. It should be also noted that the CSI required at relays for RS is in the form of link strengths, i.e. only the amplitude of fading channel is relevant, therefore phase estimation is not required. Further details on this demonstration including the format of signaling frame, specifications of the utilized hardware, etc. can be further found in [39].

For the reader interested in practical implementation aspects of cooperative diversity, we also suggest the work by Valentin *et al.* [40] which provides guidelines for RS, cooperation-aware resource allocation, medium access control (MAC) layer protocols for cooperative communications.

It is evident that more experimental studies would appear as cooperative communication finds along its way into wireless cellular standards and becomes an integral part of WSNs and *ad hoc* networks. IEEE 802.16's Relay Task Group is currently developing a draft under the P802.16j project named as "Air Interface for Fixed and Mobile Broadband Wireless Access Systems — Multihop Relay Specification." The project was initiated in 2006 and is expected to complete and approved in 2008. The first phase of 802.16j is expected to be restricted to infrastructure relay stations that extend coverage of 802.16e base stations without changing the subscriber station specifications. These relay stations will operate seamlessly with existing 802.16e subscribers. Due to backward compatibility constraints, 802.16j in fact is unable to fully exploit cooperation. Another future standard (802.16 m), which is currently in its infancy, is expected to incorporate and fully take the advantages of multi-user and cooperative techniques.

14.6. Directions for Future Work

Cooperative communication is an emerging field in which significant progress has been accomplished in recent years. The integration of cooperative communication in next generation wireless standards will lead to the design of an efficient and reliable fully distributed wireless network. However, there remain various technical challenges and open issues to be resolved before this promising concept turns into a fundamental technology for the future wireless networking. The pioneering works in this area have addressed mainly information-theoretic aspects and are, in general, limited to some idealistic assumptions such as frequency-flat channel model, availability of perfect CSI, perfect synchronization, etc. which are highly unrealistic for practical applications.

A major design challenge in high-speed broadband communication is intersymbol interference which is a result of the dispersive nature of underlying frequency-selective fading channels. The existing literature on cooperative communication has so far mainly focused on the frequency-flat fading channel model which fails to provide an accurate modeling for broadband wireless channels (e.g. video-enabled wireless surveillance networks). Although the concept of broadband cooperative communication has attracted some attention recently (see e.g. [41, 42]), research in this area is still in its infancy. Analysis, design, and optimization of broadband cooperative communication systems are therefore open problems to address. This venue of research involves the investigation of single-carrier and multi-carrier transmission solutions integrated with cooperation protocols and relaying modes.

A common assumption in the literature on cooperative diversity is the availability of perfect CSI. Under the assumption of coherent detection, the fading channel coefficients need to be first estimated and then used in the detection process. In DaF relaying, both relay and destination require a reliable channel estimate. In AaF relaying, knowledge of CSI is required at the destination terminal and may be required at the relay as well depending on the adopted scaling factor [26]. The quality of channel estimates inevitably affects the overall performance of relay-assisted transmission and might become a performance limiting factor. Development of robust and low-complexity channel estimation algorithms is, therefore, a practical open problem to investigate. Channel estimation problem in the context of DaF relaying basically consists of individual estimation of source-to-relay and relay-to-destination channels. On the other hand, in AF relaying, a cascaded channel from source-to-destination needs to be estimated. A possible alternative is to disintegrate it into individual channel estimations (i.e. separate estimations of source-to-relay and relay-to-destination channels) through the injection of a "clean" pilot symbol at relay. Besides coherent detection, differential and non-coherent detection are of potential interest. Some initial work in the area of channel estimation for cooperative communication can be found in [43, 44].

Although work in [15–17] deals with asynchronous cooperative schemes, most of the existing literature on cooperative diversity is based on the assumption of perfect synchronization among cooperating nodes. Unlike a conventional space-time coded system where multiple antennas are co-located and fed by a local oscillator, the relay nodes are geographically dispersed and each of them relies on its local oscillator. Therefore, cooperative schemes need to be implemented taking into account this asynchronous nature. Development of robust timing and frequency synchronization algorithms for cooperative communications are critical for practical implementation. Such algorithms will provide accurate synchronization information and will ensure the theoretical coding and diversity gains predicted for ideal (synchronous) cooperative systems within a practical wireless scenario.

Adaptive transmission takes advantage of the CSI available at the transmitting nodes and promises significant performance improvements particularly for fixed wireless access applications where reliable feedback is possible. The basic premise of adaptive transmission in the context of cooperative communication is to estimate relay-to-destination and source-to-destination channels at the destination terminal and feed these estimates back to the source and relay nodes to adapt the transmission according to the channel conditions. It would be of potential interest to investigate power and rate adaptation for broadband cooperative communication systems.

As indicated earlier, cooperative communications can be applied to various wireless applications including infrastructure-based networks such as cellular systems, WLANs, WMANs and infrastructure-less networks such as MANETs, VANETs, and WSNs. Obviously, each networking application comes with its own challenges and unique requirements which need to be taken into account for its integration with cooperative communication. For example, consider VANETs. These networks involve vehicle-to-vehicle

(V2V) and vehicle-to-road (V2R) communications enabling a vehicle to communicate with other vehicles and sensors/access-points installed along the road. Since vehicles have a powerful and rechargeable source of energy, power efficiency is not as important for VANETs as it is for WSNs on which majority of the current research on cooperative diversity is targeted. The underlying channel characteristics such as cascaded Rayleigh fading channel (unlike conventional Rayleigh model), the presence of impulsive noise, and high-mobility environment are also unique to VANETs. Analysis, design, and optimization of cooperative schemes specifically tailored for inter-vehicular communication scenarios are, therefore, open problems to address.

Finally, we note that our chapter has mainly focused on the physical layer aspects of cooperative communication. To realize the full potential of a cooperative network, research at the physical layer should be coupled with higher layers of the protocol stack, in particular, MAC and network layers. Potential open problems can be listed as cross-layer designs, resource allocation and management for cooperative communications, cooperative routing and scheduling for sensor networks and *ad hoc* networks.

14.7. Conclusions

The ever-increasing demand for more efficient, reliable, and cost-effective wireless communication systems motivates the study of network architectures which deviate from conventional designs. Cooperative communications is a novel communication technique and offers a fundamental paradigm shift to the design of wireless networks to achieve the increasing demands of future wireless applications. Different from conventional point-to-point communications, cooperative communication allows nodes in a wireless network to share their resources through distributed transmission/processing. This brings significant performance gains in terms of link reliability, spectral efficiency, system capacity, and transmission range. This chapter has provided an overview of this burgeoning field summarizing the fundamental ideas behind cooperative communication and presenting state-of-the-art cooperative communication techniques. Several performance comparisons have been demonstrated to provide a better insight into competing schemes. Practical issues and requirements on system design have been discussed along with potential areas of future work. The already rich literature and growing interest from the industry are indicative of a bright future for cooperative communications and networking.

14.8. Questions

Question 1. Consider a cooperative diversity system with a single relay R which assists the source node S to send information to destination node D. Perform a Monte Carlo simulation to estimate the BER performance of this single-relay network. In your simulations assume blind AaF RD protocol and consider a scenario where the relay node is close to the destination. Compare the performance of this cooperative system with that of MRC with two co-located antennas.

Question 2. Consider the cooperative transmission model given in Fig. P14.1. The nodes S and R cooperate with each other using RD protocol given in Table 14.1. In the first phase, S uses a fraction of total power P, i.e. $K_S P$, while in the second phase R uses the remaining fraction of power i.e. $(1-K_S)P$. Assume blind AaF is employed at the relay terminal. Write three equations to express the received signals at R and D during the two phases (similar to Eqs. (14.1), (14.2), (14.5)) including power allocation parameters.

Question 3. For the transmission model in Question 2, calculate instantaneous SNR for each of the underlying links.

Question 4. Pairwise error probability (PEP) is defined as the probability of deciding in favor of codeword $\hat{\mathbf{X}}$ instead of the original codeword $\mathbf{X}$. Consider the transmission model in Question 2. Assume that the destination combines the signals received during two phases of cooperation using MRC. Obtain the conditional PEP expression for a given realization of channel. Provide an upper bound for this PEP using Chernoff bound.

Question 5. Assume Rayleigh fading channel among three nodes for the transmission model given in Question 2. Show that the unconditional PEP upper bound can obtained as

$$P(\mathbf{X} \to \hat{\mathbf{X}}) \le \delta_2[1 + \beta_2 \exp(\lambda_2)\Gamma(0, \lambda_2)], \tag{P14.1}$$

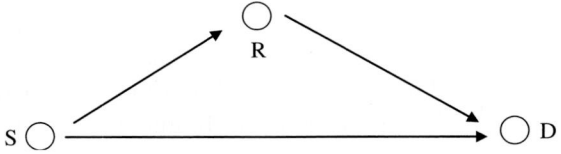

Fig. P14.1.   Cooperative transmission model.

where

$$\delta_2 = \left(1 + \frac{\text{SNR}}{2}K_S\chi_2\right)^{-1}\left(1 + \frac{\text{SNR}}{2}G_{\text{SR}}K_S\chi_2\right)^{-1},$$

$$\lambda_2 = B_D \bigg/ \left(1 + \frac{\text{SNR}}{2}G_{\text{SR}}K_S\chi_2\right),$$

$$\beta_2 = B_D - \lambda_2,$$

$$B_D = [1 + 2G_{\text{SR}}K_S\text{SNR}]/[2G_{\text{RD}}(1 - K_S)\text{SNR}].$$

Plot PEP versus power allocation parameter for $G_{\text{SR}}/G_{\text{RD}} = \{-30, 0, 30\}$ dB assuming SNR $= 10\,\text{dB}$.

Question 6. The optimum power allocation parameters can be obtained by finding the minimum of PEP given in (P14.1). Under the assumption of $G_{\text{SR}}/G_{\text{RD}} = 0\,\text{dB}$ (i.e. relay is equidistant from source and destination), show that optimum value of K_S is

$$K_S = \frac{2\text{SNR} - 1 + \sqrt{1.18 + 3.5\text{SNR} + 4\text{SNR}^2}}{6\text{SNR}}. \qquad \text{(P14.2)}$$

Using numerical optimization techniques, calculate optimum value of K_S for SNR of $30\,\text{dB}$ minimizing PEP given in (P14.1) with respect to K_S. Compare this value of K_S with that calculated through (P14.2).

Question 7. Consider a single relay-assisted communication system. Assume that the source is equipped with single antenna while the destination has N receive antennas. For CSI-assisted AaF cooperation scheme and RD protocol, show that symbol error rate is given by

$$P = \int_0^{\frac{(M-1)\pi}{M}} \left(1 + \frac{\sin^2(\pi/M)}{\sin^2\theta}\text{SNR}\right)^{-2N} d\theta. \qquad \text{(P14.3)}$$

Calculate an upper bound for (P14.3) inserting $\theta = \pi/2$ and assuming large SNR. Find the diversity order and confirm it through plots on a log–log scale.

Question 8. Assume a cooperative transmission model with four nodes: a source, a destination, and two relays. Calculate the throughput and orthogonal time slots required for transmission of one information block for each of the following schemes:

(1) All participants with RD protocol.
(2) Relay selection with RD protocol.

(3) Source and the two relay nodes cooperate using 3-antenna real-valued OSTBC [12].

Question 9. Consider a relay network based on RD protocol in DaF mode. The ML decoder is given by (14.8). Extending this idea, write an equation for ML decision for a cooperative system which uses TD protocol with distributed Alamouti space-time block code in DaF mode.

Question 10. Consider two different DaF selection relaying systems with three nodes. In the first system, the relay is allowed to participate in cooperation phase if the transmission frame of 512-symbols is correctly received at relay. In the second system, the relay participates if the instantaneous SNR in source–relay link is better than that of source–destination link. Assuming BPSK modulation and the SNR in source–relay and source–destination link is 5 dB, calculate probability of relay participation in cooperation phase for both systems.

14.9. Key Answers

Answer 1.

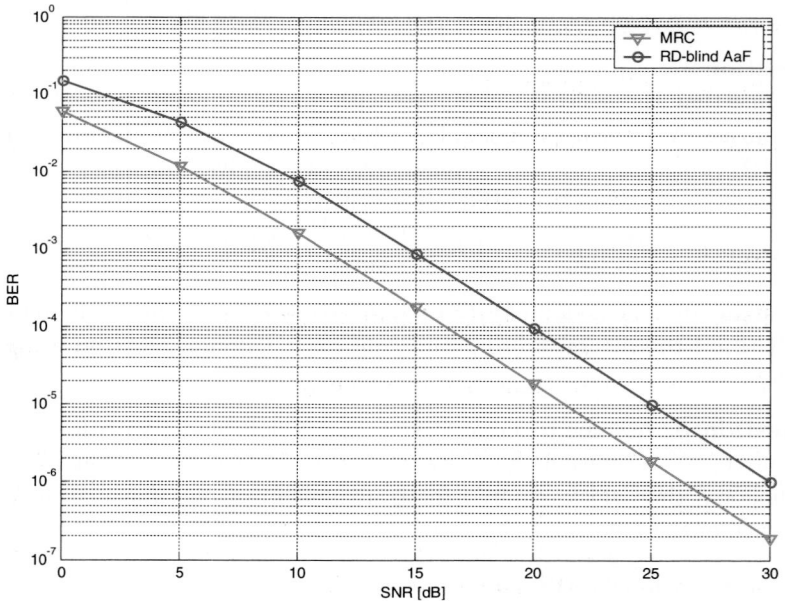

Answer 2.

$$y_{D1} = \sqrt{K_S P} h_{\text{SD}} x + n_{D1}, \tag{A14.1}$$

$$y_R = \sqrt{K_S P G_{\text{SR}}} h_{\text{SR}} x + n_R, \tag{A14.2}$$

$$y_{D2} = \sqrt{\frac{K_S(1 - K_S) G_{\text{SR}} G_{\text{RD}} P^2}{K_S G_{\text{SR}} P + N_0}} h_{\text{SR}} h_{\text{RD}} x$$

$$+ \sqrt{\frac{(1 - K_S) P G_{\text{RD}}}{K_S G_{\text{SR}} E + N_0}} h_{\text{RD}} n_R + n_{D2}, \tag{A14.3}$$

Answer 3. Define $\text{SNR} = P/N_0$.

$$\text{SNR}_{\text{SD}} = K_S |h_{\text{SD}}|^2 \text{SNR}, \tag{A14.4}$$

$$\text{SNR}_{\text{SR}} = K_S G_{\text{SR}} |h_{\text{SR}}|^2 \text{SNR}, \tag{A14.5}$$

$$\text{SNR}_{\text{RD}} = \frac{K_S(1 - K_S) G_{\text{SR}} G_{\text{RD}} \text{SNR}}{K_S G_{\text{SR}} \text{SNR} + (1 - K_S) G_{\text{RD}} |h_{\text{RD}}|^2 \text{SNR} + 1}$$

$$\times |h_{\text{SR}}|^2 |h_{\text{RD}}|^2 \text{SNR}. \tag{A14.6}$$

Answer 4. See [34] provided in the chapter.

Answer 5. See [34] provided in the chapter.

Answer 6. See [34] provided in the chapter.

Answer 7. See [26] provided in the chapter.

Answer 8.

(1) Three time slots for transmission of one symbol.
    Throughput $= 1/3$.
(2) Two time slots for transmission of one symbol.
    Throughput $= 1/2$.
(3) Four time slots for source to relay transmission (broadcast phase), and another four time slots for relaying phase using three antenna real OSTBC.

   The total time slots is, therefore, 8. Since the number of symbols transmitted is 4, the throughput is 1/2.

Answer 9.

$$
\begin{bmatrix} r_{D1} \\ r_{D2} \\ r_{D3} \\ r_{D4}^* \end{bmatrix} = \begin{bmatrix} h_{SD} & 0 \\ h_{RD}\theta_1 & h_{SD} \\ h_{SD} & 0 \\ -h_{SD}^* & h_{RD}^*\theta_2 \end{bmatrix} \begin{bmatrix} x_1 \\ x_2 \end{bmatrix} + \begin{bmatrix} n_{D1} \\ n_{D2} \\ n_{D3} \\ n_{D4} \end{bmatrix}
$$

$$
\mathbf{r} = \mathbf{H}(\theta_1, \theta_2)\mathbf{x} + \mathbf{n}
$$

$$
\hat{\mathbf{x}}_D^{ML} = \arg\max_{\mathbf{x} \in \mathbf{X}} \{ f_{\mathbf{r}|\mathbf{x}}(r \mid x) \}
$$

$$
f_{\mathbf{r}|\mathbf{x}}(r \mid x) = \frac{1}{\pi^2 N_0^2} \times \left\{ [1 - P_{SR}(\gamma_{SR})]^2 \exp\left(-\frac{|\mathbf{r} - \mathbf{H}(1,1)\mathbf{x}|^2}{2N_0}\right) \right.
$$

$$
+ P_{SR}(\gamma_{SR})[1 - P_{SR}(\gamma_{SR})] \exp\left(-\frac{|\mathbf{r} - \mathbf{H}(1,-1)\mathbf{x}|^2}{2N_0}\right)
$$

$$
+ [1 - P_{SR}(\gamma_{SR})] P_{SR}(\gamma_{SR}) \exp\left(-\frac{|\mathbf{r} - \mathbf{H}(-1,1)\mathbf{x}|^2}{2N_0}\right)
$$

$$
\left. \times [P_{SR}(\gamma_{SR})]^2 \exp\left(-\frac{|\mathbf{r} - \mathbf{H}(-1,-1)\mathbf{x}|^2}{2N_0}\right) \right\}.
$$

Answer 10. Probability of error in one bit is

$$
P_1(\text{SNR}) = \frac{1}{2}\left[1 - \sqrt{\frac{\text{SNR}}{1 + \text{SNR}}} \right].
$$

Probability of error free block of length 512 is

$$
P = [1 - P_1(\text{SNR})]^{512} = [1 - P_1(3.16)]^{512} = 1.7413 \times 10^{-15}
$$

$$
P = \Pr(\gamma_{SR} > \gamma_{SD}) = \frac{\Gamma_{SR}}{\Gamma_{SR} + \Gamma_{SD}} = 0.5.
$$

14.10. Keywords

Spatial Diversity

The use of multiple antennas at the transmitter and/or receiver side to improve the transmission performance. It takes advantage of the availability of more than one independently faded version of the transmitted signal.

Multiple-input multiple-output (MIMO) System

Wireless communication system with multiple antennas at both transmitter and receiver side.

Cooperative Diversity

A technique to reap the benefits of MIMO communications in a wireless scenario with single-antenna terminals. Cooperative communication takes advantage of the broadcast nature of wireless transmission and creates a *virtual antenna array* through cooperating nodes.

Broadcasting Phase

First phase of cooperative communication when the source node transmits (broadcasts) information towards the destination and the relay nodes.

Relaying Phase

Second phase of cooperative communication when the relay nodes retransmit information received in the broadcast phase.

Amplify-and-Forward Relaying

A type of relaying technique in which the relay nodes retransmits an amplified/scaled version of the signal received during the broadcast phase.

Decode-and-Forward Relaying

A type of relaying technique in which the relay node fully decodes, re-encodes (possibly with a different codebook), and retransmits the source node's message.

Coded Cooperation

The concept of user cooperation through the distributed implementation of existing channel coding methods such as convolutional and turbo codes.

Fixed Relaying

A type of relaying technique in which relay always forwards the message that it receives from source.

Selection Relaying

A type of relaying technique in which relay participates in cooperation phase only when source-to-relay link is good, otherwise the source reverts to non-cooperation mode.

VIII. Exercise

Question 1: Consider a cooperative diversity system with a single relay R which assists the source node S to send information to destination node D. Perform a Monte Carlo simulation to estimate the bit error rate performance of this single-relay network. In your simulations assume blind AaF RD protocol and consider a scenario where the relay node is close to the destination. Compare the performance of this cooperative system with that of MRC with two co-located antennas.

Question 2: Consider the cooperative transmission model given in Fig. P-1. The nodes S and R cooperate with each other using RD protocol given in Table 1. In the first phase, S uses a fraction of total power P, i.e. $K_S P$, while in the second phase R uses the remaining fraction of power i.e. $(1 - K_S)P$. Assume blind AaF is employed at the relay terminal. Write three equations to express the received signals at R and D during the two phases (similar to Eqs. 1,2,5) including power allocation parameters.

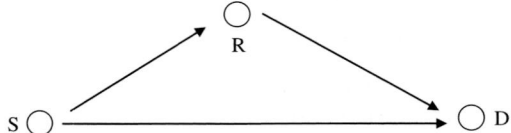

Fig. P-1.   Cooperative transmission model.

Question 3: For the transmission model in Question 2, calculate instantaneous signal-to-noise ratio for each of the underlying links.

Question 4: Pairwise error probability (PEP) is defined as the probability of deciding in favour of codeword $\hat{\mathbf{X}}$ instead of the original codeword $\mathbf{X}$. Consider the transmission model in Question 2. Assume that the destination combines the signals received during two phases of cooperation using MRC. Obtain the conditional PEP expression for a given realization of channel. Provide an upper bound for this PEP using Chernoff bound.

Question 5: Assume Rayleigh fading channel among three nodes for the transmission model given in Question 2. Show that the unconditional PEP upper bound can obtained as

$$P(\mathbf{X} \to \hat{\mathbf{X}}) \le \delta_2[1 + \beta_2 \exp(\lambda_2)\Gamma(0, \lambda_2)], \qquad \text{(P-1)}$$

where

$$\delta_2 = \left(1 + \frac{SNR}{2}K_S\chi_2\right)^{-1}\left(1 + \frac{SNR}{2}G_{SR}K_S\chi_2\right)^{-1}$$

$$\lambda_2 = B_D \Big/ \left(1 + \frac{SNR}{2}G_{SR}K_S\chi_2\right)$$

$$\beta_2 = B_D - \lambda_2$$

$$B_D = [1 + 2G_{SR}K_S SNR]/[2G_{RD}(1 - K_S)SNR]$$

Plot PEP versus power allocation parameter for $G_{SR}/G_{RD} = \{-30, 0, 30\}$ dB assuming $SNR = 10$ dB.

Question 6: The optimum power allocation parameters can be obtained by finding the minimum of PEP given in (P-1). Under the assumption of $G_{SR}/G_{RD} = 0$ dB (i.e., relay is equidistant from source and destination), show that optimum value of K_S is

$$K_S = \frac{2SNR - 1 + \sqrt{1.18 + 3.5SNR + 4SNR^2}}{6SNR}. \tag{P-2}$$

Using numerical optimization techniques, calculate optimum value of K_S for SNR of 30 dB minimizing PEP given in (P-1) with respect to K_S. Compare this value of K_S with that calculated through (P-2).

Question 7: Consider a single relay-assisted communication system. Assume that the source is equipped with single antenna while the destination has N receive antennas. For CSI-assisted AaF cooperation scheme and RD protocol, show that symbol error rate (SER) is given by

$$P = \int_0^{\frac{(M-1)\pi}{M}} \left(1 + \frac{\sin^2(\pi/M)}{\sin^2\theta} SNR\right)^{-2N} d\theta. \tag{P-3}$$

Calculate an upper bound for (P-3) inserting $\theta = \pi/2$ and assuming large SNR. Find the diversity order and confirm it through plots on a log-log scale.

Question 8: Assume a cooperative transmission model with four nodes: a source, a destination and two relays. Calculate the throughput and orthogonal time slots required for transmission of one information block for each of the following schemes:

1. All participants with RD protocol.
2. Relay selection with RD protocol.
3. Source and the two relay nodes cooperate using 3-antenna real OSTBC.

Question 9: Consider a relay network based on RD protocol in DaF mode. The ML decoder is given by (1). Extending this idea, write an equation for ML decision for a cooperative system which uses TD protocol with distributed Alamouti space-time block code in DaF mode.

Question 10: Consider two different DaF selection relaying systems with three nodes. In the first system, the relay is allowed to participate in cooperation phase if the transmission frame of 512-symbols is correctly received at relay. In the second system, the relay participates if the instantaneous SNR in source-relay link is better than that of source-destination link. Assuming BPSK modulation and the SNR in source-relay and source-destination link is 5 dB, calculate probability of relay participation in cooperation phase for both systems.

Key Answers

Answer 1: Performance comparison of collocated-MRC (BPSK) with RD-blind AaF (QPSK).

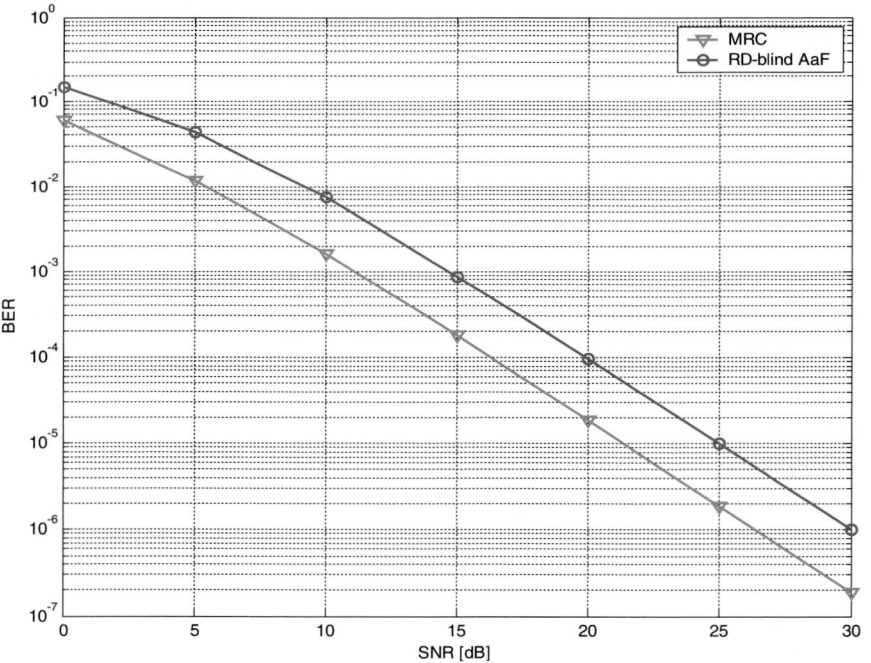

Answer 2:

$$y_{D1} = \sqrt{K_S P} h_{SD} x + n_{D1}, \tag{A-1}$$

$$y_R = \sqrt{K_S P G_{SR}} h_{SR} x + n_R, \tag{A-2}$$

$$y_{D2} = \sqrt{\frac{K_S(1 - K_S)G_{SR}G_{RD}P^2}{K_S G_{SR}P + N_0}} h_{SR} h_{RD} x$$

$$+ \sqrt{\frac{(1 - K_S)P G_{RD}}{K_S G_{SR}E + N_0}} h_{RD} n_R + n_{D2}, \tag{A-3}$$

Answer 3: Define $SNR = P/N_0$.

$$SNR_{SD} = K_S |h_{SD}|^2 \overset{\circ}{S}NR, \tag{A-4}$$

$$SNR_{SR} = K_S G_{SR} |h_{SR}|^2 SNR, \tag{A-5}$$

$$SNR_{RD} = \frac{K_S(1 - K_S)G_{SR}G_{RD}SNR}{K_S G_{SR}SNR + (1 - K_S)G_{RD}|h_{RD}|^2 SNR + 1}$$

$$\times |h_{SR}|^2 |h_{RD}|^2 SNR, \tag{A-6}$$

Answer 4: See reference [31] provided in the chapter.

Answer 5: See reference [31] provided in the chapter.

Answer 6: See reference [31] provided in the chapter.

Answer 7: See reference [24] provided in the chapter.

Answer 8:

1. Three time slots for transmission of one symbol. Throughput $= 1/3$.
2. Two time slots for transmission of one symbol. Throughput $= 1/2$.
3. Four time slots for source to relay transmission (broadcast phase), and another four time slots for relaying phase using 3 antenna real OSTBC. The total time slots is therefore 8. Since the number of symbols transmitted is 4, the throughput is $1/2$.

Answer 9:

$$\begin{bmatrix} r_{D1} \\ r_{D2} \\ r_{D3} \\ r_{D4}^* \end{bmatrix} = \begin{bmatrix} h_{SD} & 0 \\ h_{RD}\theta_1 & h_{SD} \\ h_{SD} & 0 \\ -h_{SD}^* & h_{RD}^*\theta_2 \end{bmatrix} \begin{bmatrix} x_1 \\ x_2 \end{bmatrix} + \begin{bmatrix} n_{D1} \\ n_{D2} \\ n_{D3} \\ n_{D4} \end{bmatrix}$$

$$\mathbf{r} = \mathbf{H}(\theta_1, \theta_2)\mathbf{x} + \mathbf{n}$$

$$\hat{\mathbf{x}}_D^{ML} = \arg\max_{\mathbf{x} \in \mathbf{X}} \{f_{\mathbf{r}|\mathbf{x}}(r|x)\}$$

$$f_{\mathbf{r}|\mathbf{x}}(r|x) = \frac{1}{\pi^2 N_0^2} \times \left\{ [1 - P_{SR}(\gamma_{SR})]^2 \exp\left(-\frac{|\mathbf{r} - \mathbf{H}(1,1)\mathbf{x}|^2}{2N_0}\right) \right.$$

$$+ P_{SR}(\gamma_{SR})[1 - P_{SR}(\gamma_{SR})] \exp\left(-\frac{|\mathbf{r} - \mathbf{H}(1,-1)\mathbf{x}|^2}{2N_0}\right)$$

$$+ [1 - P_{SR}(\gamma_{SR})]P_{SR}(\gamma_{SR}) \exp\left(-\frac{|\mathbf{r} - \mathbf{H}(-1,1)\mathbf{x}|^2}{2N_0}\right)$$

$$\left. \times [P_{SR}(\gamma_{SR})]^2 \exp\left(-\frac{|\mathbf{r} - \mathbf{H}(-1,-1)\mathbf{x}|^2}{2N_0}\right) \right\}$$

Answer 10: Probability of error in one bit is

$$P_1(SNR) = \frac{1}{2}\left[1 - \sqrt{\frac{SNR}{1+SNR}}\right].$$

Probability of error free block of length 512 is

$$P = [1 - P_1(SNR)]^{512} = [1 - P_1(3.16)]^{512} = 1.7413 \times 10^{-15}$$

$$P = \Pr(\gamma_{SR} > \gamma_{SD}) = \frac{\Gamma_{SR}}{\Gamma_{SR} + \Gamma_{SD}} = 0.5.$$

References

1. D. Tse and P. Viswanath, *Fundamentals of Wireless Communication* (Cambridge University Press, 2005).
2. S. Benedetto and E. Biglieri, *Principles of Digital Transmission with Wireless Applications* (Kluwer Academic/Plenum Publishers, New York, NY, 1999).
3. J. N. Laneman, Cooperative diversity in wireless networks: Algorithms and architectures, PhD Dissertation, Massachusetts Institute of Technology, Cambridge, MA, (August 2002).

4. J. N. Laneman, D. N. C. Tse and G. W. Wornell, Cooperative diversity in wireless networks: Efficient protocols and outage behaviour, *IEEE Trans. Inf. Theory* **50**, 12 (2004), 3062–3080.

5. J. N. Laneman and G. W. Wornell, Distributed space-time coded protocols for exploiting cooperative diversity in wireless networks, *IEEE Trans. Inf. Theory* **49**, 10 (2003), 2415–2525.

6. A. Sendonaris, E. Erkip and B. Aazhang, User cooperation diversity — Part I: System description, *IEEE Trans. Commun.* **51** (2003), 1927–1938.

7. A. Sendonaris, E. Erkip and B. Aazhang, User cooperation diversity — Part II: Implementation, aspects and performance analysis, *IEEE Trans. Commun.* **51** (2003), 1939–1948.

8. A. Nosratinia, T. Hunter and A. Hedayat, Cooperative communication in wireless networks, *IEEE Commun. Mag.* **42**, 10 (2004), 68–73.

9. R. Pabst, B. H. Walke, D. C. Schultz, P. Herhold, H. Yanikomeroglu, S. Mukherjee, H. Viswanathan, M. Lott, W. Zirwas, M. Dohler, H. Aghvami, D. Falconer and G. P. Fettweis, Relay-based deployment concepts for wireless and mobile broadband cellular radio, *IEEE Commun. Mag.* **42**, 9 (2004), 80–89.

10. E. C. Van der Meulen, Three-terminal communication channels, *Adv. Appl. Prob.* **3** (1971), 120–154.

11. T. M. Cover and A. A. El Gamal, Capacity theorems for the relay channel, *IEEE Trans. Inf. Theory* **25**, 5 (1979), 572–584.

12. V. Tarokh, H. Jafarkhani and A.R. Calderbank, Space-time block coding from orthogonal designs, *IEEE Trans. Inf. Theory* **45** (1999), 1456–1467.

13. Y. Jing and B. Hassibi, Distributed space-time coding in wireless relay networks, *IEEE Trans. Wireless Commun.* **5** (2006), 3524–3536.

14. S. Yang and J.-C. Belfiore, Towards the optimal amplify-and-forward cooperative diversity scheme, *IEEE Trans. Inf. Theory* **53**, 9 (2007), 3114–3126.

15. Y. Li and X.-G. Xia, A family of distributed space-time trellis codes with asynchronous cooperative diversity, *IEEE Trans. Commun.* **55** (2007), 790–800.

16. F. Oggier and B. Hassibi, A coding strategy for wireless networks with no channel information, *Proc. of Allerton Conference* (2006).

17. T. Kiran and B. Sundar Rajan, Partially-coherent distributed space-time codes with di erential encoder and decoder, *IEEE J. Sel. Areas Commun.: Special issue on Cooperative Communications and Networking* **25**, 2 (2007), 426–433.

18. G. K. Karagiannidis, T. A. Tsiftsis and R. K. Mallik, Bounds for multihop relayed communications in Nakagami-m fading, *IEEE Trans. Commun.* **54**, 1 (2006), 18–22.

19. T. A. Tsiftsis, G. K. Karagiannidis, S. A. Kotsopoulos and F.-N. Pavlidou, BER analysis of collaborative dual-hop wireless transmissions, *IEE Electron. Lett.* **40**, 11 (2004), 1732–1745.

20. T. E. Hunter and Aria Nosratinia, Diversity through coded cooperation, *IEEE Trans. Wireless Commun.* **5**, 2 (2006), 283–289.

21. R. U. Nabar, H. Bölcskei and F. W. Kneubühler, Fading relay channels: Performance limits and space-time signal design, *IEEE J. Sel. Areas Commun.* **22**, 6 (2004), 1099–1109.

22. H. Ochiai, P. Mitran and V. Tarokh, Variable rate two phase collaborative communication protocols for wireless networks, *IEEE Trans. Inf. Theory* **52**, 9 (2006), 4299–4313.

23. K. Azarian, H. E. Gamal and P. Schniter, On the achievable diversity-multiplexing tradeoff in half duplex cooperative channels, *IEEE Trans. Inf. Theory* **51**, 12 (2005), 4152–4172.

24. O. Canpolat, M. Uysal and M. M. Fareed, Analysis and design of distributed space-time trellis codes with amplify-and-forward relaying, *IEEE Trans. Veh. Technol.* **56**, 4 Part 1 (2007), 1649–1660.

25. J. W. Mark and W. Zhuang, *Wireless Communications and Networking* (Prentice Hall, ISBN 0-13-040905-7, 2003).

26. H. Mheidat and M. Uysal, Impact of receive diversity on the performance of amplify-and-forward relaying under APS and IPS power constraints, *IEEE Commun. Lett.* **10**, 6 (2006), 468–470.

27. Y. Zhao, R. S. Adve and T. J. Lim, Improving amplify-and-forward relay networks: Optimal power allocation versus selection, *IEEE Trans. Wireless Commun.* **6**, 8 (2007), 3114–3123.

28. A. Bletsas, A. Khisti, D. P. Reed and A. Lippman, A simple cooperative diversity method based on network path selection, *IEEE J. Select. Areas Commun.* (Special Issue on 4G Wireless Systems), **24**, 9 (2006), 659–672.

29. S. M. Alamouti, A simple transmit diversity technique for wireless communications, *IEEE J. Sel. Areas Commun.* **16**, 8 (1998), 1451–1458.

30. H. Wang and X.-G. Xia, Upper bounds of rates of complex orthogonal space-time block codes, *IEEE Trans. Inf. Theory* **49**, 10 (2003), 2788–2796.

31. A. Host-Madsen and J. Zhang, Capacity bounds and power allocation in wireless relay channel, *IEEE Trans. Inf. Theory* **51**, 6 (2005), 2020–2040.

32. N. Ahmed and B. Aazhang, Outage minimization with limited feedback for the fading relay channel, *IEEE Trans. Commun.* **54**, 4 (2006), 659–669.

33. M. O. Hasna and M.-S. Alouini, Optimal power allocation for relayed transmissions over Raleigh-fading channels, *IEEE Trans. Wireless Commun.* **3**, 6 (2004), 1999–2004.

34. M. M. Fareed and M. Uysal, BER-optimized power allocation for fading relay channels, *IEEE Trans. Wireless Commun.* **7**, 6 (2008), 2350–2359.

35. T. Wang, A. Cano, G. B. Giannakis and J. N. Laneman, High-performance cooperative demodulation with Decode-and-Forward relays, *IEEE Trans. Commun.* **55**, 7 (2007), 1427–1438.

36. I. S. Reed and X. Chen, *Error-Control Coding For Data Networks* (Kluwer Academic Publishers, 1999).

37. A. S. Ibrahim, A. K. Sadek, W. Su and K. J. R. Liu, Cooperative communications with channel state information: When to cooperate?, *Proc. of IEEE Global Telecommunications Conference*, Vol. 5, pp. 3068–3072, St. Louis, MO, (November 28–December 2, 2005).

38. M. M. Fareed and M. Uysal, On decode-and-forward cooperative networks with relay selection, to appear in *IEEE Transactions on Wireless Communications.*

39. A. Bletsas and A. Lippman, Implementing cooperative diversity antenna arrays with commodity hardware, *IEEE Commun. Mag.* **44**, 12 (2006), 33–40.

40. S. Valentin, H. S. Lichte, H. Karl, S. Simoens, G. Vivier, J. Vidal and A. Agustin, Implementing cooperative wireless networks — Towards feasibility and deployment, in *Cognitive Wireless Networks: Concepts, Methodologies and Visions,* (eds.) F. H. P. Fitzek and M. D. Katz (Springer, October 2007).

41. H. Mheidat, M. Uysal and N. Al-Dhahir, Equalization techniques for distributed space-time block codes with amplify-and-forward relaying, *IEEE Trans. Signal Process.* **55**, 5 Part 1 (2007), 1839–1852.

42. O.-S. Shin, A. M. Chan, H. T. Kung and V. Tarokh, Design of an OFDM cooperative space-time diversity system, *IEEE Trans. Veh. Technol.* **56**, 4 (2007), 2203–2215.

43. C. S. Patel and G. L. Stuber, Channel estimation for amplify and forward relay based cooperation diversity systems, *IEEE Trans. Wireless Commun.* **6**, 6 (2007), 2348–2356.

44. H. Mheidat and M. Uysal, Non-coherent and mismatched-coherent receivers for distributed STBCs with amplify-and-forward relaying, *IEEE Trans. Wireless Commun.* **6**, 11 (2007), 4060–4070.

Chapter 15

PUBLIC KEY CRYPTOGRAPHY AND CODING THEORY

PASCAL VÉRON

Institut de Mathématiques de Toulon,
Université du Sud Toulon-Var, France
veron@univ-tln.fr

For a long time, cryptography was only concerned by message confidentiality (how to convert a text in an incomprehensible form for non-authorized readers). Nowadays, with the development of network communications, cryptography is devoted to the study of numerous problems concerning security: authentication, message integrity checking, digital signatures, secure computation, ... The security of basically all real-life protocols is actually based on hard problems coming from number theory. The aim of this chapter is to show that algebraic coding theory offers an alternative way to define secure cryptographic primitives.

15.1. Introduction

Historically, cryptography is the science of secret writing. It has been employed since antiquity by statesmen, ambassadors, and the army. The main problem is the following: how to scramble a message before sending it in order to make it unintelligible to any outsider. In the classical model of symmetric cryptography, the message is enciphered with a function e and deciphered using a function d. These two functions depend upon a parameter k called the secret key such that for all messages m, $d(e(m,k),k) = m$. This introduces the following problem: the recipient of the message must be in possession of k. Hence the secret key must be communicated to the recipient via a secure channel. In practice, this may be very difficult to achieve. Moreover, if you want to communicate with multiple recipients, you have to distribute k to each recipient. This is not secure if one of them is dishonest since he can decipher each message encrypted with k (as it is the case for the WEP protocol). In this context, it is necessary to generate $k(k-1)/2$ distinct keys and distribute them.

In 1976, Diffie and Hellman [1] laid the foundation for public key cryptography asking the following question: is it possible to use a pair of keys (k, ℓ) such that only k be necessary for encryption, while ℓ would be necessary for decryption? For such a protocol, d and e must satisfy for all messages m, $d(e(m, k), \ell) = m$. A cryptosystem devised in this way is called a public key cryptosystem since k can be made public to all users. Indeed, there is no longer any reason to keep it secret since an attacker cannot use it in order to decipher a message (obviously, it should be computationally infeasible to determine ℓ from k). A public key cryptosystem has to verify four fundamental constraints.

(1) the cryptogram $e(m, k)$ can be computed in polynomial time from m and k,
(2) it is computationally infeasible to determine m from $e(m, k)$ and k,
(3) m can be computed from $e(m, k)$ and ℓ in polynomial time,
(4) the pair (k, ℓ) can be computed in polynomial time.

A function e which satisfies (1) and (2) is called a *one-way* function. If it satisfies also (3), then it is called a *trapdoor* function.

Since the beginning of public key cryptography, basically all real-life protocols are based on hard problems from number theory and relied on arithmetic operations over large numbers. Moreover, a lot of these protocols depend (dangerously) on only two problems: integer factoring and discrete logarithm. At this time, no one knows an efficient algorithm in order to solve these problems in a reasonable time although numerous researchers make good progress in this area. Moreover if no classical computers can factor in polynomial time, quantum computers are able to do it using Shor's algorithm [2].

Algebraic coding theory offers an alternative to number theory-based cryptosystem. Indeed, remember that the aim of algebraic coding theory is to restore a message m sent via a channel disrupted by some natural perturbation and that the goal of cryptography is to intentionally scramble a message m before sending it, so that it becomes unintelligible except for its recipient (hence this implies that the latter owns a method in order to restore the disrupted message). Obviously, there are some links between these two fields. Security of coding-based cryptographic primitives depends upon a problem which in its general form is a well-known NP-complete problem: the Syndrome Decoding Problem. Generally, these protocols are easier to implement and use only basic operations over the two element field.

15.2. Background

Only very few notions on coding theory and computational complexity theory are needed to understand the sequel of this chapter.

15.2.1. *Coding Theory*

We will assume that the reader knows basic definitions and properties of binary linear codes, otherwise a good reference on the topic is [3]. For the sequel, we will focus our attention on the decoding problem. First we recall two important results.

Proposition 15.1. *A binary linear code C of length n can correct t errors if for any $x, y \in C$ $(x \neq y)$, $B(x,t) \cap B(y,t) = \emptyset$ where $B(x,t) = \{y \in \{0,1\}^n \mid d(x,y) \leq t\}$ and $d(x,y)$ denotes the Hamming distance.*

Proposition 15.2. *A binary linear code $C(n,k)$ whose minimal distance is d can correct $\lfloor (d-1)/2 \rfloor$ errors.*

Let C be a binary $[n, k, d]$ code. Let us consider a word c' such that $c' = c_0 + e$ where $c_0 \in C$ and e is what is called an error vector. Let H be a parity check matrix of C and let s be the syndrome of c', i.e. $s = H^t c'$.

Notice that the 2^k solutions x which satisfy the equation

$$H^t x = s, \tag{15.1}$$

are given by the set $\{u + e, u \in C\}$ (remember that $\forall u \in C, H^t u = 0$). If the Hamming weight of e (i.e. the number of non-zero bits of e) satisfies

$$\forall u \in C \setminus \{0\}, \quad w(e) < w(u + e), \tag{15.2}$$

then the error e is the minimum weight solution of (15.1).

Remark 15.1. If $w(e) \leq \lfloor (d-1)/2 \rfloor$, then e satisfies Eq. (15.2).

Hence, without any extra information on the code, to decode c' one has to solve an optimization problem. Notice that searching for the minimum weight word which satisfies Eq. (15.1) is equivalent to search for the closest codeword from c'. Indeed, it is easy to see that Eq. (15.2) is equivalent to (see Question 15.1):

$$\forall u \in C \setminus \{c_0\}, \quad d(c_0, c') < d(u, c'). \tag{15.3}$$

One goal of coding theory is to find codes for which the minimum weight solution of (15.1) can be computed in polynomial time without constraints on the size of H (see next section for the definition of a polynomial time algorithm).

Among the class of codes which can be decoded in polynomial time, one of them is widely used in code-based cryptography: the Goppa codes class.

15.2.2. *Goppa codes*

In 1970, Goppa [4] introduced a new class of linear error-correcting codes which asymptotically meet the Varshamov–Gilbert bound: the so-called $\Gamma(L, g)$ codes. Binary Goppa codes can be specified through a polynomial $g(z)$ over $\mathbb{F}_{2^m}$ and a set $L \subset \mathbb{F}_{2^m}$ whose elements are not roots of $g(z)$.

Definition 15.1. Let $g(z) \in \mathbb{F}_{2^m}[z]$, $L = \{\alpha_1, \ldots, \alpha_n\} \subset \mathbb{F}_{2^m}$ such that $\forall i, g(\alpha_i) \neq 0$. The Goppa code $\Gamma(L, g)$, of length n over $\mathbb{F}_2$, is the set of codewords, i.e. *n-tuples* $(c_1, \ldots, c_n) \in \mathbb{F}_2^n$, satisfying

$$\sum_{i=1}^{n} \frac{c_i}{z - \alpha_i} \equiv 0 \pmod{g(z)}.$$

Proposition 15.3. *The dimension k of $\Gamma(L, g)$ and its minimal distance d satisfy*

$$k \geq n - m \deg g(z),$$
$$d \geq \deg \bar{g}(z) + 1.$$

where $\bar{g}(z)$ is the lowest degree perfect square which is divisible by $g(z)$.

Remark 15.2. For irreducible Goppa codes (i.e. codes for which $g(z)$ is irreducible), we deduce that the minimum distance satisfies $d \geq 2 \deg g(z) + 1$.

Proposition 15.4. *Let $L = \{\alpha_1, \ldots, \alpha_n\} \subset \mathbb{F}_{2^m}$, and $g(z) \in \mathbb{F}_{2^m}[z]$, a parity check matrix of the Goppa code $\Gamma(L, g)$ is given by*

$$H = \begin{pmatrix} g(\alpha_1)^{-1} & \cdots & g(\alpha_n)^{-1} \\ \alpha_1 g(\alpha_1)^{-1} & \cdots & \alpha_n g(\alpha_n)^{-1} \\ \vdots & \vdots & \vdots \\ \alpha_1^{\deg g(z)-1} g(\alpha_1)^{-1} & \cdots & \alpha_n^{\deg g(z)-1} g(\alpha_n)^{-1} \end{pmatrix}.$$

Although this matrix satisfies

$$\forall c \in \mathbb{F}_2^n, \quad c \in \Gamma(L, g) \Leftrightarrow H^t c = 0,$$

it is not a "real" parity check matrix since its rows belong to $\mathbb{F}_{2^m}$ so they do not generate $\Gamma(L, g)^\perp$. This matrix is the parity check matrix of a Generalized Reed–Solomon code whose restriction over $\mathbb{F}_2$ is exactly the Goppa code $\Gamma(L, g)$.

A parity check matrix $\tilde{H}$ over $\mathbb{F}_2$ can be computed from H by replacing each element of H by the corresponding column vector (of length m) when using any basis of $\mathbb{F}_{2^m}$ over $\mathbb{F}_2$.

$$H = \begin{pmatrix} g(\alpha_1)^{-1} & \cdots & g(\alpha_n)^{-1} \\ \alpha_1 g(\alpha_1)^{-1} & \cdots & \alpha_n g(\alpha_n)^{-1} \\ \vdots & & \vdots \\ \alpha_1^{r-1} g(\alpha_1)^{-1} & \cdots & \alpha_n^{r-1} g(\alpha_n)^{-1} \end{pmatrix}$$

$$\tilde{H} = \begin{pmatrix} 1 & \cdots & \cdots \\ 0 & \cdots & \cdots \\ \vdots & & \\ 1 & \cdots & \cdots \\ & & \\ \vdots & \vdots & \vdots \\ \vdots & \vdots & \vdots \end{pmatrix}$$

$\tilde{H}$ owns $m \deg g(z)$ rows which generate $\Gamma(L, g)^\perp$. Since these rows are not necessarily independent, we easily deduce that the dimension k of $\Gamma(L, g)$ satisfies $k \geq n - m \deg g(z)$. Other basic definitions and properties of Goppa codes are to be found in [3].

Patterson's algorithm [5] can be used in order to decode Goppa codes in about $\mathcal{O}(tn + t^2)$ operations over $\mathbb{F}_{2^m}$ where $t = \deg g(z)$. The algorithm only needs as input the set L and the polynomial $g(z)$. If the code is irreducible it is possible to correct up to exactly t errors.

15.2.3. *Computational Complexity Theory*

Unformally speaking, one goal of computational complexity theory is to classify problems. Usually all problems can be stated as decision problems. A decision problem is described by a name, some input and a question whose answer is yes or no.

Example 15.1. The square-root decision problem:
   Name   :   Square-root
   Input   : a and n two integers.
   Question :   Is a a square modulo n?

Another goal of complexity theory is to evaluate the ressources needed by an algorithm in order to solve a problem. An important criteria concerns the evolution of the running time of the algorithm when the size of the input grows. To analyze the running time of an algorithm, a cost is assigned to each instruction and the total cost is computed as a function f depending on the size n of the input. This function is then classified by comparing it with classical functions like: n, n^2, 2^n, $\log_2(n)$, ...

Definition 15.2. Let $g(n)$ be a non-negative function, the class $\mathcal{O}(g(n))$ is the set of non-negative functions $f(n)$ which satisfy

$$\exists B \in \mathbb{R}^+\backslash\{0\}, \quad \exists n_0 \in \mathbb{N}, \quad \text{such that} \quad \forall n \geq n_0, \quad 0 \leq f(n) \leq Bg(n).$$

Usually, we write $f(n) = \mathcal{O}(g(n))$ instead of $f(n) \in \mathcal{O}(g(n))$. Using the big $\mathcal{O}$ notation to classify an algorithm allows to obtain an upperbound on its running time according to the size of the input.

Example 15.2. Let us consider the algorithm which computes the scalar product over $\mathbb{F}_2$ of two binary vectors of size n. The size of the input is $N = 2n$. The computation of $x_i y_i$ is done with a logical AND and adding $x_i y_i$ to $x_{i+1} y_{i+1}$ is done with the XOR operator. We can thus suppose that these two operations are computed in constant time t. Hence the total running time is given by $nt + (n-1)t$ which is bounded by tN. So the running time of this algorithm is in $\mathcal{O}(N)$.

Definition 15.3. A polynomial time algorithm is an algorithm for which there exists an integer d such that its running time is in $\mathcal{O}(n^d)$, n being the size of the input.

   Decision problems are also classified. We will only mentioned in this section three important classes.

Definition 15.4. A decision problem is in the class P if it can be solved by a deterministic polynomial time algorithm when the answer is yes.

Definition 15.5. A decision problem is in the class NP if it can be solved by a non-deterministic polynomial time algorithm when the answer is yes.

Unformally speaking, a non-deterministic polynomial time algorithm is an algorithm which have to perform a sequence of bit operations in order to find the solution. The length of this sequence is polynomial in the size of the input.

Before introducing the third class, we need to present the notion of reduction. Sometimes an algorithm developed to solve a problem P_1 can be turned into an algorithm to solve a problem P_2.

Definition 15.6. We will say that a problem P_1 can be reduced to a problem P_2 if each instance of P_1 can be transformed into an instance of P_2.

Inside the NP class, there is an important subclass: the NP-complete problems.

Definition 15.7. A decision problem Π is NP-complete if:

(a) it is in NP,
(b) every problem in NP can be reduced in polynomial time to Π.

Nowadays, we have the following relation: $P \subset NP \subset NP$-complete. It is still not known if $P = NP$. From the previous definition, it should be the case if one could find a polynomial time deterministic algorithm which solves one of the problem known to be NP-complete.

Example 15.3. Finding the minimum distance of a random linear binary code is NP-complete [6].

Name	:	Minimum distance;
Input	:	H a $r \times n$ parity check matrix, p an integer > 0;
Question	:	Does there exists a non-zero codeword x such that $H^t x = 0$ and $w(x) \leq p$?

15.3. The Syndrome Decoding Problem

The Syndrome Decoding Problem is a decision problem which can be stated as follows:

Name	:	SD;
Input	:	$H(r, n)$ a binary matrix , s a binary column vector with r coordinates, p an integer;
Question	:	Is there a binary vector e of length n such that $H^t e = s$ and $w(e) \leq p$?

In the context of coding theory, if H is a parity check matrix, this means that the problem to decide whether there exists or not a word of given weight and syndrome is NP-complete.

This decision problem is linked to the optimization problem induced by maximum likelihood decoding. Indeed, searching for the closest codeword of a received word x is equivalent to find the minimum weight solution e of the equation $H^t e = H^t x$. Now, let (H, s, p) be an instance of the SD problem, the vector e exists if and only if the minimum weight solution of $H^t x = s$ is less or equal than p. On the other hand, if one knows a polynomial time algorithm to solve SD, then it can be turned into a polynomial time algorithm to compute the minimal weight of a solution of the system $H^t x = s$. In 1978, Berlekamp, McEliece and Van Tilborg [7] proved that this problem is NP-complete reducing it to the THREE-DIMENSIONAL MATCHING problem [8].

Remark 15.3. The problem still remains NP-complete if:

* the matrix H is full rank (as it is the case for a parity check matrix),
* we ask for an s with exactly p 1's.

The SD problem can be stated in terms of the generator matrix since one can go from the parity check matrix to the generator matrix (or vice-versa) in polynomial time:

Name	:	G-SD;
Input	:	$G(k, n)$ a generator matrix of a binary (n, k) code $\mathcal{C}$, $x \in \{0, 1\}^n$ and $p > 0$ an integer;
Question	:	Is there a vector e of length n and weight p such that $x + e \in \mathcal{C}$?

While the SD problem is NP-complete, there exists weak matrices for which an efficient algorithm can be developed. Hence, one can alternatively define algebraic coding theory as the science whose one goal is to build easy instances of the SD problem, in order to set up polynomial time algorithms for decoding. However for a random matrix H, it is necessary to know for which parameters (n, r, p) the problem seems to be difficult to solve.

15.4. Algorithms for the SD Problem

Nowadays, there exists five probabilistic algorithms to compute a solution to the SD problem: Lee and Brickell's algorithm [9], Leon's algorithm [10], Stern's algorithm [11], the toolbox of Canteaut and Chabaud [12], and

Johansson and Jönsonn's algorithm [13]. All these algorithms are devoted to search a word of small weight in a random code.

Proposition 15.5. *SD problem is equivalent to the following problem:*

> **Input** : $H(k, n)$ *a binary matrix of rank* k, $p > 0$ *an integer.*
> **Question** : *Is there a vector* $x \in \{0, 1\}^n$ *such that* $H^t x = 0$, $w(x) \le p$ *and* $x_n = 1$?

Proof. See exercises. $\qquad\qquad\qquad\qquad\qquad\qquad\qquad\qquad\square$

All these algorithms are based on the notion of information set decoding introduced by Prange [14].

Definition 15.8. Let G be a generator matrix of an $[n, k]$ code and $c = mG$ be a codeword. Let us denote by G_i the ith column of G and let $I = \{i_1, \ldots, i_k\}$ such that $G_I = (G_{i_1}, \ldots, G_{i_k})$ be a $k \times k$ invertible submatrix. Then these k coordinates uniquely determine the vector m, since $m = (c_{i_1}, \ldots, c_{i_k})G_I^{-1}$. The set I is called an information set.

Now suppose that a received word $x = (c+e)$ is such that no errors occur in the information set I. The error pattern e can be recovered by computing $(x_{i_1}, \ldots, x_{i_k})G_I^{-1} + x$. Hence, the main idea used in all the algorithms is to select random information sets from the generator matrix (or the parity check matrix for Stern's scheme) until the support of the error does not meet the selected set which leads to a probability of success of:

$$\frac{\binom{n-p}{k}}{\binom{n}{k}}. \tag{15.4}$$

Using the usual binomial approximation this gives the following probability of success:

$$P_{\text{succ}} = \mathcal{O}(1) \cdot 2^{-nH_2(p/n) - (1-k)H_2(p/(n-k))}, \tag{15.5}$$

where $H_2(x)$ is the classical entropy function. Hence, the work factor (number of operations) needed to compute a solution for the SD problem can be roughly estimated by

$$\frac{\text{Inv}(k)}{P_{\text{succ}}}, \tag{15.6}$$

where $\text{Inv}(k)$ is the cost for inverting a $k \times k$ matrix. Usually this operation needs k^3 binary operations (notice that in order to be more precised, we should have take into account the probability for a random $k \times k$ matrix to be invertible). The algorithms of Lee and Brickell, Leon and Stern use some heuristic in order to minimize the call to the inverse procedure by taking into account information set which contains a small part of the support of the error pattern. Canteaut and Chabaud combine these heuristics with a trick (proposed by Van Tilburg [15] and latter by Chabanne and Courteau [16]) in order to reduce the cost of the inverse procedure. Let I be the current information set for which the algorithm did not succeed, instead of randomly select k new columns, they exchange one column whose index is in I with a column whose index is in $\{1, \ldots, n\} \backslash I$ which decreases the cost of the Gaussian elimination. Interested readers can find a complete description and analysis of the first four algorithms in [17, 12]. It follows from the study of [12] that the modified version of Stern's algorithm is the best one to solve the SD problem. Another important result is that hard instances of the SD problem are obtained when the weight of the vector e is near from the theoretical minimal distance d of the code which is given by the Gilbert–Varshamov bound:

$$H_2(d/n) \simeq 1 - k/n. \tag{15.7}$$

Table 15.1 gives an example of the work factor required to find a word of weight p in a $[n, k]$ random binary code.

Remark 15.4. The asymptotic complexity of the algorithm of Canteaut–Chabaud has been empirically estimated to $(n/ \log_2(n))^{\lambda t - c}$ where c and λ are two constants such that $1 \leq c \leq 2$ and $\lambda \approx 1$.

The algorithm of Johansson and Jönsson is slightly different from the other one. The input is a list of received words and the goal is to try to decode one of them. Since the algorithm works with information sets, all the tricks used in the other algorithms can be used in order to optimize it. The probability of success grows with the size of the initial list. When this list is to small, the performances are not better than those of the other algorithms (see Table 15.2).

Table 15.1.   Computational effort required to find a word of weight p in a (n, k) code.

Code	(256,128)	(512,256)	(512,256)	(768,384)	(1024,524)	(2048,1608)
p	14	28	56	42	50	81
work factor	$2^{26.51}$	$2^{40.48}$	$2^{69.8}$	$2^{54.55}$	$2^{64.2}$	2^{100}

Table 15.2. Workfactor of Johansson and Jönsson algorithm.

Size of list	$n = 1024,\ k = 524,\ p = 50$	$n = 512,\ k = 256,\ p = 56$
1	$2^{68.1}$	$2^{72.2}$
2^5	$2^{63.7}$	$2^{68.9}$
2^{10}	$2^{59.5}$	$2^{65.9}$
2^{15}	$2^{56.2}$	$2^{64.1}$
2^{30}	$2^{50.2}$	2^{60}

15.5. A Pseudo-Random Generator

Informally speaking, a pseudo-random generator could be seen as a "blackbox" which outputs from a seed a train of bits which "looks like" random. Using complexity theory, Blum and Micali have formalized the notion of random looking [18]. An algorithm is a pseudo-random generator if its output distribution is polynomial time indistinguishable from a truly random distribution. Actually, it has been proved that the existence of a pseudo-random generator is equivalent to the existence of a one-way function. As always, the first pseudo-random generators with a "proven security" were based on the difficulty of well-known problems coming from number theory (factorization, discrete logarithm) and involved complex operations. As an example, outputting a single bit needs cubic time for RSA-based generators. A first attempt to devise a pseudo-random generator not related to number theory is due to Impagliazzo and Naor [19]. Unfortunately, security of their scheme was based on the subset sum problem which is definitively not secure.

In 1998, Fischer and Stern described an efficient pseudo-random generator based on the SD problem [20]. They first define a general collection of functions related to the SD problem:

Definition 15.9. Let $\rho \in]0,1[$, $\delta \in]0,1/2[$, and $D_n = \{(M, x) \mid M$ is a $\lfloor \rho n \rfloor \times n$ matrix, $x \in \{0,1\}^n$, $w_H(x) = \lfloor \delta n \rfloor\}$. The $SD(\rho, \delta)$ collection is the set of functions $\{f_n\}_n$ such that

$$f_n : D_n \to \{0,1\}^{\lfloor \rho n \rfloor (n+1)},$$
$$(M, x) \mapsto (M, Mx).$$

Remember that informally speaking, a one-way function f is such that it is computationally infeasible to find x from $f(x)$. For $f \in SD(\rho, \delta)$, it consists in solving an instance of the SD problem. Hence, if we choose n, ρ and δ such that the collection $SD(\rho, \delta)$ fits with hard instances of the SD problem, then it can be considered as a collection of one-way functions. Using good parameters, one can thus admit the following intractability assumption:

Intractability assumption: Let $\rho \in\ ]0,1[$, $\forall\, \delta \in\ ]0,1/2[$ such that $H_2(\delta) < \rho$, the $SD(\rho,\delta)$ collection of functions is strongly one-way $(H_2(x) = -x \log_2 x - (1-x)\log_2(1-x))$.

Notice that since $H_2(\delta) < \rho$, the functions f_n expand their inputs as it is required in the formal definition of a pseudo-random generator. Indeed $|D_n| = 2^{\lfloor \rho n \rfloor n}\binom{n}{\delta n}$, the size of the image set is $2^{\lfloor \rho n \rfloor n}2^{\lfloor \rho n \rfloor}$ and there exists ε such that $(1+\varepsilon)\log_2\binom{n}{\delta n} = \lfloor \rho n \rfloor$. Therefore, the expansion is linear. Table 15.3 gives the description of the pseudo-random generator. The algorithm A computes a vector of size n and weight $\lfloor \delta n \rfloor$ from a $\lceil \log_2\binom{n}{\delta n}\rceil$ bits number [21, 22]. In their paper, the authors proves formally that if the $SD(\rho,\delta)$ collection is one-way then the proposed algorithm is a pseudo-random generator.

Performances: If we omit the last step of the algorithm, all operations are done over bits using classical logical operators. Moreover, as suggested by the authors, the matrix can be defined as the output of a congruential generator (different seeds for each line) so that its size remains small. The main problem comes from the last step where an n-bit vector of weight $\lfloor \delta n \rfloor$ must be computed from a $\lceil \log_2\binom{n}{\delta n}\rceil$ bit number. The authors used an algorithm given by Guillot (see Table 15.4) which needs the computation of a binomial coefficient and involves classical operations over the integers. All of this can be reduced to only comparisons and substractions over $\mathbb{N}$, if one chooses to store (in a precomputation step) some binomial coefficients

Table 15.3.   SD-based random generator.

```
Input: - (M, x) ∈ Dₙ
begin
    repeat
        y ← Mx
        split y in two bit strings y₁ and y₂, such that:
            · y₁ consists in the first ⌈log₂ (ⁿ_δn)⌉ bits of y,
            · y₂ consists in the remaining bits.
        output y₂
        x ← A(y₁)
    until no more bit is needed.
end
```

Table 15.4. An algorithm to compute the ith word of weight p and length n.

```
Input: - n, p, i < (n\p) three integers
begin
     c ← (n\p)
     while n > 0 do
          c' ← c(n-p)/n /*c' ← (n-1\p)*/
          if i < c' then
               output 0
               c ← c'
          else
               output 1
               i ← i - c'
               c ← c·p/n /*c ← (n-1\p-1) */
               p ← p - 1
          endif
          n ← n - 1
     endw
end
```

in a table. For $n = 512$, $k = 256$, and $p = 55$, the table takes 880-Kbytes of memory and the algorithm generates from a 248 bits vector $(\lceil \log_2 \binom{512}{55} \rceil)$ a 256 bits vector, i.e. it generates one byte of random output. The authors claim that on a SUN Sparc10 station, the scheme outputs 3500-bits/sec as compared to an RSA-based generator (512 bits modulus) which outputs 1800-bits/sec (RSA-based generators output $\log_2 n$ pseudorandom bits by one modular multiplication with a modulus N that is n bits long).

15.6. The G-SD Identification Scheme

15.6.1. *Introduction*

An identification scheme is a cryptographic protocol which enables party A (called the "prover") to prove his identity (by means of an on-line communication) polynomially many times to party B (called the "verifier") without enabling B to misrepresent himself as A to someone else. In 1985, Goldwasser, Micali, and Rackoff described a very nice solution to this problem with zero-knowledge proofs [23], where a user convinces with a non-negligible probability an entity that he knows the solution s of a public instance of a "difficult" problem without giving any information on s (see [24] for a nice introduction to zero-knowledge). In 1986, Fiat and Shamir proved the practical significance of zero-knowledge proofs for public key identification [25]. Their scheme relies on the difficulty of factoring. Notice that, from a practical point of view, the prover may be identified to a smart card, hence it is supposed that he has reduced computational power and a small amount of memory. Since 1988, there were several attempts to build identification schemes which did not rely on number theory and use only very simple operations so as to minimize computing load. The idea to use error-correcting codes for identification is due to Harari [26], unfortunately his scheme was not zero-knowledge and not really practical due to its heavy communication load. Moreover, the scheme has been proved to be insecure in [27]. Another scheme proposed by Girault [28] has been cryptanalyzed in [29].

15.6.2. *The G-SD Scheme*

The first truly practical scheme using error-correcting codes is due to Stern [30]. The scheme uses a fixed binary (k, n) parity check matrix H which is common to all users. In 1995, a dual version of Stern's scheme has been defined: the G-SD identification scheme [31]. This version improves the communication complexity (number of bits exchanged during the protocol) for exactly the same level of security as those of Stern's scheme.

Table 15.5 lists the secret and public data used in this protocol. The pair (x, p) is the public identification of the prover. His data can be computed by a certification center having the confidence of all users or the prover can choose his secret keys and the center certifies the corresponding public keys. The principle of the protocol is the following: the prover (Alice) knows the pair (m, e) which satisfies $x + e = mG$ and $p = w(e)$. Bob (the verifier) asks Alice a series of questions. If Alice really knows (m, e), she can answer all the questions correctly. If she does not, she has a probability q of answering correctly. After r successful iterations of the protocol, Bob will be convinced that Alice knows s with probability $1 - q^r$.

Table 15.5. Public and secret data in the G-SD identification scheme.

Common public data	:	$G(k, n)$ a full rank binary matrix , a hash function denoted by $\langle . \rangle$.
Prover's secret data	:	$m \in \{0, 1\}^k$, mG and $e \in \{0, 1\}^n$.
Prover's public data	:	$x = mG + e$ and $p = w(e)$.

The identification scheme relies on the notion of commitment. Commitment is a protocol between Alice and Bob which operates in three stages:

- Stage 1: Alice hides a sequence u of bits and sends it to Bob. The hidden function is public and hard to invert.
- Stage 2: Alice and Bob execute some protocol,
- Stage 3: Alice reveals u, Bob checks the validity of the hidden value received during stage 1.

From a practical point of view, u is hidden via a cryptographic public hash function. Hence Alice sends to Bob the image $\langle u \rangle$ of u. The hash function must be collision-free (i.e. it should be "infeasible" to compute $u' \neq u$ such that $\langle u' \rangle = \langle u \rangle$). Discussion on the length of the hash value $\langle u \rangle$ can be found in [32]. Let us denote by $x \cdot y$ the concatenation of the binary strings x and y and by $y\sigma$ the image of $y \in \{0,1\}^n$ under the permutation σ of $\{1, \ldots, n\}$, the G-SD scheme includes r rounds each of these being performed as described in Table 15.6.

Table 15.6.   A round of the G-SD scheme.

- A randomly computes:

  — $u \in \{0,1\}^k$,
  — σ a permutation of $\{1, \ldots, n\}$.

  and send to B three commitments:

$$c_1 = \langle \sigma \rangle, \ c_2 = \langle (u+m)G\sigma \rangle, c_3 = \langle (uG+x)\sigma \rangle$$

- B sends a random element $b \in \{0,1,2\}$ (challenge).
- if $b = 0$,

  — A reveals $u + m$ and σ,
  — B checks the value of c_1 and c_2.

- if $b = 1$,

  — A reveals $(u+m)G\sigma$ and $e\sigma$,
  — B checks the value of c_2 and c_3 and verifies that $w(e\sigma) = p$.

- if $b = 2$,

  — A reveals σ and u,
  — B checks the value of c_1 and c_3.

15.6.3. *Security and Performances*

It can be proved that:

- The scheme is zero-knowledge, i.e. informally speaking, during the protocol the transactions contain no information on (m, e) (more formally one can construct a polynomial time machine S which outputs a communication tape having the same probability distribution as a real communication).
- A cheater can bypass the protocol with a probability bounded by $(2/3)^r$, otherwise one can construct a polynomial time probabilistic machine which either outputs a valid pair (m, e) or finds collision for the public hash function (see question 15.6).

Practical security of the scheme is linked to the parameters n, k, p, and r. Let H be a parity check matrix of the code $\mathcal{C}$ defined by G. In order to impersonate A, an intruder has to be able to compute a word e' of weight p whose image under H is $H^t x$ (this is the SD problem). If p is chosen slightly below the value of the theoretical minimum distance of $\mathcal{C}$ then the probability that there exists a word $e' \neq e$ of weight p such that $x + e'$ belongs to $\mathcal{C}$ is very low. Hence by choosing

$$n = 512, \quad k = 256, \quad p = 56,$$

searching the vector e with the probabilistic algorithms described in Sec. 15.4 needs around 2^{70} operations. Moreover taking $r = 35$, the probability of success of a cheater is bounded by 10^{-6}.

If we envisage the prover as a smart card, essentially three parameters are to be taken into account: the communication complexity (number of bits exchanged during the protocol), the complexity of the computations done by the prover, and the storage capacity needed by the prover. The G-SD identification scheme uses only very simple operations over the two element field (i.e. over bits) and can be implemented in hardware in a quite efficient way. One drawback is the size of the matrix G which must be stored by the prover. Another one is the communication complexity since at least 35 rounds are needed in order to achieve a reasonable level of security while for the same level (from a dishonest prover point of view) identification schemes based on number theory can be performed in only few rounds (4 rounds for Fiat–Shamir's scheme). Table 15.7 sums up the performances of Stern's scheme, G-SD scheme, and Fiat–Shamir's scheme (1024-bits version) giving for each one: the number of rounds needed to

Table 15.7.   SD schemes versus Fiat–Shamir scheme.

	SD	G-SD	Fiat–Shamir
Rounds	35	35	4
ROM	66048	66816	5120
Computation complexity	$2^{22.13}$	$2^{22.5}$	$2^{25.4}$
Communication complexity	40133	34160	4628

achieve a probability of success of 10^{-6} for a dishonest prover, the total communication complexity, the size of the ROM (number of bits stored by the prover), the total prover's computation complexity (number of binary operations performed by the prover during the whole protocol).

Remark 15.5. Taking into account the result of Table 15.2, Sec. 15.4, in a network of 2^{15} users, if one can distribute Johansson and Jönsson's algorithm on 2^{13} computers he can then find back one secret data in less than one week. For this very special context, the G-SD scheme has to be used with $n = 1024$, $k = 512$, and $p = 112$.

15.7. The McEliece's Public Key Cryptosystem

Despite, McEliece's cryptosystem be the first code based scheme, we decide to not describe it first because its security does not directly rely on the SD problem.

Soon after Diffie–Helmman's paper on public key cryptography, Rivest, Shamir, and Adleman exhibited such a system: the well-known RSA cryptosystem based on the factorization of integers [33]. Merkle and Hellman [34] proposed another cryptosystem based on the difficulty of the integer packing "knapsack" problem. There were several variants around this latter, but the development of the LLL algorithm made most of them insecure.

In 1978, McEliece defined the first public key cryptosystem using algebraic coding theory [35]. The basic idea is quite simple: use as a secret key a code $\mathcal{C}$ which belongs to a family of codes for which a polynomial time decoding algorithm exists and give as a public key an equivalent code $\mathcal{C}'$ which masks the algebraic structure of $\mathcal{C}$, so that $\mathcal{C}'$ looks like a random binary linear code. Table 15.8 describes the general protocol. Of course, one important parameter of this protocol is the code $\mathcal{C}$ to use:

• For n, k, and d fixed, $\mathcal{C}$ must belong to a large family of codes so that it is impossible to find it via an exhaustive search. Notice that it is

Table 15.8.   A code based public key cryptosystem.

Secret Key:

- G a generator matrix of a binary linear $[n, k, d]$ code $\mathcal{C}$ for which a polynomial time decoding algorithm $\mathcal{A}$ is known,
- S a non-singular random $k \times k$ binary matrix,
- P a random binary $n \times n$ permutation matrix.

Public Key: $G' = SGP$ and $t = \lfloor (d-1)/2 \rfloor$.

Encryption:

- Message: $m \in \{0, 1\}^k$,
- Cryptogram: $c = mG' + e$ where $e \in \{0, 1\}^n$ satisfies $w(e) = t$.

Decryption: Since $w(eP^{-1}) = w(e)$, successively compute:

- $mS = \mathcal{A}(cP^{-1}) = \mathcal{A}((mS)G + eP^{-1})$,
- $m = (mS)S^{-1}$.

enough to find an equivalent code to the public one using an algorithm due to Sendrier [29] which can determine if two generator matrices define equivalent codes and can find back the permutation.

- A polynomial time decoding algorithm must exist for $\mathcal{C}$.
- No information about the code $\mathcal{C}$ can be obtained from the generator matrix G'.

The third condition eliminates some classes of well-known "decodable" codes such as generalized Reed–Solomon codes (as shown by Sidelnikov and Shestakov [36]), and concatenated codes (as shown by Sendrier [37]). The class of binary Goppa codes as suggested by McEliece seems to satisfy these three conditions.

15.7.1. *Cryptanalysis*

McEliece recommended using an irreducible binary Goppa code of length 1024 with $L = \mathbb{F}_{2^{10}}$ and $g(z)$ an irreducible polynomial of degree 50 (see Remark 15.2). Since the number of monic irreducible polynomials of degree 50 over $\mathbb{F}_{2^{10}}$ is given by $(\sum_{d|50} \mu(d) 2^{500/d})/50$ (where μ is the Möbius function), this gives about 2^{500} candidates which clearly prevents any exhaustive search. However, two other kind of attacks can be envisaged against McEliece's cryptosystem

- a structural attack,
- a generic attack.

15.7.1.1. *A structural attack*

A structural attack against McEliece's cryptosystem consists in studying the algebraic structure of the public code $\mathcal{C}$ in order to build a decoder (or at least to find some parameters of the hidden code). Remember that L and $g(z)$ are the two essential parameters for the decoding algorithm. Until know, there does not exist any algorithm which takes as input a generator matrix of a Goppa code and which outputs these two data. However, as pointed out by Gibson, if a generator matrix G of a binary Goppa code and L are known, it is then possible to find back the polynomial $g(z)$ [38]. Hence one can devise a cryptanalysis in three steps.

(1) fix a permutation of $\mathbb{F}_{2^m}$ say $\bar{L} = \{\beta_1, \ldots, \beta_{2^m}\}$,
(2) search for a permutation π of the columns of G' which transforms the public matrix into the generator matrix $\bar{G}$ of a $\Gamma(\bar{L}, \bar{g})$ Goppa code,
(3) compute $\bar{g}$ from $\bar{G}$ and $\bar{L}$ and use the decoder of $\Gamma(\bar{L}, \bar{g})$ to decode the public code $\mathcal{C}$.

In [39], Adams and Meijer claim that there is no more than one permutation which satisfies step 2 of the cryptanalysis. This is not true, as proved by Gibson [38], who showed that there exists at least $m2^m(2^m - 1)$ such permutations. Unfortunately for $m = 10$, this represents less than $2^{-8713}\%$ of all the permutations!

Nevertheless, Loidreau and Sendrier developed a nice attack when the polynomial $g(z)$ has only binary coefficients [40]. They use the support splitting algorithm [29] which is able to decide if two linear codes are equivalent and outputs the permutation. Their structural attack uses the fact that Goppa codes defined from a binary polynomial have a non-trivial automorphism group (and so the automorphism group of the corresponding public code is also non-trivial). This cryptanalysis brings out weak keys in McEliece's cryptosystem even if their number is negligible as compared to the number of possible keys. A "real" structural attack to date necessitates a proper classification of Goppa codes.

15.7.1.2. *A generic attack*

Without the knowledge of L and g, it seems that it is computationally hard to make the difference between a random matrix and the generator matrix of a Goppa code. This is the Goppa code distinguishing problem:

> **Name**   :   GD,
> **Input**   : $G(k, n)$ a binary matrix,
> **Question** :   Is G a generator matrix of a $\Gamma(L, g)$ code?

Since there does not exist any suitable algorithm which uses the underlying Goppa code structure of McEliece's cryptosystem, cryptanalysis of the system boils down to the general problem of the decoding of a random binary linear code (the G-SD problem). In fact, cryptanalysis of McEliece's cryptosystem relies on a variant of the G-SD problem. Indeed, the weight t of the error is linked to the parameters of the code. Let $n = 2^m$, it seems that for irreducible Goppa codes the dimension k always satisfies $k = n - mt$, hence $t = (n - k)/\log_2(n)$. The underlying problem to solve is then the following:

> **Name**   :   GPBD (Goppa Parametrized Bounded Decoding)
> **Input**   : G a fullrank binary matrix $k \times n$, $y \in \{0,1\}^n$
> **Question** :   Does there exists $e \in \{0,1\}^n$ such that $y + e$ be a linear combination of rows from G and $w(e) \leq (n - k)/\log_2 n$?

This problem is NP-complete [41].

McEliece's cryptosystem with its original parameters can be cryptanalyzed in $2^{64.2}$ binary operations using the algorithms to solve the SD problem [42]. Johansonn and Jönsson algorithm can output a cleartext from a list of 1024 cryptogram in $2^{59.5}$ operations. In order to obtain a security level of 2^{80} the parameters to use are

$$m = 11, \quad n = 2048, \quad k = 1685, \quad t = 33.$$

Remark 15.6. In its original form, the cryptosystem is vulnerable to active attacks where an intruder modifies the cryptogram and uses as an oracle a deciphering machine. The protocol is also vulnerable to message replay. That is to say that an intruder is able to distinguish the fact that two cryptogram come from the same plaintext and in this context he can devise an attack which can recover the message in less than eight iterations for the original parameters (see Questions 15.7 and 15.8). In 2001,

Kobara and Imai [43] proved that the cryptosystem can be modified in order to be semantically secured against IND-CCA2 attacks.

15.7.2. *Niederreiter's Variant*

In 1986, Niederreiter [44] defined the dual version of McEliece's cryptosystem using the parity check matrix of the code instead of the generator matrix (see Table 15.9). From a security point of view Niederreiter's cryptosystem and McEliece's cryptosystem are equivalent (if used with exactly the same parameters [45]). However they differ from a practical point of view. Unlike McEliece's cryptosystem, it is not necessary to use a pseudo-random generator for encryption process. Notice, however that the plaintext is a n-binary word of weight t, hence we need a practical algorithm which maps the integers between 1 and $\binom{n}{t}$ to the set of words of weight t and length n and vice-versa. Such algorithms can be found in [20, 22].

Niederreiter's cryptosystem allows to reduce by a factor of 2 the size of the public key. Indeed, the matrix H can be expressed as $H = (I_{n-k} \mid M)$, hence it is enough to store the $(n - k) \times n$ matrix M. Such a trick is impossible in McEliece's cryptosystem since if $G' = (I_k \mid M)$ and the original message is not random, the cryptogram $c = mG' + e$ would reveal a part of the plaintext.

Table 15.9. Niederreiter's cryptosystem.

Secret key:

- A binary linear code $\mathcal{C}[n, k, d]$ for which there exists a polynomial algorithm $\mathcal{A}$ able to correct $t \leq \lfloor (d - 1)/2 \rfloor$ errors,
- $S(n - k, n - k)$ an invertible matrix,
- $P(n, n)$ a permutation matrix.

Public key: $(H' = SHP, t)$ where H is a parity check matrix of $\mathcal{C}$.

Encryption:

- Message: $m \in \{0, 1\}^n$ of weight t,
- Cryptogram: $c = H'^t m$.

Decryption:

- Compute $S^{-1}c = HP^t m$,
- Since $w(P^t m) \leq t$, apply $\mathcal{A}$ to find back $P^t m$,
- Compute $m = {}^t(P^{-1}P^t m)$.

Table 15.10. A comparison between McEliece, Niederreiter, and RSA cryptosystems.

	McEliece $(2048, 1718, t = 30)$	Niederreiter $(2048, 1718, t = 30)$	RSA-2048 $e = 2^{16} + 1$
Public key size (Kbytes)	429.5	69.2	0.5
Transmission rate	83.9%	67.3%	100%
Encryption complexity	1025	46.63	40555
Decryption complexity	2311	8450	6557176, 5

Since the public key in Niederreiter's cryptosystem is smaller and the plaintext is a word of small weight, this implies that the number of operations involved during the encryption process is less than what is done in McEliece's cryptosystem. Finally, depending on the parameters, the transmission rate (number of information symbols/ number of transmitted symbols) which is equal to $\log_2 \binom{n}{t}/(n - k)$ can be better or worst that those of McEliece (k/n).

Table 15.10 sum, up these differences and makes a comparison with the RSA cryptosystem when used with a 2048 modulus and a public exponent e equal to $2^{16} + 1$ as in openssl toolbox (the complexity is given as the number of binary operations to perform per information bit):

Remark 15.7. Notice that in his original paper, Niederreiter suggested using either a binary $[104, 24, 32]$ code (obtained by concatenation of other binary codes) or a $[30, 12, 19]$ Reed–Solomon code over $\mathbb{F}_{31}$. These two codes were verified as insecure by Brickell and Odlyzko [46] using the LLL algorithm.

15.8. The CFS Signature Scheme

A signature scheme is a protocol where the recipient of a message M can check its integrity, the sender's identity and such that the sender cannot refute that he sent M. The following general algorithm can be used in order to turn a public key cryptosystem into a signature scheme:

Step 1: Compute $h(m)$ from the message m and the hash function h,

Step 2: Apply the deciphering algorithm to $h(m)$ to get a value s,

Step 3: Use s as the signature of the message m.

Now, to obtain a formal proof of security of a signature scheme the hash function must be random and independent of the cipher and decipher functions. These constraints are not compatible with the fact that in order to use the preceding algorithm with McEliece's cryptosystem the hash values must belong to the set of decodable words which is clearly dependent of the underlying code used.

Hence if one suppose that the hash values belong to the set of random n-bits words, the main problem is to know whether it is possible to decode them. Using Niederreiter's cryptosystem instead of McEliece's one, hash values lie in the set of syndromes and must match the syndrome of an error of weight t in order to apply the deciphering function. This is the idea used by Courtois, Finiasz, and Sendrier in 2001 [47].

15.8.1. *The Scheme*

Let $\mathcal{C}$ be an irreducible $[2^m, k]$ Goppa code which can correct t errors. Then the probability for a syndrome to be a "decodable" syndrome is given by

$$\frac{\binom{2^m}{t}}{2^{n-k}}. \tag{15.8}$$

Using the fact that $k = 2^m - mt$, this can be roughly estimated as $1/t!$ Hence, instead of directly compute the value $h(m)$ the idea is to successively compute $h(m||i)$ where i is a counter which is increased by 1 until a decodable syndrome be obtained ($||$ is the concatenating operator). Table 15.11 sums up the signature scheme process and Table 15.12 its asymptotic characteristics depending on m and t.

15.8.2. *Security and Performances*

To forge a signature s for a given message M, an intruder has to find a word s of weight t such that $H_{\mathrm{pub}}{}^t s \in \{h(M||i), i \in \mathbb{N}\}$. In the formal model of the random oracle, h is random and so is this set. Hence, the only way for the intruder to forge a valid signature is too randomly choose a syndrome and try to decode it. That is to say that he has to solve an instance of the GPBD problem (see Sec. 15.7.1.2). Now suppose that an intruder can build an algorithm $\mathcal{A}$ which takes as input μ hash values and can output with probability ε a valid pair (m, s) if the underlying code used is a Goppa code. Let H be a random $(n - k, n)$ matrix. This algorithm can be used to decide whether this matrix is the parity check matrix of a Goppa code. Apply $\mathcal{A}$ to $1/\varepsilon$ set of μ hash values and check if a valid pair (m, s) has been generated. This way, we have an algorithm to solve the Goppa code indistinguishability problem (see Subsec. 15.7.1.2).

Table 15.11.   CFS signature scheme.

Public:

- $H_{\mathrm{pub}} = SHP$, the parity check matrix of a $[2^m, 2^m - mt, t]$ Goppa code,
- h a hash function from $\{0,1\}^*$ to $\{0,1\}^{n-k}$,

Secret:

- the Goppa code whose H is a parity check matrix.

Signature generation:

Let M the text to sign.

- Find the first integer i_0 such that $h(M||i_0)$ be decodable.
- Compute $s = (e||i_0)$ where e is the word of weight t which satisfies $H_{\mathrm{pub}}{}^t e = h(M||i_0)$.

Signature verification:

Let $(M, e||i_0)$ a message and its signature.

- Check that $H_{\mathrm{pub}}{}^t e = h(M||i_0)$ and $w(e) = t$.

Table 15.12.   CFS asymptotic characteristics.

Public key	$tm2^m$ bits
Signature cost	$t! t^2 m^3$ binary op.
Signature length	tm bits
Verification cost	$(t-1)tm$ binary op.
Decoding attack	$(2^m/m)^{\lambda t}$ binary op.
Structural attack	$2^{mt}/mt$ binary op.

Hence, the formal security of the scheme depends upon two difficult problems.

Now, as always there is a tradeoff to find between security and performances of the scheme. The parameters have too be well chosen so as to resist to the different attacks (decoding attack, structural attack and birthday attack) and such that the signature generation be fast (around $t!$ iterations to find the integer i_0). Since t must be small, the authors suggest to use long Goppa codes in order to obtain a secure scheme. Suggested parameters are $m = 16$ and $t = 9$.

The code used is a $(65536, 65392)$ Goppa code which can correct nine errors. A syndrome is 144 bits length which gives around 2^{85} operations in

Table 15.13.   CFS in practice.

$m = 16$ $t = 9$	
Public key	1.12 Mbytes
Signature length	145 bits
Birthday attack	2^{85}
Structural attack	2^{137}
Decoding attack	2^{80}
Signature verification	$< 1\mu\,$s

order to apply birthday attack. The complexity of decoding attack is about 2^{80} and those of structural attack 2^{137}. The signature is the concatenation of a word e of weight 9 and size 2^{16} and a counter whose value is bounded by 9!. Using an index to represent e, this one can be coded on 126 bits and 19 bits are needed for the counter, this leads to a signature length of 145 bits. The time to verify signature is less than $1\mu\,$s. Table 15.13 sums up these parameters.

Traditionally in a signature scheme there is a tradeoff between signature length and cost of the verification. Three variants of the CFS signature scheme wisely use this fact and one of them allows to obtain a signature length of 81 bits for 1 s of verification.

Remark 15.8. This scheme has been implemented in hardware on a Field Programmable Gate Array giving a signature time of 0.86 s [48].

15.9. Secret Sharing Schemes

A (k, n) secret sharing scheme is a protocol where a secret S is split into n pieces, each one being distributed to n users. If strictly fewer than k users meet together, they must not be able to compute S. Any assembly of k (or more) users can retrieve S. This problem was first considered by Shamir and he gives a solution using interpolation of polynomials over $\mathbb{Z}_p$, the secret being the constant term of a polynomial f of degree $k - 1$. Each participant owns a pair $(i, f(i))$ $(i \in \mathbb{Z}_p^{\times})$ and using Lagrange's formulas, any k users can compute f and deduce its constant term [49].

McEliece and Sarwate show that this scheme can be generalized using Reed-Solomon codes [50] for which a polynomial time decoding algorithm is known. Let $\{\alpha_1, \ldots, \alpha_n\}$ be the non-zero elements of the field $\mathbb{F}$ and $\mathcal{C}$ an $[n, k]$ RS code over $\mathbb{F}$, then each word $(m_0, \ldots, m_{k-1})$ can be encoded into the codeword $c = (c_1, \ldots, c_n)$ such that $c_i = m(\alpha_i)$ where

Table 15.14.   A code-based secret sharing scheme.

Secret: $m_0 \in \mathbb{F}_q$

Secret sharing:

· Compute the codeword $c = (c_1, \ldots, c_n)$ from the information symbols $(m_0, \ldots, m_{k-1})$, $(m_1, \ldots, m_{k-1})$ being randomly generated.

· Each user receives a pair (i, c_i).

Secret recovering:

· From $r(\geq k)$ pairs $(i_1, c_{i_1}), \ldots, (i_r, c_{i_r})$, build an n bits word c' such that $c'_i = c_i$ if $i \in \{i_1, \ldots, i_r\}$, $c'_i = 0$ otherwise.

· Use the erasure decoding algorithm to compute c and then m_0.

$m(x) = \sum_{j=0}^{k-1} m_j x^j$ (Shamir's scheme corresponds to the case where $n+1$ is prime and $\alpha_i = i$). The secret to be shared is the information symbol m_0. Table 15.14 describes the protocol. When r users meet together, they know r symbols (and their positions) of the whole codeword c. The remaining $n - r$ symbols are called *erasures*: simply replace them with 0 and they become special errors whose positions are known.

Remark 15.9. Notice that since the protocol is used over $\mathbb{F}_q$ we have $n = q - 1$.

Proposition 15.6. *Reed–Solomon codes can polynomially decode n_e errors and n_ε erasures provided that $2n_e + n_\varepsilon < n - k + 1$.*

In our case, we have $n_e = 0$ and $n_\varepsilon = n - r$, thus if $r \geq k$, every assembly of r users can compute the whole codeword c using the decoding algorithm of RS codes and deduce m_0.

Remark 15.10. Notice that $m_0 = -\sum_{i=1}^{n} c_i$ (see Question 15.10). Moreover, the encoding of RS code can be done in an efficient way without the generator matrix of the code. Hence in this protocol, there is no need to store this matrix.

This protocol has a non-negligible advantage as compared to Shamir's scheme. Suppose that a dishonest party want to denied access to the secret to legitimate users by tampering some of the pieces c_i (or being less paranoiac, just envisage that some c_i's have been tampered with some "natural" phenomena). Let t be the number of invalid c_i. Suppose r users meet together and t of them have corrupted pieces, the whole codeword c can be computed if $2t + n - r < n - k + 1$, i.e. $r \geq k + 2t$. Hence, if some pieces are damaged, it is still possible to retrieve the secret. On the other hand, since there are n users, the opponent has to alter more than $\lfloor (n - k)/2 \rfloor$ pieces to ensure that the secret be inaccessible.

A more general situation is to specify some users who have greater privileges of access to the secret than to others. An access structure consists of all subsets of participants that should be able to compute the secret, but that contains no proper subset that also could determine it. Massey proposed to treat this problem using linear codes and the notion of "minimal" codewords [51, 52].

15.10. Regular Words and Quasi-cyclic codes

There are essentially two drawbacks in code-based cryptography. First, generation of constant weight word for a scheme is not an easy problem and involves computation which slow down the whole process. Next, all the schemes depends upon a public matrix whose size is greater than the usual public data used in number theory-based cryptography.

15.10.1. *Regular Words*

An issue to the first problem is to use regular words [41] instead of constant weight words.

Definition 15.10. Let consider a binary word of size n as n/t consecutive blocks of size t. A (n, t) regular word is a word which has exactly one non-zero coordinate in each block.

From an algorithmic point of view, the generation of (n, t) regular words is obviously easier than the one of constant weight words (see Table 15.15). The regular word version of the SD problem can be stated as follows:

Name : RSD (Regular Syndrome Decoding);
Input : H a fullrank $r \times n$ binary matrix , an integer t and a syndrome y,
question : Does there exists a (n, t) regular word $e \in \{0, 1\}^n$ such that $H^t e = y$?

This problem remains NP-complete [41].

Table 15.15.   A (n, t) regular word random generator.

```
Input: - m a binary word of t log₂(n/t) bits
begin
     w ← 0 (binary word of size n)
     for i = 0 to t − 1
          extract the log₂(n/t) right bits of m
          convert those bits in an integer j ∈ [0, n/t − 1]
          w_{i n/t + j} ← 1
          shift m to the right by log₂(n/t) bits
     end for
end
Output: - w a (n, t) regular word.
```

15.10.2. *Quasi-Cyclic Codes*

Concerning the size of the matrix, this drawback can be avoided by the use of quasi-cyclic codes.

Definition 15.11. A code of length n is called quasi-cyclic of order s, for n a multiple of s, if every cyclic shift of a codeword by s coordinates is again a codeword.

The particularity of such codes is that the whole generator matrix can be derived from the knowledge of few rows. Hence, it is enough to publish these few rows (a kind of compressed version of the public matrix) instead of the whole matrix. In order to apply this trick to McEliece's cryptosystem, one has to find a large class of polynomial decodable quasi-cyclic codes. This issue has been addressed in 2005 by Gaborit [53] by using set of s quasi-cyclic subcodes of a given BCH$[2^m − 1, k, 2t + 1]$ code.

A particular class of quasi-cyclic codes is those whose generator matrix is obtained by concatenation of circulant matrix.

Definition 15.12. A $r \times r$ circulant matrix is such that the $r − 1$ latest rows are obtained by cyclic shifts of the first row.

It was shown in [54] that, if one admits a small constraint on the size n of the code then such codes behave like purely random codes (in particular they satisfy the Gilbert–Varshamov bound). Hence, they are well suited to be used in code base schemes for which a random matrix is needed. Although all classical algorithms used to find a word of given weight in a code do not give better results when applied to quasi-cyclic codes, nowadays it is not known if the decoding of a random quasi-cyclic code is an NP-complete problem.

In 2007, a modification of Stern's identification scheme has been proposed using as public matrix H, the concatenation of two $k \times k$ circulant matrices (the identity matrix and a random one) [55]. This way, the public matrix can only be described from the first line of the random matrix which in particular decreases the size of the data which must be stored by the prover. The underlying difficult problem upon which the security of the scheme is linked can be stated as follows:

Name	:	Syndrome Decoding of Double Circulant Linear Codes
Input	:	$H(k, 2k)$ a double binary circulant matrix , s a binary column vector with r coordinates, p an integer.
Question	:	Is there a binary vector e of length n such that $H^t e = s$ and $w(e) \leq p$?

Nowadays, it is not known if this problem is NP-complete.

Using this same trick and the regular words Laudauroux, Gaborit, and Sendrier have defined in 2007 a modified version of Fischer–Stern's algorithm in order to speed the output of the generator: the SYND pseudo-random generator [56]. They obtain this way a pseudo-random generator as fast as AES in counter mode [57] with few memory requirement (around 1Kbytes). Moreover, the scheme has a formal proof of security.

We sum up in Table 15.16 the characteristics of this different improvements.

Remark 15.11. For the modified version of Stern's identification scheme, there exists a variant in which the secret key is embedded in the public one. This allows to reduce again the size of the public and private data, but increases the complexity computation and the global transmission rate (see [55] for more details).

Table 15.16. Some characteristics of the improved schemes.

McEliece		Identification scheme		Pseudo-Random generator	
Code Param.	$(2047, 617)$	Code Param.	$(634, 317)$	Code Param.	$(8192, 256)$
Min. Dist.	63	Min. dist	69	Min. dist.	32
Public key	1.5 ko	Public data	634 bits	Stored data	1.03 Kbytes
QC version		Private data	951 bits	Output speed	1Gbits/s
		Transmission load	40096 bits		
		QC version		QC version + Regular words	

15.11. Other Related Works

Due to the need to hold this chapter to a reasonable length, we could not make an exhaustive and detailed survey on code-based cryptography. We summarize in this section other works and give references for interested readers.

15.11.1. *Code-based Hash Function*

At Mcrypt 2005, a provably collision resistant family of hash functions have been proposed by Augot, Finiasz, and Sendrier [58]. The Fast Syndrome-Based Hash function is based on the Merkle–Damgård design [59] which consists in iterating a compression function $\mathcal{F}$. This function takes as input a word of s bits, maps it to a word of length n and weight t and computes its syndrome from a given $r \times n$ parity check matrix (with $r < s$). The mapping is done using regular words in order to speed up the process. Inverting the function consists in solving an instance of RSD problem (see Sec. 15.10). Finding a collision (two words with same hash value) will require to solve an instance of 2-RSD problem which is also NP-complete [41]:

Name   :   2-RNSD (2-Regular Null Syndrome Decoding)
Input   : H a full rank $r \times n$ binary matrix r, p an integer,
Question :   Does there exists a 2-regular (n, p) word e such that
                 $H^t e = 0$?

Remark 15.12. A 2-regular (n, p) word is a word of length n such that each of the p consecutive blocks of size n/p contains either zero or two one.

Depending on the value of n, r, and t, the hash function can be cryptanalyzed using decoding algorithms or Wagner's generalized birthday technique [60]. Wagner's attack uses 2^a list of r bits strings as input and find a solution to RSD or 2-RNSD problem in $\mathcal{O}(2^{r/(a+1)})$ if a satisfies some constraints depending on n, r, and t. Parameters used in order to minimize the cost of the compression function always allow to select $a = 4$ which implies that r must be greater than or equal to 400. Taking into account this two kind of attacks, the size of the output functions must be of at least 5ℓ bits for a security level of 2^ℓ. The proposed scheme has two main drawbacks:

- r being large, the size of the matrix H will be large too (around 1-Mbytes for the parameters suggested in [58]). Paradoxically, the speed of the compression function can be improved with larger n while keeping a constant security level of 2^{80}.

- Usually the security of a hash function must be half its output size.

In 2007, an improvement of this scheme has been proposed by Finiasz, Gaborit, and Sendrier [61]. They first add a final compression function in order to fit the security level with the output length. Next they use quasi-cyclic codes and changed the constant weight encoder (by using a "mix" between optimal encoding and regular word encoding) in order to dramatically reduce the size of the matrix H. This way they obtain a function which hashes only two times slower than SHA-256 for the same level of security with a matrix H of 32 Kbytes. Since the recent attacks (2004) on SHA-1, the need for new hash functions with provable security is a real challenge and the improved version of FSBH is an alternative to be considered.

15.11.2. *Rank Distance Codes*

The Syndrome Decoding problem can be reconsidered using another kind of metric. Such an issue has been first addressed by Gabidulin, Paramonov, and Tretjakov [62] using rank distance defined in 1985 by Gabidulin [63].

Definition 15.13. Let $\{\gamma_1, \ldots, \gamma_m\}$ be a basis of $\mathbb{F}_{q^m}$ over $\mathbb{F}_q$ and let $a = (a_1, \ldots, a_n)$ be an element of $\mathbb{F}_{q^m}^n$. For $i \in [1, m]$, let $a_i = \sum_{j=1}^{m} a_{ij}\gamma_j$. The rank of a, denoted $\mathrm{rk}(a)$, is the rank of the matrix

$$\begin{pmatrix} a_{11} & a_{21} & \cdots & a_{n1} \\ a_{12} & a_{22} & \cdots & a_{n2} \\ \vdots & \vdots & \vdots & \vdots \\ a_{1m} & a_{2m} & \vdots & a_{nm} \end{pmatrix}.$$

Proposition 15.7. *Let a and b be two elements of $\mathbb{F}_{q^m}^n$, the mapping d_{rg} from $\mathbb{F}_{q^m}^n \times \mathbb{F}_{q^m}^n$ in $\mathbb{N}$ defined by $d_{rg}(a, b) = rg(a - b)$ is a distance.*

The SD problem can be redefined using rank distance:

Name	:	Rank-SD;
Input	:	H a full rank $r \times n$ matrix over $\mathbb{F}_{q^m}$, p an integer, s a column vector of size r over $\mathbb{F}_{q^m}$,
Question	:	Does there exist $x \in \mathbb{F}_{q^m}^n$ such that $H^t x = s$ and $\mathrm{rk}(x) \leq p$?

Unfortunately, it is not known whether this problem is NP-complete still it is believed that it is a hard problem. Using another metric, the classical decoding algorithms are no more efficient. This way the size of the public key can be reduced. However, there exist two algorithms for Rank-SD due to Ourivski and Johannson [64] whose complexity are, respectively, $\mathcal{O}((pm)^3 2^{(p-1)(r+1)})$ and $\mathcal{O}((r+p)^3 p^3 2^{(p-1)(m-p)})$.

Remark 15.13. For all $a \in \mathbb{F}_{q^m}^n$, rk$(a) \leq w(a)$. Hence, the minimal rank distance $d_{\mathrm{rk}} = \min_{a \in \mathcal{C}} \mathrm{rg}(a)$ of a code $\mathcal{C}$ defined over $\mathbb{F}_{q^m}$ satisfies $d_{\mathrm{rk}} \leq n - k + 1$.

Definition 15.14. A maximum rank distance code (MRD code) is a $\mathcal{C}[n, k]$ code whose minimal rank distance is $n - k + 1$.

In 1985, Gabidulin proposed a method in order to build MRD codes for $n \leq m$ and for which there exists a polynomial time decoding algorithm. These codes (traditionally called Gabidulin codes) are used in the GPT cryptosystem [62] whose initial version is described in Table 15.17. The cryptosystem has been cryptanalyzed by Gibson [65, 66] for the original parameters suggested for which the size of the public key were 600 bytes. Modified versions using subcodes and a Niederreiter variant has been proposed by Loidreau and Berger [67] and recently cryptanalyzed by Overbek [68, 69]. Finally to resist to this different attacks, new parameters

Table 15.17.   The GPT cryptosystem.

Secret key:

- A $\mathcal{C}(n, k)$ Gabidulin code over $\mathbb{F}_{q^m}$ which corrects $t = \lfloor (d-1)/2 \rfloor$,
- $S(k, k)$ an invertible matrix over $\mathbb{F}_{q^m}$,
- $X(k, n)$ a matrix over $\mathbb{F}_{q^m}$ such that $\forall \mu \in \mathbb{F}_{q^m}^k$, rg$(\mu X) \leq t_1$, where $t_1 < t$.

Public key: $(G' = SG + X, t_1)$, where G is a generator matrix of $\mathcal{C}$.

Encryption:

- Message: $m \in \mathbb{F}_{q^m}^k$,
- Cryptogram: $c = mG' + e$ where $e \in \mathbb{F}_{q^m}^n$ satisfies rk$(e) \leq t - t_1$.

Decryption: Since rk$(mX + e) \leq$ rk$(mX) +$ rk$(e) \leq t$, compute:

- $mS = \mathcal{A}(c) = \mathcal{A}((mS)G + mX + e)$ (where $\mathcal{A}$ is the polynomial decoding algorithm),
- $m = (mS)S^{-1}$.

have been suggested for which the size of the public key is now 2.25 Kbytes and the transmission rate is about 0.3. A good survey on rank distance and its applications in cryptography can be found in [70, 71].

15.11.3. *An Identity-based Identification Scheme*

The main problem in "real life" public key cryptography is to establish a link between a public key and its owner's identity. Without this link, anyone could change Bob's public key by its own key and will be able to send signed message as if they were generated by Bob since the verification step will make use of the modified version of Bob's public key. In practice, a public key infrastructure is used in order to manage the authenticity of the public keys. A trusted center produces certificates for public keys and people can check their validity from public key's trusted center.

In 1984, Shamir introduced the notion of identity-based public key cryptography [72]. The concept make use of a trusted third party: the KGC (Key Generation Center). This one has a master public key and a master secret key. From an identity i and the master public key, any one can derive the public key linked to i. The user whose identity is i contacts the KGC which computes its private key from i and the master secret key. Unfortunately in concrete ID-based systems, the public key of i depends upon its identity and an additional public data. In the context of identification, this value will be transmitted during the identification process.

In 2004, Bellare, Neven, and Namprempre described a generic method to derive an identity base identification scheme from a standard authentication scheme [73]. As usual this concept has only been applied to number theory schemes. In 2007 [74], Cayrel, Gaborit, and Girault considered the combination of two code base schemes (CFS signature scheme and Stern's identification scheme) in order to produce the first identity-based identification scheme using error-correcting codes (see Table 15.18). The generation of Alice's parameters is obtained from an execution of the CFS signature's scheme. Hence in order to prevent an intruder to be able to compute Alice's secret key from her identity, one has to consider the parameters that guarantee the security of the CFS scheme. The drawback is that the CFS scheme uses very long Goppa codes while Stern's scheme uses shorter ones. Since the same matrix has to be used by the KGC and by the identification process, this will overload the communication complexity. As an example, for $m = 16$ and $t = 9$ (which

Table 15.18.   A code-based IBI scheme.

Master Public Key:

- $H_{\mathrm{pub}} = SHP$, the parity check matrix of a $[2^m, 2^m - mt, 2t + 1]$ Goppa code,
- h a hash function from $\{0, 1\}^*$ to $\{0, 1\}^{n-k}$,

Master Secret Key:

- the Goppa code whose H is a parity check matrix.

Generation of Alice's Parameters:
   Input: $\mathrm{id}_A \rightarrow$ Alice's identity.
          · Find the first integer i_0 such that $h(\mathrm{id}_A + i_0)$ be decodable.
   Output:
          · Alice's secret key: s the word of weight t which satisfies $H_{\mathrm{pub}}{}^t s = h(\mathrm{id}_A + i_0)$.
          · the integer i_0.

Identification process:
   A randomly computes:
          · $y \in \{0, 1\}^n$,
          · σ a permutation of $\{1, \dots, n\}$.
   and send to B:

$$c_1 = \langle \sigma, H_{\mathrm{pub}}{}^t y \rangle, \; c_2 = \langle y\sigma \rangle, c_3 = \langle (y + s)\sigma \rangle, i_0$$

   B sends a random element $b \in \{0, 1, 2\}$ (challenge).
          · if $b = 0$,
                 · A reveals y and σ,
                 · B checks the value of c_1 and c_2.
          · if $b = 1$,
                 · A reveals $y + s$ and σ,
                 · B checks the value of c_1 and c_3.
                 Notice that $H_{\mathrm{pub}}{}^t y = H_{\mathrm{pub}}{}^t(y + s) + h(\mathrm{id}_A + i_0)$.
          · if $b = 2$,
                 · A reveals $y\sigma$ and $s\sigma$,
                 · B checks the value of c_2 and c_3 and verifies that $w(s\sigma) = t$.

Repeat the identification process until the expected security level be reached.

guarantee a security level of 2^{80}), the communication cost will be greater than 500 Kbytes.

15.12. Conclusions

While code base schemes use only elementary operations over the two elements field, they were not really considered by cryptographic community because of the size of the public data. Since these last years, numerous works have been developed in order to enhance the performance of coding-based cryptography leading to realistic alternatives to number-based theory schemes even in constrained environments such as smart cards or RFID tags. Nowadays, coding base cryptography has to be considered as a real alternative to number theory-based cryptography especially since:

- best cryptanalysis against the Syndrome Decoding problem is still exponential whereas it is subexponential for factoring,
- there does not exist a quantum algorithm which can polynomially solve the SD problem while Shor's algorithm can factor an integer N in $\mathcal{O}((\log N)^3)$ operations on a quantum computer.

15.13. Questions

Question 15.1. *Let $\mathcal{C}$ be an $[n, k, d]$ code and $c' = c_0 + e$ where $c_0 \in \mathcal{C}$ and e is a binary vector of size n. Show that*

$$\forall u \in \mathcal{C} \backslash \{0\}, \quad w(e) < w(u + e) \Leftrightarrow \forall u \in \mathcal{C} \backslash \{c_0\}, \quad d(c_0, c') < d(u, c').$$

Answer. Use the facts that $w(e) = w(c_0 + c') = d(c_0, c')$ and that $\{w(u + e), u \in \mathcal{C}, u \neq 0\} = \{w(u + c_0 + e), u \in \mathcal{C}, u \neq c_0\}$.

Question 15.2. *Let $\mathcal{C}$ be an $[n, k, d]$ code and let $t = \lfloor (d - 1)/2 \rfloor$. Show that*

$$\forall s \in \{0, 1\}^{n-k}, \quad \#\{e \in \{0, 1\}^n \mid w(e) \leq t \text{ and } H^t e = s\} \leq 1.$$

Answer. Suppose that there exists $s \in \{0, 1\}^{n-k}$ such that $\#\{e \in \{0, 1\}^n \mid w(e) \leq t \text{ and } H^t e = s\} \geq 2$. Show that this implies that there exists a codeword of weight less than or equal to $2t$ and notice that $d \geq 2t + 1$.

Question 15.3. *Show that the SD problem still remains NP-complete if H is fullrank.*

Answer. Suppose that there exists a polynomial time algorithm $\mathcal{A}$ to solve the SD problem for any instance (H, s, p) when H is fullrank. Let (H', s, p) be an instance for a random matrix H'. From H' it is easy to extract in polynomial time a fullrank submatrix H. Extract from s the corresponding coordinates, let s' be the associated vector. Use $\mathcal{A}$ to solve the instance (H', s', p). Since (H, s, p) has a solution if and only if (H', s', p) has a solution, then we have obtained this way a polynomial time algorithm to solve the SD problem which would imply that $P = NP$.

Question 15.4. *Show that the SD problem is equivalent to the following problem:*

> Input   : $H(k, n)$ *a binary matrix of rank* k, $p > 0$ *an integer.*
> Question :   *Is there a vector* $x \in \{0,1\}^n$ *such that* $H^t x = 0$, $w(x) \le p$ *and* $x_n = 1$?

Answer. Let (H, p, s) be an input of the SD Problem. Let $H' = (H \mid s)$ and let x' be a solution of the above problem for the input $(H', p + 1)$. The vector $e = (x'_1, \ldots, x'_{n-1})$ is a solution of SD for the input (H, p, s). Conversely, let (H', p') be a solution of the above problem. Consider the submatrix H obtained from the first $n - 1$ columns of H'. Denote by s the last column of H'. If e is a solution of the SD problem for the input $(H, s, p' - 1)$, then the vector $x = (e_1, \ldots, e_{n-1}, 1)$ is a solution of the above problem for the input (H', p').

Question 15.5. *Lee and Brickell algorithm proceeds as follows in order to decode a word* $x = mG + e$ *where* $w(e) \le t$:

(1) *Select an information set* $I = \{i_1, \ldots, i_k\}$ *and check if* $w(x_I G_I^{-1} G + x) \le t$ *where* $x_I = (x_{i_1}, \ldots, x_{i_k})$ *and* G_I *are the columns of* G *whose index is in* I. *In this case,* $e = x_I G_I^{-1} G + x$.
(2) *If* $w(x_I G_I^{-1} G + x) > t$, *try to find a word* μ *of length* k *and weight at most* p *such that* $w((x_I + \mu) G_I^{-1} G + x) \le t$.
(3) *If step 2 fails go to step 1.*

Their main idea is instead to try to find errors free information sets to accept to use information sets which contains at most p *errors where* p *is small. What is the probability of success of such an algorithm?*

Answer. "Good" information sets are those which contains at most p indexes which correspond to error positions, hence the probability of success is given by

$$\sum_{i=0}^{p} \binom{n - t}{k - i} \binom{t}{i} \bigg/ \binom{n}{k}.$$

Question 15.6. *Show that if an intruder is able to correctly answer to any of the three questions of the G-SD identification scheme for a fixed round,* *then he has found either a collision for the hash function or a solution for the instance of the G-SD problem.*

Answer. Let us denote by $\tilde{A}$ the intruder and let δ_1, δ_2, and δ_3 the three commitments sent by $\tilde{A}$ during a round. If $\tilde{A}$ can answer any of the three questions, this means that he can produce:

- if $b = 0$, a couple $(\tilde{\mu}, \tilde{\sigma})$ such that $\delta_1 = \langle\tilde{\sigma}\rangle$ and $\delta_2 = \langle\tilde{\mu}G\tilde{\sigma}\rangle$,
- if $b = 1$, a couple $(\tilde{y}, \tilde{e})$ such that $\delta_2 = \langle\tilde{y}\rangle$, $\delta_3 = \langle\tilde{y}+\tilde{e}\rangle$, and $w(\tilde{e}) = p$,
- if $b = 2$, a couple $(\tilde{\pi}, \tilde{u})$ such that $\delta_1 = \langle\tilde{\pi}\rangle$ and $\delta_3 = \langle(\tilde{u}G + x)\tilde{\pi}\rangle$.

This gives the following system:

$$\langle\tilde{\sigma}\rangle = \langle\tilde{\pi}\rangle,$$
$$\langle\tilde{\mu}G\tilde{\sigma}\rangle = \langle\tilde{y}\rangle,$$
$$\langle(\tilde{u}G + x)\tilde{\pi}\rangle = \langle\tilde{y}+\tilde{e}\rangle,$$
$$w(\tilde{e}) = p.$$

Hence either $\tilde{A}$ can produce collision for the hash function or from the previous system he can deduce that $\tilde{e}$ satisfies $w(\tilde{e}) = p$ and $x + \tilde{e}\sigma^{-1} = (\tilde{\mu} + \tilde{u})G$ which implies that $(\tilde{\mu} + \tilde{u}, \tilde{e})$ is a solution to the underlying G-SD problem.

Question 15.7. *If we admit that the weight distribution of a Goppa code acts like the one of a random code (i.e. we consider that $Pr(w(c) = i) = \binom{n}{i}/2^n$) show that it is possible in McEliece's cryptosystem to decide if two distinct cryptogram c_1 and c_2 come from the same plaintext m.*

Answer. Let m_1 and m_2 be two plaintext and let $c_1 = m_1G + e_1$ and $c_2 = m_2G + e_2$ where G is the public matrix. First show that $w(c_1 + c_2) \leq 2t \Rightarrow w((m_1 + m_2)G) \leq 4t$. Now if $m_1 = m_2$ then $w(c_1 + c_2) \leq 2t$. Hence if $w(c_1 + c_2) \leq 2t$ then either $m_1 = m_2$, or $m_1 \neq m_2$ and $w((m_1 + m_2)G) \leq 4t$. To finish the proof, show that for the parameters used in McEliece's cryptosystem the probability that the weight of a codeword be less than or equal to $4t$ is negligible.

Question 15.8. *Show that in McEliece's cryptosystem, if two cryptogram come from the same plaintext m, than one can devise an algorithm to compute m whose probability of success can be estimated to $\dfrac{\binom{n-2t+t^2/n}{k}}{\binom{n-2t+2t^2/n}{k}}$.*

Answer. We know from the preceding question that if $w(c_1 + c_2) \leq 2t$, then c_1 and c_2 come from the same plaintext m. Choose an information set I among the positions where $c_1 + c_2$ is zero. If $w(c_1 + c_{1,I}G_I^{-1}G) \leq t$ then $m = c_{1,I}G_I^{-1}G$ else choose another information set I. The total number of information sets is $\binom{n-w(e_1+e_2)}{k}$. "Good" information sets are positions for which e_1 and e_2 are zero. Now since e_1 and e_2 are independent, the probability that coordinate i of e_1 and e_2 be equal to 1 is t^2/n^2. Hence, on average there are t^2/n common non-zero positions in e_1 and e_2. Thus, the number of "good" information set can be estimated to $\binom{n-w(e_1+e_2)-t^2/n}{k}$. To end the proof show that $w(e_1 + e_2)$ is about $2t + 2t^2/n$.

Question 15.9. *Could you explain why in the CFS signature scheme the asymptotic size of the signature length is tm?*

Answer. The signature is the concatenation of a word of length 2^m and weight t (which can be represented as an integer in the range $[1, \binom{2^m}{t}]$) with an integer bounded by $t!$. To end the proof consider that $\binom{2^m}{t} \simeq 2^{tm}/t!$.

Question 15.10. *Let α be a primitive element of $\mathbb{F}_q$, Reed–Solomon codes can be defined as $Im\phi$ where:*

$$\phi : \mathbb{F}_q^k \to \mathbb{F}_q^{q-1}$$
$$m \mapsto \phi(m) = (m(1), m(\alpha), \dots, m(\alpha^{q-2})),$$

and $m(x) = \sum_{i=0}^{k-1} m_i x^i$. Show that $m_0 = -\sum_{i=0}^{q-2} c_i$.

Answer. Prove that $\forall j \in [1, k-1]$, $\sum_{i=0}^{q-2}(\alpha^j)^i = 0$.

15.14. Keywords

$(\alpha t, t)$ regular word

A binary word of size αt such that there is only one non-zero bit in each of the t consecutive blocks of size α.

Birthday attack

An attack whose goal is to find collision for a hash function. The name is derived from the mathematical birthday paradox problem which states that for a function f and a set H of n elements, the probability to find two elements which have the same image under f is in $\mathcal{O}(\sqrt{n})$.

Circulant matrix

A matrix whose each row is a circular shift of the first row.

Collision

Two elements x and y which have same hash values.

Commitment

Public values hidden by a hash function before the beginning of a protocol and disclosed during of after the protocol.

Communication complexity

Number of bits exchanged during a protocol between two entities.

Decision problem

A problem for which the answer is yes or no.

Erasure

An error in a codeword whose position is known.

Hash function

A function which takes as input a binary sequence of any length and output a fixed length sequence.

IND CCA2 attack

Indistinguishability under adaptive chosen ciphertext attack. A model in which the cryptanalyst has to solve the following challenge: giving two plaintext and one cryptogram, find which plaintext has been encrypted. The cryptanalyst can access a decryption oracle before and after the challenge to decrypt any message except the cryptogram.

Information set

A set of k coordinates of an n bit word which are errors free.

NP problem

A problem which can be solved by a non-deterministic polynomial time algorithm.

NP complete problem

A problem Π which is in NP such that every other problem in NP can be reduced in polynomial time to it.

One-way function

Unformally speaking a one-way function is a function such that $f(x)$ can be computed in polynomial time whereas it is "hard" to invert .

Polynomial time algorithm

An algorithm whose running time is bounded by An^d, where A and d are some constants and n is the size of the input.

Pseudo-random generator

A deterministic polynomial time algorithm which outputs a "random" sequence of bits from a fixed input.

Random oracle model

A theoretical model in which hash functions act as real random functions.

Syndrome

In coding theory, the syndrome of the word x is $H^t x$ where H is a parity check matrix.

Zero-knowledge proof

An interactive protocol in which someone proves that he knows a secret s without revealing any information on it.

References

1. W. Diffie and M. E. Hellman, New directions in cryptography, *IEEE Trans. Inform. Theory* **IT-22**, 6 (1976), 644–654.
2. P. W. Shor, Polynomial-time algorithms for prime factorization and discrete logarithms on a quantum computer, *Proceedings of the 35th Annual Symposium on Foundations of Computer Science*, (1994), 20–22.
3. F. J. MacWilliams and N. J. A. Sloane, *The Theory of Error-Correcting Code* (North-Holland, 1977).
4. V. D. Goppa, A new class of linear error correcting codes, *Probl. Pered. Inform.* (1970), 24–30.
5. N. Patterson, Algebraic decoding of goppa codes, *IEEE Transactions on Information Theory* **21**, 2 (1975), 203–207.
6. A. Vardy, The intractability of computing the minimum distance of a code, *IEEE Trans. Inform. Theory* **43**, 6 (1997), 1757–1766.
7. E. R. Berlekamp, R. J. McEliece and H. C. A. van Tilborg, On the intractability of certain coding problems, *IEEE Trans. Inform. Theory* **24**, 3 (1978), 384–386.
8. M. R. Garey and D. S. Johnson, *Computers and Intractability, A Guide to the Theory of NP-Completeness* (W. H. Freeman and Company, New York, 1979).
9. P. J. Lee and E. F. Brickell, An observation on the security of mceliece's public-key cryptosystem, *Advances in Cryptology — EUROCRYPT '88, Lecture Notes in Computer Science*, Vol. 330 (Springer-Verlag, 1988), pp. 275–280.

10. J. S. Leon, A probabilistic algorithm for computing minimum weights of large error-correcting codes, *IEEE Trans. Inform. Theory* **34**, 5 (1988), 1354–1359.

11. J. Stern, A method for finding codewords of small weight, *Coding Theory and Applications*, *Lecture Notes in Computer Science*, Vol. 388 (Springer-Verlag, 1988), pp. 106–113.

12. A. Canteaut and F. Chabaud, A new algorithm for finding minimum-weight words in a linear code: Application to mceliece's cryptosystem and to narrow-sense bch codes of length 511, *IEEE Trans. Inform. Theory* **44**, 1, (1998), 367–378.

13. T. Johansson and F. Jönsson, On the complexity of some cryptographic problems based on the general decoding problem, *IEEE Trans. Inform. Theory* **48**, 10 (2002), 2669–2678.

14. E. Prange, The use of information sets in decoding cyclic codes, *IRE Trans.* **IT-8** (1962), 85–89.

15. J. van Tilburg, On the mceliece public-key cryptosystem, in *Advances in Cryptology — CRYPTO '88*, *Lecture Notes in Computer Science*, Vol. 403 (Springer-Verlag, 1988), pp. 119–131.

16. H. Chabanne and B. Courteau, Application de la méthode de décodage itérative d'omura à la cryptanalyse du système de mc eliece, *Rapport de Recherche 122*, Université de Sherbrooke (October, 1993).

17. A. Canteaut, *Attaques de cryptosystèmes à mots de poids faible et construction de fonctions t-résilientes*, PhD Thesis, Université Paris VI, (1996).

18. M. Blum, How to prove a theorem so no one else can claim it, *Proc. Int. Congress of Mathematicians*, Berkeley, (1986), CA, pp. 1444–1451.

19. R. Impagliazzo and M. Naor, Efficient cryptographic schemes provably as secure as subset sum, *30th Symp. Foundations of Computing Science*, (1989), pp. 236–241.

20. J.-B. Fischer and J. Stern, An efficient pseudo-random generator provably as secure as syndrome decoding, *Advances in Cryptology — EUROCRYPT '96*, *Lecture Notes in Computer Science*, Vol. 1070 (Springer-Verlag, 1996), pp. 245–255.

21. P. Guillot, Algorithme pour le codage à poids constant.

22. N. Sendrier, Efficient generation of binary words of given weight, *Cryptography and Coding — 5th IMA Conference*, *Lecture Notes in Computer Science*, Vol. 1025 (Springer-Verlag, 1995), pp. 184–187.

23. S. Goldwasser, S. Micali and C. Rackoff, The knowledge complexity of interactive proof systems, *SIAM, J. Comput.* **18** (1989), 186–208.

24. J.-J. Quisquater and L. Guillou, How to explain zero-knowledge protocols to your children, *Advances in Cryptology-Crypto '89*, *Lecture Notes in Computer Science*, Vol. 435 (Springer-Verlag, 1990), pp. 628–631.

25. A. Fiat and A. Shamir, How to prove yourself: Practical solutions to identification and signature problems, *Advances in Cryptology — CRYPTO '86*, *Lecture Notes in Computer Science*, Vol. 263 (Springer-Verlag, 1987), pp. 186–194.

26. S. Harari, A new authentication algorithm, *Coding Theory and Applications, Lecture Notes in Computer Science*, Vol. 388 (Springer-Verlag, 1988), pp. 91–105.

27. P. Véron, Cryptanalysis of Harari's identification scheme, in *Cryptography and Coding, 5th IMA Conference, Lecture Notes in Computer Science*, Vol. 1025 (Springer-Verlag, 1995), pp. 264–269.

28. M. Girault, A (non-practical) three-pass identification protocol using coding theory, *Advances in Cryptology, Auscrypt'90, Lecture Notes in Computer Science*, Vol. 453 (Springer-Verlag, 1990), pp. 265–272.

29. N. Sendrier, Finding the permutation between equivalent linear codes: The support splitting algorithm, *IEEE Trans. Inform. Theory* **46**, 4 (2000), 1193–1203.

30. J. Stern, A new identification scheme based on syndrome decoding, *Advances in Cryptology — CRYPTO '93, Lecture Notes in Computer Science*, Vol. 773 (Springer-Verlag, 1993), pp. 13–21.

31. P. Véron, Improved identification schemes based on error-correcting codes, *Appl. Algebra Eng. Commun. Comput.* **8**, 1 (1996), 57–69.

32. M. Girault and J. Stern, On the length of cryptographic hash-values used in identification schemes, *Advances in Cryptology — CRYPTO '94, Lecture Notes in Computer Science*, Vol. 839 (Springer-Verlag, 1994), pp. 202–215.

33. R. L. Rivest, A. Shamir and L. Adleman, A method for obtaining digital signatures and public-key cryptosystems, *Commun. ACM.* **26**, 1 (1983), 96–99, ISSN 0001-0782. doi: http://doi.acm.org/10.1145/357980.358017.

34. R. Merkle and M. Hellman, Hiding information and signatures in trapdoor knapsacks, *IEEE Trans. Inform. Theory* **24** (1978), 525–530.

35. R. J. McEliece, A public-key cryptosystem based on algebraic coding theory, *JPL DSN Progress Report* (1978), 114–116.

36. V. Sidelnikov and S. Shestakov, On cryptosystems based on generalized Reed–Solomon codes, *Diskretnaya Math.* **4** (1992), 57–63.

37. N. Sendrier, On the structure of a randomly permuted concateneted code, *EUROCODE '94*, pp. 169–173. Inria (1994).

38. J. K. Gibson, Equivalent goppa codes and trapdoors to mceliece's public key cryptosystem, in *Advances in Cryptology — EUROCRYPT '91, Lecture Notes in Computer Science*, Vol. 547 (Springer-Verlag, 1991), pp. 517–521.

39. C. Adams and H. Meijer, Security-related comments regarding mceliece's public-key cryptosystem, *IEEE Trans. Inform. Theory* **35** (1989), 454–455.

40. P. Loidreau and N. Sendrier, Weak keys in the mceliece public-key cryptosystem, *IEEE Trans. Inform. Theory* **47**, 3 (2001), 1207–1211.

41. M. Finiasz, *Nouvelles constructions utilisant des codes correcteurs d'erreurs en cryptographie à clé publique*, PhD Thesis, Ecole Polytechnique, (2004).

42. A. Canteaut and N. Sendrier, Cryptanalysis of the original mceliece cryptosystem, *Advances in Cryptology — ASIACRYPT '98, Lecture Notes in Computer Science*, Vol. 1514 (Springer-Verlag, 1998), pp. 187–199.

43. K. Kobara and H. Imai, Semantically secure mceliece public-key cryptosystems-conversions for mceliece pkc, *4th International Workshop on*

Practice and Theory in Public Key Cryptography, PKC 2001, Lecture Notes in Computer Science, Vol. 1992 (Springer-Verlag, 2001), pp. 19–35.

44. H. Niederreiter, Knapsack-type cryptosystems and algebraic coding theory, *Problems Control Inform. Theory* **15**, 2 (1986), 159–166.

45. Y. X. Li, R. H. Deng and X. mei Wang, On the equivalence of mceliece's and niederreiter's public-key cryptosystems, *IEEE Trans. Inform. Theory* **40**, 1 (1994), 271.

46. E. Brickell and A. Odlyzko, *Cryptanalysis: A Survey of Recent Results* (1992), 501–540.

47. N. Courtois, M. Finiasz and N. Sendrier, How to achieve a mceliece-based digital signature scheme, *Advances in Cryptology — ASIACRYPT 2001, Lecture Notes in Computer Science*, Vol. 2248 (Springer-Verlag, 2001), pp. 157–174.

48. J.-L. Beuchat, N. Sendrier, A. Tisserand and G. Villard, Fpga implementation of a recently published signature scheme, Tech. Rep. 5158, Inria (March, 2004).

49. A. Shamir, How to share a secret, *Communications of the ACM* **22**, 11 (1979) 612–613.

50. R. J. McEliece and D. V. Sarwate, On sharing secrets and Reed–Solomon codes, *Communications of the ACM* **24**, 9 (1981), 583–584.

51. M. J.-L. Minimal, Codewords and secret sharing, in *6th Joint Swedish–Russian Workshop on Information Theory* (1993), pp. 276–279.

52. A. E. Ashikhmin and A. Barg, Minimal vectors in linear codes, *IEEE Trans. Inform. Theory* **44**, 5 (1998), 2010–2017.

53. P. Gaborit, Shorter keys for code based cryptography, in *Proceeedings of WCC'05* (2005), pp. 81–90.

54. P. Gaborit and G. Zémor, Asymptotic improvement of the Gilbert–Varshamov bound for linear codes, in *Proceeedings of ISIT'06* (2006), pp. 287–291.

55. P. Gaborit and M. Girault, Lightweight code-based identification and signature, in *Proceeedings of ISIT'07* (2007).

56. P. Gaborit, C. Laudauroux and N. Sendrier, Synd: A fast code-based stream cipher with a security reduction, *Proceeedings of ISIT'07* (2007).

57. R. Housley, Using advanced encryption standard (aes) counter mode with ipsec encapsulating security payload (esp). RFC 3686, Network Working Group (January, 2004).

58. D. Augot, M. Finiasz and N. Sendrier, A family of fast syndrome based cryptographic hash functions, eds. E. Dawson and S. Vaudenay, *Mycrypt, Lecture Notes in Computer Science*, Vol. 3715 (Springer, 2005), pp. 64–83.

59. I. Damgård, A design principle for hash functions, *Advances in Cryptology — CRYPTO '89, Lecture Notes in Computer Science*, Vol. 435 (Springer-Verlag, 1990), pp. 416–427.

60. D. Wagner, A generalized birthday problem, *Advances in Cryptology — CRYPTO '02, Lecture Notes in Computer Science*, Vol. 2442 (Springer-Verlag, 2002), pp. 288–304.

P. Véron

61. M. Finiasz, P. Gaborit and N. Sendrier, Improved fast syndrome based cryptographic hash function, *ECRYPT Hash Workshop 2007* (2007).
62. E. M. Gabidulin, A. V. Paramonov and O. V. Tretjakov, Ideals over a non-commutative ring and thier applications in cryptology, *Advances in Cryptology — EUROCRYPT '91, Lecture Notes in Computer Science*, Vol. 547 (Springer-Verlag, 1991), pp. 482–489.
63. E. Gabidulin, Theory of codes with maximal rank distance, *Problemy Peredachi Informatsii* **21** (1985), 1–12.
64. A. Ourivski and T. Johansson, New technique for decoding codes in the rank metric and its cryptography applications, *Probl. Inform. Transm.* **38**, 3 (2002), 237–246.
65. J. K. Gibson, Severely denting the gabidulin version of the mceliece public key cryptosystem, *Design Codes and Cryptography* **6**, 1 (1995), 37–45.
66. J. K. Gibson, The security of the gabidulin public-key cryptosystem, *Advances in Cryptology — EUROCRYPT '96, Lecture Notes in Computer Science*, Vol. 1070 (Springer-Verlag, 1996), pp. 212–223.
67. T. Berger and P. Loidreau, How to mask the structure of codes for a cryptographic use, *Designs Codes and Cryptography* **35** (2005), 63–79.
68. R. Overbeck, A new structural attack for gpt and variants, eds. E. Dawson and S. Vaudenay, *Mycrypt, Lecture Notes in Computer Science*, Vol. 3715 (Springer, 2005), pp. 50–63.
69. R. Overbeck, Extending gibson's attacks on the gpt cryptosystem, ed. Ø. Ytrehus, *WCC, Lecture Notes in Computer Science*, Vol. 3969 (Springer, 2005), pp. 178–188.
70. P. Loidreau, *Etude et optimisation de cryptosystèmes à clé publique fondé sur la théorie des codes correcteurs*, PhD Thesis, Ecole Polytechnique (2001).
71. P. Loidreau, *Metrique rang et cryptographie* (Université Pierre et Marie Curie, Paris VI, 2007).
72. A. Shamir, Identity-based cryptosystems and signature schemes, *CRYPTO'84* (1984), 47–53.
73. M. Bellare, C. Namprempre and G. Neven, Security proofs for identity-based identification and signature schemes, *EUROCRYPT, Lecture Notes in Computer Science*, Vol. 3027 (Springer, 2004), pp. 268–286.
74. P.-L. Cayrel, P. Gaborit and M. Girault, Identity-based identification and signature schemes using correcting codes, eds. D. Augot, N. Sendrier and J.-P. Tillich, *WCC 2007* (INRIA, 2007).